CARBOHYDRATES IN DRUG DESIGN

CARBOHYDRATES IN DRUG DESIGN

edited by
Zbigniew J. Witczak
Karl A. Nieforth
University of Connecticut
Storrs, Connecticut

MARCEL DEKKER

MARCEL DEKKER, INC. NEW YORK • BASEL • HONG KONG

Library of Congress Cataloging-in-Publication Data

Carbohydrates in drug design/ edited by Zbigniew J. Witczak, Karl A.
Nieforth.
p. cm.
Includes index.
ISBN 0-8247-9982-8 (hardcover: alk. paper)
1. Carbohydrate drugs. I. Witczak, Zbigniew J.
II. Nieforth, Karl A.
[DNLM: 1. Carbohydrates—chemistry. 2. Carbohydrates—therapeutic
use. 3. Carbohydrates—pharmacokinetics. 4. Drug Design. QU 75
C26443 1997]
615'.3—dc21
DNLM/DLC
for Library of Congress

97-12543
CIP

The publisher offers discounts on this book when ordered in bulk quantities. For more information, write to Special Sales/Professional Marketing at the address below.

This book is printed on acid-free paper.

MARCEL DEKKER, INC.
270 Madison Avenue, New York, New York 10016
http://www.dekker.com

Current printing (last digit):
10 9 8 7 6 5 4 3 2 1

PRINTED IN THE UNITED STATES OF AMERICA

Preface

In the last four decades, carbohydrates, one of the most abundant natural product classes, have received considerably less attention from the majority of mainstream, synthetic chemists than other fields of natural product chemistry. Recently, this situation has changed and carbohydrate research is becoming a significant focus for synthetic chemists, particularly in medicinal and pharmaceutical chemistry. Within these disciplines, exciting new developments in the isolation and study of glycoproteins, glycopeptides and oligosaccharides have contributed to dramatic advances in our knowledge of the critical biological roles of these compounds.

Significant progress in this new field, glycobiology, and new discoveries of unique biological functions of carbohydrates have had a direct impact on a diverse group of biological studies from cancer to diabetes and cell-cell adhesion in inflammation and metastasis. Moreover, we have witnessed explosive developments in synthetic carbohydrate chemistry as efforts have been to expand our basic knowledge of relatively simple molecules such as KDO and sialyl Lewis^x, glycosylphosphatidylinositol (GPI) membrane anchors, Acarbose and other new antidiabetic agents as well as heparin oligosaccharides. Some of these new developments have major clinical implications and are the subject of chapters in this book. Other areas which are not part of this text, including macrolide antibiotics and modified aminoglycoside antibiotics, were not ignored; they were omitted because of limited space and scope of the book.

Carbohydrates have a long history of medicinal use as nutritional supplements, anticoagulants, plasma expanders, macrolide antibiotics, etc. and have recently been introduced as antidiabetics. The first of these, Acarbose, was introduced by Bayer in 1990 and approved by the FDA in 1995 for the USA market. The second was Voglibose, which was introduced in 1994 by Takeda Industries, and marketed only in Japan. A few other carbohydrate therapeutics based on oligosaccharides (heparin derivatives) and polysaccharides (glycans) are in the final phases of clinical trials or waiting approval.

The following pages offer stimulating accounts of the rapidly developing and challenging field of the chemistry of carbohydrates presented by internationally recognized scholars. Significant developments in this fascinating arena review state-of-the-art technologies used to unravel mechanisms of actions, study conformational and steric effects, compare and contrast synthetic and enzymatic approaches and finally develop new leads in emerging fields of glycobiology.

Each of the authors has drawn heavily on existing literature and his or her own personal experiences to trace the discovery of new leads and the subsequent development of these leads. The reader is provided with an insight into the factors which led to these new, potentially therapeutic carbohydrates that might be applied in future discoveries. Every chapter provides an illustrative overview of the highly diverse, crucial functions played by mono-, oligo- and polysaccharides in biological systems, and the means by which these observations may be exploited by the imaginative biological chemist. The reader is afforded a glimpse into the bright future of carbohydrate therapeutics as discussions unfold on auranofin, Sucraflate, SLe^x, heparin pentasaccharide, gangliosides, β-glucans, new analogs of anthracycline antibiotics and pyrimidine nucleosides.

This volume is intended for all researchers and graduate students in carbohydrate chemistry, medicinal chemistry, biochemistry and glycobiology. We hope that this volume is found informative and useful to the community of researchers in both academic and industrial laboratories as well as other scientists involved in drug discovery and interested in staying abreast of the progress in the carbohydrate field.

We would like to thank the authors for their well-written articles and their patience during the editing process. We would also like to thank our colleagues who served as external referees for their very helpful suggestions and advice to improve the text. We hope that this book will be an important stimulus for new ideas in this exciting field of research. We believe that this is a challenging subject with a broad universe of interest which, because of its strong commercial potential, will quickly become a major interest of the chemical and pharmaceutical industries. Finally, we take this opportunity to make special mention of Professor Akira Hasegawa's chapter (Chapter 4, p. 137). Professor Hasegawa passed away on October 10, 1996, and we are privileged to have his contribution in this book.

Zbigniew J. Witczak
Karl A. Nieforth

Contents

Contributors

Osman Achmatowicz, Ph.D. Professor, Pharmaceutical Research Institute, Warsaw, Poland

Susanne Alban, Ph.D. Research Associate, Institute of Pharmacy, University of Regensburg, Regensburg, Germany

J. Warren Beach Assistant Professor, Department of Medicinal Chemistry, College of Pharmacy, University of Georgia, Athens, Georgia

Karol S. Bruzik, Ph.D. Assistant Professor, Department of Medicinal Chemistry and Pharmacognosy, College of Pharmacy, University of Illinois, Chicago, Illinois

Stanislas Czernecki, Ph.D. Professor, Laboratory of Carbohydrates, Université Pierre et Marie Curie, Paris, France

Eric De Clercq, Ph.D. Professor, Rega Institute, Catholic University of Leuven, Leuven, Belgium

Akira Hasegawa, Ph.D.[†] Professor, Department of Applied Bio-organic Chemistry, Gifu University, Gifu, Japan

Mariko Kakuta, Ph.D. Professor, Department of Home Economics, Koman Women's University, Osaka, Japan

Milton J. Kiefel, Ph.D. Research Fellow, Department of Medicinal Chemistry, Victorian College of Pharmacy, Monash University, Parkville, Victoria, Australia

[†] Deceased.

Makato Kiso, Ph.D. Professor, Department of Applied Bio-organic Chemistry, Gifu University, Gifu, Japan

Ling Lei, Ph.D. Postdoctoral Research Associate, Faculty of Engineering, Ehime University, Matsuyama, Japan

Robert J. Linhardt, Ph.D. Professor, Department of Medicinal Chemistry and Natural Products Chemistry, College of Pharmacy, University of Iowa, Iowa City, Iowa

Akira Misaki, Ph.D. Emeritus Professor, Faculty of Human Life Sciences, Osaka City University, Osaka, Japan

Seiichiro Ogawa, Ph.D. Professor, Department of Applied Chemistry, Keio University, Yokohama, Japan

Shoichiro Ozaki, Ph.D. Professor, Faculty of Engineering, Ehime University, Matsuyama, Japan

Christophe Pannecouque, Ph. D. Research Associate, Rega Institute, Catholic University of Leuven, Leuven, Belgium

Roman Perez-Soler, Ph.D. Professor, M.D. Anderson Cancer Center, University of Texas, Houston, Texas

Waldemar A. Priebe, Ph.D. Associate Professor, M.D. Anderson Cancer Center, University of Texas, Houston, Texas

René Roy, Ph.D. Professor, Department of Chemistry, University of Ottawa, Ottawa, Ontario, Canada

Barbara Szechner, Ph.D. Research Associate, Pharmaceutical Research Institute, Warsaw, Poland

Toshihiko Toida, Ph.D. Professor, Faculty of Pharmaceutical Sciences, Chiba University, Chiba, Japan

Jean M. Valery, Ph.D. Assistant Professor, Laboratory of Carbohydrates, Université Pierre et Marie Curie, Paris, France

Leon A. G. M. van den Broek, Ph.D. Senior Research Scientist, Department of Medicinal Chemistry II, N.V. Organon, Oss, The Netherlands

Mark von Itzstein, Ph.D. Professor, Department of Medicinal Chemistry, Victorian College of Pharmacy, Monash University, Parkville, Victoria, Australia

Zbigniew J. Witczak, Ph.D. Assistant Professor, Department of Pharmaceutical Sciences, School of Pharmacy, University of Connecticut, Storrs, Connecticut

Myriam Witvrouw, Ph.D. Research Associate, Rega Institute, Catholic University of Leuven, Leuven, Belgium

1
Carbohydrates as New and Old Targets for Future Drug Design

Zbigniew J. Witczak
University of Connecticut
Storrs, Connecticut

1. INTRODUCTION

The long standing neglect of carbohydrates by the pharmaceutical and academic communities was most likely due to the underestimation of the biological importance of this class of natural compounds. Early considerations of carbohydrates as energy sources, structural components, and protective agents have been slowly refocused on the fact that they have significant diagnostic and therapeutic potential. New scientific evidence has suggested a relationship between carbohydrate structure and many biological functions. Recent developments show that the diversity and complexity of carbohydrates permit these molecules to carry out a wide range of functions. Studies continue unabated as do important, new developments with significant consequences in medicinal chemistry. Research on carbohydrates is now undergoing considerable growth and promises to be a major source of drug discovery leads [1-9]. One feature of the isolation of new carbohydrates from natural sources is that many belong to what, hitherto, were considered to be rare classes of sugars, i.e. amino-sugars, branched-chain sugars, deoxy sugars, thio-sugars, cyclitols including *myo*-inositols and higher sugars. However, it is remarkable how some of the unique carbohydrate properties can evoke new concepts and technologies. The diversity and functional specialization of biologically active carbohydrate molecules seem now to be a hallmark and can be expected to bring about similar revolutions in clinical treatment. Such investigations have led to the discovery of new biosynthetic reactions and enzyme control mechanisms, and contribute to an understanding of certain fundamental processes, for example, in the interaction of cells with their environment and with other cells. The roles of carbohydrates in biological processes and diseases are becoming better understood.

Carbohydrate-containing biomolecules are found on all cell surfaces and because of their inherent structural diversity, many oligosaccharides are information carriers and recognition molecules through linkages with other components such as lipids and proteins. Many studies have revealed that carbohydrates provide signals for protein targeting and serve as receptors for binding toxins, viruses and hormones. They control vital events in fertilization and early development, regulate many critical immune system recognition events and target aging cells for destruction. Cell-cell interactions, such as antigen-antibody interactions and virus-host interactions are classical examples of the aforementioned biochemical functions. Another extremely important, recent discovery in cell-cell adhesion in inflammatory responses is the role of sialyl Lewis^x, (sLe^x) a terminal tetrasaccharide of glycolipids. Sialyl Lewis^x is displayed on the surface of white blood cells and is responsible for the repair of injured tissues. This particular discovery has significant potential in the development of new nonsteroidaal antiinflamatory drugs, as well as anticancer drugs designed to prevent the spread of cancer cells: metastasis.

In metastasis cell surface carbohydrates change upon malignant transformation [10] and are responsible for the significant differences in surface properties between metastatic and nonmetastatic cells [10-12]. It is also well documented that the total and neuraminidase-releasable sialic acid contents of tumor cell surfaces are closely related to the metastatic potential of the tumor cells [13-18]. These important new discoveries are excellent, logical leads as key steps for rational design of anti-cancer agents for the treatment of metastatic tumors. Particularly, the development of specific cancer vaccines to induce an anticancer immune response now appears more feasible. This might also offer alternative treatments to chemotherapy and radiation therapy for metastatic cancers as well.

The ultimate goal of a cancer vaccine design is the generation of antigen (carbohydrates as targets) specific vaccines (active specific immunotherapy ASI) by using chemically well-characterized synthetic antigens as active immunogens. The fact that particular antigens might be selective or, ideally, specific for cancer cells could prove very useful in stimulating antibody production and promoting active immunity against cancers.

Recent advances in monoclonal antibody technology together with rapid progress in synthetic and structural chemistry, have identified and characterized a number of tumor-associated antigens [19-23]. These new important discoveries in molecular biology, immunology, and synthetic carbohydrate chemistry offer great potential for further development of new and diverse cancer vaccines. The immunological and clinical aspects of cancer vaccine developments were reviewed [24].

Presently, there are several established carbohydratc-based products with "biopharmaceutical" applications as well as other new products with potential applications in medicine. Recently purified polysaccharides of bacterial origins, for example have been prepared for use as antigenic vaccines against pneumococcal and meningococcal infections. Due to the ability of certain polysaccharides to cross react with other antisera, they may also provide immunity against other infections.

Other artificial antigens, glycosylated recombinant proteins and immunoadjuvants are emerging as new areas of interest and extensive research.

Other glycobiology-related areas such as neurobiology, neuropathology, and neuropharmacology have investigated the important function of carbohydrates in the understanding of neurobiology. In recent years, neurobiotechnology companies have initiated rapid development of trophic factors and monoclonal antibodies based on recombinant techniques. In this context studies have been initiated of carbohydrates and other glycosylated compounds such as gangliosides, glycosaminoglycans, glycolipids, and lecitins, that may play a significant role in neuropathology or act as therapeutic agents in a wide range of disorders. Gangliosides are one of the leading examples of fast track development. Carbohydrate research in neurobiology may well soon give rise to interesting developments in the treatment of neurological disorders.

These new trends and directions in basic and applied research in multidisciplinary fields associated with the emerging field of glycobiology will provide tremendous opportunities for the development of new carbohydrate therapeutics. Present work suggests that these will include not only complex carbohydrate molecules but relatively simple small monosaccharide molecules as well.

2. CLASSIFICATION

Because of the multifunctionality of carbohydrates and for overall clarity, it is useful to classify biologically active carbohydrates according to their therapeutic activity such as antiinflammatory, anticancer, antidiabetic, anticonvulsant, antibiotic as well as antiviral. Traditional classification would divide the existing drugs or new analogs being developed as potential therapeutics into the following chemical classes of derivatives: mono- and disaccharides, oligosaccharides, and polysaccharides.

3. ANTICONVULSANT AGENTS

An example of a simple sugar with strong biological activity is a new prototype of an antiepileptic drug based on fructose, being developed by R.W. Johnson Pharmaceutical Institute under the generic name **Topiramate**.

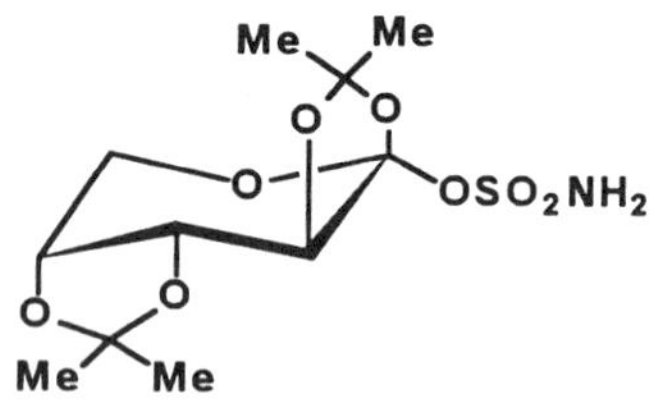

Fig. 1 Chemical structure of Topiramate.

Topiramate, a novel fructose sulfamate, is presently in late-phase clinical trials and has proven to be an effective anti-epileptic drug [25]. The anti-convulsant properties of Topiramate were discovered through biological screening. Recently, a carbocyclic analogue of Topiramate was also synthesized [26]. However, no biological data, were reported.

4. ANTIVIRAL AGENTS

The potent and selective influenza neuraminidase inhibitor, **4-Guanidino Neu-5Ac2en** (K_i viral = 0.1 nm; K_i human = 100 nm) inhibits the replication of both

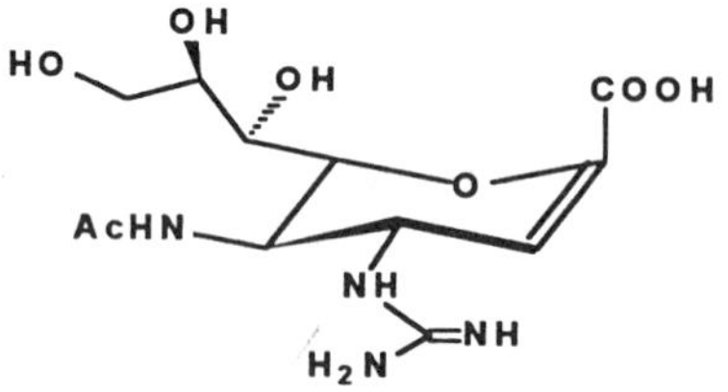

Fig. 2 Chemical structure of **4-Guanidino-NEU5AC2en**.

influenza A and B viruses in cell culture and is efficacious in animal models when administered intranasally [27-28]. A group of researchers headed by von Itzstein [29-31] developed this synthetic derivative currently being tested by Glaxo. Most importantly, the design of this compound is based on protein structure data from a complex of influenza sialidase with an inhibitor [32]. (See Chapter 2 by von Izstein and Kiefel). C-Glycoside analogues of sialic acid previously synthesized by two research groups [33-34] have been recently tested together with new analogs as potential influenza inhibitors [35]. The results are very encouraging for predicting the active site of the receptor as well as biologically active conformations capable of receptor binding. Another category of simple monosaccharides with potentially, multiple biological activities are the azasugars.

Azasugars (see Chapter 11 by van den Broek) belong to the polyhydroxypiperidine and polyhydroxypyrrolidine groups. This class of sugar analogues is represented by a well known family of

Fig.3 Chemical structures of Deoxyfuconojirimycin (left) and α-Homofuconojirimycin (right).

specific glucosidase inhibitors derived from nojirimycin and its reduced form, 1-deoxynoijrimycin, with potential as antidiabetic [36] and antitumor [37] agents. Because some derivatives have exhibited activity against the human immunodeficiency virus (HIV) [38-41], there has been a tremendous effort recently in the search for new azasugars and their analogues [42-51]. Among other azasugars, piperidine analogs with the same secondary hydroxyl groups absolute configuration as fucose are very powerful inhibitors of most fucosidases [52].

Deoxyfuconojirimycin [53-54] and **α -Homofuconojirimycin** [55] both have K_i of 10^{-8} mol dm^3 or less against a number of fucosidases and are the most potent fucosidase inhibitors yet reported. The research group from Organon Int. [56] reported an extensive review of the chemistry of azasugars and their biological activities as potential inhibitors of N-glycoprotein processing glycosidases and HIV-I infections.

The interest in this group of derivatives has been further illustrated in research reports from many active industrial laboratories [57].

Two of the representatives of this class are **Castanospermine** and **Swainsonine** which are drug candidates for treating cancer and viral infections [58-61]. The first one was isolated in 1981 from the seeds of the Australian tree *Castanospermum australe*, and inhibits a number of glycosidases. It was also isolated (1988) from the dried pods of *Alexa leiopetala*. The first synthesis of castanospermine [58] and determination of absolute configuration was reported by Bernotas and Ganem [62]. Castanospermine is also is a strong inhibitor of various intestinal glucosidases.

Fig. 4 Chemical structures of (-) Swainsonine (left) and (+) Castanospermine (right).

The hydroxyl group with the (S)-configuration on C-1 of castanospermine is essential for optimal glucosidase activity just as is the hydroxyl group on C-6 of deoxynojirimycin. The inhibitory effects of various castanospermine derivatives on glucosidases, fucosidases, and mannosidases have demonstrated that alterations of configuration in the piperidine ring as well as removal of the hydroxyl groups lead to significantly weaker α-, and β-glycosidase inhibition. Because of its toxicity, which is probably due to nonspecific inhibition of various glucosidases, castanospermine is unsuitable for therapeutic use in diabetes.

The second representative of bicyclic pyrrolidine derivatives is swainsonine, a powerful inhibitor of α-D-mannosidases which are involved in the biosynthesis of glycoproteins [63]. It may be utilized in cancer chemotheraphy [64-65]. Surprisingly, glucosidases are not inhibited by swainsonine. It also was shown that swainsonine effectively inhibits human B cell development .

Recently, new sulfated polysaccharides from the lentinan and curdlan groups were reported to have high anti-HIV activity and low anticoagulant activity [66]. Similarly, ribofuranan and dextran sulfates also had high anti-HIV activity, however, with simultaneously high anticoagulant activity. It was assumed that this particular effect may be attributed to the flexible backbone containing oxymethylene groups between the glucopyranose rings. It now seems unlikely that sulfated polysaccharides with potent anticoagulant side effects will not be used as future anti-AIDS drugs. New prototypes of semisynthetic oligosaccharides as glycoconjugate vaccines [67] offer potential therapy for many, presently incurable diseases.

5. ANTIDIABETIC AGENTS

5.1. Glucosidase Inhibitors

The discovery of the glucosidase inhibitory activity of 1-deoxynojirimycin has stimulated interest in the development of specific glycosidase inhibitors for study and potential treatment of metabolic disorders such as diabetes [68-70] and lysosomal storage diseases [71-73]. An excellent review of oral antidiabetics including α-glucosidase inhibitors has been recently published [74]. Despite its potent glycohydrolase inhibitory activity, castanospermine is no longer under development as an oral antidiabetic.

α-glucosidase inhibitors [74] are also effective in lowering the post-prandial glycemic rise after carbohydrate ingestion in normal and diabetic patients.They reduce insulin release or insulin requirements and some reduce plasma lipids.They represent an adjunct to dietary therapy of diabetes, either alone or in combination with insulin or other oral hypoglycemic agents. A large number of α-glycosidase inhibitors were originally isolated from cultures of various species of *Streptomyces* and strains of the genus *Bacillus*, but are now synthesized. One of the first synthesized compounds is **Miglitol** (Fig. 5), a hydroxyethyl derivative of deoxynojiramycin, (1,5-dideoxy-1,5-[2-hydroxyethylimino]-D-glucitol, Bay m 1099) which has been used in clinical studies of type II diabetics. It is an effective α-glucosidase inhibitor with a broad spectrum of substrate activity and is fully absorbed and totally excreted by the kidneys. This gives it an advantage over Acarbose, another antidiabetic drug from the pseudooligosaccharides family, which is poorly absorbed. Further long-term phase III studies are underway to establish full benefits of treatments with Miglitol [75-76],

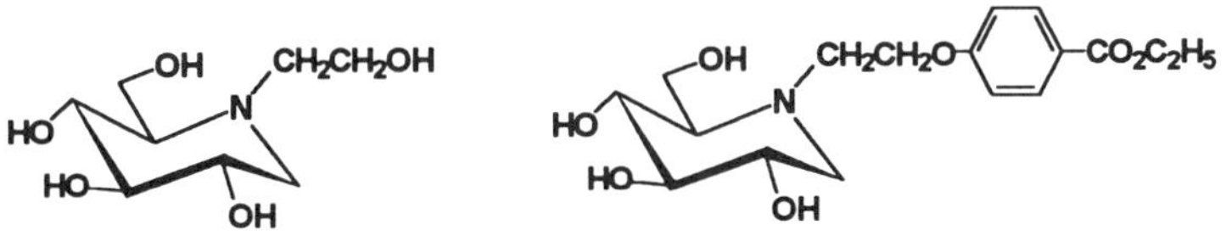

Fig. 5 Chemical structure of Miglitol (left) and Emiglitate (right).

especially in the treatment of refractory type II diabetics. Miglitol is distributed rapidly and heterogeneously to organs and tissues of rats. Miglitol is excreted quantitatively unchanged via urine both after intravenous and oral administration. Most importantly no biotransformation of Miglitol has been observed in rats, dogs or humans.

Another analogue of Miglitol is **Emiglitate** (1,5-dideoxy-1,5-[2-4-ethoxycarbonylphenoxy]ethylimino-D-glucitol) (Bay o 1248). It is more lipophilic than MIGLITOL, however its inhibitory profile is not as useful. The drug is also less well tolerated. Development of the drug after phase II studies was terminated.

Voglibose (A0-128) introduced in 1994 by Takeda Inc., is the first α-glucosidase inhibitor marketed in Japan. Voglibose is obtained by reductive alkylation of valiolamine with hydroxyacetone (Fig. 6). The alternative approach proceeds by reductive amination of the synthetically available inosose with 2-aminopropane-1,3-diol, and subsequent debenzylation. Voglibose, a very simple valiolamine derivative is an extremely potent sucrase (IC_{50} 4.6 x 10^{-9} M) and maltase (IC_{50} 1.5 x 10^{-8} M) inhibitor.

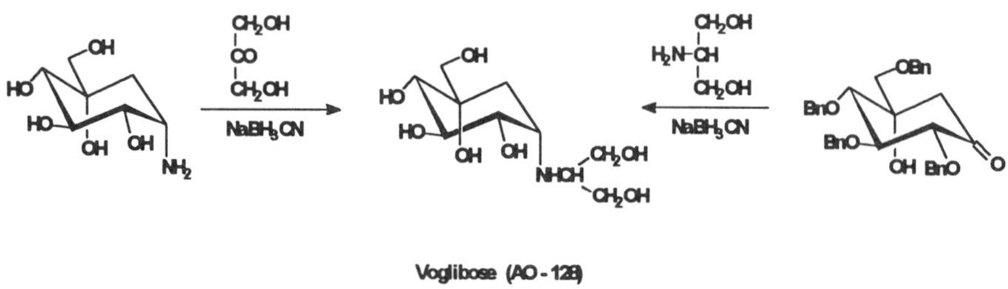

Fig. 6 Synthesis of Voglibose.

The C-disaccharide analogue of nojirimycin, **MDL 73945,** synthesized at Merrell Dow (Strasbourg, France) is a highly potent and selective inhibitor of intestinal α-glucohydrolases that effectively reduces glycemic and insulin responses to a carbohydrate load and has a long duration of action [77].

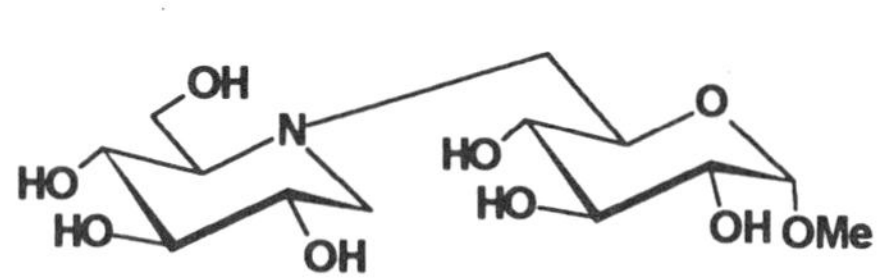

Fig. 7 Chemical structure of new antidiabetic agent MDL 73945.

Because of its uniqueness among the known intestinal α-glucohydrolase inhibitors, this new derivative warrants further evaluation as a drug for reducing post-prandial hyperglycemia in both insulin-dependent and non-insulin dependent diabetes mellitus

(NIDDM).

Among disaccharides and nonclassified carbohydrates with potential antidiabetic activity, one particular class of derivatives is worthy of mention because of its specificity in the inhibition of glucohydrolases. These are relatively simple structures which contain nitrogen in the sugar ring. These azadisaccharides provide an attractive approach to design new leads to antidiabetic agents, especially for the treatment of NIDDM. This class of relatively simple derivatives of aza-sugars are **Glucopyranosylmoranolines** [78] synthesized in Japan in 1978.

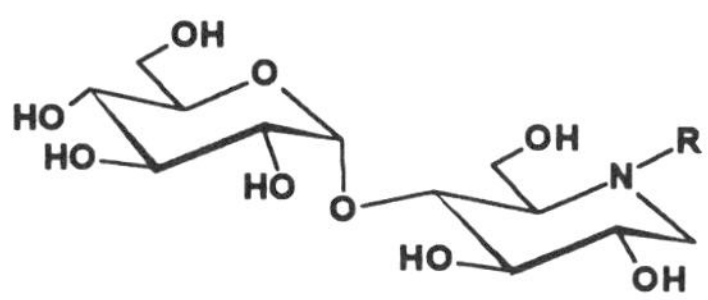

R = $-H$, $-CH_3$, $-C_2H_5$, $-C_3H_7$, $-C_4H_9$,

$BrC_6H_4CH_2-$, $CH_3C_6H_4CH_2-$,

Fig. 8 Chemical structures of Glucopyranosyl moranolines.

The synthesis of new analogs of **Valienamine** derivatives is also reported [79].

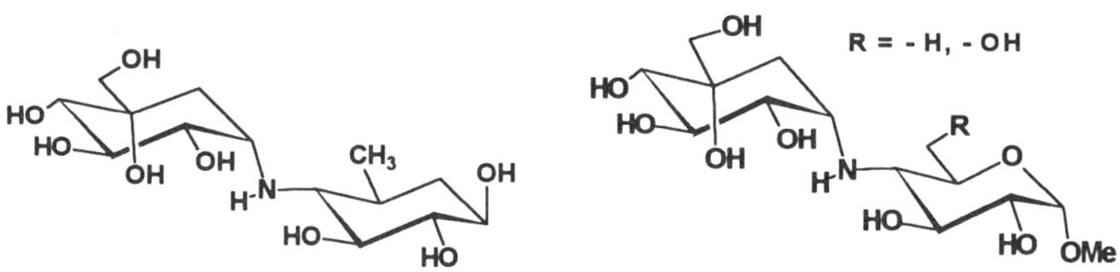

Fig. 9 Chemical structures of Valienamine derivatives.

These amino bridged disaccharides resemble and are similar in structure to well known aminoglycoside antibiotics containing valienamine, a natural carbocyclic amine.

Among the oligosaccharide classes of therapeutics the first α-glucosidase inhibitor **Acarbose, (Precose)** was introduced in 1990 by Bayer AG, Germany under the trade name **Glucobay**. It is the first of a new class of drugs for treating NIDDM and has been recently (1995) approved in U.S.A. by the Food and Drug Administration for patients whose NIDDM can not be managed by diet alone. Acarbose, a complex oligosaccharide isolated from *Actinoplanes*, is reportedly useful as an adjuvant therapy in diabetes [80-82]. The first total synthesis of Acarbose was achieved by Ogawa (see Chapter 10). Acarbose provides physicians with the first new therapeutic class for the treatment of diabetes in nearly 40 years.By inhibiting α-glucosidase,Acarbose,delays carbohydrate metabolism in the gastrointestinal tract and modulates changes in food induced blood sugar levels. Acarbose is a potent, fully competitive inhibitor of glucoamylase, sucrase, maltase and isomaltase activity in the small intestine of various species. Particularly, it reversibly and competitively inhibits two enzymes involved in carbohydrate digestion i.e. pancreatic a-amylase and membrane bound intestinal α-glucosidase. Pancreatic α-amylase hydrolyzes complex starches to oligosaccharides in the lumen of the small intestine, whereas a-glucosidase hydrolyzes oligosaccharides, trisaccharides, and disaccharides to glucose and other monosaccharides in the brush border of the small intestine. Most importantly, Acarbose does not inhibit lactase, so it is not expected to cause lactose intolerance.

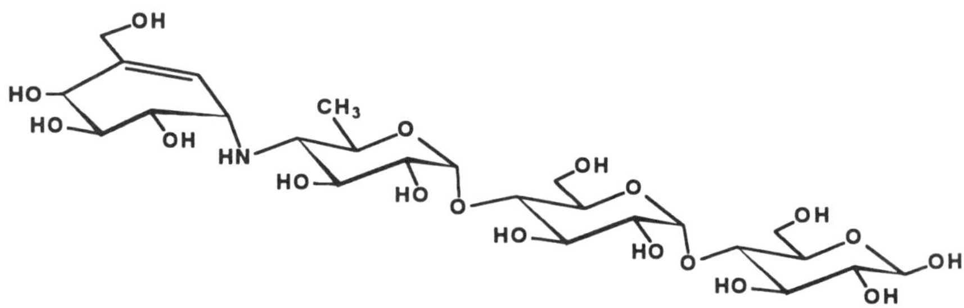

Fig. 10 Chemical structure of Acarbose.

Acarbose may be used as single therapy or in combination with a sulfonylurea when diet plus either Acarbose or a sulfonylurea do not result in adequate glycemic control.

The evidence from the clinical evaluation of α-glucosidase inhibitors, much of which is based on Acarbose, shows that they can be effective therapeutic agents in both type I and especially in type II diabetes. A number of β-glucosides of Acarbose have been prepared which are hardly more effective than Acarbose itself.

5.2. Higher Pseudo-Oligosaccharides

The most common representatives of this class of potential antidiabetics are derivatives of Acarviosin formed as by-products of the fermentation of Acarbose. Other examples are the Terestatines A, B, C and the inhibitor A-5662 of which Acarviosin is the central unit. However, none of these high-molecular inhibitors have reached clinical studies or market.

5.3. Carbohydrate Biguanides

An interesting group of potential antidiabetic drugs is being developed by the R.W. Johnson Pharmaceutical Research Institute [83-84]. Monosaccharides containing a biguanide functionality exhibit significant hypoglycemic activity, however they do not lower fasting blood glucose as compared to the reference drug Metformin. Two of the most active compounds are glucose analogs with a C-6 biguanidine group in which the hydroxyls are substituted with alkyl moieties. This is a classical example of important biological activity resulting from the incorporation of a guanidine moiety into various types of sugars.

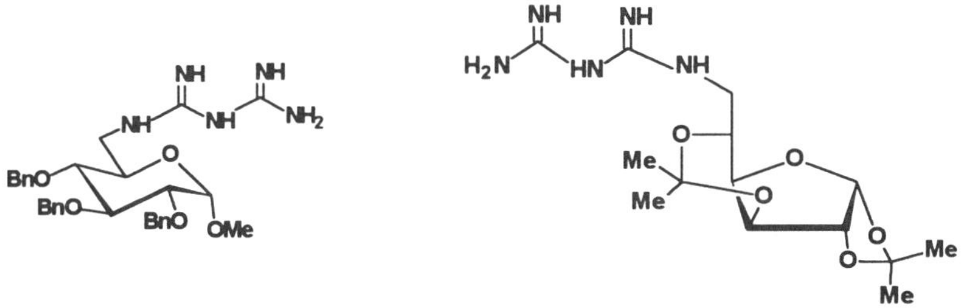

Fig. 11 Chemical structures of new biguanidine carbohydrate antidiabetic agents.

Interestingly, this group of natural and modified carbohydrates is one of the most important leads in the development of carbohydrate based drugs. Clinical studies are underway to verify and evaluate the expected paharmacological activities.

5.4. Inositol Phosphates

Recent advances in molecular and cellular biology have demonstrated that the receptor-controlled hydrolysis of phospholipids leading to intracellular second messengers is a fundamental and widespread mechanism for the transmission of signals across cell membranes, i.e., signal transduction. It is now clear that inositol phosphate esters play a key role in the linkage of large numbers of surface glycoproteins and cell membranes [85] through the glycophosphatidyl inositol (GPI) anchor molecule. Moreover, phosphooligosaccharides containing an inositol phosphate moiety of similar structure to the GPI may have insulin mimetic activity and could be the intracellular second messengers of insulin [86], possibly derived from the GPI. Moreover, inositol phosphate glycan fragments produced from GPI anchors can reproduce the actions of insulin and the insulin sensitive inositol glycan on fat and liver cells. Considering that the exact structure of the GPI anchor is known, the design and synthesis of insulin-mimetic compounds are logical steps to exploit the above class of compounds. This discovery has stimulated research for antidiabetic drugs structurally related to GPI for the treatment of non-insulin dependent Diabetes Mellitus (NIDDM). Two excellent, recently published books [87-88] and an extensive review [89] deal with synthetic and biological aspects of inositol phosphates. Two chapters in this book review the specialized study of chiral phosphates of inositol (Chapter 8 by Ozaki and Lei) and phospahatidylmyo-inositol (Chapter 9 by Bruzik).

6. ANTITUMOR AGENTS

One important growth factor signaling pathway also utilizes inositol phosphates to convey information inside the cell. When an oncogene is over-expressed or constitutively activated through mutation, the cell receives a continuous signal to grow. This mechanism suggests the possibility of controlling oncogenesis by inhibiting the signaling pathways and mediating the effects of activated oncogenes. This is a new and mechanistically based approach to develop new classes of drugs that possibly can control cancer cell growth [90-91]. The race is on to develop a simple analog of inositol phosphate derivatives with the ability to prevent formation and growth of cancer cells of any form. Interestingly, derivatives of *myo*-inositol substituted at the 3-position selectively inhibited the growth of cells in the absence of physiological concentrations of *myo*-inositol [92]. Two potential inhibitors of cancer cell growth *in vitro,* 3-deoxy-3-fluoro-*myo*-inositol and 3-deoxy-3-fluoro-phosphatidylinositol [93] (Fig. 12) clearly demonstrate this approach to the control of cell proliferation.

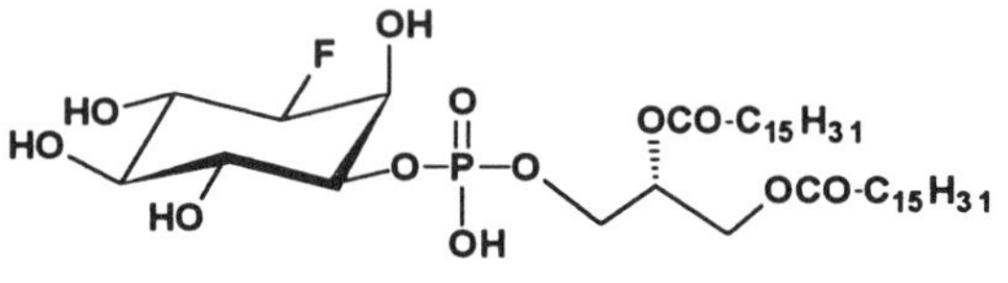

Fig. 12 Chemical structure of 3-deoxy-3-fluoro-phosphatidyl inositol.

Among the simple monosaccharides, two important aminosugars with antibiotic activity are well known, i.e. **Streptozotocin** and **Prumycin**.

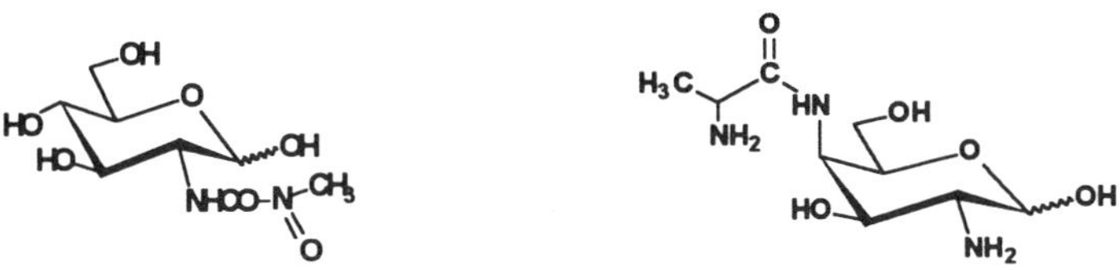

Fig. 13 Chemical structures of Streptozotocin (left) and Prumycin (right).

Streptozotocin [94] is important only as an antitumor and diabetogenic agent, owing to its specific toxicity to the b-cells of the islets of Langerhans. It is clinically useful in the therapy of malignant insulinomas and Hodgkins' disease. However, this antineoplastic antibiotic is highly toxic when administered orally. Prumycin has antitumor activity, but is less active than Streptozocin. Both antibiotics, however are used only in rare critical cases when alternative theraphy can not be administered.

6.1. Sialyl Lewis^x (sLe^x) mimics

The trisaccharide determinant 3-O-a-fucosyl-N-acetyllactosamine known as the Lewis^x antigen, has been found both in glycoprotein and in membrane glycolipids of mammalian cells [95]. The glycan with an α-(2-3) or α-(2-6)-sialylated galactose moiety known as sialyl Lewis^x (sLe^x) is a widely distributed ligand for glycoprotein receptors (lectins) known as selectins L, E, P. The biochemical mechanism of the cell-adhesion process involves recognition and binding of carbohydrate ligands like sLe^x. It is located on leukocyte surfaces and binds to glycoproteins which are adhesion molecules (selectins L, E, P) expressed on the endothelial cell surface during the inflammation process. One of the ligands recognized by E-selectin is sLe^x. Two excellent chapters discuss the potential utility of sialic acid (see Chapter 2 by von Itzstein) and sLe^X (see Chapter 3, by Roy).

Potential therapeutic approaches include using sLe^x itself or simplified structures based on sLe^x to bind to endothelial cell receptors and thereby prevent leukocyte adhesion. The chemical [96-99] and enzymatic [99-101] synthesis of sialyl Lewis^x was recently reported. Conformational studies of sLe^x by NMR in an aqueous solution [102] revealed three-dimensional geometry by complete assignment of ^{1}H and ^{13}C resonances. This was further supported by MM2 calculations. The sulfate containing an oligosaccharide ligand for the human E-selectin molecule has also been described [103-104]. The total synthesis of sulfated Le^x and Le^a type oligosaccharide selectin ligands was recently reported [105].

Two novel disaccharides composed of sialic acid and L-fucose linked *via* C-2-4 have been recently synthesized [106]. Both blocked the adhesion of a sialyl Le^x expressing myeloid cell line U937 to activated endothelial cell cultures at a 25 and 30 fold higher concentration than sLe^x.Interestingly, a bivalent sLe^x anchored on a galactose residue *via* α-1,3- and α-1,6-linkages prepared from chemically synthesized trisaccharide and enzymatically glycosylated, was about 5-fold better than sLe^x [107].

This particular discovery also indicates that the sLe^x moiety could be replaced by relatively simple synthetic analogs. Another important characteristic of the E selectin-sialyl Le^x interaction is its multivalency, which stimulated to the synthesis of N-glycopeptide clusters with $Lewis^x$ antigen side chains [108].

A long-range target is the replacement of the sLe^x moiety with mimetics that are effective when taken orally, have better pharmacokinetics, and of course are metabolically more stable than the previously developed analogs. Finally, since sLe^x induced leukocyte adhesion could also be blocked by inhibiting the corresponding enzymes responsible for the biosynthesis of this oligosaccharide group, intensive efforts are underway to develop such inhibitors [109-124]. One of the approaches used recombinant human 1,3-fucosyltransferase (α-1,3FucT) [109] for the preparation of a new family of analogs of fucosylated oligosaccharides as potential 1,3-fucosyltransferase inhibitors. When tested in the presence of five-member pyrollidine analogs of fucose, a profound synergistic activity of inhibition of 1,3-fucosyltransferase was observed in the presence of guanosine diphosphate (GDP) and these azasugars analogs of fucose. Normally, pyrrolidine analogs of fucose are generally less potent inhibitors of fucosidases than the piperidine analogues [113-114]. This discovery will be useful in the design of further fucosidase inhibitors and antitumor therapeutics in general.

Crucial to the new development of this new family of carbohydrate mimetics is the determination of the biologically active conformation of the carbohydrate ligand when bound to selectin. From recently reported data [107], it seems that the structural elements of the sLe^x unit which are essential for recognition by E selectin include the carboxyl function of sialic acid, two hydroxyl functions C-4 and C-6 of the galactose residue and the three hydroxyl functions of the fucose moiety. Interestingly, the methyl group at C-5 of this L-sugar unit is not essential for the activity and can be replaced with an arabinose unit as proposed by the authors [107]. This finding, together with the modification of the fucose moiety, has been explored by the synthesis of the carbocyclic analog of GDP- fucose [110], GMP fucose derivatives [112], the C-trisaccharide of Le^x [115] and bicyclic α-L-homofucose [116]. Enzyme inhibition studies of a-L-homofucose showed no significant inhibition effect on all tested human liver fucosyltransferases. The thio analog of L-fucose [112], one of a few very powerful inhibitors of α-L-fucosidase, is a promising target for further development of the above class of inhibitors. A new class of (1-4)-linked, S-thiodisaccharides

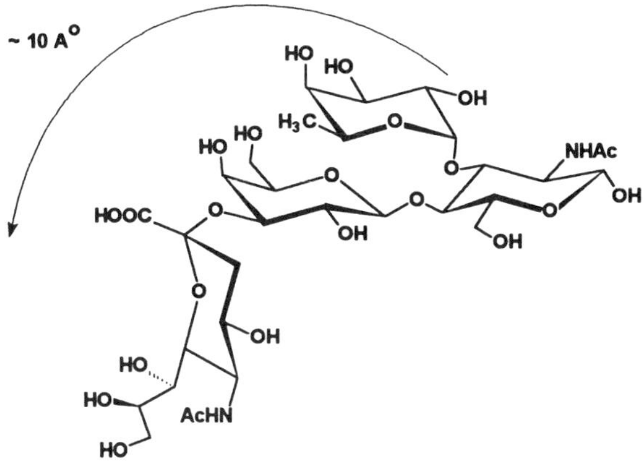

Fig. 14 Active binding site and conformation of sLe^x.

containing 5-thio-L-fucose and L-fucose units [123-125] were synthesized and tested against L-fucosidase from bovine kidney and bovine epididymis revealing a mixed mode of inhibition at K_i = 4.9 mM and K_i = 3.5mM respectively. Preliminary data [124] reported that a 5-thio analog of (1-2)-linked thiodisaccharides showed a potent competitive inhibition of K_i = 30mM against α-L-fucosidase from bovine epididymis.

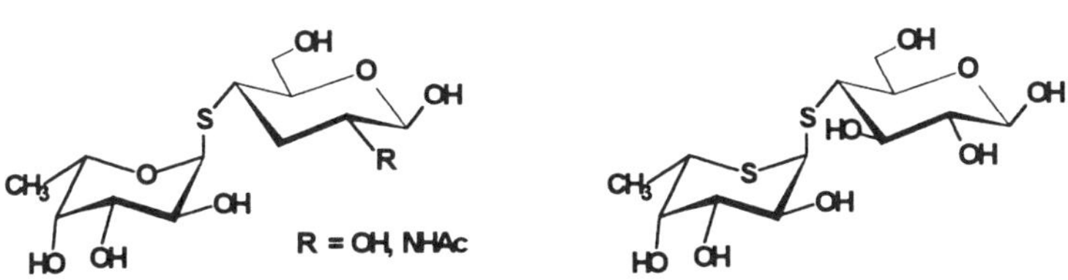

Fig. 15 Chemical structures of new thio-disaccharides as potential inhibitors of L-fucosidase.

6.2. Polysaccharides

Although the polysaccharides are one of the most important classes of natural products, their significance as purely therapeutics is limited. Only some nontoxic antitumor polysaccharides, derived from bacteria, fungi, and algae have demonstrated importance. Good antitumor activity is observed in the case of (1-3)-β-D-glucans having a α-D-glucopyranosyl group linked (1-6) to every 3rd or 4th residue of the main chain. In some cases the (1-4) linkages also exist. Lichen glucans generally have β-D-(1-6) linkages only. Glucomannans, and mannans have both β-D-(1-6) and β-D-(1-2) linkages. The molecular weights of the glucans vary from 10,000 to 1,000,000, with averages of 50,000 to 100,000.

Some polysaccharides have antibacterial and/or antifungal activity. Generally, most of the polysaccharides of this group have host-mediated antitumor activities. These types of polysaccharides do not affect the tumor cells directly, but after administration they stimulate the reticuloendothelial system of the host to produce antitumor substances indirectly, resulting in a host-mediated action. Moreover, because of the lack of any cytotoxicity, they are nontoxic to the host. The nontoxic β-glucan polysaccharides with antitumor activity were reviewed [126].

One important polysaccharide from the above β-glucan family is **Schizophylan** introduced in 1985 by Kaken, Japan under the trade name **Sonifilan**. Schizophylan is an immunostimulant related to Lentinan, isolated from the culture filtrate of *Schizophyllum commune*. It is useful in combination with other antineoplastic treatments in the managements of carcinomas of the lungs, stomach, uterus and breasts. Lentinan, a purified β-D-glucan polysaccharide of 500.000 M.W. extracted from *Lentinus edodus,* is composed of a linear β-(1-3)-glucan with β-(1-6)-linked branches (DB = 1/12.5). There is also some structural evidence of branches of greater than one glucosyl unit and some internal β-(1-6)-linkages within the main chain. Lentinan is used in combination with other chemotherapeutic agents for breast, colorectal, gastric, prostate and pancreatic cancers.

Other antitumor fungal (1-3)-glucans related to Schizophyllan were reviewed [127]. Interestingly, branched (1-3)-β-D-glucans with D-arabinosyl or D-mannosyl branches exhibited high antitumor activity, whereas branched (1-4)-D-glucans with similar side chains had no antitumor activity [128].

Among, the most important *in vivo* effects of β-D-glucans, are stimulation of antitumor responses, stimulation of cytokines, stimulation of wound healing, stimulation of resistance to microbial challenge and stimulation of hematopoesis. Whereas studies of mechanisms of glucan interactions with the immune system and *in vivo* efficacy are in progress, the precise structural features which interact with the immune system receptor(s) and the receptor(s) to which β-glucans bind have not been defined. Long term phase III clinical studies of a genetically modified *Saccharomyces cervisiae*, β-(1-3)-glucan (**PGG-glucan Betafectin**) developed by Alpha-Beta Technology are underway. The targeted clinical application of Betafectin is its specific efficacy in preventing postoperative infections in patients undergoing major thoracic or abdominal surgery. PPG-glucan, also enhances the antimicrobial activity of human peripheral blood monocytes and neutrophils.

Another class of complex carbohydrates is represented by **Dextran,** with a molecular weight lower than 70,000 which has been used extensively as a plasma expander. Plasma expanders are non-antigenic and are degraded *in vivo* by means of a dextranase found in the liver. These aqueous polymer solutions are capable only of replacing the blood volume and cannot carry oxygen.

An additional polysaccharide, starch transformed into the water soluble hydroxyethyl starch (HES) derivative, has attracted interest as a plasma substitute, since it can have a significant intravascular half-time depending on the substitution ratio in hydroxyethyl groups that hampers α-amylase attack. HES has been shown to be even better tolerated than **Dextran**.

The other classical representative of the above class of polysaccharides is **Sodium Cellulose Phosphate** developed for the treatment of type I absorptive hypercalciuria.

Glycosaminoglycans

Among these derivatives, some glycosaminoglycans have been associated with neuropathology, whereas others may be used as therapeutic agents. One of the classical neuropathological examples is amyloid plaques, which are protein deposits with a highly unique peptide structure and sequence that are found in a number of diseases such as Alzheimer's disease, Down's syndrome, and diabetes. One therapeutic class of the above derivatives is the sulfated glycosaminoglycans such as heparan sulfate (see Chapter 6 by Alban and Chapter 7 by Linhardt and Toida) which may influence the specific amyliod conformation that determines the final location of the deposit.

The glycosaminoglycan known as **ADCON-L,** which is a component of a large proteoglycan is being developed by Gliatech, a Cleveland based company. The glycan is normally secreted by glial (oligodendrocytes, astrocytes, and ependymal) cells after injury to nerve tissue. Although it does not affect healing, the glycan component prevents the formation of specific scar tissue after disc surgery. The process of the formation of scar tissues causes a painful disease known as epidural fibrosis. The potential application of **ADCON-L** is in the prevention of adhesion in abdominal, pelvic and heart surgeries.

Gangliosides

Gangliosides are naturally occurring sialic acid containing glycosphingolipids structurally defined as hematosides, lacto-, neolacto-, globo-, and ganglio-series gangliosides. They are found in the tissues of the central nervous system as principal membrane constituents of vertebrate cells. Their functional significance still remains obscure. However, numerous experimental data suggest that apart from their basic functions as structural membrane components, these glycolipids play an important role in receptor-mediated signal transduction, as well as direct cell/cell and cell/extracellular matrix interactions. These natural products appear to have anti-excitotoxic effects that may be helpful in protecting against damage caused by stroke, trauma, and epilepsy. Gangliosides also demonstrate pro-neurotrophic effects that may help in the treatment of diabetic neuropathy, Alzheimer's as well as Parkinson's diseases. Recently, several gangliosides have been studied and evaluated in clinical trials. Fidia Company, in Padua, Italy is developing preparations of GM-1 ganglioside for cerebrovascular accidents. Interestingly, the functionality and potential of the targeted GM-1 ganglioside are very promising. Clinical trials demonstrate that GM-1 considerably improves the function of patients with previously paralyzed muscles if administered within 72 hours of injury. There is some evidence to suggest that GM-1 could be effective if administered even years after injury. The ganglioside family offers promising opportunities for new therapies and is currently under fast track development.

7. ANTIBIOTICS

This class of useful and potential therapeutics has gained attention with the appearance of complex nucleoside antibiotics which exhibit a variety of biological activities including anthelmintic, antifungal, antiviral, antitumor, herbicidal and insecticidal. The above class of natural products has been reviewed by Garner [129] and more recently by Knapp [130].

An interesting example of the thioglycoside moiety occurring in the linkosaminide family of antibiotics [131] is represented by **Lincomicin**. Linkosaminide antibiotics are characterized by an alkyl 6-amino-6,8-dideoxy-1-thio-α-D-galacto-octa-pyranoside joined with a proline ring by an amide linkage. 1962 marked the discovery of biological properties of the above family of antibiotics [132]. **Clindamycin** and **Celesticetin** are two other representatives. Modification of the sugar moiety of Linkomycin by the introduction of a chlorine at C-7 in the reaction with thionyl chloride produced the new analog Clindamycin [133]. Celesticetin is the third analogue with a modified thiosugar moiety at C-1. Clindamycin is utilized in the treatment of common infections caused by gram-positive cocci. It is also highly efficacious in the treatment of anaerobic infections including actinomycosis. Chapter 15 by Achmatowicz and Szechner deals exclusively with the chemistry of these derivatives.

Another representative of natural thiosugars is a new family of eneydiyne antibiotics (isolated in 1980) consisting of Esperamycin, Calichemicin, and Dynamycin A. Enediyne antibiotics [134] are among the most potent antitumor agents known displaying *in vitro* and/or *in vivo* activity at a ng/ml level. The natural thiosugar moiety

moiety present in the representative Esperamicin A_1 from this family is chemically 2, 4, 6,-trideoxy-4-methylthio-α-D-ribo-hexopyranose.

7.1. Aminoglycoside Antibiotics

The aminoglycoside class of antibiotics is one of the most important and useful among traditionally recognized carbohydrate therapeutics [135-137]. A majority of the aminoglycosides are produced by *Actinomycetes.* A few new representative members from this class have been recently introduced to the market, among them: **Arbekacin** by Meiji Seika, Japan, under the trade name **Habekacin**. Abekacin is a semi-synthetic derivative of dibekacin useful in the treatment of bacterial infections [138-140].

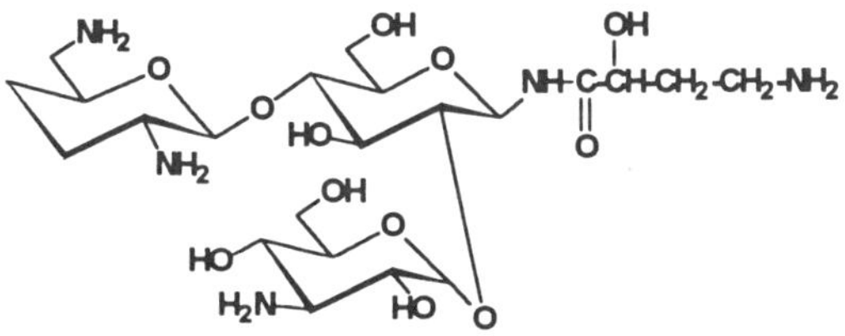

Fig. 16 Chemical structure of Arbekacin (Habekacin).

This aminoglycoside is active against a broad spectrum of bacteria, including some of the gentamycin, kanamycin, and tobramycin resistant pathogens. Compared to well known Amikacin and Dibekacin, ototoxicity is reportedly milder.

7.2 Macrolide Antibiotics

This class of antibiotics is represented by new analogs possessing relatively the same or similar sugar moieties as previously reported representatives such as Erythromycin. Macrolide antibiotics are produced as secondary metabolites of soil microorganisms. The majority have been produced by various strains of *Streptomyces.* Macrolide antibiotics are classified according to the size of the macrocyclic lactone ring forming the aglycone, either as 12-, 14-, or 16- membered ring macrolides. They are polyfunctional molecules and the majority of macrolides contain at least one aminosugar moiety.

One of the new macrolide antibiotics is **Clarithrmycin** originating in Japan and introduced by Abbott under the trade name **Klacid**. It is an acid stable antibiotic

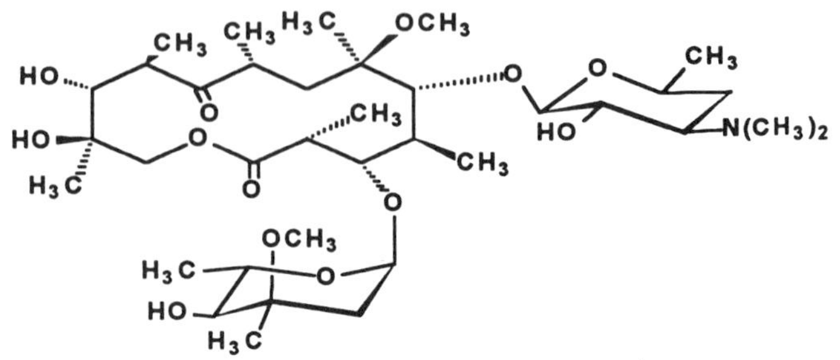

Fig. 17 Chemical structure of macrolide antibiotic Clarithromycin.

indicated for use in the treatment of skin, urinary, and respiratory tract infections [141-143]. Compared to erythromycin, Clarithromycin exhibits almost the same *in vitro* activity against conventional pathogens, but is far better tolerated due to fewer gastrointestinal problems.

7.3. Avermectins Antibiotics

These types of natural products have excellent insecticidal activity [144]. The complex structural diversity of Avermectins is an important factor for further development of new analogues of this particularly interesting class of antibiotics [145].

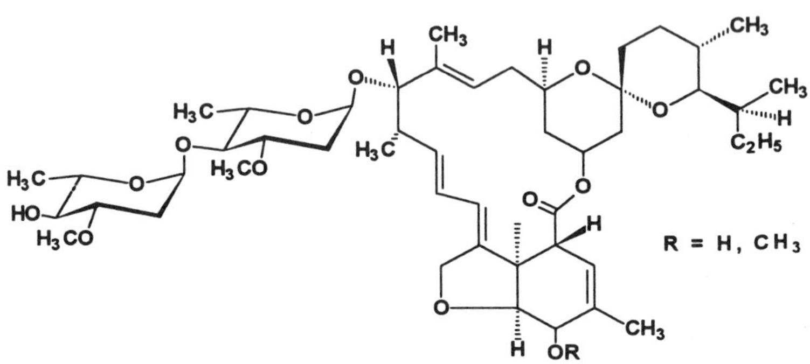

Fig. 18 Chemical structure of Avermectins.

New synthetic and biochemical studies revealed that introduction of an amino or aminomethyl function in C-4" of L-oleandrose, significantly increased insecticidial activity.

7.4. Oligosaccharide Antibiotics

This class of carbohydrates with very complex structures and many asymmetric centers is classified into four categories: **Everninomycins, Curamycins, Avilamycins** and **Falmbamycins** [146] (Fig. **19**). Everninomycins are the first among this class of antibiotics to have their precise structures determined and are classified as B, C, D, and D1. Everninomycin B and D are two main oligosaccharides of the antibiotic complex produced by *Micromonospora carbonacea.* They are active against a variety of strains of *Staphylococcus, Streptococcus, Bacillus* and *Mycobacteria,* but have no activity against *Enterobacteriace* or *Pseudomonas.* Flambamycin is produced by *Streptomyces hygroscopicus.*

The structure-activity relationships in this particular group of unusual antibiotics are not completely understood. The presence of the free phenolic hydroxyl group and ortho ester linkages, however, are essential for antibiotic activity.

8. ANTICOAGULANT AGENTS (HEPARIN ANALOGUES)

This group of carbohydrate therapeutics is a constant source of attention and research and new developments continue to be reported. Two excellent chapters (Chapter 6 by Alban and Chapter 7 by Linhardt and Toida) discuss this class of therapeutics with emphasis on currently used agents as well as future possibilities. Among significant new discoveries it is worthy to mention the improvement of biological properties, especially prolongation of the half-life in plasma, as well as reinforcement of the specific activity of new O-acylated analogues [147-148]. The highly functional heparin pentasaccharide that binds antithrombin is another example of fast track developments. New methods of chemical and enzymatic depolymerization and modification of heparin [149] are particularly promising in the development of highly functional analogues of pentasaccharide as new generations of anticoagulant agents.

9. MISCELLANEOUS AGENTS

Among the large number of unclassified carbohydrate therapeutics (mainly mono- and disaccharides) with a wide spread of biological activities one particular class of antirheumatic agents should be mentioned. An interesting representative of this class

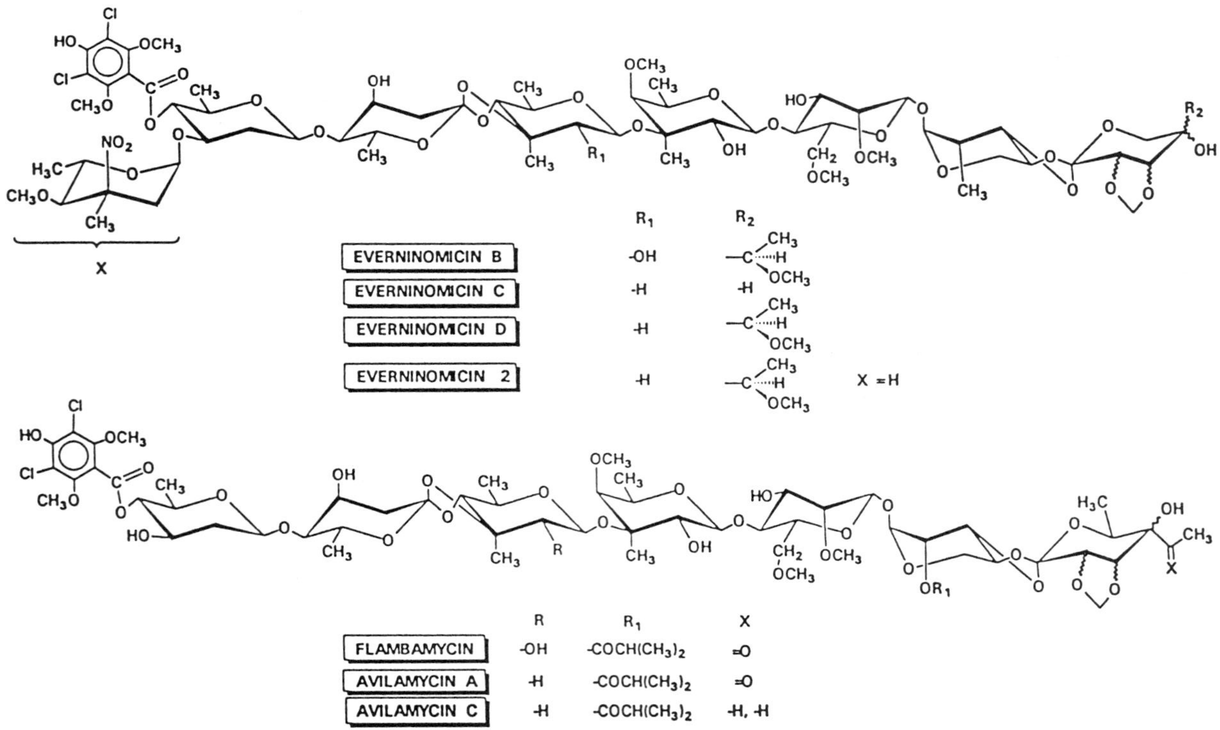

Fig. 19 Chemical structures of Oligosaccharide Antibiotics: Avilamycins, Flambamycin, and Everninomicins.

of derivatives is **Auranofin** (Fig. 20), used clinically for the treatment of chronic rheumatoid conditions such as rheumatoid arthritis. This compound has been reported to inhibit IL-1 synthesis [150]. It was introduced by Smith, Kline & French in 1983 under trade name **Ridaura** [151]. It is the first orally effective derivative of gold to

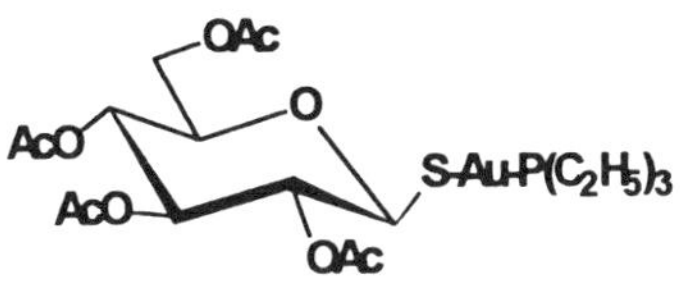

Fig. 20 Chemical structure of Auranofin (Ridaura).

be marketed. It is better tolerated and more convenient than gold sodium thiomalate which is administered intramuscularly. Aurothioglucose, is also produced by Schering Corporation under the trade name Solganal. This a-thiocompound must be given by injection to be effective.

Other examples of synthetic carbohydrates are Amiprilose and its hydrochloride salt, Therafectin. Both are known to exhibit antiproliferative and antiinflammatory activities [152]. Amiprilose acts as an immunomodultor [153] and therefore has a therapeutic effect on autoimmune disorders such as arthritis, psoriasis, eczema, and systemic lupus erythematosus [154-156]. It has low toxicity and no serious side effects [157-158] but is required in large doses for effective therapy. This presents a problem, especially for oral administration because treatment of inflammatory or autoimmune disorders is often chronic. A second Terafectin generation compound for the treatment of rheumatioid arthritis, GW 80126 is being developed by Greenwich Pharmaceuticals and has entered its six month Phase II studies.

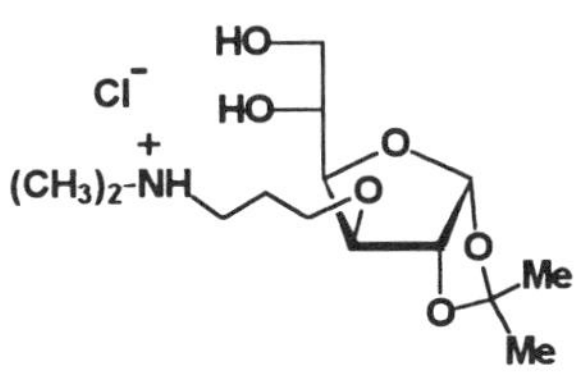

Fig. 21 Chemical structure of Amiprilose (Therafectin).

The recent reports [159-160] on the synthesis of new 6-thio analogs of Amiprilose showed that they are approximately 5-1000 times more potent than the parent compound.

One of the earliest thrapeutically useful disaccharides, still in use as a therapeutic, is **Lactulose**. Lactulose is used in the treatment of hepatic coma and chronic constipation.

Another classical example of the derivatives from the above class of sugars is **Sucralfate,** an aluminum hydroxide complex of sucrose sulfate (Fig. 22) currently

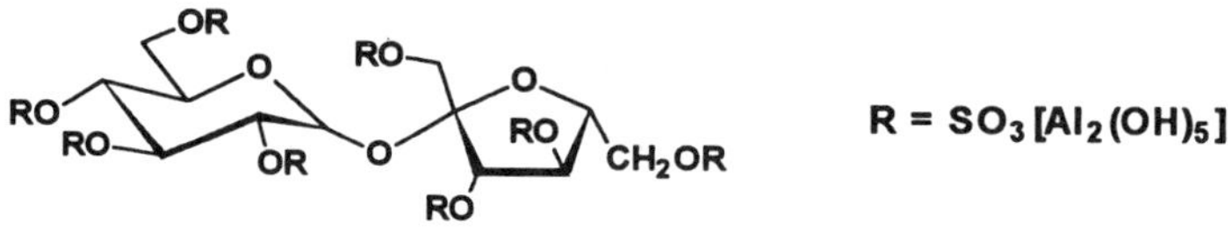

Fig. 22 Chemical structure of Sucralfate.

being used in the therapy of duodenal ulcers [161]. Introduced in 1970 in Japan, it acts by selectively binding to necrotic ulcer tissue, thus counteracting further destruction to the mucosa.

10. CONCLUSION

In this short overview, it is quite apparent that carbohydrates play diverse and crucial roles in a wide variety of biological systems. As the technology for the synthesis and analysis of simple and complex carbohydrates advances, the number of examples of oligosaccharides which play specific functional roles in biology will surely increase.

Modern synthetic medicinal chemistry has clearly made a major contribution toward improved synthetic or enzymatic production of many complex and simple carbohydrate molecules for their evaluation as potential therapeutics. Moreover, advancements in selective glycoside bond formation, especially enzymatically *via* cloned glycosyltransferases will definitely play an extremely important role in the search for new agents. This and other anticipated discoveries will definitely speed up new leads in carbohydrate based therapeutics. Lead candidates are already apparent in a few cases such as sialyl Le^x analogues as potential antiinflammatory and anticancer agents and sialyl glycosides as potential antiviral agents against the influenza virus or new analogues of azasugars as anti-HIV, antiinflammatory, and antitumor agents.

Another class of potential candidates, are nontoxic analogues of (1-3)-glucan polysaccharides with considerable immunostimulant activity. Also neuropharmaceuticals from the gangliosides family, the most promising being GM-1 (for cerebrovascular accidents) are being investigated in clinical trials.

New insights from this field will increase our understanding of many biological processes and suggest new approaches in the diagnostics and treatment of human diseases, including AIDS, Alzheimer's disease, diabetes and metastatic cancer. With favorable clinical results for many of these new, potential carbohydrate therapeutics already heralding the advent of new clinical studies, the future of carbohydrate based therapeutics appears full of challenges and enormous promise.

REFERENCES

1. Bock K, Clausen H, eds. Complex Carbohydrates in Drug Research Structural and Functional Aspects. Alfred Benzon Symposium 36, Copenhagen: Munksgaard, 1994.
2. Garegg PJ, Lindberg AA, ed. Carbohydrate Antigens, ACS Symposium Series No. 519, ACS. Washington, D.C. 1993.
3. Esen A, ed. β-Glucosidases: Biochemistry and Molecular Biology. ACS Symposium Series No. 533, ACS Washington, D. C. 1993.
4. Ogura H. Hasegawa A. Suami T, eds. Carbohydrates-Synthetic Methods and Application in Medicinal Chemistry. Kodansha, Tokyo, & VCH Weinheim, 1992.
5. Varki A. Biological Roles of Oligosaccharides: All of the Theories are Correct? Glycobiology. 1993; 3: 97-130.
6. Schnaar RL. Complex Carbohydrates in Drug Development. In: Adv. in Pharmacology, 1992; 23: 35-84.
7. Musser JH. Carbohydrates as Drug Discovery Leads. In: Ann. Rep. in Med. Chem. 1992; 27: 301-310.
8. Musser JH. Trends in New Lead Identification: A structural Approach. In: Wermuth CG, ed. Medicinal Chemistry for the 21st Century. Blackwell Scientific Publications, 1992: 25-38.
9. van Boeckel CAA. Some Recent Application of carbohydrates and Their Derivatives in the Pharmaceutical Industry. Recl Trav Chim Pays-Bas. 1986; 105: 35-48.
10. Alhadeff JA. Malignant Cell Glycoproteins and Glycolipids. In: Crit Rev Onc Hemt. 1989; 9: 37- 58.
11. Matsusako T, Muramatsu H, Shirahama T, Muramatsu T, Ohi Y. Expression of Carbohydrate Signal, Sialyl Dimeric Le^x Antigen, is Associated with Metastatic Potential of Transitional Cell Carcinoma of the Human Urinary Bladder. Biochem Biophys Res Commun. 1991; 181: 1218-1222.
12. Dennis JW, Laferte C, Waghorne ML, Breitman ML, Krebel RS. a-1-6 Branching of Asn-Linked Oligosaccharidesis Directly Associated with

Metastasis. Science. 1987; 236: 582-585.
13. Bosmann HB, Bieber GF, Brown AE, Case KR, Gersten DM, Kimmerer TW, Lione A. Biochemical Parameters Correlated with Tumor Cell Implantation. Nature. 1983; 246: 487-489.
14. Yogeeswaran G, Stein BS, Sebastian H. Altered Cell Surface Organization of Gangliosides and Sialoglycoproteins of Mouse Metastatic Melanoma Variant Lines Selected in Vivo for Enhanced Lung Implantation. Cancer Res. 1978; 38: 1336-1344.
15. Tao TW, Burger MM, Non-metastasising Variants Selected from Metastasising Melanoma Cells. Nature. 1977: 270: 437-438.
16. Yogeeswaran G, Salk P. Metastatic Potential Is Positively Correlated with Cell Surface Silylation of Cultured Murine Tumor Cell Lines,. Science. 1981; 212: 1514-1516.
17. Matsumoto H, Muramatsu H, Muramatsu T, Shimazu H. Carbohydrate Profiles Shown by a Lecitin and a Monoclonal Antibody Correlate with Metastatic Potential and Prognosis of Human lung Carcinomas. Cancer. 1992; 69: 2084-2090.
18. Miyake M, Taki T, Hitomi S, Hakomori S. Correlation of Expression of H/Le^y /Le^b Antigens with Survival in Patients with Carcinoma of the Lung. N Engl J Med. 1992; 327: 14-18.
19. Muramatsu T. Carbohydrate Signals in Metastasis and Prognosis of Human Carcinomas. Glycobiology, 1993; 3: 294-296.
20. Fukuda M. Carbohydrate-Dependent Cell Adhesion. Biorg & Med Chem. 1995; 3: 207-215.
21. Lowe JB. Carbohydrate Recognition in Cell-Cell Interaction. In: Fukuda M, Hindsgaul O, eds. Molecular Glycobiology. IRL Press. Oxford, 1994: 163-205.
22. Hakomori S. Abberant Glycosylation in Tumors and Tumor-Associated Carbohydrate Antigens. Adv Cancer Res. 1989; 52: 257-331.
23. Toyokuni T. Singhal AK. Synthetic Carbohydrate Vaccines Based on Tumor-Associated Antigens. Chemical Soc Rev. 1995; 231-242.
24. MacLean GD, Longenecker BM. Can J Oncol. 1994; 4: 249-258. Specific Immunotherapy of Cancer with Vaccines. Bystryn J-C, Ferrone S, Livingston P, eds. The New York Academy of Sciences New York, 1993.
25. Maryanoff BE, Nortey SO, Gardocki JF, Shank RP, Dodgson SP. Anticonvulsant O-Alkyl Sulfamates. 2,3:4,5 Bis-O-(1-methylethylidene)-β-D-fructopyranose Sulfamate and Related Compounds. J Med Chem. 1987; 30: 880-887.
26. McComsey DF, Maryanoff BE. Improved Synthesis of Pseudo-β-D-

Fructopyranose, a Carbocyclic Monosaccharide from (-)-Quinic Acid. J Org Chem. 1994; 59: 2652-2654.
27. von Itzstein M, Wu WY, Kok GB, Peg MS, Dyason J C, Jin B, Van Phan T, Smythe ML, White HF, Oliver SW, Colman PM, Varghese JN, Ryan DM, Woods JM, Bethell RC, Hotman VJ, Cameron JM, Penn CR. Rational Design of Potent Sialidase Based Inhibitors of Influenza Virus Replication. Nature. 1993; 363: 418-423.
28. von Itzstein M, Wu W-Y, Jin B. The Synthesis of 2,3-didehydro-2,4-dideoxy-4-guanidinyl-N-acetylneuraminic acid: A Potent Influenza Virus Sialidase Inhibitor. Carbohydr Res. 1994; 259: 301-305.
29. Holzer CT, von Itzsten M, Jin, B, Pegg MS, Stewart WP, Wu WY. Inhibition of Sialidases from Viral, Bacterial and Mammalian Sources by Analogues of 2-deoxy,2,3-dihydro-N-acetylneuraminic acid Modified at C-4 Position. Glycoconj J. 1993; 10: 40-44.
30. Chong AKJ, Pegg MS, von Itzstein M. Influenza Virus Sialidase: Effect of Calcium on Steady-State Kinetic Parameters.Biochem Biophys Acta. 1991; 1077: 65-71.
31. Chong AKJ, Pegg MS, Taylor NT, von Itzstein M. Evidence for Sialosyl Cation Transition State Complex in the Reaction of Sialidase from Influenza Virus. Eur J Biochem. 1992; 207: 335-343.
32. Unverzagt C. Got the Flu? Try a Designer Agent Derived from a Sugar. Angew Chem Int Ed Engl. 1993; 32: 1691-1693.
33. Paulsen H, Matschulat P. Synthesis of C-glycosides of N-Acetylneuraminic Acid And Other Derivatives. Liebigs Ann Chem. 1991; 487-495.
34. Walliman K, Vasella A. C-Glycosides of N-Acetylneuraminic Acid. Synthesis and Inhibition of *Vibrio cholerae* Sialidase. Helv Chim Acta. 1991; 74: 1520-1532.
35. Nagy JO, Wang P, Gilbert JH, Schaefer ME, Hill TG, Callstrom MR, Bednarski MD. Carbohydrate Materials Bearing Neuraminidase-Resistant C-Glycosides of Sialic Acid Strongly Inhibit the in Vitro Infectivity of Influenza Virus. J Med Chem. 1992; 35: 4501-4502.
36. Rhinehart L, Robinson KM, King CH, Liu PS. Castanospermine-Glucosides as Selective Disaccharide Inhibitors. Biochem Pharmacol. 1990; 39: 1537-1543.
37. Sunkara PS. Bowlin TL. Liu PS. Sojerdsma A. Antiretroviral Activity of Castanospermine and Deoxynojirimycin Specific Inhibitors of Glycoprotein Processing. Biochem Biophys Res Commun. 1987; 148: 206-200.
38. Gruters RA, Neefjes JJ, Teresmette M, de Goede Rey, Tulp A, Huisman HG, Miedema, F, Plogh HL. Interference with HIV-induced Syncytium Formation and Viral Infectivity by Inhibitors of Trimming Glucosidase.

Nature. 1987; 330: 74-77.

39. Fleet GW J, Karpas A, Dwek RA, Fellows LE, Tyms AS, Petursson S, Namgoong SK, N. G. Ramsden, P. W. Smith JC, Son F, Wilson DR, Witty Jacob, GS, Rademacher, ThW. Inhibition of HIV Replication by Amino Sugar Derivatives. Febs Lett. 1988; 237: 128-132.

40. Karpas A, Fleet GWJ, Dwek RA, Petursson S. Namgoong K, Ramsden NG, Jacob GS, Rademacher ThW. Aminosugar derivatives As Potential Anti-human Immunodeficiency Virus Agents. Proc Natl Acad Sci USA. 1988; 85: 9229-9233.

41. Tyms AS, Taylor DL, Sincere PS, Kang MS. In Design of Anti-AIDS Drugs. De Clerq E, ed. Elsevier, Amsterdam, 1990: 257-318.

42. Casiraghi G, Rassu G, Spanu P, Pinna L. N-(tert-Butoxycarbonyl)-2-(tert-butyldimethylsilyloxy)pyrrole:A Promising Compound for Synthesis of Nonracemic Hydroxylated Pyrrolidine Derivatives. J Org Chem. 1992; 57: 3760-3763.

43. Ina H, Kibayashi C, Total Syntheses of (+)-Castanospermine and (+)-1-Epicastanospermine and Their 1-O-Acyl Deivatives from a Common Chiral Building Block. J Org Chem. 1993; 58: 52-61.

44. Casiraghi G, Rassu G, Spanu P, Pinna L, Ulgheri F. Total Synthesis of (+)-2,8,8a-Tri-epi-swainsonine and (-)-1-epi-Swainsonine. J Org Chem. 1993; 58: 3397-3400.

45. Zhou P, Mohd Salleh H, Honek J. Facile Chemoenzymatic Synthesis of 3-(Hydroxymethyl)-6-Epicastanospermine, J. Org. Chem. 1993; 58: 264-266.

46. Hudlicky T, Rouden J, Luna H. Rational Design of Aza Sugars via Biocatalysis: Mannojirimycin and Other Glycosidase Inhibitors. J Org Chem. 1993; 58: 985-987.

47. Arnone A, Bravo P, Donadelli A, Resnatii G. Asymetric Synthesis of Fluorinated Analogues of 1-Dexynijirimycin. J Chem Soc Chem Commun. 1993; 984-985.

48. Furneaux RH, Lynch GP, Way G, Winchester B. Synthesis of -Acetamido-1,2,4-Trideoxy-1,4-Imino-D-Galactitol,A New Hexosaminidase Inhibitor. Tetrahedron Lett. 1993; 3477-3480.

49. Kefalas P, Grierson DS. Synthesis of Pyrrolo-Castanospermine Analogs. A Novel Fluoride Ion Induced Rearrangement of 3-Trialkyl Substituted Dihydro-2H-1,2-Oxazine to A Pyrrole. Tetrahedron Lett. 1993; 3555-3558.

50. Bruce, I.; Fleet, GW J, Cenci di Bello I, Winchester B. Iminoheptitols as Glycosidase Inhibitors: Synthesis of Homomannonojirimycin, 6-epi-α-Homomannonojirimycin and of a Highly Substituted Pipecolic Acid. Tetrahedron. 1992; 48: 10191-10200.

51. Dondoni A, Merino P, Perrone D. Totally Chemical Synthesis of Azasugars *via* Thiazole Intermediates I. Stereodivergent Routes to (-) Nojirimycin, (-)-Mannonojirimycin and Their 3-Deoxy Derivatives from Serine. Tetrahedron. 1993; 49: 2939-2956.
52. Winchester B, Barker C, Baines S, Jacob GS, Namgoong SK, Fleet GWJ.Inhibition of a-L-Fucosidase by Derivatives of Deoxyfuconojirimycin and Deoxymannojirimycin. Biochem J. 1990; 265: 277-282.
53. Fleet GW, Petursson S, Campbell AL, Mueller RA, Behling JR, Babiak KA, Ng JS, Scaros MG. Short Efficient Synthesis of the α-L-Fucosidase Inhibitor Deoxyfuconojirimycin [1,5-Dideoxy-1,5-Imino-L-fucitol] from D-lyxonolactone, J Chem Soc Perkin Trans. 1. 1989; 665-666.
54. Hughes AB, Rudge AJ. Deoxynojirimycin: Synthesis and Biological Activity. Natural Product Rep. 1994; 135-162.
55. Andrews DM, Bird MI, Cunningham MM, Ward P. Synthesis of and Glycosidase Inhibition by α-L-Homofuconojirimycin, Bioorg Med Chem Lett. 1993; 3: 2533-2536.
56. van den Broek LAGM, Vermaas D J, Heskamp BM, van Boeckel CA A, Tan MC AA, Bolscher JGM, Ploegh HL, van Kemenade FJ, deGoede R.EY, Miedema F. Chemical Modification of Azasugars, Inhibitors of N-Glycoprotein-processing Glycosidases and of HIV Infection. Rec Trav Chim Pays-Bas. 1993; 112: 82-94.
57. Jacobs G. Aminosugar Inhibitors As Anti-HIV Agents, In Carbohydrate and Carbohydrate Polymers. M. In: Yalpani M. Ed. ATL Press, 1993: 74-86.
58. Burgess K. Hemderson I. Synthetic Aproaches to Stereoisomers and Analogues of Castanospermine. Tetrahedron. 1992; 48: 4045-4066.
59. Humphries MJ, Matsumoto K, White SL, Olden K. Inhibition of Experimental Metastasis by Castanospermine in Mice: Blockage of Two Distinct Stages of Tumor Colonization by Oligosaccharide Processing Inhibitors. Cancer Res. 1986; 46: 5215-5222.
60. Johnson VA, Walker BD, Barlow MA, Paradis TJ, Chou TC, Hirsch MS. Synergistic Inhibition of Human Immunodeficiency Virus Type 1 and Type 2 Replicationm *in vitro* by Castanospermine and 3'-Azido-3-' Deoxythymidine, Antimicrobial Agents and Chemother. 1989; 33: 53-57.
61. Tulsiani DRP, Broquist HP, Touster O. Marked Difference in the Swainsonine Inhibition of Rat Liver Lysosomal α-D-Mannosidase, Rat Liver Golgi Mannosidase II and Jack Bean α-D-Mannosidase, Arch Biochem Biophysics. 1985; 236: 427-434.
62. Bernotas RC, Ganem B. Total Synthesies of (+) Castanospermine and (+) Deoxynoijrimycin. Tetrahedron Lett. 1984; 25: 165-168.

63. Dennis JW. Effects of Swainsonine and Polyinosinic: Polycitidic Acid on Murine Tumor Cell Growth and Metastasis. Cancer Res. 1986; 46: 5131-5136.
64. Hino M, Nakayama O, Tsurumi K, Adachi K, Shibata T, Terano H, Kohasaka M, Aoki H, Imanaka H. Studies of an Immunomodulator Swainsonine I. Enhancement of immune Response by Swainsonine *in vitro*. J Antibiot. 1985; 38: 926-935.
65. Kino T, Inamura N, Nakahara K, Kiyoto S, Goto Y, Terano H, Kohsaka M, Aoki H, Imanaka H. Studies of an Immunomodulator Swainsonine II. Effect of Swainsonine on Mouse Immunodeficient System and Experimental Murine Tumor. J Antibiot. 1985; 38: 936-940.
66. Uryu, T.; Kaneko, Y.; Yoshida, T.; Mihara, R.; Shoji. T.; Katsuraya, K.; Nakashima, H.; Yamamoto, N., Synthesis of Anti-HIV Active Sulfated Polysaccharides and Sulfate Alkyl Oligosaccharides, In Carbohydrates and Carbohydrate Polymers. Yalpani M, ed. ATL Press, 1993, pp.102-115.
67. Egan, W. The Development and Current Status of Glycoconjugate Vaccines. In: Ann Rep in Med Chem. 1992; 28: 257-265.
68. Liu PS, Total Synthesis of 2,6-Dideoxy-2,6-imino-7-O-β-D-*glycero*-L-*gulo*-heptitol Hydrochloride: A Potent Inhibitor of α-Glucosideses, J Org Chem. 1987; 52: 4717-4721.
69. Elmers BR, Rhinehart BL, Robinson KM. Castanospermine: An Apparent Tight-binding Inhibitor of Hapatic Lysosomal α-Glucosidase. Biochem Pharmacol. 1987; 36: 2381-2385.
70. Yoshikuni Y, Ezure Y, Aoyagi Y, Enomoto H. Inhibition of Intestinal a-Glucosidase and Postprandial Hyperglycemia by N-substituted Moranoline Derivatives. J Pharmacobio-Dyn. 1988; 11: 356-362.
71. Rhinehart BL. Begovic ME, Robinson KM. Quantitative Relationship of Lysosomal Glycogen Accumulation to Lysosomal α-Glucosidase Inhibition in Castanospermine-Treated Rats. Biochem Pharmacol. 1991; 41: 223-228.
72. Grace ME, Graves PN, Smith FI, Grabowski GA. Analysis of Catalytic Activity and Inhibitor Binding of Human Acid β-Glucosidase by Site-directed Mutagenesis. J Biol Chem. 1990; 265: 6827-6835.
73. Greenberg P, Merrill AH, Liotta DC, Grabowski GA. Human Acid a-Glucosidase: Use of Sphingosyl and N-Aalkyl-glucosamine Inhibitors to investigate the Properties of the Active Site. Biochim Biophys Acta. 1990; 1039: 12 -20.
74. Truscheit E, Frommer W, Junge B, Müller L, Schmidt DD, Wingender W. Chemistry and Biochemistry of Microbial α-glucosidase Inhibitors. Angew. Chem Int Ed Engl. 1981; 20: 744-761.

75. Kuhlmann J, Puls, Eds. Handbook of Experimental Pharmacology. Vol 119. Oral Antidiabetics. Springer-Verlag Berlin Heidelberg, 1996.
76. Joubert PH, Foukaridis GN, Bopape ML. Miglitol May Have a Blood Glucose Lowering Effect Unrelated to Inhibition of Alpha-Glucosidase. Eur J Clin Pharmacol. 1987; 31: 723-724.
77. Robinson KM, Begovic ME, Rhinehart BL, Heineke EW, Ducep JB, Kastner PR, Marshall FN, Danzin C. New Potent α-Glucohydrolase Inhibitor MDL73945 with Long Duration of Action in Rats. Diabetes. 1991; 40: 825-830.
78. Yoshikuni Y, Ezure Y, Seto T, Mori K, Watanabe K, Enomoto H. Synthesis and α-Glucosidase Inhibiting Activity of a New a-Glucosidase Inhibitor, 4-O-α-D-Glucopyranosylmoranoline and Its N-Substituted Derivatives. Chem Pharm Bull. 1989; 37: 106-109.
79. Hori S, Fukase H, Matsuo T, Kameda Y, Asano N, Matsui K. Synthesis and α-D-Glucosidase Inhibitory Activity of N-Substituted Valiolamine Derivatives as Potential Oral Antidiabetic Agents. J Med Chem. 1986; 29: 1038-1046.
80. Joubert PH, Venter HL, Foukaridis GN. The Effect of Miglitol and Acarbose After An Oral Glucose Load: A Novel Hypoglycaemic Mechanism.Br J Clin Pharmacol. 1990; 30: 391-396.
81. Junge B, Boshagen H, Stoltefuz J, Muller L. Derivatives of Acarbose and Their Inhibitory Effects on α-glucosidases. In: Brodbeck U. Ed. Enzyme Inhibitors. Verlag Chemie Weinheim 1980; pp123-137.
82. Chakrabarti S, Cherian PV, Sima AAF. The Effect of Acarbose on Diabetes-and Age -Related Basement Membrane Thickening in Retinal Capillaries of the BB/W-Rat. Diabetes Res Clin Pract . 1993; 20: 123-128.
83. Reitz AB, Tuman RW. Carbohydrate Biguanide as Potential Antidiabetic Drugs. In New Antidiabetic Drugs, Bailey, C. J.; Flatt, PR. Eds., Smith-Gordon, 1990, pp. 171-176.
84. Reitz AB, Tuman RW, Marchione CS, Jordan Jr., AD, Bowden CR, Maryanoff BE, Carbohydrate Biguanides as Potential Hypoglycemic Agents. J Med Chem. 1989; 32: 2110-2116.
85. Ferguson MAJ, Williams AF. Cell-Surface Anchoring of Proteins *via* Glycosylphosphatidylinositol Structures. Ann Rev Biochem. 1988; 57: 285-320.
86. Saltiel AR, Cuatrecasas P, Insulin Stimulates the Generation from Hapatic Plasma Membranes of Modulators Derived from Inositol Glycolipid. Proc Nat Acad Sci USA. 1986; 83: 5793-5797.
87. Billington DC. The Inositol Phosphates Chemical Synthesis and Biological Significance, VCH Verlag, Berlin, 1993.

88. Reitz AB, ed. Inositol Phosphates and Derivatives Synthesis, Biochemistry, and Therapeutic Potential. ACS Symposium Series, 463, ACS, Washington D.C. 1991.
89. Potter BVL, Lampe D. Chemistry of Inositol Lipid Mediated Cellular Signaling. Angew Chem Int Ed Engl. 1995; 34: 1933-1972.
90. Powis G, Signaling Targets for Anticancer Drug Development, Trends Pharmacol Sci. 1991; 12: 188-198.
91. Powis G, Kozikowski AP, Inhibitors of *myo*-Inositol Signaling, In New Targets for Cancer Chemotheraphy, Ker, D. J.; Workman P., Eds. CRC Press; Boca Raton, 1992.
92. Kozikowski AP, Faug AH, Malaska MJ, Tückmantel W,Ognyanov VI, Powis G. Inositol Analogs as Potential Anticancer Therapeutics. Curr Med Chem. 1994; 1: 1-12.
93. Kozikowski AP, Tückmantel W, Powis G. Synthesis and Biological Activity of D-3-Deoxy-3-fluorophosphatidylinositol- A New Direction in the Design of Non-DNA Targeted Anticancer Agents. Angew Chem Intl. Ed. Engl. 1992; 31: 1379-1381.
94. Wiley PF, Jahnke HK, Chidester CG, Mizsak SA, Spaulding LB, Argoudelis AD, Streptozocin: Structure and Chemistry. J Org Chem. 1979; 44: 9-16.
95. Ichikawa Y, Halcomb RL, Wong C-H. Sticky Solutions. Chem. Brit. 1994; 117-121.
96. Kameyama A, Ishida C, Kiso M, Hasegawa A. Total Synthesis of Sialyl Lewis X. Carbohydr Res. 1991; 209: C1-C4.
97. Nicolau KC, Hummel CW, Backovich NG, Wong C-H. Stereocontrolled Synthesis of Silayl Le^x , the Oligosaccharide Binding Ligand to ELAM-1 (Sialyl = N-acetylneuramin), J Chem Soc Chem Commun. 1991; 870-872.
98. Danishefsky SJ, Gervay J, Peterson JM, McDonald FE, Koseki, K., Oriyama T, Griffith DA, Wong C-H, Dumas DP.Remarkable Regioselectivity in the Chemical Glycosylation of Glycal Acceptors: A Concise Solution to the Synthesis of Sialyl-Lewis X Glucal. J Am Chem Soc. 1992; 114: 8329-8331.
99. Danishefsky SJ Koseki K, Griffith DA, Gervay J, Peterson JM, McDonald FE, Oriyama T. Azaglycosylation of Complex Stannyl Alkoxides with Glycal-Derives Iodosulfonamides: A Straightforward Synthesis of Sialyl-Lewis X Antigen and Other Oligosacccharide Domains. J Am Chem Soc. 1992; 114: 8331-8333.
100. Dumas DP, Ichikawa Y, Wong C-H. Lowe JB, Nair RP. Enzymatic Synthesis of Sialyl Le^x and Derivatives Based on A Recombinant

Fucosyltransferase. Bioorg Med Chem Lett. 1991; 1: 425-429.
101. Ball GE, O'Neil RA, Schultz JE, Lowe JB, Weston BW, Nagy JO, Brown EG, Hobbs CJ, Bednarski MD. Synthesis and Structural Analysis Using 2-D NMR of Sialyl Lewis X (SLe^x) and Lewis (Le^a) Oligosaccharides: Ligands Related to E-Selectin [ELAM-1] Binding. J Am Chem Soc. 1992; 114: 5449-5451.
102. Lin YC, Hummel CW, Huang DH, Ichikawa Y, Nicolau KC, Wong C-H, Conformational Studies of Sialyl Lewis X in Aqueous Solution. J Am Chem Soc. 1992; 114: 5452-5454.
103. Nelson RM, Dolich S, Aruffo A, Cecconi O, Bevilacqua MP. Higher-Affinity Oligosaccharide Ligands for E-Selectin. J Clin Invest. 1993; 91: 1157-1166.
104. Yuen C-T, Lawson AM, Chai W, Larkin M, Stoll MS, Stuart AC, Sullivan FX, Ahern TJ, Feizi T. Novel Sulfated Ligands for the Cell Adhesion Molecule E-Selectin Revealed by the Neoglycolipid Technology among O-Linked Oligosaccharides on an Ovarian Cystadenoma Glycoprotein. Biochemistry. 1992; 31: 9126-9131.
105. Nicolau KC, Bockovich NJ Carcanague DR. Total Synthesis of Sulfated Le^x and Le^a -Type Oligosaccharide Selectin Ligands. J Am Chem Soc. 1993; 115: 8843-8444.
106. Allanson NM, Davidson AH, Martin FM. A Novel Mimis of the Sialyl Lewis X Determinant, Tetrahedron Lett. 1993; 34: 3945-3948.
107. DeFrees SA, Gaeta F CA, Lin Y-C, Ichikawa Y, Wong C-H. Ligand Recognition by E-Selectin: Analysis of Conformation and Activity of Synthetic Monomeric and Bivalent Sialyl Lewis X Analogs. J Am Chem Soc. 1993; 115: 7549-7550.
108. Von dem Burch K, Kunz H. Synthesis of N-Glycopeptide Clusters with Lewis^x Antigen Side Chains and Their Coupling to Carrier Proteins. Angew Chem Int Ed Engl. 1994; 33: 101-103.
109. Wong C-H, Dumas DP, Ichikawa Y, Koseki K, Danishefsky SJ, Weston BW, Lowe JB. Specificity, Inhibition and Synthetic Utility of a Recombinant Human α-1,3-Fucosyltransferase. J Am Chem Soc. 1992; 114: 7321-7322.
110. Cai S, Stroud MR, Hakomori S, Toyokuni T. Synthesis of Carbocyclic Analgues of Guanosine 5'-(α-L-Fucopyranosyl diphosphate) (GDP-Fucose) as Potential Inhibitors of Fucosyltransferases. J Org Chem. 1992; 57: 6693-6696.
111. Plewe M, Sandhoff K, Schmidt RR. Synthesis of 4-Epoxy-4-C-methyleneglycosylceramides, Potential Glycosyltransferase Inhibitors, Liebigs Ann Chem. 1992; 699-708.

112. Schmidt RR, Frische K. A New Galactosyl Transferase Inhibitor. Bioorg Med Chem Lett. 1993; 3: 1747-1750.
113. Frische K, Schmidt RR. Glycal-1-ylmethylphosphonates-Precursors of Glycosyltrasnsferase Inhibitors. Liebigs Ann Chem. 1994; 297-303.
114. Dumas DP, Kajimoto T, Lie K. K-C, Wong, C-H, Berkowitz DB, Danishefsky SJ. Azasugar and Glycal Inhibitors of α-L Fucosidase. Bioorg Med Chem Lett. 1992; 2: 33-36.
115. Witczak ZJ. C-Trisaccharide of Lewis^x from Levoglucosenone. XVII International Carbohydrate Symposium, Ottawa, Ontario Jul 17-22. 1994. Abstract B.1-96: Witczak ZJ, Sun J, Li Y. 208th ACS National Meeting, Washington, DC. Aug 21-25, 1994. Abstract Carb-58.
116. Beacham, AR, Biggadike K, Taylor HE, Hackert L, Winchester B. G, Watkin DJ, Fleet GWJ. Short Synthesis of a Bicyclic Mimics of α-L-fucose. J Chem Soc Chem Commun. 1994; 2001-2002.
117. Scheffler K, Ernst B, Katopodis A, Magnani JL, Wang W-T, Weisemann R, Peters T. Determination of the Bioactive Conformation of the Carbohydrate Ligand in the E-Selectin/Sialyl Lewis^x Complex. Angew Chem Int Ed Engl. 1995; 34: 1841-1844.
118. Huang, H., Wong, C-H, Synthesis of Biologically Active Sialyl Lewis X Mimetics. J Org Chem. 1995; 60: 3100-3106.
119. Kaila N, H-A, Yu Y, Xiang. Design and Synthesis of Novel Sialyl Lewis X Mimics. Tetrahedron Lett. 1995; 36: 5503-5506.
120. Manning DD, Bertozi CR, Pohl NL, Rosen SD, Kiessling LL. Selectin-Saccharide Interactions: Revealing Structure-Function Relationships with Chemical Synthesis. J Org Chem. 1995; 60: 6254-6255.
121. Hashimoto H, Fujimori T, Yuasa H. Synthesis of 5-thio-L-fucose and Its Inhibitory Effect on Fucosidase. J Carbohydr Chem. 1990; 9: 683-694.
122. Hashimoto H, Hayakawa M. Synthesis of a New Fucosidase Inhibitor 1,5-Dideoxy-1,5-imino-L-talitol, via cyanotrimethylsilanolisys of α,β-ribofuranoside and Its Inhibitory Activities. Chem Lett. 1989; 1881-1884.
123. Hashimoto H, Izumi M. Efficient and Stereoselective 1,2-cis Glycoside Formation of 5-Thio-aldopyranoses: Glycosylation with Peracetylated 5-Thio-D-arabinopyranosyl and 5-Thio-L-fucopyranosyl Trichloro-acetimidates. Tetrahedron Lett. 1993; 34: 4949-4952.
124. Hashimoto H, Shimada K, Horito S. Synthesis of α-L-Fucopyranosyl Disaccharides with Thioglycosidic Linkages and Characterization of a-Fucosidases from Bovine Kidney and Epididiymis by their Inhibitory Activities. Tetrahedron: Asymmetry. 1994; 5: 2351-2366.
125. Witczak ZJ, Sun J, Mielguj R. Synthesis of L-fucopyranosyl-4-thiodisaccharides from Levoglucosenone and their Inhibitory activity on

α-L-Fucosidase. Bioorg Med Chem Lett. 1995; 5: 2169-2174.

126. Witczak ZJ, Whistler RL, Structure and Antitumor Activity of Polysaccharides, In Industrial Polysaccharides-The Impact of Biotechnology and Advanced Methodologies. Stivala SS, Crescenzi V, Dea ICM, eds. Gordon & Breach, 1987, pp. 157-173.
127. Misaki A, Kishida E, Kakuta M, Tabata K. In Carbohydrates and Carbohydrate Polymers. Yalpani M, ed. ATL Press, 1993, pp.116-129.
128. Matsuzaki K, Sato T, Enomoto K, Yamamoto I, Ohshima R, Hatanaka. K.; Uryu T, Eaku H, Sone Y, Misaki A. Synthesis of Water Soluble, Branched Polysaccharides having D-mannopyranose, D-arabinofuranose, or Oligo-D-arabinofuranose side-chains and their Antitumor Activity. Carbohydr Res. 1986; 157: 171-182.
129. Garner, P., Synthetic Approaches to Complex Nucleoside Antibiotics, In Studies in Natural Products Chemistry, Atta-ur-Rhaman, ed. Elsevier Amsterdam, 1988, Part A Vol. 1. p.397.
130. Knapp S. Synthesis of Complex Nucleoside Antibiotics. Chem Rev. 1995; 95: 1859-1876.
131. Mason DJ, Dietz A, Deboer C. Linkomycin, A New Antibiotic I. Discovery and Biological Properties. Antimicrob Agents Chemother. 1962; 554-559.
132. Birkenmeyer RD, Kagan F. Lincomycin XI. Synthesis and Structure of Clindamycin, A Potent Antibacterial Agent. J Med Chem. 1970; 13: 616-619.
133. Fass RJ. Saslaw S. Clindamycin Clinical and Laboratory Evaluation of Parenteral Tharaphy. Am J. Med Sci. 1972; 263: 369-382.
134. Borders DB, Doyle TW, eds. Enediyne Antibiotics As Antitumor Agents. Marcel Dekker Inc. New York, 1995.
135. Berdy J, Aszalos A, Bostian M, McNitt KL, eds. Handbook of Antibiotic Compounds, Vol. I. Carbohydrate Antibiotics, CRC Press, Boca Raton, 1980.
136. Umezawa H, Hoeper IR, eds. Handbook of Experimental Pharm. Vol. 80. Aminoglycoside Antibiotics. Springer Verlag, Berlin, 1982.
137. Ristuccia AM. Aminoglycosides. In: Antimicrobial Theraphy. Ristuccia AM. Cunha BA, eds. Raven Press, New York 1984, pp 305-327.
138. Fillastre JP, Leroy A, Humbert G, Moulin HB, Beruadet P, Josse S. Pharmacokinetics of Habekacin in Patients with Renal Insufficiency. Antimicrob Agents Chemother. 1987; 31: 575-577.
139. Hill CD. Counts GW, Turck M. *In vitro* Comparison of Dibekacin and Gentamicin Derivatives. Antimicrob Agents Chemother. 1981; 19: 190-192.

140. Leroy A, Humbert G, Fillastre JP. Pharmacokinetics of Dibekacin in Normal Subjects and in Patients with Renal Filure.J AntimicrobChemother. 1980; 6: 113-120.
141. Fernandes PB, Bailer R, Swanson R, Hanson CW, McDonald E, Ramer N, Hardy D, Shipkowitz N, Bower RR, Gade E. In vitro and in vivo Evaluation of A 56268 (TE-031): A New Macrolide. Antimicrob Agents Chemother. 1986; 30: 865-873.
142. Ferero JL, Bopp BA, Marsh KG, Quigley S C, Johnson MJ, Anderson DL, Lamm JE, Tolman KG, Sanders SW, Cavanaugh JH. Metabolism and Disposition of Clarithromycin in Man. Drug Metab Dispos. 1990; 18: 441-446.
143. Hodinka RL, Jack-Wait K, Gilligan PH. Comparative *in vitro* Activity of A 56268 (TE-031): A New Macrolide Antibiotic. Eur J Clin Microbiol. 1987; 6: 103-108.
144. Blizzard T, Fisher MH, Mrozik H, Shih TL. In: Recent Progress in the Chemical Synthesis of Antibiotics. Lucacs G, Ohno M, eds., Springer-Verlag, Berlin, 1990, pp.66-83.
145. Krohn K, Kirst HA, Maag H, eds. Antibiotics and Antiviral Compounds Chemical Synthesis and Modification. VCH, Weinheim, 1993.
146. Ganguly AK, Sarre OZ, Greeves D, Motron J. Structure of Everninomicin D[1.] J Am Chem Soc. 1975; 97: 1982-1985.
147. Sinaÿ P, Jacquinet JC, Petitou M, Duchaussoy P, Lederman I, Choay J, Torri G. Total Synthesis of Pentasacchride fragment Having High Affinity for Antithrombin III. Carbohydr. Res., 1984; 132: C5-C9.
148. Saivin S, Petitou M, Lormeau JC, Dupouy D, Sie P, Caranobe C, Houin G, Boneu B. Pharmacologic Properties of an Unfractioned Heparin Butyryl derivative with long-lasting Effects. J Lab Clin Med. 1992; 119: 189-196.
149. Linhardt R J, Chemical and Enzymatic Methods for the Depolymerization and Modification of Heparin, In: Carbohydrates-Synthetic Methods and Applications in Medicinal Chemistry.Ogura H, Hasegawa A, Suami T, eds. Kodansha, Tokyo, & VCH Weinheim. 1992, pp. 385-401.
150. Sutton BM, McGusty E, Walz DT, DiMartino MJ. Oral Gold. Antiarthritic Properties of Alkylphosphinegold Coordination Complexes. J Med Chem. 1972; 15: 1095-1098.
151. Brown DH, Smith WE. The Chemistry of the Gold Drugs Used in the Treatment of Rheumatoid Arthritis. Chem Soc Rev 1980; 9: 217-240.
152. Linhardt RJ, Baenziger NC, Ronsen B. Structure of Amiprilose Hydrochloride a Novel Anti-inflammatory Agent. J Pharm Sci. 1990; 79: 158-162.
153. Harris ED, Sporn MB. Proliferative Diseases. American J Medicine. 1981;

70: 1231-1236.

154. Brickerhoff CE. Decreased Cell Proloferation and PGE_2 Production by Fibroblasts Treated with a Modified Hexose Sugar Amiprilose Hydrochloride. Agents Actions. 1990; 30: 322-328.

155. Morrison CJ, Gordon P, Hashimoto T. Enhanced Killing of Candida albicans by Cultured Peritoneal Exudate Cells Treated with SM-1213, A Synthetic Immunomodulator. Antim Agents Chemother. 1984; 26: 74-77.

156 Kieval RI, Young CT, Prohazka D, Brickerhoff CE, Trentham DE. Evaluation of a Modified Hexose Sugar Amiprilose Hydrochloride in Experimental Models of Synovitis. J Rheumatol. 1989; 16: 67-74.

157 Garrett ER, Van Peer A. Pharmacokinetics of Intravenous and Oral 1,2-O-isopropylidene-3-O-3'-(N',N'-dimethylamino-n-propyl)-D-glucofuranose Hydrochloride in Dog as A Function of Dose and Characterization of Metabolites. J Pharm Sci. 1983; 72: 1045-1057.

158. Weinblatt, ME, Fraser PA, Anderson R, Coblyn JS, Trentham DE. Diminution of the T8 Subset by Amiprilose hydrochloride in Refractory Rheumatoid Arthritis. J Rheumatol. 1987; 14: 859-860.

159. Arora SK, Whistler RL, Thomas AV. WO Patent . 1992; 92/04359; Chem Abstr. 1992; 117: 90700q.

160. Ronsen B. Arora SK, Thomas AV. EP Patent. 1990; 379,397; Chem Abstr. 1991; 114: 229286t.

161. Ronsen B, Arora SK, Thomas AV. EP Patent. 1990; 404,136; Chem Abstr. 1991; 114: 247656t.

162. Hollander D. Tytgat GNJ, eds. Sucralfate From Basic Science to the Bedside. Plenum Publishing Co., New York, 1995.

2
Sialic Acid Analogues as Potential Antimicrobial Agents

Mark von Itzstein and Milton J. Kiefel
Monash University
Parkville, Victoria, Australia

1. INTRODUCTION

The dawn of a post-antibiotic era is almost upon us. New challenges for the control of disease caused by pathogenic microorganisms are being presented. The tolerance of many pathogenic microbes to our present day antibiotics and the re-emergence of devastating diseases such as tuberculosis, leads one to conclude that new strategies to combat these disease-causing microorganisms must be developed with a great deal of urgency. The family of carbohydrates collectively known as the sialic acids and the enzymes that metabolise them have been implicated in the pathogenicity of a number of infectious microorganisms [1-3]. This has prompted much research on the sialic acid family and in the following sections we have attempted to summarise the recent developments in sialic acid chemistry and their prospects as drugs of the future.

2. ASPECTS OF THE CHEMISTRY OF SIALIC ACIDS

The importance of sialic acids as α-ketosidically-linked terminal units of many glycoproteins and glycolipids [1, 4] and the increased realisation of the significant roles such compounds play in the pathogenesis of some microorganisms and various disease states [4-8], has resulted in a substantial increase in the reported syntheses of structurally-modified sialic acids over recent years. In particular, the recent articles by Zbiral [9] and DeNinno [10] as well as earlier accounts [4, 11, 12] give an excellent insight into the chemistry of sialic acids and their analogues. Given the comprehensive nature of these accounts, this section will focus on some of the advances in the chemistry of sialic acids over the last five years.

2.1 Preparations of Sialic Acids

The biosynthesis of sialic acids is highly complex, and involves one of three known pathways starting from *N*-acetylmannosamine [10, 13] and condensation with some form of pyruvate. Laboratory preparations of sialic acids are complicated by virtue of the structural complexity of sialic acids, and may be conveniently divided into either enzymatic or chemical syntheses. By using either of these approaches, or by a combination of both, one of the most well-known sialic acids, 5-acetamido-3,5-dideoxy-D-*glycero*-D-*galacto*-2-nonulo-pyranosonic acid (*N*-acetylneuraminic acid, Neu5Ac, **1**) has been prepared with modifications at every carbon atom [9].

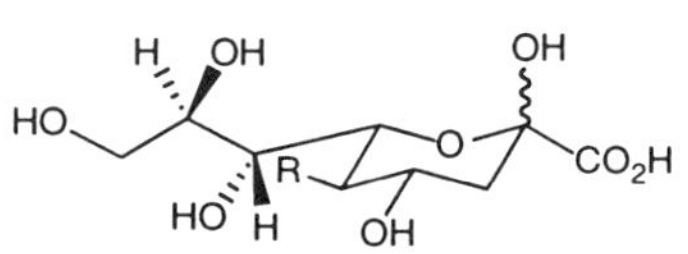

(**1**) R = NHAc

2.1.1 Enzymatic methods

The most common approach to the enzymatic synthesis of Neu5Ac analogues involves the chemical preparation of modified *N*-acetylmannosamine (ManNAc) derivatives and their conversion to the corresponding Neu5Ac analogues using Neu5Ac aldolase (*N*-acylneuraminate pyruvate lyase; EC 4.1.3.3) in the presence of pyruvate (Scheme 1). For synthetic purposes the equilibrium can be shifted to favour the aldol condensation product using an excess of pyruvate [14]. The increased interest in Neu5Ac analogues as biological probes over recent years has prompted several studies on the substrate specificity of Neu5Ac aldolase. In general terms, modifications of ManNAc at C-2 or C-6 are readily tolerated by this enzyme and provide Neu5Ac analogues modified at C-5 and C-9, respectively [14-16]. Importantly, mannose, together with several structural analogues, is readily accepted as a substrate of Neu5Ac aldolase. Using these starting substrates, the preparation of 3-deoxy-D-*glycero*-D-*galacto*-2-nonulopyranosonic acid (KDN) (**2**) and various analogues has been achieved[14, 15, 17]. These compounds themselves are of interest as biological probes.

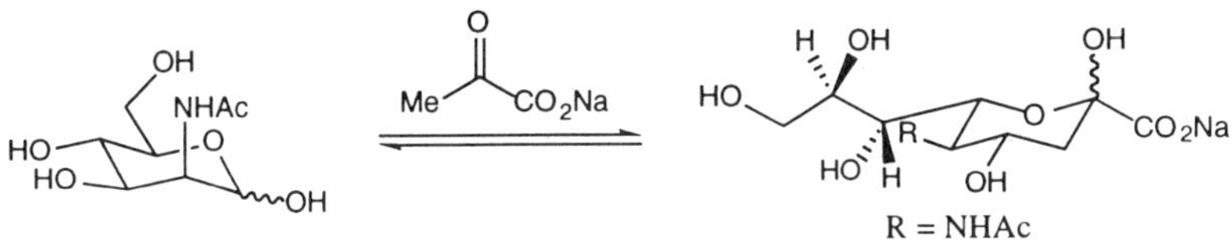

R = NHAc

Scheme 1

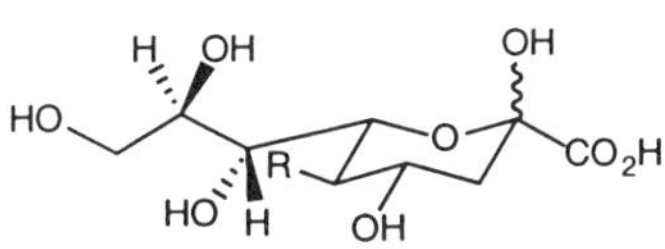

(**2**) R = OH

Such studies on the substrate specificity of Neu5Ac aldolase have resulted in some excellent reports detailing the tolerance of this enzyme towards structural modifications of both Neu5Ac [18, 19] and ManNAc [14, 15] substrates. In particular, the detailed reports by Wong [14, 15] provide a good overview of the permissible modifications to the ManNAc substrate. Accordingly, this account will concentrate on the more recent reports of the synthesis of Neu5Ac analogues from structurally-modified ManNAc derivatives using Neu5Ac aldolase.

Modifications to the *N*-acetamido group of ManNAc have resulted in the synthesis of 5-azido- and 5-carbobenzoxy-Neu5Ac derivatives **3** and **4**, respectively [20]. These compounds were subsequently chemically modified to the 5-amino derivative **5** which was *N*-acylated to provide several *N*-acylneuraminic acid analogues which were evaluated for their inhibition of influenza virus adhesion [20]. 5-*N*-Thioacylneuraminic acid derivatives (eg. **6**) have similarly been prepared from ManNAc [21]. The novel C-5 homologated neuraminic acid derivative **7** has been synthesised in 75% yield from the corresponding C-2 homologated ManNAc compound **8** [22]. Compound **8** was itself prepared from the 2-oxo derivative **9** *via* the 2-*C*-nitromethyl-mannopyranoside **10** [22].

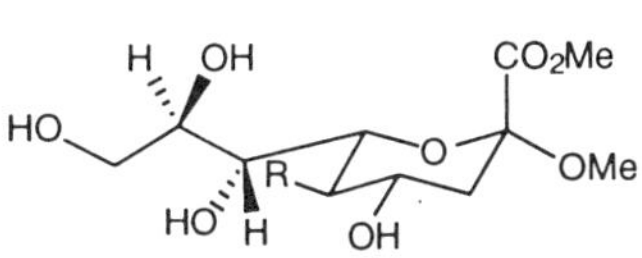

(**3**) R = N_3
(**4**) R = NHCBz
(**5**) R = NH_2

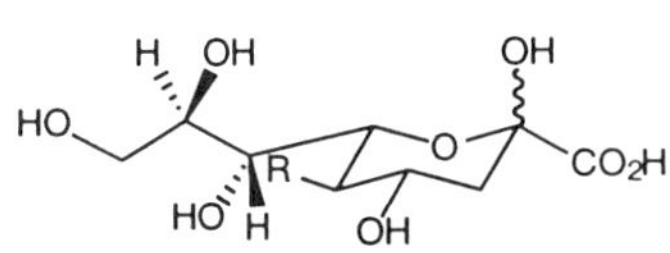

(**6**) R = NHC(S)R'
(**7**) R = CH_2NHAc

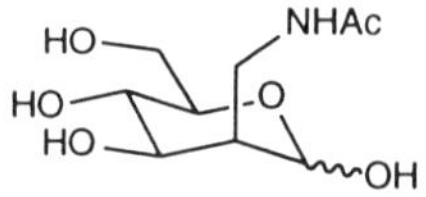

(**8**)

(**9**)

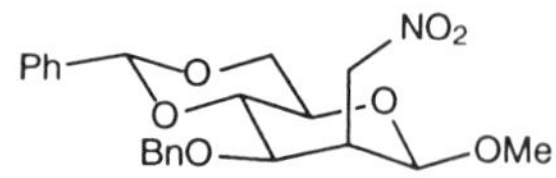

(**10**)

Modifications at C-3 of either ManNAc or mannose results in compounds which are typically not accepted as substrates by Neu5Ac aldolase [15]. It has been reported however that exposure of 3-deoxymannose to Neu5Ac aldolase provided a mixture of four acids, tentatively identified as the furanose **11** of the two nonulosonic acids epimeric at C-4 [23].

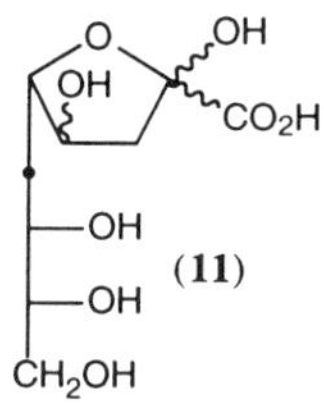

(**11**)

Very few reports have appeared concerning the use of C-4 or C-5 modified ManNAc substrates. Deoxygenation or methylation at C-4 of ManNAc furnishes the corresponding 7-deoxy-Neu5Ac (**12**) [24] and 7-*O*-methyl-Neu5Ac (**13**) [16, 24], respectively, after exposure to Neu5Ac aldolase. The synthesis of 7-azido-7-deoxy-Neu5Ac (**14**) from 4-azido-4-deoxy-ManNAc (**15**), itself prepared by treatment of the 1,6:3,4-dianhydro derivative (**16**) with lithium azide [25], represents the only report of a Neu5Ac analogue with a nitrogen substituent at C-7 [26, 27]. Indeed, since the chemical modification of Neu5Ac itself at C-7 is particularly difficult (see for example [28]) these enzymatic approaches towards such compounds represent a practical solution to this problem.

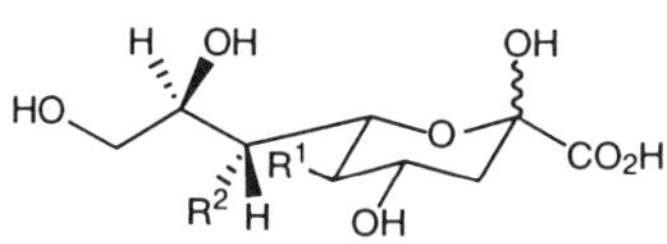

(**12**) R^1 = NHAc, R^2 = H
(**13**) R^1 = NHAc, R^2 = OMe
(**14**) R^1 = NHAc, R^2 = N_3

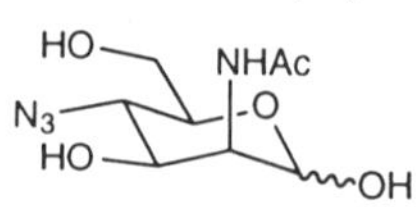

(**15**)

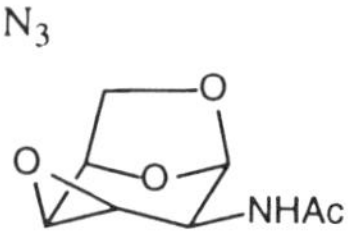

(**16**)

5-*O*-Methyl-ManNAc is a poor substrate for Neu5Ac aldolase, giving only trace amounts of the corresponding 8-*O*-methyl-Neu5Ac compound **17** even after extended reaction times [16]. This result may however be due to steric hindrance, since it has recently been shown that both 5-deoxy-ManNAc and 5-deoxy-mannose are readily converted into 8-deoxy-Neu5Ac (**18**) and 8-deoxy-KDN (**19**), respectively, by Neu5Ac aldolase [23].

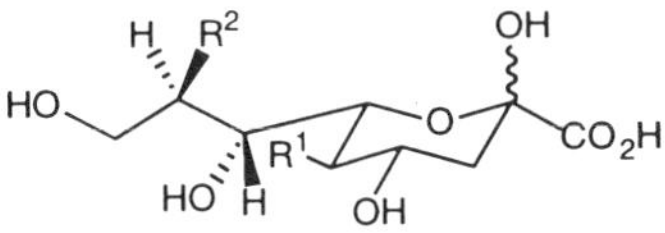

(**17**) R^1 = NHAc, R^2 = OMe
(**18**) R^1 = NHAc, R^2 = H

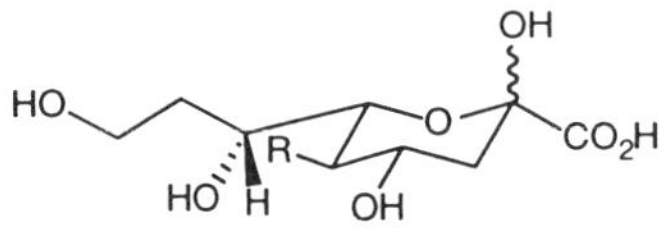

(**19**) R = OH

The value of fluorescence labelling of Neu5Ac analogues in order to study various sialic acid-recognition phenomena, has led to the synthesis of the fluorescent sialic acid analogue **20** from the 9-BOC-glycyl derivative **21** [29]. Compound **21** was obtained *via* the aldol condensation between the ManNAc derivative **22** and pyruvate mediated by Neu5Ac aldolase. Whilst most 6-substituted ManNAc derivatives are accepted as substrates for Neu5Ac aldolase, **22** contains the most sterically demanding C-6 substituent yet reported [29].

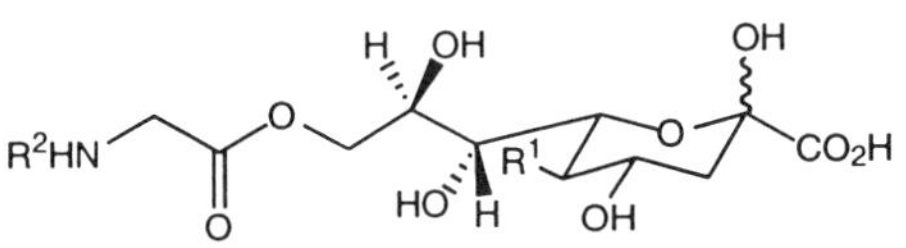

(**20**) R^1 = NHAc, R^2 = $-SO_2-$

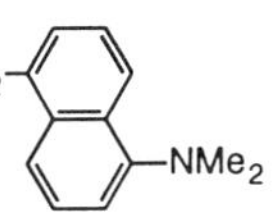

(**21**) R^1 = NHAc, R^2 = BOC

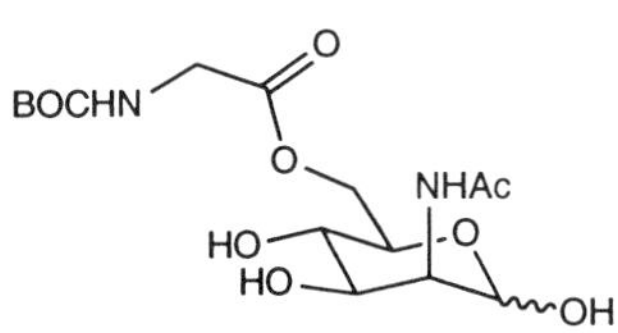

(**22**)

Carbohydrates apart from ManNAc and mannose derivatives are also accepted as substrates by Neu5Ac aldolase. Such carbohydrates include gulose, allose, rhamnose, ribose, xylose and arabinose, as well as some L-sugars [15, 17, 30], and provide avenues for the synthesis of various analogues of KDN (**2**) and 3-deoxy-D-*manno*-2-octulosonic acid (KDO, **23**) compounds which, like Neu5Ac (**1**), are becoming increasingly recognised as valuable biological probes.

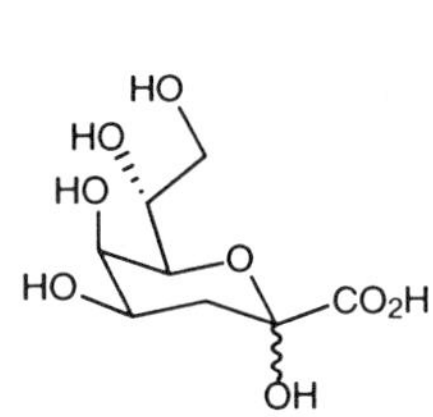

(**23**)

2.1.2 Chemical methods

The first chemical synthesis of Neu5Ac (**1**) was reported in 1958 [31] and involved the condensation of *N*-acetylglucosamine with oxobutanedioic acid in the presence of base to give **1** in 1–2% yield. Since that time more efficient syntheses of Neu5Ac have been developed, one in particular being the condensation of ManNAc (**24**) with the potassium salt of di-*tert*-butyl oxobutanedioate (Scheme 2) followed by a base-induced decarboxylation to afford **1** in 34% yield [32].

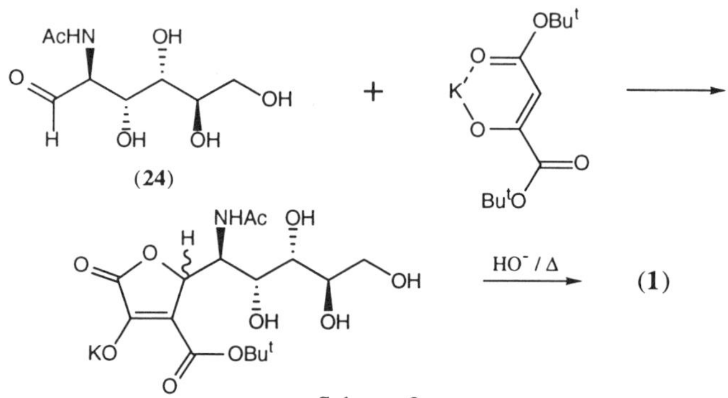

Scheme 2

Whilst there have been several improvements made to these early syntheses, the general synthetic strategy is still relevant today [10]. Recently, a short and highly efficient route to sialic acids has been reported wherein an indium-mediated allylation of ManNAc with either ethyl α-(bromomethyl) acrylate [33] or α-(bromomethyl)acrylic acid [34] in acidic aqueous ethanol furnished the corresponding enoate **25** (threo:erythro = 4:1) or enoic acid **26** (threo:erythro = 3:1), respectively, in high yield. Ozonolysis of the diastereomeric enoates **25** followed by oxidative workup and acetylation gave the protected Neu5Ac derivative **27** in 51% yield [33]. Similarly, ozonolysis of **26** gave Neu5Ac (and its C-4 epimer) in 74% yield [34]. Both KDN (**2**) [34] and KDO (**23**) [35] have also been prepared using this indium-mediated allylation strategy. Other metal catalysts which have been used in such condensation reactions include nickel (see for example [36]), zinc and tin.

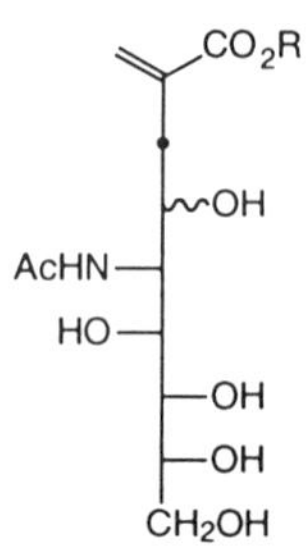

(**25**) R = Et
(**26**) R = H

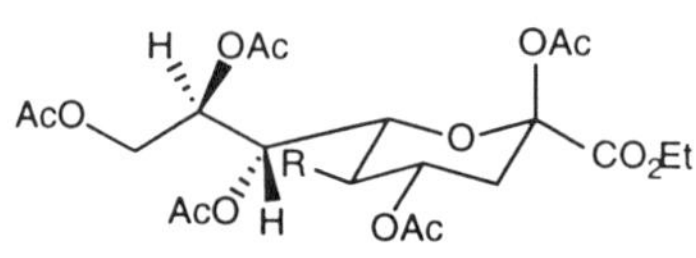

(**27**) R = NHAc

In another recent synthesis of KDN (**2**), also commencing from mannose, the strategy centred on the Wittig condensation between the aldehyde **28** and the thiazole ylide **29** to give the enone **30**, followed by the conjugate addition of benzyl oxide to **30**, to give predominantly the *syn*-adduct **31** [37]. Similarly, conjugate addition of benzyl amine to **30** gave the adduct **32** which was elaborated into 4-acetamido-4-deoxy-KDN (**33**) [38].

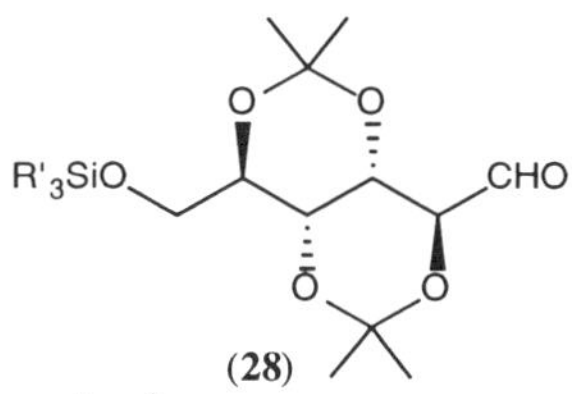

(**28**)

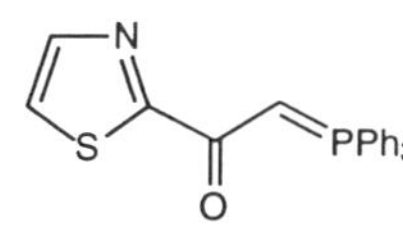

(**29**)

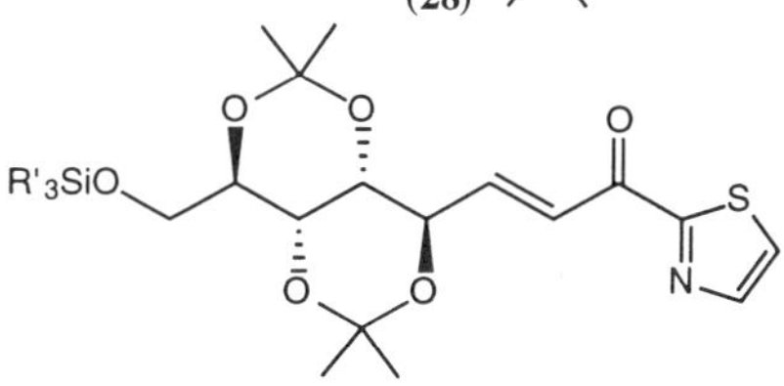

(**30**)

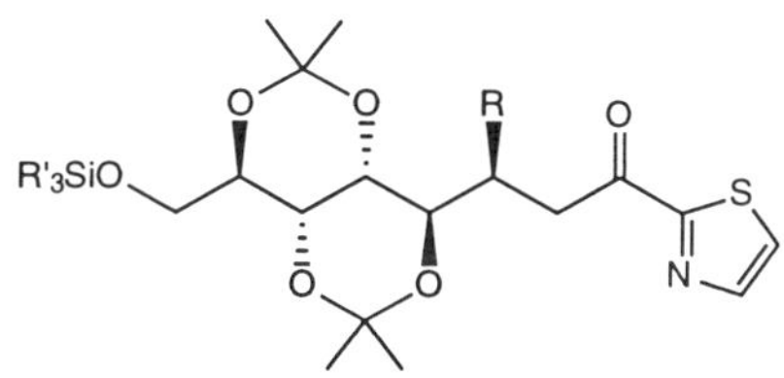

(**31**) R = OBn
(**32**) R = N(Ac)Bn

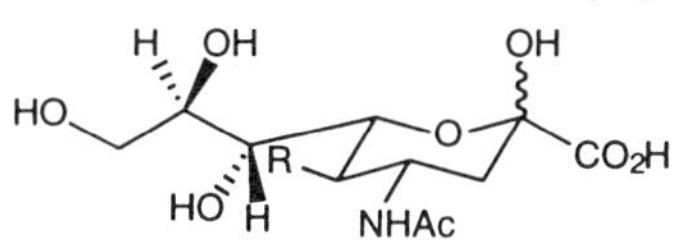

(**33**) R = OH

To date there has only been one report of the total synthesis of Neu5Ac (**1**) from non-carbohydrate precursors [39]. The key steps in this synthesis are presented in Scheme 3. The recent report of a new acyl anion equivalent, *viz*: *tert*-butyl 2-alkoxy-2-cyanoacetate [40], may prove valuable in more efficient syntheses of sialic acids and related compounds.

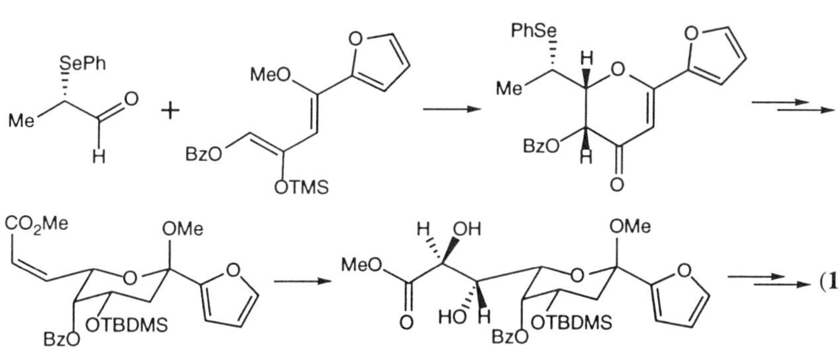

Scheme 3

As already discussed, enzymatic syntheses of Neu5Ac analogues involve the chemical modification of ManNAc and then exposure of the resultant ManNAc derivative to Neu5Ac aldolase. Similarly, chemical syntheses of Neu5Ac analogues invariably commence from Neu5Ac itself, and in this way modifications have been made at every carbon of Neu5Ac. An excellent article by Zbiral [9] provides a detailed overview of this aspect of sialic acid chemistry, and therefore this account will generally concentrate on the most recent developments in this field. Some analogues of Neu5Ac, including 2-deoxy- and 2,3-didehydro-derivatives, have resulted in compounds which are of particular biological interest and these will be presented later.

C-2 modifications of Neu5Ac, apart from 2-deoxy- and 2,3-didehydro-derivatives (*vide infra*), typically involve the synthesis of various glycosides. The preparation of α-sialosides is of particular importance given that sialic acids occur naturally as the terminal units of many glycoproteins and glycolipids [1, 4]. Consequently, there is considerable literature devoted to their preparation ([9, 10, 41-44]. In general terms, the glycosidation of Neu5Ac is more difficult than with other carbohydrates due to the presence of the carboxylic acid function at C-2 (which disfavours the formation of an intermediate oxocarbocation) and the lack of a "directing" substituent at C-3 [10]. Nevertheless, glycosides of Neu5Ac have been prepared using Koenigs-Knorr methodology involving the condensation between an activated Neu5Ac compound such as **34** (X = leaving group) and an alcohol in the presence of a promoter [41, 42]. In many instances the leaving group in **34** is a β-chloro substituent (**34a**; X = β-Cl). Recent modifications of this general strategy include the use of a Neu5Ac derivative with an acetoxy or phenoxycarbonyloxy leaving group (**34b**; X = β-OC(O)R) [45], or a diethylphosphite leaving group (**34c**; X = β-$OP(OEt)_2$) [46]. Sialyl donors with sulfur-containing leaving groups (**34d**; X = α-SMe [47]: **34e**; X = α-SC(S)OEt [48, 49] (see also Section 2.2.2) or **35**, which also bears a participating group at C-3 ([50]; for a similar strategy see refs. [51] & [43]), provide sialosides in high chemical yield and excellent stereoselectivity.

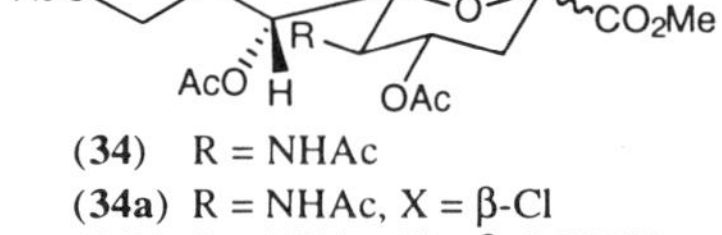

(**34**) R = NHAc
(**34a**) R = NHAc, X = β-Cl
(**34b**) R = NHAc, X = β-OC(O)R
(**34c**) R = NHAc, X = β-$OP(OEt)_2$
(**34d**) R = NHAc, X = α-SMe
(**34e**) R = NHAc, X = α-SC(S)OEt

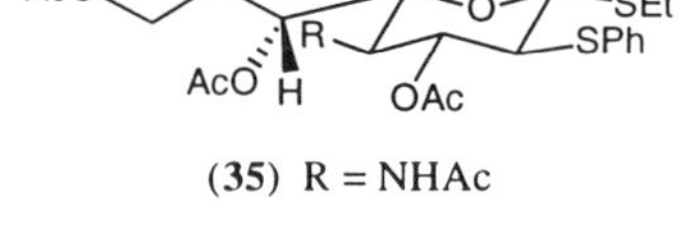

(**35**) R = NHAc

The use of a participating substituent at C-3 in an attempt to increase the efficiency of the glycosidation of Neu5Ac derivatives, first reported in 1986 [52], has been thoroughly discussed elsewhere [41].

Other modifications of the C-2 position of Neu5Ac include the synthesis of the C-2 homologated Neu5Ac analogue **36** [36], along with its C-5 and C-6 epimers, *via* a nickel-mediated aldol condensation reaction between the ManNAc derivative **37** and disodium acetonedicarboxylate. Interestingly, compound **36** (but not its C-5 or C-6 epimers) was found to be unstable, readily decarboxylating to give the C-2 methyl derivative **38**.

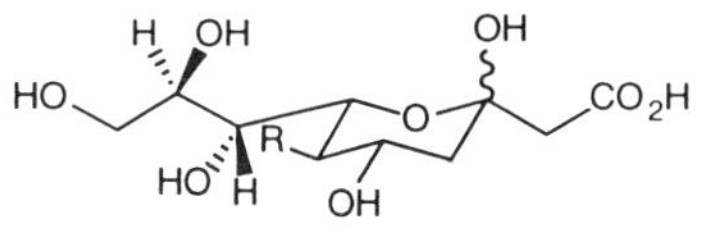

(**36**) R = NHAc

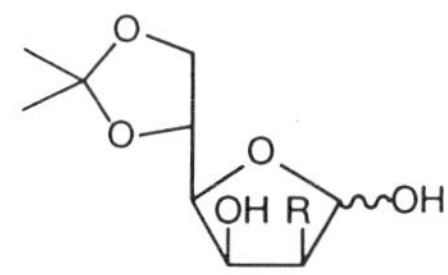

(**37**) R = NHAc

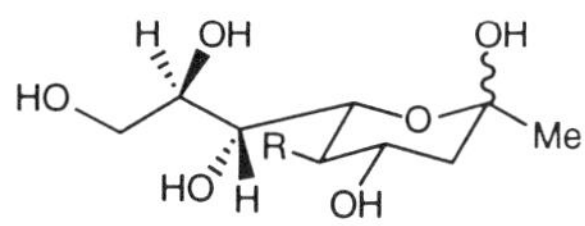

(**38**) R = NHAc

The *cis*-selective Wittig condensation between the dialdose **39** and the Wittig salt **40** to furnish the C_9 adduct **41** was used as a key step in the synthesis of the 3,5-dideoxy-2-nonulose derivative **42** [53]. Compound **42** can also be prepared directly from Neu5Ac1,β2Me_2 (**43**) *via* reduction with sodium borohydride [54].

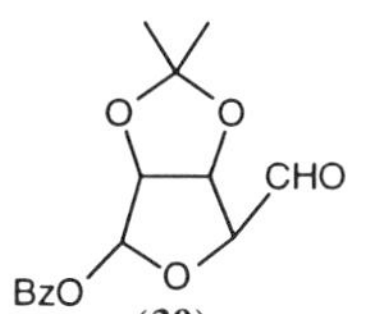

(**39**)

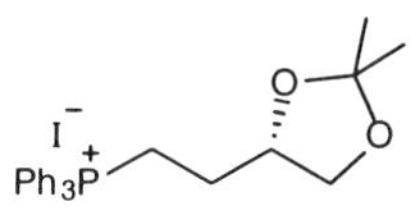

(**40**)

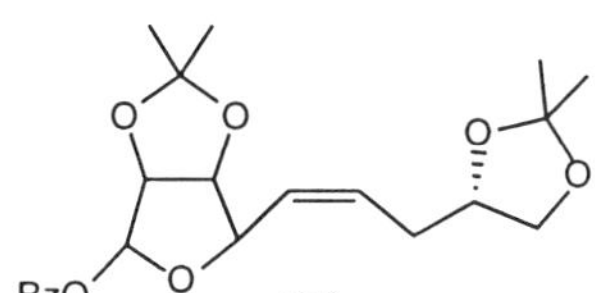

(**41**)

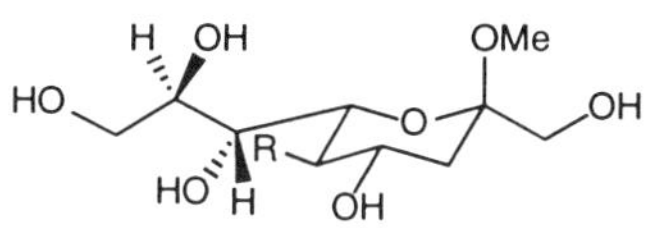

(**42**) R = NHAc

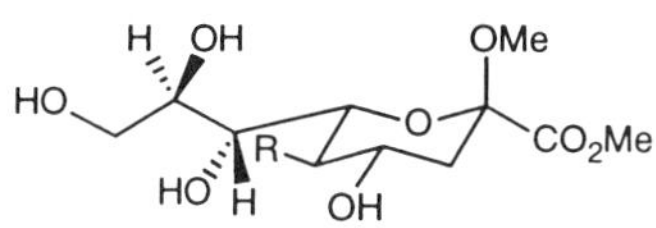

(**43**) R = NHAc

Analogues of Neu5Ac structurally varied at C-3, apart from those for improving glycosidation reactions [41], generally involve exchange of one of the protons with a hydroxyl group or halogen [9]. For example, some C-3 fluorinated derivatives of Neu5Ac have been prepared as potential inhibitors of influenza virus sialidase [55, 56] (see Section 3.2.2). It must be remembered that both C-3 protons can be readily exchanged by deuterium, $H3_{ax}$ in the range pD 6.5–9.0 and $H3_{eq}$ at pD above 12.5 [57] This phenomenon has been used to prepare specifically C-3 labelled Neu5Ac analogues for studying enzymatic transformations [58, 59].

The discovery that 4-*O*-acetylsialosyl oligosaccharides are stable to sialidases [60] has prompted the synthesis of several C-4 modified sialic acids as potential probes for structure-activity relationships [9]. Indeed, along with C-9, modifications at C-4 are the most commonly reported analogues of Neu5Ac [9].

Preparations of various C-4 hydroxy analogues of Neu5Ac, including *O*-acetyl and *O*-methyl derivatives and their *epi-* counterparts, are adequately covered elsewhere [9]. The synthesis of the 4-oxo derivative **44**, *via* the selective oxidation of the corresponding 4,7-diol [61] has allowed the preparation of a variety of C-4 modified Neu5Ac analogues.

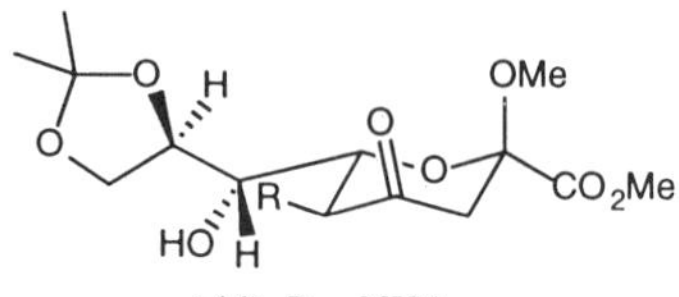

(**44**) R = NHAc

Treatment of **44** with tetramethylzirconium in THF gave a 3:2 mixture of the epimeric 4-*C*-methyl compounds **45** and **46**, respectively [62], whilst exposure of **44** to tributoxymethylzirconium furnished exclusively **45**. Similarly, methylenation of **44** using the system zirconocene dichloride/zinc/diiodomethane furnished the 4-methylene derivative **47**, which under hydrogenation conditions gave the two epimeric 4-deoxy-4-*C*-methyl Neu5Ac analogues **48** and **49** [62, 63]. Interestingly, the deprotected 4-*C*-methylene-β-methylketoside **50** was found to be a competitive inhibitor of CMP-sialate synthase [62].

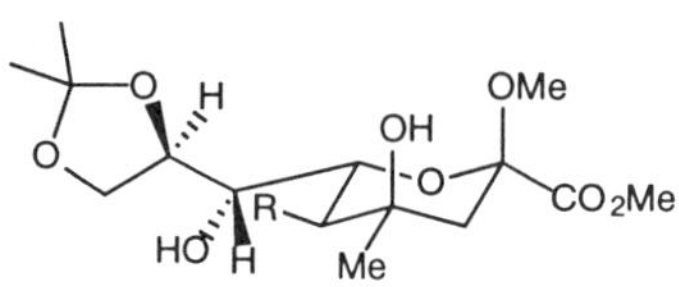

(**45**) R = NHAc

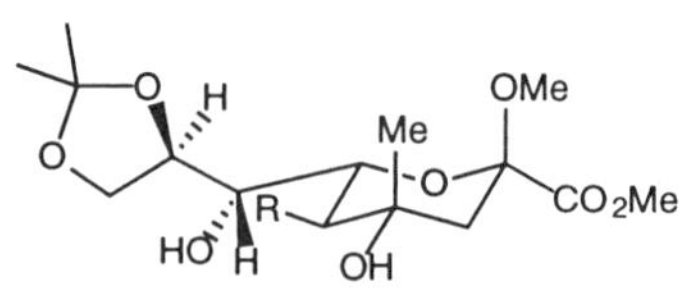

(**46**) R = NHAc

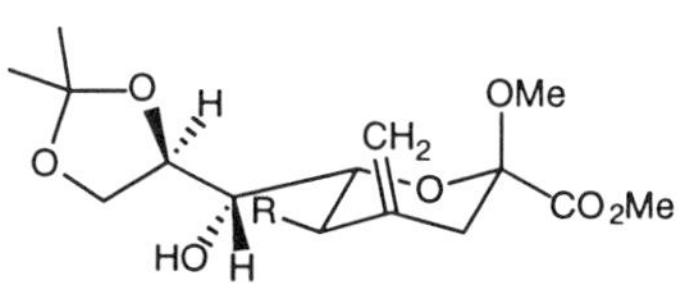

(**47**) R = NHAc

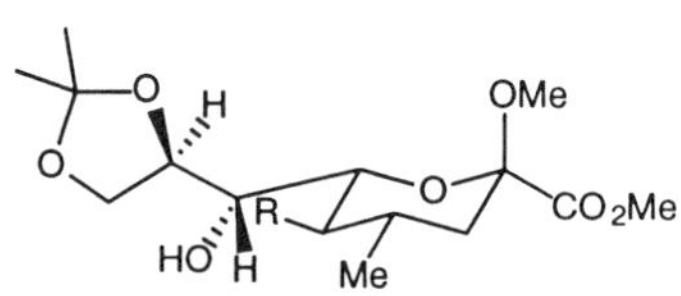

(**48**) R = NHAc

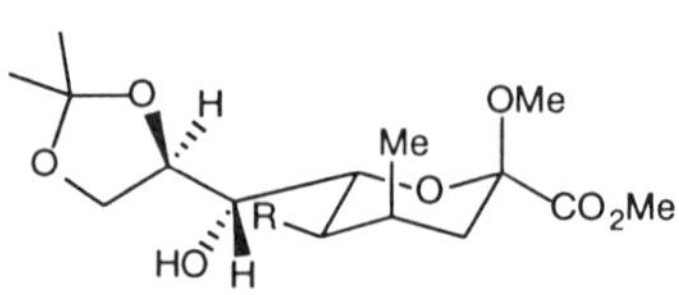

(**49**) R = NHAc

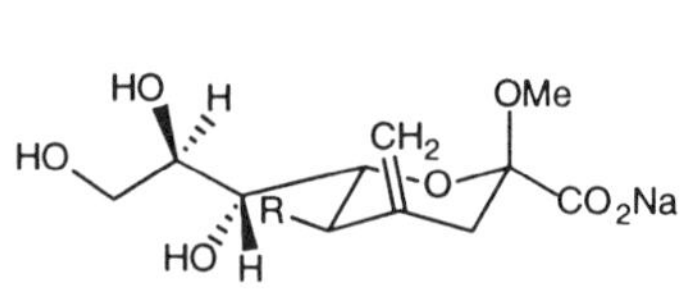

(**50**) R = NHAc

The 4-oxo-derivative **44** has also served as a synthetic precursor to the novel 4-oxime derivative **51**, prepared in 87% yield as the single oxime shown by treatment of **44** with methoxylamine [64].

4-Deoxy-Neu5Ac (**52**) has been prepared in a synthesis commencing from ManNAc [65]. A key feature of this 8-step synthesis was the Wittig-Wadsworth-

Emmons reaction between the aldehyde **53** and the ylide prepared from ethyl 2-(*tert*-butyldimethyl)silyloxy-2-(dimethylphosphono)acetate to give the adduct **54**, which after de-*O*-acetylation and saponification with concomitant de-*O*-silylation gave **52** [65].

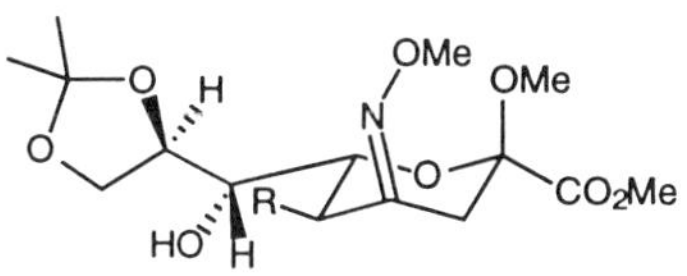

(**51**) R = NHAc

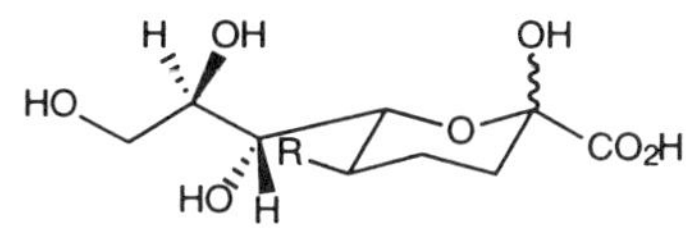

(**52**) R = NHAc

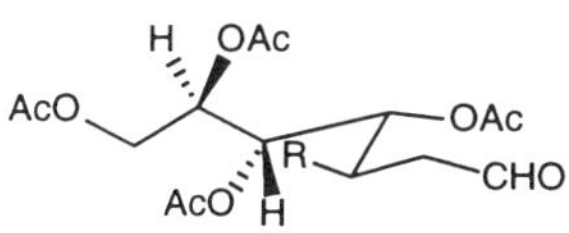

(**53**) R = NHAc

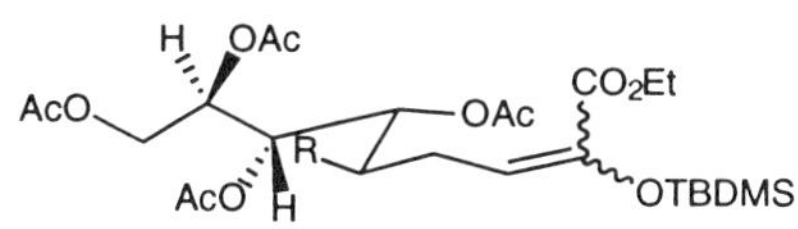

(**54**) R = NHAc

In a simple synthesis of 4-deoxy-4-guanidino-Neu5Acα2Me (**55**), a potential inhibitor of influenza virus haemagglutinin (see Section 3.2.1), treatment of the 4-azido-4-deoxy-Neu5Ac2en derivative **56** (for the preparation of **56** and related compounds see Section 2.2.3) with *N*-bromosuccinimide in methanol gave the 4-azido bromide **57**, together with its diastereomer **58** [66]. Debromination and concomitant azide reduction of the desired α-methylketoside **57** yielded the 4-amino hydrochloride salt **59**, which was elaborated to **55** under standard conditions.

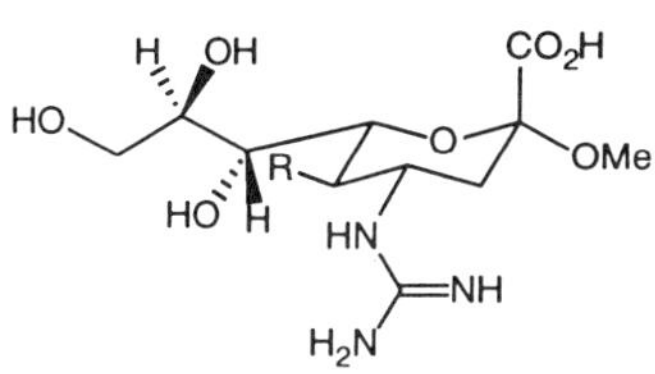

(**55**) R = NHAc

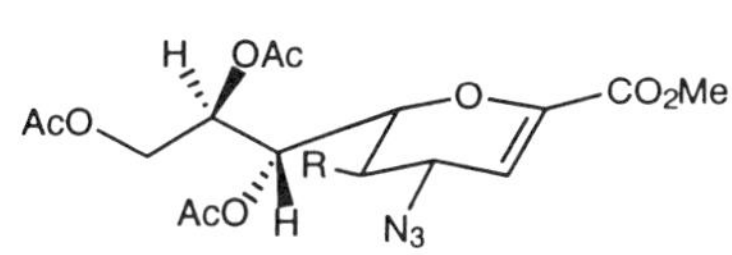

(**56**) R = NHAc

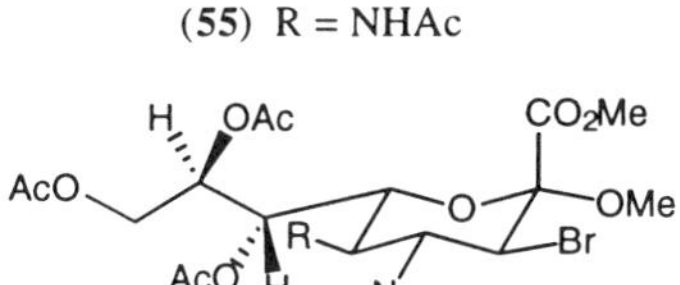

(**57**) R = NHAc

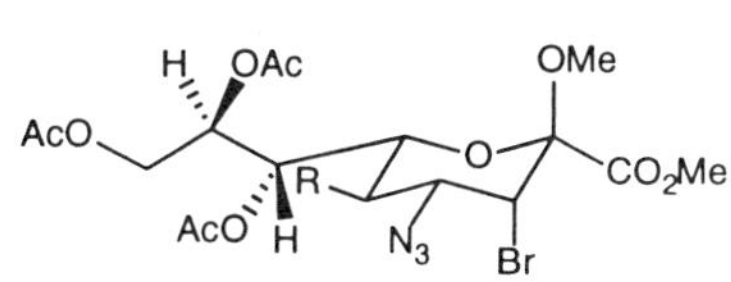

(**58**) R = NHAc

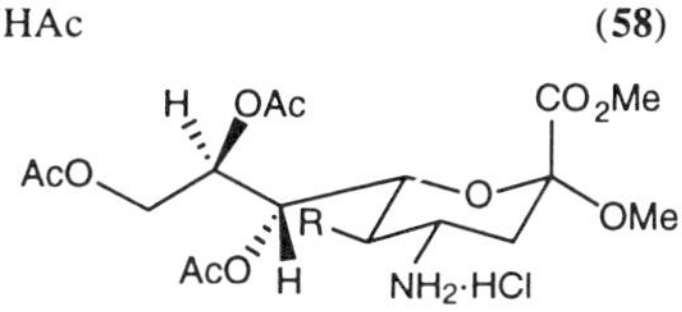

(**59**) R = NHAc

Structural variations at C-5 of Neu5Ac fall into two broad groups, *viz*: those with various *N*-acyl groups and those in which the *N*-acetamido group has been replaced with a different functionality. For those compounds in the latter group it must be remembered that sialic acid analogues not bearing an *N*-acyl substituent at C-5 are perhaps better described as KDN analogues.

The reported syntheses of modified *N*-acyl analogues of Neu5Ac, including *N*-glycolyl [11, 67], haloacetamido [11, 67] and substituted alkylacetamido [11, 20], are too numerous to mention here [11]. Apart from the preparation of such derivatives using Neu5Ac aldolase with the appropriate ManNAc substrate (see Section 2.1.1), the most common chemical syntheses involve the acylation of neuraminic acid (**60**) itself. Neuraminic acid can be prepared directly from Neu5Ac *via* hydrolysis of the *N*-acetamido group [11, 21, 67, 68] or by hydrogenation of a 5-azido or 5-carbobenzoxy [20] group in Neu5Ac. Recently, this general strategy has been used in the preparation of the two anomers of the *N*-trifluoroacetyl benzyl glycoside **61** [69], several *N*-acyl derivatives including the *N*-cyclopropanoyl analogue **62** [20] and *N*-thioacyl derivatives such as **63** [21].

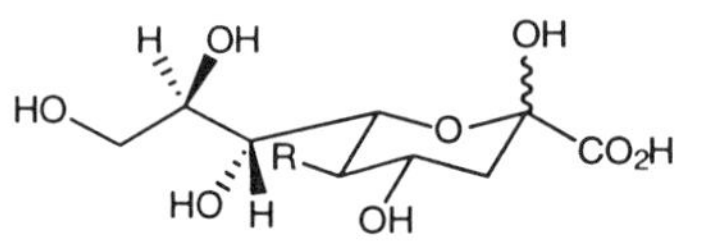

(**60**) $R = NH_2$

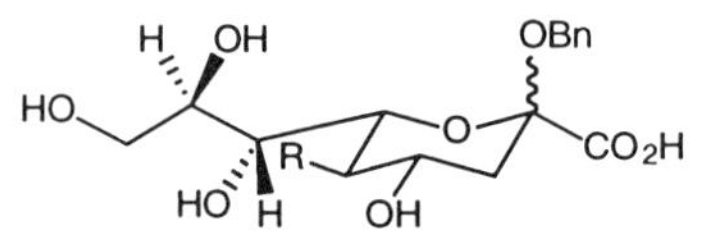

(**61**) $R = NHC(O)CF_3$

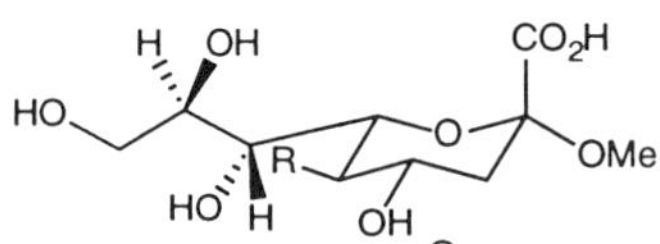

(**62**) R =
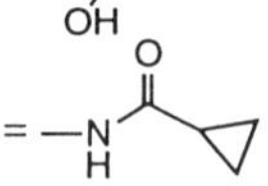

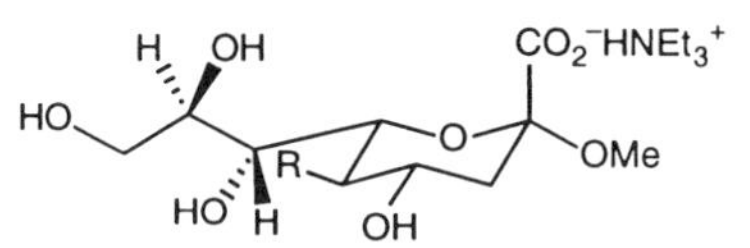

(**63**) R = NHC(S)Et

In an elegant synthesis of several C-5 modified Neu5Ac analogues the unstable 5-diazo derivative **64**, which was prepared *in situ* from Neu5Ac *via* the nitroso derivative **65**, is a pivotal intermediate [70]. Thus, treatment of **64** with tetrabutylammonium acetate in acetic acid gave the per-acetylated derivative **66** of KDN, whilst exposure of **64** to ammonia in toluene furnished 5-azido-5-deoxy-KDN (**67**). 5-Deoxy-KDN (**68**) was also prepared from the 5-nitroso derivative **65** [70].

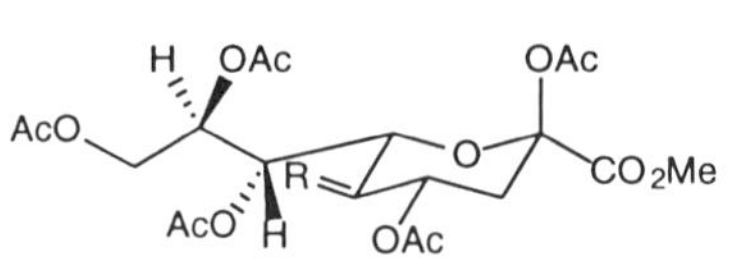

(**64**) $R = N_2$

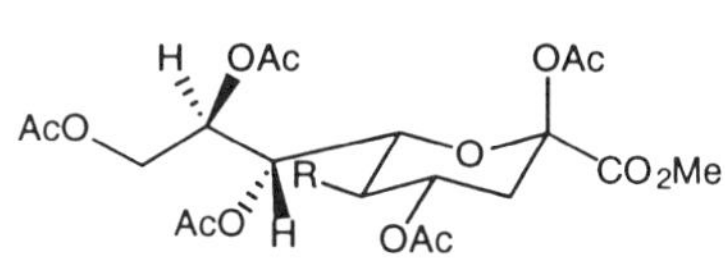

(**65**) R = N(Ac)NO
(**66**) R = OAc
(**67**) $R = N_3$
(**68**) R = H

There are few reports on structural modifications at C-6 of Neu5Ac. In a synthesis of 6-*C*-methyl-Neu5Ac (**69**) a key step was the Kornblum reaction of the nitropyranose **70** to give exclusively **71** in 94% yield. The nitromethyl group in **71** was then transformed to a methyl group, and subsequent oxidation and protecting group manipulation led to **72** which was elaborated to **69** [71]. Unlike Neu5Ac, compound **69** exists exclusively as the β-anomer shown. The synthesis of 6-amino-2,6-dideoxy-Neu5Ac (**73**), and related compounds, *via* the chain elongation of the nitromannosamine derivative **74** to ultimately give **75** has also been reported by these workers [72]. A series of glycerol side-chain shortened analogues of **73** have also been prepared [73]. 6-Amino-2,6-dideoxy-Neu5Ac (**73**) along with its C-4 epimer were found to be competitive inhibitors of viral sialidases [72].

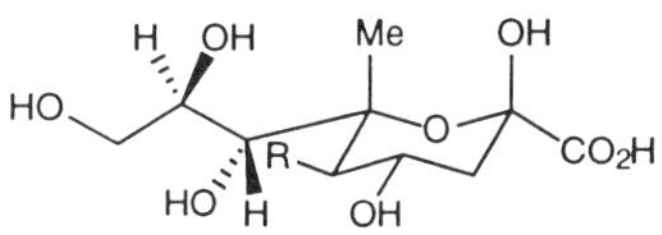

(**69**) R = NHAc

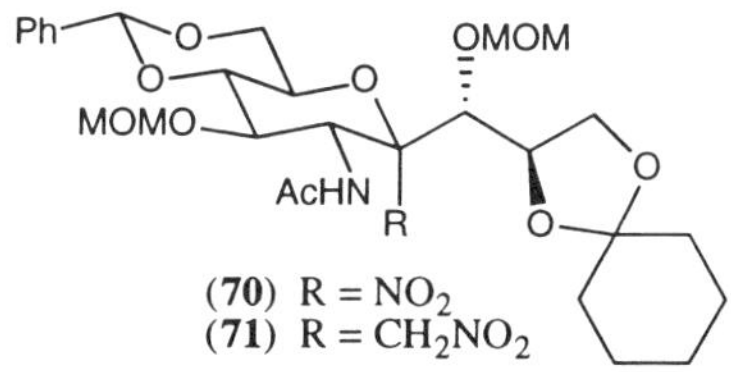

(**70**) R = NO_2
(**71**) R = CH_2NO_2

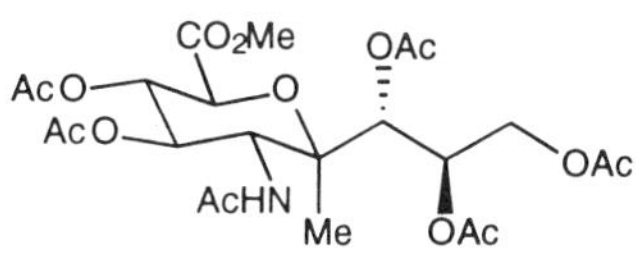

(**72**)

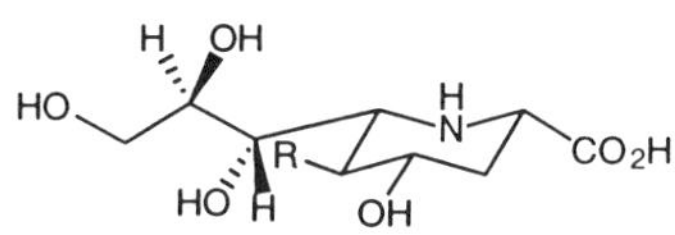

(**73**) R = NHAc

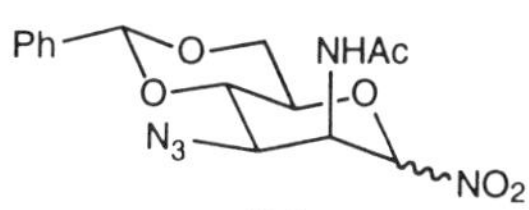

(**74**)

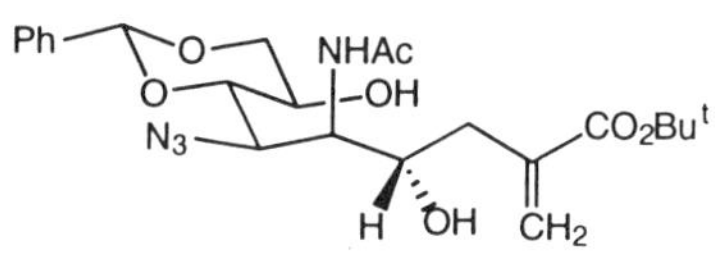

(**75**)

The condensation of D-glucose with oxalacetic acid to give the furanose derivative **76** has been used to prepare the 6-acetamido-nonulofuranosonic acid **77**, and its C-4 epimer **78** [74]. Both **77** and, to a lesser extent, **78** were found to inhibit influenza virus sialidase [74].

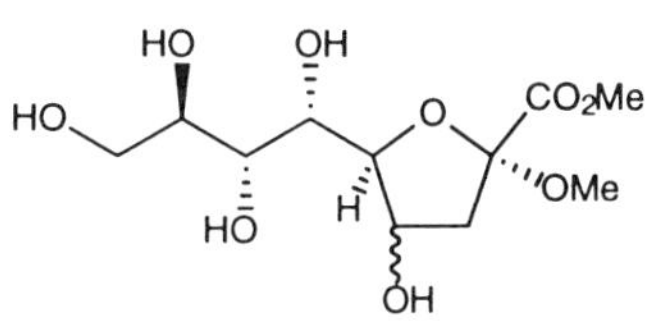

(**76**)

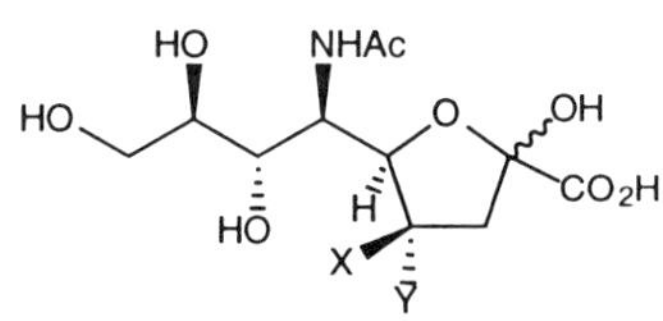

(**77**) X = H, Y = OH
(**78**) X = OH, Y = H

The carbocyclic analogues **79** and **80** of Neu5Ac and KDO, respectively, have been synthesised in several steps from the Diels-Alder *endo*-adduct **81** of furan and acrylic acid [75]. Compound **79** was found to be a weak inhibitor of sialidase from *Streptococcus sp.* [75].

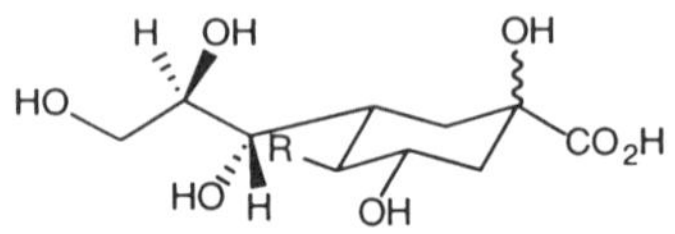

(**79**) R = NHAc

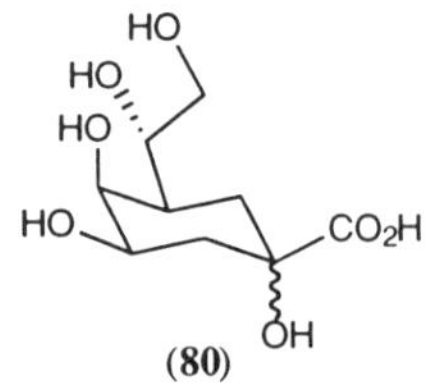

(**80**)

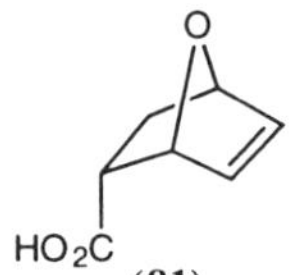

(**81**)

Several modifications have been made to the glycerol side-chain of Neu5Ac, including chain-shortened analogues and various epimeric or deoxy derivatives. Such compounds are discussed elsewhere [9]. The synthesis of the 7-oxo- and 8-oxo-Neu5Ac derivatives **82** and **83**, respectively, has been achieved *via* the oxidation of the selectively protected Neu5Ac derivative **84** [28]. Interestingly, the 7-oxo- derivative **82** can be directly converted to the 8-oxo- derivative **83** under well-defined conditions [28].

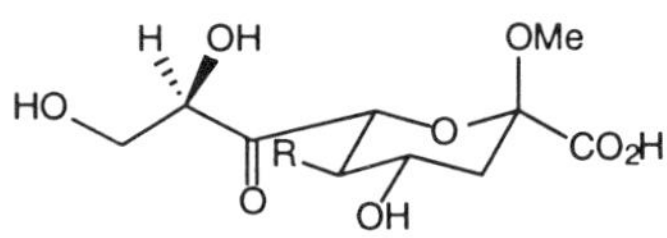

(**82**) R = NHAc

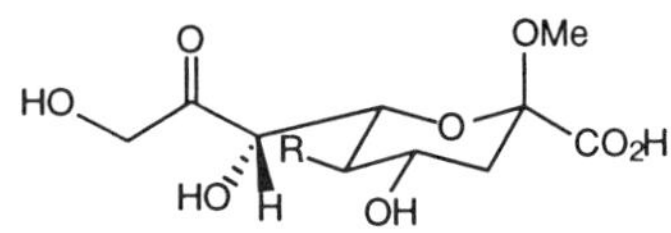

(**83**) R = NHAc

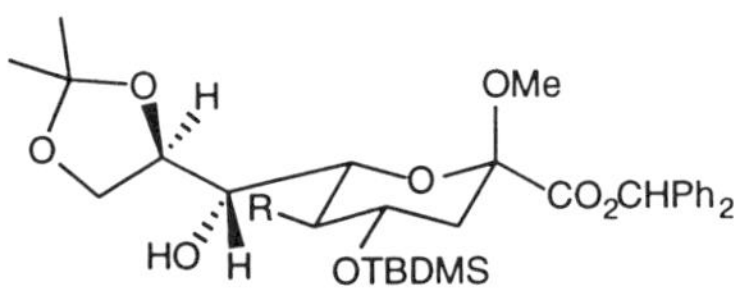

(**84**) R = NHAc

Syntheses of C-9 modified Neu5Ac derivatives, particularly those involving exchange of the hydroxyl group with a nitrogen-based functionality, are generally achieved *via* the enzymatic conversion of the corresponding C-6 mannosamine derivative and are discussed elsewhere [9, 14]. 9-Deoxy-9-fluoro-Neu5Ac (**85**) has been prepared directly from Neu5Ac *via* the 9-*O*-trityl derivative **86**, which was de-*O*-tritylated and treated with diethylamino-sulfur trifluoride in order to introduce the fluorine functionality [76].

A series of 9-deoxy-9-*N*-substituted Neu5Ac derivatives have been prepared from the 9-*O*-tosyl derivative **87**. Displacement of the tosyl group with azide gave 9-azido-9-deoxy-Neu5Ac (**88**), which was then elaborated into the 9-amino-9-deoxy derivative **89** and the 9-deoxy-9-thioacetamido analogue **90** [21].

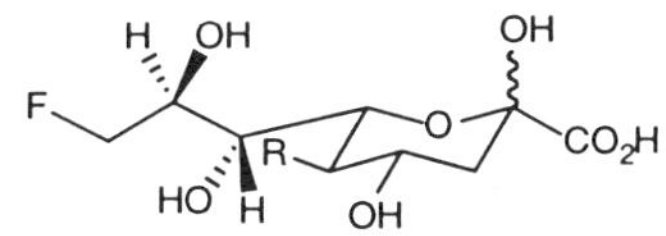

(**85**) R = NHAc

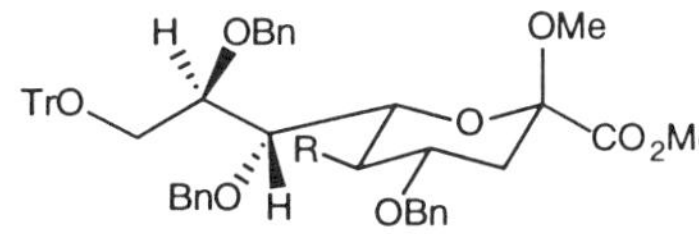

(**86**) R = NHAc

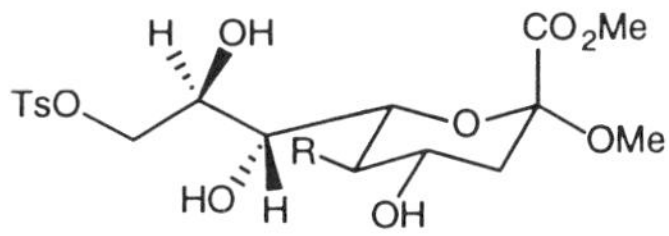

(**87**) R = NHAc

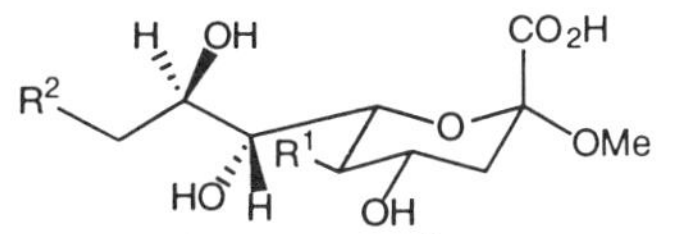

(**88**) R^1 = NHAc, R^2 = N_3
(**89**) R^1 = NHAc, R^2 = NH_2
(**90**) R^1 = NHAc, R^2 = NHC(S)Me

In a recent report detailing the preparation of glycerol chain-extended analogues of Neu5Ac, Swern oxidation and subsequent Wittig olefination of the 9-hydroxy Neu5Ac derivative **91** led to the C_{11} α,β-unsaturated ester **92**, the C_{11} saturated ester **93** and the C_{11} α,β-unsaturated aldehyde **94**, all in good overall yield [77].

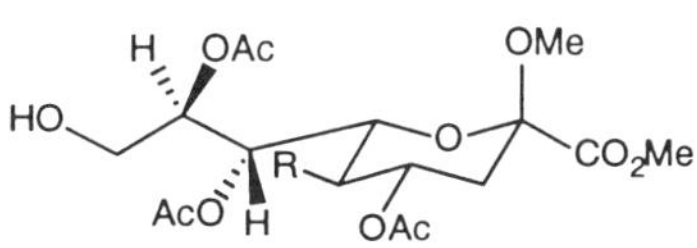

(**91**) R = NHAc

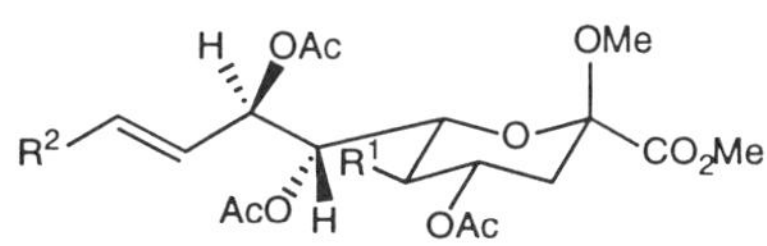

(**92**) R^1 = NHAc, R^2 = CO_2Me
(**94**) R^1 = NHAc, R^2 = CHO

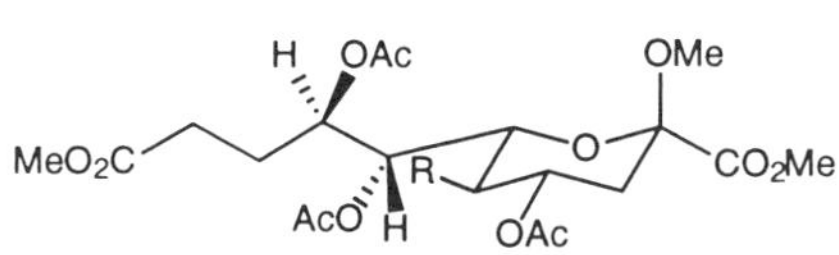

(**93**) R = NHAc

In many of the chemical syntheses of modified sialic acids presented above, an important consideration is the ability to prepare analogues of Neu5Ac in which various hydroxyl groups can be selectively manipulated. The preparation of partially acetylated derivatives is of particular interest. In this regard, the synthesis of several 9-*O*-acyl Neu5Ac derivatives, directly from Neu5Ac, using ortho esters [78], and the preparation of various mono-, di- and tri-*O*-acetylated derivatives [79] are valuable contributions. These reports, along with the selective syntheses of Neu4,5Ac_2 [78], Neu5,7Ac_2 [80] and Neu4,5,9Ac_3 derivatives [81], give an insight into the chemical transformations necessary to achieve manipulation at a given position of Neu5Ac.

2.2 Synthesis of Sialic Acid–Based Biological Probes

2.2.1 2-Deoxy derivatives

The importance of 2-deoxy-Neu5Ac derivatives as probes for structure–activity relationships with proteins associated with Neu5Ac recognition and metabolism cannot be underestimated. In particular, the discovery that an axially–oriented carboxylate group is vital for the binding of Neu5Ac to the X-31 haemagglutinin of influenza virus, whereas the α-glycosidic oxygen is not involved in any recognition process [82, 83] has prompted the synthesis of several 2-deoxy-Neu5Ac derivatives [9].

2-Deoxy-$2H_{eq}$-Neu5Ac (**95**) and 2-deoxy-$2H_{ax}$-Neu5Ac (**96**) have been prepared as a 9:2 mixture, respectively, *via* catalytic hydrogenation of the 2-chloro-2-deoxy-β-Neu5Ac derivative **34a** followed by saponification [84], whilst treatment of **34a** with tributyltin hydride gives exclusively **96** [84]. In an alternative approach towards the synthesis of 2-deoxy-Neu5Ac and 2-deoxy-KDO analogues, a simple and efficient one-pot anomeric deoxygenation of the corresponding peracetylated methyl esters of Neu5Ac and KDO using SmI_2 gave **95** and **97**, respectively, in excellent yield after saponification [85].

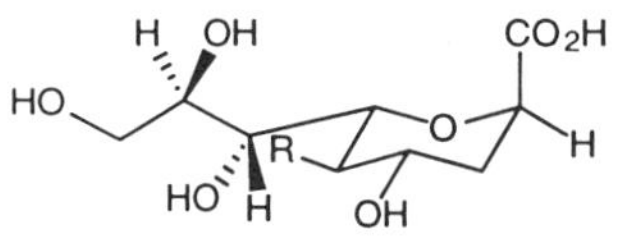

(**95**) R = NHAc

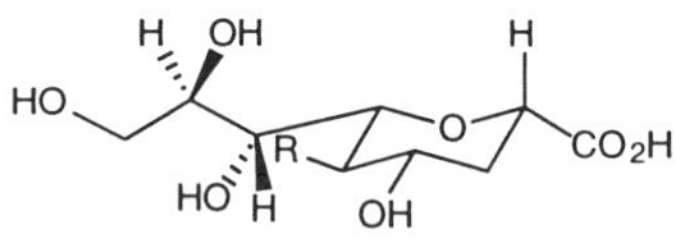

(**96**) R = NHAc

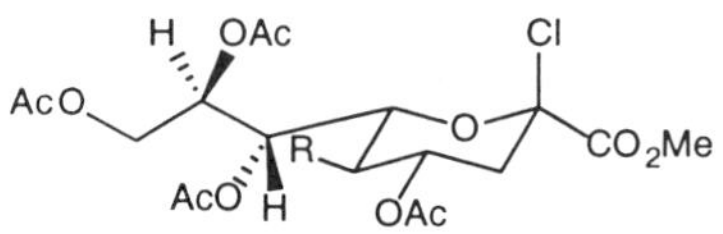

(**34a**) R = NHAc

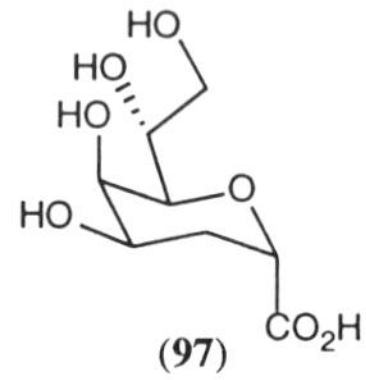

(**97**)

Related 2-deoxy-Neu5Ac analogues have been prepared in similar ways to **95**, or by using **95** itself as starting material [9]. Both **95** and **96** have been found to possess useful biological properties, with **95** showing inhibition of influenza virus haemagglutinin [86] whilst its isomer **96** is weakly recognised by *Vibrio cholerae* sialidase [87]. Several C-4 modified 2-deoxy-$2H_{eq}$-Neu5Ac derivatives, analogous to compounds **44** to **50** discussed earlier (Section 2.1.2), have been synthesised *via* the 4-oxo-derivative **98** [86].

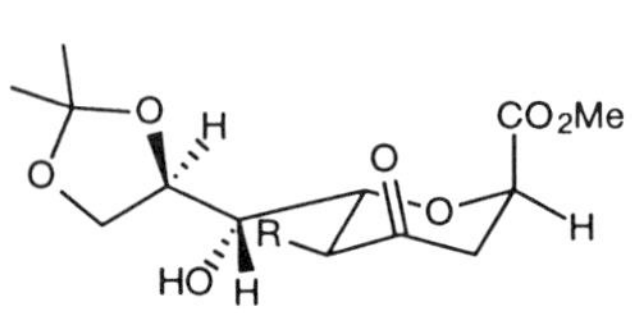

(**98**) R = NHAc

The 2-chloro-2-deoxy-β-Neu5Ac derivative **34a**, a valuable intermediate for glycoside syntheses (Section 2.1.2), has also been used as a precursor of 2-deoxy-2-substituted-Neu5Ac derivatives. In one example, exposure of **34a** to sodium azide under phase transfer catalysed conditions furnished the 2-azido-2-deoxy-α-Neu5Ac derivative **99** in excellent yield [88].

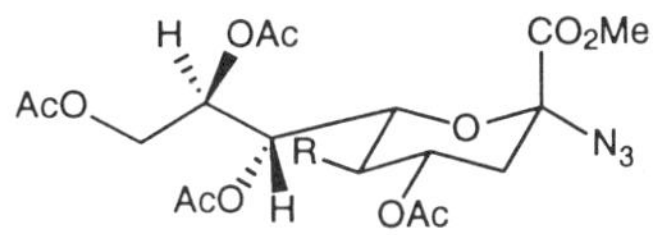

(**99**) R = NHAc

The finding that the phosphonic acid **100**, a pyrrolidine analogue of Neu5Ac, is a notably stronger inhibitor of *Vibrio cholerae* sialidase than the corresponding carboxylic acid [89] has prompted the synthesis of the phosphonic acid analogues **101** and **102** of 2-deoxy-Neu5Ac. Thus, Neu5Ac2en1Me (**103**) was elaborated to the acid **104** and subsequent oxidative decarboxylation gave predominantly **105** [87]. Phosphonylation of **105** then gave a mixture of **101** and **102** after deprotection. The synthesis of a similar derivative to **105**, *viz:* **106**, together with the α,β-unsaturated lactone **107**, has been achieved by the lead tetraacetate mediated decarboxylation of Neu4,5,7,8,9Ac$_5$ [90].

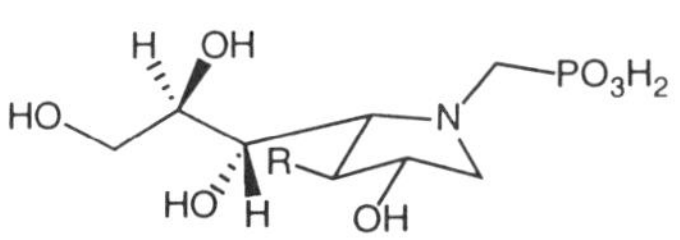
(**100**) R = NHAc

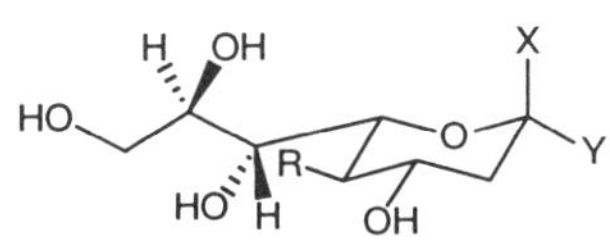
(**101**) R = NHAc, X = PO_3H_2, Y = H
(**102**) R = NHAc, X = H, Y = PO_3H_2

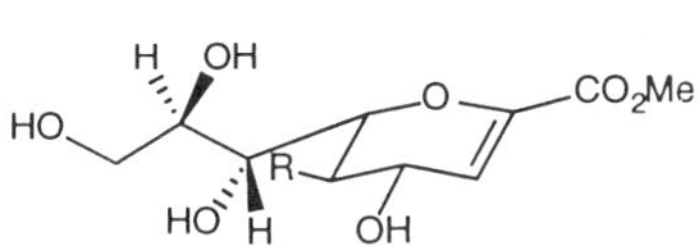
(**103**) R = NHAc

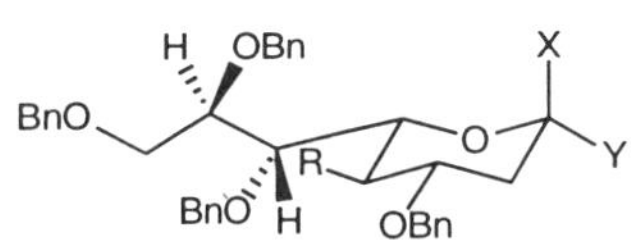
(**104**) R = NHAc, X = H, Y = CO_2H
(**105**) R = NHAc, X = OAc, Y = H

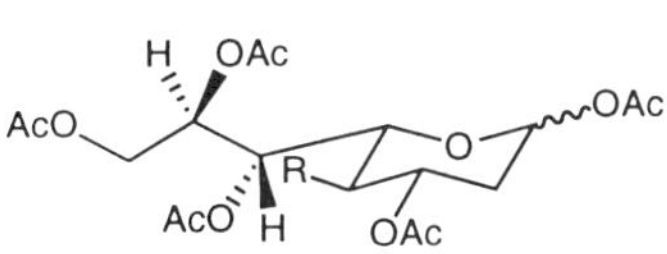
(**106**) R = NHAc

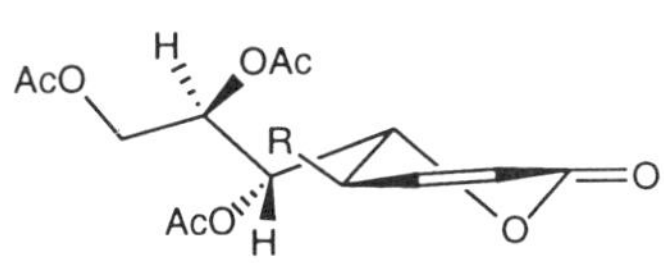
(**107**) R = NHAc

The phosphonic acids **101** and **102** are stronger inhibitors of *Vibrio cholerae* sialidase than the corresponding Neu5Ac compounds, and whereas only that 2-deoxy-Neu5Ac analogue with an equatorial carboxylic acid group (ie. **96**) inhibits *Vibrio cholerae* sialidase, both phosphonic acids are inhibitors [87]. Furthermore, both **101** and **102** have been found to inhibit sialidase from influenza virus (type A, N9 and N2), although only the α-anomer **102** is a better inhibitor than Neu5Ac (see also Section 3.2.2) [91]. It is believed that the increased inhibition of this sialidase by **102**, compared to that of Neu5Ac, is due to a stronger electrostatic interaction of the phosphonyl group in the active site as well as a lower distortion energy requirement for binding [91].

2.2.2 Thioketosides

Thioketosides of Neu5Ac are of interest not only as glycosyl donors [43, 51, 92-95], but also as biological probes for sialic acid-recognising proteins [96, 97] given that the thioglycosidic linkage is resistant to enzyme degradation. Simple alkyl or aryl 2-thioketosides of Neu5Ac, eg. **108**, have been prepared in a variety of ways. Early efforts at preparing such thioketosides required several steps from Neu5Ac ([98] & refs therein). Subsequently it has been found that treatment of peracetylated Neu5Ac1Me with thiols in the presence of boron trifluoride etherate gave **108** as predominantly the β-ketoside [48], whilst exposure of the same substrate to (methylthio)trimethylsilane in the presence of trimethylsilyl triflate afforded the methylthioketoside of Neu5Ac (**108**, R = Me) as a 1:1 mixture of anomers [47]. Similarly, reaction of the 2-chloro-2-deoxy-β-Neu5Ac derivative **34a** with sulfur nucleophiles afforded the α-thioketoside of **108** [48, 97, 99, 100]. The 2-chloro-2-deoxy-β-Neu5Ac derivative **34a** has also been used as a precursor to various α-arylthioketosides of Neu5Ac using phase transfer catalysed conditions [79, 95]. In this way compounds such as **109** [79, 95] and **110** [95] have been prepared in high yield.

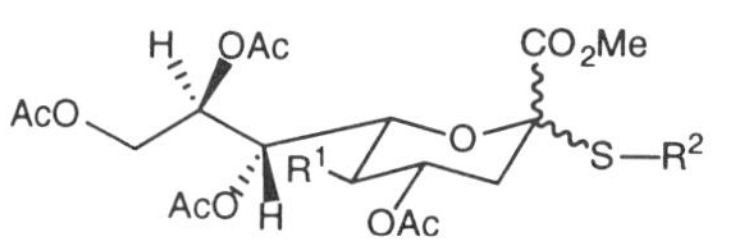

(**108**) R^1 = NHAc, R^2 = alkyl, aryl

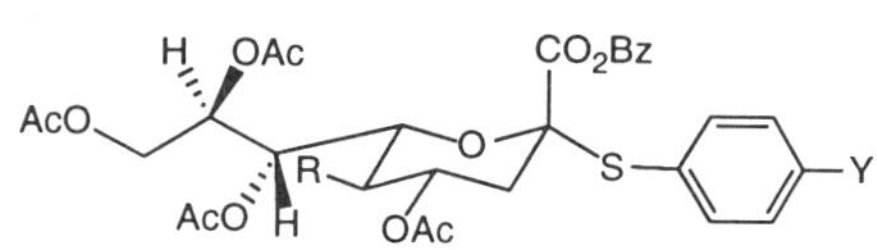

(**109**) R = NHAc, Y = NH_2,NO_2

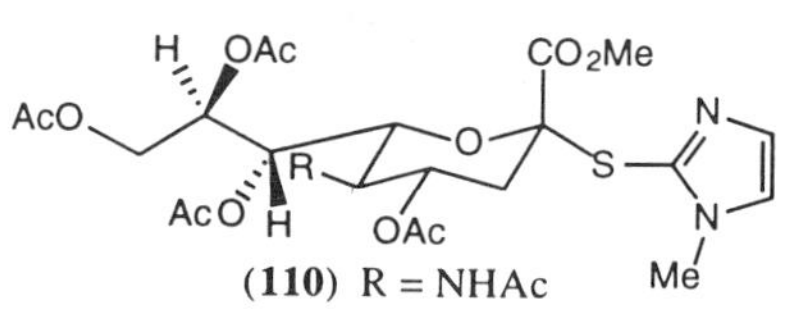

(**110**) R = NHAc

The synthesis of the 4-, 7-, 8- and 9-deoxy-Neu5Ac-2-methylthioketosides (eg. **111**), as a way of probing the role of the hydroxyl groups of Neu5Ac in the biological function of gangliosides, has been achieved *via* the reduction of the corresponding *O*-(phenoxy)thiocarbonyl derivative (eg. **112**) followed by introduction of the thiomethyl group by replacement of an anomeric acetate [93]. Various partially acetylated derivatives of **113**, including mono-, di- and tri-*O*-acetylated analogues, have been prepared using either selective acetylating agents (trimethyl orthoacetate or dimethylacetamide dimethyl acetal) or *via* a partial Zemplén de-*O*-acetylation of **109** [79].

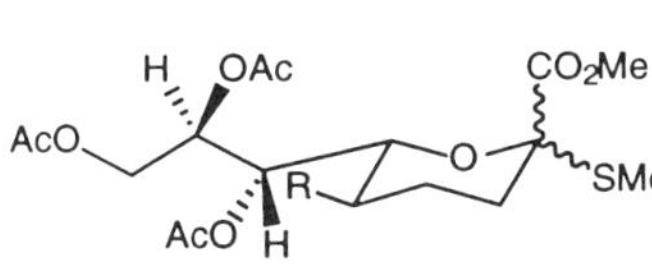

(**111**) R = NHAc

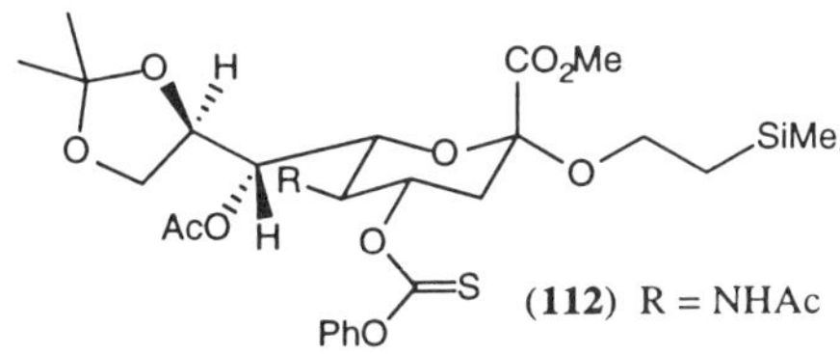

(**112**) R = NHAc

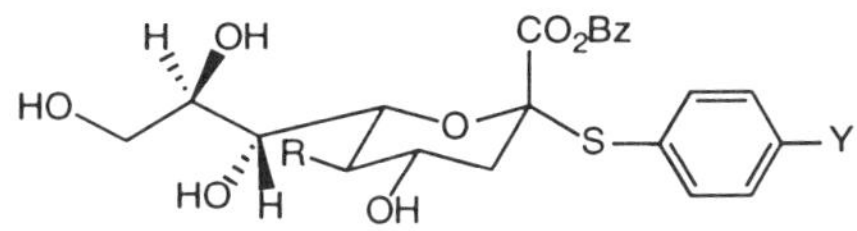

(**113**) R = NHAc, Y = NH_2,NO_2

It has been found that when simple thioketosides such as **108** (R = alkyl, aryl) act as glycosyl donors the glycosidation proceeds in high yield and affords predominantly the α-glycosides, even if **108** is a mixture of anomers [43]. This methodology has been successfully applied to the synthesis of several α-glycosides of Neu5Ac, including α-(2→3)- and α-(2→6)-sialosyl glycosides, gangliosides and lactosyl ceramides, and is discussed in detail elsewhere [43].

More complex α-thiosialosides (eg. **114**, R^2 = glycosides, oligosaccharides, fluorescent probes) have been prepared as biological probes for sialic acid-recognising proteins, given the stability of the thioketosidic linkage towards sialidases [101]. Indeed, the discovery that thioglycosidic analogues of gangliosides are not hydrolysed by influenza virus sialidase [102] has prompted the synthesis of several thioglycosidic analogues of glycosphingolipids [43].

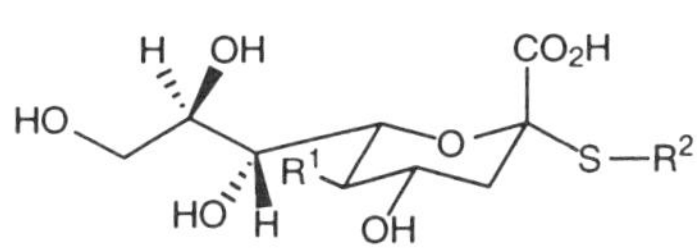

(**114**) R^1 = NHAc

The general strategy for the synthesis of thioglycosidic analogues of these biologically important compounds centers on the 2-thioacetyl derivative **115** Neu5Ac. Compound **115** is typically prepared in high yield by reaction of the 2-chloro-2-deoxy-β-Neu5Ac derivative **34a** with potassium thioacetate [43, 99]. De-thioacetylation of **115** and then coupling with an activated glycosyl acceptor is the commonly used method for gaining access to various α-linked thiosialosides [43]. In such thioglycosidation reactions the glycosyl acceptors are usually activated using a halogen, tosylate or triflate. Initially, the de-thioacetylation of **115** was achieved by using an equimolar amount of sodium methoxide at low temperature [97, 99], but partial de-*O*-acetylation under such conditions [103, 104] prompted the development of a method for the selective thiodeacetylation of **115** using diethylamine in *N,N*-dimethylformamide [105]. This latter method has been successfully applied to the synthesis of α-(2→6)-

thioglycosides of Neu5Ac (eg. **116**) [105], α-(2→3)- and α-(2→4)-thiosialosides [106] and the Neu5Ac-α-5'-thiocytidine derivative **117** [107].

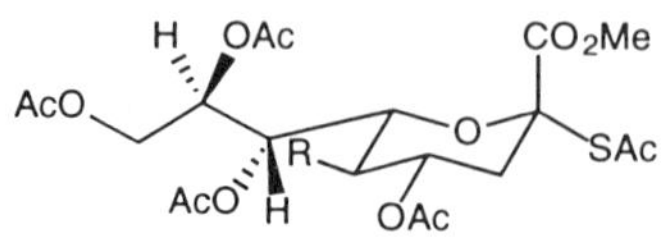

(**115**) R = NHAc

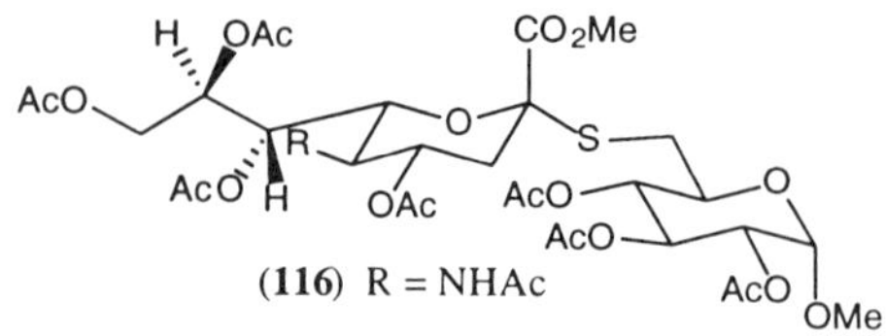

(**116**) R = NHAc

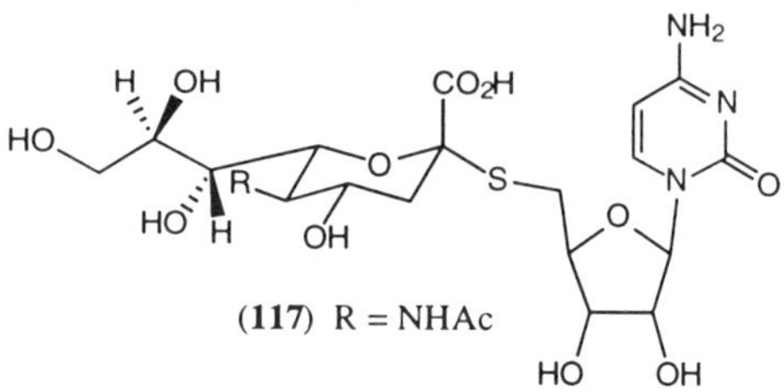

(**117**) R = NHAc

The use of thioglycosidic derivatives of Neu5Ac as potential inhibitors of sialic acid-recognising proteins has not yet been fully realised. However, thiosialosides such as **117** appear to be candidates for the inhibition of cell-surface sialylation and subsequent pulmonary metastasis [108, 109], whilst we have shown that α-(2→6)-thiosialosides such as **118**, which bears a 9-*O*-acetyl group, show a significant increase in rotaviral (bovine, NCDV) inhibition compared to the corresponding non-acetylated thiosialosides [106].

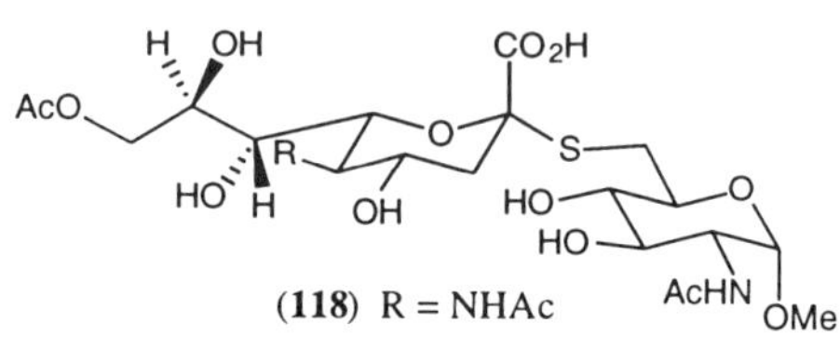

(**118**) R = NHAc

Other thioglycosides of Neu5Ac also exhibit useful biological properties. Thioglycosidic analogues of gangliosides [43] exhibit significant inhibition of influenza virus sialidases [102]. Interestingly, fluorescence-labelled thioketosides such as **119** [79] which, like **118**, also bears a 9-*O*-acetyl group, and **120** [110] which contains a 4-*O*-acetyl group have been synthesised. These compounds, together with the azidoaryl thioketoside derivative **121** [97], have all been prepared as probes for the analysis of sialic acid-recognising proteins.

2.2.3 2,3-Unsaturated derivatives

The discovery in 1969 that 2,3-didehydro-2,3-dideoxy-*N*-acetylneuraminic acid (Neu5Ac2en, **122**) is one of the strongest competitive inhibitors of sialidases [111] has prompted the synthesis of numerous analogues [9, 42].

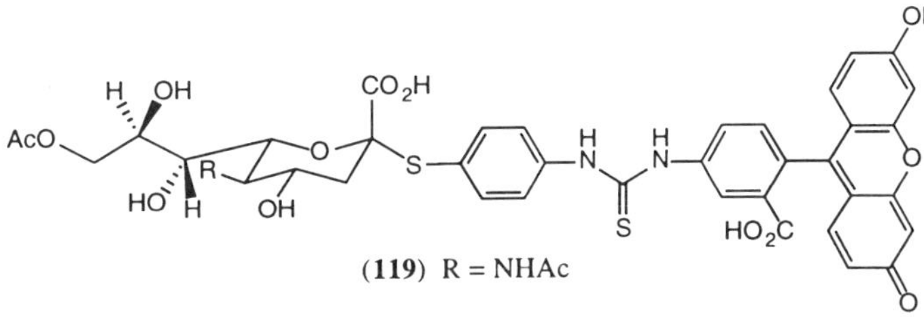

(**119**) R = NHAc

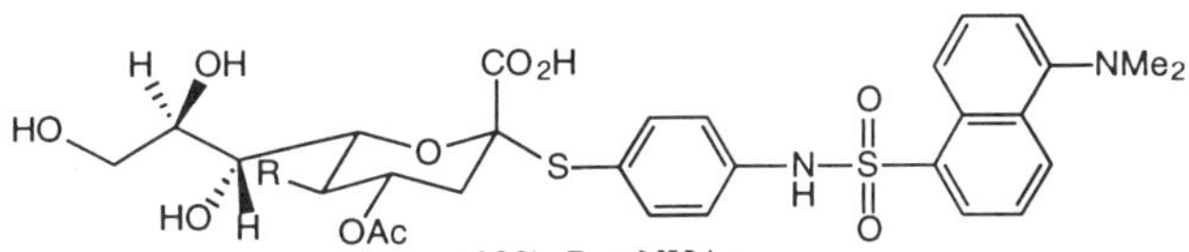

(**120**) R = NHAc

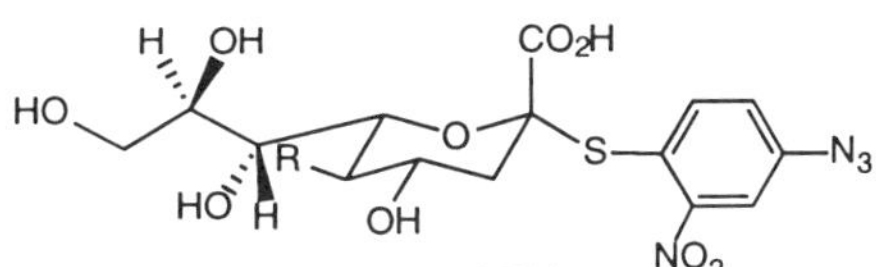

(**121**) R = NHAc

The effects that simple structural modifications of Neu5Ac2en have on the strength of inhibition of sialidases have been described elsewhere [9]. In the case of *Vibrio cholerae* sialidase the most important structural features for favourable interaction appear to be the carboxyl group, the axially-oriented 7-hydroxyl group, the quasi-equatorial 4-hydroxyl group, and the carbonyl moiety of the *N*-acetyl group [9]. Changes to any of these structural features has a significant effect on sialidase inhibition [9, 11]. More recent findings detailing the effects of structurally-modified Neu5Ac2en derivatives, especially C-4 analogues, on influenza virus sialidase inhibition are presented in the following section (Section 3.2.2).

The synthesis of Neu5Ac2en (**122**) itself is usually achieved by exposure of the 2-chloro-2-deoxy-β-Neu5Ac derivative **34a** to a base such as triethylamine or silver carbonate [42, 111] followed by deprotection. Other methods for the preparation of **122** include the treatment of Neu5Ac1Me with acetic anhydride and sulfuric acid to give Neu5Ac2en1Me and 4-*epi*-Neu5Ac2en1Me in equal amounts [112], whilst heating Neu4,5,7,8,9Ac$_5$1Me at 90° in dioxan also gives **122** after deprotection [111]. 2,3-Unsaturated Neu5Ac derivatives are also frequently formed as by-products from glycosidation reactions [10, 41].

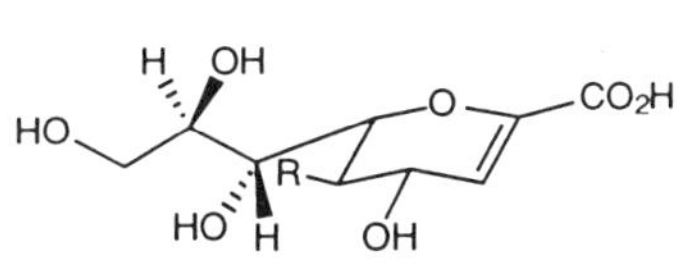

(**122**) R = NHAc

Apart from the synthesis of various *epi-* [113], deoxy- [114] and C-5 modified [115, 116] Neu5Ac2en derivatives, most studies in recent times have been directed towards the preparation of C-4 modified Neu5Ac2en derivatives (*vide infra*). The synthesis of both 4-azido-4-deoxy-Neu5Ac2en (**123**) and 4-*epi*-4-azido-4-deoxy-Neu5Ac2en (**124**) was achieved from Neu4,5,7,8,9Ac$_5$1Me *via* the oxazoline **125**. Exposure of **125** to trifluoroacetic acid in aqueous THF gave the 4-*epi*-Neu5Ac2en derivative **126** which, under Mitsunobu conditions, gave the epimeric 4-azido compounds **123** and **124** [117]. The 4-azido derivative **123** was further elaborated into 4-amino-4-deoxy-Neu5Ac2en, 4-deoxy-4-formamido-Neu5Ac2en and 4-acetamido-4-deoxy-Neu5Ac2en [117]. 4-Deoxy-Neu5Ac2en (**127**), prepared in several steps from *N*-acetylmannosamine [118], shows similar inhibitory activity as Neu5Ac2en towards influenza virus sialidase.

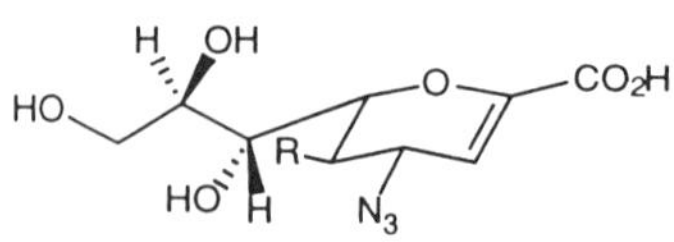

(**123**) R = NHAc

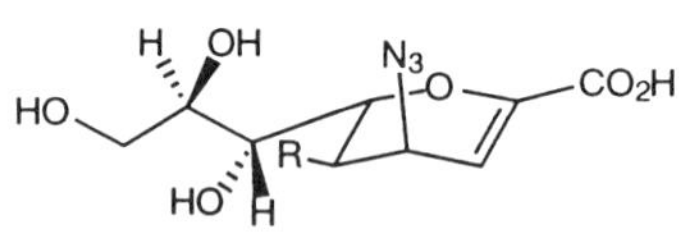

(**124**) R = NHAc

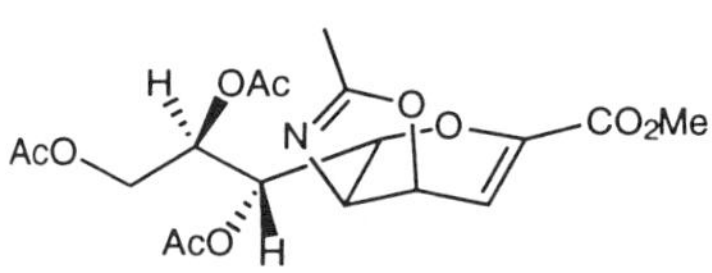

(**125**)

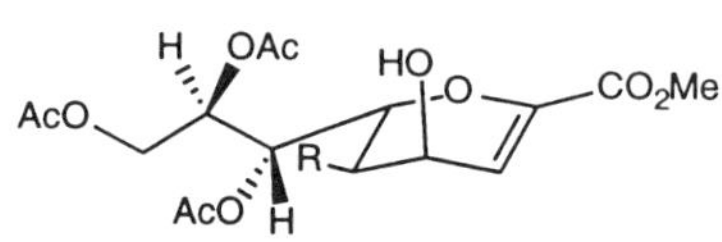

(**126**) R = NHAc

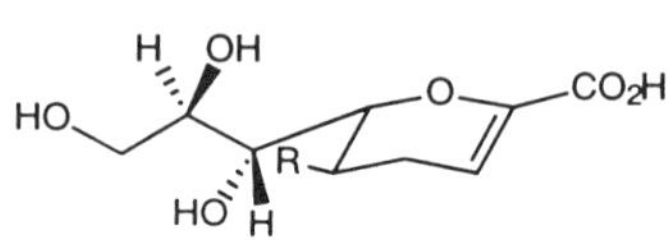

(**127**) R = NHAc

Recently a simple, efficient and stereoselective method for the introduction of a substituent at C-4 has been reported in which the oxazoline derivative **125**, prepared by exposure of Neu4,5,7,8,9Ac$_5$2en1Me to boron trifluoride etherate, is the key intermediate [119]. Treatment of **125** with lithium azide under acidic catalysis gave the 4-azido derivative **56** in >90% yield, whilst exposure of **125** to thiolacetic acid in anhydrous *N,N*-dimethylformamide gave the 4-thioacetate derivative **128** [119].

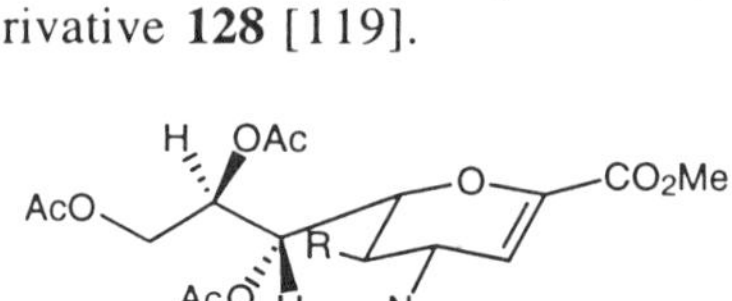

(**56**) R = NHAc

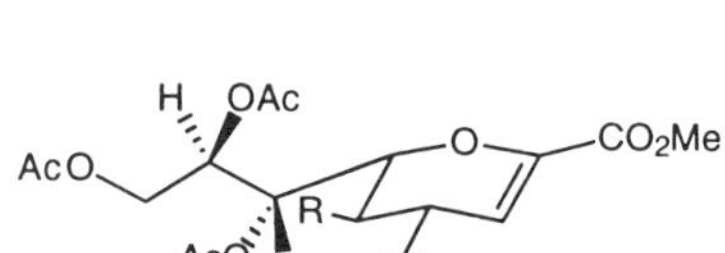

(**128**) R = NHAc

The 4-azido-4-deoxy-Neu5Ac2en derivative **56** has proved a versatile intermediate in the synthesis of potent influenza virus sialidase inhibitors. Thus, reduction of **56** under hydrogenation conditions gave 4-amino-4-deoxy-Neu5Ac2en (**129**) after saponification [120]. Exposure of **129** to aminoimino-methanesulfonic acid in the presence of base gave 4-deoxy-4-guanidino-Neu5Ac2en (**130**) [120], a compound which has been found to be a potential anti-influenza drug (see Section 3.2.2).

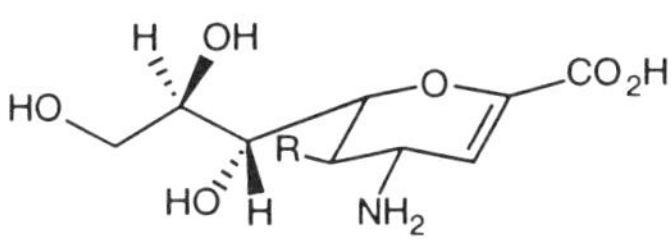

(**129**) R = NHAc

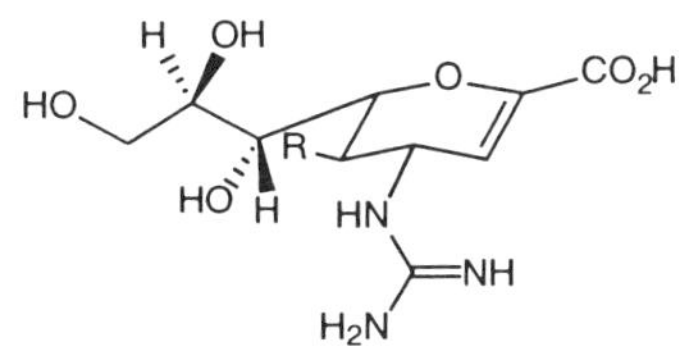

(**130**) R = NHAc

The discovery that both **129** and **130**, but in particular **130**, are potent inhibitors of influenza virus sialidase [121] (see also Section 3.2.2) has prompted the synthesis of a variety of analogues of **130**. A series of 4-deoxy-4-guanidino modified analogues have been prepared (eg. **131**; R = nitro, ethoxycarbonyl, methyl, amino, hydroxy) although none of these derivatives have been shown to inhibit influenza virus sialidase as effectively as **130** [122]. Similarly, carbocyclic analogues such as **132** [123] and chain-shortened derivatives like **133** [124] have been prepared. Interestingly, **133** shows similar influenza virus sialidase inhibition to Neu5Ac2en [124].

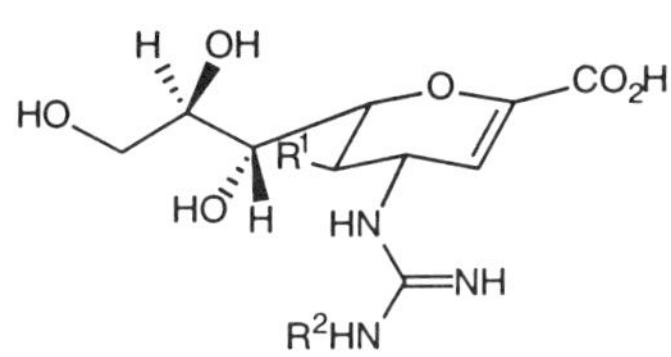

(**131**) R^1 = NHAc, R^2 = NO_2, EtO_2C, Me, NH_2, OH

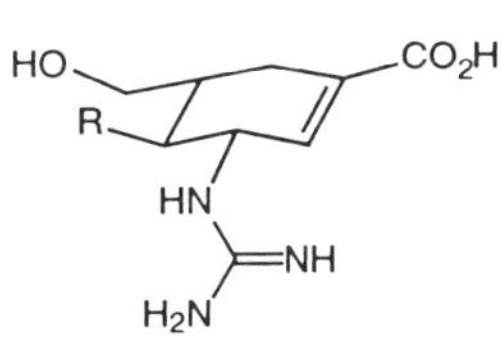

(**132**) R = NHAc

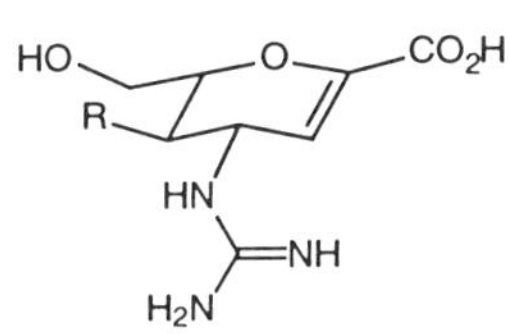

(**133**) R = NHAc

Several C-5 modified 4-deoxy-4-guanidino-Neu5Ac2en analogues have also been prepared, including the 5-desacetamido derivative **134** [125], and various amide (eg. **135**) and sulfonamide (eg. **136**) C-5 substituted compounds [126]. Changes to the acetamido group at C-5 have an effect on the inhibitory activity of these compounds towards influenza virus sialidase, with **134** showing very poor activity [125] whilst both **135** and **136** show comparable activity to **130** [126].

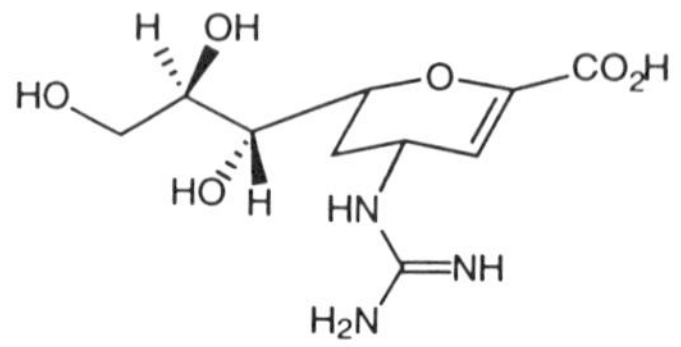

(**134**)

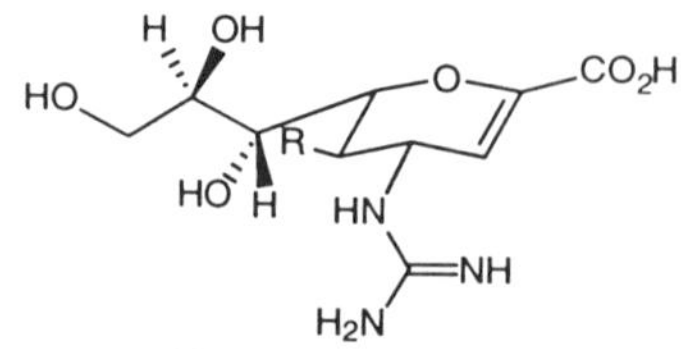

(**135**) R = NHC(O)CF_3
(**136**) R = $NHSO_2Me$

The finding that phosphonic acid analogues of 2-deoxy-Neu5Ac (*viz.* **101** and **102**; Section 2.2.1) are stronger inhibitors of sialidases than the corresponding parent Neu5Ac compounds [87] has prompted an elegant synthesis of the phosphonic acid analogue **137** of Neu5Ac2en [127]. Some of the key transformations in the synthesis of **137** are shown in Scheme 4. Thus, the oxidative decarboxylation of **138** gave the anomeric acetate derivative **139**, whilst phosphonylation of the imidate **140** ultimately gave **137** *via* a photobromination-reductive elimination sequence [127]. Whilst **137** was found to be a strong inhibitor of *Vibrio cholerae* sialidase, the expected increase in activity compared to Neu5Ac2en was not observed [127].

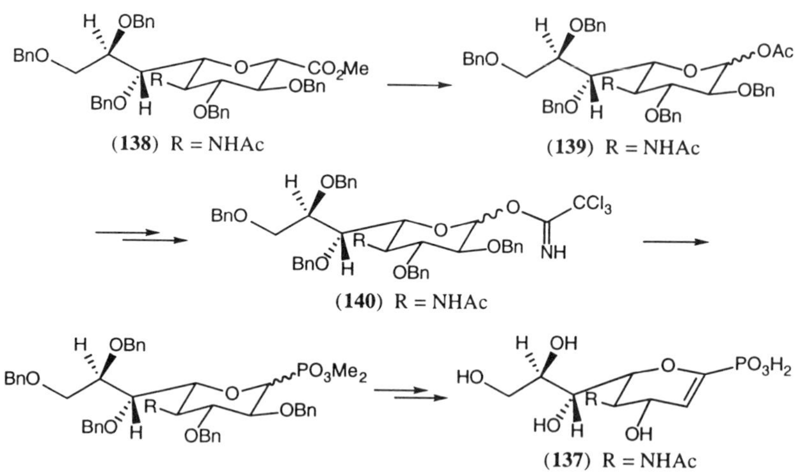

Scheme 4

The synthesis of the 6-*C*-methyl-Neu5Ac derivative **69** by Vasella and Wyler [71] (Section 2.1.2) also resulted in the preparation of the 6-*C*-methyl- and 6-*C*-hydroxymethyl-Neu5Ac2en derivatives **141** and **142**, respectively [71]. Both compounds **141** and **142** are weak inhibitors of *Vibrio cholerae* sialidase [71].

The preparation of the Neu5Ac3en methylketosides **143** and **144** [128] represents the only report to date of the synthesis of unsaturated Neu5Ac analogues which do not contain a 2,3-unsaturated functionality. Compounds

143 and **144** were prepared *via* selenoxide elimination from **145** and **146**, respectively, followed by saponification. The α-methylketoside **143** is a weak inhibitor of influenza virus sialidase [128].

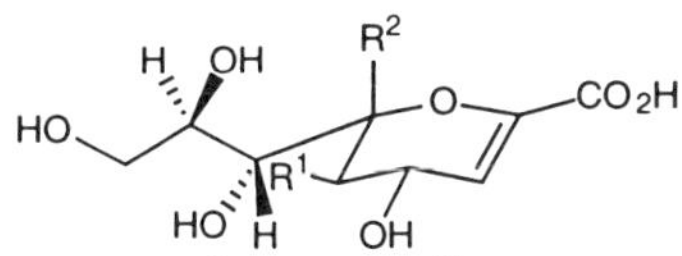

(**141**) R^1 = NHAc, R^2 = Me
(**142**) R^1 = NHAc, R^2 = CH_2OH

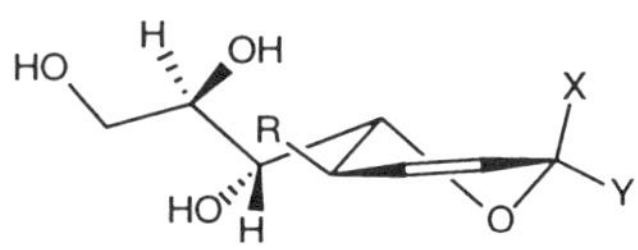

(**143**) R = NHAc, X = CO_2Me, Y = OMe
(**144**) R = NHAc, X = OMe, Y = CO_2Me

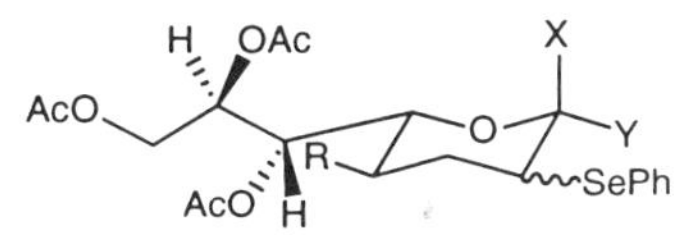

(**145**) R = NHAc, X = CO_2Me, Y = OMe
(**146**) R = NHAc, X = OMe, Y = CO_2Me

3. SIALIC ACIDS AS POTENTIAL ANTI-INFLUENZA DRUGS; A CASE STUDY

The search for a comprehensive prophylactic and/or therapeutic drug for influenza has been long and relatively intense. The World Health Organisation (WHO) considers influenza a serious socio-economic disease and closely monitors, on a worldwide basis, the appearance and impact of the virus. Even in recent times influenza has caused substantial loss of life and these events have further fuelled the search for its control by an effective and safe treatment [129].

3.1 Influenza as a Disease

Influenza has probably been around for much of humanity's history. The 1918-1919 influenza pandemic claimed the lives, directly or indirectly, of some 20 million people and put this disease on the world stage. This particular pandemic put in train a substantial research effort to find a preventative agent and/or cure for this deadly disease. Interestingly, it wasn't until 1933 that a virus was identified as the causative agent of influenza [130].

Influenza viruses are members of the orthomyxoviridae family which are further classified, on serological differences, into three distinct types, A, B and C. The human population appears to be most affected by types A and B, as these cause severe widespread epidemics (types A and B), pandemics (type A) as well as regional, more localised, epidemics (type B) [131]. Of particular concern, and most vulnerable, are the immunocompromised (high risk group) such as the elderly, the very young and those with existing conditions such as chronic pulmonary and heart disease. An added dimension to the severity of this disease is its economic impact. A major epidemic can cost employers millions of dollars through absenteeism and hence loss of productivity.

3.1.1 The virus particle

The influenza virus is spherical in shape and is approximately 1 x 10^3 Å in diameter. Eight separate single-stranded RNA segments make up the genome of the virus. Two of the gene products, haemagglutinin and sialidase, are membrane-bound glycoproteins. Haemagglutinin is a trimer of identical subunits, which recognises α-ketosides of *N*-acetylneuraminic acid and acts, initially, as the virus' biological glue, facilitating the adhesion of the virus particle to the host cell. Influenza virus sialidase (EC 3.2.1.18; neuraminidase; *N*-acetylneuraminate glycohydrolase), a tetrameric glycoprotein consisting of four identical disulfide-linked subunits (M_r 60 kDa), is considered to act as a pair of biological scissors; facilitating the release of freshly synthesised virion progeny. These glycoproteins are seen as spike-like projections extending out from the surface of the virus particle [132]. The x-ray crystal structures of both influenza virus haemagglutinin (Figure 1) and sialidase (Figure 2) have been long known [82, 133, 134, 135].

3.2 Influenza Virus Glycoproteins as Drug Design Targets

Both of the influenza virus surface glycoproteins are thought to be useful drug design targets and have been the focus of a number of research groups throughout the world. The potential interruption of the adhesion process of the virus particle to its target cell has enjoyed significant ideological popularity, whilst, at the present moment, the inhibition of virion progeny release has provided the most optimistic anti-influenza drug candidate.

3.2.1 Haemagglutinin

Attempts to design and synthesise potential anti-influenza drugs based on the inhibition of the influenza virus haemagglutinin glycoprotein have been previously reviewed [9, 131, 136]. One strategy of this research effort has been to take advantage of the polyvalent nature of the interaction between haemagglutinin and the target cell surface sialoglycoconjugates. This has led to the synthesis by some elegant chemistry of polyvalent 'cluster' or 'dendritic' sialosides which appear to have a significantly increased affinity over monomeric sialosides [131, 137-140]. Recently, a useful synthesis of some *C*-sialoside polymers, which are sialidase resistant and presumably are competitive ligands for haemagglutinin, has been reported. These polymers have shown significant inhibition of *in vitro* replication of influenza virus [141]. More recently Watowich *et al* have designed a number of potential monomeric 'functionalised'

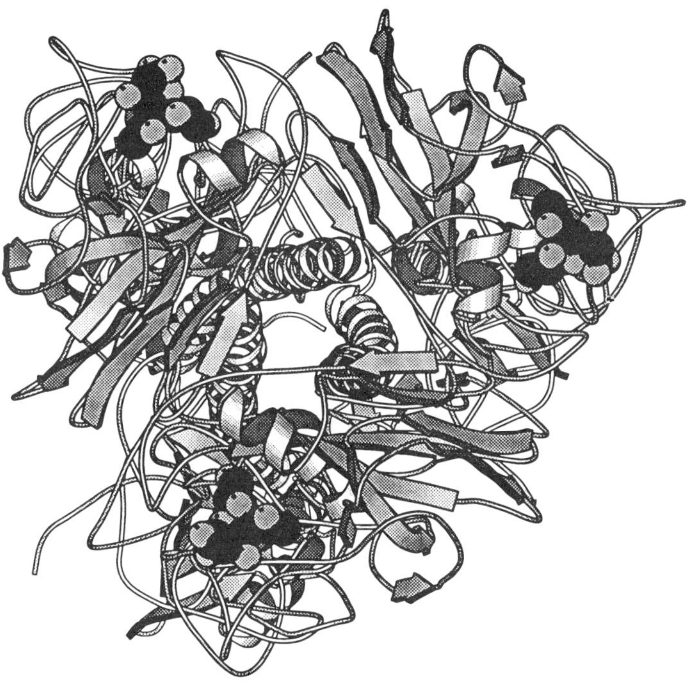

Figure 1: A view of influenza virus haemagglutinin trimer with Neu5Ac bound in each of the three binding sites.

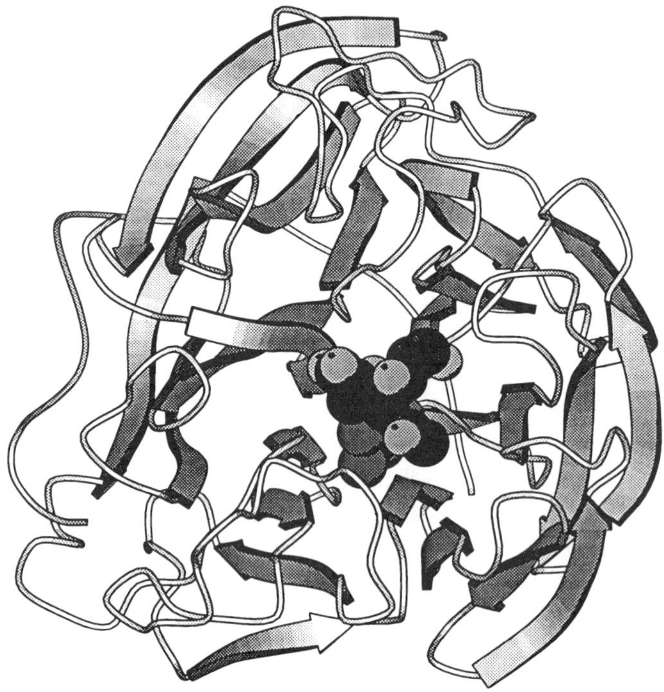

Figure 2: A monomer of influenza virus sialidase bound with 4-deoxy-4-guanidino-Neu5Ac2en (**130**).

sialosides which are proposed to enhance the affinity of these ligands for haemagglutinin [142].

The synthesis of one of these modified sialosides has been recently published [66]. The target compound, methyl 4-deoxy-4-guanidino-α-D-*N*-acetylneuraminic acid (4-deoxy-4-guanidino-Neu5Acα2Me, **55**), was successfully prepared from a 2,3-unsaturated precursor as described in Section 2.1.2. This compound is presently under further investigation and preliminary crystallographic data suggests that it binds in the expected manner [143].

3.2.2 Sialidase

Influenza virus sialidase has been implicated as being essential in the release of virion progeny away from infected cells [144]. At the molecular level sialidase cleaves terminal sialic acid residues that are α-ketosidically linked to various glycoconjugates (Scheme 5).

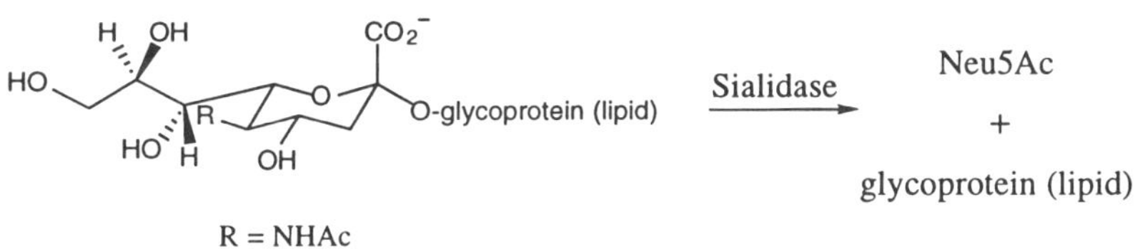

Scheme 5

As a direct result of post-translational sialoglycosylation of both haemagglutinin and sialidase, the virion progeny tend to clump, specifically through the haemagglutinin-sialic acid interaction. Given sialidases release the aggregated progeny, inhibition of this process is envisaged to provide the immune system with enough time to appropriately respond to the newly synthesised virion progeny before it is released from the infected cell's surface. Potent competitive *N*-acetylneuraminic acid-based inhibitors of influenza virus sialidase have been known for nearly three decades [145]. These 2,3-unsaturated derivatives of *N*-acetylneuraminic acid, described in Section 2.2.3 are likely transition state analogues [118, 146]. More recent evidence in support of this notion has been published [59, 147, 148]. Through a series of detailed kinetic isotope effect studies using the synthetic substrate 4-methylumbelliferyl-*N*-acetyl-α-D-neuraminic acid (**147**) and its C-3 perdeuterated counterpart (**148**) [59], computational chemistry [59, 147] and ^{1}H-NMR spectroscopy experiments [59] it has been proposed that influenza virus sialidase catalyses the release of α-*N*-acetylneuraminic acid *via* a putative sialosyl cation intermediate (**149**) which is significantly stabilised by the general negatively-charged environment within the active site of the enzyme. Interestingly, the use of an alternative synthetic sialoside, *p*-nitrophenyl-*N*-acetyl-α-D-neuraminic acid (**150**) and its C-3 perdeuterated analogue (**151**) did not support the putative sialosyl cation (**149**) as a likely transition state [149]. An explanation for this discrepancy has been recently reported, in which it would appear that the leaving group properties of the sialoside's aglycon moiety may have significant impact on secondary isotope effects through varying degrees of cation stabilisation [148].

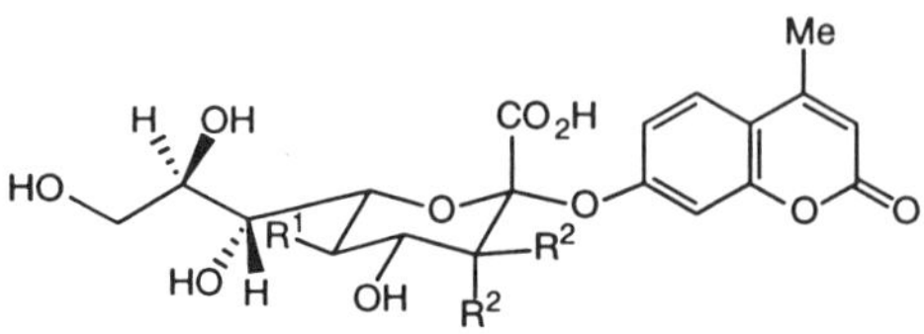

(147) R^1 = NHAc, R^2 = H
(148) R^1 = NHAc, R^2 = D

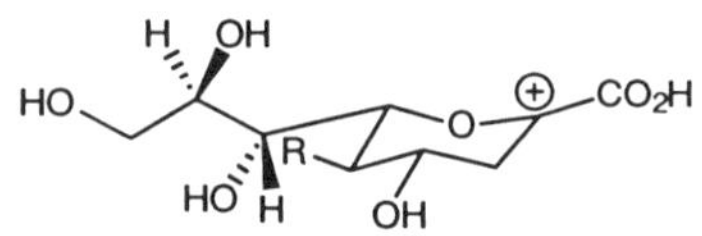

(149) R = NHAc

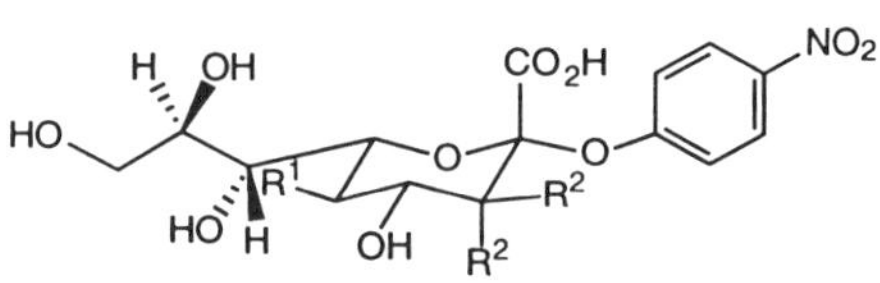

(150) R^1 = NHAc, R^2 = H
(151) R^1 = NHAc, R^2 = D

Recently, the design and synthesis of potent influenza virus sialidase inhibitors, based on the crystal structure of the enzyme [135], has been achieved [120, 121, 150]. The most potent of these compounds, 4-deoxy-4-guanidino-Neu5Ac2en (**130**), has shown good efficacy in a human experimental infection study [151] and is presently under further investigation in human clinical trials. Initially, this compound was thought to be a slow-binding inhibitor of sialidase from only influenza virus A with an apparent K_i of 3 x 10^{-11} M [152]. However, it has been recently shown that it acts as a slow-binding inhibitor of sialidase from both influenza virus A and B [153]. The clinical significance, if any, of this slow inhibition by 4-deoxy-4-guanidino-Neu5Ac2en (**130**) is not known.

At a molecular level 4-deoxy-4-guanidino-Neu5Ac2en (**130**) has been predicted to bind influenza virus sialidase with a number of significant charge-charge interactions (Figure 3) [147]. The majority of the total binding energy has been attributed to these interactions and is primarily associated with the introduced guanidinyl moiety at C-4 on Neu5Ac2en (**122**) and acidic residues within the sialidase active site as shown in Figure 3 [147].

One of the most salient and interesting features of 4-deoxy-4-guanidino-Neu5Ac2en (**130**) is the degree of selectivity for influenza virus sialidase [154]. A range of bacterial sialidases, other viral sialidases and mammalian sialidases have been found to be resistant to this potent influenza virus sialidase inhibitor [121, 154]. An explanation for this excellent selectivity has been postulated [154] and it is clear that only sialidase from influenza virus can tolerate relatively bulky basic moieties in the Neu5Ac C-4 binding pocket.

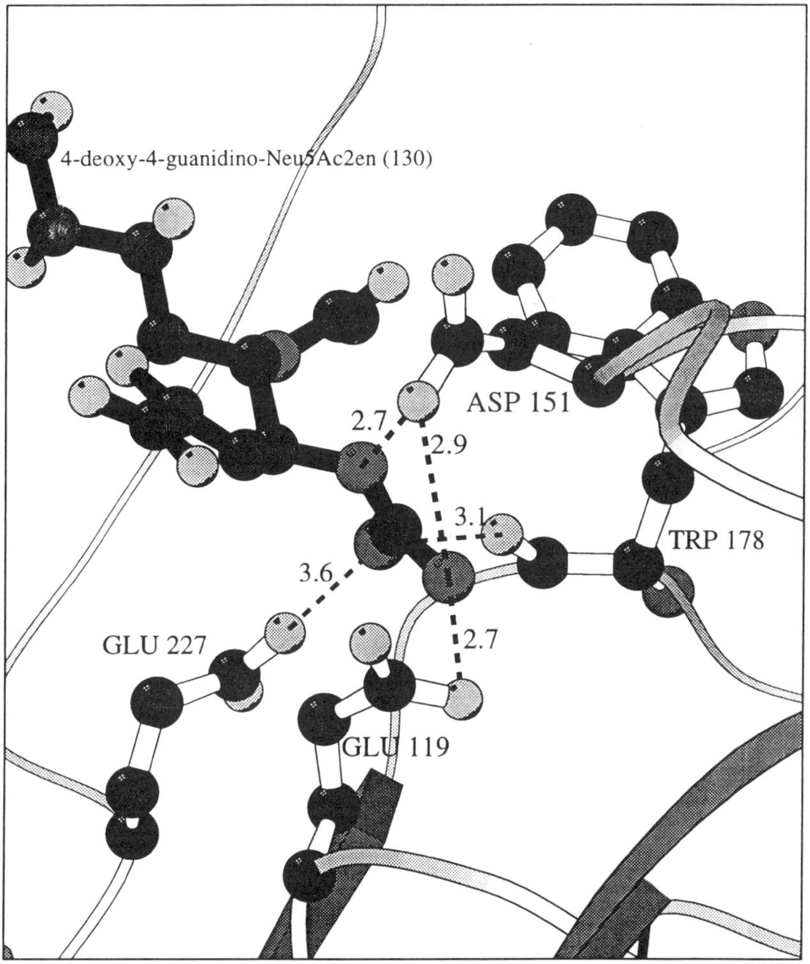

Figure 3: 4-Deoxy-4-guanidino-Neu5Ac2en (**130**) bound into the active site pocket interacting with several conserved amino acid residues.

The perennial question of the possible emergence of drug resistant influenza virus mutants to 4-deoxy-4-guanidino-Neu5Ac2en (**130**) is presently under investigation by several groups, worldwide.

Other *N*-acetylneuraminic acid derivatives have been prepared that also appear to inhibit, to varying degrees, influenza virus sialidase [55, 91]. The former work [55] reports the evaluation of two previously synthesised [56] fluorinated *N*-acetylneuraminic acid derivatives (**152**) and (**153**). The latter study [91] details the evaluation of the binding of phosphonate analogues of *N*-acetylneuraminic acid to influenza virus sialidase using x-ray crystallography and a standard sialidase assay. These phosphonate analogues (**101**) and (**102**) proved to have 10 to 100 times higher affinity for influenza virus sialidase when compared with *N*-acetylneuraminic acid (**1**). Further examples of Neu5Ac-based influenza virus sialidase inhibitors have been discussed in Sections 2.1.2 and 2.2.

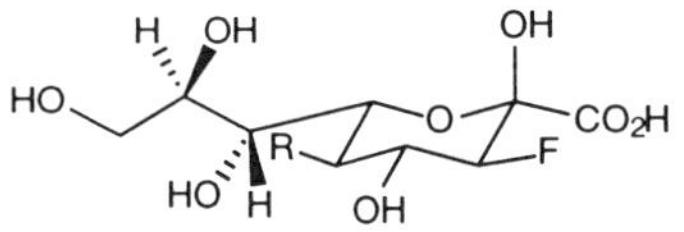

(**152**) R = NHAc

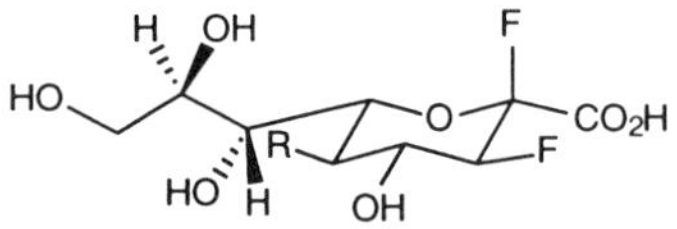

(**153**) R = NHAc

4. CONCLUSION AND FUTURE PROSPECTS

The potential of sialic acids as anti-microbials is clearly enormous. The understanding of the roles that these highly-functionalised carbohydrates play in the pathogenicity of infectious microorganisms is crucial to the development of these new-wave drugs. As we move into a post-antibiotic era the discovery of novel classes of anti-microbials becomes of paramount importance. The sialic acid family appears to be one such class and will provide many researchers with numerous challenges over the next decade.

Acknowledgements

The authors would like to thank Jeff Dyason for his much-appreciated assistance in the preparation of the figures in this chapter. We would also wish to thank Ms Fiona Crone for her excellent contribution in the preparation and editing of this chapter.

REFERENCES

1. Schauer R. Chemistry, metabolism and biological functions of sialic acids. Adv Carbohydr Chem Biochem 1982; 40: 131-234.
2. Schauer R. Sialic acids and their role as biological masks. TIBS 1985; 10: 357.
3. Corfield T. Bacterial sialidases - roles in pathogenicity and nutrition. Glycobiology 1992; 2: 509-521.
4. Schauer R, ed. Sialic Acids - Chemistry, Metabolism & Function. Cell Biology Monographs. Vol 10, Wien, New York: Springer Verlag, 1982.
5. Jeanloz RW and Codington JF. In: Biological Roles of Sialic Acid. Rosenberg A and Schengrund CL, eds. New York: Plenum Press, 1976: 201-227.
6. Gottschalk A, ed. Glycoproteins, their Composition, Structure and Function. 2nd ed. Amsterdam-London-New York: Elsevier, 1972.
7. Varki A. Diversity in the sialic acids. Glycobiology 1992; 2: 25-40.
8. Troy IInd FA. Polysialylation: from bacteria to brains. Glycobiology 1992; 2: 5-23.
9. Zbiral E. Synthesis of sialic acid analogs and their behaviour towards the enzymes of sialic acid metabolism and haemagglutinin X-31 of Influenza A-virus. In: Carbohydrates: Synthetic Methods and Applications in Medicinal Chemistry. Ogura H, Hasegawa A and Suami T, eds. New York: VCH, 1992: 304-339.
10. DeNinno MP. The synthesis and glycosidation of *N*-acetylneuraminic acid. Synthesis 1991; 583-593.
11. Tuppy H and Gottschalk A. The structure of sialic acids and their quantitation. In: Glycoproteins: Their Composition, Structure and Function. Gottschalk A, ed. Amsterdam-London-New York: Elsevier, 1972: 403-449.
12. Holmquist L. Synthesis of *N*-acetylneuraminic acid derivatives and studies of their interaction with *Vibrio cholerae* neuraminidase. FOA Reports 1975; 9: 1-19.
13. McGuire EJ. In: Biological Roles of Sialic Acid. Rosenberg A and Schengrund CL, eds. New York: Plenum Press, 1976: 123-158.
14. Wong C-H and Whitesides GM, eds. Enzymes in Synthetic Organic Chemistry. Tetrahedron Organic Chemistry Series, Baldwin JE and Magnus PD, eds, Vol 12, Oxford: Elsevier, 1994.
15. Fitz W, Schwark J-R and Wong C-H. Aldotetroses and C(3)-modified aldohexoses as substrates for *N*-acetylneuraminic acid aldolase: A model for the explanation of the normal and the inversed stereoselectivity. J Org Chem 1995; 60: 3663-3670.
16. Augé C, David S, Gautheron C, Malleron A and Cavayé B. Preparation of six naturally occuring sialic acids with immobilized acylneuraminate pyruvate lyase. New J Chem 1988; 12: 733-744.

17. Augé C, Gautheron C, David S, Malleron A, Cavayé B and Bouxom B. Sialyl aldolase in organic synthesis: from the trout egg acid, 3-deoxy-D-*glycero*-D-*galacto*-2-nonulosonic acid (KDN), to branched-chain higher ketoses as possible new chirons. Tetrahedron 1990; 46: 201-214.

18. Zbiral E, Kleineidam RG, Schreiner E, Hartmann M, Christian R and Schauer R. Elucidation of the topological parameters of *N*-acetylneuraminic acid and some analogues involved in their interaction with the *N*-acetylneuraminate lyase from *Clostridium perfringens*. Biochem J 1992; 282: 511-516.

19. Schauer R, Stoll S, Zbiral E, Schreiner E, Brandstetter NH, Vasella A and Baumberger F. Interaction of *N*-acetyl-4-,7-, 8- or 9-deoxyneuraminic acids and *N*-acetyl-4-, 7- or 8-mono-epi-and -7,8-di-epineuraminic acids with *N*-acetylneuraminate lyase. Glycoconjugate J 1987; 4: 361-369.

20. Sparks MA, Williams KW, Lukacs C, Schrell A, Priebe G, Spaltenstein A and Whitesides GM. Synthesis of potential inhibitors of hemagglutination by influenza virus: Chemoenzymic preparation of N-5 analogs of *N*-acetylneuraminic acid. Tetrahedron 1993; 49: 1-12.

21. Isecke R and Brossmer R. Synthesis of 5-*N*- and 9-*N*-thioacylated sialic acids. Tetrahedron 1994; 50: 7445-7460.

22. Koppert K and Brossmer R. Synthesis of the C-5 homologue of *N*-acetylneuraminic acid by enzymatic chain elongation of 2-C-acetamidomethyl-2-deoxy-D-mannose. Tetrahedron Lett 1992; 33: 8031-8034.

23. David S, Malleron A and Cavayé B. Aldolases in organic synthesis: acylneuraminate-pyruvate lyase accepts furanoses as substrates. New J Chem 1992; 16: 751-755.

24. Halcomb RL, Fitz W and Wong C-H. Enzymatic synthesis of 7-deoxy-*N*-acetylneuraminic acid and 7-*O*-methyl-*N*-acetylneuraminic acid. Tetrahedron - Asymmetry 1994; 5: 2437-2442.

25. Thomson R and von Itzstein M. Synthesis of 4-substituted-2-acetamido-2,4-dideoxy-mannopyranoses using 1,6-anhydro sugar chemistry. Carbohydr Res 1995; 274: 29-44.

26. Kong DCM and von Itzstein M. The first synthesis of a C-7 nitrogen-containing sialic acid analogue, 5-acetamido-7-azido-3,5,7-trideoxy-D-*glycero*-D-*galacto*-2-nonulopyranosonic acid (7-azido-7-deoxy-Neu5Ac). Tetrahedron Lett 1995; 36: 957-960.

27. Kong DCM and von Itzstein M. Chemoenzymatic synthesis of a novel sialic acid analogue using membrane-enclosed enzyme catalysis. In XVIIth International Carbohydrate Symposium. 17-22 July, 1994. Ottawa, Canada.

28. Hartmann M, Christian R and Zbiral E. Synthesis of the β-methyl ketosides of 4-*oxo*-, 7-*oxo*-, and 8-*oxo*-*N*-acetylneuraminic acid and the corresponding 7,7- and 8,8-dimethoxy derivatives. Their behaviour towards CMP-sialate synthase. Liebigs Ann Chem 1990; 83-91.

29. Fitz W and Wong C-H. Combined use of subtilisin and *N*-acetylneuraminic acid aldolase for the synthesis of a fluorescent sialic acid. J Org Chem 1994; 59: 8279-8280.
30. Kragl U, Gödde A, Wandrey C, Lubin N and Augé C. New synthetic applications of sialic acid aldolase, a useful catalyst for KDO synthesis. Relation between substrate conformation and enzyme stereoselectivity. J Chem Soc Perkin Trans 1 1994; 119-124.
31. Cornforth JW, Firth ME and Gottschalk A. The synthesis of *N*-acetylneuraminic acid. J Biol Chem 1958; 68: 57-61.
32. Kuhn R and Baschang G. Aminozucker-Synthesen. XXV. Synthese der lactaminsäure. Liebigs Ann Chem 1962; 659: 156-163.
33. Gordon DM and Whitesides GM. Indium-mediated allylations of unprotected carbohydrates in aqueous media: A short synthesis of sialic acid. J Org Chem 1993; 58: 7937-7938.
34. Chan T-H and Lee M-C. Indium-mediated coupling of α-(bromomethyl)acrylic acid with carbonyl compounds in aqueous media. Concise syntheses of (+)-3-deoxy-D-*glycero*-D-*galacto*-nonulosonic acid and *N*-acetylneuraminic acid. J Org Chem 1995; 60: 4228-4232.
35. Gao J, Harter R, Gordon DM and Whitesides GM. Synthesis of KDO using indium-mediated allylation of 2,3:4,5-di-*O*-isopropylidene-D-arabinose in aqueous media. J Org Chem 1994; 59: 3714-3715.
36. Hagedorn H-W, Merten H and Brossmer R. Synthesis of the deculopyranosonic acid analog of *N*-acetylneuraminic acid, its 5-epimer and 6-epimer, and of 5-acetamido-1,3,5-trideoxy-D-*glycero*-D-*galacto*-non-2-ulopyranose. Carbohydr Res 1992; 236: 89-96.
37. Dondoni A, Marra A and Merino P. Installation of the pyruvate unit in glycidic aldehydes via a Wittig olefination-Michael addition sequence utilizing a thiazole-armed carbonyl ylid. A new stereoselective route to 3-deoxy-2-ulosonic acids and the total synthesis of DAH, KDN, and 4-*epi*-KDN. J Amer Chem Soc 1994; 116: 3324-3336.
38. Dondoni A, Boscarato A and Marra A. Synthesis of a sialic acid analog with the acetamido group at C-4. Tetrahedron - Asymmetry 1994; 5: 2209-2212.
39. Danishefsky SJ, DeNinno MP and Chen S-H. Stereoselective total syntheses of the naturally occurring enantiomers of *N*-acetylneuraminic acid and 3-deoxy-D-*manno*-2-octulosonic acid. A new and stereospecific approach to sialo and 3-deoxy-D-*manno*- 2-octulosonic acid conjugates. J Amer Chem Soc 1988; 110: 3929-3940.
40. Takahashi T, Okano T, Harada T, Imamura K and Yamada H. Reactions of a new acyl anion equivalent; $^{-}COCO_2R$, and its use to preparation of α-ketoesters. Synlett 1994; 121-122.
41. Okamoto K and Goto T. Glycosidation of sialic acid. Tetrahedron 1990; 46: 5835-5857.
42. Vliegenthart JFG and Kamerling JP. Synthesis of sialic acids and sialic acid derivatives. In: Sialic Acids: Chemistry, Metabolism and Function.

Cell Biology Monographs. Schauer R, ed. Wien, New York: Springer Verlag, 1982: 59-76.

43. Hasegawa A and Kiso M. Systematic synthesis of gangliosides toward the elucidation and biomedical application of their biological functions. In: Carbohydrates: Synthetic Methods and Applications in Medicinal Chemistry. Ogura H, Hasegawa A and Suami T, eds. New York: VCH, 1992: 243-266.

44. Ito Y, Gaudino JJ and Paulson JC. Synthesis of bioactive sialosides. Pure Appl Chem 1993; 65: 753-762.

45. Mukaiyama T, Sasaki T, Iwashita E and Matsubara K. Stereoselective synthesis of *N*-acetyl-α-neuraminosyl-galactose disaccharide. Chem Lett 1995; 455-456.

46. Martin TJ, Brescello R, Toepfer A and Schmidt RR. Synthesis of phosphites and phosphates of neuraminic acid and their glycosyl donor properties – convenient synthesis of GM_3. Glycoconjugate J 1993; 10: 16-25.

47. Hasegawa A, Ohki H, Nagahama T, Ishida H and Kiso M. A facile, large-scale preparation of the methyl 2-thioglycoside of *N*-acetylneuraminic acid, and its usefulness for the α-stereoselective synthesis of sialoglycosides. Carbohydr Res 1991; 212: 277-281.

48. Marra A and Sinaÿ P. Stereoselective synthesis of 2-thioglycosides of *N*-acetylneuraminic acid. Carbohydr Res 1989; 187: 35-42.

49. Liebe B and Kunz H. Synthesis of sialyl-Tn antigen. Regioselective sialylation of a galactosamine threonine conjugate unblocked in the carbohydrate portion. Tetrahedron Lett 1994; 35: 8777-8778.

50. Ercegovic T and Magnusson G. A new *N*-acetylneuraminic acid donor for highly stereoselective α-sialylation. J Chem Soc Chem Commun 1994; 831-832.

51. Ito Y and Ogawa T. Highly stereoselective glycosylation of sialic acid aided by stereocontroling auxiliaries. Tetrahedron 1990; 46: 89-102.

52. Okamoto K, Kondo T and Goto T. A stereospecific synthesis of β-glycosides of *N*-acetylneuraminic acid and secondary alcohols. Chem Lett 1986; 1449-1452.

53. Bodenmüller A and Schmidt RR. Synthesis of *N*-acetylneuraminic acid derivatives and structural analogues. Liebigs Ann Chem 1994; 541-548.

54. Ogura H, Furuhata K, Itoh M and Shitori Y. Syntheses of 2-*O*-glycosyl derivatives of *N*-acetyl-D-neuraminic acid. Carbohydr Res 1986; 158: 37-51.

55. Hagiwara T, Kijima-Suda I, Ido T, Ohrui H and Tomita K. Inhibition of bacterial and viral sialidases by 3-fluoro-*N*-acetylneuraminic acid. Carbohydr Res 1994; 263: 1067-172.

56. Nakajima T, Hori H, Ohrui H, Meguro H and Ido T. Synthesis of *N*-acetyl-3-fluoro-neuraminic acids. Agri Biol Chem 1988; 52: 1209-1215.

57. Vliegenthart JFG, Dorland L, van Halbeek H and Haverkamp J. NMR spectroscopy of sialic acids. In: Sialic Acids: Chemistry, Metabolism and

Function. Cell Biology Monographs. Schauer R, ed. Wien, New York: Springer Verlag, 1982: 127-171.

58. Dorland L, Haverkamp J, Schauer R, Veldink GA and Vliegenthart JFG. The preparation of specifically deuterium- or tritium-labeled *N*-acetylneuraminic acid and cytidine-5'-monophospho-β-*N*-acetylneuraminic acid as precursors for glycoconjugate synthesis. Biochem Biophys Res Commun 1982; 104: 1114-1119.
59. Chong AKJ, Pegg MS, Taylor NR and von Itzstein M. Evidence for a sialosyl cation transition-state complex in the reaction of sialidase from influenza virus. Eur J Biochem 1992; 207: 335-343.
60. Schauer R. Chemie und Biologie der Acylneuraminsäuren. Angew Chem 1973; 85: 128-140. Angew Chem Int Ed Engl 1973; 12: 127-138.
61. Hartmann M and Zbiral E. Structural variations of *N*-acetylneuraminic acid, 12: a new useful approach to 4-epi- and 8-epi-*N*-acetylneuraminic acid. Monatsh Chem 1989; 120: 899-906.
62. Hartmann M, Christian R and Zbiral E. Structural variations of *N*-acetylneuraminic acid, Part 19. Synthesis of both epimeric pairs of the 4-C-methyl- and 4-deoxy-4-C-methyl- as well as of the β-methylketoside of 4-deoxy-4-C-methylene-*N*-acetylneuraminic acid. Behaviour towards CMP-sialate synthase. Monatsh Chem 1991; 122: 111-125.
63. Hartmann M and Zbiral E. Synthesis of the epimeric pair of 4-deoxy-4-(R)- and 4-deoxy-4-(S)-C-methyl-*N*-acetylneuraminic acid. Tetrahedron Lett 1990; 31: 2875-2878.
64. Groves DR, Wilson JC and von Itzstein M. The synthesis and NMR investigation of a methyloxime derivative of *N*-acetylneuraminic acid. Aust J Chem 1995; 48: 1217-1220.
65. Estenne G, Saroli A and Doutheau A. A new synthesis of *N*-acetyl-4-deoxy-D-neuraminic acid. J Carbohydr Chem 1991; 10: 181-195.
66. Ciccotosto S and von Itzstein M. Synthesis of methyl 5-acetamido-3,4,5-trideoxy-4-guanidinyl-D-*glycero*-D-*galacto*-2-nonulopyranosidonic acid (4-deoxy-4-guanidino-Neu5Acα2Me). Tetrahedron Lett 1995; 36: 5405-5408.
67. Roy R and LaFerriére CA. Synthesis of protein conjugates and analogues of *N*-acetylneuraminic acid. Can J Chem 1990; 68: 2045-2054.
68. Ogura H and Furuhata K. Studies on sialic acids I. Determination of anomeric configuration of neuraminic acid derivatives by circular dichroism. Tetrahedron Lett 1981; 22: 4265-4268.
69. Byramova NE, Tuzikov AB and Bovin NV. Studies on the synthesis of sialosides and sialic acid analogs 2. A simple procedure for the synthesis of the methyl and benzyl glycosides of Neu5Ac and 4-*epi*-Neu5Ac. Conversion of the benzyl and methyl glycosides of Neu5Ac into *N*-trifluoroacetylneuraminic acid benzyl glycosides. Carbohydr Res 1992; 237: 161-175.
70. Schreiner E and Zbiral E. Structural variations of *N*-acetylneuraminic acid. 16. A convenient approach to 3-deoxy-D-*glycero*-D-*galacto*-

nonulosonic acid (KDN), 5-azido-5-deoxy-KDN and 5-deoxy-KDN, and their 4-methylumbelliferyl 2α-glycosides. Liebigs Ann Chem 1990; 581-586.
71. Vasella A and Wyler R. Synthesis of the 6-*C*-methyl and 6-*C*-(hydroxymethyl) analogues of *N*-acetylneuraminic acid and of *N*-acetyl-2,3-didehydro-2-deoxyneuraminic acid. Helv Chim Acta 1990; 73: 1742-1763.
72. Baumberger F, Vasella A and Schauer R. Synthesis of new sialidase inhibitors, 6-amino-6-deoxysialic acids. Helv Chim Acta 1988; 71: 429-445.
73. Glänzer BI, Györgydeák Z, Bernet B and Vasella A. Analogues of sialic acids as potential sialidase inhibitors. Synthesis of C_6 and C_7 analogues of *N*-acetyl-6-amino-2,6-dideoxyneuraminic acid. Helv Chim Acta 1991; 74: 343-369.
74. Yamamoto T, Kumazawa H, Inami K, Teshima T and Shiba T. Syntheses of sialic acid isomers with inhibitory activity against neuraminidase. Tetrahedron Lett 1992; 33: 5791-5794.
75. Ogawa S, Yoshikawa M, Taki T, Yokoi S and Chida N. Synthesis of carbocyclic analogues of 3-deoxy-D-*manno*-2-octulosonic acid and *N*-acetylneuraminic acid. Carbohydr Res 1995; 269: 53-78.
76. Sharma M, Petrie IIIrd CR and Korytnyk W. General methods for modification of sialic acid at C-9. Synthesis of *N*-acetyl-9-deoxy-9-fluoroneuraminic acid. Carbohydr Res 1988; 175: 25-34.
77. Kiefel M, Bennett S and von Itzstein M. Synthesis of C_{11} chain-extended analogues of *N*-acetylneuraminic acid. J Chem Soc Perkin Trans 1 1996; 439-442.
78. Ogura H, Furuhata K, Sato S, Anazawa K, Itoh M and Shitori Y. Synthesis of 9-*O*-acyl- and 4-*O*-acetyl-sialic acids. Carbohydr Res 1987; 167: 77-86.
79. Reinhard B and Faillard H. Regioselective acetylations of sialic acid α-ketosides. Liebigs Ann Chem 1994; 193-203.
80. Anazawa K, Furuhata K and Ogura H. Synthesis of 7-*O*-acetyl-*N*-acetylneuraminic acid derivative. Chem Pharm Bull 1988; 36: 4976-4979.
81. Haverkamp J, Schauer R, Wember M, Kamerling JP and Vliegenthart JFG. Synthesis of 9-*O*-acetyl and 4,9-di-*O*-acetyl derivatives of the methyl ester of *N*-acetyl-β-D-neuraminic acid methyl glycoside. Hoppe-Seyler's Z Physiol Chem 1975; 1575-1583.
82. Weis W, Brown JH, Cusack S, Paulson JC, Skehel JJ and Wiley DC. Structure of the influenza virus haemagglutinin complexed with its receptor, sialic acid. Nature 1988; 333: 426-431.
83. Sauter NK, Bednarski MD, Wurzburg BA, Hanson JE, Whitesides GM, Skehel JJ and Wiley DC. Hemagglutinins from two influenza virus variants bind to sialic acid derivatives with millimolar dissociation

constants: a 500-MHz proton Nuclear Magnetic Resonance study. Biochemistry 1989; 28: 8388-8396.

84. Schmid W, Christian R and Zbiral E. Synthesis of both epimeric 2-deoxy-*N*-acetylneuraminic acids and their behaviour towards CMP-sialate synthetase - A comparison with 2-β-methylketoside of *N*-acetylneuraminic acid. Tetrahedron Lett 1988; 29: 3643-3646.

85. Hanessian S and Girard C. Anomeric deoxygenation of 2-ulosonic acids using SmI_2: rapid access to 2-deoxy-KDO and 2-deoxy-NANA. Synlett 1994; 863-864.

86. Hartmann M and Zbiral E. Synthesis of 2-deoxy-2Heq - sialic acid analogues structurally varied at C-4. Liebigs Ann Chem 1991; 795-801.

87. Wallimann K and Vasella A. Phosphonic-acid analogues of the *N*-acetyl-2-deoxyneuraminic acids: synthesis and inhibition of *Vibrio cholerae* sialidase. Helv Chim Acta 1990; 73: 1359-1372.

88. Tropper FD, Andersson FO, Braun S and Roy R. Phase transfer catalysis as a general and stereoselective entry into glycosyl azides from glycosyl halides. Synthesis 1992; 618-620.

89. Czollner L, Kuszmann J and Vasella A. Synthesis of pyrrolidine analogues of *N*-acetylneuraminic acid as potential sialidase inhibitors. Helv Chim Acta 1990; 73: 1338-1358.

90. Potter JJ and von Itzstein M. Decarboxylation of a sialic acid derivative. Carbohydr Res 1996; 282: 181-187.

91. White CL, Janakiraman MN, Laver WG, Philippon C, Vasella A, Air GM and Luo M. A sialic acid-derived phophonate analog inhibits different strains of influenza virus neuraminidase with different efficiencies. J Mol Biol 1995; 245: 623-634.

92. Kanie O, Kiso M and Hasegawa A. Glycosylation using methylthioglycosides of *N*-acetylneuraminic acid and dimethyl (methylthio)sulfonium triflate. J Carbohydr Chem 1988; 7: 501-506.

93. Hasegawa A, Adachi K, Yoshida M and Kiso M. Synthesis of the methyl thioglycosides of deoxy-*N*-acetylneuraminic acids for use as glycosyl donors. Carbohydr Res 1992; 230: 257-272.

94. Marra A and Sinaÿ P. A novel stereoselective synthesis of *N*-acetyl-α-neuraminosyl-galactose disaccharide derivatives, using anomeric *S*-glycosyl xanthates. Carbohydr Res 1990; 195: 303-308.

95. Cao S, Meunier SJ, Andersson FO, Letellier M and Roy R. Mild stereoselective syntheses of thioglycosides under PTC conditions and their use as active and latent glycosyl donors. Tetrahedron - Asymmetry 1994; 5: 2303-2312.

96. Khorlin AY, Privalova IM, Zakstelskaya LY, Molibog EV and Evstigneeva NA. Synthetic inhibitors of *Vibrio cholerae* neuraminidase and neuraminidases of some influenza virus strains. FEBS Lett 1970; 8: 17-19.

97. Warner TG and Lee LA. An azidoaryl thioglycoside of sialic acid. A potential photoaffinity probe of sialidases and sialic acid-binding proteins. Carbohydr Res 1988; 176: 211-218.
98. Murase T, Kameyama A, Kartha KPR, Ishida H, Kiso M and Hasegawa A. Synthetic studies on sialglycoconjugates 5. Facile, regio- and stereoselective synthesis of ganglioside GM_4 and its position isomer. J Carbohydr Chem 1989; 8: 265-283.
99. Hasegawa A, Nakamura J and Kiso M. Studies on the thioglycosides of *N*-acetylneuraminic acid 1. Synthesis of alkyl α-glycosides of 2-thio-*N*-acetylneuraminic acid. J Carbohydr Chem 1986; 5: 11-19.
100. Lubineau A and Le Gallic J. Stereoselective syntheses of alkyl- and alkyl-2-thio-α-sialosides. J Carbohydr Chem 1991; 10: 263-268.
101. Rothermel J and Faillard H. Phase-transfer-catalysed synthesis of *N*-acetylneuraminic acid α-thioketosides and inhibition studies with *Clostridium perfringens* sialidase. Biol Chem Hoppe-Seyler 1989; 370: 1077-1084.
102. Suzuki Y, Sato K, Kiso M and Hasegawa A. New ganglioside analogs that inhibit influenza virus sialidase. Glycoconjugate J 1990; 7: 349-356.
103. Hasegawa A, Terada T, Ogawa H and Kiso M. Studies on the thioglycosides of *N*-acetylneuraminic acid 10: synthesis of *S*-(α-Sialosyl)-(2→6)-*O*-2-acetamido-2-deoxy-β-D-hexopyranosyl ceramide and its related compounds. J Carbohydr Chem 1992; 11: 319 - 331.
104. Hasegawa A, Morita M, Ito Y, Ishida H and Kiso M. Studies on the thioglycosides of *N*-acetyneuraminic acid 7: synthesis of *S*-(α-sialyl)-(2→6)-β-D-hexopyranosyl and -(2→6)-β-D-lactosyl ceramides and their biological activity. J Carbohydr Chem 1990; 9: 369-392.
105. Bennett S, von Itzstein M and Kiefel MJ. A simple method for the preparation of thioglycosides of *N*-acetylneuraminic acid. Carbohydr Res 1994; 259: 293-299.
106. Kiefel MJ, Beisner B, Bennett S, Holmes ID and von Itzstein M. Synthesis and biological evaluation of *N*-acetylneuraminic acid-based rotavirus inhibitors. J Med Chem 1996; 39: 1314-1320.
107. Smalec B and von Itzstein M. An improved synthesis of an important *S*-sialylnucleoside analogue. Carbohydr Res 1995; 266: 269-272.
108. Kijima-Suda I, Toyoshima S, Itoh M, Furuhata K, Ogura H and Osawa T. Inhibition of sialyltransferase of murine lymphocytes by disaccharide nucleosides. Chem Pharm Bull 1985; 33: 730-739.
109. Harvey BE and Thomas P. Inhibition of CMP-sialic acid transport in human liver and colorectal cancer cell lines by a sialic acid nucleoside conjugate (K1-8110). Biochem Biophys Res Commun 1993; 190: 571-575.
110. Roth A and Faillard H. Synthesis of fluorescent 7,8,9-tri-*O*-acetyl-*N*-acetyl- and 4-*O*-acetyl-*N*-acetylneuraminic acid α-thioketosides. Liebigs Ann Chem 1993; 485-489.

111. Meindl P and Tuppy H. Ber 2-deoxy-2,3-dehydro-sialinsäuren. 1. Mitt: Synthese und Eigenschaften von 2-Deoxy-2,3-dehydro-*N*-acylneuraminsäuren und deren Methylestern. Monatsh Chem 1969; 100: 1296-1306.

112. Kumar V, Kessler J, Scott ME, Patwardhan BH, Tanenbaum SW and Flashner M. 2,3-dehydro-4-epi-*N*-acetylneuraminic acid; a neuraminidase inhibitor. Carbohydr Res 1981; 94: 123-130.

113. Zbiral E, Brandstetter HH, Christian R and Schauer R. Synthesis of the C-7-, C-8-, and C-7,8-side chain epimers of 2-deoxy-2,3-didehydro-*N*-acetylneuraminic acid and their behaviour towards sialidase from *Vibrio cholerae*. Liebigs Ann Chem 1987; 781-786.

114. Zbiral E, Schreiner E, Kleineidam RG and Schauer R. Structural variations of *N*-acetylneuraminic acid 10. Synthesis of 2,7-, 2,8- and 2,9-dideoxy- and 2,4,7-trideoxy-2,3-didehydro-*N*-acetylneuraminic acid and their behaviour towards sialidase from *Vibrio cholerae*. Liebigs Ann Chem 1989; 159-165.

115. Meindl P and Tuppy H. 2-Deoxy-2,3-dehydrosialic acids. III. Synthesis and properties of 2-deoxy-2,3-dehydroneuraminic acid and of new *N*-acyl derivative. Monatsh Chem 1973; 104: 402-414.

116. Schreiner E, Zbiral E, Kleineidam RG and Schauer R. Structural variations on *N*-acetylneuraminic acid, Part 21: 2,3-Didehydro-2-deoxysialic acids structurally varied at C-5 and their behaviour towards sialidase from *Vibrio cholerae*. Carbohydr Res 1991; 216: 61-66.

117. Schreiner E, Zbiral E, Kleineidam RG and Schauer R. Synthesis of some 2,3-didehydro-2-deoxysialic acids structurally varied at C-4 and their behaviour towards sialidase from *Vibrio cholerae*. Liebigs Ann Chem 1991; 129-134.

118. Driguez P-A, Barrère B, Quash G and Doutheau A. Synthesis of transition-state analogues as potential inhibitors of sialidase from *Influenza* virus. Carbohydr Res 1994; 262: 297-310.

119. von Itzstein M, Jin B, Wu W-Y and Chandler M. A convenient method for the introduction of nitrogen and sulfur at C-4 on a sialic acid analogue. Carbohydr Res 1993; 244: 181-185.

120. von Itzstein M, Wu W-Y and Jin B. The synthesis of 2,3-didehydro-2,4-dideoxy-4-guanidinyl-*N*-acetylneuraminic acid: A potent influenza virus sialidase inhibitor. Carbohydr Res 1994; 259: 301-305.

121. von Itzstein M, Wu W-Y, Kok GB, Pegg MS, Dyason JC, Jin B, Phan TV, Smythe ML, White HF, Oliver SW, Colman PM, Varghese JN, Ryan DM, Woods JM, Bethell RC, Hotham VJ, Cameron JM and Penn CR. Rational design of potent sialidase-based inhibitors of influenza virus replication. Nature 1993; 363: 418-423.

122. Chandler M, Bamford MJ, Conroy R, Lamont B, Patel B, Patel VK, Steeples IP, Storer R, Weir NG, Wright M and Williamson C. Synthesis of the potent influenza neuraminidase inhibitor 4-guanidino Neu5Ac2en. X-ray molecular structure of 5-acetamido-4-amino-2,6-anhydro-3,4,5-

trideoxy-D-*erythro*-L-*gluco*-nononic acid. J Chem Soc Perkin Trans 1 1995; 1173-1180.

123. Chandler M, Conroy R, Cooper AWJ, Lamont RB, Scicinski JJ, Smart JE, Storer R, Weir NG, Wilson RD and Wyatt PG. Approaches to carbocyclic analogues of the potent neuraminidase inhibitor 4-guanidino-Neu5Ac2en. X-ray molecular structure of *N*-[(1S,2S,6R)-2-azido-6-benzyloxymethyl-4-formylcyclohex-3-enyl]acetamide. J Chem Soc Perkin Trans 1 1995; 1189-1197.

124. Bamford MJ, Pichel JC, Husman W, Patel B, Storer R and Weir NG. Synthesis of 6-,7- and 8-carbon sugar analogues of potent anti-influenza 2,3-didehydro-2,3-dideoxy-*N*-acetylneuraminic acid derivatives. J Chem Soc Perkin Trans 1 1995; 1181-1187.

125. Starkey ID, Mahmoudian M, Noble D, Smith PW, Cherry PC, Howes PD and Sollis SL. Synthesis and influenza virus sialidase inhibitory activity of the 5-desacetamido analogue of 2,3-didehydro-2,4-dideoxy-4-guanidinyl-*N*-acetylneuraminic acid (GG167). Tetrahedron Lett 1995; 36: 299-302.

126. Smith PW, Starkey ID, Howes PD, Sollis SL, Keeling SP, Cherry PC, von Itzstein M, Wu W-Y and Jin B. Synthesis and influenza virus sialidase inhibitory activity of analogues of 4-guanidino-Neu5Ac2en (GG167) with modified 5-substitutents. Eur J Med Chem 1996; 31: 143-150.

127. Vasella A and Wyler R. Synthesis of a phosphonic acid analogue of *N*-acetyl-2,3-didehydro-2-deoxyneuraminic acid, an inhibitor of *Vibrio cholerae* sialidase. Helv Chim Acta 1991; 74: 451-463.

128. Maudrin J, Barrère B, Chantegrel B, Deshayes C, Quash G and Doutheau A. Synthesis of 5-acetamido-3,4,5-trideoxy-D-*manno*-non-3-en-2-ulosonic acid α and β-methyl ketoside and their behavior towards the sialidase from influenza virus. Bull Soc Chim Fr 1994; 131: 400-406.

129. Curwen M, Dunnell K and Ashley J. Hidden influenza deaths. BMJ 1990; 300: 896.

130. Smith WC, Andrews CH and Laidlaw PP. A virus obtained from influenza patients. Lancet 1933; 2: 66-68.

131. Whittington A and Bethell R. Anti-infectives: recent developments in the antiviral therapy of influenza. Exp Opin Ther Patents 1995; 5: 793-803.

132. Krug RM, ed. The Influenza Viruses. The Viruses, Fraenkel-Conrat H and Wagner RR, eds, New York & London: Plenum Press, 1989.

133. Wilson IA, Skehel JJ and Wiley DC. Structure of the haemagglutinin membrane glycoprotein of influenza virus at 3 Å resolution. Nature 1981; 289: 366-373.

134. Rogers GN, Paulson JC, Daniels RS, Skehel JJ, Wilson IA and Wiley DC. Single amino acid substitutions in influenza haemagglutinin change receptor binding specificity. Nature 1983; 304: 76-78.

135. Colman PM, Varghese JN and Laver WG. Structure of the catalytic and antigenic sites in influenza virus neuraminidase. Nature 1983; 303: 41-44.

136. von Itzstein M, Barry JG and Chong AKJ. Anti-infectives overview: the development of potential anti-influenza drugs. Curr Opin Thera Patents 1993; 3: 1755-1762.

137. Spaltenstein A and Whitesides GM. Polyacrylamides bearing pendant α-sialoside groups strongly inhibit agglutination of erythrocytes by influenza virus. J Amer Chem Soc 1991; 113: 686-687.

138. Mochalova LV, Tuzikov AB, Marinina VP, Gambaryan AS, Byramova NE, Bovin NV and Matrosovich MN. Synthetic polymeric inhibitors of influenza virus receptor-binding activity suppress virus replication. Antiviral Res 1994; 23: 179-190.

139. Sabesan S, Duus JO, Neira S, Domaille P, Kelm S, Paulson C and Bock K. Cluster sialoside inhibitors for influenza virus - synthesis, NMR, and biological studies. J Amer Chem Soc 1992; 114: 8363-8375.

140. Roy R, Zanini D, Meunier SJ and Romanowska A. Solid-phase synthesis of dendritic sialoside inhibitors of Influenza A Virus haemagglutinin. J Chem Soc Chem Commun 1993; 1869-1872.

141. Nagy JO, Wang Peng GJH, Schaefer ME, Hill TG, Callstrom MR and Bednarski MD. Carbohydrate materials bearing neuraminidase-resistant C-glycosides of sialic acid strongly inhibit the *in vitro* infectivity of influenza virus. J Med Chem 1992; 35: 4501-4502.

142. Watowich SJ, Skehel JJ and Wiley DC. Crystal structure of influenza virus hemagglutinin in complex high-affinity receptor analogs. Structure 1994; 2: 719-731.

143. Watowich SJ, 1995, personal communication.

144. Liu C, Eichelberger MC, Compans RW and Air GM. Influenza type A virus neuraminidase does not play a role in viral entry, replication, assembly, or budding. J Virol 1995; 69: 1099 - 1106.

145. Meindl P and Tuppy H (Thomae Dr Karl, GMBH). 2-Deoxy-2,3-dehydro-*N*-acylneuraminic acids. 1972: patent Ger 1493249. Chem Abs 78: 30153e.

146. Flashner M, Kessler J and Tanenbaum SW. The interaction of *N*-acetylneuraminic acid and its analogs with neuraminidase. In: Drug Action and Design: Mechanism-Based Enzyme-Inhibitors. Kalman I, ed. Amsterdam: Elsevier, 1979: 27-44.

147. Taylor NR and von Itzstein M. Molecular modeling studies on ligand binding to sialidase from influenza virus and the mechanism of catalysis. J Med Chem 1994; 37: 616-624.

148. Tiralongo J, Pegg MS and von Itzstein M. Effect of substrate aglycon on enzyme mechanism in the reaction of sialidase from influenza virus. FEBS Lett 1995; 372: 148-150.

149. Guo X, Laver WG, Vimr E and Sinnott ML. Catalysis by two sialidases with the same protein fold but different stereochemical courses: a

mechanistic comparison of the enzymes from Inflenza A Virus and *Salmonella typhurimurium.* J Amer Chem Soc 1994; 116: 5572-5578.

150. von Itzstein M, Dyason JC, Oliver SW, White HF, Wu W-Y, Kok GB and Pegg MS. A study of the active site of influenza virus sialidase: an approach to the rational design of novel anti-influenza drugs. J Med Chem 1996; 39: 388-391.

151. Hayden FG, Treanor JJ, Betts RF, Lobo M, Esinhart JD and Hussey EK. Safety and efficacy of the neuraminidase inhibitor GG167 in experimental human influenza. JAMA 1996; 275: 295-299.

152. Pegg MS and von Itzstein M. Slow-binding inhibition of sialidase from influenza virus. Biochemistry and Molecular Biology International 1994; 32: 851-858.

153. Hart GJ and Bethell RC. 2,3-Didehydro-2,4-dideoxy-4-guanidino-*N*-acetyl-D-neuraminic acid (4-guanidino-Neu5Ac2en) is a slow-binding inhibitor of sialidase from both Influenza A and Influenza B virus. Biochem Mol Biol Int 1995; 36: 695-703.

154. Holzer CT, von Itzstein M, Jin B, Pegg MS, Stewart WP and Wu W-Y. Inhibition of sialidases from viral, bacterial and mammalian sources by analogues of 2-deoxy-2,3-didehydro-*N*-acetylneuraminic acid modified at the C-4 position. Glycoconjugate J 1993; 10: 40-44.

3
Sialoside Mimetics and Conjugates as Anti-inflammatory Agents and Inhibitors of Flu Virus Infections

René Roy
University of Ottawa
Ottawa, Ontario, Canada

1. INTRODUCTION

Sialic acid terminated glycans (sialyloligosaccharides) present on cell surface glycoproteins and glycosphingolipids are involved in a large variety of biological events [1]. As boundary residues (capping sequence), sialic acids (Fig. 1) are ideally positioned to participate in numerous carbohydrate-protein interactions mediating cell surface recognition phenomena. Collectively, this family of sialylated oligosaccharides is known to serve as ligands for microbial toxins [2], bacterial and viral adhesins [3,4], and for mammalian lectins responsible for cell-cell adhesion [5,6]. Sialic acids and sialyloligosaccharides also play other functions that span from trivial: structural, physical, stabilizing, protective to very sophisticated ones such as cell signal mediators [7]. Surprisingly, sialic acids can also act as "masking" antigens where they cover other important carbohydrate antigens [8].

Most recent interests by the pharmaceutical industry toward sialosides rely on their involvement in inflammation processes and in flu virus infections. In the first case, various sialyloligosaccharides have been identified as ligands for the cell adhesion molecules implicated in the cascade of events leading to inflammation. In the case of flu virus infections, sialic acid plays two critical roles. In one instance, it constitutes the key monosaccharidic host receptor for the viral hemagglutinins. In the second instance, it acts as substrate for the viral sialidase (neuraminidase) which is believed to be essential for the release of viron particles from the infected cells.

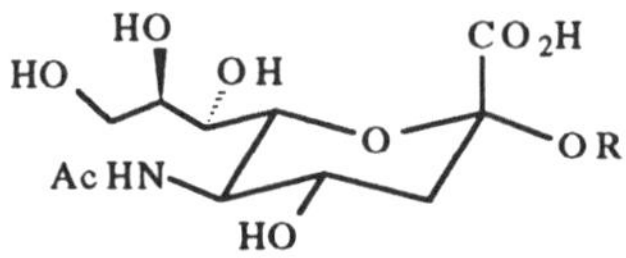

Fig. 1 α-Sialosides as they appear on most natural cell surface gangliosides and sialylated glycoproteins.

This chapter will focus around two areas of intensive research activities related to sialoconjugates. The first part will illustrate how a sialylated tetrasaccharide known as sialyl Lewis^x (**1**, sLe^x) having the structure NeuAcα-(2,3)-Galβ-(1,4)-[Fucα-(1,3)]-GlcNAcβ-(1-OR) has become a lead compound for the pharmaceutical industry interested in drug development against inflammatory diseases. Sialyl Lewis^x and related structures, which occur as partial structural elements of glycosphingolipids (gangliosides) and glycoproteins on cell membranes, have been clearly established as key actors in the blockage of a well coordinated sequence of events leading to inflammation. Carbohydrate-mediated cell adhesion is an important event initiated by tissue injury or infection. Intercellular adhesions, which are at the origin of the migration of white blood cells to the site of infection, are mediated by a family of sialoconjugate-binding molecules collectively known as selectins. The above sialyl Lewis^x and/or related oligosaccharides function as ligands for all three selectins. Over-recruitment of these cells at the site of injury or infection is responsible for acute and/or chronic diseases such as myocardial shock (ischemia-reperfusion injury), stroke, thrombosis, encephalitis, rheumatism, psoriasis, dermatitis, acute respiratory distress syndrome (ARDS), and even metastasis. This section will describe the structures of many sLe^x glycomimetics designed to become potential antagonists involved in inflammation.

The second part of this chapter will describe how low affinity sialic acid derivatives have been transformed into potent multivalent inhibitors of adherence of influenza virus to human erythrocytes. Influenza virus particles are covered with many copies of their surface glycoproteins. One of which, N-acetyl-neuraminidase, is related to sialic acid hydrolases. The second one serves as receptor for the viral attachment to host ligands. The implications of these numerous receptors on the virions, coupled with their low affinity toward individual sialosides, suggested inhibition approaches based on molecules also presenting multiple copies of the sialic acid ligand. This chapter will be devoted to the design of such multivalent inhibitors. A particular emphasis will be given to a new family of "glycopolymers" and "glycodendrimers" containing sialic acid residues.

2. SIALYL LEWIS^X AS SELECTIN ANTAGONISTS

The three selectins (L, P, and E) responsible for the adhesion of leukocytes to vascular endothelium in the early cascade of events leading to inflammation processes are a family of transmembrane glycoproteins [reviewed in 9-11]. L-Selectins (homing receptors) are constitutively expressed on various leukocytes, polymorphonuclear neutrophils (PMN), and monocytes. They are involved in the trafficking of lymphocytes to high endothelial venules of peripheral lymph nodes. They also mediate binding to cytokine-activated endothelium. P-Selectins are temporarily stored in alpha and dense granules of platelets and in Weibel-Palade bodies of endothelial cells. They are rapidly (<10 min) redistributed to the cell surface after activation by thrombin, histamine, phorbol esters, and other mediators [reviewed in 12]. E-Selectins are the latest of the adhesion molecules to appear on vascular endothelial cells (2-6 h). They are up-regulated after stimulation by bacterial endotoxins (LPS) and by cytokines associated with inflammation, including interleukin-1 (IL-1) and tumor necrosis factor-α (TNF-α). In turn, IL-1 stimulates phospholipase A_2 (PLA_2) thereby inducing prostacyclin (PGI_2) production by endothelial cells. E-Selectins also mediate the adhesion of PMN, monocytes and a subpopulation of T-lymphocytes to inflamed endothelium. Each selectin contains a signal sequence, a C-type (calcium dependent) lectin domain similar to the carbohydrate-recognition domain (CRD) of asialoglycoprotein hepatic receptor, an epidermal growth factor-like domain (EGF) surmounting short consensus repeat of complement binding protein, a transmembrane anchoring motif, and a short cytoplasmic tail [11].

The mechanism of inflammation can be summarized as follows (Scheme 1). Low-affinity (carbohydrate-based) binding of sLe^x expressed on the surface of endothelial cells is initially recognized by L-selectin on leukocytes. This first event allows capture and rolling of the leukocytes on the endothelium. The ensuing firm leukocyte adhesion is then secured by a combination of P- and E-selectins assisted by integrin bindings. The next step (protein-based) is mediated by leukocyte integrins such as CD11/CD18 which bind with high affinity to intercellular adhesion molecules (ICAM-1) belonging to integrin-Ig superfamily of molecules and also expressed on activated endothelial cells [13].

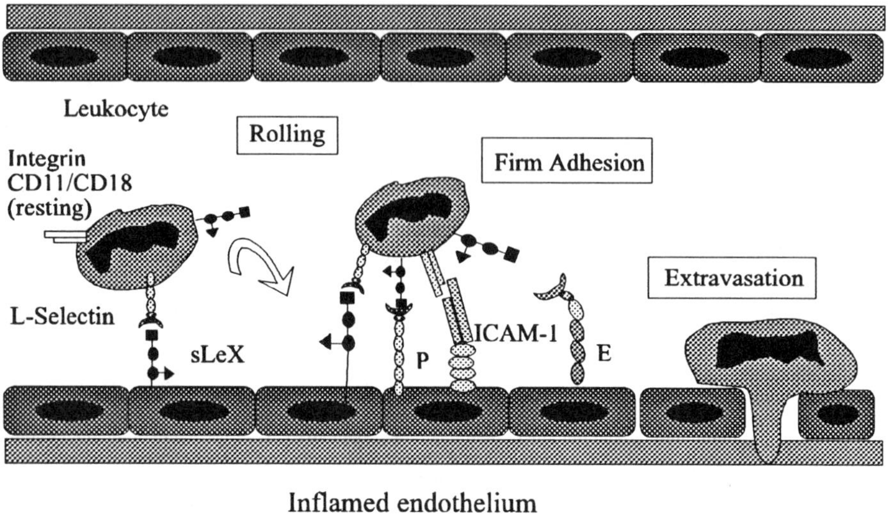

Scheme 1. Mechanism of leukocytes extravasation (diapedesis). Constitutively expressed L-selectins on leukocytes bind to endothelial $sLe^{x/a}$ (or other natural ligands such as sialomucins) causing slow down and rolling. Following tissue injury or infection, P-selectins are up-regulated (< 10 min) and help firm adhesion of leukocytes through sLe^{x} binding. Further tight adhesion occurs between activated integrins (CD11/CD18) on leukocytes and ICAM-1 activated endothelium. E-Selectins participate to the binding after 2-6 h of injury or infection

Several useful pharmaceutical strategies can be envisaged to prevent selectin-mediated leukocyte recruitment and extravasation (diapedesis) during inflammation. A summary is listed below:

- Inhibition of binding by anti-selectin antibodies
- Inhibition of selectin expression
- Inhibition of carbohydrate ligand biosyntheses
- Inhibition of binding with peptides and antibodies against sLe^{x}
- Inhibition of binding by selectin ligands

Monoclonal antibodies to all three selectins have proved useful in demonstrating their involvement with a diversity of acute and chronic inflammatory conditions [reviewed in 11]. Anti-sense nucleotides directed at inhibition of expression of ICAM-1 (ISIS Pharmaceuticals) are in clinical Phase II stage of development [14]. Other conventional anti-inflammatory drugs are also the subject of renewed activities. The involvement of sialyl $Lewis^{x/a}$ as low affinity ligands for all three selectins and the strong implication of the L-fucose residue in the binding interactions has triggered much attention in the design of specific and potent fucosyl transferase inhibitors. The most likely candidate seems to be fucosyl-transferase VII (Fuc-T VII) because it appears to endow a diversity of cells with the ability to bind to E-selectin. Strong evidences suggest that Fuc-T VII is the myeloid-specific (natural ligand) fucosyl-transferase involved with the production of sLe^x [15].

Although all four strategies are the focus of intensive research, the latter has attracted considerable appeal from the pharmaceutical industry because it is amenable to small drug molecule discoveries acting as antagonists. As mentioned above, sLe^x has been the lead molecule that has revived wide interest for carbohydrate-based therapeutics.

2.1. Sialyl Lewis^x Mimetics

Binding of selectins to their carbohydrate ligands has been unequivocally demonstrated to be required for neutrophil rolling and extravasation and to play a major role in lymphocyte recirculation and platelet adhesion [16]. The biological implications of the selectins strongly suggest that compounds that block their functions would serve as anti-inflammatory agents. Given the facts that the identity of the natural carbohydrate ligands for the selectins is still a debatable issue and that sLe^x (**1**) constitutes a valuable lead compound with demonstrated activity *in vivo* [17-19], the daunting task of designing potent inhibitors remains. In the case of carbohydrate-selectin interactions, the remarkable specificity of binding immediately suggests the traditional approach of using structure-activity relationship (SAR) studies. This is the approach that this section of the chapter will summarize.

Since the early discovery of sLe^x (**1**) as a putative ligand in E-selectin (ELAM-1) mediated cell adhesion [5,16], tremendous activities have been devoted to this topic [12, 20-25]. The indication that sLe^x is biologically active in animal models including myocardial reperfusion injury [17], protection of lung injury [18] and liver inflammation [19], further supports its use as an anti-inflammatory agent.

Several chemical [25-31], as well as chemo-enzymatic [32, 33], syntheses of sLe^x (**1**) have been reported. Numerous reports have also described the solution conformations of **1** by NMR spectroscopy [32, 34-42] and by molecular dynamics simulations [38, 39]. Although it has been previously claimed that the conformation in the bound state was identical with that in solution [37], the information could not be reconciled with more recent investigations using E-selectin [42]. In the latest study, it was demonstrated that the orientations of both the sialic acid and the fucose residues in the bound state lead to transfer NOEs that differ considerably from those of free sLe^x. E-selectin complexes exclusively a conformation of sLe^x from an ensemble of free solution conformations in which the sialic acid has an orientation corresponding approximately to conformation A of Rutherford *et al.* [39] or conformation C of Ichikawa *et al.* [32].

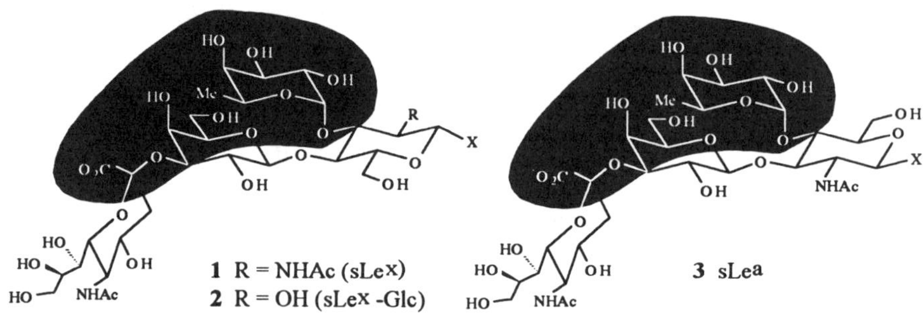

The association constant of sLe^x to E-selectin has been adequately measured by fluorescence polarization [43]. In the above investigation, it was shown that sLe^x, like most other carbohydrate ligands, binds with relatively small affinity (K_D 120 ±31 μM, *i.e.* ~ 0.1 mM) to its receptor.

The selectins have calcium dependent carbohydrate binding domains of the C-type lectins [11]. The similarities between the L-fucose epitope and D-mannose (see Scheme 2) was used to postulate that the active sites of the selectins might be modeled onto those of mannose-binding proteins (MBP) [44-47]. These early assumptions were fully rewarded with the recent publication of the X-ray data of E-selectin at 2.0 Å resolution [48].

Using synthetic as well as artificial neoglycoconjugate ligands, a number of investigators have demonstrated that the regioisomeric sLe^a (**3**) also constituted a ligand for L- and E-selectins [49, 50], and presumably P-selectin. The findings that the analog sLe^x-Glc (**2**) showed even higher affinities for the selectins [12, 20, 51] combined with a number of structure-activity studies [25,51-53], allows a contour map indicating the likely active topography for **1-3** to be established (see shadowed structures above).

More recent studies have identified glycosaminoglycans that bind to selectins. These findings may help identify the natural ligands for the selectins [11, 21, 22, 54, 55]. Some of the glycoproteins belong to the family of sialomucins, thus further supporting sLe^x as a lead structure. However, highly sulfated ligands have also been identified in these studies. In fact 3'-sulfo-Le^x (**4**) is also active *in vivo* [reviewed in 6]. These investigations have shed some doubt about the real structure of the biologically active ligands. Additionally, a number of anionic saccharides such as polyphosphomannan, fucoidan, sulfatide, heparin fragments, keratan and chondroitin sulfates showed some low level of binding to L- and P-selectins but not E-selectin [22, 55].

2.1.1 Sialic acid isostere

Based on cumulated evidences [21, 22, 25, 53-55] with both natural [21, 22, 55] as well as synthetic [25, 53, 54] ligands, it was recognized that the sialic acid residue could be easily replaced by a negatively charged group. For instance, new E-selectin ligands containing a sulfate group have been identified by screening neoglycolipids derived from a human ovarian cystadenoma [56]. It was found that 3'-sulfo-Le^{a} (**3a**) and 3'-sulfo-Le^{x} (**4**) bound to E-selectin as strongly as sLe^{x} (**1**) [54]. However, the sulfated analogs were better inhibitors for L- and P-selectins [53-55] (Table 1). According to a recent study [54], the order of ligands affinity for E-selectin is: 3'sulfo-Le^{a} (**3a**)> sLe^{a} (**3**)> 3'sulfo-Le^{x} (**4**)> sLe^{x} (**1**) and hexasaccharides> penta-> tetra> tri-saccharides, with **3a** binding 15 fold better than (1).

Synthetic analogs of sLe^{x} having the sialic acid residue modified in numerous ways have provided the basis for simpler glycomimetics in which only the negative charge of the sialic acid was deemed essential, while OH-2, OH-3, and OH-4 of fucose, and OH-6 of galactose were essential for E- and L-selectins [12, 20, 25, 51-53]. P-selectin, however, requires only OH-3 of fucose. Thus, the three selectins differ slightly in their binding properties. The structures of **3a-5** with the corresponding topographical overlaps to those of sLe^{x} (**1**) and sLe^{a} (**3**) are illustrated. Therefore, replacement of sialic acid by a sulfate group or by other negatively charged residues was the next most logical modification. These early observations have formed the basis for the synthesis of glycomimetic analogs also described in this section.

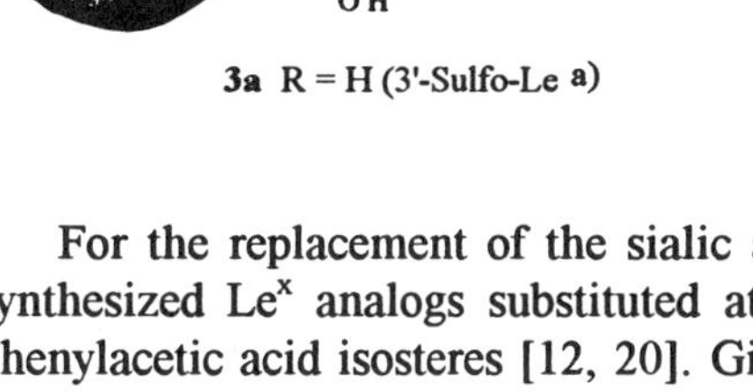

3a R = H (3'-Sulfo-Le a)

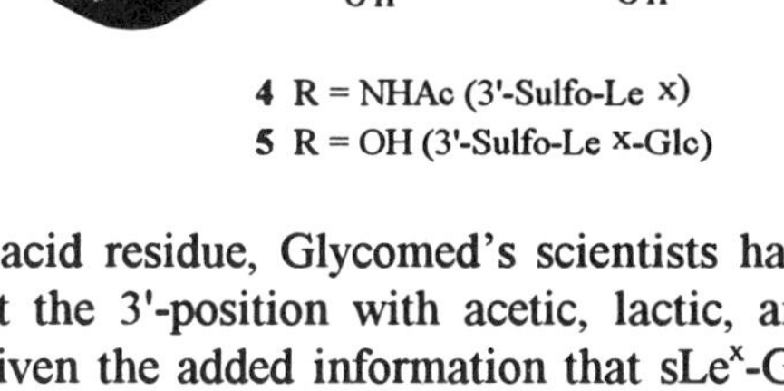

4 R = NHAc (3'-Sulfo-Le x)
5 R = OH (3'-Sulfo-Le x-Glc)

For the replacement of the sialic acid residue, Glycomed's scientists have synthesized Le^{x} analogs substituted at the 3'-position with acetic, lactic, and phenylacetic acid isosteres [12, 20]. Given the added information that sLe^{x}-Glc (**2**) was a better ligand for all selectins (IC_{50} 0.1 mM (**2**, X = OH) with E-selectin), they scaffolded their isosteric sialic acid group on 3-fucosylated lactose. A series of more or less potent analogs **6-9** were obtained. Except for the analog containing acetic acid side chain (**6**, IC_{50} > 4mM for L-, P-, and E-), all other derivatives showed binding almost equipotent toward E-selectin (**7-9**, IC_{50} < 0.75 mM). Alternatively, compounds **7-8** showed weak binding properties to L- and P-selectins. Glycomimetic **9** on the other hand, with its lipophilic aglycone and its (S)-lactic acid isostere, exhibited improved binding affinities to all three selectin receptors (IC_{50} < 1 mM) (Table 1).

In an attempt to further explore the bioactive conformation of sLe^x, scientists from CIBA have synthesized an analog in which the sialic acid residue was replaced by a carboxylic acid function fixed in the equatorial position by a 1,3-dioxane acetal fused to C-4 of galactose [57]. Equatorial vinylation of a 4-ketogalactoside intermediate, followed by reductive ozonolysis provided a 4-hydroxymethyl arm which was subsequently acetalized to OH-3 with furfuraldehyde serving as carboxylic acid group precursor (oxidative ozonolysis). The resulting C-4 branched analogue **10**, having the axial hydroxyl group of galactose still intact, was found inactive in a competitive E-selectin binding assay up to a concentration of 10 mM. It was concluded that the position of the carboxylic acid in the bound conformation differed decisively from one of the preferred solution conformations [32, 34-42]. This conclusion was recently confirmed by transfer NOE experiments using sLe^x bound to E-selectin [42]. Thus the negative charge of the sLe^x ligand must be exactly positioned to best fit the complementary function (s) in the E-selectin active site with very little if any orientational flexibility.

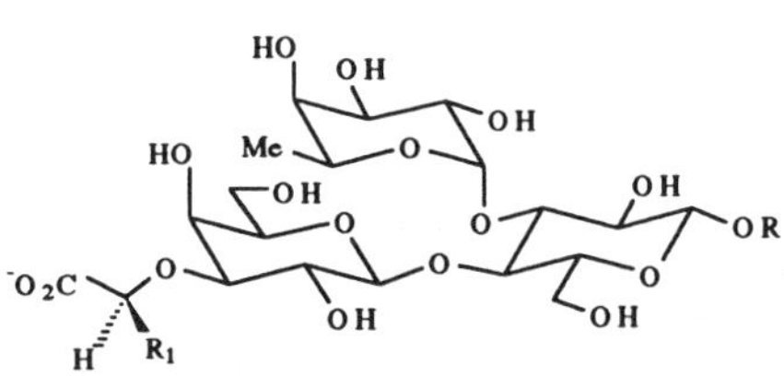

6 R = R_1=H
7 R = H, R_1= Ph (R/S)
8 R = H, R_1= Me
9 R = $(CH_2)_6OH$, R_1 = Me

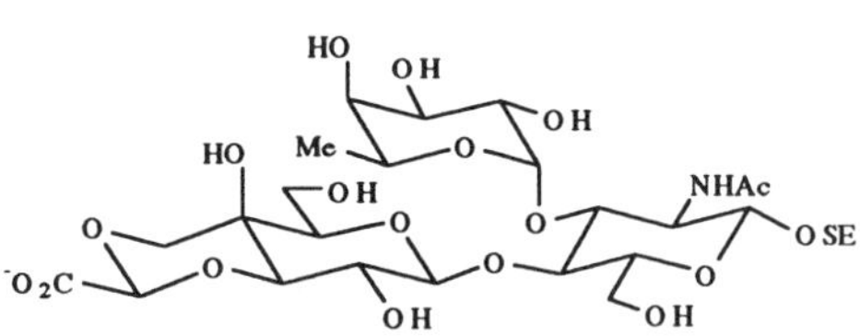

10

2.1.2 N-Acetylglucosamine isostere

As mentioned above, the regioisomeric sLe^x (**1**) and sLe^a (**3**) have been shown to bind to E-[49, 50, 54] and L-selectin [55] with compound **3** binding slightly better than **1**. This information, together with the fact that computer modeling using hard sphere exo-anomeric (HSEA) calculations [49] demonstrating substantial overlap between NeuAc/Fuc/Gal residues, suggested that the N-acetylglucosamine residue (GlcNAc) may only act as a bridging element tethering fucose to the sialylgalactose moieties. These observations were further substantiated with sLe^x-Glc (**2**) in which the GlcNAc residue was substituted for glucose (Glc) [51]. In this last case, the only difference was the replacement of the C-2 N-acetyl by a hydroxyl group. The resulting analog **2** showed slightly better affinity toward all three selectins in competitive binding assays [12, 51].

Further isosteric replacement of the GlcNAc residue has involved D-glucal to provide sLe^x-glucal **11** [41]. Using the same inhibition of binding experiments described above (E-selectin/HL-60 cells), compound **11** was shown to be as active as sLe^x **1** itself (IC_{50} 2.1 mM). Conformational analysis of **11** by NMR indicated that the solution conformation of **1-3**, and **11** were identical. Similar successful modifications involving replacement of GlcNAc residue by deoxynojirimycin derivatives have also been described (not shown) [25].

Using related grounds, Toepfer *et al.*[58] have replaced GlcNAc residue by (1R, 2R)-*trans*-1,2-cyclohexanediol to provide **12**. The C_2-symmetry of the latter permits mimicking both sLe^x (**1**) and sLe^a (**3**) topostructures. Compound **12** has IC_{50}s against E- and P-selectins of 0.4 and 0.6 mM, respectively.

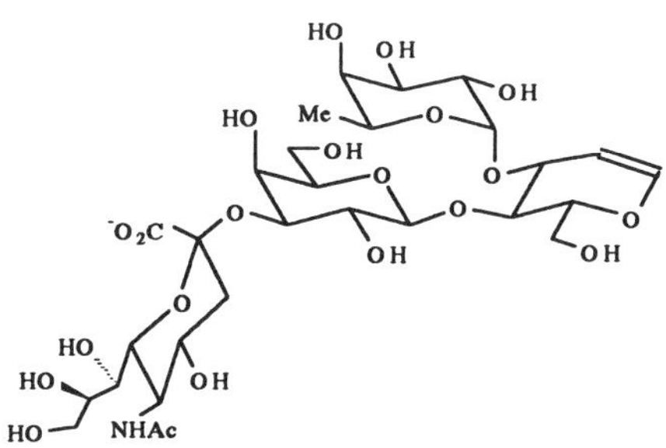

11 sLe^x -Glycal

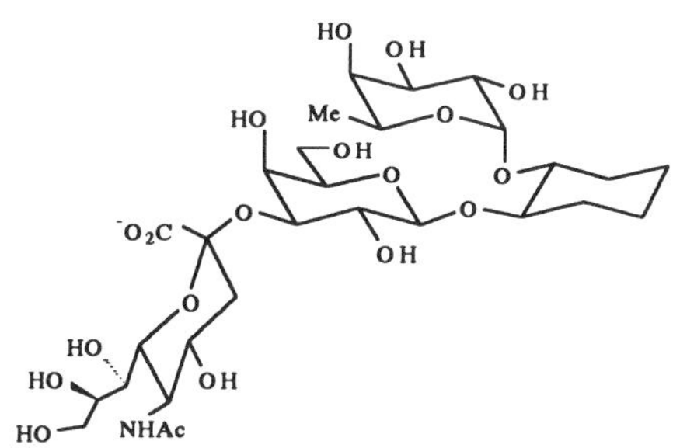

12

The above arguments may be however over-simplistic since its has also been shown that incorporating an 8-methoxycarbonyloctyl aglycone to either sLe^x (IC_{50} 380 μM) or sLe^a (IC_{50} 130 μM) (**1**, **3**: X= β-$O(CH_2)_8CO_2Me$) increased their binding activity nearly two fold against E-selectin [59]. Similarly, replacement of the C-2 N-acetyl group (GlcNAc) by an amino or azido group resulted in substantial increases in activity, with the most potent inhibitor being amino substituted sLe^a (**3**), which was 36-fold more active (IC_{50} 21 μM) than the reducing tetrasaccharide sLe^x [59]. Alternatively, replacement of the N-acetyl by an N-propionyl group abolished binding. None of these modifications proved useful in the case of P-selectin. Taken together, these data suggest that replacement of the N-acetyl group with a smaller moiety or one capable of ionic bond participation can strengthen binding to E-selectin. The lipophilic aglycone may contribute to increased binding by lesser entropic lost.

In the pursuit of smaller and simpler analogs of sLe^x (**1**) and given the fact mentioned above, the group at Glycomed [60] has made early modifications involving tethering the essential fucose residue to the sialylgalactose moiety through a dimethylene bridge (not shown, see **16**). The resulting analog lacked the GlcNAc residue completely and yet it showed good activity *in vivo* (mouse peritonitis model).

2.1.3 Galactose isostere

Previous work by Hasegawa *et al.* [25] and by Stahl *et al.* [61] using 4-deoxy galactose analogs showed poor binding properties with E- and L-selectins. P-selectin however did not show any substantial decrease in binding using the same 4-deoxy Gal analog. Alternatively, the 6-deoxy analog was not recognized by any of the three selectins, suggesting that OH-6 is critical for binding. These observations correlate well with the proposed implication of OH-6 in the chelation of the calcium ions [62].

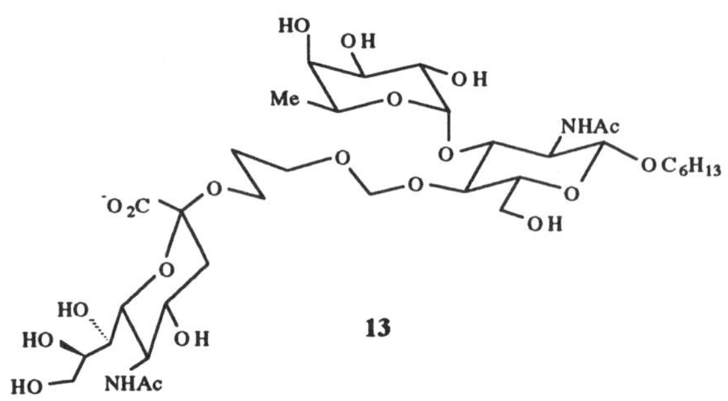

13

Another potential antagonist of E-selectin binding in which the galactose moiety was replaced by a six-atom bridge (oxypropoxymethoxy) was recently published [63]. Although no biological activity was described for **13**, the synthesis described an interesting preparation of a methylene acetal linkage which was hoped to adopt a conformation similar to that of the galactose residue. The procedure necessitated the activation of a methylthiomethyl ether (MTM, $ROCH_2SCH_3$) preinstalled as an α-linked aglycone on sialic acid [2-O-3-(methylthiomethoxy)propyl)]. The mixed acetal bridge between the sialoside aglycone and OH-4 of the GlcNAc residue was effected using an N-iodosuccinimide-triflic acid activation (NIS-TfOH) similar to that used in glycosylation reactions. In the same paper, a simpler analog of **13** missing the fucose residue was also prepared in order to have access to an inhibitor of α-(1-

3)-fucosyltransferase. Preliminary studies showed that the resulting analog did not function as an acceptor of Fuc-T III.

2.1.4 Fucose isostere

The L-fucose residue was early recognized as an essential functionality for binding to all three selectins [25, 53, 64]. It has been reasoned that because selectins are members of the family of calcium-dependent lectins (C-type), the fucose residue would have similar calcium-binding properties to that of D-mannose in analogous mannose-binding proteins (MBP). In fact, before the X-ray data became available for E-selectin [44-47], it was argued that MBP would constitute a good selectin binding model. It later appeared that these presumptions were validated [48]. These arguments can be fully supported by close examination of the relative orientations of the three hydroxyl groups of L-fucose and D-mannose (Scheme 2). Of particular interest is the relative *cis* orientation of OH-3,4 (Fuc) and OH-2,3 (Man) which is suitable for calcium chelation. Furthermore, these two hydroxyl groups are essential for binding as shown with a series of deoxy and fluoro deoxy derivatives [25, 64]. Indeed, such topographical mimicry has been previously accounted for in previous lectin binding studies and with other monosaccharide mimics like GlcNAc→NeuAc and Gal→Rhamnose.

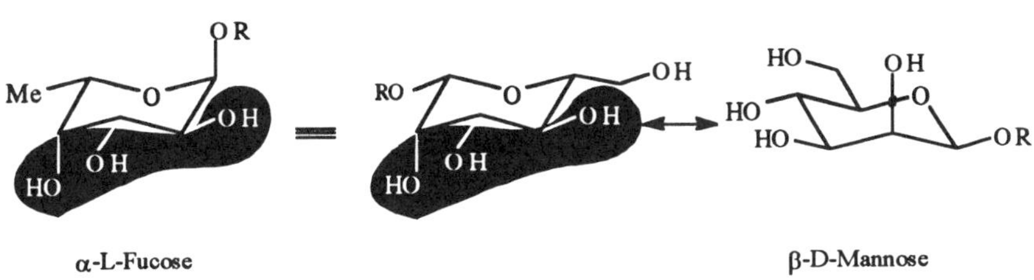

Scheme 2. Potential topostructure overlaps showing the mimicry between α-L-fucose and β-D-mannose for the calcium binding sites of selectins and mannose binding protein (MBP).

Additionally, results from electrospray mass spectrometry on a series of deoxy fucose analogs, in conjunction with computer-assisted modeling of sLe^x, suggest that calcium ion coordinates ring oxygen atoms from the Gal-GlcNAc moieties and the OH-2 of fucose [62]. A D-arabinose substitute lacking the C-6 methyl group of fucose, was also found to be 5-fold less potent (IC_{50} 5 mM) than the parent sLe^x derivative (**1**) with human recombinant E-selectin (HL-60 cells) [64]. These data strongly indicate that any modifications of fucose on sLe^x are generally deleterious toward E-selectin binding. Other results showed that all three hydroxyls are required for recognition by E- and L-selectin while only the C-3 hydroxyl is essential for P-selectin [53].

Using more drastic functional group alterations, researchers from Glycomed synthesized glycomimetics **14** and **15** in which the fucose residue was replaced by a glycerol side chain (no chirality indicated) on the sLe^x-Glc framework. It was anticipated that the glycerol functionality could still act as a calcium chelator. They found that while **14** was totally inactive in inhibition assays on all three selectins (IC_{50} >5mM), compound **15** showed slightly better inhibition toward L- (IC_{50} 2.1 mM) and P-selectins (IC_{50} 1 mM). E-selectin binding was also totally abrogated [12, 20].

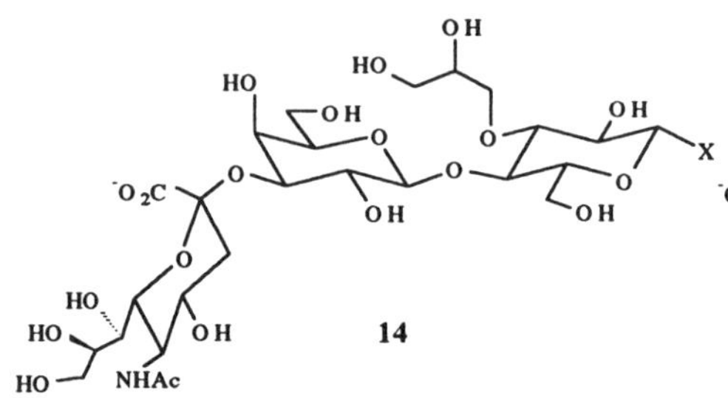

14

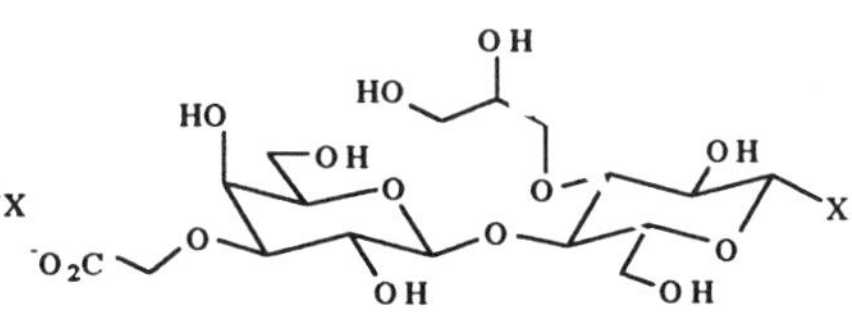

15

2.1.5 NeuAc/GlcNAc isosteres

All of the evidence accumulated so far has pointed toward the implication of the sialic acid carboxylate, the hydroxyl groups of the galactose and fucose residues, and likely the ring oxygens of the Gal/GlcNAc residues. These results have stimulated the syntheses of analogs in which both sialic acid and N-acetylglucosamine were altered. Based on these cumulated information, glycomimetics **16-19** were recently prepared [65-67]. Acetic acid was used as sialic acid isostere as discussed above and the remaining Gal-Fuc moieties, judged essential for binding, were simply tethered using four atom spacers as bridging units. Synthesis of the first glycosidic bond with galactose using a monoprotected ethylene glycol derivative and further glycosylation with perbenzylated fucosyl fluoride, provided disaccharide **16** after suitable functional group manipulations [66]. As expected (see 2.1.2), glycomimetic **16** showed promising antagonist activity with E-selectin (IC_{50} 1.5 mM). Replacement of the two glycosidic oxygens by carbon atoms provided analogs **17-18**, prepared from their corresponding C-allyl glycosides and Wittig coupling [65]. Both analogs showed very weak binding activities (>15 mM) against E-selectin. Compounds **16-18** were resistant to α-fucosidase and β-galactosidase.

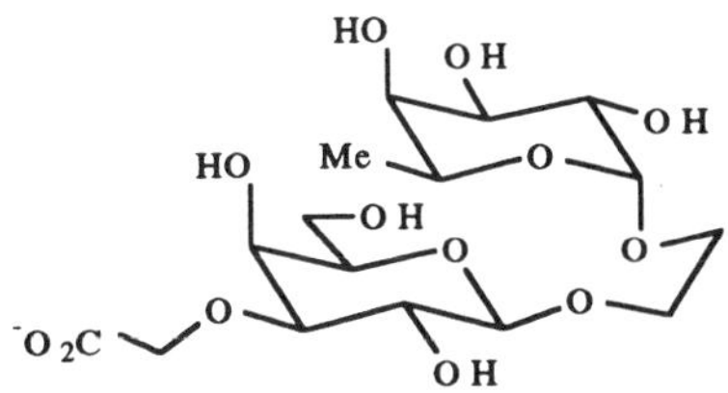

16

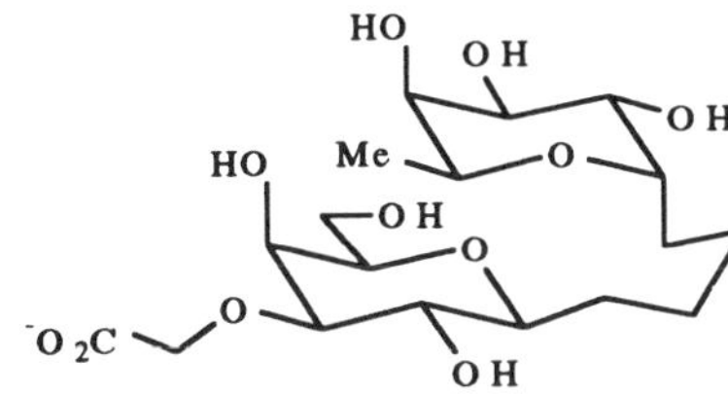

17

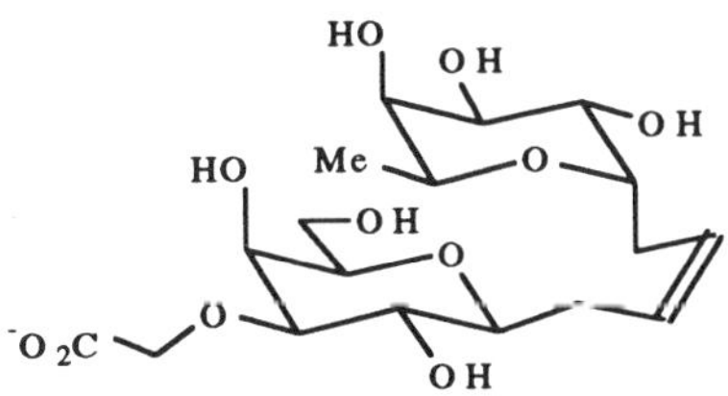

18

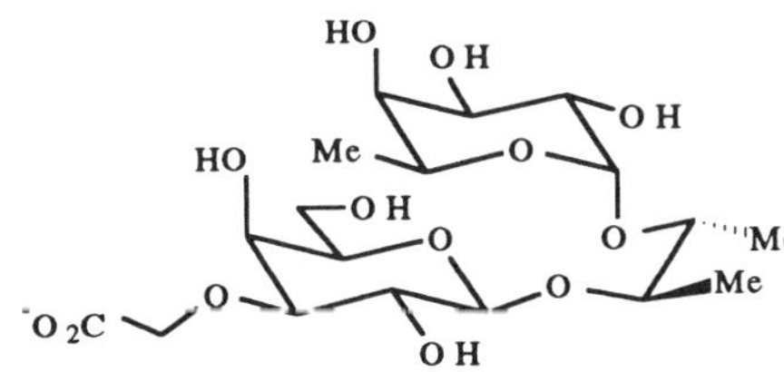

19

The poor binding properties of **17-18** compared to **16** are likely the result of higher conformational flexibility and the lack of essential oxygens. In order to provide analogs with more rigid conformation while keeping oxygen atoms, Prodger *et al.* [67] synthesized glycomimetic **19** using chiral (2R, 3R)-(-)-2,3-butanediol as bridging element. Sequential glycosylation with perbenzylated methylthio fucosyl donor and peracetylated galactosyl fluoride provided the key structural disaccharide in poor overall yield. Interestingly, as opposed to previous syntheses of the acetic acid isostere using O-alkylation with bromoacetates, the authors performed ozonolysis and oxidation of a 3'-O-allyl derivative.

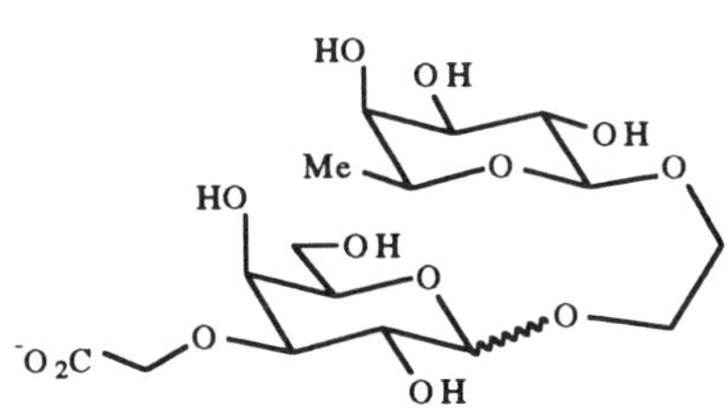

20

Using an approach similar to that described for **16**, Ragan and Cooper [68] prepared analog **20** as an intractable mixture of galactoside anomers. They also used a 3'-O-allyl ether as precursor of the isosteric acetic acid. Compound **20** was totally devoid of activity toward E-selectin.

2.1.6 Gal/GlcNAc isosteres

Before the topographical requirements of the presumed selectin receptors were fully appreciated, few groups have unsuccessfully attempted the substitution of both galactose and N-acetylglucosamine residues. Attempts at tethering fucose to sialic acid resulted in complete or partial loss of activity as judged by inhibition of binding to E-selectin (Table 1). Compound **21** (R=H or Me) was claimed to block the adhesion of an sLe^x myeloid cell line U937 to activated endothelial cell cultures at 25-30-fold higher concentrations than sLe^x [69, 70]. It was prepared by a strategy involving Wadsworth-Emmons coupling of a C-4 branched fucosyl ketophosphonate with sialic acid aldehydes of various lengths. Longer spacers between the sialic acid and the fucosyl residues resulted in further loss of activity.

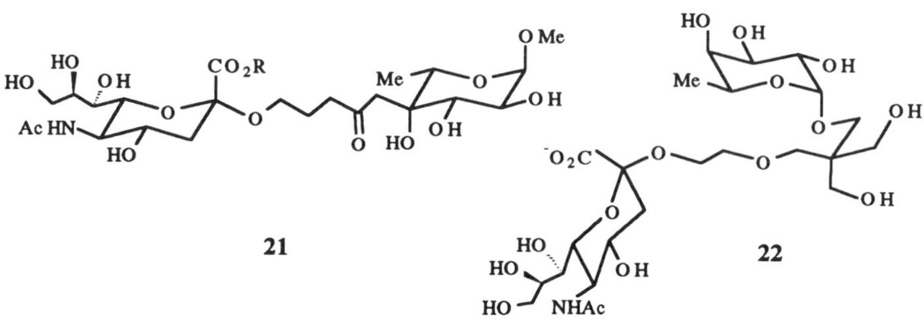

21 **22**

Using pentaerythritol as linker and an ethylenedioxy unit to mimic the galactose unit, compound **22** and its three anomerically different variants were prepared. They were also deprived of biological activity against E-selectin [71]. Interestingly, glycomimetic **23**, built by sequential glycosylation of 1,6-hexanediol, was also inactive against P- and E-selectins (IC_{50} > 4mM) but showed some affinity to L-selectin (IC_{50} 1-3 mM) [12]. The same analog with an acetic acid isostere in place of the sialic acid had an IC_{50} of 0.5 mM against L-selectin [12]. This simple example reemphasized the different ligand requirements between the selectins.

In order to provide a less flexible spacer and to fix the distance and orientation between the two tethered sugars, Kaila *et al.* [72] prepared analogs **24-26** using 1,4- and 1,3-benzenedimethanol. Conformational analyses of **24-26** were examined with the Quanta/CHARMM modeling package using energy minimization. The results illustrated that the intramolecular distances between the two carbohydrate residues could be positioned near that of sLe^x (6.8 Å between the glycosidic oxygens of sialic acid and fucose). A 2-deoxy analog **24** was also prepared because of earlier findings indicating that only the OH-3 group is necessary for P-selectin, as opposed to L- and E-selectins [53].

Compounds **24-26** were shown to be 20 fold less active than sLe^x in a static inhibition assay using human recombinant α-1 acid glycoprotein (AGP) having five residues of sLe^x and E-selectin (15% inhibition at 0.5 mM).

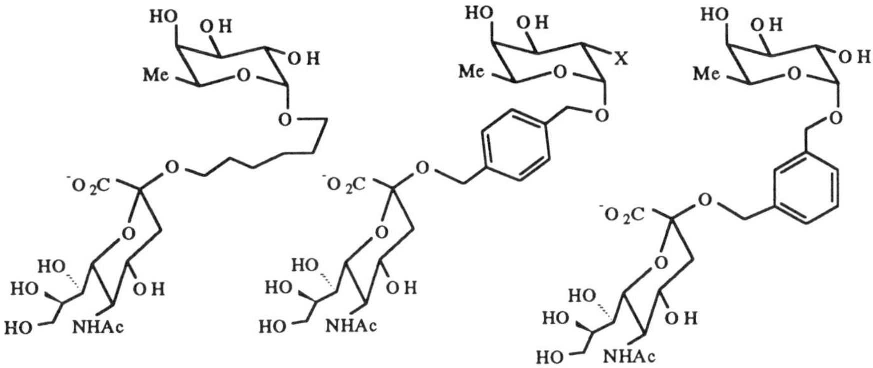

23 **24** X = H **25** X = OH **26**

By an extension of the strategy previously described for the preparation of **12**, Toepfer *et al.* [58] synthesized sLe^x mimetics **27-31**. The GlcNAc domain was substituted by a (1R, 2R)-trans-1,2-cyclohexanediol, while the Gal domain was simulated by a polyol moiety. Surprisingly, analog **27** having a 1,3-propanediol and a β-sialoside linkage showed similar receptor affinities to those of sLe^x for both E- (IC_{50} 1.6 mM) and especially P-selectins (IC_{50} 1.0 mM). Analogs with more hydroxyl groups **28-29** or longer intra-sugar residue distances **30-31** were less satisfactory. Using a similar approach, inactive compound **32** (IC_{50} 10 mM) which is lacking the essential hydroxyl groups of galactose was prepared from D-tartaric acid and a six carbon spacer containing a carboxylate group [66].

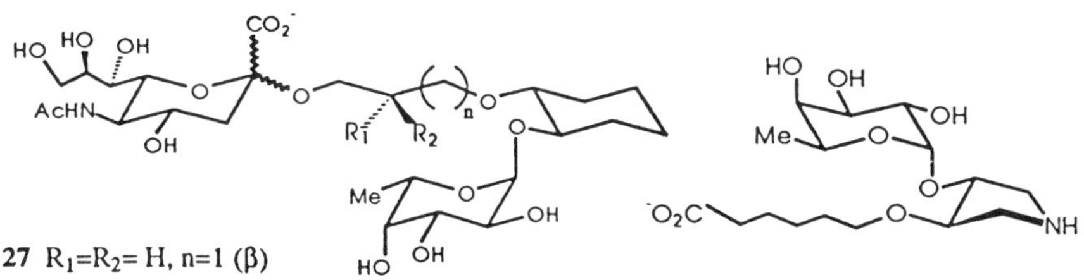

27 R_1=R_2= H, n=1 (β)
28 R_1= H, R_2=OH, n=1 (α,β)
29 R_1=R_2= CH_2OH, n=1 (α)
30 R_1=R_2= H, n=2 (β)
31 R_1=R_2= H, n=3 (α)

32

Table 1 Relative Inhibitory Potency of Sialyl Lewis^x Mimetics Having Carbohydrate Isosteres in Selectin-Ig Chimera Competitive Binding Assays

Compound	IC$_{50}$ (mM)			Ref.
	L-Selectin	P-Selectin	E-Selectin	
1 (sLe^x)	2.5	2.5	0.75-2.5[a]	53, 59
2 (sLe^x-Glc)	2.3	1.3	0.1	53
3 (sLe^a)	-	-	0.22-2.1[a]	59
4 (3'-SO_3-Le^x)	2.3	1.5	>5	12
5 (3'-SO_3-Le^x-Glc)	>5	>4	2.3	12
6	Weak	Weak	~4	12
7	3-4	3-4	0.75	12
8	>4-5	>4-5	0.75	12
9	<0.5-1	0.75-1	0.5-0.7	12
10	-	-	>10	57
11	-	-	2.1	41
12	-	0.6	0.4	58
13	-	-	-	63
14	>5	>5	>5	12, 20
15	2.1	1	>5	12, 20
16	-	-	1.5	65, 66
17	-	-	20	65
18	-	-	15	65
19	-	-	-	67
20	-	-	>40	68
21	-	-	>25[b]	69, 70
22	-	-	>5	71
23	1-3	>4	>4	12
24	-	-	>20[c]	72
25	-	-	>20	72
26	-	-	>20	72
27	-	1.0	1.6	58
28	-	2.5	2.4	58
29	-	2.0	-	58
30	-	2.6	-	58
31	-	2.7	-	58
32	-	-	10	66

[a] The lower values were obtained using sLe$^{x/a}$-BSA conjugates as standard [59].
[b] Cell assays.
[c] Using asialoglycophorin as standard.

2.2 Glycopeptidomimetics

The level of sophistication now achieved in solid-phase glycopeptide synthesis make it a procedure of choice for the construction of glycopeptide libraries dedicated to the search of improved selectin antagonists. This is exactly the approach followed by researchers from the Calsberg Laboratory (Denmark) [73]. In their initial investigations, they linked the essential fucose residue to N-acetyl L-serine as an α-linked glycoside. Further attachment to other amino acid sequences lead to glycopeptides **33** and **34** having interesting IC_{50} values of < 0.25 mM against L- and P-selectins in the usual competitive ELISA experiments (Table 2). Presumably, in these two examples, the carbonyl groups must be suitably oriented to accommodate the complementary active sites of the selectins.

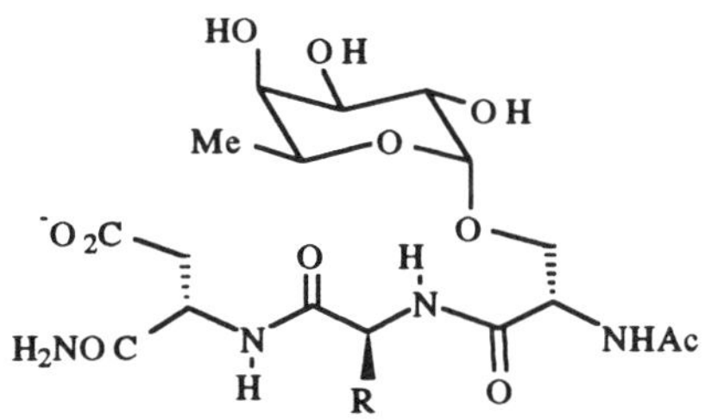

33 R = H, Gly
34 R = CH_2CONH_2, Asn

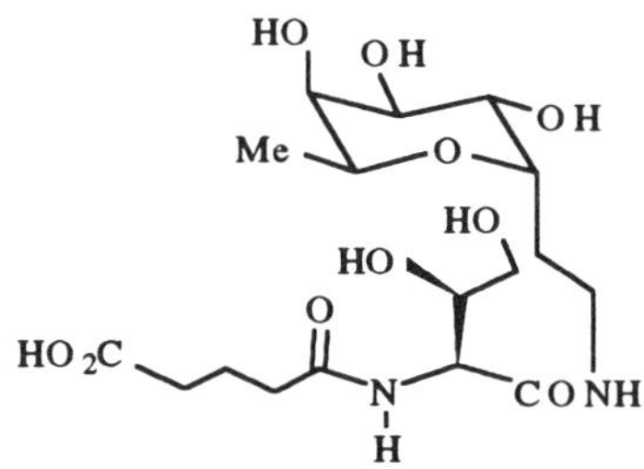

35

Likewise, an analogous glycopeptide strategy using solution chemistry was recently reported by a group from the Scripps Research Institute [65, 74]. The synthesis of a C-linked fucopeptide analog **35** resulted in an active compound (IC_{50} of 1.3 mM) against E-selectin [65]. No detail was mentioned about the inhibition of the other selectins. However, combined together, the above two facts suggest that it is possible to provide sLe^x glycomimetics having the capacity to inhibit all three selectins.

Using α-O-L-fucosyl-L-threonine as a key functional element with potentially less flexibility, Wong *et al.* [74] also prepared the series of analogs **36-40**. The first three members of this new class of analogs showed promising activities against E-selectin with IC_{50}s in the lower 0.55-1.3 mM range. The more rigid derivative **40** was essentially inactive (IC_{50} > 3 mM).

When (1R, 2R)-2-aminocyclohexanol was used in place of L-threonine as a scaffolding structure for α-O-L-fucoside, the analogs **41-48** were generated [74]. Compound **41** was slightly less active than its L-threonine counterpart (IC_{50} 1.5 mM, E-selectin assay). Moreover, the one carbon homologue containing glutaric acid analogue (**42**) lost binding affinity (IC_{50} 7 mM), reiterating the sensitivity of the spatial requirement for the carboxylate group. Alternatively, compound **43** having an N-Boc- protected L-glutamic acid residue in the vicinity of the carboxylate still showed activity similar to that of sLe^x, while its deprotected amine derivative **44** was inactive (IC_{50} 10 mM). Reversing the L-glutamic acid linkage to its natural C-1 position afforded analogs **45** (IC_{50} 6 mM) and **46** (IC_{50} > 10 mM) deprived of inhibitory activities.

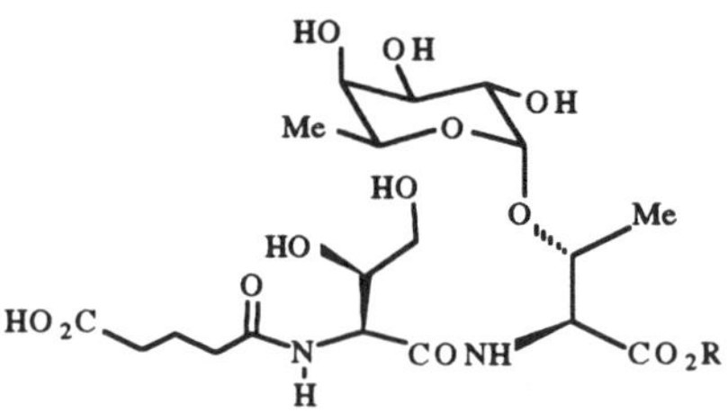

36 R = Et
37 R = H

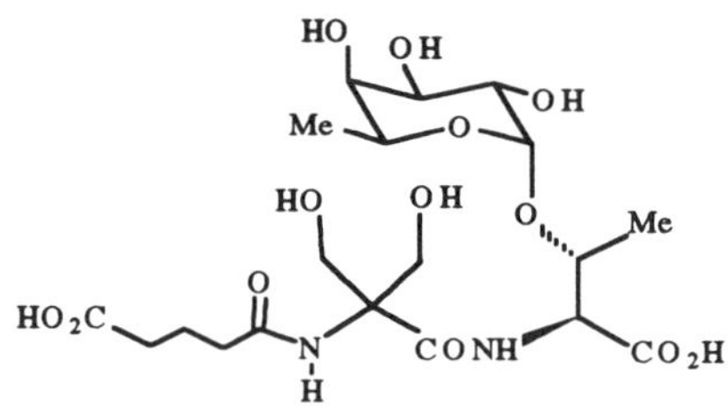

38

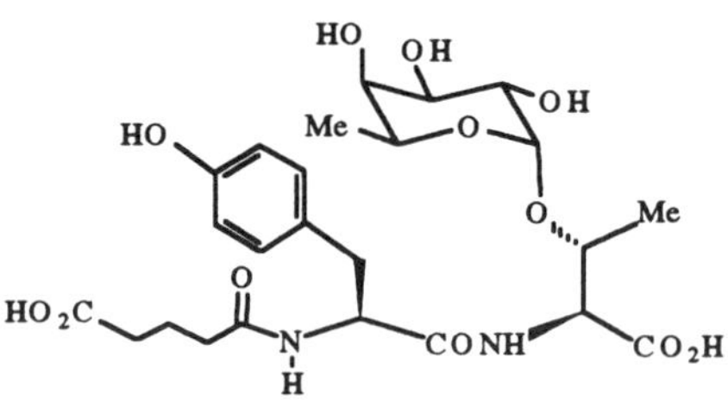

39

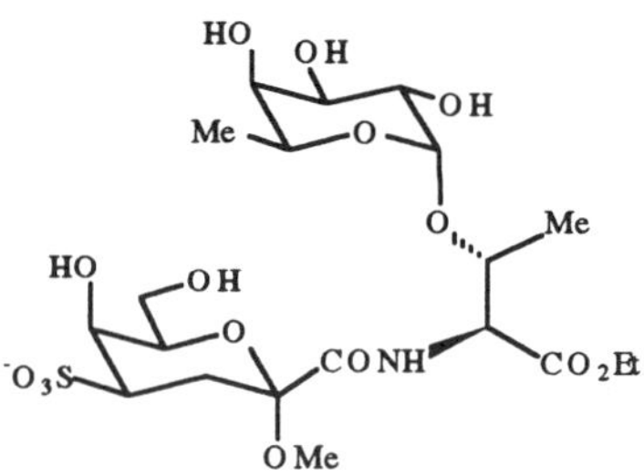

40

As opposed to tyrosine-containing analog **39** described above, compound **47** was inactive in the cyclohexanic series. C-branched O-3-acetic acid-linked 2-deoxy galactoside derivative **48** was also shown to be inactive (IC_{50} > 3 mM), presumably because of the misorientation of its critical hydroxyl and carboxylate functionalities. It is worth mentioning that the C-1 branched galactoside core was enzymatically prepared using sialic acid aldolase.

2.2.1 Chimeric glycopeptide mimetics

The firm adhesion of leukocytes on activated endothelium follows the weak carbohydrate-dependent (sLe^x) rolling event and relies on stronger protein-protein interactions imparted by the many adhesion molecules present in extracellular matrices [10, 11]. These adhesion molecules contain the tripeptide arginine-glycine-aspartic acid (RGD) as their recognition site. These proteins include fibronectin, collagens, fibrinogen, von Willebrand factor, and few other proteins [75]. The RGD sequence is recognized by a family of structurally related receptors, integrins, including GpIIb/IIIa expressed on platelets. It is therefore possible that the leukocyte integrins CD11/CD18 interact with the endothelium-attached ICAM-1 through RGD interactions.

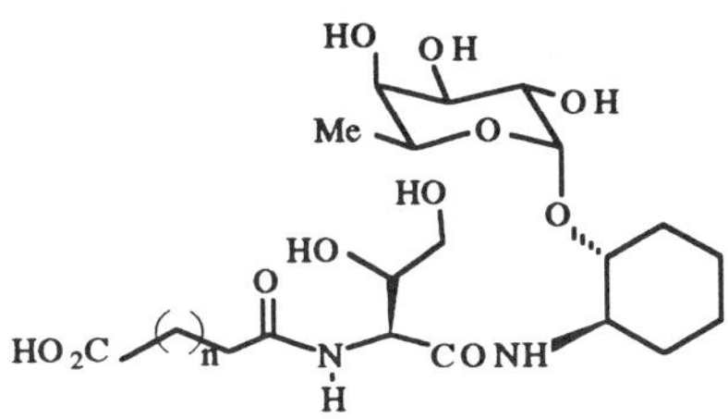

41 n = 1
42 n = 2

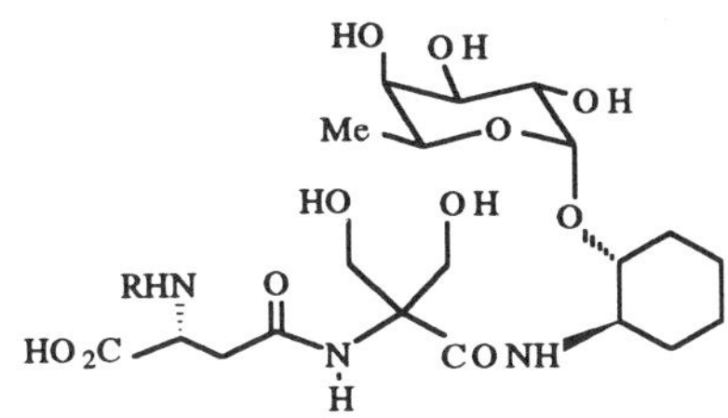

43 R = Boc
44 R = H

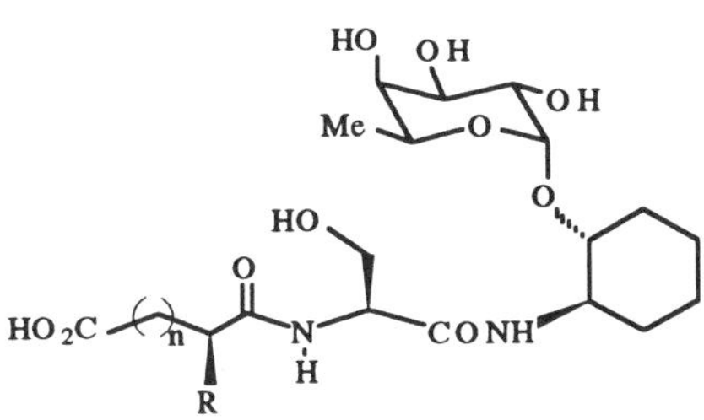

45 R = H, n = 2
46 R = NHBoc, n = 1

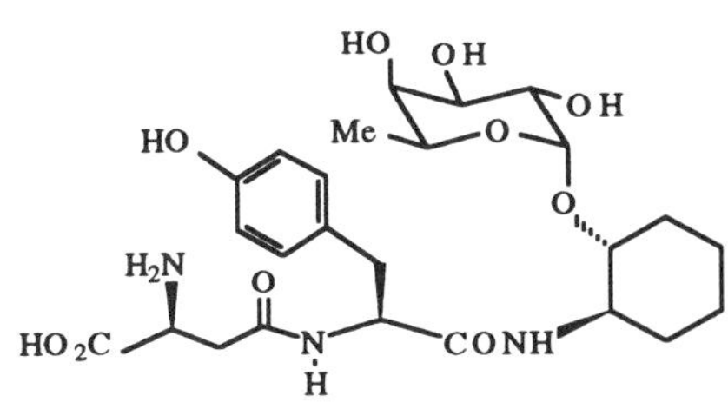

47

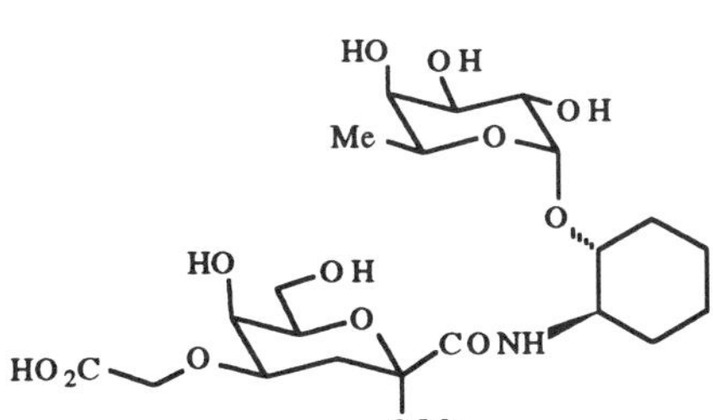

48

Small peptides containing the sequence RGD have been shown to be potent antagonists of the fibrinogen receptor and are particularly effective when the RGD sequence is conformationally constrained [76]. These results have formed the basis for intensive activity toward the design of antiadhesive peptide mimetics [77]. Based on these premises, a group at Hoechst AG (Germany) has initiated the synthesis of a hybrid glycopeptide (**49**) which combines the structural elements of the RGD motif with those of sLe^x [78]. Convergent synthesis of both fragments followed by amide bond coupling reaction between the protected peptide and a glycosylamine sLe^x derivative afforded **49**.This chimeric analog was found particularly potent in a cell adhesion test in which competitive binding of **49** was measured on recombinant human P-selectin immunoglobulin fusion protein (P-selectin-Ig) relative to the usual tumor cell lines HL60 (IC50 26 OM). A non-glycosylated RGD analog used as control gave IC_{50} value greater than 0.8 mM. Thus, **49** constitutes the most active inhibitor so far.

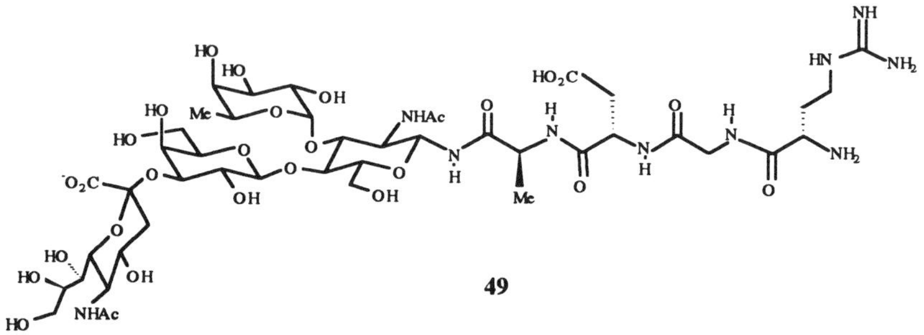

49

Table 2 Relative Inhibitory Potency of Sialyl Lewis^x Glycopeptidomimetics in Selectin-Ig Chimera Competitive Binding Assays

Compound	IC_{50} (mM)			Ref.
	L-Selectin	P-Selectin	E-Selectin	
33	<0.25	<0.25	-	73
34	<0.25	<0.25	-	73
35	-	-	1.3	65
36	-	-	0.55	74
37	-	-	1.0	74
38	-	-	1.0	74
39	-	-	1.3	74
40	-	-	>3	74
41	-	-	1.5	74
42	-	-	7	74
43	-	-	1.0	74
44	-	-	10	74
45	-	-	6	74
46	-	-	>10	74
47	-	-	>10	74
48	-	-	>3	74
49	-	0.026	-	78

2.3 Pharmacophores

Another fruitful approach used by researchers from Glycomed has provided different lead compounds having activities in both *in vitro* and *in vivo* experiments [79]. Using conformational energy computations, minimum solution conformation by high field NMR, and structure-activity relationships (SAR) cumulated thus far, it was possible to identify pharmacophores from the Fine Chemicals Directory (FCD 91.1) database. The search in the data base was made using MACCS-3D software (Molecular Design, Ltd). The database searches were performed both in the 2-D mode for initial screening and in the 3-D mode for the final selections. From their studies, it was found that glycyrrhizin (**51**), a natural triterpene glycoside from licorice already used in Chinese herbal medicines as an anti-inflammatory agent, could effectively block selectin recognition of sLe^x (**1**) in a dose-dependent manner.

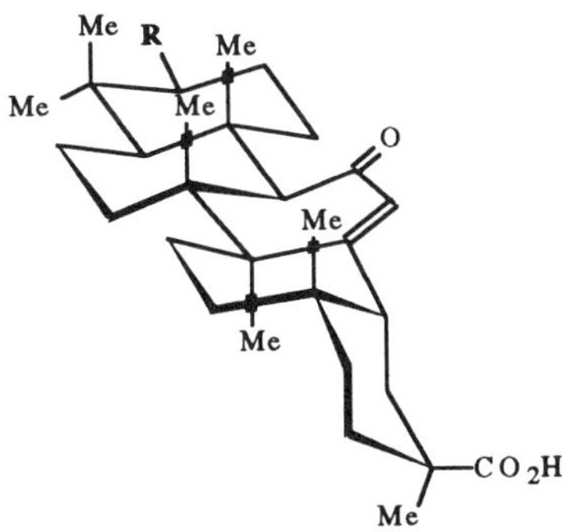

51 R = 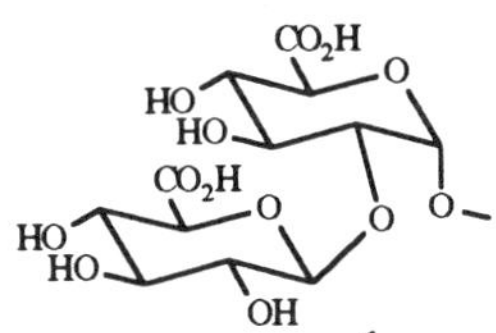

52 R = 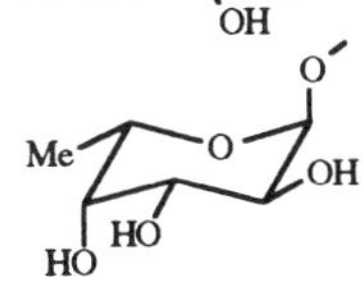

53 R = 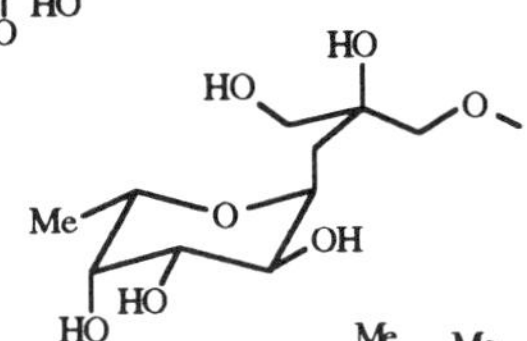

50 R = OH (18-β-Glycyrrhetinic Acid)

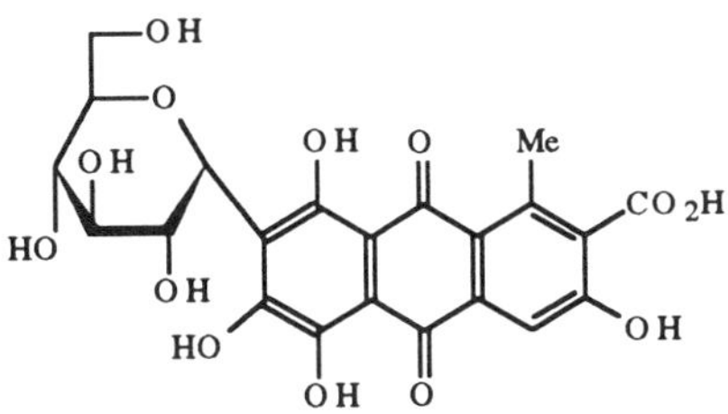

54

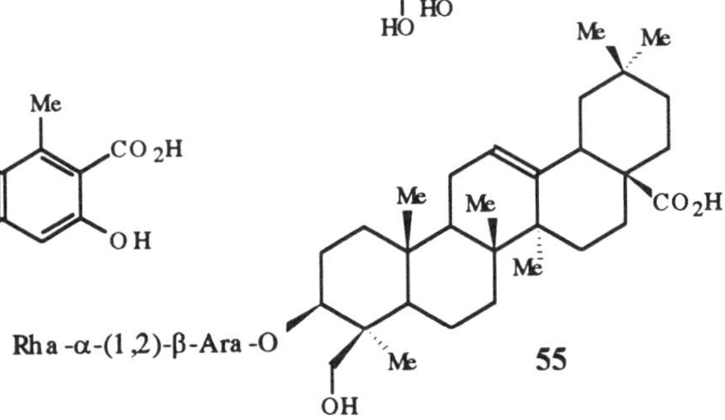

55

In fact, glycyrrhizin blocked L- and P-selectin binding to sLe^x in the micromolar range (IC_{50} < 0.5 mM) more effectively than any monovalent oligosaccharide tested so far. E-selectin binding was least affected, suggesting again that it is least tolerant of variations from the basic sLe^x structure.

Glycyrrhetinic acid (**50**), the aglycone of glycyrrhizin was relatively ineffective in the inhibition assays, indicating that the glucuronic acid residues were important. The search has also identified carminic acid (**54**) and α-hederin (**55**) as other active compounds with the latter (IC_{50} < 0.5 mM) being sensibly more potent than the former (IC_{50} > 2 mM) against L- and P-selectins.

In attempts to further improve the biological activity, the two glucuronic acid residues of glycyrrhizin (**51**) were replaced by other uncharged sugars (glucose, galactose, maltose, and fucose, not shown). Substitution by O-fucoside (**52**) retained full activity against L-selectin (IC_{50} 0.6 mM) and partial activity against P-selectin (IC_{50} 2 mM). Of significance was the finding that **52** demonstrated an increased ability, relative to **51**, to block swelling and neutrophil influx in an arachidonic acid-stimulated inflammation model in mice.

Further replacement of O-fucoside by a C-fucoside analog **53** enhanced the activity for the three selectins (IC_{50} < 0.5 mM). Compound **53** was designed to be stable to fucosidases and to more closely mimic the 10-12 Å distance between the carboxylate and fucose in sLe^x. It was also active *in vivo* in the above mice experiments.

Table 3 Relative Inhibitory Potency of Sialyl $Lewis^x$ Pharmacophores in Selectin-Ig Chimera Competitive Binding Assays

Compound	IC^{50} (mM)			Ref.
	L-Selectin	P-Selectin	E-Selectin	
50	>2	>2	>2	79
51	0.5	<0.5	>2	79
52	0.6	2	>2	79
53	<0.5	<0.5	<0.5	79
54	2-3	2.3	2.3	79
55	~0.5	<0.5	>2	79

2.4 Multivalent sLe^x Conjugates and Analogues

The generally low *in vitro* IC_{50}s (~ 0.5 mM) encountered for all the analogs illustrated so far do not render justice to the much better IC_{50} (1 μM) of sLe^x (**1**) shown in *in vivo* experiments for lung injury protection in rats [18]. A likely reason to explain these results may originate from the fact that, in most cases, divalent selectin-Ig chimera were used as receptors, while multivalent high-density sLe^x antigens were used (HL-60 cell lines or neoglycoproteins) as ligands. Obviously, the binding interactions of such high affinity ligands are more difficult to inhibit. Given that the biological forms of the selectins [11, 21,

22,80] and the natural ligand themselves, *i.e.* GLYCAM-1/Sgp50, CD34/Sgp90, MAdCAM-1 for L-selectin, and PSGL-1 for P-selectin [22, 81], are likely to have clustered arrangements, it seemed a logical step to synthesize sLe^x conjugates as multivalent structures.

Multivalent binding is expected to enhance attractive forces through cooperative avidity effects (additive). This effect has been well characterized for asialoglycoprotein hepatic receptors [82] and for the inhibition of influenza virus attachment to host sialylated receptors (see next section). This phenomena has accounted for up to several thousand-fold increases in binding associations when the ligands are presented in polymeric forms ("glycopolymers") [83, 84].

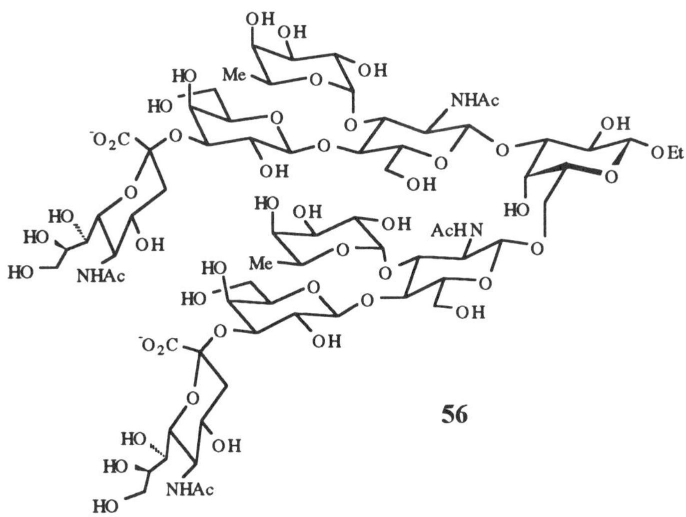

56

Employing an approach commensurate to that utilized in the preparation of divalent influenza virus hemagglutinin inhibitors (see next section), scientists from Cytel and Scripps have made divalent sLe^x inhibitors by a chemoenzymatic strategy [40, 41]. The two sLe^x ligands were scaffolded on an ethyl β-D-galactoside core to provide nonasaccharides such as **56**. Initial attachment of the N-acetylglucosamine residues at any two combinations of the galactose hydroxyl groups afforded divalent structures with different spatial arrangements. It was further established that each sLe^x moiety had a conformation similar to that of the parent monomer. Inhibition of HL-60 cell adhesion to recombinant soluble E-selectin showed IC_{50}s ranging from 0.2 mM (**56**) to 0.45 mM for positional isomers. A general trend of inhibition was found depending on the relative position of the two sLe^x substituents: 3,6-linked > 2,3

$\geq$ 4,6 $\geq$ 2,6 > monomer. In this assay, the value for the pentasaccharide monomer was 0.8 mM and the tetrasaccharide monomer was 1.2 mM.

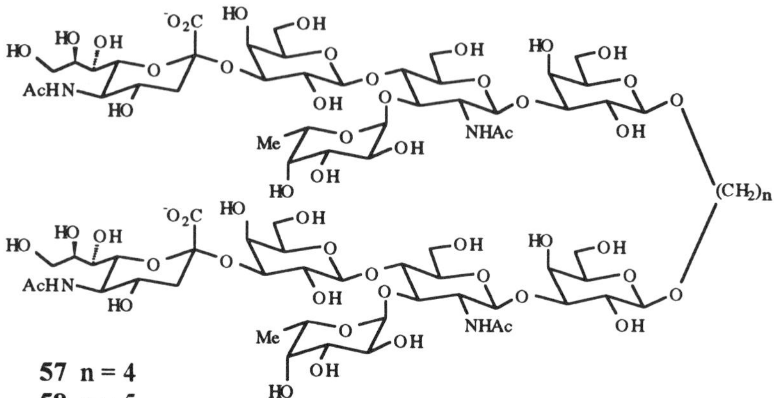

57 n = 4
58 n = 5

In addition, when 1,4-butanediol or 1,5-pentanediol were used as tethering elements, divalent structures **57** and **58** were obtained, respectively. Interestingly, they showed the same level of activity as that of the pentasaccharide. It was argued that the discrepancy in activity between the two types of divalent conjugates **56** versus **57/58** most likely arises from the relative differences in the orientation and distances between the monomeric domains.

Using a blockwise approach, Öhrlein (CIBA, Basel) [85] has performed a chemoenzymatic synthesis of oligomeric sLe^x (**59**) starting from an N-acetyl-glucosamine-linked asparagine building block. To take into account possible multiple interactions with the selectins, the inter-sLe^x distances were controlled by tethering each monomeric sLe^x unit with ω-amino acids of various lengths. In this manner, molecules containing up to three sLe^x moieties were prepared and tested in a static competitive assay using chimeric E-selectin-Ig (GlycoTech Corp., Maryland). While monomer **59** had an IC_{50} of 0.60 mM, trimer **59b** (m= 1, n= 4) showed an activity of 0.13 mM based on each sLe^x unit. This represents a nearly 5 fold improvement over that of the monomer. The same trimer with one more methylene unit (n= 5), lost some binding efficiency (IC_{50} 0.39 mM/ sLe^x) supporting the idea that the intermolecular distances between the sLe^x units might be important for tight binding.

The synthetic strategy was however cumbersome since each of the sLe^x units of the oligomers was scaffolded individually, therefore unduly lengthening the chemoenzymatic process. A monomeric sLe^x core structure, similar to that used in the preparation of trimeric clustered structures such as **60** [86], would have been less demanding even if chemoenzymatically prepared.

A recent report has described the syntheses of trimeric sLe^x clusters like **60** using either an sLe^x precursor having an O-linked six carbon aglycone terminated with an amine group or the same glycosylaminated sLe^x derivative used in the preparation of the RGD-sLe^x glycopeptide **49** described above [78].

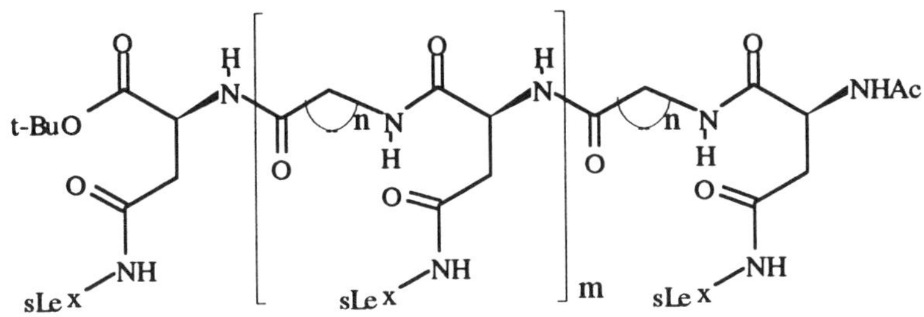

59 Monomer (m = 0; α–NHAc)
59a Dimer (m = 0; α–NHAc; n = 1-
59a Dimer (m = 0; α–NHAc; n = 1-)

The approach provided high yield access to a family of conjugates containing short to long-chain spacers. When used in inhibition of binding of E- and P-selectins-Ig fusion proteins to HL-60 cells, compound **60** and related shorter or longer homologs (not shown), showed IC_{50}s ranging from 0.13 mM to > 1mM (Table 4). Trivalent sLe^x conjugates were each significantly more potent than their monomeric standard in blocking adhesion to E-selectin, the trend being **60** > shorter spacered-**60** > longer-spacered-**60**. The same trend applied to P-selectin. It was also observed that the P-selectin receptor was slightly more sensitive to the spacer length, presumably because of a more pronounced multivalency effect. The multivalency effect was however not reproduced in *in vivo* experiments using LPS-induced leukocyte adhesion in rat mesenteric venules (10 mg/kg, i.v.) [86].

In further attempts to fix the relative orientation of the sLe^x residues, while keeping the multivalency effect operational, Sprengard *et al.* [87] have synthesized cyclic peptides such as **61** in which the sequence was: $Ala_{(1)}$-$Asp_{(3)}$-Ser-$Asp_{(2)}$-D-$Ala_{(2)}$-$Asp_{(1)}$-Gly-OH. The three aspartic acid residues were then coupled with aminated sLe^x precursor described above. Compounds like **61** showed IC_{50}s in the range of 0.35-0.6 mM when tested as inhibitors against E-selectin-Ig.

A few more analogs having monomeric sLe^x residues and succinic acid, fluorescein, biotin, or 4-(N-maleimidomethyl)cyclohexane-1 carboxylate (not shown) were similarly prepared by the same group [88]. The former showed no significant change in binding. However, the latter compounds, with relatively lipophilic groups, demonstrated noticeably improved affinities, the cyclohexane-1 carboxylate derivative showing 5-fold and 3-fold improvement toward E- (IC_{50} 0.28 mM) and P-selectin (IC_{50} 0.38 mM), respectively.

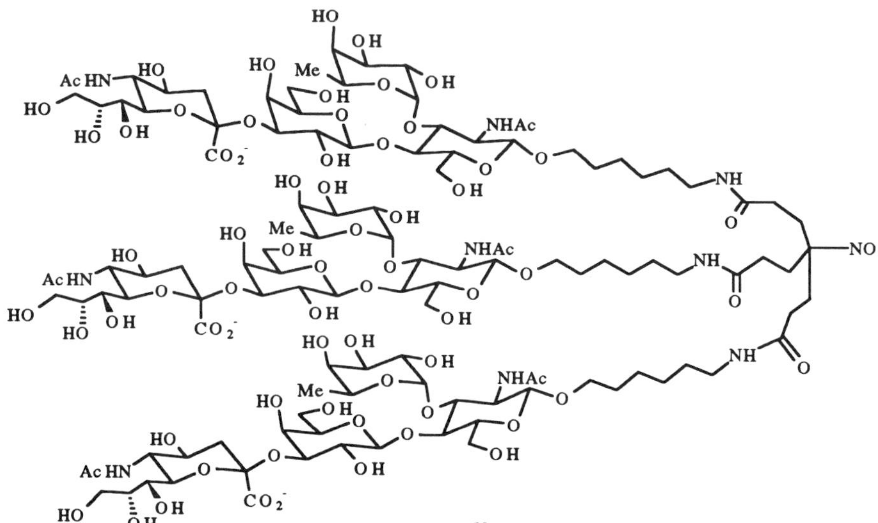

60

In order to gain more insight toward the understanding of multivalent effects in selectin adhesion and to examine the interplay of polysulfated structures believed to be the natural ligands, Roy *et al.* [89] synthesized a series of dendritic clusters similar to **62** using 3'-sulfo-Le^x-Glc (**5**) as the basic structural element. A convergent approach was used in order to provide access to "glycodendrimers" [90] containing other potential ligands. This strategy has proved very useful in inhibition of adhesion of influenza viruses to human erythrocytes [91, 92] and in other studies related to carbohydrate-protein interactions [93-96].

The process of making trivalent 3'-sulfo-Le^x-Glc (**62**) involved using gallic acid onto which three suitable oligoethylene glycol spacer arms were attached. The ethylene glycol bridging units allowed water solubility and careful adjustment of the intermolecular distances between the ligand moieties. Following appropriate functional groups manipulation, a thiolated derivative of **5** was condensed with the corresponding N-chloroacetylated dendritic core according to a strategy already described for other sialosides [90-92]. Using a similar approach, thiolated **5** was also coupled to dendritic L-lysine core structures to provide di- (**62a**), tetra-(**62b**), and octameric (**62c**) dendrimers (not shown) closely related to structure **86** (see next section 3.4).

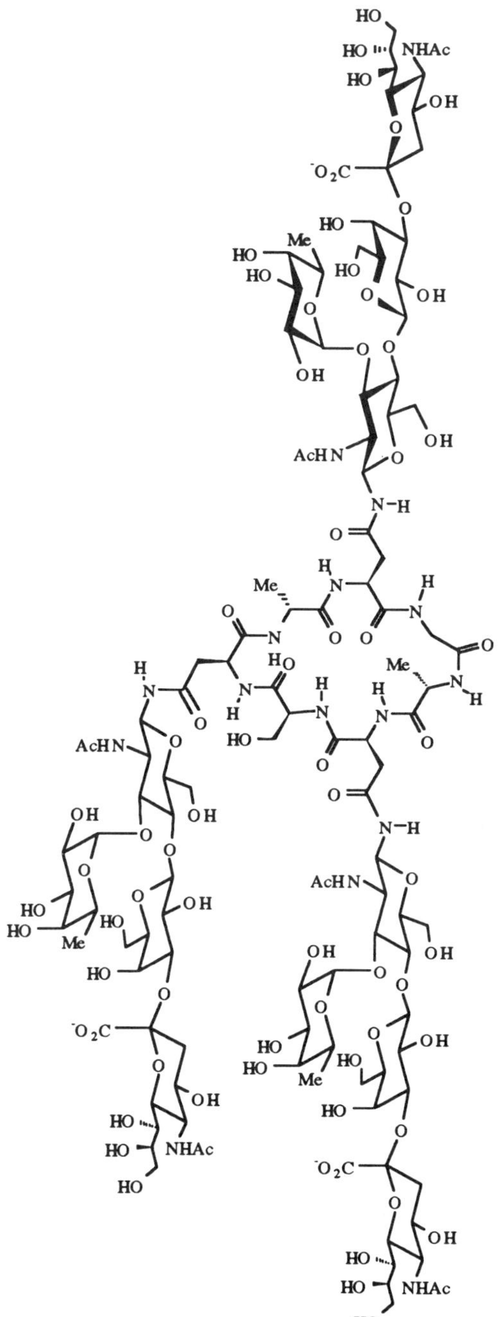
HO
HO
NHAc
HO
OH
O
$^{-}O_2C$
O
HO
Me
HO
HO
HO
O
OH
O
O
O
OH
OH
AcHN
O
N-H
O
H
N
Me
H
N
O
N
O
O
O
Me
N
H
H
N
O
H
N
H
N
O
HO
O
AcHN
O
OH
OH
O
O
O
O
OH
HO
HO
Me
HO
HO
O
$^{-}O_2C$
O
OH
HO
HO
NHAc
HO
O
N-H
AcHN
O
OH
OH
O
O
O
O
OH
HO
HO
Me
HO
HO
O
$^{-}O_2C$
O
OH
HO
HO
NHAc
HO

61

Preliminary inhibition data with L- and E-selectins using chimeric-Ig fusion proteins and hexasaccharide glycolipid as reference ligand, illustrated that compounds of this new class of glycoconjugates can inhibit both selectins at μM concentrations with **62** having IC_{50}s of 0.15 mM for E-selectin and 5 μM for L-selectin. The analogous dimer to octamer **62a-c** had IC_{50}s of 30, 7, and 2 OM with E-selectin and 10, 2, and 1 OM respectively for L-selectin (Table 4). As demonstrated in previous studies [90-96], unspecific binding due to the core structures of the dendrimers are unlikely since glycodendrimers with unrelated carbohydrate ligands did not show any cross-reactivity.

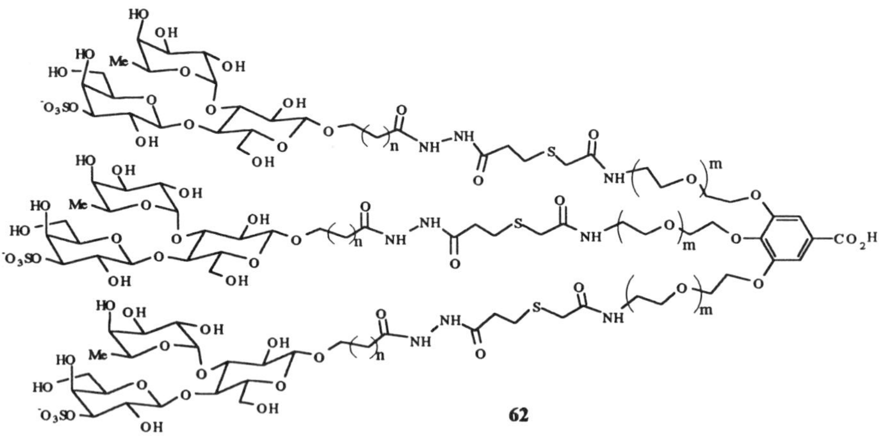

62

The successes previously encountered in using antigenic water-soluble polymers having covalently attached carbohydrate residues [reviewed in 83, 84] have also prompted the syntheses of selectin ligands in polymeric forms [89, 97-100]. Through their intrinsic numerous repeating units, glycopolymers are the quintessential molecules expressing multivalency. In the study described above [89], the synthesis of copolyacrylamide **62d** was also described. The same precursor used for the synthesis of the dendrimers was used in the preparation of **62d**. The reaction simply involved synthesizing the selectin ligand as a monomer possessing an N-acryloylated functionality. The 3'-sulfo-Le^x-Glc containing monomer was then copolymerized with acrylamide in water. Preliminary inhibition data with E- and L-selectins showed it to be a very powerful inhibitor [89]. Other investigators have similarly prepared polymeric forms of the selectin ligand using biocompatible polymers [97]. Other strategies involved preparing the ligands with amine functionalities in the aglycones which can be directly attached to premade polymers having active ester functionality [98-100] or by using combined chemoenzymatic syntheses of either the monomers or the polymers [101].

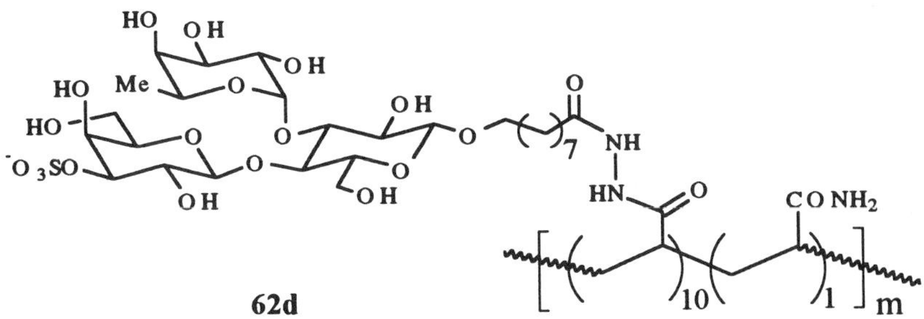

62d

The inhibition data for the respective multivalent ligands are listed in Table 4 below. These data strongly suggest that, when properly scaffolded, the selectin ligands can present very powerful binding properties. Thus, by combining multivalency, heterobifunctional ligands, and glycomimetic strategies, it is possible to obtain better insights into the interplay of molecules involved in cellular adhesion.

Table 4 Relative Inhibitory Potency of Multivalent Sialyl Lewis^x Mimetics in Competitive Binding Assays with Selectin-Ig Chimera

Compound	IC50 (mM)			Ref.
	L-Selectin	P-Selectin	E-Selectin	
56	-	-	0.4	40, 41
57	~5	~1	~1	40
58	~5	~1	~1	40
59	-	-	0.60	85
59a (n=1)	-	-	0.46	85
59a (n=2)	-	-	-	85
59a (n=3)	-	-	0.64	85
59a (n=4)	-	-	1.2	85
59a (n=5)	-	-	0.47	85
59a (n=6)	-	-	0.80	85
59a (n= 7)	-	-	0.48	85
59b (n=4)	-	-	0.14	85
59b (n=5)	-	-	0.39	85
60	-	0.14	0.13	86
61	-	-	0.35-0.60	87
62	0.005	-	0.15	89
62a	0.01	-	0.30	89
62b	0.002	-	0.07	89
62c	0.001	-	0.02	89

2.5 Concluding Remarks

The observations accumulated to date indicate that, while L- and P-selectins showed more distinct binding patterns than E-selectin toward the sLe^x motif, each of the selectins demonstrated subtle affinity differences between themselves. These data also suggest that sLe^a constitutes a better lead compound for E-selectin, while 3'-sulfo-Le^x appears to be a better candidate for L- and P-selectins. The implications of sulfated forms of $sLe^{x/a}$ known to occur on L-selectin mucin ligands GlyCAM-1 and other naturally occurring O-linked glycoproteins ligands [22] have stimulated numerous synthetic investigations toward the syntheses of sulfated $sLe^{x/a}$ analogs [102-110] reported for the first time in 1992 [111]. Recent findings showed that 3'-sulfo-Le^a-Glc was more effective than the related 3'-sulfo-Le^x (**4**) in blocking E- and L-selectin and as effective for P-selectin. Interestingly, synthetically prepared 6'-sulfo-Le^a-Glc [107] and enzymatically prepared 6-sulfo-Le^x (*i.e.* at the GlcNAc residue) [108] were both found compatible to L-selectin binding. In fact, the latter analog was four times more active than sLe^x [108] while the former completely abolished E-selectin binding [107]. The enzymatic synthesis of 6-sulfo-Le^x [108] is of importance because it raises the issue of cautious examination of fucosyl transferase substrates as the means to detect potential natural ligands. Alternatively, it was also found that 3',6'-disulfo-Le^a-Glc analog abolished binding to all three selectins [107]; a fact somewhat surprising based on the observations that polyanionic polysaccharides were found to bind to L- and P-selectins.

The increased binding properties of some of the glycomimetics described above: sulfated $Le^{x/a}$, fucosylated peptides, chimeric sLe^x-RGD analogs, and multivalent ligands in particular, are likely to afford antagonists with desired high binding properties. The pharmacophore approach has also provided the opportunity to discover new lead compounds. Natural herbs and plants extracts used as anti-inflammatory preparations by folk medicine should similarly be considered.

However, it is likely inevitable that the next generation of leads will originate from peptides alone since it has already been found that synthetic peptides, based on a conserved region (^{48}YYWIGIRK^{55}-NH2) of the lectin domain of E-, L-, and P-selectins blocked both *in vitro* and *in vivo* selectin-mediated adhesion [112]. Moreover, another synthetic peptide sequence (SPYGRC) taken from pertussis toxin was shown to bind directly to sialylated oligosaccharides including $sLe^{x/a}$ [113]. Once refined, these peptides will themselves become leads to metabolically stable peptidomimetics of high affinities.

As opposed to the above findings, the next section will describe a carbohydrate-protein interaction related to influenza flu virus hemagglutinin in which the entire sialic acid moiety was found critical for binding.

3. MULTIVALENT SIALIC ACID CONJUGATES AS INHIBITORS OF INFLUENZA VIRUS HEMAGGLUTININS

The initial step in the cascade of events leading to influenza virus infections is mediated by the tight attachment of virus particles to host cell receptors. These receptors are constituted of cell surface sialylated glycoproteins and glycolipids (gangliosides). The following steps involve receptor-mediated endocytosis where the acidic pH is postulated to trigger viral membrane conformational change which allows its fusion to that of the host with ensuing release of the viral nucleoplasmid. The first event responsible for the receptor-virus interaction is therefore an attractive target for antiviral intervention [114].

Influenza virus particles are spheroidal and approximately 100 nm in diameter. The outer-membrane envelope contains ~500 copies of rod-shaped spikes composed of hemagglutinin (HA) trimers and ~100 copies of mushroom-shaped spikes composed of neuraminidase tetramers. The hemagglutinin constitutes the receptors for α-sialoside ligands. Its X-ray structure has been refined at 3 Å resolution [115] including one report in which the HA has been co-crystallized with sialyllactose [116]. The X-ray analysis showed the sialic acid binding pockets to be ~40Å apart. The neuraminidase catalyzes the hydrolysis of α-sialoside linkages from the host sialyloligosaccharides. It facilitates the transport of the virus to and from the infection site by permitting its passage through the sialic acid-rich mucins and presumably from the infected cells where it helps releasing new-born virion progeny. The X-ray structure of this glycoprotein has also been refined at 2.9 Å resolution [117]. Pictorial diagrams of both influenza virus glycoproteins are depicted in the accompanying chapter of this book by von Itzstein and Kiefel [118]. The chapter describes in length the chemistry of sialic acids and the complementary strategy of antiviral intervention based on the inhibition of neuraminidase actions. This strategy has lead to the discovery of the most potent (nM) anti-influenza drug. The approach was based on the rational design of transition state analogues and proper docking of the substrates into the enzyme's active site by molecular modeling.

Numerous researchers have attempted to map the hemagglutinin combining site through the syntheses of sialic acid analogs [119-121]. As opposed to sLe^x-selectin interactions discussed in the first section, most of the sialic acid functionalities appeared essential for favorable binding interactions. Here again, the carboxylate group constitutes the critical site of attachment. A summary of the key binding interactions is illustrated below for Influenza A virus (Scheme 3). Consequently, and except for few exceptions [122-123], there has been no monovalent sialoside analogs shown to bind stronger than sialic acid. However, the analogs described above used modifications on the aglycone or at the OH-4 positions which are not directly inside the active site since they are known to be exposed to the outside aqueous environment.

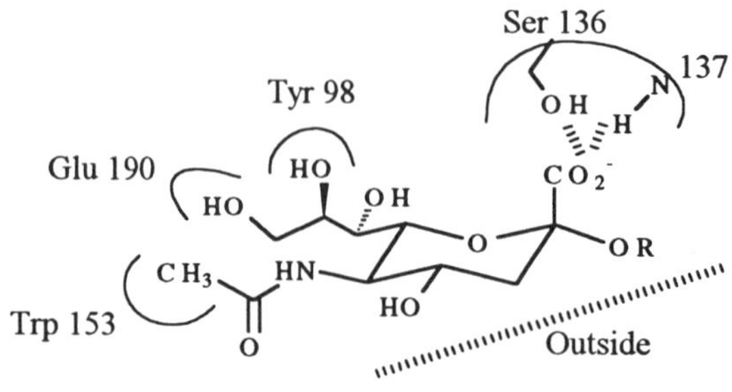

Scheme 3 Schematic diagram showing the binding interactions of an α-sialoside in the binding site of influenza A (X-3, H3N2) virus hemagglutinin.

As mentioned above, carbohydrate protein interactions are generally of low affinity and sialic acid-influenza virus interactions are no exception. Detailed ^{1}H-NMR experiments with α-sialosides and bromealin-released hemagglutinin (BHA) have revealed an intrinsic weak dissociation constant of 2.8 mM [124]. In fact, it has been known for a long time that influenza virus was bound to its ligand with low affinity since naturally occurring sialomucins of high sialic acid contents were a thousand-fold better inhibitors of human erythrocytes hemagglutination than monosaccharides or mucins of low sialic acid contents [125].

Evidence gained toward the requirements for multivalency in tight binding interactions have been cumulating. This phenomenon has been recognized as the key step for many pathogenic infections reported to date and has been the subject of a postulate [126]. Even before being generally accepted, it has triggered the syntheses of a wide range of glycoconjugates having covalently attached sialic acids. A brief description of synthetic conjugates bearing multiple copies of sialic acid residue follows. These new sialoconjugates have been designed to improve the inhibitory potential of sialic acid toward flu virus infections and to gain insight in our understanding of multivalency. The following discussion describes sialoconjugates of increasing complexity. Some of the new conjugates discussed below have also demonstrated good potential for screening and diagnosis of anti-sialic acid antibodies [127]. Additionally, they raised the issue of multiple intra- or intermolecular interactions since the hemagglutinin itself appears as a trimeric unit each possessing its own sialic acid binding site which appears as a shallow grove near the surface. Some of the above conjugates have been determinant in defining subtle structural variations which exist between the different strains of influenza viruses. Of the most studied strains, classified serologically as A, B, and C, the former two appeared to be more virulent to humans.

3.1 Clusters

The synthesis of bidentate α-sialoside ligands such as **63** and **64** has been reported [128]. It was found that, while none of these clusters showed improved binding to isolated hemagglutinin (HBA), both demonstrated evidence of increased affinity with the intact virus particles as measured by inhibition of hemagglutination of human erythrocytes. It was postulated that the divalent sialosides bind intermolecularly to adjacent HA trimers on the viral surface. By varying the distances between the two sialic acid residues, it was estimated that the distance between the binding sites on adjacent HA trimers was ≈ 55 Å. It was also observed that the more rigid series of bidentate ligands having oligoglycine spacers (**64**) was more potent than the oligoethylene glycol series (**63**). In the inhibitions experiments, **64** (n = 3) was two-fold better than **63** (n = 1), and **64** (n = 4) was five-fold better than **63** (n = 2) and 100-fold better than the monosaccharide. Moreover, the dependence of binding affinity on linker length indicated a defined distance between the sialic acids binding *loci*. Alternatively, bidentate ligands of longer spacer length showed decreased binding affinity, supporting the lack of cross-linking between two virus particles.

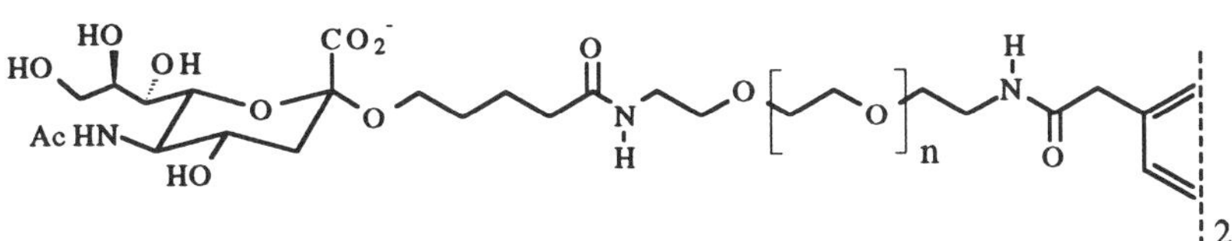

63

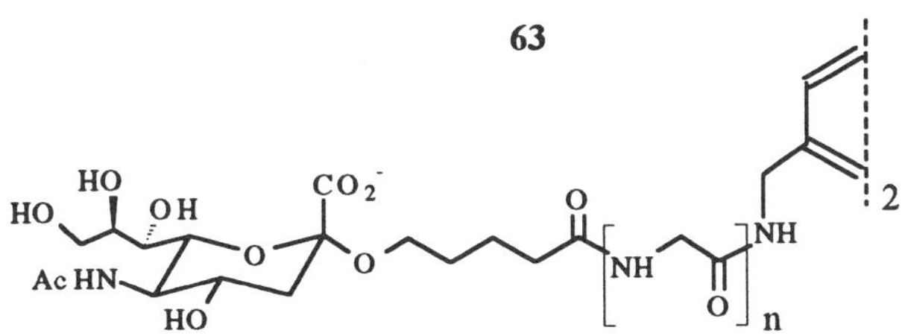

64

A similar investigation was conducted by Sabesan *et al.* [129, 130] using five heptasaccharides containing two α-D-NeuAc(2,6)-β-D-Gal(1,4)-β-D-GlcNAc fragments anchored at two different hydroxyl groups of β-D-galactoside used as tethering element. It was anticipated that the conformational rigidity and the relative orientation of the oligosaccharides would assist binding by favorable entropic factors when compared to the more flexible ligands described above. The heptasaccharides with O-2/O-3, O-2/O-4, O-2/O-6, O-4/O-6, or O-3/O-6 (**65**) linkages were prepared by combined chemo-enzymatic strategy.

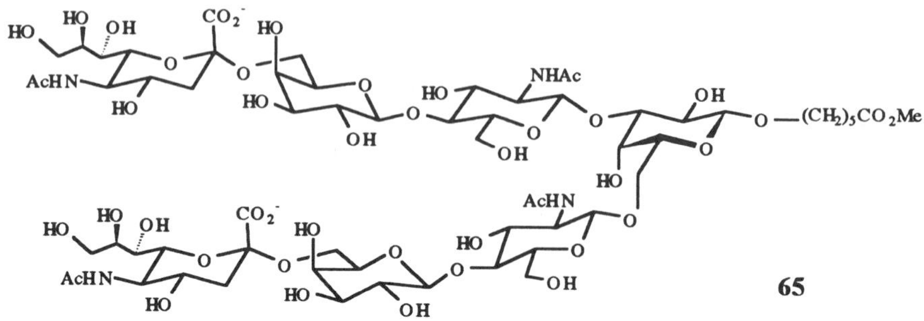

65

Using one-dimensional and two-dimensional NOESY NMR experiments together with Monte Carlo simulations, the authors calculated that the end C-2 to C-2 inter sialic acid distances of the bivalent ligands were ~19 Å in 2,4-heptasaccharide to ~9 Å in 3,6- and 4,6-heptasaccharides, with the others falling within these two distances. Inhibition of hemagglutination experiments showed that, relative to methyl α-sialoside, the 3,6- (**65**) and 4,6-disialosides with a 9 Å intramolecular distance exhibited 10- and 8.4-fold better inhibitory activity, respectively. In addition, the 2,4-disialoside with a ~19 Å distance had no improved inhibitory potency. Based on the above discussion, the intramolecular distance between the two sialic acid moieties of the best divalent ligand described here would be too short to allow intermolecular binding to adjacent hemagglutinins of the influenza virus. However, the two sialic acids are well positioned to interact intramolecularly with two binding sites within the hemagglutinin trimer. The same authors have also proposed the synthesis of a neoglycoprotein conjugate, prepared by coupling **65** to bovine serum albumin (BSA). The conjugate was shown to have an IC_{50} of less than 0.25 mM [130]. The use of protein carriers to increase carbohydrate ligand multivalency is however a practice which should be completely avoided since the resulting conjugates will possess immunogenic properties, *i.e* they would act as vaccines [127].

In a more recent study, a large number of divalent linear, as well as branched, glycopeptide clusters similar to **66** and **67** were prepared using a combined chemical and enzymatic approach [131]. The two copies of the target ligand made of NeuAcα-(2,6)-Gal-β-D-(1,4)-β-D-GlcNAc-Asn were scaffolded enzymatically on preformed β-D-GlcNAc-Asn linked residues interspaced by various oligoglycine (**66**) or oligo-L-proline spacers (**67**) mounted by solid-phase synthesis on Pam-resin using Boc strategy. The branched glycopeptide (**67**) was obtained using Nα, Nε$(Boc)_2$-protected L-lysine from which two simultaneous oligo-L-proline spacers were synthesized and capped with an AcAsn(β-D-GlcNAc)-glycine residue. This strategy has allowed the preparation of influenza virus ligands having different geometries and distances between the sialic acids.

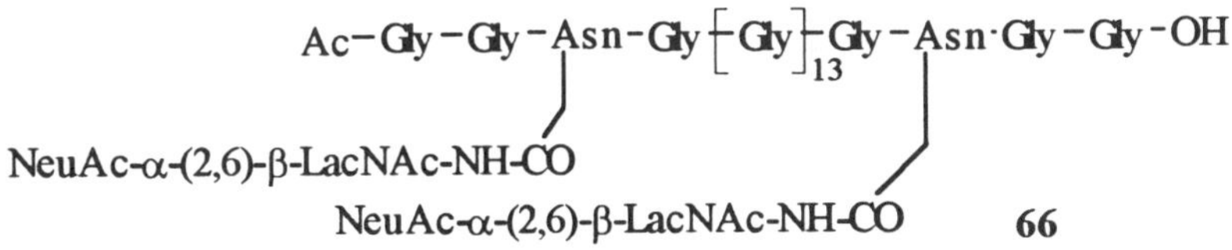

66

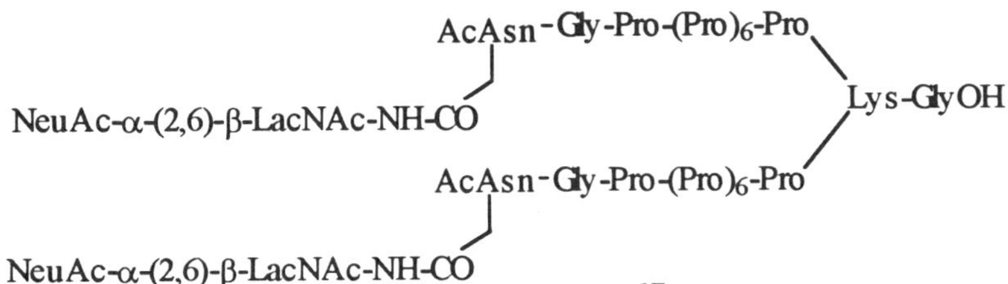

67

The study showed that the inhibitory properties of these divalent sialosides are also critically dependent on the inter sialoside distances and the geometry imposed by the peptide backbones. The more flexible oligoglycine spacers provided compounds which were eight-fold more potent than the parent monosialoside. Alternatively, the more rigid proline rich spacers with some helical conformation showed only moderate increase of binding. The conclusions were consistent with intramolecular cross-linking of two binding sites within the same HA trimer for shorter clusters and possibly bridging of two viruses for longer ones.

3.2 Polymers

By virtue of their repeating units, polymers and particularly carbohydrate containing polymers [reviewed in 83, 84] constitute excellent opportunities to study carbohydrate protein interactions where multivalency is critical. Owing to the low dissociation constants observed in most of the above interactions, glycopolymers made of sialic acid have offered the first real breakthrough in dramatically improving the inhibitory properties of sialoconjugates. The first example of such a polymer (77) was reported for the first time in 1987 [132]. Since then, the literature has witnessed intense activity in this area.

Sialylated glycopolymers have been prepared by essentially two procedures. In the first one, both sialic acid monomers and backbone monomers are copolymerized in aqueous solutions using radical initiation chemistry. In this manner, O- [132-136], N- [137], S- [138], and C- [139] linked sialopolymers have been prepared. In the second approach, suitably functionalized sialosides are grafted onto preformed polymers bearing reactive functionalities [140-144].

In the first sialopolymer synthesis, an O-allyl α-sialoside monomer (**68**) was prepared using classical Koenigs-Knorr chemistry [132]. Direct co-polymerization with acrylamide furnished copolyacrylamide **77**. Although the resulting copolymer showed strong binding properties with wheat germ agglutinin (WGA), a plant lectin known to bind sialosides, it lacked affinity for influenza A virus HA [135]. These results were attributed to the short spacer which did not allowed sufficient penetration in the HA active site. In a later version, the same precursor **68** was extended with cysteamine under radical conditions. After N-acryloylation, the new monomer **70** produced copolyacrylamide **78** having a spacer of ~11 Å which was sufficient for tight binding to the influenza A virus HA. A very similar approach was then followed for the synthesis of another copolyacrylamide resulting from monomer **72** [136].

In order to prepare polymers in which the sialic acid linkage would be simultaneously resistant to the action of the viral neuraminidase, a C-glycoside version of **70** (**71**) was synthesized starting from the C-allyl α-sialoside **69** [139, 143]. Compound **69** was prepared as an α/β mixture (1:1) using allyl tributyltin and AIBN [145] or a catalytic amount of bis(tributyltin) under photolytic conditions [146]. Further transformation of **69** into amine **73** and grafting to polymers bearing pendant reducing sugars afforded grafted copolymer **83** [143].

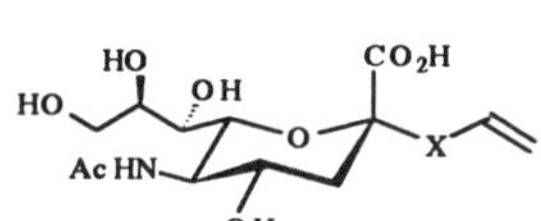

68 X = OCH_2
69 X = CH_2

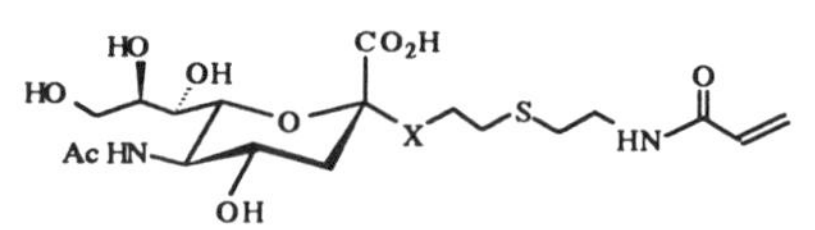

70 X = OCH_2
71 X = CH_2

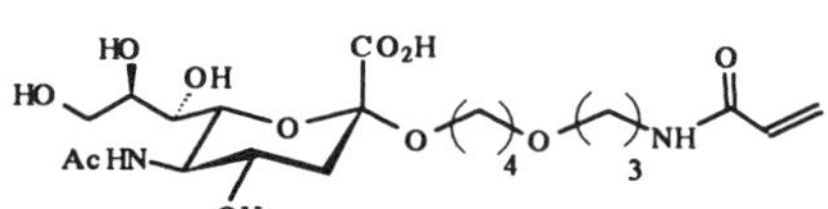

72

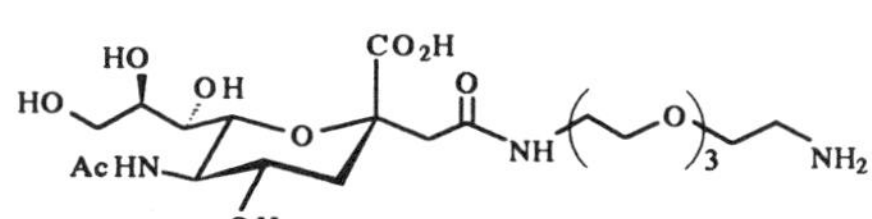

73

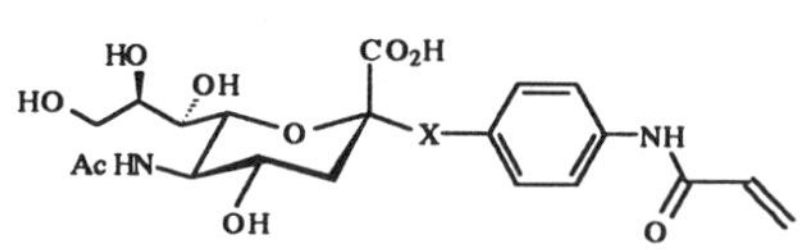

74 X = O
75 X = S

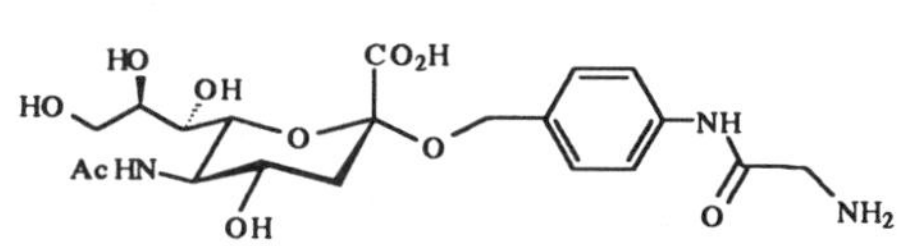

76

Using a strategy which was designed to produce copolymers stable to the deleterious action of the viral neuraminidase together with higher affinity [120] aromatic sialoside ligands which could bind to both all three strains of influenza virus A, B, and C (the ligand is a 9-O-acetyl sialoside), O- and S-linked α-sialosides **74** and **75** were prepared [134, 138]. After copolymerization with acrylamide as before, glycopolymers **80** and **81** were produced. The thioglycoside containing copolymer **81** was subsequently treated with trimethyl orthoacetate and, following mild acid hydrolysis, afforded a polymer in which the sialic acid residues were regioselectively O-acetylated at O-9 corresponding to the C-strain ligand. Interestingly, the 9-O-acetylated polymer corresponding to **81** was able to inhibit human erythrocytes hemagglutination against the C-strain as planned. However the non-acetylated version (**81**) failed inhibiting the A strain, suggesting that the binding pocket for the latter is deeper than that of the former.

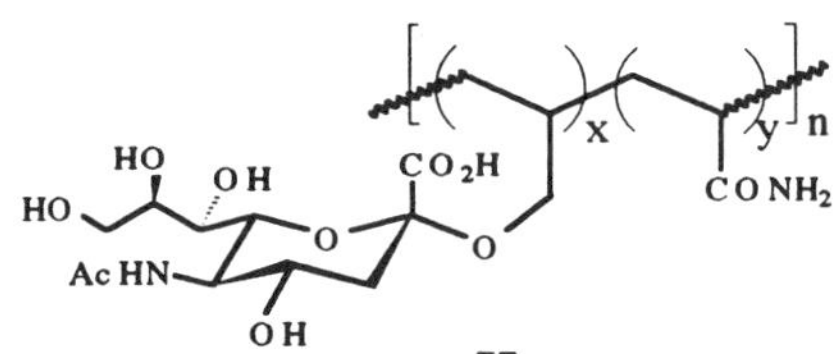

77

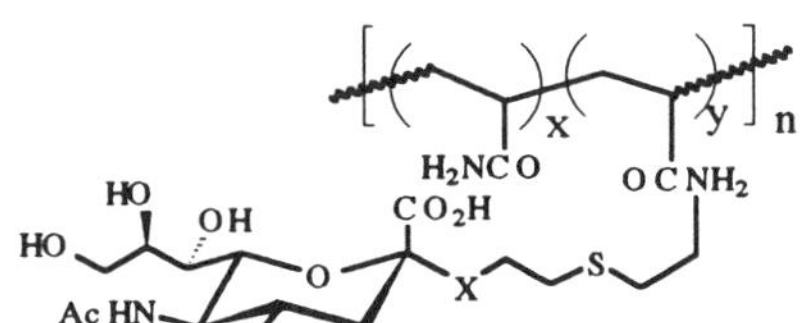

78 X = OCH_2
79 X = CH_2

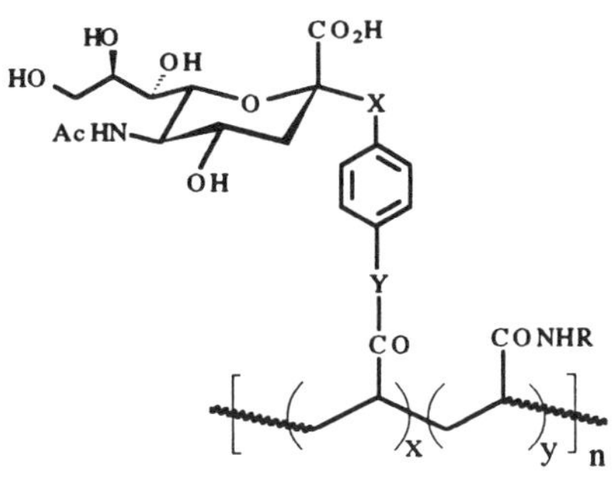

80 X = O, Y = NH
81 X = S, Y = NH
82 X = OCH_2, Y = $NHCOCH_2NH$

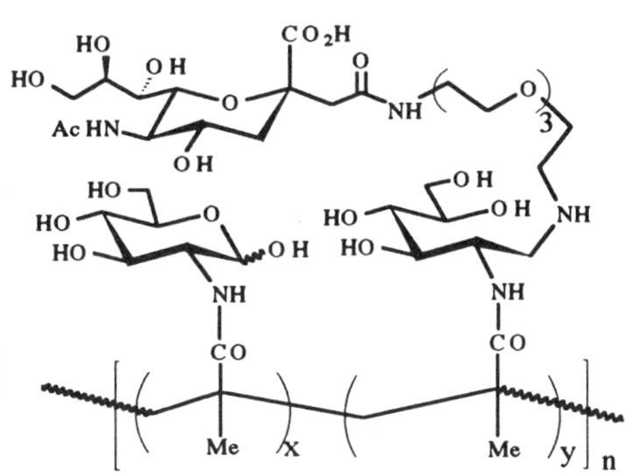

83

Using a complementary strategy, Russian workers have produced a large number of sialopolymers such as **82** with different backbone properties [140-142]. The corresponding monomer **76** was also synthesized with an aromatic aglycone. After extension with a more nucleophilic glycine spacer, **76** was grafted on preformed poly(4-nitrophenyl acrylate) of fixed molecular weight. This approach has the advantage of producing a large variety of copolymers having the same molecular weight.

All the sialopolymers described above showed greatly improved inhibitory properties toward the influenza virus hemagglutinin when compared to monosaccharides or divalent sialosides. They showed up to 50,000-fold increased affinity. There are many factors which can explain the cooperative effect of multivalent sialosides. As pointed earlier, it is likely that these copolymers, which have molecular weights in the order of 100 kDa and above, can equally cross-link both the entire virus particles and the hemagglutinin trimers on the viral surface. It is also possible that single strands of copolymers cover large surfaces of the virions, therefore interfering further with their adhesion to erythrocytes by steric hindrance.

3.3 Liposomes

An interesting variant to the solution of multivalent interactions include the synthesis of liposomes [147-149]. Fluid bilayered liposomes having sialic acid residues exposed at their surfaces have the ability to interact effectively with HA molecules on the surface of the viruses. The fluidity of the lipid bilayer offers the advantage of allowing the sialosides to diffuse laterally, therefore optimizing the number of productive interactions. Moreover, they are, in principle, more easily amenable to detailed theoretical analysis than their sialopolymer counterparts. In a way, they represent a synthetic extension to naturally occurring gangliosides.

In one approach, an O-linked sialoside (**84**) was prepared from the same precursor used in the preparation of monomer **72** [147]. The amine functionality was further extended with a bridging glutaryl diamide unit which was linked to L-lysine. Attachment of the lipid tail was effected through L-cysteine possessing a tetradecylamide and a thiodecane chain. The Nα-lysine amine was used to anchor a fluorescent tag [5-(dimethylamino)naphthalene-1-sulfonyl] (Dans) group for quantitative measurements. The liposomes were then produced by standard technique using egg phosphatidylcholine and cholesterol. Small 25-100 nm unilamellar liposomes were thus assembled.

The biochemical activity was measured by inhibition of hemagglutination of chicken red blood cells against influenza A virus. The calculated concentration necessary for complete inhibition was 20 nM in sialic acid residues. As previously observed with sialopolymers, an optimum concentration was observed below and above which the inhibitory properties decreased. In the above liposomes, the optimum was reached at 2.5-18 mole percent. For comparison purposes, an analog of **84** missing the lipid tails, and therefore incapable of liposome formation, had an inhibitory concentration of ~200 μM representing a factor of 10,000 higher than the polyvalent liposomes. These results indicated that the liposomes were moderately more effective in inhibiting agglutination than are sialosides on water-soluble polymer supports. Additionally, they were as good or better than the best-known natural sialomucins and macroglobulins [125]. Expectedly however, the major drawback of these liposomes is that they initiated lysis at concentrations above 10 μM. It is still not known if these liposomes can also inhibit infectivity *in vivo*.

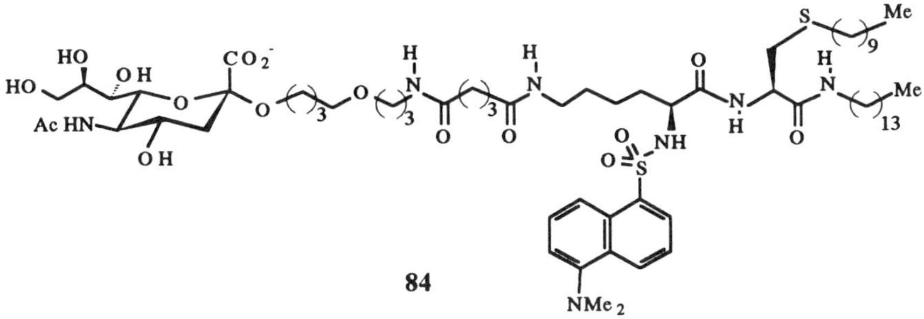

84

In an attempt to provide more metabolically and physically stable sialylated liposomes, polymerized liposomes containing α-C-linked sialosides were synthesized [148, 149]. To this end, C-allyl sialoside (**69**) was oxidatively ozonolyzed and further transformed into the amino derivative **73**. Coupling of amine **73** with a long chain diacetylenic acid furnished the required lipid precursor. Mixing the diacetylenic acid with the corresponding diacetylenic sialic acid residue and polymerization by irradiation at 254 nm gave polymeric liposome **85** having a conjugated polydiacetylenic backbone of alternating ene-yne groups. Transmission electron microscopy revealed ellipsoid structures averaging on the order of 40 nm in length and 15 nm in width. Liposomes with varying levels of sialoside content were thus obtained (0-60%) [148]. Liposomes having 5% and 10% sialic acid content showed IC_{50}s which were 30,000 times lower than that of a monosaccharide taken as reference. Interestingly, as observed previously, liposomes having a higher level of sialoside incorporations (30% and 60%) failed to inhibit erythrocytes agglutination by the viruses, while liposomes with only 1% sialosides also failed inhibiting. Moreover, the new biomaterial was shown to inhibit infectivity *in vivo* when using low content (1%) polymerized sialosides. The results indicated that even if no inhibition of hemagglutination occurs, there is still the possibility of inhibition of infectivity.

These interesting structures were also used for direct colorimetric detection of virus adhesion using the known chromatic properties of polydiacetylene bilayer assembled on glass slides [149]. The bilayer was composed of a self-assembled inner octadecylsilane monolayer and a surmounting Langmuir-Blodgett monolayer of the polydiacetylene. The method was used in quantitative measurements by observations of the color transition from blue to red which occurs as a result of conformational changes in the conjugated backbone. The bilayer assembly showed visible absorptions with maximum at 620 nm and a weaker absorption at 550 nm and thus appears as a blue film. When the film was incubated with influenza A virus (X-31), the absorption at 550 nm increased with a concomitant decrease from the maximum at 620 nm, resulting in a red-colored film visible to the naked eye.

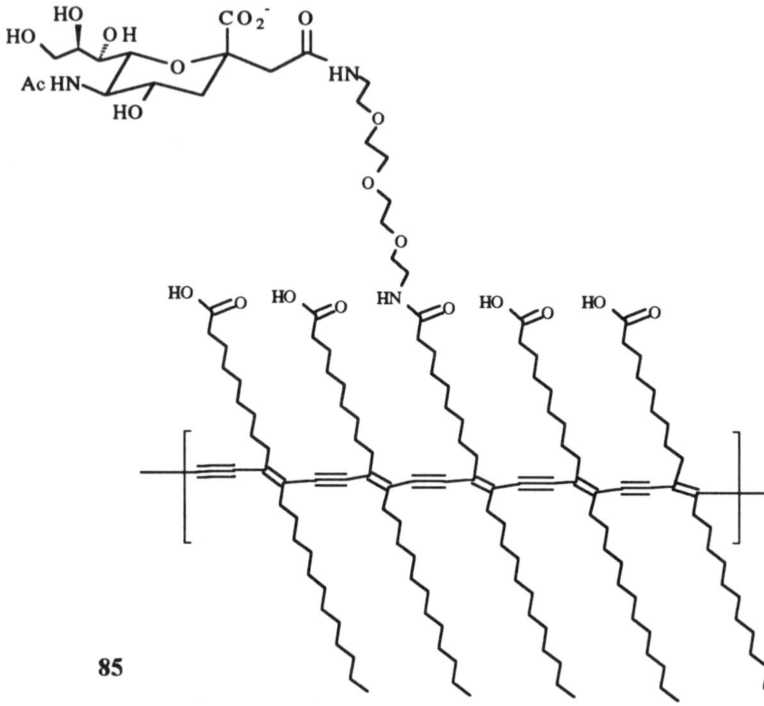

85

3.4 Dendrimers

Many of the disadvantages of the above clusters, polymers, and liposomes were anticipated to be levied with new forms of multivalent sialosides which would better mimic the shapes of multi-antennary glycoproteins [90]. It was anticipated that a new class of biopolymers prepared in dendritic shapes would result in useful inhibitors by virtue of their well characterized carbohydrate valency, shapes, and orientation. These goals were fully achieved with sialylated dendrimers such as **86** [90-92] and many others synthesized with different carbohydrate residues [93-96], including the sLe^x analog discussed above [89].

Solid-phase peptide synthesis was used to assemble a multi-branched lysine core unto which was grafted the desired number of α-thio-sialosides. In this way, dendrimers with 2, 4, 8, (**86**) and 16 thio-sialoside residues were obtained. In inhibition of hemagglutination experiments with influenza A virus, the dendrimers showed IC_{50} values which exponentially decreased as valencies increased. The dendrimer with 16 sialosides (IC_{50} 19 μM) was as potent as previously described polymers. This new generation of glycoconjugates is likely to provide lead structures for new potent inhibitors and drug carriers.

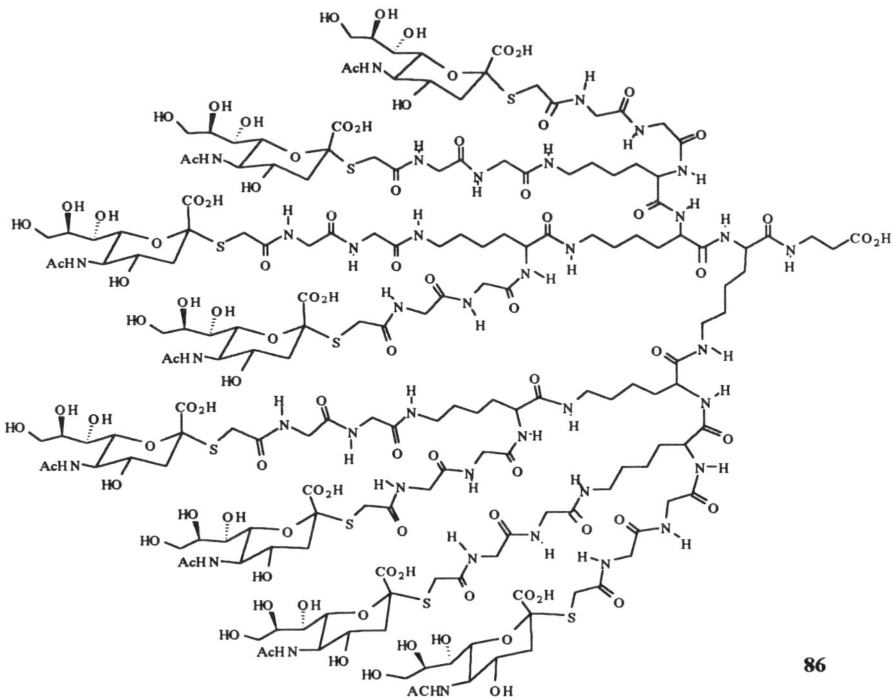

86

4. CONCLUSIONS

Many of the glycomimetics and multivalent glycoconjugates described here have opened the way for potentially new therapeutics. Medicinal carbohydrate chemistry is thus witnessing rejuvenated interest from the pharmaceutical industry and from young researchers. It is also likely that, with the advent of combinatorial chemistry, combined with the tremendous progress recently made in glycobiology, even better refined and new biologically relevant structures will be unraveled. As mentioned above, chimeric glycoconjugates may also be used to better understand inflammation mechanisms and the cellular adhesion processes at their origin. A number of other pathogenic infections, also depending on carbohydrate interactions, will benefit from the above findings.

Acknowledgments

I am truly indebted to the many active researchers in the field who provided me invaluable information before publication of their work. Many thanks to K. Bock, H. Kunz, R. Öhrlein, and C.-H. Wong.

REFERENCES

1. Schauer, R, ed. Sialic Acids-Chemistry, Metabolism & Functions. Cell Biology Monographs. Vol 10, Wien: Springer Verlag, 1992.
2. Karlsson K-A. Microbial Recognition of Target-Cell Glycoconjugates. Curr Opin Struct Biol 1995; 5:622-635.
3. Ito Y, Gaudino JJ, Paulson JC. Synthesis of Bioactive Sialosides. Pure Appl Chem 1993; 65:753-762.
4. Paulson JC. In: Conn M, ed. The Receptors. Vol 2. New York: Academic Press, 1985, 131-219.
5. Springer TA, Larsky LA. Sticky Sugars for Selectins. Nature 1991; 349:196-197.
6. Paulson JC. Selectin/Carbohydrate-Mediated Adhesion of Leukocytes. In: Harlan JM, Liu DY, eds. Adhesion: Its Role in Inflammatory Diseases. New York: WH Freeman. 1992:19-42.
7. Varki A. Biological Roles of Oligosaccharides: All of the Theories are Correct. Glycobiology 1993; 3:97-130.
8. Schauer R. Sialic Acids as Antigenic Determinants of Complex Carbohydrates. In: Wu AM, ed. The Molecular Immunology of Complex Carbohydrates. Advances in Experimental Medicine and Biology. Vol 28. New York: Plenum Press, 1988, 47-72.
9. Larsky LA. Selectins: Interpreters of Cell-Specific Carbohydrate Information During Inflammation. Science 1992; 258:964-969.
10. Springer TA. Traffic Signals for Lymphocyte Recirculation and Leukocyte Emigration: The Multistep Paradigm. Cell 1994; 76:301-314.
11. Larsky LA. Selectin-carbohydrate Interactions and the Initiation of the Inflammatory Response. Ann Rev Biochem 1995; 64:113-139.
12. Dasgupta F, Rao BNN. Anti-adhesive Therapeutics: a New Class of Anti-Inflammatory Agents. Exp Opin Invest Drugs 1994; 3:709-724.
13. Kotovuori P, Tontti E, Pigott R, Shepherd M, Kiso M, Hasegawa A, Renkonen R, Nortamo P, Altieri DC, Gahmberg CG. The Vascular E-Selectin Binds to the Leukocyte Integrins CD11/CD18. Glycobiology 1993; 3:131-136.
14. Bennett CF. Inhibition of Cell Adhesion Molecule Expression with Antisense Oligonucleotides: Activity *in vitro* and *in vivo*. Clin Chem 1994; 40:644-645.
15. Sasaki K, Kurata K, Funayama K, Nagata M, Watanabe E. Ohta S, Hanai N, Nishi T. Expression Cloning of a Novel α-1,3-Fucosyltransferase that is Involved in Biosynthesis of the Sialyl Lewis X Carbohydrate Determinants in Leukocytes. J Biol Chem 1994; 269:14730-14737.

16. Phillips ML, Nudelman E, Gaeta FCA, Perez M, Singhal AK, Hakomori S-I, Paulson JC. ELAM-1 Mediates Cell Adhesion by Recognition of a Carbohydrate Ligand, Sialyl-Le^x. Science 1990; 250:1130-1135.
17. Buerke M, Weyrich AS, Zheng Z, Gaeta FCA, Forrest MJ, Lefer AM. Sialyl $Lewis^x$-containing Oligosaccharide Attenuates Myocardial Reperfusion Injury in Cats. J Clin Invest 1994; 93:1140-1148.
18. Mulligan MS, Paulson JC, DeFrees S, Zheng Z-L, Lowe JB, Ward PA. Protective Effects of Oligosaccharides in P-Selectin-Dependent Lung Injury. Nature 1993; 363:149-151.
19. Kajihara J-I, Guoji Y, Kato K, Suzuki Y. Sulfatide, a Specific Sugar Ligand for L-selectin, Blocks CCl_4-Induced Liver Imflammation in Rats. Biosc Biotech Biochem 1995; 59:155-157.
20. Musser JH, Rao N, Nashed M, Dasgupta F, Abbas S, Nemetalla A, Date V, Foxall C, Asa D, James P, Tyrell D, Brandley BK. Structure-Activity Studies Based on the Sialyl Lewis X Epitope. In: Claassen, V, ed. Trends in Receptors Research. Elsevier, Amsterdam. 1993; 33-40.
21. Feizi T. Oligosacharides that Mediate Mammalian Cell-Cell Adhesion. Curr Opin Struct Biol 1993; 3:701-710.
22. Rosen SD, Bertozzi CR. The Selectins and their Ligands. Curr Opin Cell Biol 1994; 6:663-673.
23. Ichikawa Y, Halcomb RL, Wong C-H. Sticky Solutions. Chem Britain 1994; 117-121.
24. Giannis A. The Sialyl $Lewis^X$ Group and its Analogues as Ligands for Selectins: Chemoenzymatic Syntheses and Biological Functions. Angew Chem Int Ed Engl 1994; 33:178-180.
25. Hasegawa A, Kiso M. Design and Synthesis of Cell Adhesion Carbohydrate Ligands/Inhibitors. Accompanying Chapter in this Book.
26. Palcic MM, Venot AP, Ratcliffe M, Hindsgaul O. Enzymatic Synthesis of Oligosaccharides Terminating in the Tumor-Associated Sialyl $Lewis^a$ Determinant. Carbohydr Res 1989; 190:1-11.
27. Kameyama A, Ishida H, Kiso M, Hasegawa A. Total Synthesis of Tumor-Associated Ganglioside, Sialyl Lewis X. J Carbohydr Chem 1991; 10:549-560.
28. Danishefsky SJ, Gervay J, Peterson JM, McDonald FE, Koseki K, Griffith DA, Oriyama T, Marsden SP. Application of Glycals to the Synthesis of Oligosaccharides: Convergent Total Syntheses of the Lewis X Trisaccharide Sialyl Lewis X Antigenic Determinant and Higher Congeners. J Am Chem Soc 1995; 117:1940-1953.
29. Nicolaou KC, Hummel CW, Iwabuchi Y. Total Synthesis of Sialyl Dimeric Le^x. J Am Chem Soc 1992; 114:3126-3128.

30. Singh K, Fernandez-Maryolas A, Martin-Lomas M. Synthesis of Oligosaccharides Structurally Related to E-Selectin Ligands. J Chem Soc, Chem Commun 1994; 775-776.
31. Iida M, Endo A, Fujita S, Numata M, Matsuzaki Y, Sugimoto M, Nunomura S, Ogawa T. A Total Synthesis of Glycononaosyl Ceramide with a Sialyl Dimeric Le^x Sequence. Carbohydr Res 1995; 270:C15-C19.
32. Ichikawa Y, Lin Y-C, Dumas DP, Shen G-J, Garcia-Junceda E, Williams MA, Bayer R, Ketcham C, Walker LE, Paulson JC, Wong C-H. Chemical-Enzymatic Synthesis and Conformational Analysis of Sialyl Lewis X and Derivatives. J Am Chem Soc 1992; 114:9283-9298.
33. Halcomb RL, Huang H, Wong C-H. Solution- and Solid-Phase Synthesis of Inhibitors of *H. pylori* Attachment and E-Selectin-Mediated Leukocyte Adhesion. J Am Chem Soc 1994; 116:11315-11322.
34. Lin Y-C, Hummel CW, Huang D-H, Ichikawa Y, Nicolaou KC, Wong C-H. Conformational Studies of Sialyl Lewis X in Aqueous Solution. J Am Chem Soc 1992; 114:5452-5454.
35. Ball GE, O'Neil RA, Schultz JE, Lowe JB, Weston BW, Nagy JO, Grown EG, Hobbs CJ, Bednarski MD. Synthesis and Structural Analysis Using 2-D NMR of Sialyl Lewis X (SLe^x) and Lewis X (Le^x) Oligosaccharides: Ligands Related to E-Selectin (ELAM-1) Binding. J Am Chem Soc 1992; 114:5449-5451.
36. Rao BNN, Moreland M, Brandley BK. The Potential of Carbohydrates as Cell Adhesion Inhibitors. Med Chem Res 1991; 1:1-8.
37. Hensley P, McDevitt PJ, Brooks I, Trill JJ, Feild JA, McNulty DE, Connor JR, Griswold DE, Kumar NV, Kopple KD, Carr SA, Dalton BJ, Johanson K. The Soluble Form of E-Selectin is an Asymmetric Monomer. J Biol Chem 1994; 269:23949-23958.
38. Kogelberg H, Rutherford TJ. Studies on the Three-Dimensional Behaviour of the Selectin Ligands $Lewis^a$ and Sulphated $Lewis^a$ Using NMR Spectroscopy and Molecular Dynamics Simulations. Glycobiology 1994; 4:49-57.
39. Rutherford TJ, Spackman DG, Simpson PJ, Homans SW. 5 Nanosecond Molecular Dynamics and NMR Study of the Conformational Transitions in the Sialyl-Lewis X Antigen. Glycobiology 1994; 4:59-68.
40. DeFrees SA, Kosch W, Way W, Paulson JC, Sabesan S, Halcomb RL, Huang D-H, Ichikawa Y, Wong C-H. Ligand Recognition by E-Selectin: Synthesis, Inhibitory Activity, and Conformational Analysis of Bivalent Sialyl Lewis X Analogs. J Am Chem Soc 1995; 117:66-79.
41. DeFrees SA, Gaeta FCA, Lin Y-C, Ichikawa Y, Wong C-H. Ligand Recognition by E-Selectin: Analysis of Conformation and Activity of

Synthetic Monomeric and Bivalent Sialyl Lewis X Analogs. J Am Chem Soc 1993; 115:7549-7550.

42. Scheffer K, Ernst B, Katopodis A, Magnani JL, Wang WT, Weisemann R, Peters T. Determination of the Bioactive Conformation of the Carbohydrate Ligand in the E-Selectin/Sialyl Lewis^x Complex. Angew Chem Int Ed Engl 1995; 34:1841-1844.
43. Jacob GS, Kirmaier C, Abbas SZ, Howard SC, Steininger CN, Welply JK, Scudder P. Binding of Sialyl Lewis X to E-Selectin as Measured by Fluorescence Polarization. Biochemistry 1995; 34:1210-1217
44. Hollenbaugh D, Bajorath J, Stenkamp R, Aruffo A. Interaction of P-Selectin (CD62) and its Cellular Ligand: Analysis of Critical Residues. Biochemistry 1993; 32:2960-2966.
45. Bajorath J, Hollenbaugh D, King G, Harte Jr W, Eustice DC, Darveau RP, Aruffo A. CD62/P-Selectin Binding Sites for Myeloid Cells and Sulfatides Are Overlapping. Biochemistry 1994; 33:1332-1339.
46. Drickamer K. Three-Dimensional View of a Selectin Cell Adhesion Molecule. Glycobiology 1994; 4:245-248.
47. Bajorath J, Stenkamp R, Aruffo A. Comparison of a Protein Model with Its X-ray Structures: The Ligand Binding Domain of E-Selectin. Bioconjugate Chem 1995; 6:3-6.
48. Graves BJ, Crowther RL, Chandran C, Rumberger JM, Li S, Huang K-S, Presky DH, Familletti PC, Wolitzky BA, Burns DK. Insight into E-Selectin/Ligand Interaction from the Crystal Structure and Mutagenesis of the Lec/EGF Domains. Nature 1994; 367:532-538.
49. Berg EL, Robinson MK, Mansson O, Butcher EC, Magnani JL. A Carbohydrate Domain Common to Both Sialyl Le^a and Sialyl Le^x is Recognized by the Endothelial Cell Leukocyte Adhesion Molecule ELAM-1. J Biol Chem 1991; 266:14869-14872.
50. Berg EL, Magnani JL, Warnock RA, Robinson MK, Butcher EC. Comparison of L-Selectin and E-Selectin Ligand Specificities: the L-Selectin Can Bind the E-Selectin Ligands Sialyl Le^x and Sialyl Le^a. Biochem Biophys Res Commun 1992; 184:1048-1055.
51. Tyrrell D, James P, Rao N, Foxall C, Abbas S, Dasgupta F, Nashed M, Hasegawa A, Kiso M, Asa D, Kidd J, Brandley BK. Structural Requirements for the Carbohydrate Ligand of E-Selectin. Proc Natl Acad Sci USA 1991; 88:10372-10376.
52. Foxall C, Watson SR, Dowbenko D, Fennie C, Larsky LA, Kiso M, Hasegawa A, Asa D, Brandley BK. The Three Members of the Selectin Receptor Family Recognize a Common Carbohydrate Epitope, the Sialyl Lewis X Oligosaccharide. J Cell Biol 1992; 117:895-902.
53. Brandley BK, Kiso M, Abbas S, Nikrad P, Srivastava O, Foxall C, Oda Y, Hasegawa A. Structure-function Studies on Selectin Carbohydrate

Ligands. Modifications to Fucose, Sialic Acid and Sulphate as a Sialic Acid Replacement. Glycobiology 1993; 3:633-639.

54. Yuen C-T, Bezouška K, O'Brien J, Stoll M, Lemoine R, Lubineau A, Kiso M, Hasegawa A, Bockovich NJ, Nicolaou KC, Feizi T. Sulfated Blood Group Lewis[a]. A Superior Oligosaccharide Ligand for Human E-Selectin. J Biol Chem 1994; 269:1595-1598.

55. Green PJ, Yuen C-T, Childs RA, Chai W, Miyasaka M, Lemoine R, Lubineau A, Smith B, Ueno H, Nicolaou KC, Feizi T. Further Studies of the Binding Specificity of the Leukocyte Adhesion Molecule, L-Selectin, Towards Sulphated Oligosaccharides. Suggestion of a Link Between the Selectin- and the Integrin-Mediated Lymphocyte Adhesion Systems. Glycobiology 1994; 4:29-38.

56. Yuen C-T, Lawson AM, Chai W, Larkin M, Stoll MS, Stuart AC, Sullivan FX, Ahern TJ, Feizi T. Novel Sulfated Ligands for the Cell Adhesion Molecule E-Selectin Revealed by the Neoglycolipid Technology Among O-Linked Oligosaccharides on an Ovarian Cystadenoma Glycoprotein. Biochemistry 1992; 31:9126-9131.

57. Thoma G, Schwarzenbach F, Duthaler RO. Synthesis of a Sialyl Lewis X Mimic with Fixed Carboxylic Acid Group: Chemical Approach Toward the Elucidation of the Bioactive Conformation of Sialyl Lewis X. J Org Chem 1996; 61:514-524.

58. Toepfer A, Kretzschmar G, Bartnik E. Synthesis of Novel Mimetics of the Sialyl Lewis X Determinant. Tetrahedron Lett 1995; 36:9161-9164.

59. Nelson RM, Dolich S, Aruffo A, Cecconi O, Bevilacqua MP. Higher-Affinity Oligosaccharide Ligands for E-Selectin. J Clin Invest 1993; 91:1157-1166.

60. Dasgupta F, Casabonne ME, Gilbert JH, Musser H, Schaeffer ME. Tethered Analogs of Sialyl Lewis X. XVIIth International Carbohydrate Symposium. Ottawa, Canada. July 17-22, 1994; D2-7.

61. Stahl W, Sprengard U, Kretzschmar G, Kunz H. Synthesis of Deoxy Sialyl Lewis-X Analogues, Potential Selectin Antagonists. Angew Chem Int Ed Engl 1994; 33:2096-2098.

62. Siuzdak G, Zheng Z-L, Rhamphal JY, Ichikawa Y, Nicolaou KC, Gaeta FCA, Chatman KS, Wong C-H. Examination of the Sialyl Lewis X-Calcium Complex by Electrospray Mass Spectrometry. Bioorg Med Chem Lett 1994; 4:2863-2866.

63. Heskamp BM, Veeneman GH, Van der Marel GA, Van Boeckel CAA, Van Boom JH. Synthesis of a SLe[x] Mimic: a Potential E-Selectin Binding Antagonist. Recl Trav Chim Pays-Bas 1995; 114:398-402.

64. Ramphal JY, Zheng Z-L, Perez C, Walker LE, DeFrees SA, Gaeta FCA. Structure-Activity Relationships of Sialyl Lewis[x]-Containing

Oligo-Saccharides. 1. Effect of Modifications of the Fucose Moiety. J Med Chem 1994; 37:3459-3463.
65. Uchiyama T, Vassilev VP, Kajimoto T, Wong W, Huang H, Lin C-C, Wong C-H. Design and Synthesis of Sialyl Lewis X Mimetics. J Am Chem Soc 1995; 117:5395-5396.
66. Huang H, Wong C-H. Synthesis of Biologically Active Sialyl Lewis X Mimetics. J Org Chem 1995; 60:3100-3106.
67. Prodger J, Bamford MJ, Gore PM, Holmes DS, Saez V, Ward P. Synthesis of a Novel Analogue of Sialyl Lewis X. Tetrahedron Lett 1995; 36:2339-2342.
68. Ragan JA, Cooper K. Synthesis of a Galactose-Fucose Disaccharide Mimic of Sialyl Lewis X. Bioorg Med Chem Lett 1994; 4:2563-2566.
69. Allanson NM, Davidson AH, Martin FM. A Novel Mimic of the Sialyl Lewis X Determinant. Tetrahedron Lett 1993; 34:3945-3948.
70. Allanson NM, Davidson AH, Floyd CD, Martin FM. The Synthesis of Novel Mimics of the Sialyl Lewis X Determinant. Tetrahedron: Asymmetry 1994; 5:2061-2076.
71. Hanessian S, Prabhanjan H. Design and Synthesis of Glycomimetic Prototypes- A Model Sialyl Lewis X Ligand for E-Selectin. Synlett 1994; 868-870.
72. Kaila N, Yu H-A, Xiang Y. Design and Synthesis of Novel Sialyl Lewis X Mimics. Tetrahedron Lett 1995; 36:5503-5506.
73. Bock K, Peters T, Meldal M, Duus JO, Hindsgaul O, Lowery T, Srivastava OP, Foxall C, Bradbury K, Rao BNN. Fucopeptides as Sialyl-Le X Mimetics for the Inhibition of Selectin Binding. Personnal communication; WO Patent No. 9510296 A1, 1995:04:20.
74. Wu S-H, Shimazaki M, Lin C-C, Qiao L, Moree WJ, Weitz-Schmidt G, Wong C-H. Synthesis of Fucopeptides as Sialyl Lewis^x Mimetics. Angew Chem Int Ed Engl 1996; 35:88-90, and personal communication.
75. Ruoslahti E, Pierschbacher MD. New Perpectives in Cell Adhesion: RGD and Integrins. Science 1987; 238:491-497.
76. McDowell RS, Gadek TR. Structural Studies of Potent Constraint RGD Peptides. J Am Chem Soc 1992; 114:9245-9253.
77. Ku TW, Miller WH, Bondinell WE, Erhard KF, Keenan RM, Nichols AJ, Peishoff CE, Samanen JM, Wong AS, Huffman WF. Potent Non-Peptide Fibrinogen Receptor Antagonists Which Present an Alternative Pharmacophore. J Med Chem 1995; 38:9-12.
78. Sprengard U, Kretzschmar G, Bartnick E, Hüls C. Synthesis of an RGD-Sialyl Lewis-X Glycoconjugates: A New Highly Active Ligand for P-Selectin. Angew Chem Int Ed Engl 1995; 34:990-993.

79. Rao BNN, Anderson MB, Musser, JH, Gilbert, JH, Schaefer ME, Foxall C, Brandley BK. Sialyl Lewis X Mimics Derived from a Pharmacophore Search Are Selectin Inhibitors with Anti-Inflammatory Activity. J Biol Chem 1994; 269:19663-19666.
80. Norgard KE, Moore KL, Diaz S, Stults NL, Ushiyama S, McEver RP, Cummings RD, Varki A. Characterization of a Specific Ligand for P-Selectin on Myeloid Cells. J Biol Chem 1993; 268:12764-12774.
81. Yoshida T, Toyama-Sorimachi N, Miyasaka M, Lee YC. A Crucial Effect of Ligand Clustering on the Inhibition of Binding by L-Selectin in Intercellular Adhesion. Biochem Biophys Res Commun 1994; 204:969-974.
82. Lee RT, Lee YC. Enhanced Biochemical Affinities of Multivalent Neoglycoconjugates. In: Lee YC, Lee RT, eds. Neoglycoconjugates: Preparation and Applications. San Diego: Academic Press, 1994, 23-50.
83. Roy R. Design and Syntheses of Glycoconjugates. In: Khan S-H, O'Neil R, eds. Modern Methods in Carbohydrate Synthesis. Amsterdam: Harwood Academic, 1996, 378-402.
84. Roy R. Blue-Prints, Synthesis and Applications of Glycopolymers. Trends Glycoscience Glycotech (TIGGS) 1996; 8:79-99.
85. Öhrlein R. Chemo-enzymatic Synthesis of Oligovalent Sialyl-Le^x. XVIIth International Carbohydrate Symposium. Ottawa, Canada. July 17-22, 1994; D1-20 and personal communication.
86. Kretzschmar G, Sprengard U, Kunz K, Bartnik E, Schmidt W, Toepfer A, Hörsch B, Krause M, Seiffge D. Oligosaccharide Recognition by Selectins: Synthesis and Biological Activity of Multivalent Sialyl Lewis-X Ligands. Tetrahedron 1995; 51:13015-13030.
87. Sprengard U, Schudock G, Schmidt W, Kretzschmar G, Kunz H. Multiple Sialyl Lewis-X N-Glycopeptides: New Effective Ligands for E-Selectin. Angew Chem Int Ed Engl 1996; 35: 321-324.
88. Sprengard U, Kunz H, Hülls C, Schmidt W, Seiffge D, Kretzschmar G. Synthesis and Biological Activity of Novel Sialyl $Lewis^x$ Conjugates. Bioorg Med Chem Lett 1996; 6: 509-514.
89. Roy R, Park WKC, Srivastava OP, Foxall C. Combined Glycomimetic and Multivalent Strategies for the Design of Potent Selectin Antagonists. Bioorg Med Chem Lett 1996; 6: in press; unpublished observations, submitted to Angew Chem Int Ed Engl.
90. Roy R. Glycodendrimers: A New Class of Biopolymers. Polymer News 1996; 21: 226-232.
91. Roy R, Zanini D, Meunier SJ, Romanowska A. Solid-phase Synthesis of Dendritic Sialoside Inhibitors of Influenza A Virus Hemagglutinin. J Chem Soc, Chem Commun 1993; 1869-1872.

92. Roy R, Zanini D, Meunier SJ, Romanowska A. Synthesis and Antigenic Properties of Sialic Acid Based Dendrimers. ACS Symposium Series 1994; 560:104-119.
93. Zanini D, Park WKC, Roy R. Synthesis of Novel Dendritic Glycosides. Tetrahedron Lett 1995; 36:7383-7386.
94. Zanini D, Park WKC, Meunier SJ, Wu Q, Aravind S, Kratzer B, Roy R. Syntheses and Biological Properties of Glycodendrimers. PMSE 1995; 73:82-83.
95. Park WKC, Kratzer B, Zanini D, Wu Q, Meunier SJ, Roy R. Glycodendrimers: New Powerful Inhibitors of Carbohydrate-protein Interactions. Glycoconjugate J 1995; 12:456.
96. Pagé D, Zanini D, Roy R. Macromolecular Recognition: Effect of Multivalency in the Inhibition of Binding of Yeast Mannan to Concanavalin A and Pea Lectins by Mannosylated Dendrimers. Bioorg Med Chem 1996; in press.
97. Stahl W, Sprengard U, Ahlers M, Hörsch B, Bartnick E, Walch A, Kretzschmar G, Kunz H. Multivalent Sialyl Lewis-X on a Biodegradable Polymer, personal communication.
98. Nifant'ev NE, Shashkov AS, Tsvetkov YE, Tuzikov AB, Abramenko IV, Gluzman DF, Bovin NV. Synthetic Probe for Study of Molecules Related to Selectin Family. ACS Symposium Series 1994; 560:267-275.
99. Zemlyanukhina TV, Nifant'ev NE, Shashkov AS, Tsvetkov YE, Bovin NV. Selectin Receptors: Synthesis of Spacer-Armed Sulfated Trisaccharides Lewis A and Lewis X and Neoglycoconjugates Thereof. Carbohydr Lett 1995; 1:277-284.
100. Zemlyanukhina TV, Nifant'ev NE, Kononov LO, Shaskhov AS, Bovin NV. Selectin Receptors 2. Synthesis of HSO_3-3'Le^a-The Sulfated Ligand of the Cell Adhesion Molecule E-Selectin. Russ J Bioorg Chem 1994; 20:315-317.
101. Roy R, Park WKC, Kosch W, Paulson JC. Unpublished results.
102. Lubineau A, Lemoine R. Regioselective Sulfation of Galactose Derivatives Through the Stannylene Procedure. New Synthesis of the 3'O-Sulfated Lewis^a Trisaccharide. Tetrahedron Lett 1994; 35:8795-8796.
103. Nicolaou KC, Bockovich NJ, Carcanague DR. Total Synthesis of Sulfated Le^x and Le^a-Type Oligosaccharide Selectin Ligands. J Am Chem Soc 1993; 115:8843-8844.
104. Vig R, Jain RK, Piskorz CF, Matta KL. Selectin Ligands: Synthesis of 3'-O-Sialyl-6'-O-Sulfo Lewis^a, NeuAcα2-3(6-O-SO_3Na)Galβ1-3(Fuc-α1-4)-GlcNAcβ-OMe. J Chem Soc, Chem Commun 1995; 2073-2074.

105. Jain RK, Vig R, Locke RD, Mohammad A, Matta KL. Selectin Ligands: 2,3,4-tri-O-acetyl-6-O-pivaloyl-α/β-galactopyranosyl Halide as Novel Glycosyl Donor for the Synthesis of 3-O-Sialyl or 3-O-Sulfo Le^x and Le^a Type Structures. Chem Commun 1996; 65-67.
106. Jain RK, Vig R, Rampal R, Chandrasekaran EV, Matta KL. Total Synthesis of 3'-O-Sialyl, 6'-O-Sulfo $Lewis^x$, NeuAcα2-3(6-O-SO_3Na)-Galβ1-4(Fucα1-3)-GlcNAcβ-OMe: A Major Capping Group of GLYCAM-1. J Am Chem Soc 1994; 116:12123-12124.
107. Manning DD, Bertozzi CR, Pohl NL, Rosen SD, Kiessling LL. Selectin-Saccharide Interactions: Revealing Structure-Function Relationships with Chemical Synthesis. J Org Chem 1995; 60:6254-6255.
108. Scudder PR, Shailubhai K, Duffin KL. Streeter PR, Jacob GS. Enzymatic Synthesis of a 6'-Sulphated Sialyl-$Lewis^x$ which is an Inhibitor of L-Selectin Binding to Peripheral Addressin. Glycobiology 1995; 5:929-932.
109. Numomura S, Iida M, Numata M, Sugimoto M, Ogawa T. Total Synthesis of Sulfated Le^x Pentaosyl Ceramide. Carbohydr Res 1994; 263:C1-C6.
110. Lubineau A, Le Gallic J, Lemoine R. First Synthesis of the 3'Sulfated $Lewis^a$ Triasaccharide, Putative Ligand for the Leucocytes Homing Receptor. J Chem Soc, Chem Commun 1993; 1419-1418.
111. Chandrasekaran EV, Jain RK, Matta KL. Ovarian Cancer α1,3-L-Fucosyltransferase. J Biol Chem 1992; 267:23806-23814.
112. Briggs JB, Oda Y, Gilbert JH, Scaefer ME, Macher BA. Peptides Inhibit Selectin-Mediated Cell Adhesion *in vitro* and Neutrophil Influx into Inflammatory Sites *in vivo*. Glycobiology 1995; 5:583-588.
113. Heerze LD, Smith RH, Wang N, Armstrong GD. Utilization of Sialic Acid-Binding Synthetic Peptide Sequences Derived from Pertussis Toxin as Novel Anti-Inflammatory Agents. Glycobiology 1995; 5:427-433.
114. Wiley DC, Skehel, JJ. The Structure and Function of the Hemagglutinin Membrane Protein of Influenza Virus. Ann Rev Biochem 1987; 56:365-394.
115. Wilson IA, Skehel JJ, Wiley DC. Structure of the Haemagglutinin Membrane Glycoprotein of Influenza Virus at 3 Å Resolution. Nature 1981; 289:366-373.
116. Weis W, Brown JH, Cusak JH, Paulson JC, Skehel JJ, Wiley DC. Structure of the Influenza Virus Hemagglutinin Complexed with its Receptor, Sialic Acid. Nature 1988; 333:426-431.

117. Varghese JN, Laver WG, Colman PM. Structure of the Influenza Virus Glycoprotein Antigen Neuraminidase at 2.9 Å Resolution. Nature 1983; 303:35-40.
118. von Itzstein M, Kiefel MJ. In: Sialic Acid Analogues as Potential Antimicrobial Agents. Accompanying Chapter 2 in this Book.
119. Zbiral E. Synthesis of Sialic Acid Analogs and their Behavior Towards the Enzymes of Sialic Acid Metabolism and Hemagglutinin X-31 of Influenza Virus. In: Ogura H, Hasegawa A, Suami T, eds. Carbohydrates. Synthetic Methods and Applications in Medicinal Chemistry. Tokyo: Kodansha, 1992:304-339.
120. Pritchett TJ, Brossmer R, Rose U, Paulson JC. Recognition of Monovalent Sialosides by Influenza Virus H3 Hemagglutinin. Virology 1987; 160:502-506.
121. Matrosovich MN, Gambaryan AS, Tuzikov AB, Byramova NE, Mochalova LV, Golbraikh AA, Shenderovich MD, Finne J, Bovin NV. Probing of the Receptor-Binding Sites of the H1 and H3 Influenza A and Influena B Virus Hemagglutinins by Synthetic and Natural Sialosides. Virology 1993; 196:111-121.
122. Toogwood PL, Galliker PK, Glick GD, Knowles JR. Monovalent Sialosides that Bind Tightly to Influenza A Virus. J Med Chem 1991; 34:3138-3140.
123. Weinhold EG, Knowles JR. Design and Evaluation of a Tightly Binding Fluorescent Ligand for Influenza A Hemagglutinin. J Am Chem Soc 1992; 114:9270-9275.
124. Sauter NK, Bednarski MD, Wurzburg BA, Hanson JE, Whitesides GM, Skehel JJ, Wiley DC. Hemagglutinins from Two Influenza Variants Bind to Sialic Acid Derivatives with Millimolar Dissociation Constants: A 500-MHz Proton Nuclear Magnetic Resonance Study. Biochemistry 1989; 28:8388-8395.
125. Pritchett TJ, Paulson JC. Basis for the Potent Inhibition of Influenza Virus Infection by Equine and Guinea Pig α_2-Macroglobulin. J Biol Chem 1989; 264:9850-9858.
126. Matrosovich MN. Towards the Development of Antimicrobial Drugs Acting by Inhibition of Pathogen Attachment to Host Cells: a Need for Polyvalency. FEBS Lett 1989; 252:1-4.
127. Roy R, Laferrière, Pon RA, Gamian A. Induction of Rabbit Immunoglobulin G Antibodies Against Synthetic Sialylated Neoglycoproteins. Methods Enzymol 1994; 247:351-361.
128. Glick GD, Knowles JR. Molecular Recognition of Bivalent Sialosides by Influenza Virus. J Am Chem Soc 1991; 113:4701-4703.

129. Sabesan S, Duus JØ, Domaille P, Kelm S, Paulson JC. Synthesis of Cluster Sialoside Inhibitors for Influenza Virus. J Am Chem Soc 1991; 113:5865-5866.
130. Sabesan S, Duus JØ, Neira S, Domaille P, Kelm S, Paulson JC, Bock K. Cluster Sialoside Inhibitors for Influenza Virus: Synthesis, NMR, and Biological Studies. J Am Chem Soc 1992; 114:8363-8375.
131. Unverzagt C, Kelm S, Paulson JC. Chemical and Enzymatic Synthesis of Multivalent Sialoglycopeptides. Carbohydr Res 1994; 251:285-301.
132. Roy R, Laferrière CA, Gamian A, Jennings HJ. N-Acetylneuraminic Acid: Neoglycoproteins and Pseudopolysaccharides. J Carbohydr Chem 1987; 6:161-165.
133. Roy R, Laferrière CA. Synthesis of Antigenic Copolymers of N-Acetylneuraminic Acid Binding to Wheat Germ Agglutinin and Antibodies. Carbohyd Res 1988; 177:C1-C4.
134. Laferrière CA, Roy R, Andersson FO. Syntheses of Water-Soluble Polyacrylamide-Containing Sialic Acid. Methods Enzymol 1994; 242:271-280.
135. Gamian A, Chomik M, Laferrière CA, Roy R. Inhibition of Influenza A Virus Hemagglutinin and Induction of Interferon by Synthetic Sialylated Glycoconjugates. Can J Microbiol 1991; 37:233-237.
136. Spaltenstein A, Whitesides GM. Polyacrylamides Bearing Pendant ♋-Sialoside Groups Strongly Inhibit Agglutination of Erythrocytes by Influenza Virus. J Am Chem Soc 1991; 113:686-687.
137. Roy R, Laferrière CA. Michael Addition as the Key Step in the Synthesis of Sialyloligosaccharide-Protein Conjugates from N-Acryloylated Glycopyranosylamines. J Chem Soc, Chem Commun 1990; 1709-1711.
138. Roy R, Andersson FO, Harms G, Kelm S, Schauer R. Synthesis of Esterase-Resistant 9-O-Acetylated Polysialoside as Inhibitor of Influenza C Virus Hemagglutinin. Angew Chem Int Ed Engl 1992; 31:1478-1481.
139. Sparks MA, Williams KW and Whitesides GM. Neuraminidase-Resistant Hemagglutinin Inhibitors: Acrylamide Copolymers Containing a C-Glycoside of N-Acetylneuraminic Acid. J Med Chem 1993; 36:778-783.
140. Byramova NE, Mochalova LV, Belyanchikov JM, Matrosovich MN, Bovin NV. Synthesis of Sialic Acid Pseudopolysaccharides by Coupling of Spacer-Connected Neu5Ac with Activated Polymer. J Carbohydr Chem 1991; 10:691-700.
141. Matrosovich MN, Mochalova LV, Marinina VP, Byramova NE, Bovin NV. Synthetic Polymeric Sialoside Inhibitors of Influenza Virus Receptor-binding Activity. FEBS Lett 1990; 272:209-212.

142. Mochalova LV, Tuzikov AB, Marinina VP, Gambaryan AS, Byramova NE, Bovin NV, Matrosovich MN. Synthetic Polymeric Inhibitors of Influenza Virus Receptor-Binding Activity Suppress Virus Replication. Antiviral Res 1994; 23:179-190.

143. Nagy JO, Wang P, Gilbert JH, Schaefer ME, Hill TG, Callstrom MR, Bednarski MD. Carbohydrate Materials Bearing Neuraminidase-Resistant C-Glycosides of Sialic Acid Strongly Inhibit the inVitro Infectivity of Influenza Virus. J Med Chem 1992; 35:4501-4502.

144. Roy R, Pon RA, Tropper FD, Andersson FO. Michael Addition of Poly-L-Lysine to N-Acryloylated Sialosides. Syntheses of Influenza A Virus Haemagglutinin Inhibitor and Group B Meningococcal Polysaccharide Vaccines. J Chem Soc, Chem Commun 1993; 264-265.

145. Paulsen H, Matschulat P. Synthesis of the C-Glycoside of N-Acetylneuramic Acid. Liebigs Ann Chem 1991; 487-495.

146. Nagy JO, Bednarski MD. The Chemical-Enzymatic Synthesis of a Carbon Glycosides of N-Acetylneuraminic Acid. Tetrahedron Lett 1991; 32:3953-3956.

147. Kingery-Wood JE, Williams KW, Sigal GB, Whitesides GM. The Agglutination of Erythrocytes by Influenza Virus is Strongly Inhibited by Liposomes Incorporating an Analog of Sialyl Ganglioside. J Am Chem Soc 1992; 114:7303-7305.

148. Spevak W, Nagy JO, Charych DH, Schaefer ME, Gilbert JH, Bednarski MD. Polymerized Liposomes Containing C-Glycosides of Sialic Acid: Potent Inhibitors of Influenza Virus *in Vitro* Infectivity. J Am Chem Soc 1993; 115:1146-1147.

149. Charych DH, Nagy JO, Spevak W, Bednarski MD. Direct Colorimetric Detection of Receptor-Ligand Interaction by a Polymerized Bilayer Assembly. Science 1993; 261:585-588.

4
Design and Synthesis of Cell Adhesion Carbohydrate Ligands/Inhibitors

Akira Hasegawa and Makato Kiso
Gifu University
Gifu, Japan

I. INTRODUCTION

Cell surface carbohydrates have long been suggested to act as cell-cell recognition molecules [1-3]. Sialic acid or sulphate-containing glycoconjugates in particular have recently been recognized to play important roles in many biological processes such as cell adhesion, cell differentiation and proliferation, microbial and viral infections, immune responses, inflammation, tumor progression, and so on. In fact the discovery of selectins [4-6] and more recently the I-type lectins (sialoadhesin family) [7-8] have demonstrated the critical importance of protein-carbohydrate binding in cell-cell interactions, providing new opportunities for development of therapeutic agents based on carbohydrates as binding inhibitors. We describe here mainly our results of the structure-function studies on selectin carbohydrate ligands/inhibitors. Some pharmacological studies in disease models are also discussed.

II. SELECTIN LIGANDS

The selectins are a family of carbohydrate-binding proteins involved in leukocyte trafficking, thrombosis, inflammation, tumor metastasis. Until recently, a variety of sialylated, sulphated and/or fucosylated carbohydrates have been reported as possible selectin ligands [9]. Since 1990, the sialyl

Lewis[x] (sialyl-Le^X) determinant, α-Neu5Ac(2→3)-β-D-Gal(1→4)[α-L-Fuc(1→3)]-β-D-GlcNAc, which is found as the terminal carbohydrate structure in both cell membrane glycolipids and glycoproteins, has been recognized [9-11] as a minimal effective structure for selectin binding.

A. Sialyl-Le^X Gangliosides and Analogs

The first total syntheses of hexasaccharidic sialyl-Le^X (**1**) [12], pentasaccharidic sialyl-Le^X (**2**) [13], and the sialyl α(2→6) positional isomer (**3**) [14] (Fig.1) have been achieved according to our established method [15]. The use of these synthetic gangliosides in both direct binding and inhibition assays revealed [11] that both sialyl-Le^X (**1**) and (**2**) were effectively recognized by E-selectin in dose-dependent manners, while the sialyl α(2→6) isomer (**3**) was not recognized.

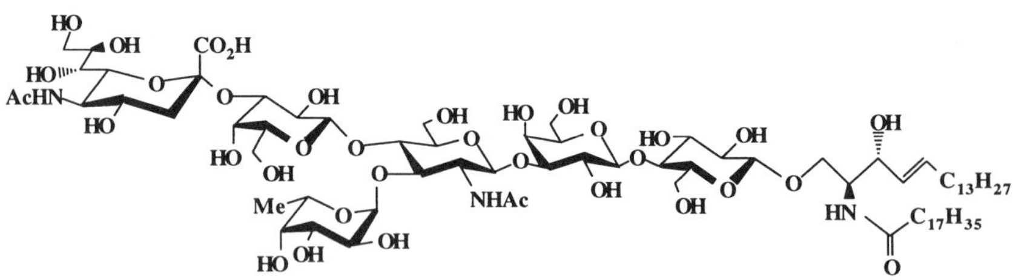

1 Sialyl Le^x (hexa)

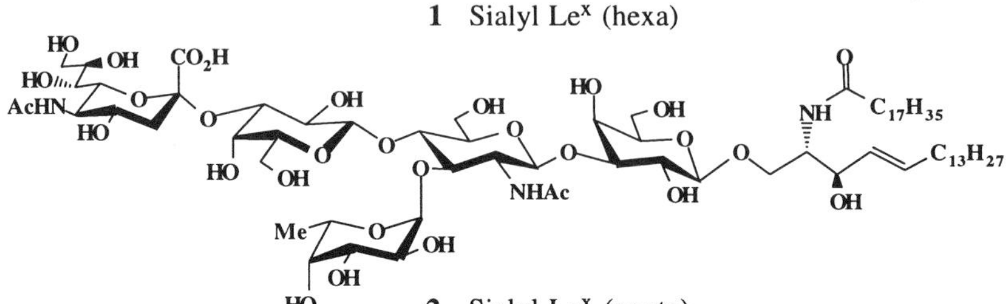

2 Sialyl Le^x (penta)

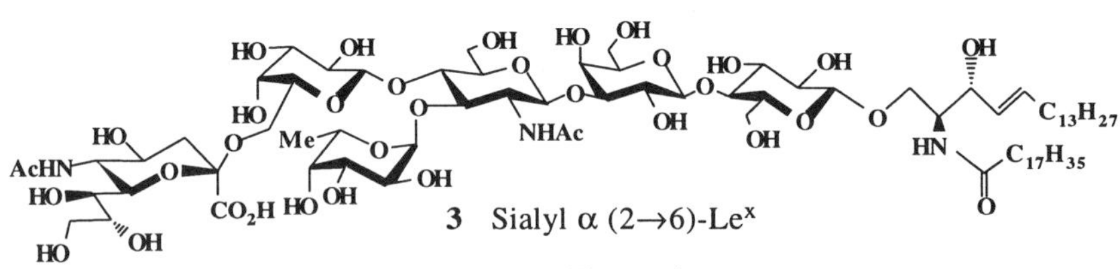

3 Sialyl α (2→6)-Le^x

Figure 1

The binding ability of **1** was obviously higher than **2** suggesting the necessity of suitable sugar-chain length.

It was also demonstrated [16] that both P-(CD 62) and E-selectin (ELAM-1) recognized the same carbohydrate epitope, sialyl-Le^X, as a high-affinity ligand by use of the synthetic ganglioside (**1**) [12]. Furthermore,

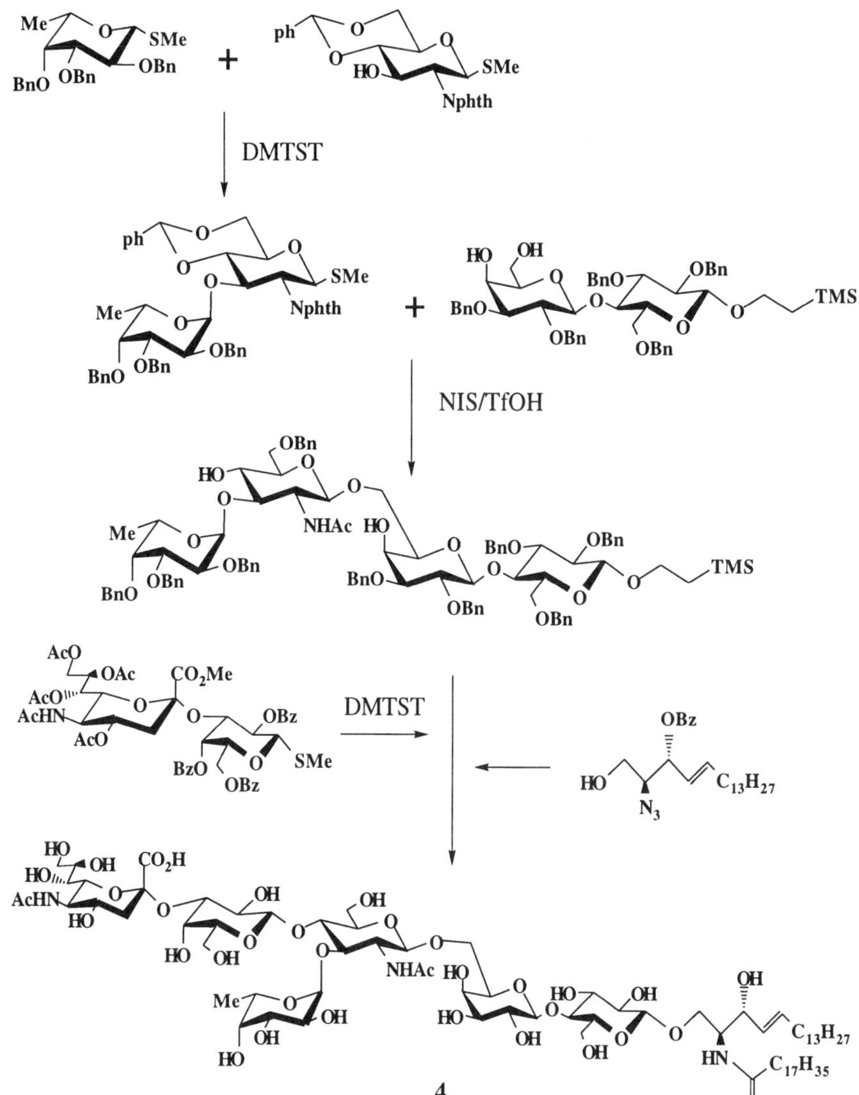

4

Figure 2

the use of ELISA assay developed with immobilized sialyl-Le^X gangliosides and selectin-IgG chimeras demonstrated [17] that three members of the selectin family

(E-, P-, and L-selectin) recognized a common carbohydrate epitope sialyl-Le^X in a calcium-dependent manner. We have also synthesized sialyl-Le^X β(1→6)-lactosyl ceramide (**4**) [18] by using four kinds of methylthioglycosides as glycosyl donors, and dimethyl(methylthio)sulfonium triflate (DMTST) [19-20] and N-iodosuccinimide-trifluoromethanesulfonic acid (NIS/TfOH) [21-23] as glycosyl promoters (Fig.2). This sialyl-Le^X isomer (**4**) was also efficiently recognized by three selectins as well. These data suggested that sialyl-Le^X oligosaccharide may form the basis of a recognition domain common to all three selectins.

These findings as well as the use of soluble inhibitors including tetrasaccharidic sialyl-Le^X, together with NMR data on the conformation (Fig.3) of oligosaccharides in solution, led to a hypothesis [11] for the binding face of the sialyl-Le^X epitope interacting with E-selectin which suggested that the carboxyl group of sialic acid, the 4- and 6-hydroxyls of galactose, and the 2-, 3-, and 4-hydroxyls of fucose may be involved in selectin-carbohydrate binding.

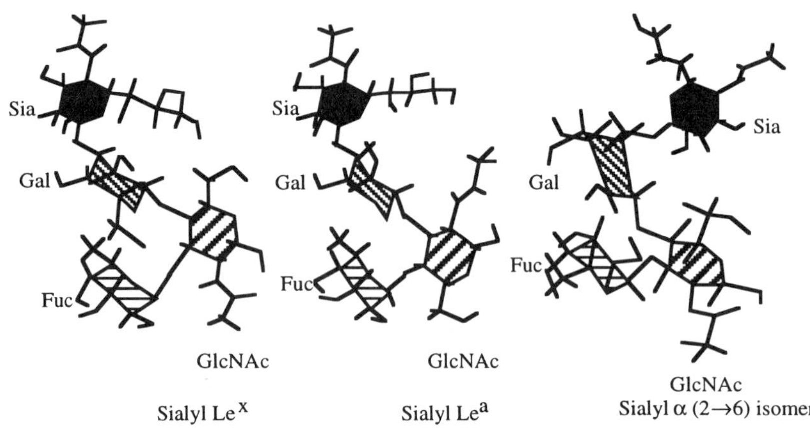

Figure 3

1. Sialic acid modification

The syntheses of sialyl-Le^X ganglioside analogs containing modified sialic acids, such as C7 (**5**)-, C8 (**6**)- and 8-epi(**7**)-N-acetylneuraminic acids [24], N-glycolylneuraminic acid (**8**) [25], and the deaminated neuraminic acid (**9**, KDN) [26] (Fig.4), have been achieved by using the suitably modified sialyl-α-(2→3)-galactose donors as described in [15]. The ELISA assay for the binding of selectin-IgG chimeras with immobilized sialyl-Le^x ganglioside analogs has shown [27] that these modifications in the sialic acid residue have little or no effect on the binding of any of the selectins. These sialic acid-modified ganglioside analogs are not only high-affinity ligands for selectins but also potent binding inhibitors.

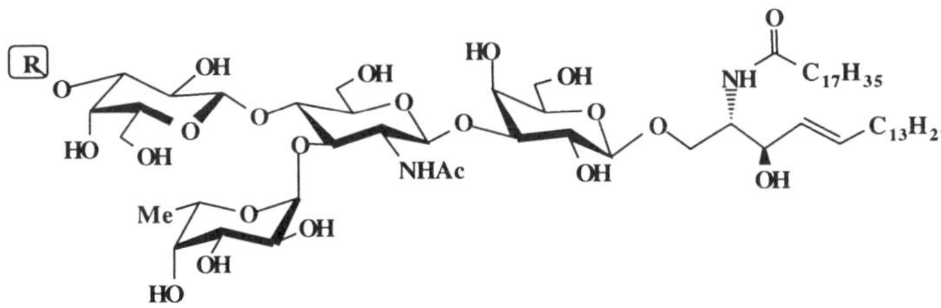

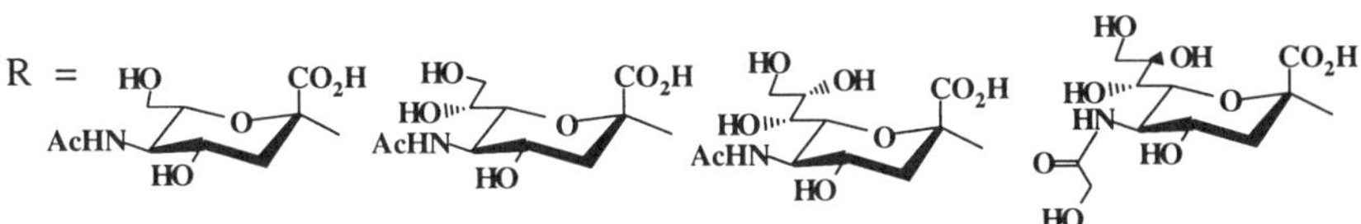

5 (C_7) 6 (C_8) 7 (8-epi) 8

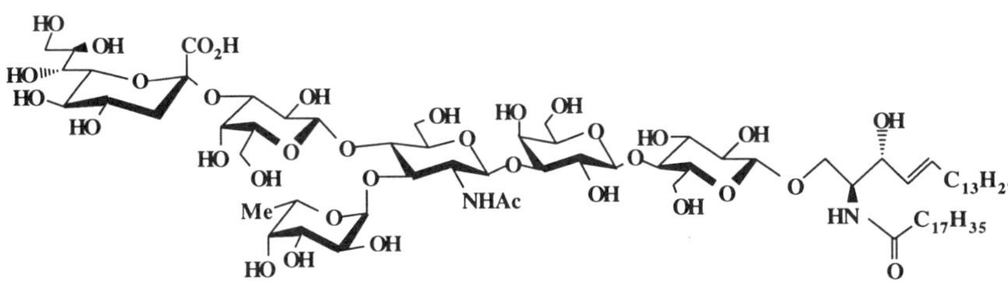

9 KDN-Le^x (hexa)

Figure 4

It is noteworthy that the IC_{50} values of the KDN derivative (**9**) against the immobilized sialyl-Le^X ligand (penta, **2**) were 0.03 μM for L-selectin, 1 μM for P-selectin, and 1.6 μM for E-selectin, respectively. This result is in disagreement with a report by Nelson et al. [28], in which KDN-Le^X was significantly less active in E-selectin assays than sialyl-Le^X. Our results show that the sialic acid residue can be replaced by KDN and N-glycolylneuraminic acid without decreasing of binding/inhibition activities. These findings, together with previous data, suggest the crucial role of the carboxyl moiety of sialic acid which is considered to be an anchor point for selectin-sialyl-Le^X interactions [29-31].

2. Fucose modification

It has been found [11,32] that binding of selectins was dependent on fucose, which is considered to be implicated in the calcium binding [29-31] critical for selectin recognition together with an ionic interaction mediated by sialic acid. In order to clarify the structural requirements of the fucose residue for selectin recognition, a variety of sialyl-Le^X analogs containing 2-deoxy (**10**)-, 3-deoxy (**11**)-, 4-deoxy (**12**)-, 2-O-methyl (**13**)-, 2-epi (**14**)-, 4-epi (**15**)-, and 2,4-di-epi (**16**)-L-fucose (Fig.5) have systematically been synthesized [33-35].

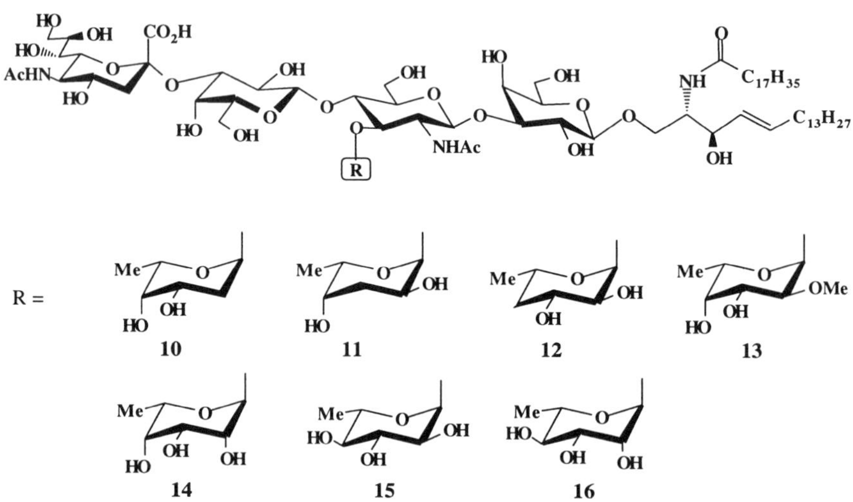

Figure 5

In the ELISA assay, the deoxyfucose-containing sialyl-Le^x ganglioside analogs (**10-12**) did not bind to E- and L-selectin at all, demonstrating [27] that for E- and L-selectin, hydroxyls on positions 2, 3 and 4 of the fucose residue are all required for recognition. P-selectin, however, required only 3-hydroxyl substitution. Removal of the 2- or 4-hydroxyl groups did not markedly affect the recognition of P-selectin. On the other hand, the inhibition assay against the immobilized sialyl-Le^X ganglioside (**2**) has shown [Kondo H et al., unpublished results] that the 2-epi (**14**) and 4-epi (**15**) analogs were potent binding inhibitors with the following IC_{50} values: **14** (3.2 μM for L-selectin, 47 μM for P-selectin, and 0.62 mM for E-selectin), and **15** (2.3 μM for L-selectin, 13 μM for P-selectin, and >1.0 mM for E-selectin). The other analogs also expressed inhibitory activities with IC_{50} values of 20~40 μM for L-selectin, and 30~90 μM for P-selectin, respectively, but had little or no activities for E-selectin.

3. Galactose modification

Chemical modification of the galactose residue in the pentasaccharidic sialyl-Le^x ganglioside (**2**) have been achieved [36] by employing sialyl-α-(2→3)-deoxygalactose donors (**17a-c**) which were coupled with suitably protected trisaccharide acceptors (**18**) (Fig.6). The resulting pentasaccharides were converted, by an established procedure [15], to the desired ganglioside analogs containing 4-deoxy(**19**), 6-deoxy(**20**), or 4,6-dideoxy(**21**) galactose.

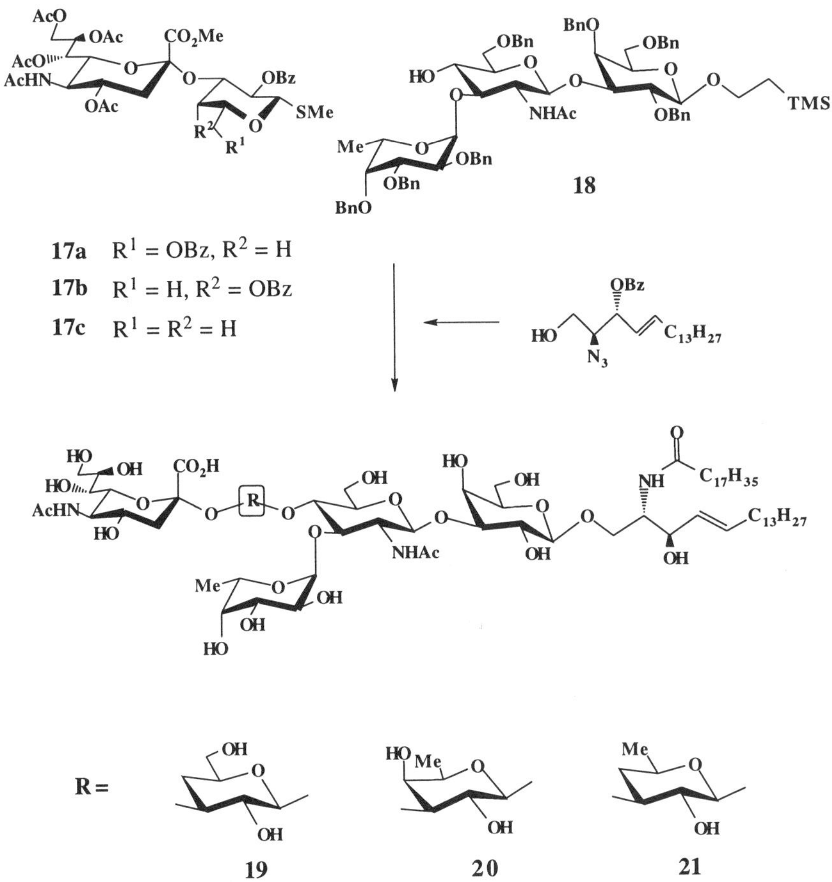

Figure 6

The ELISA assay for selectin binding [27] revealed that the 4-deoxy analog (**19**) was not recognized by E- and L-selectin, respectively. For P-selectin, however, **19** was recognized in almost the same order as the parent sialyl-Le^X ganglioside. On the other hand, the 6-deoxy analog (**20**) was not recognized at all by all three selectins, suggesting that the 6-hydroxyl of galactose might be more crucial for the selectin binding. In the inhibition assay, however, both **19** and **20** showed relatively strong activities for L- and P-selectin (IC_{50}, 3~4 μM for L-selectin and 20~30 μM for P-selectin), but no activity for E-selectin. Very recently, similar results have also been reported for the sialyl-Le^X tetrasaccharide epitope analogs [37].

4. Sulphate as a sialic acid substitute

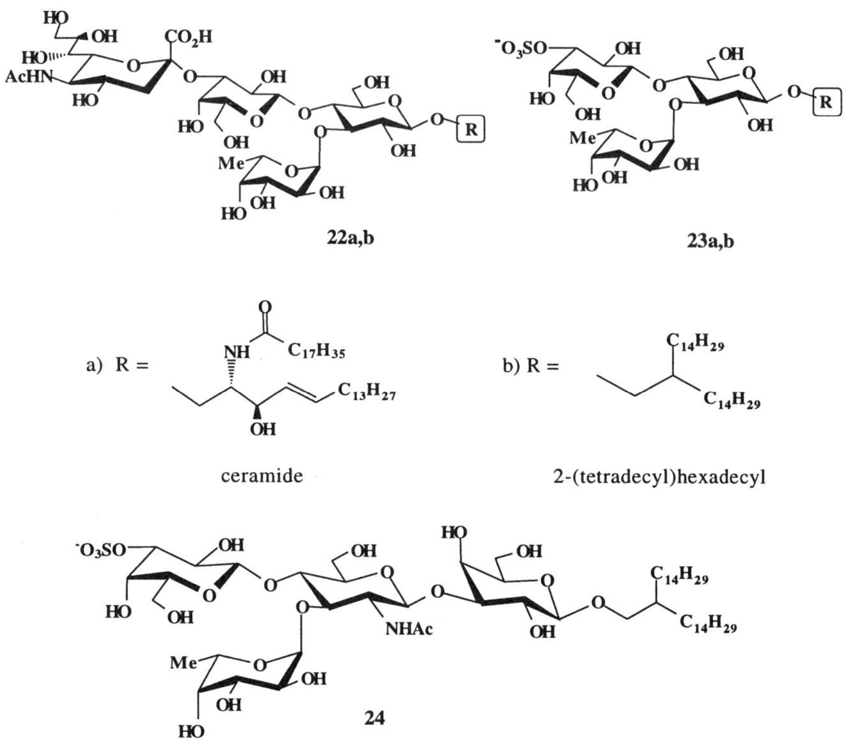

Figure 7

It has been reported that E- and L-selectin bind to the sulphated Le^X/Le^a structures [38-40], while L- and P-selectin bind to sulphatides and HNK-1 (sulphoglucuronylneolacto lipid, SGNL) epitope [41]. We have synthesized [r42] several, novel sialyl-Le^X (Glc)-(**22a,b**) and sulpho-Le^X (Glc)-lipids (**23a,b**) (Fig.7) for testing the binding ability to selectins. Even though the synthetic sulpho-Le^X (Glc)-lipids (**23a,b**) were significantly shorter (three sugar residues) than the sialyl-Le^X (hexa, **1**), it was bound with equal (E-selectin) or greater (L- and P-selectin) efficiency than **1** [27]. Since E-selectin seems to prefer a longer sugar core as previously described, these data may suggest that the sulpho-Le^X epitope is more avidly recognized by E-selectin as well. It is of interest that the recognition of **23a,b** for E-selectin was calcium dependent, while for L- and P-selectin, it was less dependent on calcium, maintaining ≥ 50% of control binding with addition of up to 5 μM EDTA. This result suggests two modes of interaction: one is Ca^{2+} dependent and one is Ca^{2+} independent. In this respect, **23a,b** are recognized by L- and P-selectin in a manner similar to sulphatides [43] as described in section B. These sialyl- and sulpho-Le^X-lipids (**22~24**) have also been found to be potent binding inhibitors for three selectins with the following IC_{50} values: 0.01~0.3 mM for E-selectin, 0.002~0.25 mM for P-selectin, and 0.004~40 μM for L-selectin, respectively.

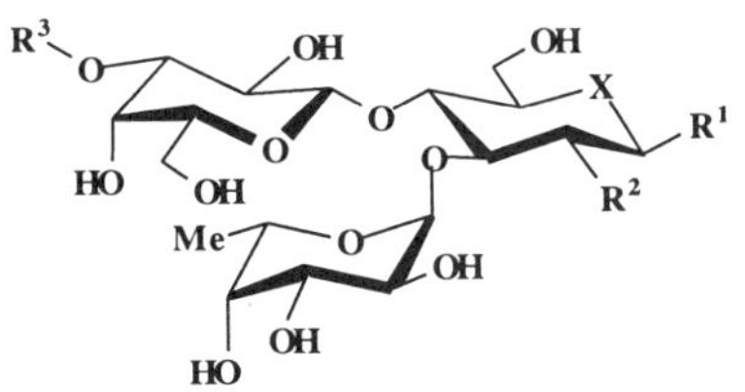

	X	R^1	R^2	R^3	
25	O	OH	NHAc	NeuAc	(sialyl Le^x)
26	O	H	NHAc	NeuAc	
27	O	H	OH	NeuAc	
28	O	H	H	NeuAc	
29	NR	H	OH	NeuAc	
30	O	H	NHAc	SO_3^-	
31	O	H	OH	SO_3^-	
32	O	H	H	SO_3^-	
33	NR	H	OH	SO_3^-	
34	O	H	H	PO_3^-	

Figure 8

5. Soluble inhibitors

A variety of sialyl-, sulpho-, and phosphono-Le^X analogs (Fig.8) containing 1-deoxy-N-acetylglucosamine (**26** and **30**) [44], 1-deoxy- and 1,2-dideoxy-glucose (**27**,**28**,**31**,**32** and **34**) [45], and several 1-deoxynojirimycin derivatives (**29** and **33**) [46] have been synthesized for evaluation of their utility as soluble inhibitors [47]. Sialyl-Le^X tetrasaccharide (**25**) was found to inhibit the binding of selectin-IgG chimeras to immobilized sialyl-Le^X (penta, **2**) with IC_{50} values of 0.6 mM for E-selectin, and > 1.0 mM for P- and L-selectin, respectively. The sialyl-Le^X derivatives (**26-29**) inhibited E-selectin binding with an IC_{50} of

0.1~0.8 mM, while the sulpho-Le^X derivatives (**30-32**) were all less potent (IC_{50}, > 1.0 mM). In contrast, sulphated compounds **31** and **32** were more potent (IC_{50}, 0.5-0.7 mM) for P-selectin binding than the sialylated compounds **27** and **28** (IC_{50}, > 1.0 mM). The inhibition potency, however, is much lower than that of glycolipid-type compounds described in the preceding section, such as **22** and **23**, indicating the crucial importance of hydrophobic interaction in the selectin blocking. It is a little surprising, however, that compound **26** showed strong inhibitory activities for the P- and L-selectin binding with IC_{50} values of 0.044 and 0.035 mM, respectively.

B. Sulphatide and Its Analogs

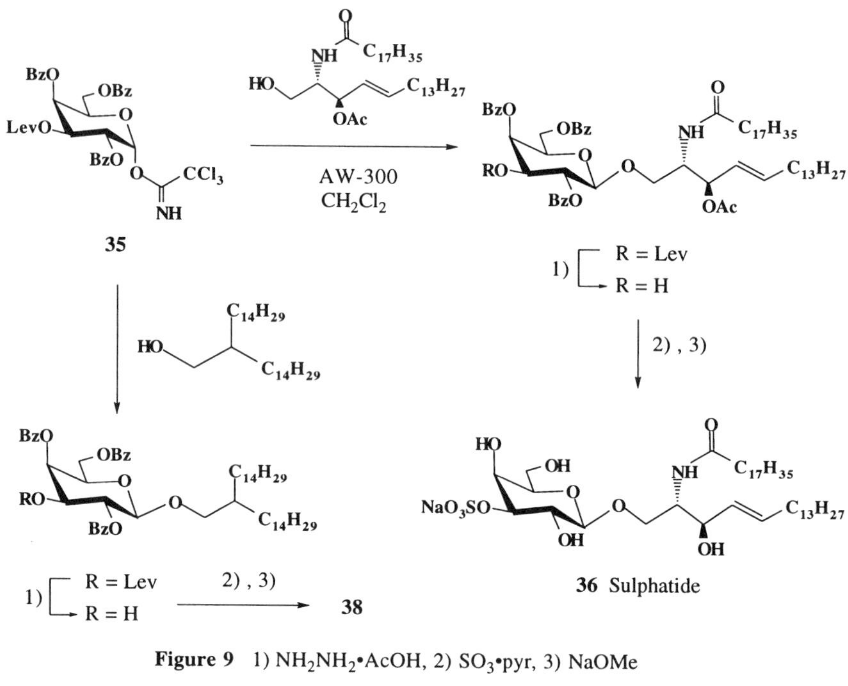

Figure 9 1) NH_2NH_2•AcOH, 2) SO_3•pyr, 3) NaOMe

Sulphated carbohydrates, including sulphatides, appear to have binding activity for selectins. The specificity of binding, in general, seems to favor L- and P-selectins. As described in section II.A.4., however, the synthetic sulpho-Le^X(Glc)-lipids (**23a,b**) were strongly bound with equal (E-selectin) or greater (L- and P-selectin) efficiency than the hexasaccharidic sialyl-Le^X ganglioside (**1**), considered to be one of the biologically relevant ligands for selectins.

Sulphatide (**36**) and its analogs (**37-44**) (Fig.9, Fig.10) have been synthesized stepwise, by the coupling of ceramide [48-49] or 2-

(tetradecyl)hexadecanol with levulinyl (Lev)-protected imidates such as **35** [44] followed by sulphation using SO_3-pyridine reagent as shown in Fig.9.

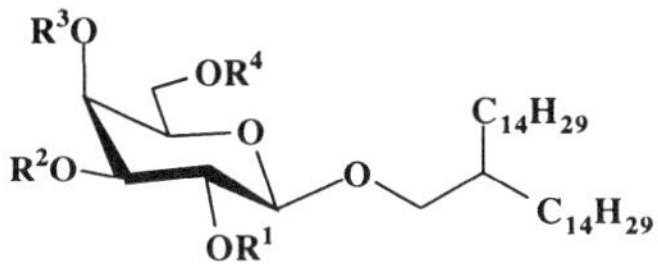

	R^1	R^2	R^3	R^4
37	SO_3Na	H	H	H
38	H	SO_3Na	H	H
39	H	H	SO_3Na	H
40	H	H	H	SO_3Na
41	H	SO_3Na	H	SO_3Na
42	H	SO_3Na	SO_3Na	SO_3Na

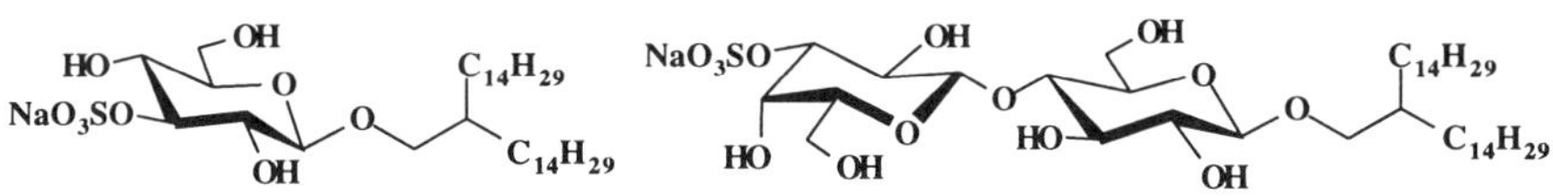

43 **44**

Figure 10

In the ELISA assay using L-selectin-IgG chimera, sulphatide showed the most active binding among the various sulphated or non-sulphated glycolipids tested [50]. All of the active glycolipids carried the non-reducing terminal galactose residue sulphated at the 3 position, suggesting that the terminal HO_3S-3Galβ1-structure is essential for the binding of L-selectin. Synthetic sulphatide analogs (**38**,**41**,**42**, and **44**) having the HO_3S-3Galβ1- structure bound more preferentially to the chimera than the others. It was also found that the reactivity of sulphatide was higher than that of sialyl-Le^X either in the presence of 1 mM Ca^{2+} or 1 mM Ca^{2+} and 5 mM EGTA, indicating that the L-selectin-sulphatide recognition is much less Ca^{2+} dependent than that of sialyl-Le^X. Similar results were obtained for the binding of L- and P-selectin chimeras to the sulphated Le^X analog **23a**,**b** as described in section II.A.4. [27].

It has also been reported [22,51] that sulphatides could block the binding not only of L-selectin to carbohydrate ligands but also of P-selectin to myeloid cells.

Therefore, sulphatides and related compounds may be useful for therapeutic applications.

C. Therapeutic Application/Glycomimetics

A variety of sialyl- and sulpho-Le^X analogs have been found to be potent inhibitors of selectin-carbohydrate binding, which mediates the early stage of adhesion of leukocytes to activated endothelial cells. Therefore, much interest has been focused on possible therapeutic applications in inflammation, reperfusion injury, tumor metastasis, and autoimmune diseases.

It has been shown [11] that the glucose analog of sialyl-Le^X had stronger inhibitory activity than sialyl-Le^X itself. Also the pharmacophore region of the sialyl-Le^X structure for E- and L-selectin was identified as the carboxyl group of sialic acid, the 2-, 3- and 4-hydroxyls of fucose, and the 4- and 6-hydroxyls of galactose by systematic structure-function studies [27].

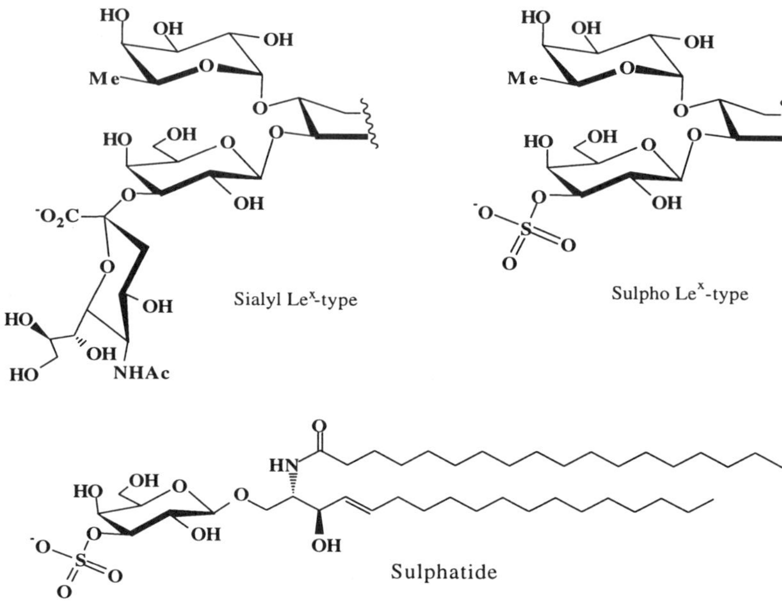

Figure 11

P-selectin, however, required only the 3-hydroxyl of fucose and the 6-hydroxyl of galactose, together with the carboxyl group of sialic acid. On the other hand, the sulpho-Le^X type compounds (Fig.11), such as **23a**,**b**, have been found to efficiently bind to E-, L- and P-selectin. In addition, it has also been found that sulphatides and the analogs [**37**-**44**] bind potently to L- and P-selectin, but not to E-selectin.

Based on these findings, a number of sialyl-Le^X derivatives and mimetics have been designed [28,52,53] to develop a new class of anti-inflammatory agents. Bivalent sialyl-Le^X derivatives [54] had higher inhibitory activity against E-selectin than did the sialyl-Le^X monomer, suggesting a possible multivalent ligand-receptor interaction. More interest has been focused on novel sialyl-Le^X

mimetics with simpler structures, higher activity, and better stability [53]. However, only a few compounds which had the six functional groups (-CO_2^- of NeuAc; 2-, 3-, and 4-OH of Fuc; 4- and 6-OH of Gal) and the suitable conformation showed inhibitory activities (IC_{50}, 1~2 mM) comparable with that of sialyl-Le^X. Computer aided, pharmacophore-based drug design is also attractive [55].

Current studies of the *in vivo* blocking of selectin-dependent inflammatory reactions have shown that sialyl-Le^X derivatives are highly protective against acute lung injury and reperfusion injuries in animal models [56]. Similarly, a sulpho-Le^X analog (**23b**) has been found to have potent anti-inflammatory activity against a rat dermatitis (skin inflammation) [Matsuo S et al., unpublished result]. It was also demonstrated that sulphatide (**36**) and its synthetic analogs (**38**,**41**,**42**) were highly protective against lung injury [56] and kidney inflammation in rats [57].

Although there are many mysteries and questions regarding selectin ligands and their binding modes, a variety of approaches for development of therapeutic agents based on carbohydrates as binding blockers are now in progress.

III. FUTURE DIRECTIONS

In addition to the selectins, a new family of cellular interaction molecules that recognize sialic acid-containing carbohydrates, has recently been identified as members of an immunoglobulin superfamily [7-8]. This family, termed "sialoadhesin family" or "I-type lectins," consists of the B-cell antigen CD22 [58], the macrophage receptor sialoadhesin [59], the myelin-associated glycoprotein (MAG) [7], and the marker of myelomonocytic cells CD33 [60]. These molecules have been shown to recognize the sialyl-α-(2→6)- and sialyl-α-(2→3)-linked oligosaccharides, but the details of their structures are still obscure.Therefore, systematic investigation for their specific carbohydrate ligands (counter-receptors) may open new perspectives not only on the biological studies of cell adhesion molecules but also on development of novel therapeutic agents.

We have recently succeeded in the first total synthesis of α-series ganglioside GQ1bα (Fig.12) [61], which is relatively minor and one of the most complex structures among the known gangliosides. This ganglioside GQ1bα has been found to be one of the highest affinity ligands of MAG [62], suggesting the crucial importance of the hybrid-type polysialoganglioside structure.On the other hand, it has also been demonstrated [63] that gangliosides GM_3 and GM_4 have strong immunosuppressive activity, by using a series of synthetic gangliosides and their analogs. Among those, a GM_3 ganglioside analog (Fig.13) was found to be a potent inhibitor of cellular immune responses [64]. This reaction is also sialic acid-dependent, including KDN. Further structure-function studies may contribute to a comprehensive understanding of carbohydrate-protein interactions.

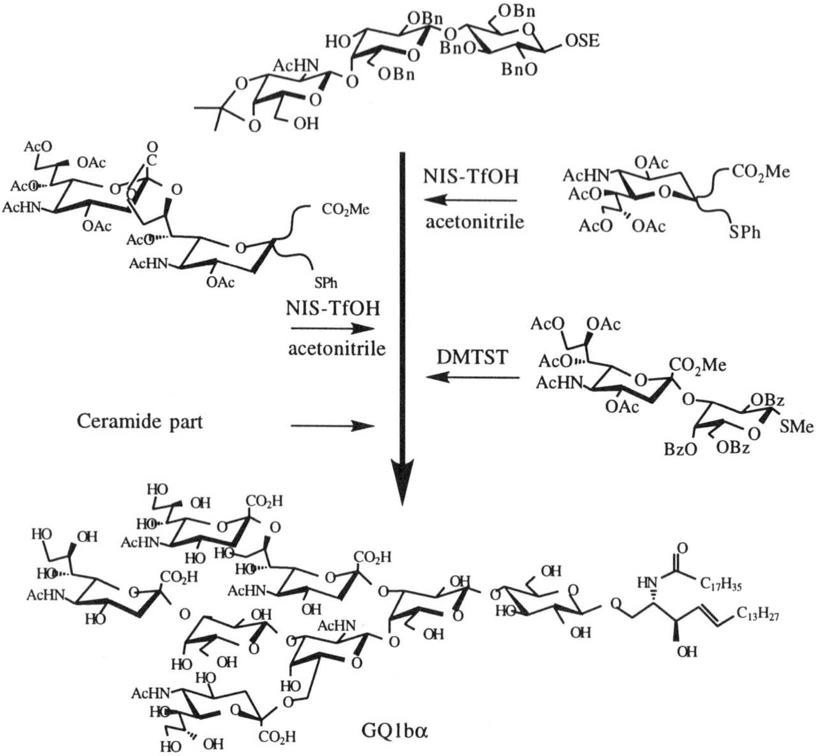

Figure 12

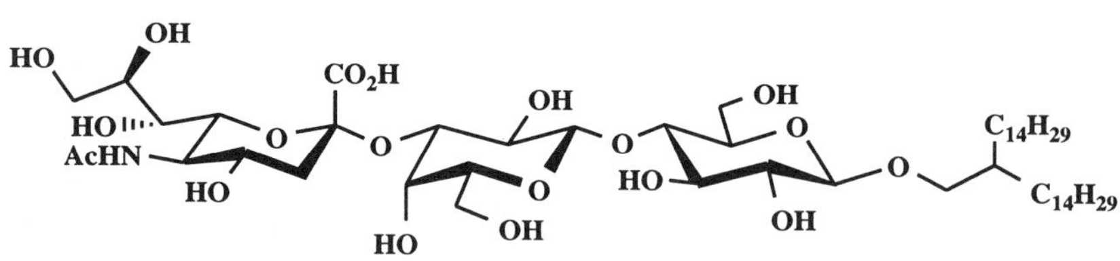

Figure 13

Acknowledgments

This work was supported in part by grants-in-aid for Scientific Research on Priority Areas and Developmental Scientific Research from the Japanese Ministry of Education, Science and Culture.

REFERENCES

1. Hakomori S. Glycosphingolipids. Sci Am 1986; 254: 32-41.
2. Sharon N, Lis H. Carbohydrate in cell recognition. Sci Am 1993; 268: 74-81.
3. Varki A. Biological roles of oligosaccharides: all of the theories are correct. Glycobiology 1993; 2: 97-130.
4. Lasky LA. Selectins: interpreters of cell-specific carbohydrate information during inflammation. Science 1992; 258: 964-969.
5. Cummings RD, Smith DF. The selectin family of carbohydrate-binding proteins: structure and importance of carbohydrate ligands for cell adhesion. BioEssays 1992;14: 849-856.
6. Bevilacqua MP, Nelson RM. Selectins. J Clin Invest 1993; 91: 379-387.
7. Kelm S, Pelz A, Schauer R, Filbin MT, Tang S, de Bellard ME, Schnaar RL, Mahoney JA, Hartnell A, Bradfield P, Crocker PR. Sialoadhesin, myelin-associated glycoprotein and CD22 define a new family of sialic acid-dependent adhesion molecules of the immunogloblin superfamily. Curr Biol 1994; 4: 965-972.
8. Powell LD, Varki A. I-type lectins. J Biol Chem 1995; 270: 14243-14246.
9. Varki A. Selectin ligands. Proc Natl Acad Sci USA 1994; 91: 7390-7397.
10. Brandley BK, Swiedler SJ, Robbins PW. Carbohydrate ligands of the LEC cell adhesion molecules. Cell 1990; 63: 861-863.
11. Tyrrell D, James P, Rao N, Foxall C, Abbas S, Dasgupta F, Nashed M, Hasegawa A, M. Kiso, Asa D, Kidd J, Brandley BK. Structural requirements for the carbohydrate ligand of E-selectin. Proc Natl Acad Sci USA 1991; 88: 10372-10376.
12. Kameyama A, Ishida H, Kiso M, Hasegawa A. Total synthesis of tumor-associated ganglioside, sialyl Lewis x. J Carbohydr Chem 1991; 10: 549-560.
13. Hasegawa A, Ando T, Kameyama A, Kiso M. Stereocontrolled synthesis of sialyl Lewis x epitope and its ceramide derivative. J Carbohydr Chem 1992; 11: 645-658.
14. Kameyama A, Ishida H, Kiso M, Hasegawa A. Synthesis of sialyl-α(2-6)-Lewis x. J Carbohydr Chem 1991; 10: 729-738.
15. a) Hasegawa A, Kiso M. Synthesis of sialyl Lewis x ganglioside and analogs. Methods in Enzymology 1994; 242: 158-173, and references cited therein. b) Hasegawa A, Kiso M. Systematic synthesis of gangliosides toward the elucidation and biomedical application of their biological functions. In: Ogura H, Hasegawa A, Tuami T, eds. Carbohydrates: Synthetic Methods and Applications in Medicinal chemistry. Tokyo: Kodansha, 1992: 243-266. c) Hasegawa A. Synthesis of sialo-oligosaccharides and their ceramide derivatives as tools for elucidation of biologic functions of gangliosides. In: Kovac P, ed. Synthetic Oligosaccharides. Indispensable Probes for the Life Sciences.

ACS Symposium Series 560. Washington DC: American Chemical Society, 1994: 184-197.

16. Polley MJ, Phillips ML, Wayner E, Nudelman E, Singhal AK, Hakomori S, Paulson JC. CD62 and endothelial cell-leukocyte adhesion molecule 1 (ELAM-1) recognize the same carbohydrate ligand, sialyl-Lewis x, Proc Natl Acad Sci USA 1991; 88: 6224-6228.
17. Foxall C, Watson SR, Dowbenko D, Fennie C, Lasky LA, Kiso M, Hasegawa A, Asa D, Brandley BK. The three members of the selectin receptor family recognize a common carbohydrate epitope, the sialyl Lewis x oligosaccharide. J Cell Biol 1992; 117: 895-902.
18. Hotta K, Itoh K, Kameyama A, Ishida H, Kiso M, Hasegawa A. Stereocontrolled synthesis of positional isomers of tumor-associated ganglioside antigens, sialyl Lewis x and sialyl paragloboside. J Carbohydr Chem 1995; 14: 115-133.
19. Fügedi P, Garegg PJ. A novel promoter for the efficient construction of 1,2-trans linkages in glycoside synthesis, using thioglycosides as glycosyl donors. Carbohydr Res 1986; 149: c9-c12.
20. Kanie O, Kiso M, Hasegawa A. Glycosylation using methylthio-glycosides of N-acetylneuraminic acid and dimethyl (methylthio) sulfonium triflate. J Carbohydr Chem 1988; 7: 501-506.
21. Konradsson P, Udodong UE, Fraser-Reid B. Iodonium promoted reactions of disarmed thioglycosides. Tetrahedron Lett 1990; 31: 4313-4316.
22. Veeneman GH, van Leeuwen SH, van Boom JH. Iodonium ion promoter reactions at the anomeric centre. II. An efficient thioglycoside mediated approach toward the formation of 1,2-trans linked glycosides and glycosidic esters. Tetrahedron Lett 1990; 31: 1331-1334.
23. Hasegawa A, Nagahama T, Ohki H, Hotta K, Ishida H, Kiso M. Reactivity of glycosyl promoters in α-glycosylation of N-acetyl-neuraminic acid with the primary and secondary hydroxyl groups in the suitably protected galactose and lactose derivatives. J Carbohydr Chem 1991; 10: 493-498.
24. Yoshida M, Uchimura A, Kiso M, Hasegawa A. Synthesis of chemically modified sialic acid-containing sialyl-Le^x ganglioside analogues recognized by the selectin family. Glycoconjugate J 1993; 10: 3-15.
25. Hasegawa A, Uchimura A, Ishida H, Kiso M. Synthesis of a sialyl Lewis x ganglioside analogue containing N-glycolyl in place of the N-acetyl group in the N-acetylneuraminic acid residue. Biosci Biotech Biochem 1995; 59: 1091-1094.
26. Terada T, Kiso M, Hasegawa A. Synthesis of KDN-lactotetraosylceramide, KDN-neolactotetraosylceramide, and KDN-Lewis x ganglioside. Carbohydr Res 1994; 259: 201-218.
27. Brandley BK, Kiso M, Abbas S, Nikrad P, Srivastava O, Foxall C, Oda Y, Hasegawa A. Structure-function studies on selectin carbohydrate

ligands. Modification to fucose, sialic acid and sulphate as a sialic acid replacement. Glycobiology 1993; 3: 633-639.
28. Nelson RM, Dolich S, Aruffo A, Cecconi O, Bevilacqua MP. Higher-affinity oligosaccharide ligands for E-selectin. J Clin Invest 1993; 91: 1157-1166.
29. Graves BJ, Crowther RL, Chandran C, Rumberger JM, Li S, Huang K, Presky DH, Familletti PC, Wolitzky BA, Burns DK. Insight into E-selectin/ligand interaction from the crystal structure and mutagenesis of the lec/EGF domains. Nature 1994; 367: 532-538.
30. Bajorath J, Hollenbaugh D, King G, Harte W, Eustice DC, Darvea RP, Aruffo A. CD62/P-selectin binding sites for myeloid cells and sulfatides are overlapping. Biochemistry 1994; 33: 1332-1339.
31. Kogan TP, Revelle BM, Tapp S, Scott D, Beck PJ. A single amino acid residue can determine the ligand specificity of E-selectin. J Biol chem 1995; 270: 14047-14055.
32. Tiemeyer M, Swiedler SJ, Ishihara M, Moreland M, Schweingruber H, Hirtzer P, Bandley BK. Carbohydrate ligands for endothelial leukocyte adhesion molecule 1. Proc Natl Acad Sci USA 1991; 88: 1138-1142.
33. Hasegawa A, Ando T, Kato M, Ishida H, Kiso M. Synthesis of deoxy-L-fucose-containing sialy Lewis x ganglioside analogues. Carbohydr Res 1994; 257: 67-80.
34. Hasegawa A, Kato M, Ando T, Ishida H, Kiso M. Synthesis of the methyl thioglycosides of four stereoisomers and the 2-methoxy derivative of L-fucose. Carbohydr Res 1995; 274: 155-163.
35. Hasegawa A, Kato M, Ando T, Ishida H, Kiso M. Synthesis of sialyl Lewis x ganglioside analogues containing modified L-fucose residues. Carbohydr Res 1995; 274: 165-181.
36. Komba S, Ishida H, Kiso M, Hasegawa A. Synthesis of deoxygalactose-containing sialyl-Le^x ganglioside analogs to elucidate the structure necessitated for selectin recognition. Glycoconjugate J 1996; 13: 1-14.
37. Stahl W, Sprengard U, Kretzschmar G, Kunz H. Synthesis of deoxy sialyl Lewis x analogues, potential selectin antagonists. Angew Chem Int Ed Engl 1994; 33: 2096-2098.
38. Green PJ, Tamatani T, Watanabe T, Miyasaka M, Hasegawa A, Kiso M, Yuen C-T, Stoll MS, Feizi T. High affinity binding of the leukocyte adhesion molecule L-selectin to 3'-sulphated-Le^a and -Le^x oligosaccharides and the predominance of sulphate in this interaction demonstrated by binding studies with a series of lipid-linked oligosaccharides. Biochem Biophys Res commun 1992; 188: 244-251.
39. Yuen C-T, Lawson AM, Chai W, Larkin M, Stoll MS, Stuart AC, Sullivan FX, Ahern TJ, Feizi T. Novel sulfated ligands for the cell adhesion molecule E-selectin revealed by the neoglycolipid technology among O-linked oligosaccharides on an ovarian cystadenoma glycoprotein. Biochemistry 1992; 31: 9126-9131.
40. Yuen C-T, Bezouska K, O'Brien J, Stoll M, Lemonie R, Lubineau A, Kiso M, Hasegawa A, J.Bockovich N, C.Nicolaou K, Feizi T. Sulfated

blood group Lewis^{a}, a superior oligosaccharide ligand for human E-selectin. J Biol Chem 1994; 269: 1595-1598.

41. Needham LK, Schnaar RL. The HNK-1 reactive sulfoglucronyl glycolipids are ligands for L-selectin and P-selectin but not E-selectin. Proc Natl Acad Sci USA 1993; 90: 1359-1363.
42. Hasegawa A, Ito K, Ishida H, Kiso M. Synthesis of sialyl and sulfo Lewis x analogs containing a ceramide or 2-(tetradecyl)hexadecyl residue. J Carbohydr Chem 1995; 14: 353-368.
43. Asa D, Gant T, Oda Y, Brandley BK. Evidence for two classes of carbohydrate binding sites on selectins. Glycobiology 1992; 2: 395-400.
44. Maeda H, Ishida H, Kiso M, Hasegawa A. Synthesis of sulfo- and sialyl-Lewis x epitope analogs containing the 1-deoxy-N-acetylglucosamine in place of N-acetylglucosamine residue. J Carbohydr Chem 1995; 14: 369-385.
45. Maeda H, Ito K, Ishida H, Kiso M, Hasegawa A. Synthesis of sulfo-, pohosphono- and sialyl-Lewis x analogs containing the 1-deoxy- and 1,2-dideoxyhexopyranoses in place of N-acetylglucosamine. J Carbohydr Chem 1995; 14: 387-406.
46. Kiso M, Furui H, Ando K, Ishida H, Hasegawa A. Systematic synthesis of N-methyl-1-deoxynojirimycin-containing, Le^{x}, Le^{a}, sialyl-Le^{x} and sialyl-Le^{a} epitopes recognized by selectins. Bioorg Med Chem 1994; 2: 1295-1308.
47. Ohmoto H, Nakamura K, Inoue T, Kondo N, Inoue Y, Yoshino K, Kondo H, Ishida H, Kiso M, Hasegawa A. Studies on selectin blocker 1. Structure-activity relationships of sialyl Lewis x analogs. J Med Chem. In press.
48. Kiso M, Nakamura A, Tomita Y, Hasegawa A. A novel route to D-erythro-sphingosine and related compounds from mono-O-isopropylidene-D-xylose or -D-galactose. Carbohydr Res 1986; 158: 101-111.
49. Kiso M, Nakamura A, Hasegawa A. Synthesis of 1-O-(N-acetyl-α-and-β-D-neuraminyl)-ceramides. J Carbohydr Chem 1987; 6: 411-422.
50. Suzuki Y, Toda Y, Tamatani T, Watanabe T, Suzuki T, Nakao T, Murase K, Kiso M, Hasegawa A, Tadano-Aritomi K, Ishizuka I, Miyasaka M. Sulfated glycolipids are ligands for a lymphocyte homing receptor, L-selectin (LECAM-1), binding epitope in sulfated sugar chain. Biochem Biophys Res Commun 1993; 190: 426-434.
51. Aruffo A, Kolanus W, Walz G, Fredman P, Seed B. CD62/P-selectin recognition of myeloid and tumor cell sulfatides. Cell 1991; 67: 35-44.
52. Dasgupta F, Rao BNN. Anti-adhesive therapeutics: a new class of anti-inflammatory agents. Exp Opin Invest Drugs 1994; 3: 709-724.
53. Uchiyama T, Vassilev VP, Kajimoto T, Wong W, Huang H, Lin CC, Wong CH. Design and synthesis of sialyl Lewis x mimetics. J Am Chem Soc 1995; 117: 5395-5396, and references cited therein.

54. DeFrees SA, Kosch W, Way W, Paulson JC, Sabesan S, Halcomb RL, Huang DH, Ichikawa Y, Wong CH. Ligand recognition by E-selectin: synthesis, inhibitory activity, and conformational analysis of bivalent sialyl Lewis x analogs. J Am Chem Soc 1995; 117: 66-79.
55. Rao BNN, Anderson MB, Musser JH, Gilbert JH, Schaefer ME, Foxall C, Brandley BK. Sialyl Lewis x mimics derived from a pharmacophore search are selectin inhibitors with anti-inflammatory activity. J Biol Chem 1994; 269: 19663-19666.
56. Mulligan MS, Miyasaka M, Suzuki Y, Kawashima H, Iizuka M, Hasegawa A, Kiso M, Warner RL, Ward PA. Anti-inflammatory effects of sulfatides in selectin-dependent acute lung injury. Int J Immunopharmac 1995; 7: 1107-1113, and references cited therein.
57. Suzuki Y. Sulfated glycolipids as potential ligands for L- and P-selectins and their role in inflammation in kidney and lung. Glycoconjugate J 1995; 12: 539.
58. Powell LD, Jain RK, Matta KL, Sabesan S, Varki A. Characterization of sialyloligosaccharide binding by recombinant soluble and native cell-associated CD22. J Biol Chem 1995; 270: 7523-7532, and references cited therein.
59. Crocker PR, Freeman S, Gordon S, Kelm S. Sialoadhesin binds preferentially to cells of the granulocytic lineage. J Clin Invest 1995; 95: 635-643.
60. Freeman SD, Kelm S, Barker EK, Crocker PR. Characterization of CD33 as a new member of the sialoadhesin family of cellular interaction molecules. Blood 1995; 85: 2005-2012.
61. Hotta K, Ishida H, Kiso M, Hasegawa A. First total synthesis of a cholinerglc neuron-specific ganglioside GQ1bα. J Carbohydr Chem 1995; 14: 491-506.
62. Yang LJS, Zeller CB, Shaper NL, Kiso M, Hasegawa A, Shapiro RE, Schnaar RL. Gangliosides are neuronal ligands for myelin-associated glycoprotein. Proc Natl Acad Sci USA. In press.
63. Ladisch S, Hasegawa A, Li R, Kiso M. Immunosuppressive activity of chemically synthesized gangliosides. Biochemistry 1995; 34: 1197-1202.
64. Ladisch S, Hasegawa A, Li R, Kiso M. A chemically synthesized sialic acid-containing glycoconjugate, 2-(tetradecylhexadecyl)-O-(5-acetamido-3,5-dideoxy-D-glycero-α-D-galacto-2-nonulopyranosylonic acid)-(2→3)-O-β-D-galactopyranosyl-(1→4)-β-D-glucopyranoside, is a potent inhibitor of cellular immune responses. Biochem Biophys Res commun 1994; 203: 1102-1109.

5
Polysulfates: Chemistry and Potential as Antiviral Drugs

Myriam Witvrouw, Christophe Pannecouque, and Eric De Clercq
Rega Institute for Medical Research, Catholic University of Leuven
Leuven, Belgium

I. SUMMARY

In the early 1960s it was demonstrated that heparin and related polyanionic substances act as herpes simplex virus (HSV) inhibitors [1-4]. Shortly after the identification of human immunodeficiency virus (HIV) as the causative agent of the Acquired Immune Deficiency Syndrome (AIDS) in 1984, heparin and other polysulfates were found to be potent and selective inhibitors of HIV-1 replication in cell culture [5-9]. Since 1988, the activity spectrum of the polysulfates has been shown to extend to various enveloped viruses (10-25), including viruses that emerge as opportunistic pathogens [e.g., herpes simplex virus (HSV), cytomegalovirus (CMV)] in immunosuppressed (e.g. AIDS) patients.

As potential anti-HIV drug candidates, polysulfates offer a number of promising features. They are able to block HIV replication in cell culture at concentrations as low as 0.01 $\mu g\ ml^{-1}$ without toxicity to the host cells at concentrations up to 2.5 mg ml^{-1}. We noted that some polysulfates show a differential inhibitory activity against different HIV strains, suggesting that marked differences exist in the target molecules with which polysulfates interact. They not only inhibit the cytopathic effect of HIV, but also prevent HIV-induced syncytium (giant cell) formation. Furthermore, experiments carried out with dextran sulfate samples of increasing molecular weight and with sulfated cyclodextrins of different degrees of sulfation have shown that antiviral activity increases with increasing molecular weight and degree of sulfation. A sugar backbone is not strictly needed for the anti-HIV activity of polysulfates since sulfated polymers composed of a carbon-carbon backbone have also proved to be highly efficient anti-HIV agents *in vitro*. Other, yet to be defined, structural features may also play an important role. As demonstrated with the sulfated cyclodextrins, polysulfates may act synergistically with other anti-HIV drugs [e.g., azidothymidine (AZT)]. They are not known to lead to virus-drug resistance development and they show activity against HIV mutants that have become resistant to reverse transcriptase inhibitors, such as AZT, tetrahydro-imidazo [4,5,1-jk] [1,4]-benzodiazepin-2(1*H*)-thione (TIBO) and others. From studies on their mechanism of action we concluded that polysulfates exert their anti-HIV activity by shielding off the positively charged sites in the V3 loop of the viral envelope glycoprotein (gp120). The V3 loop is necessary for virus attachment to cell surface heparan sulfate, a primary binding site, before more specific binding occurs to the CD4 receptor of $CD4^+$ cells. This general mechanism also explains the broad antiviral activity of polysulfates against enveloped viruses. Variations in the viral envelope glycoprotein region may result in differences in the susceptibility of different enveloped viruses to compounds that interact with their envelope glycoproteins.

Polysulfates suffer from a number of disadvantages. They are very poorly absorbed following oral administration (po) as demonstrated in rabbits, although oral bioavailability may be enhanced by appropriate chemical modifications (e.g., substitution with benzyl groups). Such a modified ß-cyclodextrin sulfate showed oral uptake in rabbits with a plasma half-life of 4 hours. When polysulfates were administered intravenously (iv) to rabbits, half-lives of 20 minutes to 4 hours were noted. We established a simple and sensitive bioassay system to monitor plasma concentrations of polysulfates based on the measurement of anti-HIV acitivity. When dextran sulfate (DS) and pentosan polysulfate (PS) were administered by continuous iv infusion in symptomatic HIV-infected patients, no anti-HIV activity was observed, and the levels of p24 and CD4 did not change [26]. Moreover, continuous iv infusion of DS and PS proved to be toxic, counteracting the coagulation process and causing thrombocytopenia, although the anticoagulant activity could be drastically reduced by appropriate chemical modifications as proven with heparin. The efficacy of polysulfates in the therapy and/or prophylaxis of retroviral infections remains to be demonstrated both in animal models and in humans.

It is important to consider not only treatment of patients who are already infected with HIV, but also prophylaxis and protection from HIV and/or other virus infections. Since (i) sexual transmission is responsible for the large majority of HIV infections worldwide; (ii) this transmission is mostly mediated via mononuclear cells that infect epithelial cells of the genital tract; and since

(iii) polysulfates effectively inhibit cell-cell adhesion, polysulfates may be considered as potentially effective in a vaginal formulation to protect from HIV infection.

II. CHEMISTRY

In the search for inhibitors of HIV replication, a great number of polyanionic biopolymers have been shown to be active *in vitro*. Heparin was one of the first sulfated polysaccharides tested and found to be active against HIV *in vitro*. The isolation and synthesis of other polyanionic and especially polysulfated polymers were initially stimulated by the interest in better anticoagulants. A specific sequence occurring in heparin is responsible for interaction with antithrombin III (AT III) in the coagulation cascade [27-28]. This anticoagulant activity makes polysulfates less attractive as antiviral drugs. Much effort has been spent on the development of synthetic and semisynthetic antiviral polysulfates which lack anticoagulant activity. The most important requirement for antiviral activity is the presence of a sufficient number of anionic groups. The simplest compound that fulfills this requirement is a carbon-carbon chain bearing hydrogenosulfate substituents such as polyvinylalcoholsulfate (PVAS), a sulfated polyvinyl alcohol polymer.

A. Dextran Sulfate

In the early 1940s dextran sulfuric acid esters were synthesized in order to investigate their inhibitory action on coagulation and their toxicity [29].

Dextran is a high-molecular polysaccharide produced from sucrose by several bacteria, but only *Leuconostoc mesenteroides* and *Leuconostoc dextranicum* are used for commercial purposes. Its backbone consists predominantly of α-(1-6) linked D- glucose units. Dextran can differ in chain length and degree of branching. The branching occurs *via* α-(1-3) and α-(1-4) branch points. Dextran fractions of different average molecular weight are obtained by partial hydrolysis.

These different fractions were sulfated in order to obtain dextran sulfates of different molecular weight. Therefore dextran of a certain molecular weight range is added in portions to a well stirred solution of an excess of chlorosulfonic acid in anhydrous pyridine. The reaction occurs under effective stirring at 65°C for 5 hours. After neutralisation the sodium salt of the ester is precipitated with ethanol. After dialysis it is precipitated one more time with ethanol.

The obtained sodium salt of dextran sulfate (DS, Fig. 1) contains two to three sulfate groups per glucose unit.

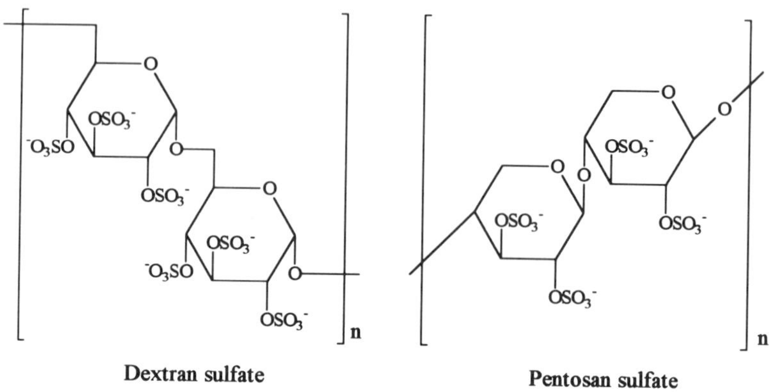

Figure 1 Structural formulae of dextran sulfate (DS) and pentosan polysulfate (PS). Structures are shown as if all hydroxyl groups are substituted by sulfates.

B. Pentosan Sulfate

In the search for synthetic polysulfated analogues of heparin, xylan sulfuric esters were synthesized. Due to the toxicity synthetic procedures were adapted in order to obtain xylan sulfuric esters of lower molecular weight [30].

Xylans are hemicelluloses which consist of chains of ß-(1-4) linked D-xylopyranosyl units.

The semisynthetic sulfated polysaccharide is obtained by adding xylan to a vigorously stirred solution of an excess of chlorosulfonic acid in anhydrous pyridine at 65°C. Then the reaction mixture is kept at 80°C for 2.5 hours. The pyridinium salt of the ester is precipitated from methanol. The product is bleached and further partly hydrolysed with hydrogen peroxide in acidic conditions at 97°C. After neutralisation the pentosan sulfate is bleached and dialyzed. The sodium salt is obtained by crystalization from a mixture of water, ethanol and acetone (2:4:4 v:v:v) by addition of sodium acetate.

Approximately 100% of the hydroxyl groups of the pentosan sulfate (PS, Fig. 1) used in our antiviral experiments are sulfated.

C. Sucralfate

Sucralfate (Fig. 2), ß-D-fructofuranosyl-α-D-glucopyranoside octakis (hydrogenosulfate) aluminum complex, was first described in 1967 by Nametaka *et al.* [31]. It is used to promote ulcer healing by binding to damaged mucosa and stimulating other mucosal protecting mechanisms. Sucrose sulfate aluminum complex is prepared by addition of an aqueous solution of $Al(OH)_2Cl$ to a dilute aqueous solution of sucrose sulfate sodium salt [31].

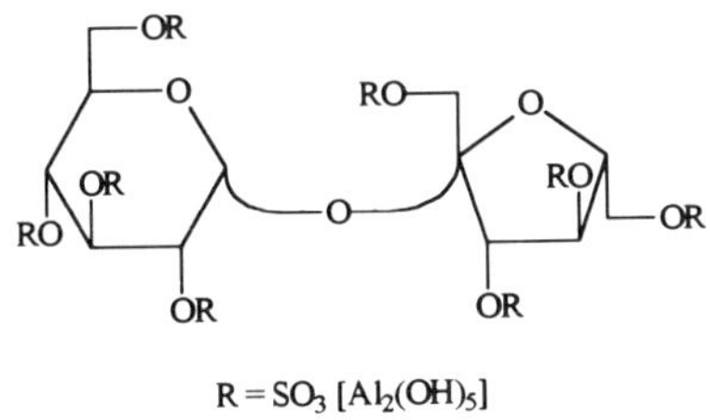

Sucralfate

Figure 2 Structural formula of sucralfate.

D. Heparin

Heparin (Fig. 3) is the most important polysulfated polysaccharide in medical therapy. It was discovered eighty years ago by McLean and Howell [32,33]. Crafoord was the first to use heparin in the treatment of coagulation diseases, 50 years ago.

As are most of the polysulfated polysaccharides, heparin is isolated from natural sources. Commercial heparin is isolated from beef lung and pork intestinal mucosa.

The name heparin comes from its first isolation from liver ("hepar"). Its biosynthesis is initiated in the rough endoplasmatic reticulum of mast cells by the synthesis of a core protein which is rich in serine-glycine repeats [34,35]. The linkage region is composed of neutral sugars attached to serine residues from which the glycosaminoglucan chains extend.

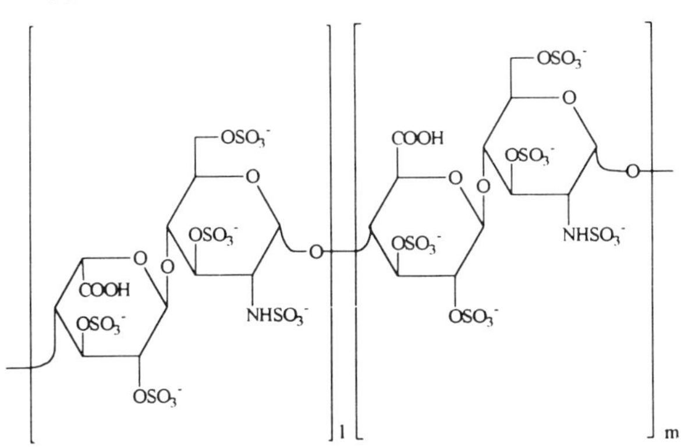

Heparin

Figure 3 Structural formula of heparin. Structure is shown as if all hydroxyl groups are substituted by sulfates.

Repeating units of 1in molecules possess a molecular weight ranging from 5,000 to 40,000 Da and are composed of alternating 1,4-linked uronic (iduronic or glucuronic) acid and glucosamine residues (Table 1). The repeating disaccharide units of heparin are dissimilar since not all of the glucuronic acid residues are epimerized to iduronic acid, the amine residues of glucosamine can remain partially acylated, and *O*- and *N*-sulfation does not necessarily occur at all positions.

E. Periodate-Oxidized Heparin Derivatives

Many oxidative and peroxidative methods have been described for the depolymerisation of heparin. Periodate oxidation occurs at unsulfated glucuronic and iduronic acid residues, containing vicinal diol groups [36]. As only a small number of unsulfated uronic acids are present in heparin, the cleavage sites are restrained. The oxidation step is followed by an elimination by base treatement and reduction using sodium borohydride, as described by Lormeau *et al.* [37].

Due to the restrained cleavage sites, the overall structure of the periodate-treated molecule is very similar to that of heparin itself (Figs. 3 and 4). The chains consist of 6 to 30 monosaccharides and are essentially built up from disaccharide units of *N*-sulfated glucosamine and 2-*O*-sulfated uronic acids.

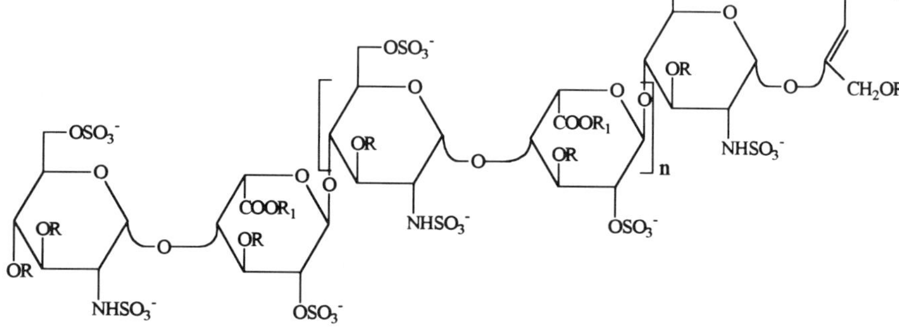

	R	R_1
Periodate oxidized heparin	H	H
O-butyrylated periodate oxidized heparin	H / $CH_3(CH_2)_2CO$	H
O-butyrylated benzyl ester of periodate oxidized heparin	H / $CH_3(CH_2)_2CO$	H / $C_6H_5CH_2$

Figure 4 Structural formulae of heparin derivatives.

Table 1 Average MW, Percentage of Sulfation, Sugar Content and Configuration of the Sugar Oligomers (i.e. Nature and Type of Interglycosidic Linkage) of Polysaccharides

Compound	Average MW (x 10^{-3})	% of sulfation	Sugar content	Configuration
1. Linear				
1.1. Chemical synthesized				
DS	from 1 to 500	81[b]	glucose	_1,6
PS	3	100[b]	xylose	ß1,4
Sucralfate	2.1[a]	100[b]	sucrose	_1,4
1.2. From natural origin				
Heparin	11	ND	uronic acid [iduronic (60%) or glucuronic (40%)]-glucosamine	ß1,4
Dermatan sulfate	ND	6[c]	iduronic acid *N*-acetyl-galactosamine	ß1,3
Xylomannan sulfate	ND	25[c]	xylose-mannose	_1,3-linked mannose (98%) having single stubs of ß1,2-xylose
Galactan sulfate	ND	28[c]	galactose	ß1,3
2. Circular				
α-cyclodextrin hexasulfate	1.7[a]	40[b]	glucose	_1,6
ß-cyclodextrin heptasulfate	2.1[a]	43[b]	glucose	_1,6
γ-cyclodextrin octasulfate	2.3[a]	40[b]	glucose	_1,6
α-cyclodextrin dodecasulfate	2.3[a]	70[b]	glucose	_1,6
ß-cyclodextrin tetradecasulfate	2.5[a]	63[b]	glucose	_1,6
γ-cyclodextrin hexadecasulfate	3.1[a]	73[b]	glucose	_1,6

Formulae of (the oligomeric units of) sulfated polysaccharides are depicted in Figs. 1-3, 5 and 6.

[a]Exact MW. [b]% of sulfation of hydroxyl groups; [c]Sulfate content on dry weight.

F. O-Acylated Heparin Derivatives

To increase the ratio of anti-HIV activity to anticoagulant activity, the hydroxyl groups of heparin have been acylated in various degrees. The sodium salt of heparin is commercially available and, as most of the reactions are performed in organic solvents, it is necessary for solubility reasons to convert this to a less hydrophilic salt such as the tributylammonium salt. This may easily be accomplished by passing the heparin sodium salt over an acidic ion exchanger resin followed by neutralization with aqueous tributylamine. The acylation is performed using acid anhydrides in dimethylformamide (DMF) with 4-dimethylaminopyridine (DMAP) as a catalyst. The degree of acylation can be controlled by the acid anhydride/heparin ratio, temperature and reaction time. Using acid anhydride, DMF and DMAP, no desulfation was observed. A possible side reaction is the formation of mixed anhydrides between the carboxylic acid of heparin and the acyl group of the acid anhydride. It is sufficient to treat these undesired derivatives with an aqueous bicarbonate solution to hydrolyze the mixed anhydride [38].

When an acyl halide is used to acylate the hydroxyl functions, acylation of the nitrogen is observed. For heparin this results in an *N,O*-acylated product (Fig. 4) that is formed by *N*-desulfation followed by *N*-acylation. As the *N*-acyl groups (carboxamides) are more stable than the *O*-acyl groups (esters) towards alkaline hydrolysis, this method can be used to synthesize selectively *N*-acylated heparin analogues [38].

G. Esterification of the Carboxylic Acid

Heparin ammonium salt is esterified by reaction with an alkyl halide in DMF (Fig. 4). Some of the described derivatives have been obtained by combination of acylation and esterification; therefore, heparin is first reacted with an alkyl halogenide, *in casu* benzyl bromide, followed by reaction with an acid anhydride, as described above [38].

H. Dermatan Sulfate

Dermatan sulfate (Fig. 5) was isolated in 1941 by Meyer and Chaffee [39]. As this product was first isolated from pig skin ("dermos") the name dermatan sulfate was introduced. Dermatan sulfate has also been isolated from beef lung after removal of heparin. Later it was detected in many other organs such as heart, brain, and spleen. As it also displayed anticoagulant activity it was called ß-heparin [40].

Dermatan sulfate is built up from alternating 1,3-linked iduronic acid and galactosamine (Table 1), the repeating sequence in dermatan sulfate being (1-3)-

O-(2-acetamido-2-deoxy-β-D-galactopyranosyl-4-sulfate)-(1-4)-O-α-L-idopyranuronosyl [41]. The chain is usually 20 to 100 monosaccharides long.

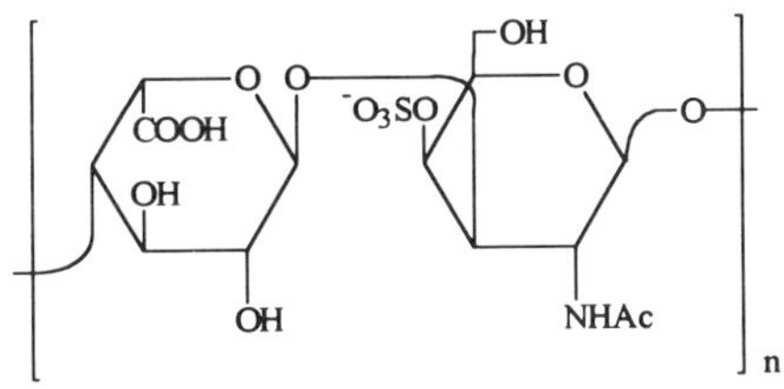

Figure 5 Structural formula of dermatan sulfate.

I. Xylomannan

This natural sulfated polysaccharide was extracted from the red seaweed *Nothogenia fastigiata*, a seaweed growing in Argentina. The dry seaweed is extracted with boiling water using the method described by Matulewicz and Cerezo [42]. The water-soluble polysaccharides were fractionated with cetrimide. The fraction that shows antiviral activity consists of 2.1 % ß-D-xylose and 97.9 % α-D-mannose. The structure of xylomannan consists of an α-D-(1,3)-linked mannan chain, which is sulfated at the 2 and 6 positions, and has single stubs of ß-(1,2)-linked D-xylose (Table 1) [42].

J. Galactan Sulfate

This polysaccharide is extracted from the red seaweed *Agardhiella tenera* growing in Venezuela. The plant is washed in running water. After filtration the polysaccharide is precipitated by addition of isopropanol and converted to its potassium salt [24,43]. This galactan contains various ratios of D- and L-galactose, 3,6-anhydro-D- and L-galactose and half-ester sulfate [44-45]. The sulfate content of our sample was 28% of the dry weight (Table 1).

K. Cyclodextrin Sulfates

Cyclodextrins (Fig. 6) are natural occurring cyclic α-(1-4) linked D-glucopyranose rings consisting of 6, 7, 8 units and are called α–, β– or γ–cyclodextrins, respectively. The α–, β– and γ–cyclodextrins are sulfated with sulfur trioxide trimethylamine complex in dimethylformamide [15, 46-47]. The rate of sulfation can be controlled by changing the reaction conditions. When 8.1 equivalent sulfur trioxide trimethylamine complex per glucose unit is used at 70°C for 24 hours, the sulfated α-, ß- or γ-cyclodextrins thus obtained contain approximately one sulfate group per glucose unit. Moreover, the sulfate groups are introduced at the most accessible hydroxyl functions, i.e. at the 6-position of the glucopyranose.

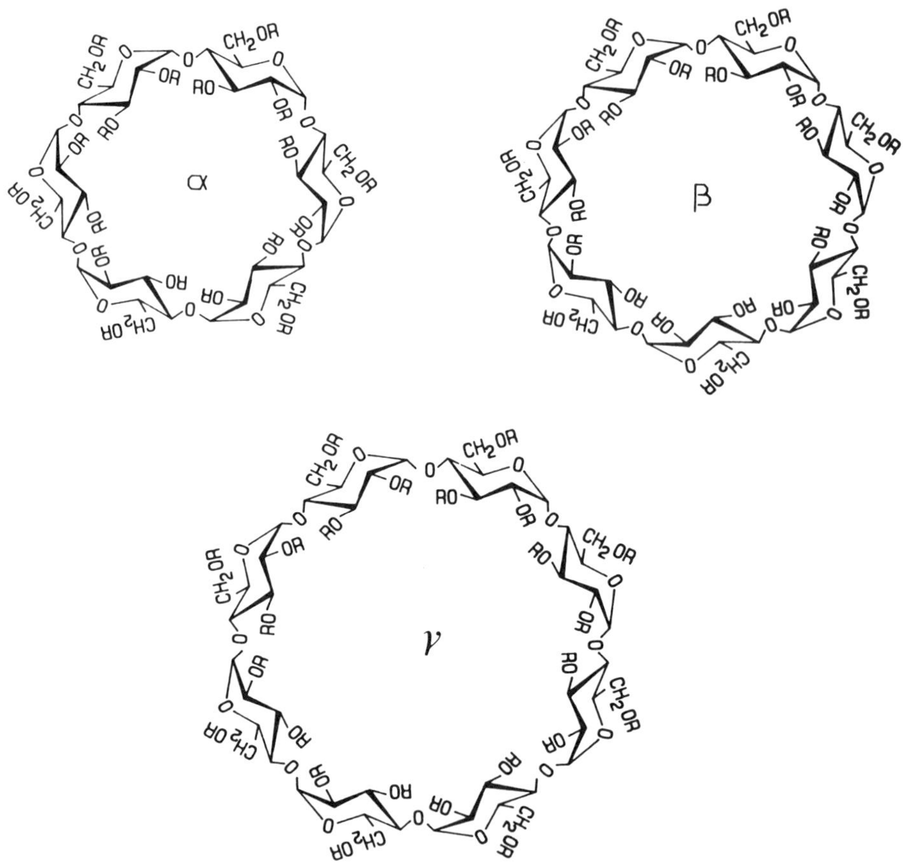

Figure 6 Structural formulae of α-, ß- and γ-cyclodextrin.

When the reaction is performed in more diluted conditions at higher temperature (110°C) with the same ratio of sulfur trioxide trimethylamine complex per cyclodextrin, the initial hexa-, hepta- or octasulfate products remain in solution and are converted to the dodeca-, tetradeca- or hexadecasulfates, respectively (Table 1). After the reaction the sulfated cyclodextrins can be precipitated as their sodium salts.

L. Modified Cyclodextrin Sulfates

In order to obtain polysulfated compounds with better pharmacological properties, a series of modifications were introduced in the cyclodextrin structures (Fig. 7). The modified cyclodextrin sulfates (mCDS) discussed here are representative of a much greater number. Important features present in most mCDS members are lipophilic substituents. It is known that polyanionic macromolecules such as dextran sulfate poorly penetrate lipophilic membranes. The small size and lipophilic character of the mCDS makes them bioavailable *via* administration per os, which is not the case for molecules such as dextran sulfate, pentosan polysulfate, and others.

The first step in the synthesis of tris(6-benzylthio-6-deoxy)-ß-cyclodextrin hexadecasulfate (mCDS 11) is partial 6-*O*-mesitylenesulfonylation of ß-cyclodextrin as described by Tsujihara *et al.* [48] In the next step the the sodium salt of benzylmercaptan is used to substitute the mesitylenesulfonyloxy groups with the formation of tris(6-benzylthio-6-deoxy)-ß-cyclodextrin. Finally, mCDS 11 is obtained by sulfation using sulfur trioxide-pyridine complex [49].

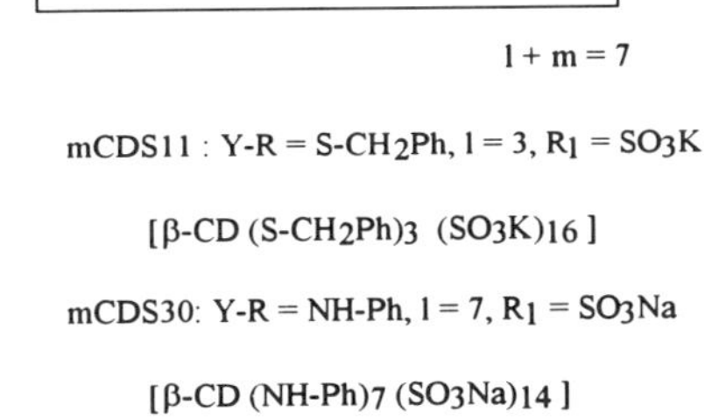

mCDS11 : Y-R = S-CH_2Ph, l = 3, R_1 = SO_3K

[β-CD (S-CH_2Ph)3 (SO_3K)16]

mCDS30: Y-R = NH-Ph, l = 7, R_1 = SO_3Na

[β-CD (NH-Ph)7 (SO_3Na)14]

Figure 7 Structural formulae for ß-cyclodextrin derivatives

For the synthesis of heptakis(6-phenylamino-6-deoxy)-ß-cyclodextrin tetradecasulfate (mCDS 30) a similar strategy is used. The seven glucopyranose units are selectively 6-*O*-mesitylenesulfonylated [48]. The phenylamino substitution reaction is performed by refluxing heptakis(6-*O*-mesitylenesulfonyl)-ß-cyclodextrin in aniline. Sulfur trioxide-pyridine complex is used for the final sulfation step [49].

M. Polyvinylalcohol Sulfate and its Copolymer with Acrylic Acid

Polyvinyl alcohol sulfate (PVAS) (Fig. 8) and the copolymer of acrylic acid with vinyl alcohol sulfate (PAVAS) (Fig. 8) were first developed for their anticoagulant properties by Machovich *et al.* [50] as synthetic analogues of heparin.

Polyvinyl alcohol (PVA) is a synthetic polymer obtained from the polymerization of vinyl acetate followed by hydrolysis of the acetate esters. It is a stable polymer which is soluble in water due to the presence of the hydroxyl groups. The copolymer of acrylic acid with vinyl alcohol (PAVA) is obtained from radical copolymerization of vinyl acetate and acrylic acid followed by subsequent basic hydrolysis of the acetate esters.

The PVA and PAVA sulfates (termed PVAS and PAVAS, respectively) are synthesized by slow addition of concentrated sulfuric acid to a 10% mass percent solution of the polymers at low temperatures (-5°C to 5°C).

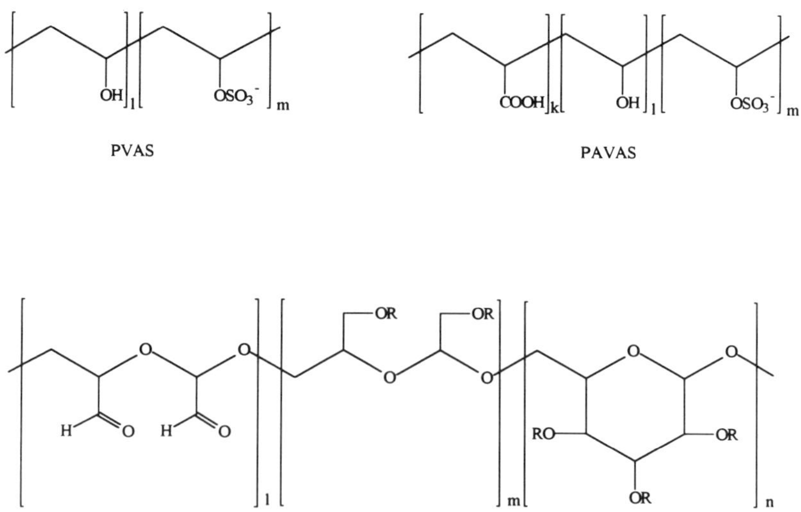

R = H or SO_3^- ; 62.2% of the hydroxyl groups are sulfated

PAPS

Figure 8 Structural formulae of polyvinylalcohol sulfate (PVAS), its co-polymer with acrylic acid (PAVAS) and polyacetal polysulfate (PAPS).

The degree of sulfation is controlled by variation of the final concentration of sulfuric acid in the mixture (60-80% mass percent). Table 2 presents the results of the analysis of the composition of tested PVAS and PAVAS congeners. The average molecular weight and the sulfation rate of the hydroxyl groups are displayed. For the PAVAS the ratio of carboxylic acid residues is included [14].

N. Polyacetal Polysulfates

Polyacetal carboxylic acids were investigated for their antiviral properties in 1970 by Claes *et al.* [51]. Part of that published synthetic procedure was used for the synthesis of the polysulfated analogues. Dextran was oxidized using periodate resulting in cleavage of the anhydropyranose units, leading to the polyaldehyde oxydextran. Analysis revealed that 86.4 % of the anhydro pyranosyl units were converted to the dialdehyde. Reaction with sodium borohydride led to reduction of the aldehydes with formation of primary hydroxyl groups. Analysis of the reduction product revealed that 8.2 % of unreacted dialdehyde remained.

Table 2 Average MW, Composition, and Percentage of Sulfation of the Non-Carbohydrate Polymers

Compound	Average MW ($\times 10^{-3}$)	% of the acrylic acid residues $100 \times (\frac{k}{k+l+m})^a$	% of sulfation[b]
1. Carbon-carbon backbone			
PVAS # 1	20	-	50
PVAS # 2	40	-	51
PVAS # 3	40	-	84
PAVAS # 1	5	10	31
PAVAS # 3	5	10	67
PAVAS # 6	20	10	31
PAVAS # 8	20	10	82
2. Polyacetal backbone			
PAPS	32	-	62

[a]The symbols k, l and m are depicted in Fig. 8.
[b]Calculated with the following formula for the PVAS and PAVAS congeners: $100 \times (\frac{m}{k+l+m})^a$

The primary hydroxyl groups were easily sulfated using pyridine-sulfur trioxide complex [22]. The obtained molecule [polyacetal polysulfate (PAPS)] (Fig. 8) has a flexible backbone consisting of a polyacetal chain in which 62.2 % of the hydroxyl groups are sulfated (Table 2). This backbone is essentially biodegradable, albeit slowly, with the result that the compound persists in tissue long enough to be useful as a drug.

III. INHIBITORY EFFECT ON VIRUS-INDUCED CYTOPATHIC EFFECT *IN VITRO* : STRUCTURE-ACTIVITY RELATIONSHIP

A. Activity Against HIV

A number of polysulfates have been examined for their inhibitory effect on HIV replication in cell cultures. DS, PS, and heparin are representatives of sulfated polysaccharides that consist of several repeating monosaccharide or disaccharide units substituted with sulfate groups (Figs. 1, 3 and Table 1).

Polysulfates (i.e. sulfated polysaccharides) show marked differences in their *in vitro* anti-HIV activity, depending on both the virus strain and cell type [21,52-54]. The sulfated polysaccharides DS and PS are significantly more inhibitory to HIV-2(ROD) than to HIV-1(III_B), HIV-1(RF), or HIV-2(EHO) (Table 3). Heparin and its derivatives are less active against the two HIV-2 strains, the differences in sensitivity being particularly striking for the periodate-oxidized heparin fragments (Table 3). In CEM cells, DS is equally active against all HIV strains tested [HIV-1(III_B), HIV-1(RF), HIV-2(ROD), HIV-2(EHO)] but heparin proved more active against the two HIV-2 strains (54). These findings reflect differences in the interaction of these polyanions with the different antigenic properties of the HIV envelope glycoproteins of different HIV strains (see section D. on Mechanism of anti-HIV action). Differences in anti-HIV activity between cell types may mean that these compounds are also able to interfere with cellular factors (i.e. CD4 receptor) involved in the entry of HIV in the cells.

DS [molecular weight (MW) 3,000-500,000], PS and heparin are highly potent and selective inhibitors of HIV-1 replication in cell cultures (Table 4). These compounds protect MT-4 cells against HIV-1-induced cytopathicity at a concentration of 0.2-0.8 μg ml^{-1}. Neither DS, nor PS or heparin reduce the viability of mock-infected MT-4 cells at concentrations up to 2.5 mg ml^{-1} [54]. The selectivity index (SI) is approximately 10,000. To determine the importance of molecular size in the anti-HIV activity of sulfated polysaccharides, various samples of DS prepared from dextran fractions with MW ranging from 1,000 to 500,000, but with an identical sulfate content (81%), were examined for their inhibitory effect on HIV-1 induced cytopathicity (Table 4).

Table 3 Comparative Antiviral Activity of Different Polysulfates Against Various Strains of HIV-1 and HIV-2

Compound	EC_{50}[a] ($\mu g\ ml^{-1}$)			
	HIV-1		HIV-2	
	III_B	RF	ROD	EHO
DS (MW 5,000)	0.9	0.4	0.08	3.8
PS	0.8	0.9	0.01	2.7
Heparin	1.4	0.7	44	19
Periodate oxidized heparin	1.5	0.3	> 250	18
O-Butyrylated periodate oxidized heparin	0.1	0.1	0.7	1.3
O-Butyrylated benzyl ester of periodate oxidized heparin	0.2	0.2	0.9	1.3

[a]50% Effective concentration, or concentration required to inhibit by 50% the cytopathicity of HIV in MT-4 cells.
Sources: refs. 38, 54.

A marked increase in anti-HIV-1 activity was observed when the MW increased from 1,000 to 5,000 or 10,000. At higher MW (10,000 - 500,000), the antiviral activity of DS tended to level off. Against HIV-2-induced cytopathicity, the 1,000 MW DS sample proved more inhibitory than the higher MW samples.

Sucralfate (Fig. 2 and Table 1) showed a completely different anti-HIV activity profile than reported for DS, PS, and heparin. Little activity was noted against HIV-1 (EC_{50} : 218 $\mu g\ ml^{-1}$), but better activity was shown against HIV-2 (EC_{50} : 7.3 $\mu g\ ml^{-1}$).

To increase the ratio of the anti-HIV activity to the anticoagulant activity, heparin samples selectively substituted at the OH and/or COOH groups were prepared (Fig. 4). *O*-Acylation (e.g. butyrylated) of the heparin fragments obtained by periodate depolymerization and their esters (e.g. benzyl ester) (Fig. 4), yielded products with 2- to 10-fold higher activity against both HIV-1- and HIV-2-induced cytopathicity (EC_{50} : 0.3 - 0.4 $\mu g\ ml^{-1}$), with little, if any, cytotoxicity, and very low anticoagulant activity *in vitro* [38].

In contrast with heparin, DS and PS, dermatan sulfate (Fig. 5 and Table 1) did not demonstrate any anti-HIV activity (Table 4). Since dermatan sulfate contains less than one sulfate group per disaccharide unit, this result indicates that a certain number of sulfate groups (at least two per monosaccharide unit) are required for anti-HIV activity. If, however, dermatan sulfate is supersulfated, it becomes as active as DS and PS [55].

The anti-HIV activity of two sulfated polysaccharides (xylomannan- and galactan-rich) extracted from seaweeds (*Nothogenia fastigiata* and *Aghardhiella tenera*, respectively) has also been investigated [23-25]. The xylomannan sulfate was only weakly active against HIV-1 and HIV-2 (EC_{50} : 13.7 and 13.4 μg ml^{-1}, respectively) (Table 4) and it was rather toxic (CC_{50} : 33 μg ml^{-1}) (data not shown). When the galactan sulfate was examined for its inhibitory effect on the cytopathicity of HIV-1(III_B) and HIV-2(ROD), the compound provided protection at much lower concentrations (EC_{50} : 0.5 μg ml^{-1}) (Table 4). No reduced viability of mock-infected MT-4 cells at concentrations up to 2.5 mg ml^{-1} was noted (data not shown).

Besides the linear sulfated polysaccharides, we have evaluated a series of sulfated cyclic oligosaccharides (cyclodextrins) containing six (α-cyclodextrin), seven (ß-cyclodextrin), or eight (γ-cyclodextrin) glucopyranose units (Fig. 6 and Table 1) for their activity against HIV. The α-, ß-, and γ-cyclodextrins containing two sulfate groups per glucose unit were more active than the corresponding cyclodextrins containing only 1 sulfate group (Table 4).

Table 4 Inhibitory Effects of Different Groups of Polysulfates on HIV-Induced Cytopathicity

Compound	EC_{50}[a] (μg ml^{-1})	
	HIV-1(III_B)	HIV-2(ROD)
1. Polysaccharides		
1.1. Linear		
1.1.1. Chemically synthesized		
DS (MW 1,000)	7.1	0.1
(MW 1,500)	2.9	0.1
(MW 3,400)	0.7	0.3
(MW 5,000)	0.5	0.3
(MW 10,000)	0.2	1.2
(MW 40,000)	0.4	2.3
(MW 70,000)	0.6	3.9
(MW 110,000)	0.6	4.1
(MW 500,000)	0.6	4.1
PS	0.8	0.01
Sucralfate	218	7.3
1.1.2. From natural origin		
Heparin	0.8	4.5
Periodate oxidized heparin	0.5	77

Table 4 Continued

O-butyrylated periodate oxidized heparin	0.3	0.3
O-butyrylated benzyl ester of periodate oxidized heparin	0.4	0.4
Dermatan sulfate	> 625	> 625
Xylomannan sulfate	13.7	13.4
Galactan sulfate	0.6	0.5
1.2 Circular		
α-cyclodextrin hexasulfate	26.2	1.9
ß-cyclodextrin heptasulfate	3.0	1.8
γ-cyclodextrin octasulfate	0.8	1.6
α-cyclodextrin dodecasulfate	6.5	0.4
ß-cyclodextrin tetradecasulfate	0.8	1.0
γ-cyclodextrin hexadecasulfate	0.2	0.4
mCDS11	0.5	0.6
mCDS30	0.1	0.8
2. Non-carbohydrate polymers		
2.1. Carbon-carbon backbone		
PVAS # 1	0.3	0.3
PVAS # 2	0.3	0.4
PVAS # 3	0.4	0.4
PAVAS # 1	0.3	0.2
PAVAS # 3	0.3	0.4
PAVAS # 6	0.3	0.3
PAVAS # 8	0.3	0.5
2.2. Polyacetal backbone		
PAPS	0.4	0.04

[a]50% Effective concentration, or concentration required to inhibit by 50% the cytopathicity of HIV-1(III_B) or HIV-2(ROD) in MT-4 cells.
Sources: refs. 14, 15, 21, 22, 24, 25, 38, 49 and 54.

Also, the anti-HIV-1 activity of the sulfated cyclodextrins (CDS) depended on the number of glucopyranose units, γ-CDS (EC_{50} : 0.2 µg ml^{-1}) being more active than ß-CDS (EC_{50} : 0.8 µg ml^{-1}), which in turn was more active than α-CDS (EC_{50} : 6.5 µg ml^{-1}) (Table 4) against HIV-1-induced cytopathicity in MT-4 cells. With respect to HIV-2 activity, all three compounds were almost equally active, although the CDS with two sulfate groups per glucose unit proved somewhat more active (EC_{50} : 0.4-1.0 µg ml^{-1}) than those containing only 1 sulfate group per glucose unit (EC_{50} : 1.6-1.9 μg ml^{-1}). The sulfated ß- and γ-cyclodextrins were more active against HIV-1 than HIV-2 (Table 3). Modified CDS (mCDS), where lipophilic groups were introduced in ß-CDS were synthesized [tris (6-benzylthio-6-deoxy)-ß-cyclodextrin hexadecasulfate (mCDS11) and heptakis (6-phenylamino-6-deoxy)-ß-cyclodextrin tetradecasulfate (mCDS30), see Fig. 7] and proved more inhibitory to the replication of HIV-1 (EC_{50}: 0.5 and 0.1 μg ml^{-1}) and HIV-2 (EC_{50} : 0.6 and 0.8 μg ml^{-1}) than unmodified ß-CDS (EC_{50} for HIV-1 and HIV-2 : 0.8 and 1.0 μg ml^{-1}, respectively) (Table 4).

A sugar backbone is not strictly needed for the anti-HIV activity of polysulfates, since sulfated non-carbohydrate polymers (Fig. 8 and Table 2) based on a polyacetal backbone, such as polyacetal polysulfate (PAPS) [22], or, simply, a carbon-carbon backbone, such as polyvinylalcohol sulfate (PVAS) and its copolymer with acrylic acid (PAVAS) [14, 56], have also proved to be highly efficient anti-HIV agents *in vitro* (Table 4). PAPS, PAVAS, and PVAS, irrespective of their MW or degree of sulfation, have proved to be as efficient, or even more efficient inhibitors of HIV-1 replication (EC_{50} : 0.2-0.4 μg ml^{-1}) than the sulfated polysaccharides. PAVAS and PVAS, again irrespective of their MW or degree of sulfation, showed the same activity against HIV-2 (0.2-0.4 μg ml^{-1}), whereas PAPS was 10-fold more potent against HIV-2 (EC_{50} : 0.04 μg ml^{-1}) (Table 4).

Polysulfates are able to block HIV replication in cell culture at concentrations as low as 0.01 μg ml^{-1} without toxicity for the host cells, thus achieving *in vitro* selectivity indices up to 250,000. The fact that some polysulfates show a different inhibitory potency against HIV-1 and HIV-2, suggest that there exist marked differences in the target molecules of HIV-1 and HIV-2 with which the polysulfates interact (see section D. Mechanism of anti-HIV action).

B. Activity Against Enveloped Viruses Other Than HIV

Polysulfates were found to be active against a wide variety of enveloped viruses [10-25] such as herpes viruses [herpes simplex virus type 1 (HSV-1), thymidine kinase-deficient (TK^-) HSV-1, herpes simplex virus type 2 (HSV-2), cytomegalovirus (CMV)], togaviruses (Sindbis virus, Semliki Forest virus), arenaviruses (Junin virus, Tacaribe virus), rhabdoviruses [vesicular stomatitis virus (VSV)], orthomyxoviruses (influenza A virus) and paramyxoviruses [respiratory syncytial virus (RSV)]. The polysulfates also showed some inhibitory activity against poxviruses [vaccinia virus (VV)] but proved not to be inhibitory to non-enveloped viruses such as picorna-, reo- and adenovirus and some enveloped myxoviruses (influenza B virus and parainfluenza virus type 3).

When DS samples derived from dextran with a MW ranging from 1,000 to 500,000 were examined for their inhibitory effects on the replication of RNA viruses other than HIV and DNA viruses, the 1,000 and 1,500 MW DS samples were much less active than the higher MW samples against Sindbis virus, Semliki forest virus, VSV, influenza A virus, RSV, HSV-1, TK^- HSV-1, HSV-2 and VV, but their activity against Junin virus and Tacaribe virus was only slightly lower than that of the higher MW samples (Table 5 and Table 6). As a rule, the antiviral activity increased with increasing MW, with an optimum at approximately MW 40,000. However, against CMV (Table 6) the 1,000 and 1,500 MW DS samples were almost as active as the higher MW samples.

PS (MW 3,000) was a more potent inhibitor of Sindbis virus, HSV and CMV replication than DS (MW 3,400). Semliki forest virus, Junin virus and Tacaribe virus proved to be less sensitive to PS than DS, whereas VSV, influenza A virus and RSV were equally sensitive to PS and DS (MW 3,400).

Heparin (MW 11,000) proved to be less effective than DS (MW 10,000) against all viruses tested. The red seaweed-derived xylomannan sulfate was more active than DS (MW 5,000) against influenza A virus and HSV-1, but less active against Junin virus, Tacaribe virus, RSV, HSV-2 and CMV. When a naturally derived galactan sulfate was examined for its inhibitory effect on RNA viruses other than HIV, the compound was found to be particularly active against Sindbis virus. It inhibited Sindbis virus at a concentration that was at least 10-fold lower than the concentration at which DS (MW 5,000) inhibited this virus. Galactan sulfate was not active against Junin virus and Tacaribe virus, slightly active against VSV, but equally active as DS (MW 5,000) against Semliki forest virus, influenza A virus and RSV. The compound was more inhibitory than DS (MW 5,000) against HSV-1, TK^- HSV-1 and VV, although HSV-2 and CMV proved less sensitive.

When the CDSs were examined for their inhibitory effect on the replication of DNA viruses, the compounds with two sulfate groups per glucose unit were more active than CDS with one sulfate group per glucose unit, and this difference was particularly striking for the AD-169 strain of CMV (Table 6). Their EC_{50} for HSV was 10- to 100-fold higher than the EC_{50} of DS (MW 5,000), when evaluated under the same conditions. These compounds showed the highest activity against the AD-169 strain of CMV with a 2- to 3-fold higher EC_{50} than for DS (MW 5,000). None of the CDSs were active against VV and all RNA viruses, other than HIV.

A similar MW-activity relationship was found when the PVAS and PAVAS congeners were examined for their inhibitory effects on the replication of Sindbis virus, Semliki forest virus, VSV, HSV-1, TK- HSV-1, HSV-2 and VV (Table 5 and Table 6). As a second rule, the antiviral activity increased with increasing degree of sulfation. However, their activity against Junin virus, Tacaribe virus, RSV and CMV was not markedly influenced by their MW and degree of sulfation (Table 5 and Table 6). The sulfated polymers with a carbon-carbon backbone (PVAS and PAVAS) were found to be markedly active against Sindbis virus and CMV. VV was also inhibited by PVAS and PAVAS, and here the degree of sulfation played a significant role: only those compounds with a

Table 5 Inhibitory Effects of Different Groups of Polysulfates on the Replication of Enveloped RNA Viruses Other Than HIV

Compound	EC_{50}[a] (μg ml^{-1})						
	Sindbis virus	Semliki Forest virus	Junin virus	Tacaribe virus	VSV	Influenza A virus	RSV
1. Polysaccharides							
1.1. Linear							
1.1.1. Chemical synthesized							
DS (MW 1,000)	300	> 400	0.2	0.2	70	200	12
(MW 1,500)	> 400	> 400	-	-	70	200	1.4
(MW 3,400)	185	100	0.2	0.4	4.5	42	0.4
(MW 5,000)	150	120	0.3	0.4	2	5.8	0.2
(MW 10,000)	15	14	0.2	0.3	0.7	1.4	0.1
(MW 40,000)	2	14	0.04	0.2	0.2	1.0	0.1
(MW 70,000)	14	14	0.1	0.2	0.5	1.4	0.1
(MW 110,000)	2	20	-	-	0.7	-	-
(MW 500,000)	7	20	-	-	0.7	-	-
PS	20	> 400	1.6	2.8	4	20	0.8
1.1.2. From natural origin							
Heparin	200	> 400	1.8	2.0	> 50	> 200	-
Xylomannan sulfate	-	-	10	7.8	-	0.2	0.9
Galactan sulfate	11	85	> 100	> 100	14	0.4	0.4

2.	Non-carbohydrate polymers							
2.1.	Carbon-carbon backbone							
	PVAS # 1	2	> 400	0.4	1	0.7	1.0	-
	PVAS # 2	1.4	40	10	10	0.6	-	3.6
	PVAS # 3	2	20	3.2	3.2	0.2	-	1.1
	PAVAS # 1	70	> 400	> 50	> 50	20	5.8	27
	PAVAS # 3	10	> 400	3.7	3.7	0.7	1.0	3
	PAVAS # 6	2	> 400	5	3.2	2	9.8	12
	PAVAS # 8	0.7	20	3.2	2	0.7	0.7	3.6
2.2.	Polyacetal backbone							
	PAPS	20	> 400	4	4	2	0.8	0.8

[a]50% Effective concentration, or concentration required to inhibit by 50% the cytopathicity of Sindbis virus, Semliki forest virus, Junin virus or Tacaribe virus in Vero cells, VSV or RSV in HeLa cells and Influenza A virus in MDCK cells.
Sources: refs. 10, 11, 14, 18, 21, 22, 24, 25 and 54.

Table 6 Inhibitory Effects of Different Groups of Polysulfates on the Replication of Enveloped DNA Viruses

Compound	EC_{50}[a] (µg ml^{-1})					
	HSV-1 (KOS)	TK⁻ HSV-1 (B2006)	HSV-2 (G)	CMV (AD-169)	CMV (Davis)	VV
1. Polysaccharides						
1.1. Linear						
1.1.1. Chemical synthesized						
DS (MW 1,000)	110	100	95	2.5	3	275
(MW 1,500)	10	20	7	0.3	0.8	400
(MW 3,400)	4.5	2	2	0.7	0.8	35
(MW 5,000)	2	2	1.4	0.7	0.7	80
(MW 10,000)	2	1.4	1.4	0.6	0.5	10
(MW 40,000)	2	0.7	0.3	0.3	0.3	4.5
(MW 70,000)	2	0.9	0.6	0.6	0.6	20
(MW 110,000)	1	1.4	0.7	0.1	0.3	20
(MW 500,000)	2	2	0.7	0.1	0.3	40
PS	0.7	1.1	0.5	0.4	0.1	20
1.1.2. From natural origin						
Heparin	3.6	5	2	11	3	> 50
Xylomannan sulfate	1.6	-	2.5	10	2.8	-
Galactan sulfate	0.7	0.7	2.5	-	4.0	7

1.2.	Circular						
	α-cyclodextrin hexasulfate	300	150	200	50	15	> 400
	ß-cyclodextrin heptasulfate	150	20	150	40	7	> 400
	γ-cyclodextrin octasulfate	150	70	70	25	20	> 400
	α-cyclodextrin dodecasulfate	150	70	70	1.5	7	> 400
	ß-cyclodextrin tetradecasulfate	70	70	40	2	9	> 400
	γ-cyclodextrin hexadecasulfate	70	20	20	1.5	10	> 400
2.	Non-carbohydrate polymers						
2.1.	Carbon-carbon backbone						
	PVAS # 1	1.4	1.5	1.4	-	0.1	14
	PVAS # 2	0.7	0.6	0.6	-	0.3	12
	PVAS # 3	0.9	0.5	0.5	0.08	0.2	9
	PAVAS # 1	7	7	7	-	0.03	> 400
	PAVAS # 3	0.7	2	1	0.04	0.03	7
	PAVAS # 6	4.5	2	11	-	0.08	> 400
	PAVAS # 8	0.7	1.3	1.3	0.08	0.07	20
2.2.	Polyacetal backbone						
	PAPS	2	2	7	0.08	0.1	10

[a]50% Effective concentration, or concentration required to inhibit by 50% the cytopathicity of HSV-1, TK^- HSV-1, HSV-2 or VV in E_6SM cells and CMV in HEL cells.
Sources: refs. 14, 21, 22, 24, 25 and 54.

sulfation percentage equal to or higher than 50% showed activity against VV (Table 6). Against Semliki forest virus only those compounds with a sulfation percentage equal to or higher than 50% and a MW of 40,000 for PVAS congeners and 10,000 for PAVAS congeners showed activity (Table 5). When PAPS was examined for its inhibitory effect on the replication of RNA viruses other than HIV, the compound was found to be more active against influenza A virus and Sindbis virus at an at least 7-fold lower EC_{50} than the EC_{50} of DS (MW 5,000), but it was less active against Junin virus, Tacaribe virus and RSV. VSV proved equally active to PAPS and DS (MW 5,000). When PAPS was examined for its inhibitory effect on the replication of DNA viruses, it proved particularly active against CMV. It inhibited the replication of this virus at a concentration of 0.1 $\mu g\ ml^{-1}$. VV proved at least 10-fold more sensitive to PAPS than DS (MW 5,000). HSV-1 and TK^- HSV-1 showed comparable sensitivity and HSV-2 was 5-fold less sensitive to PAPS than DS (MW 5,000).

Carbohydrate polysulfates must fulfill certain structural requirements to be effective against HIV and other enveloped viruses. A similar structure-activity relationship was found for all enveloped viruses with regard to the MW and degree of sulfation of the sulfated polysaccharides: the antiviral activity increased with increasing MW or degree of sulfation. These requirements proved less stringent for the sulfated synthetic polymers. These compounds were highly efficient against some viruses (i.e. HIV-1, HIV-2, Junin virus, Tacaribe virus, RSV and CMV), irrespective of their MW and degree of sulfation. While the MW and degree of sulfation certainly play an important role in the antiviral activity of polysulfates, other yet still to be defined, structural features must be involved as well. For instance, PAVAS # 3 is a more potent inhibitor of CMV infection than is DS (MW 5,000) although they have a similar MW and comparable degree of sulfation. The importance of the carbohydrate backbone has been previously demonstrated for the plant lectins [57,58]. Only *N*-acetylglucosamine-specific and mannose-specific plant lectins proved active against HIV, HSV and CMV. *N*-acetylgalactosamine-specific plant lectins proved only effective against HSV. These findings are in agreement with the nature of the sugars present in the HIV, HSV and CMV envelope glycoproteins. The oligosaccharides of HSV and CMV envelope glycoproteins have not been described in the same detail as for HIV, but HSV is believed to contain additional *O*-linked oligosaccharides (acetyl-galactosamine), known to be present in a localized cluster in gC (the heparin-binding moiety) of HSV. This coincides with the finding that enzymatic digestion of heparan sulfate (containing *N*-acetylglucosamine residues), but not chondroitin sulfate (containing *N*-acetylgalacosamine residues), influences HIV infectivity [59] (see section D: Mechanism of anti HIV-action).

C. Inhibition of Syncytium Formation Between HIV-Infected Cells and Uninfected Cells

Syncytium (giant cell) formation induced by the interaction between the gp120 glycoprotein expressed on the surface of cells infected with HIV and the CD4 receptor of uninfected $CD4^+$ cells may play an important role in the depletion of T4 lymphocytes in AIDS patients [60]. Schols *et al.* [61] have demonstrated by using a flow cytometric method that this syncytium formation leads to a selective destruction of the uninfected $CD4^+$ cells. In this assay the cells are stained with specific monoclonal antibodies (mAbs) that react with antigens expressed on either HIV-1-infected HUT-78 cells [HUT-78/HIV-1(III_B) cells] or uninfected MOLT-4 cells (Fig. 9).

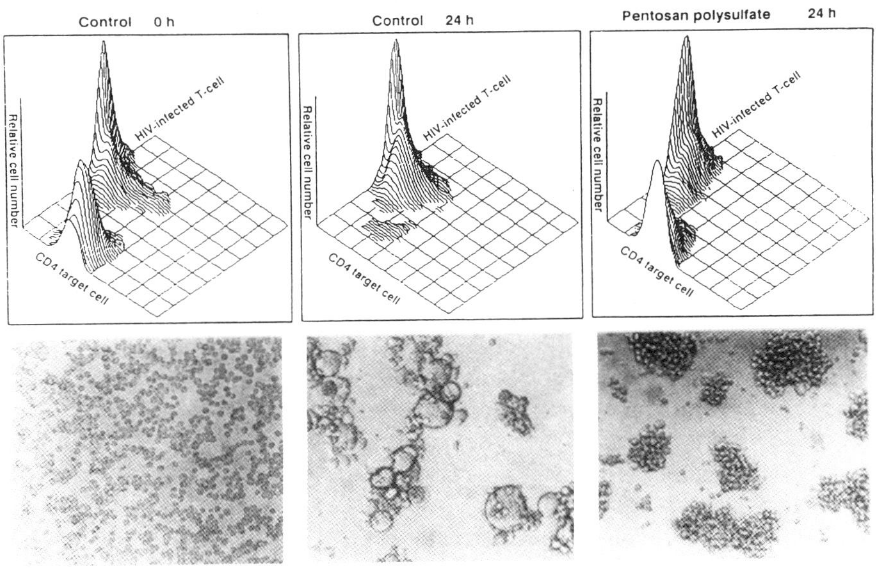

Figure 9 Three-dimensional presentation of flow cytometric analysis of cocultures of HUT-78/HIV-1(III_B) and MOLT-4 cells (upper panels) and microscopic appearance of these cocultures (lower panels). MOLT-4 cells were cultured with an equal number of HUT-78/HIV-1(III_B) cells in the absence (left and middle panels) or presence of PS (100 μg ml^{-1}) (right panels). Immediately after cocultivation (left panels) or at 24 hours after cocultivation (middle and right panels), the cells were harvested and stained with two different fluorescent mAb (anti-HLA-DR-PE and anti-Leu2a-FITC). In upper panels, the x and y axes indicate the intensity of green (FITC) and red (PE) fluorescence, respectively. The z axis corresponds to the cell number. Therefore, the peaks along the x and y axes represent the cells that express Leu2a antigen (MOLT-4 cells) and HLA-DR antigen [HUT-78/HIV-1(III_B) cells], respectively.

In the presence of PS (100 µg ml^{-1}), no giant cell formation is observed and the uninfected MOLT-4 cells are protected against the destructive effect of the persistently HIV-1-infected HUT-78 cells.

Except for the low-MW DS samples, sucralfate, periodate oxidized heparin, dermatan sulfate and CDS, all polysulfates that are inhibitory to HIV-1 replication are able to block HIV-1-induced giant cell formation. Yet, higher concentrations are required for inhibition of giant cell formation than for inhibition of HIV-1 replication (Table 4 and Table 7). A similar correlation was found between the MW of the DS samples and their inhibitory effect on HIV-1-induced giant cell formation as mentioned above for their inhibitory effect on HIV-1-induced cytopathicity. A marked increase in anti-HIV activity was observed when the MW increased from 1,000 to 5,000 or 10,000. At higher MW (10,000-500,000), the activity of DS against HIV-1 tended to level off. For HIV-2-induced syncytium formation, the 1,000-MW sample proved more inhibitory than the higher MW samples. In agreement with the activity of sucralfate against HIV-induced cytopathicity, it only proved active against HIV-2-induced giant cell formation (EC_{50} : 2 µg ml^{-1}). The CDSs differ markedly from DS (MW 5,000) in their inhibitory effect on HIV-1-induced giant cell formation. While not inhibiting HIV-1-induced giant cell formation, they inhibit HIV-2-induced giant cell formation at a concentration of approximately 1 µg ml^{-1}, which is 30-fold lower than the concentration at which DS (MW 5,000) inhibits HIV-2-induced giant cell formation. Like the CDSs, low-MW DS samples are unable to inhibit HIV-1-induced syncytium formation (Table 7). Yet, the CDS and low-MW DS samples are active in inhibiting the cytopathic effect of HIV-1. This means that the MW requirements are more stringent for inhibition of HIV-1-induced giant cell formation than for inhibition of HIV-1-induced cytopathicity, but, in contrast, the MW requirements for inhibition of HIV-2-induced syncytium formation are less stringent than for inhibition of HIV-2-induced cytopathicity.

Polysulfates are known to interact with the V3 region of the viral envelope gp120 glycoprotein (see section D: Mechanism of anti-HIV action). This would explain why they inhibit fusion (syncytium formation) of HIV-infected cells with uninfected $CD4^+$ cells. The fact that CDSs inhibit HIV-2-induced, but not HIV-1-induced, syncytium formation, suggests that there exist marked differences in those parts of the HIV-1 and HIV-2 gp120 glycoproteins that are involved in fusion. Again, modified ß-CDS's proved more inhibitory to HIV-1-induced syncytium formation than unmodified ß-CDS (EC_{50} : > 500 µg ml^{-1}), whereas their activity against HIV-2-induced giant cell formation was less pronounced (EC_{50} : 3-7 µg ml^{-1}) in comparison with the unmodified ß-CDS (EC_{50} : 1 µg ml^{-1}).

The sulfated non-carbohydrate polymers PVAS, PAVAS and PAPS achieved 50% reduction in HIV-1- and HIV-2-induced syncytium formation at a concentration of 1-2 μg ml^{-1} (Table 7), irrespective of their MW or degree of sulfation.

In conclusion, most of the polysulfates which have been found to inhibit HIV replication, also inhibit syncytium formation, although the structural requirements with respect to MW and number of sulfate groups seem to be more stringent for inhibition of syncytium formation than for inhibition of HIV replication (cytopathicity) (Table 4, Table 7) [62,63].

Table 7 Inhibitory Effects of Different Groups of Polysulfates on HIV-Induced Syncytium Formation

Compound	EC_{50}[a] (μg ml^{-1})	
	HIV-1(III_B)	HIV-2(ROD)
1. Polysaccharides		
1.1. Linear		
1.1.1. Chemically synthesized		
DS (MW 1,000)	500	1.8
(MW 1,500)	200	12
(MW 3,400)	3.4	11
(MW 5,000)	2.2	30
(MW 10,000)	1.2	40
(MW 40,000)	1.2	30
(MW 70,000)	2.0	40
(MW 110,000)	3.0	40
(MW 500,000)	3.0	28
PS	14	6
Sucralfate	> 250	2
1.1.2. From natural origin		
Heparin	12.3	-
Periodate oxidized heparin	> 200	> 200
O-butyrylated periodate oxidized heparin	1.5	-
O-butyrylated benzyl ester of periodate oxidized heparin	1.0	1.0
Dermatan sulfate	> 500	-
Xylomannan sulfate	200	200
Galactan sulfate	> 5	> 5
1.2 Circular		
α-cyclodextrin hexasulfate	> 500	-
ß-cyclodextrin heptasulfate	> 500	-
γ-cyclodextrin octasulfate	> 500	1
α-cyclodextrin dodecasulfate	> 500	1

Table 7 Continued

ß-cyclodextrin tetradecasulfate	> 500	1
γ-cyclodextrin hexadecasulfate	> 500	1
mCDS11	7	7
mCDS30	1	3
2. Non-carbohydrate polymers		
2.1. Carbon-carbon backbone		
PVAS # 1	2	2
PVAS # 2	2	1
PVAS # 3	2	2
PAVAS # 1	1	1
PAVAS # 3	1	1
PAVAS # 6	2	1
PAVAS # 8	2	2
2.2. Polyacetal backbone		
PAPS	1	1

[a]50% Effective concentration, or concentration required to inhibit by 50% syncytium formation between MOLT-4 cells and HIV-1(III_B)- or HIV-2(ROD)-infected HUT-78 cells.
Sources: refs. 14, 15, 21, 22, 24, 25, 38, 49 and 54.

For all polysulfates tested the concentration required to block HIV replication (Table 4) was consistently lower than the concentration required to block syncytium formation (Table 7).

D. Mechanism of Anti-HIV Action

As mentioned above, the mechanism by which the polysulfates inhibit HIV replication can be attributed to the inhibition of the binding of the virions to $CD4^+$ cells [64,66] and subsequent syncytium formation [9,62,66]. The inhibition of HIV attachment to $CD4^+$ cells can be demonstrated by monitoring virus-cell binding with radiolabeled HIV particles [7,9], a radioimmunoassay (67), flow cytometry (Fig. 10) [14,15,21,22,24,49,54,65] and a p24 ELISA assay (Table 8) [24]. Dose-dependent inhibition of virus adsorption is observed with polysulfates, as demonstrated for DS (Fig. 10 and Table 8). The inhibitory effect on giant cell (syncytium) formation can be demonstrated following cocultivation of persistently HIV-1 or HIV-2-infected HUT-78 cells with MOLT-4 cells [62] (see section C: Inhibition of syncytium formation between HIV-infected cells and uninfected cells). However, conflicting results have been published as to the target molecule(s) for this action.

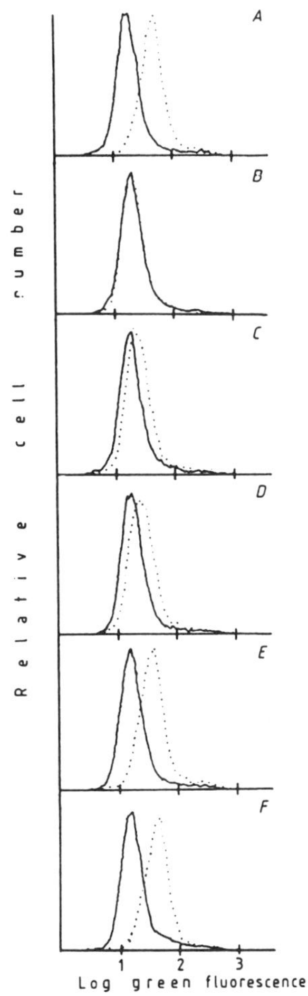

Figure 10 Effect of various concentrations of DS on HIV-1(III_B) binding to MT-4 cells. The solid line histograms represent the nonspecific fluorescence in the absence of DS (A) or in the presence of DS at 25 μg ml^{-1} (B); 5 μg ml^{-1} (C), 1 μg ml^{-1} (D), 0.2 μg ml^{-1} (E), and 0.04 μg ml^{-1} (F). The dotted line histograms represent the specific cellular HIV-1 fluorescence in the absence of DS (A) or in the presence of DS at 25 μg ml^{-1} (B), 5 μg ml^{-1} (C), 1 μg ml^{-1} (D), 0.2 μg ml^{-1} (E), and 0.04 μg ml^{-1} (F). The mean fluoresence values were as follows: 22 (A), 25 (B), 22 (C), 23 (D), 21 (E), and 22 (F) for the nonspecific fluorescence and 52 (A), 25 (B), 28 (C), 30 (D), 45 (E), and 51 (F) for the specific fluorescence.

Table 8 The Inhibitory Indexes of Dextran Sulfate (MW 5,000) for HIV-1(III_B) Binding (III_B) and Cytopathic Effect (II_{CPE}) in MT-4 Cells

Concentration (μg ml^{-1})	II_{VB}		II_{CPE}
	Immunofluorescence (FACS)	p24 content ELISA)	
25	1.00	0.97	1.00
5	0.80	0.75	0.86
1	0.77	0.69	0.64
0.2	0.20	0.57	0.40
0.04	0.03	0.37	0.29

The inhibitory index for virus binding (II_{VB}) determined by FACS analysis was calculated according to the following formula: $1\text{-}(MF_{VC}\text{-}MF_{CC})/(MF_V\text{-}MF_C)$, whereby MF_{VC} is the mean fluorescence (MF) with a given concentration of the compound in HIV-1 inoculated cells, MF_{CC} is the mean fluorescence for the control cells (not exposed to HIV-1) treated with compound, MF_V is the mean fluorescence for the HIV-1-inoculated cells (not treated with any compound), and MF_C is the mean fluoresence for the control cells (not exposed to HIV-1 and not treated with any compound). The inhibitory index for virus binding (II_{VB}) determined by p24 ELISA was calculated according to the following formula: $1\text{-}(OD_T)_{HIV}\text{-}(OD_C)_{mock}/(OD_C)_{HIV}\text{-}(OD_C)_{mock}$, the inhibitory index for cytopathic effect (II_{CPE}) was calculated according to the following formula: $1\text{-}(OD_T)_{HIV}\text{-}(OD_C)_{HIV}/(OD_C)_{mock}\text{-}(OD_C)_{HIV}$, whereby $(OD_T)_{HIV}$ is the optical density measured with a given concentration of the test compound in HIV-1-infected cells; $(OD_C)_{HIV}$ is the optical density measured for the control untreated HIV-1-infected cells; $(OD_C)_{mock}$ is the optical density measured for the control, untreated, mock-infected cells.

DS and other polysulfates do not interfere with the binding of OKT4A/Leu3a mAbs to the $CD4^+$ target cells [7,9,24,66,68,69] (OKT4A/Leu3a mAbs bind at the same site as HIV and can block HIV infection), although DS has been shown to interact with the CD4 molecule at an epitope distinct from the OKT4A epitope [68]. The latter authors also developed a solid-phase assay to study the influence of DS on the interaction of rCD4 and gp120 in the absence of any other cellular or viral components. The interaction of rCD4 with rgp120 was inhibited by DS but not dextran. Pretreatment experiments indicated that DS interferes predominantly with rCD4. This interaction seems to be relatively irreversible (70). Thiele *et al.* [71] and Parish *et al.* [72] have demonstrated that DS MW 500,000, but not DS MW 8,000 or PS, seems to modulate detection of CD4 receptors (also the OKT4A epitope), but only at a very high concentration, on PBL cells. The modulation of CD4 by DS MW 500,000 is highly temperature-dependent suggesting an active, energy-dependent mode of action. Although the low-MW sulfated polysaccharides DS MW 5,000 and PS do not generate this cellular response, they retain their high antiviral activity even at a much lower concentration. A molecular weight of 5,000 to 10,000 seems optimal for the inhibitory effects of DS on HIV-1 replication and giant cell formation [21,62].

Polysulfates interact with positively charged amino acids concentrated in the principal neutralizing domain of gp120 (V3) [73-75]. They interfere with the binding of a specific mAb (9284) to the V3 loop of gp120 of HIV-1 (Fig. 11) [23,24,76] but not with the binding of mAb 9305, a neutralizing anti-gp120/V3 mAb that binds to an adjacent epitope [70]. Taken together, these data suggest that polysulfates interact with the V3 loop of gp120 as has been previously demonstrated with heparin fractions [77,78] and inhibit the binding of certain anti-V3 loop mAbs but also reveal that the portion of the V3 loop recognized by 9305 is distinct from the polysulfate binding site on gp120. The inhibitory effect of DS on gp120/V3 detection by Flow Activated Cell Sorter (FACS) analysis appears to be reversible [70,76]. When the treatment period with DS was extended, there was no further decrease of gp120 detection [76]. DS, PS, and heparin are only slightly inhibitory to the binding of specific mAbs to gp41 and gp160/41, and to the binding of polyclonal anti-HIV-1 serum to HIV-1 infected cells (inhibition obtained with DS at 100 $\mu g\ ml^{-1}$ was less than 10% for the anti-gp41 mAb, 15% for the anti-gp160/41 mAb and 12% for the polyclonal anti-HIV-1 serum) [65]. Thus the inhibitory effect of DS and its congeners on the interaction of HIV gp120 with the cellular CD4 receptor can be ascribed to a specific "shielding" of gp120/V3.

These findings indicate that the mechanism by which polysulfates inhibit HIV replication appears to be related to an interaction with the epitope of the V3 loop of gp120 recognized by the mAb 9284. This interaction is probably mediated by the binding of negatively charged sulfate groups to positively charged amino acids side chains in the V3 loop. We postulate that polysulfate binding to gp120 disrupts two (related) processes critical to HIV infection, i.e. inhibition of HIV binding to the cell surface, followed by interference with the virus-cell fusion step. The initial virus-binding event depends on the interaction with the CD4 membrane receptor, but additionally has to overcome the electrostatic repulsive forces between the cell membrane and the virion envelope. The early inhibitory effect of polysulfates can be the result of a disruption of ionic interactions between charged regions of viral surface glycoproteins, including gp120, and membrane phospholipids. Enveloped viruses that bud from cells carry part of the cell membrane and therefore may interact more effectively with the negatively charged cell surface if the outer part of the envelope protein

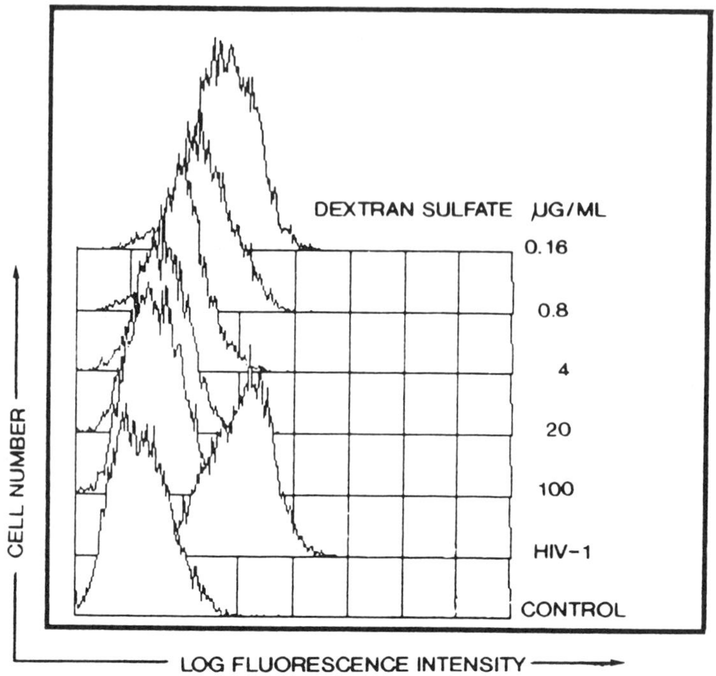

Figure 11 Inhibitory effect of DS (MW 5,000) on the binding of anti-gp120 mAb, recognizing the V3 loop of gp120, to persistently HIV-1-infected HUT-78 cells (strain III_B). The first histogram represents the control fluorescence of uninfected HUT-78 cells incubated with anti-gp120 mAb and RaM-IgG-(Fab')$_2$-PE. The second histogram represents the specific fluorescence of HIV-1-infected HUT-78 cells incubated with the anti-gp120 mAb and Ram-IgG-(FAb')$_2$-PE. The third, fourth, fifth, sixth and seventh histogram represent the binding of anti-gp120 mAb, but now the cells were pretreated for 15 minutes with DS at concentrations varying from 0.16 µg ml^{-1} (back) to 100 µg ml^{-1} (front). The mean fluorescence values were (back to front) 81, 74, 63, 52, 47, 96 and 40, respectively.

contains regions of high positive charge density [79]. The V3 loop in HIV may provide this function and the binding of polysulfates to the loop not only abolishes positive charge but also adds an additional negative potential, thereby electrostatically preventing the interaction between the virion and the cell. This mechanism may also explain the broad antiviral activity of polysulfates against enveloped viruses.

Enveloped viruses, e.g. HSV, CMV, and pseudorabies virus (PRV) have been shown to interact with cell surface heparan sulfate as an initial binding site [80-83]. They are also inhibited by the polysulfates. By analogy, heparan sulfate has recently been reported to participate in HIV-cell attachment and virus entry [59]. The importance of the V3 loop in the attachment to heparan sulfate molecules at the cell surface is in agreement with features previously reported for the heparin-binding-moieties of e.g. CMV gC-II and PRV gIII [81,84,85]. The binding of gC-II to immobilized heparin depends on structures maintained by disulfide bonds. Moreover, the PRV gIII glycoprotein also contains an arginine-rich region as previously documented for the HIV V3 loop. The function of heparan sulfate depends on sulfation of these molecules, again suggesting that ionic interactions contribute to virus binding to heparan sulfate. Based on previous data showing a potent inhibitory effect of polysulfates on binding of mAb 9284 to the gp120 V3 loop, positively charged amino acids in the V3 domain may bind to cell-surface heparan sulfate. The antiviral activity of heparin (which is chemically very similar to heparan sulfate) [86] may be achieved by occupying the sites on the virion envelope that are necessary for attachment of the virus particles to cell surface heparan sulfate. It is likely that molecules that are related to heparin, such as DS, PS, CDS, PVAS, PAVAS and PAPS, achieve their antiviral activity in a similar way as heparin.

The most likely role of the V3 loop is to assist in membrane fusion after gp120 binds to CD4. Once again, the high positive charge density could be essential to this function presumably by bringing two negatively charged membranes together [79]. Specifically, this could be accomplished by binding to a negatively charged region on the cell membrane, thereby orienting the fusiogenic portion of gp41 and the viral membrane so that fusion occurs. The V3 region may be physically close to gp41, since mutations in the loop have been reported to lead to gp41-gp120 dissociation [64]. Proteolysis has also been shown to occur within the V3 loop and could be involved in inducing fusion [87]. Binding of the polysulfate to the V3 region may neutralize the local positive charge density and thus disturb binding of the V3 region to the CD4 receptor, proteolysis, and fusion. On the other hand, antibodies directed against the V3 domain may not inhibit virion binding [88], because they do not eliminate the positive charge density but still physically block an essential interaction between the V3 loop and either a cell surface protease or the cell surface membrane.

Moreover, the fact that the polysulfates are inhibitory to some enveloped viruses (see section B. Activity against enveloped viruses other than HIV) seems to depend on the composition of the amino acid sequences of the viral envelope glycoproteins that are involved in virus-cell fusion [18,89]. The viruses sensitive to the polysulfates share a tripeptide segment (Phe-Leu-Gly). This tripeptide segment may be involved either directly (as a target sequence) or indirectly (*via* its influence on the overall conformation) in the inhibitory effects of these sulfated compounds. Polysulfates have been reported to inhibit reverse transcriptase activity of various retroviruses [7, 90-93]. However, this inhibitory effect cannot be considered as specific, since the compounds also interfere with host cellular DNA polymerases [9]. Should reverse transcriptase act as a target for the anti-HIV activity of polysulfates, it would be mandatory that the compounds are taken up into the cells. Given their polyanionic character, such uptake may not readily occur. Also, inhibition of reverse transcriptase by polysulfates is achieved only at concentrations which are far in excess of those required for inhibition of virus replication, which, in turn, coincides quite well with those that have been found to be inhibitory to virus adsorption. It seems justified, therefore, to attribute the anti-HIV-1 activity of the polysulfates to their inhibitory effects on virus adsorption and fusion.

In conclusion, polysulfates are assumed to exert their anti-HIV activity by occupying the positively charged sites of the V3 loop of gp120 that are necessary for attachment of HIV to cell surface heparan sulfate molecules. Polysulfates may interact with several viral epitopes, i.e. not only with gp120, but also with gp41 epitopes. Since the gp120 V3 loop and gp41 are believed to be involved in HIV attachment and fusion to $CD4^+$ cells, it is clear that inhibition of these functions by polysulfates should result in the inhibition of virus binding to $CD4^+$ cells and syncytium formation between HIV-infected cells and $CD4^+$ uninfected cells.

E. *In Vitro* Anti-HIV Activity in Combination with Nucleoside Analogues

Numerous antivirals with various mechanisms of action against HIV, the causative agent of AIDS, have been identified. At present, the only drugs approved by the US Food and Drug Administration (FDA) for the treatment of AIDS, are targeted at the viral reverse transcriptase. The nucleoside analogue 3'-azido-3'-deoxythymidine (AZT) was the first one approved for clinical use. The active intracellular metabolite, AZT-triphosphate (AZT-TP), inhibits reverse transcriptase (RT) and acts as a chain terminator. The second nucleoside analogue, 2',3'-dideoxyinosine (ddI), that has been approved, acts at the RT level via its active metabolite ddATP. The two other nucleoside analogues that have been approved are ddC and d4T, which act at the RT level via their 5'-triphosphates ddCTP and d4T-TP, respectively. The significant toxicity associated with administration of these inhibitors (myelosuppression, neuropathy and pancreatitis) and the appearance of drug-resistant HIV strains [94,95] emphasize the critical need to identify new classes of anti-HIV agents to be used alone or in combination with the existing antiviral agents. Combination of different antiviral drugs has a triple aim:
(i) it may prevent or delay the emergence of drug-resistant virus strains. (ii) it may lead to increased activity (synergism) and extend the antiviral spectrum to include viruses which would otherwise not be affected by the individual compounds and (iii) it may allow reducing the individual dosages and thus decrease drug toxicity.

In 1987, Ueno and Kuno [6] reported synergism between DS and AZT. This observation was confirmed by Hayashi *et al.* [96] and Busso and Resnick [52]. Whereas the purine nucleoside analogues ddA and ddI act clearly synergistically with ß-CDS, the pyrimidine nucleoside analogues AZT, ddC and D4T show a sub-synergistic interaction with ß-CDS (Fig. 12). The synergistic action may result from the capturing of the nucleoside analogues within the central cavity of the cyclodextrins. The higher synergistic activity achieved by the purine nucleoside analogues may suggest that they are captured more easily than the pyrimidine nucleoside analogues. Anand *et al.* [97,98] also reported a profound antiviral synergism with three polyanionic compounds (α-CDS, ß-CDS, PS) and AZT *in vitro*. Various other drugs, i.e. soluble CD4 [99], castanospermine [100], PMEA [101], phosphonoformate (foscarnet) [102], TIBO derivatives [103] α-interferon [104] have been previously reported to act synergistically with AZT *in vitro*.

Optimal AIDS therapy may require the use of combinations of agents that exhibit synergistic antiviral effects (e.g. polysulfates and nucleoside analogues). Ideally, such combinations should target different sites in the HIV replicative cycle (e.g. virus attachment and reverse transcriptase), inhibit viral replication in a broad range of cell types, and yet not display accrued toxicity. Combination therapies may help preventing the emergence of drug-resistant HIV mutants and may also allow the use of the individual drugs below their toxicity threshold.

F. Pharmacokinetic Profile

Because of its marked *in vitro* activity against HIV and other enveloped viruses, DS [53,105-107] and PS [53,108-110] have been subjected to pharmacokinetic studies. Initial studies were conducted with oral DS (MW 8,000) in patients with AIDS and AIDS-related complex [105]. The compound appeared to be well tolerated, but did not show much benefit. It was later ascertained that, after oral administration, DS (MW 8,000) is very poorly absorbed in humans [26], as has also been demonstrated in rats [106]. There are several assay methods to monitor blood drug levels achieved by polysulfates. MacGregor *et al.* examined the plasma concentration of PS using a competitive binding assay, in which radiolabeled PS competed with plasma samples for binding to polybrene-sepharose [111]. The plasma half-life and peak levels of PS in humans following intravenous (iv) injection of 100 mg PS are 50 to 58 min and 16 to 24.5 μg ml^{-1}, respectively [111]. Moreover, after an intravenous dose of 225 mg DS in humans, peak plasma concentrations by competitive binding assay were 26 to 35 μg ml^{-1}. DS disappeared from the plasma with a half-life of 84 to 120 minutes.

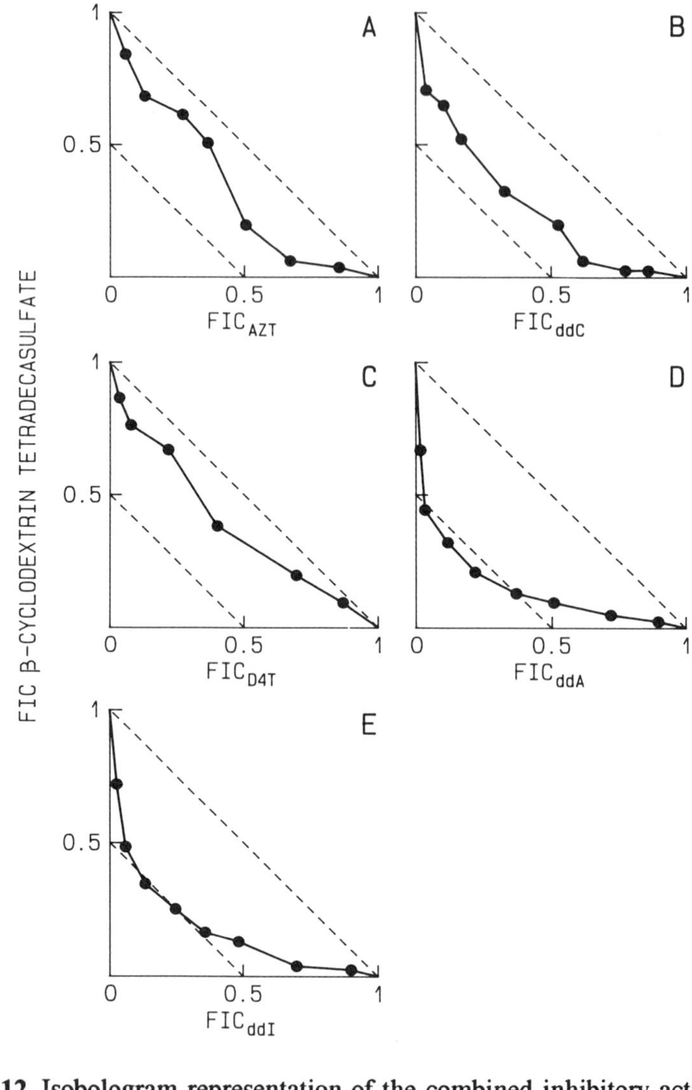

Figure 12 Isobologram representation of the combined inhibitory actions of ß-cyclodextrin tetradecasulfate and (AZT) (a), ddC (b), D4T (c), ddA (d) and ddI (e) on the cytopathic effect of HIV-1(III_B) in MT-4 cells. Broken lines represent the unity lines for FIC equal to 1 and 0.5, respectively. Mean results from two separate experiments are shown.

Only 25% of the dose was recovered in the urine of healthy volunteers with normal renal function, indicating that most of the drug was cleared by a non-renal mechanism [26]. This chemical method suffers from the drawback that it may also detect degradation products that are no longer active as anti-HIV agents.

Since polysulfates are primarily known for their anticoagulant activity and are used clinically for this purpose, a bioassay could, in principle, be based on the measurement of this anticoagulant activity [112]. Based on this procedure, the plasma half-life of heparin in humans is approximately 35 min (after iv injection of 23 mg heparin) [112]. However, the anti-HIV activity and anticoagulant effects of polysulfates are not necessary correlated [38,62].

The ideal bioassay for monitoring the concentration of these compounds in biological specimens should be based on the measurement of anti-HIV activity. We have established a simple and sensitive bioassay system to measure serum concentrations of polysulfates [53]. This assay is based on the inhibitory effect of the serum samples (containing polysulfates) on HIV-2 replication in cell culture. As mentioned above, DS and PS inhibit HIV-2 replication at a markedly lower concentration than HIV-1 replication. In addition, HIV-2 replication should not be influenced by the presence of anti-HIV-1 neutralizing antibodies in the plasma samples from AIDS patients. Fig. 13 shows the serum concentrations following iv administration of PS at 0.1 and 0.5 g kg^{-1} to rabbits. This method permits the detection in (rabbit) serum samples of DS and PS concentrations as low as 3.0 and 0.5 μg ml^{-1}, respectively. The serum concentrations of iv administered DS and PS declined biphasically with an initial half-life of approximately 84 min. [53].

This anti-HIV bioassay system was used to monitor serum concentrations of PS following iv administration at different doses (600, 300 and 100 mg) to 6 patients with HIV infection. One hour after administration, serum concentrations reached peak values of 113, 54 and 28 μg ml^{-1}, respectively. The serum levels of PS decreased rapidly, but the compound remained detectable 48, 24 and 4 hours after injection, respectively. The estimated half-life of PS in humans was approximately 2 hours [110]. This result confirmed a previous reported half-life of PS in humans of 1 hour 14 minutes [113].

When several polysulfates were administered intravenously to rabbits at a dose of 0.1 g kg^{-1}, serum peak concentrations ranging from 0.3 mg ml^{-1} for PS to 1.5 mg ml^{-1} for PVAS and PAVAS were reached (Table 9). When DS and PS were injected intravenously in rabbits at 0.5 g kg^{-1}, serum peak levels of 2.2 to 3.0 mg ml^{-1} were reached. Concentration vs. time data were fitted to a monoexponential function to calculate the elimination rate constant, serum half-life, distribution volume, total body clearance and area under the curve (AUC) of polysulfates.

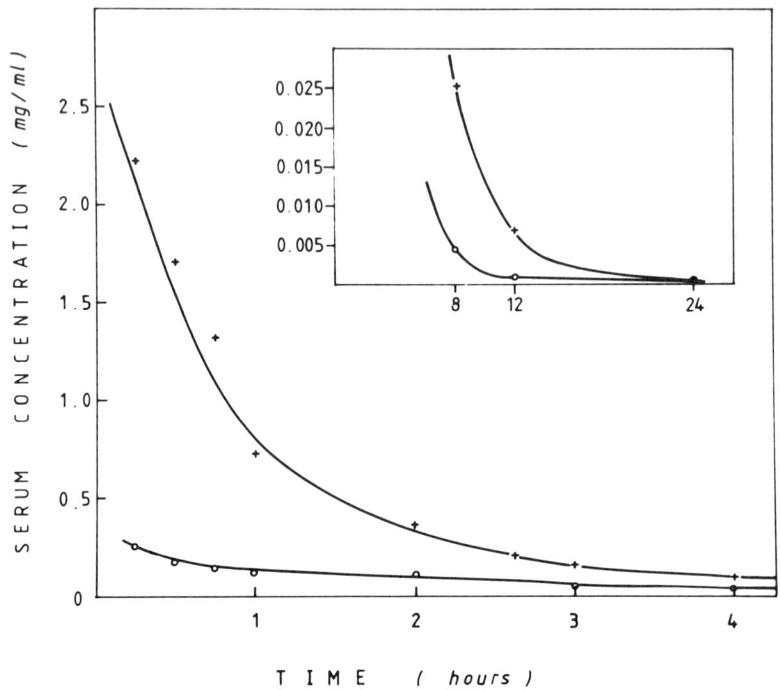

Figure 13 Serum concentrations following intravenous administration of PS [0.1 g kg^{-1} (O) or 0.5 g kg^{-1} (+)] to rabbits.

Polysulfates disappeared from the serum with half-lives of 20 min for γ-cyclodextrin hexadecasulfate up to 2 hours for PVAS and PAVAS, with total body clearances of 0.11 and 0.02 l/h/kg, respectively. Clearance values for the different DS and PS doses (0.1 and 0.5 g kg^{-1}) were not found to be different (0.11-0.12 and 0.24-0.18 l/h/kg, respectively). Thus the pharmacokinetics of DS and PS upon iv bolus administration was dose independent within the dose range of 100 to 500 mg kg^{-1}. For each iv administered polysulfate, data fitted well to monoexponential functions, the correlation coefficient of the linear regression curves being higher than 0.939.

After oral administration of polysulfates via gastric intubation only the modified ß-cyclodextrin hexadecasulfate bearing three benzylthio groups as lipophilic moieties (mCDS11) gave concentrations above the minimum detectable concentration (Table 9). The serum drug concentration at 1 hour after oral administration of mCDS11 at a dose of 0.25 g kg^{-1} was 0.5 mg ml^{-1}, thus indicating high oral bioavailability. Its half-life was 4 hours (Table 9). Similar findings were also obtained following oral administration of mCDS11 to rats [114]. Toxicity was not observed after oral administration of mCDS11 to mice at a dose of 3 g kg^{-1} [114] and at doses of 0.5 and 2 g/kg/day for 5 days [49]. Recently, Moriya *et al.* [115] carried out a detailed investigation of modified CDS. A series of C-2 modified CDS (instead of C-6 modified) analogues was synthesized in attempts to develop a more effective, orally bioavailable anti-HIV agent. The plasma peak level of the modified ß-CDS mCDS71, determined by the anti-HIV bioassay, after oral administration to rats at 1 g kg^{-1}, was 0.32 mg ml^{-1} [115]. This and similar compounds should be pursued for their potential as oral drug candidates in the prophylaxis and/or therapy of HIV infections.

In conclusion, polysulfates are very poorly absorbed following oral administration. Only a modified ß-CDS gave concentrations above the minimum detectable concentration with a half-life of 4 hours.

G. *In Vivo* and Clinical Efficacy

In marked contrast with the wealth of *in vitro* antiviral data is the paucity of antiviral data of polysulfates in animal models. Beneficial effects have only been reported for PS [116], sulfoevernan [117] and Viva Natural, a dietary seaweed extract [118] in the murine leukemia virus (MLV) system. It would seem mandatory to examine whether other polysulfates are equally effective as the previously reported in the MLV model and/or whether the remarkable efficacy of these compounds, especially of sulfoevernan, extends to other *in vivo* models. Attempts to demonstrate activity with DS in cats infected with feline leukemia virus [119] and ducklings infected with duck hepatitis B virus [120] failed. In the cat model, DS proved to be prohibitively toxic, and in the ducklings it was ineffective against the hepatitis B virus infection.

Because polysulfates are potent and selective inhibitors of HIV-1 and HIV-2 *in vitro* (Table 3), DS (MW 8,000) and PS have been introduced in the clinic for the treatment of AIDS patients. Serum drug concentrations achieved by these sulfated polymers were highly dependent on the route of administration. Clinical trials of oral DS [105] and PS [109] were terminated soon after the drug was shown to be poorly absorbed (bioavailability < 1%) [26]. No significant changes in surrogate markers (p24) or clinical endpoints (CD4), suggesting anti-HIV activity, were recorded from those trials. Not only are sulfated polyanions poorly absorbed when given perorally, their oral administration may also cause intestinal overgrowth of facultative bacilli, accompanied by histological alterations in the intestinal mucosa [121].

Since DS and PS appear to be poorly absorbed through the gastro-intestinal tract, parenteral administration of these drugs seems mandatory. DS has been administered as a continuous iv infusion to symptomatic HIV patients, at the maximally tolerated dose (i.e. the initial DS infusion rate was 22.5 mg h^{-1}), to produce the greatest acceptable increase in thrombolytic activity (activated partial thromboplastin time), thus achieving the greatest possible exposure to drug over time and maximizing any potential effect on the virus [106]. Although plasma peak concentrations far exceeded (200-fold) the EC_{50} (0.5 μg ml^{-1}) at which DS inhibits HIV-1 replication *in vitro*, circulating HIV p24 levels increased in all subjects who received the drug for more than 3 days. Continuous iv infusion of DS proved to be toxic, producing reversible thrombocytopenia [107].

Recently, Pluda and colleagues from the U.S. National Institutes of Health (NIH) administered PS to 16 HIV-positive patients with Kaposi's sarcoma (KS) in a Phase I clinical trial. Since PS inhibits fibroblast growth factor (FGF), it was the goal to prevent or block the neovascularization thought to play a role in the pathogenesis of KS. Administration of PS, as a continuous iv infusion, yielded no significant tumor response or anti-HIV activity. However, plasma PS levels of at least 7 μg ml^{-1} were reached. The EC_{50} at which PS inhibits HIV-1 replication *in vitro* is 0.8 μg ml^{-1}.

A possible explanation for the failure of DS and PS to achieve a clinical response in these studies is that effective levels of these sulfated compounds were lower in tissues than could be predicted from serum or plasma concentrations, either because of poor penetration or because of extensive binding to plasma proteins. Also, given the MW and strong anionic nature of sulfated compounds, they might be expected to remain confined to the intravascular space and not attain therapeutic tissue levels.

In conclusion, in some animal models for retroviral infection there was evidence for the efficacy of polysulfates whereas in other animal models, no such evidence could be obtained. Also when DS and PS were introduced in clinical trials by continuous iv infusion to symptomatic HIV patients, no beneficial effects on surrogate markers were noted. Continuous iv infusion of DS or PS produced unwelcome toxicity. The main dose-limiting side effects were anticoagulation and thrombocytopenia. Because of their toxicity and lack of beneficial effect on surrogate markers, DS, PS and other polysulfates are unlikely to have a practical role in parenteral treatment of symptomatic HIV infection, unless they have been modified to such extent that they lose their undesirable features.

H. Future Perspectives

As a retrovirus, HIV characteristically inserts its genome into host cell chromosomes, leading to a persistently infected state. The problems encountered in developing agents that could specifically excise the integrated copies of the HIV genome from all infected host cells of a patient suggests that an "absolute cure" may be exceedingly difficult to achieve. Nonetheless, it is important to consider not only treatment of patients who are already infected with the virus, but also prophylaxis and protection from new infections. The awareness of the necessity for chemo-prophylaxis is more recent.

Since heterosexual transmission is now responsible for the large majority of infections worldwide [122], it is of paramount importance to develop effective prevention strategies to combat further spread of the epidemic *via* this route. Various groups and organizations have, based on epidemiological and sociological research, pointed to the importance of female-controlled methods to prevent or reduce the risk of sexually acquired HIV. The urgent need for methods to prevent sexual transmission of HIV that are under the control of women was addressed at a meeting convened by the World Health Organization (WHO) in Geneva on 11-13 November 1993 [123]. It led to the recommendation to encourage the development of vaginal microbicides active against HIV, ideally, preparations that women could use without the knowledge of their sexual partners. Although their use should not be seen as a substitute for, but as a complement to, other approaches to stop the sexual spread of HIV, it would provide women with a means of protection when they could not persuade their partner to use a condom.

Considerable evidence suggests that sexual transmission of HIV is mediated *via* mononuclear cells that can infect epithelia of the genital tract. Pearce-Pratt and Phillips [124] described and employed an *in vitro* model to examine the mechanism of cell-to-cell transmission of this virus and to identify agents that may be effective in a vaginal formulation to prevent HIV transmission *via* sexual contact. They found that seminal fluid significantly increases the number of lymphocytes adhering to epithelia. On the other hand, sulfated polysaccharides, or polysulfates in general, effectively inhibit cell-cell adhesion. Polysulfates have yet other properties that make them an attractive candidate for a vaginal anti-HIV formulation. They are inexpensive, water-soluble, and poorly absorbed [26,125] and, in addition to their anti-HIV activity, they are also endowed with activity against various other enveloped viruses (such as HSV). These results suggest that polysulfates may be effective in a vaginal formulation to reduce the probability of sexually transmitted HIV infection and other virus infections. Our recent data obtained with PVAS and PAVAS in the topical prevention of HSV-2 infection in mice [126] point to the feasibility of such an approach.

REFERENCES

1. Nahmias AJ, Kibrick S. Inhibitory effect of heparin on herpes simplex virus. J Bacteriol 1964; 87:1060-1066.
2. Nahmias AJ, Kibrick S, Bernfeld P. Effect of synthetic and biological polyanions on herpes simplex virus. Proc Soc Exp Biol Med 1964; 115:993-996.
3. Takemoto KK, Fabisch P. Inhibition of herpes virus by natural and synthetic acid polysaccharides. Proc Soc Exp Biol Med 1964; 116:140-144.

4. Vaheri A. Heparin and related polyionic substances as virus inhibitors. Acta Pathol Microbiol Scand 1964; Suppl 171: 7.
5. Ito M, Baba M, Sato A, Pauwels R, De Clercq E, Shigeta S. Inhibitory effect of dextran sulfate and heparin on the replication of human immunodeficiency virus (HIV) in vitro. Antiviral Res 1987; 7:361-367.
6. Ueno R, Kuno S. Dextran sulfate, a potent anti-HIV agent in vitro having synergism with zidovudine. Lancet 1987; i:1379.
7. Baba M, Pauwels R, Balzarini J, Arnout J, Desmyter J, De Clercq E. Mechanism of inhibitory effect of dextran sulfate and heparin on replication of human immunodeficiency virus in vitro. Proc Natl Acad Sci USA 1988; 85:6132-6136.
8. Baba M, Nakajima M, Schols D, Pauwels R, Balzarini J, De Clercq E. Pentosan polysulfate, a sulfated oligosaccharide, is a potent and selective anti-HIV agent in vitro. Antiviral Res 1988; 9:335-343.
9. Mitsuya H, Looney DJ, Kuno S, Ueno R, Wong-Staal F, Broder S. Dextran sulfate suppression of viruses in the HIV family: inhibition of virion binding to CD4+ cells. Science 1988; 240:646-649.
10. Baba M, Snoeck R, Pauwels R., De Clercq, E. Sulfated polysaccharides are potent and selective inhibitors of various enveloped viruses, including herpes simplex virus, cytomegalovirus, vesicular stomatitis virus and human immunodeficiency virus. Antimicrob Agents Chemother 1988; 32:1742-1745.
11. Andrei G, De Clercq E. Inhibitory effect of selected antiviral compounds on arenavirus replication in vitro. Antiviral Res 1990; 14:287-300.
12. Andrei G, Snoeck R, Goubau P, Desmyter J, De Clercq, E. Comparative activity of selected antiviral compounds against clinical isolates of human cytomegalo-virus. Eur J Microbiol Infect Dis 1991; 10:1026-1033.
13. Andrei G, De Clercq, E. Molecular approaches for the treatment of hemorrhagic fever virus infections. Antiviral Res 1993; 22:45-75.
14. Schols D, De Clercq E, Balzarini J, Baba M, Witvrouw M, Hosoya M, Andrei G, Snoeck R, Neyts J, Pauwels R, Nagy M, Györgyi-Edelényi J, Machovich R, Horváth I, Löw M, Görög S. Sulphated polymers are potent and selective inhibitors of various enveloped viruses, including herpes simplex virus, cytomegalovirus, vesicular stomatitis virus, respiratory syncytial virus, and toga-, arena- and retroviruses. Antiviral Chem. Chemother 1990; 1:233-240.
15. Schols D, De Clercq E, Witvrouw M, Nakashima H, Snoeck R, Pauwels R, Van Schepdael A, Claes P. Sulphated cyclodextrins are potent anti-HIV agents acting synergistically with 2',3'-dideoxynucleoside analogues. Antiviral Chem Chemother 1991; 2:45-53.
16. Lüscher-Mattli M, Glück R. Dextran sulfate inhibits the fusion of influenza virus with model membranes, and suppress influenza virus replication in vivo. Antiviral Res 1990; 14:39-50.

17. Lüscher-Mattli M, Glück R, Kempf C, Zanoni-Grassi M. A comparative study of the effect of dextran sulfate on the fusion and the in vitro replication of influenza A and B, Semliki Forest, vesicular stomatitis, rabies, Sendai, and mumps virus. Arch. Virol 1993; 130:317-326.
18. Hosoya M, Balzarini, J, Shigeta S, De Clercq E. Differential inhibitory effects of sulfated polysaccharides and polymers on the replication of various myxoviruses and retroviruses, depending on the composition of the target amino acid sequences of the viral envelope glycoproteins. Antimicrob Agents Chemother 1991; 35:2515-2520.
19. Mastromarino P, Conti C, Petruzziello R, Lapadula R, Orsi N. Effect of polyions on the early events of Sindbis virus infection of Vero cells. Arch Virol 1991; 121:19-27.
20. Mastromarino P, Seganti L, Petruzziello R, Gabrieli R, Divizia M, Panà A, Orsi N. Influence of polyanions on the early steps of enterovirus infection. J Chemother 1991; 3:203-208.
21. Witvrouw M, Schols D, Andrei G, Snoeck R, Hosoya M, Pauwels R, Balzarini J, De Clercq E. Antiviral activity of low-MW dextran sulphate (derived from dextran MW 1000) compared to dextran sulphate samples of higher MW. Antiviral Chem Chemother 1991; 2:171-179.
22. Witvrouw M, Schols D, Andrei G, Snoeck R, Ikeda S, Pauwels R, Van Schepdael A, Arnout J, Claes P, Desmyter J, De Clercq E. New polyacetal polysulphate active against human immuodeficiency virus and other enveloped viruses. Antiviral Chem Chemother 1992; 3:351-360.
23. Witvrouw M, Desmyter J, De Clercq E. Polysulfates as inhibitors of HIV and other enveloped viruses. Antiviral Chem Chemother 1994; 5:345-359.
24. Witvrouw M, Este JA, Quinones Mateu ME, Reymen D, Andrei G, Snoeck R, Ikeda S, Pauwels R, Vittori Bianchini N, Desmyter J, De Clercq E. Activity of a sulfated polysaccharide extracted from the red seaweed Aghardhiella tenera against human immunodeficiency virus and other enveloped viruses. Antiviral Chem Chemother 1994; 5:297-303.
25. Damonte E, Neyts J, Pujol CA, Snoeck R, Andrei G, Ikeda S, Witvrouw M, Haines H, Matulewicz MC, Cerezo A, Coto CE, De Clercq E. Antiviral activity of a sulfated polysaccharide from the red seaweed Nothogenia fastigiata. Biochem Pharmacol 1994; 47:2187-2192.
26. Lorentsen KJ, Hendrix CW, Collins JM, Kornhauser DM, Petty BG, Klecker RW, Flexner C, Eckel RH, Lietman PS. Dextran sulfate is poorly absorbed after oral administration. Ann Intern Med 1989; 111:561-566.
27. Lindahl U, Backstrom G, Thunberg L. The antithrombin-binding sequence in heparin. J Biol Chem 1983; 258:9826.
28. Atha DH, Stephens AW, Rosenberg RD. Evaluation of critical groups required for binding of heparin to antithrombin. Proc Natl Acad Sci USA 1984; 81:1030.

29. Grönwall A, Ingelman B, Mosimann H. A dextran sulphuric acid ester with heparin activity. Uppsala Lakareförcn. Forh. 1945; 50:397-404.
30. Swiss pat. 293,566 (1953 to Wander). Sulfuric acid esters of xylan. C.A. 1955; 49:1787h.
31. Nametaka M, Tanaka T, Sakamoto N, Moro K. Oligosaccharide sulfates and monosaccharide polysulfates for medical purposes. III. antiulcerogenic properties of sucrose sulfate-aluminum complex. Yakugaku Zasshi 1967; 87:889-893.
32. McLean J. The thromboplastic action of cephalin. Amer J Physiol 1916; 41:250-257.
33. Howell WH. Havey Lectures, Ser 12:1916-1917.
34. Lindahl U, Feingold DS, Rodén L. Biosynthesis of heparin. Trends in Biological Science 1986; 11:221.
35. Lindahl U. Heparin: Chemical and biological properties, Clinical Applications, pp 159-189, Eds DA Lane, U Lindahl, CRC Press Inc, Boca Raton, Florida, 1989.
36. Cifonelli JA, King JA. Structural characteristics of heparan sulfates with varying sulfate contents. Biochemistry 1977; 16:2137.
37. Lormeau J-C, Petitou M, Choay J. Low molecular weight heparins with a regular structure, their preparation and biological uses. Eur. Pat. Appl 1987, 2874477; Chem Abstr 1989; 11:66.
38. Bârzu T, Level M, Petitou M, Lormeau J-C, Choay J, Schols D, Baba M, Pauwels R, Witvrouw M, De Clercq E. Preparation and anti-HIV activity of O-acylated heparin and dermatan sulfate derivatives with low anticoagulant effect. J Med Chem 1993; 36:3546-3555.
39. Meyer K, Chaffee E. The mucopolysaccharides of skin. J Biol Chem 1941; 138:491.
40. Marbet R, Winterstein A. Probleme der Blutgerinnung. ß-Heparin, ein neuer, blutgerinnungshemmender Mucoitinschwefelsäureester. Helv Chim Acta 1951; 34:2311.
41. Jeanloz RW, Stoffyn PJ, Trémège M. Chemical structure of ß-heparin. Fed Proc 1957; 16:201.
42. Matulewicz MC, Cerezo AS. Water-soluble sulfated polysaccharides from the red seaweed Chaetangium fatigiatum. Analysis of the system and structures of the α-D-(1-3)-linked mannans. Carbohydrate Pol 1978; 7: 121-132.
43. DiNinno V, McCandless EL. The chemistry and immunochemistry of carrageenans from Eucheuma and related algal species. Carbohydrate Res 1978; 66:85-95.
44. Percival E, McDowel RH. Chemistry and Enzymology of Marine Algal Polysaccharides. New York: Academic Press, 1967.
45. Rees, D.A. Carbohydrate sulfates. Ann Rep Chem Soc (London) 1965; 469.

46. Berger L, Lee J. US patent No. 2,923,704, 1960.
47. Folkman J, Weisz PB, Joullié, MM, Li WW, Ewing WR. Control of Angiogenesis with synthetic heparin substitutes. Science 1989; 243:1490-1493.
48. Tsujihara K, Kurita H, Kawazu M. The highly selective sulfonylation of cycloheptaamylose and syntheses of its pure amino derivatives. Bull Chem Soc Jpn 1977; 50:1567-1571.
49. Otake T, Schols D, Witvrouw M, Naesens L, Nakashima H, Moriya T, Kurita H, Matsumoto K, Ueba N, De Clercq E. Modified cyclodextrin sulfates (mCDS11) have potent inhibitory activity against HIV and high oral bioavailability. Antiviral Chem Chemother 1993; 5:155-161.
50. Machovich R, Nagy M, Györgyi-Edelényi J, Csomor K, Horváth I. Anticoagulant effect of sulphated poly/vinyl alcohol-acrylic acid/copolymers. Thromb Haemostas 1986; 56:397-400.
51. Claes P, Billiau A, De Clercq E, Desmyter J, Schonne E, Vanderhaeghe H, De Somer P. Polyacetal carboxylic acids: a new group of antiviral polyanions. J Virol 1970; 5:313-320.
52. Busso M, Resnick L. Anti-human immunodeficiency virus effects of dextran sulfate are strain dependent and synergistic or antagonistic when dextran sulfate is given in combination with dideoxynucleosides. Antimicrob Agents Chemother 1990; 34:1991-1995.
53. Witvrouw M, Baba, M, Balzarini J, Pauwels R, De Clercq E. Establishment of a bioassay to determine serum levels of dextran sulfate and pentosan polysulfate, two potent inhibitors of human immunodeficiency virus. J Acquir Immun Defic Syndr 1990; 3:343-347.
54. Schols D, Pauwels R, Witvrouw M, Desmyter J, De Clercq E. Differential activity of polyanionic compounds and castanospermine against HIV replication and HIV-induced syncytium formation depending on virus strain and cell type. Antiviral Chem Chemother 1992; 3:23-29.
55. Jurkiewicz E, Panse P, Jentsch K-D, Hartmann H, Hunsmann G. In vitro anti-HIV-1 activity of chondroitin polysulfate. AIDS 1989; 3:423-427.
56. Baba M, Schols D, De Clercq E, Pauwels R, Nagy M, Györgyi-Edelényi J, Löw M, Görög S. Novel sulfated polymers are highly potent and selective inhibitors of human immunodeficiency virus replication and giant cell formation. Antimicrob Agents Chemother 1990; 34:134-138.
57. Stig Hansen J-E, Nielsen C, Vestergaard BF. Inhibition of human immunodeficiency virus 1 (HIV-1) and herpes simplex virus 1 (HSV-1) infectivity with a broad range of lectins. Scand J Infect Dis 1991; 23:425-430.

58. Balzarini J, Neyts J, Schols D, Hosoya M, Van Damme E, Peumans W, De Clercq E. The mannose-specific plant lectins from Cymbidium hybrid and Epipactis helleborine and the (N-acetylglucosamine)n-specific plant lectin from Urtica dioica are potent and selective inhibitors of human immunodeficiency virus and cytomegalovirus replication in vitro. Antiviral Res 1992; 18:191-207.
59. Patel M, Yanagishita M, Roderiquez G, Bou-Habib DC, Oravecz T, Hascall VC, Norcross MA. Cell-surface heparan sulfate proteoglycan mediates HIV-1 infection of T-cell lines. AIDS Res Hum Retrovir 1993; 9:167-174.
60. Haseltine WA. Replication and pathogenesis of the AIDS virus. J Acquir Immune Defic Syndr 1988; 1:217-240.
61. Schols D, Pauwels R, Baba M, Desmyter J, De Clercq E. Syncytium formation and destruction of bystander CD4+ cells cocultured with T cells persistently infected with human immunodeficiency virus as demonstrated by flow cytometry. J Gen Virol 1989; 70:2397-2408.
62. Baba M, Schols D, Pauwels R, Nakashima H, De Clercq E. Sulfated polysaccharides as potent inhibitors of HIV-induced syncytium formation: a new strategy towards AIDS chemotherapy. J Acq Immune Defic Syndr 1990; 3:493-499.
63. Montefiori DC, Robison WE, Modliszewski A, Rowland JM, Schuffman SS, Mitchell WM. Differential inhibition of HIV-1 cell binding and HIV-1-induced syncytium formation by low molecular weight sulphated polysaccharides. J Antimicrob. Chemother 1990; 25:313-318.
64. Kowalski H, Potz J, Basiripour L, Dorfman T, Goh WC, Terwilliger E, Dayton A, Rosen C, Haseltine W, Sodroski J. Functional regions of the envelope glycoprotein of human immunodeficiency virus type 1. Science 1987; 237:1351-1355.
65. Schols D, Baba M, Pauwels R, De Clercq E. Flow cytometric method to demonstrate whether anti-HIV-1 agents inhibit virion binding to T4+ cells. J Acq Immune Def Syndr 1989; 2:10-15.
66. Bagasra O, Lischner HW. Activity of dextran sulfate and other polyanionic polysaccharides against human immunodeficiency virus. J Infect Dis 1988; 158:1084-1087.
67. Nakashima H, Yoshida O, Baba M, De Clercq E, Yamamoto N. Anti-HIV activity of dextran sulfate as determined under different experimental conditions. Antiviral Res 1989; 11:233-246.
68. Lederman S, Gulick R, Chess L. Dextran sulfate and heparin interact with CD4 molecules to inhibit the binding of coat protein (gp120) of HIV. J Immunol 1989; 143:1149-1154.
69. Schols D, Baba, M, Pauwels R, Desmyter J, De Clercq E. Specific interaction of aurintricarboxlic acid with the human immunodeficiency virus/CD4 cell receptor. Proc Natl Acad Sci USA 1989; 86:3322-3326.

70. Lederman S, Bergmann JE, Cleary AM, Yellin MJ, Fusco PJ, Chess L. Sulfated polyester interactions with the CD4 molecule and with the third variable loop domain (V3) of gp120 are chemically distinct. AIDS Res Hum Retrovir 1992; 8:1599-1610.
71. Thiele B, Braig HR, Ehm I, Kunze R, Ruf B. Influence of sulfated carbohydrates on the accessibility of CD4 and other CD molecules on the cell surface and implications for human immunodeficiency virus infection. Eur J Immunol 1989; 19:1161-1164.
72. Parish CR, Low L, Warren HS, Cunningham AL. A polyanion binding site on the CD4 molecule. Proximity to the HIV-gp120 binding region. J Immunol 1990; 145:1188-1195.
73. Callahan LN, Phelan M, Mallinson M, Norcross MA. Dextran sulfate blocks antibody binding to the principal neutralizing domain of human immunodeficiency virus type 1 without interfering with gp120-CD4 interactions. J Virol 1991; 65:1543-1550.
74. Batinic D, Robey FA. The V3 region of the envelope glycoprotein of human immunodeficiency virus type 1 binds sulfated polysaccharides and CD4-derived synthetic peptides. J Biol Chem 1992; 267:6664-6671.
75. Mbemba E, Chams V, Gluckman J-C, Klatzmann D, Gattegno L. Molecular interactions between HIV-1 major envelope glycoprotein and dextran sulfate. Biochim Biophys Acta 1992; 1138:62-67.
76. Schols D, Pauwels R, Desmyter J, De Clercq E. Dextran sulfate and other polyanionic anti-HIV compounds specifically interact with the viral gp120 glycoprotein expressed by T-cells persistently infected with HIV-1. Virology 1990; 175:556-561.
77. Cardin AD, Taylor DL, Krstenansky JL, Tyms AS, Jackson RL. A heparin which binds to the envelope glycoprotein gp120 inhibits human immunodeficiency virus replication. Trans Assoc Am Physicians 1989; 102:101-109.
78. Jackson RL, Busch SJ, Cardin AD. Glycosaminoglycans: molecular properties, protein interactions, and role in physiological processes. Physiol Rev 1991; 71:481-539.
79. Stegmann T, Doms RW, Helenius A. Protein-mediated membrane fusion. Annu Rev Biophys Chem 1989; 18:187-211.
80. WuDunn D, Spear PG. Initial interaction of herpes simplex virus with cells is binding to heparan sulfate. J Virol 1989; 63:52-58.
81. Sawitzky D, Hampl H, Habermehl K-O. Comparison of heparin-sensitive attachment of pseudorabies virus (PRV) and herpes simplex virus type 1 and identification of heparin-binding PRV glycoproteins. J Gen Virol 1990; 71:1221-1225.
82. Lycke E, Johansson M, Svennerholm B, Lindahl U. Binding of herpes simplex virus to cellular heparan sulfate, an initial step in the adsorption process. J Gen Virol 1991; 72:1131-1137.

83. Neyts J, Snoeck R, Schols D, Balzarini J, Esko JD, Van Schepdael A, De Clercq E. Sulfated polymers inhibit the interaction of human cytomegalovirus with cell surface heparan sulfate. Virology 1992; 189:48-58.
84. Robbins AK, Watson RJ, Whealy ME, Hays WW, Enquist LW. Characterization of a pseudorabies virus glycoprotein gene with homology to herpes simplex virus type 1 and type 2 glycoprotein C. J Virol 1986; 58:339-347.
85. Kari B, Gehrz R. Structure, composition and heparin binding properties of a human cytomegalovirus glycoprotein complex designated gC-II. J Gen Virol 1993; 74:255-264.
86. Kjellén L, Lindahl U. Proteoglycans: Structures and interactions. Annu Rev Biochem 1991; 60:443-475.
87. Hattori T, Koito A, Takatsuki K, Kido H, Katunuma N. Involvement of tryptase-related cellular protease(s) in human immunodeficiency virus type 1 infection. FEBS Lett 1989; 248:48-52.
88. Skinner MA, Ting R, Langlois AJ, Weinhold KJ, Lyerly HK, Javaherian K, Matthews TJ. Characteristics of a neutralizing monoclonal antibody to the HIV envelope glycoprotein. AIDS Res Hum Retrovir 1988; 4:187-197.
89. Gallaher WR. Detection of a fusion peptide sequence in the transmembrane protein of human immunodeficiency virus. Cell 1987; 50:327-328.
90. Nakashima H, Kido N, Kobayashi N, Motoki Y, Neushul M, Yamomoto N. Purification and characterization of an avian myeloblastosis and human immunodeficiency virus reverse transcriptase inhibitor, sulfated polysaccharides extracted form sea algae. Antimicrob Agents Chemother 1987; 31:1524-1528.
91. Sydow G, Klöcking H-P. Effect of pentosan polysulfate (SP54) on the reverse transcriptase activity of several retroviruses. Biomed Biochim Acta 1987; 46:527-530.
92. Moelling K, Schulze T, Diringer H. Inhibition of human immunodeficiency virus type 1 RNaseH by sulfated polyanions. J Virol 1989; 63:5489-5491.
93. Cronn RC, Whitmer JD, North TW. RNaseH activity associated with reverse transcriptase from feline immunodeficiency virus. J Virol 1992; 66:1215-1218.
94. Larder BA, Darby G, Richman DD. HIV with reduced sensitivity to zidovudine (AZT) isolated during prolonged therapy. Science 1989; 243:1731-1734.

95. St. Clair MH, Martin JL, Tudor-Williams G, Bach MC, Vavro CL, King DM, Kellam P, Kemp SD, Larder BA. Resistance to ddI and sensitivity to AZT induced by a mutation in HIV-1 reverse transcriptase. Science 1991; 245:1557-1559.
96. Hayashi S, Fine RL, Chou T-C, Currens MJ, Broder S, Mitsuya H. In vitro inhibition of the infectivity and replication of human immunodeficiency virus type 1 by combination of anti-retroviral 2',3'-dideoxynucleosides and virus-binding inhibitors. Antimicrob Agents Chemother 1990; 34:82-88.
97. Anand R, Nayyar S, Galvin TA, Merril CR, Bigelow LB. Sodium pentosan polysulfate (SPP), as anti-HIV agent also exhibits synergism with AZT, lymphoproliferative activity, and virus enhancement. AIDS Res Hum Retrovir 1990; 6:679-689.
98. Anand R, Nayyar S, Pitha J, Merril CR. Sulfated sugar alfa-cyclodextrin sulfate, a uniquely potent anti-HIV agent, also exhibits marked synergism with AZT, and lymphoproliferative activity. Antiviral Chem Chemother 1990; 1:41-46.
99. Johnson VA, Barlow MA, Chou TC, Fisher RA, Walker BD, Hirsch, MS, Schooley RT. Synergistic inhibition of human immunodeficiency virus type 1 (HIV-1) replication in vitro by recombinant soluble CD4 and 3'-azido-3'-deoxythymidine. J Infect Dis 1989; 159:837-844.
100. Johnson VA, Walker BD, Barlow MA, Paradis TJ, Chou TC, Hirsch MS. Synergistic inhibition of human immunodeficiency virus type 1 and type 2 replication in vitro by castanospermine and 3'-azido-3'-deoxythymidine. Antimicrob Agents Chemother 1989; 33:53-57.
101. Smith MS, Brian EL, De Clercq E, Pagano JS. Susceptibility of human immunodeficiency virus type 1 replication in vitro to acyclic adenosine analogs and synergy of the analogs with 3'-azido-3'-deoxythymidine. Antimicrob Agents Chemother 1989; 33:1482-1486.
102. Eriksson BFH, Schinazi RF. Combinations of 3'-azido-3'-deoxythymidine (zidovudine) and phosphonoformate (foscarnet) against human immunodeficiency virus type 1 and cytomegalovirus in vitro. Antimicrob Agents Chemother 1989; 33:663-669.
103. Buckheit RW Jr, Germany-Decker J, Hollingshead MG, Allen LB, Shannon WM, Janssen PAJ, Chirigos MA. Differential antiviral activity of two TIBO derivatives against the human immunodeficiency and murine leukemia viruses alone and in combination with other anti-HIV agents. AIDS Res Hum Retrovir 1993; 9:1097-1106.
104. Hartshorn KL, Vogt MW, Chou TC, Blumberg RS, Byington R, Schooley RT, Hirsch MS. Synergistic inhibition of human immunodeficiency virus in vitro by azidothymidine and recombinant alpha A interferon. Antimicrob Agents Chemother 1987; 31:168-172.

105. Abrams DI, Kuno S, Wong R, Jeffords K, Nash M, Molaghan JB, Gorter R, Ueno R. Oral dextran sulfate (UA001) in the treatment of the acquired immunodeficiency syndrome (AIDS) and AIDS-related-complex. Ann Intern Med 1989; 110:183-188.
106. Hartman NR, Johns DG, Mitsuya H. Pharmacokinetic analysis of dextran sulfate in rats as pertains to its clinical usefulness for therapy of HIV infection. AIDS Res Hum Retrovir 1990; 6:805-812.
107. Flexner C, Barditch-Crovo PA, Kornhauser DM, Farzadegan H, Nerhood LJ, Chaisson RE, Bell KM, Lorentsen KJ, Hendrix CW, Petty BG, Lietman PS. Pharmacokinetics, toxicity, and activity of intravenous dextran sulfate in human immunodeficiency virus infection. Antimicrob Agents Chemother 1991; 35:2544-2550.
108. Maffrand J-P, Herbert J-M, Bernat A, Defreyn G, Delebassee D, Savi P, Pinot J-J, Sampol J. Experimental and clinical pharmacology of pentosan polysulfate. Seminars Thromb Hemost 1991; 17:186-198.
109. Peters M, Ruf B, Pohle HD. Intravenous versus oral application of pentosan polysulfate (HOE/BAY 946). Vth International Conference on AIDS, Montreal, Canada, June 1989 (abstract WBP299).
110. Peters M, Witvrouw M, De Clercq E, Ruf B. Pharmacokinetics of intravenous pentosan polysulphate (HOE/BAY 946) in HIV-positive patients. AIDS 1991; 5:1534-1535.
111. MacGregor IR, Dawes J, Pepper DS, Prowse CV, Stocks J. Metabolism of sodium pentosan polysulfate in man measured by a new competitive binding assay for sulfated polysaccharides - comparison with effects upon anticoagulant activity, lipolysis and platelet α-granule proteins. Thromb Haemost 1985; 53:411- 414.
112. Dawes J, Bara L, Billaud E, Samama M. Relationship between biological activity and concentration of a low-molecular-weight heparin (PK 10169) and unfractionated heparin after intravenous and subcutaneous administration. Haemostasis 1986; 16:116-122.
113. Gerard M, Weerts D, De Wit S, Sommereyns B, Witvrouw M, Schols D, Desmyter J, De Clercq E. Pentosan polysulfate (PPS) in HIV patients (P): a phase 1 study. 30th ICAAC, Atlanta, Georgia, October 1990.
114. Moriya T, Kurita H, Matsumoto K, Otake T, Mori H, Morimoto M, Ueba N, Kunita N. Potent inhibitory effect of a series of modified cyclodextrin sulfates (mCDS) on the replication of HIV-1 in vitro. J Med Chem 1991; 34:2301-2304.
115. Moriya T, Saito K, Kurita H, Matsumoto K, Otake T, Mori H, Morimoto M, Ueba N, Kunita N. A new candidate for an anti-HIV-1 agent: Modified cyclodextrin sulfate (mCDS71). J Med Chem 1993; 36:1674-1677.

116. Biesert L, Suhartono H, Winkler I, Meichsner C, Helsberg M, Hewlett G, Klimetzek V, Molling K, Schlumberger HD, Schrinner E, Brade H-D, Rübsamen-Waigmann H. Inhibition of HIV and virus replication by polysulfated polyxylan: HOE/BAY 946, a new antiviral compound. AIDS 1988; 2:449-457.

117. Weiler BE, Schroder HC, Stefanovich V, Stewart D, Forrest JM, Allen LB, Bowden BJ, Kreuter MH, Voth R, Muller WE. Sulphoevernan, a polyanionic polysaccharide, and the narcissus lectin potently inhibit human immunodeficiency virus infection by binding to viral envelope protein. J Gen Virol 1990; 71:1957-1963.

118. Furusawa E, Furusawa S, Chou SC. Antileukemic activity of Viva-Natural, a dietary seaweed extract, on Rauscher murine leukemia in comparison with anti-HIV agents, azidothymidine, dextran sulfate and pentosan polysulfate. Cancer Lett 1991; 56:197-205.

119. Mathes LE, Hayes KA, Swenson CL, Polas PJ, Weisbrode SE, Kociba GJ. Evaluation of antiviral activity and toxicity of dextran sulfate in feline leukemia virus-infected cats. Antimicrob Agents Chemother 1991; 35:2147-2150.

120. Offensperger WB, Offensperger S, Walter E, Blum HE, Gerok W. Inhibition of duck hepatitis B virus infection by lysosomotropic agents. Virology 1991; 183:415-418.

121. Wells CL, Rhame FS. Effect of oral dextran sulfate on the mouse intestinal tract. J Acquir Immune Defic Syndr 1990; 3:361-365.

122. Nkowane BM. Prevalence and incidence of HIV infection in Africa: a review of data published in 1990. AIDS 1991; 5:S7-S16.

123. Lange JMA, Karam M, Piot P. Boost for vaginal microbicides against HIV. Lancet 1993; 341:1356.

124. Pearce-Pratt R, Phillips DM. Studies of adhesion of lymphotropic cells: implications for sexual transmission of HIV. Biol Reprod 1993; 48:431-445.

125. Yarchoan RH, Mitsuya H, Myers CE, Broder S. Clinical pharmacology of 3'-azido-2',3'-dideoxythymidine (zidovudine) and related dideoxynucleo-sides. N Engl J Med 1989; 321:726-738.

126. Neyts J, De Clercq E. Effect of polyanionic compounds on intracutaneous and intravaginal herpesvirus infection in mice: impact on the search for vaginal microbicides with anti-HIV activity. J Acquir Immun Defic Syndr Human Retrovir 1995; 10:8-12.

6
Carbohydrates with Anticoagulant and Antithrombotic Properties

Susanne Alban
University of Regensburg
Regensburg, Germany

I. INTRODUCTION

After a long period in which carbohydrates in biology and medicine were more or less neglected, a resurgence in interest has occurred in the consideration of carbohydrates as molecules which play an important role in biological processes. Carbohydrates as a class of biomolecules have important functions in biological systems; therefore it is not surprising that several drugs currently available contain defined carbohydrate structures. Through discoveries in the new discipline glycobiology, this class of therapeutics may gain even more importance in the future. Nevertheless, carbohydrate containing drugs have already a long tradition beginning with the use of cardiotonic glycosides in the ancient Egypt [1]. The second biomolecule introduced in therapy after the polypeptide insulin was the polysaccharide heparin which has been used in the prophylaxis and treatment of thrombosis since 1935. Consequently, from a historical point of view, the use of carbohydrates as anticoagulants and antithrombotics is especially remarkable.

Although to date heparin is the drug of choice in prophylaxis and treatment of thrombosis, there are some related problems with the consequence that considerable research efforts are now underway in the search for alternatives to traditional heparin. There are several different strategies in the development of newer antithrombotics [2], such as antiplatelet drugs, endothelial lining modulators, viscosity modulators, biotechnology-based proteins, synthetic peptides, synthetic antithrombin nonpeptide agents, peptide conjugates, enzymes, venoms and finally glycosaminoglycan (GAG) derived drugs including sulfated polysaccharides, which shall be the subject of this chapter.

This carbohydrate group of potential therapeutics is currently the topic of intensive research and developments in this area continue to be reported [3]. Besides the chemical variants of heparin itself, sulfated polysaccharides generally represent a large pool of posssible antithrombotic drugs. Among these so-called heparinoids are included the natural GAG occurring in vertebrates, the sulfated polysaccharides of marine or bacterial origin, and the all- or semi-synthetically produced sulfated oligo- and polysaccharides.

II. HEPARIN AND LOW MOLECULAR WEIGHT HEPARINS

A. Heparin

1. History

In 1916, a medical student Jay McLean isolated an anticoagulating substance from dog liver which consequently was called 'heparin' by William H. Howell. Shortly afterwards, heparin was also found in other organs like lung and intestine [4]. On the basis of the original mode of isolation heparin was initially assumed to be a phospholipid. In 1935 Jorpes succeeded in establishing heparin as a complex polysaccharide, namely a GAG consisting of uronic acid and hexosamine residues. Due to the high content of sulfate groups, Jorpes identified heparin as the most highly charged polyanion of natural origin [5]. Concurrently, the use of heparin in the prevention of postoperative thrombosis was initiated. Even today, heparin is the drug of choice in the prophylaxis as well as in the therapy of thrombosis.

2. Occurrence, physiological significance, pharmaceutical production

Heparin occurs in the tissue of vertebrates, especially in the lung, liver, intestine mucosa, thymus, spleen, heart, lymph, vascular endothelium and plasma, where it is stored exclusively in the granules of basophilic granulocytes and mast cells, respectively [6]. It forms complexes with histamine and proteins. Its turnover is very slow and it is released only after cell lysis. Most likely, heparin does not act as an anticoagulant in the organism, but possibly has an important function for cell protection by binding of proteinases [7].

Since heparin cannot be totally synthesized due to its structural complexity, for industrial-scale production it has to be extracted from mammalian tissue, mainly bovine lung and porcine or bovine intestinal mucosa.The extraction of 1 kg starting material followed by complex formation, fractionating precipitation, alkaline treatment and bleaching results in only 150 mg heparin, the dose for two to three days of thrombosis prophylaxis for one patient.

Individual manufacturers use different sources for their heparins, which respectively show certain structural differences [8], and in addition, they use various methods of isolation and purification. Since the establishment of an internationally accepted standardization has failed so far, a multitude of different commercial heparin preparations exist with large variations in their chemical and physiological properties.

3. Biosynthesis

In addition to the multiplicity of heparin sources and the technology of production, a further reason for the lack of comparability of different heparin preparations is the chemical complexity of this natural product. Heparin is not a homogeneous substance consisting of an unique sequence of carbohydrate residues, but a mixture of different GAG chains. Its unbranched carbohydrate backbone is basically simple and consists of uronic acids, i.e. D-glucuronic (GlcA) or L-iduronic acid (IdoA), and D-glucosamine (GlcN) units, joined in an alternating sequence through 1,4-glycosidic linkages. The variable location of N-acetyl, N-sulfate and O-sulfate groups on these three units gives rise to at least 10 differing monosaccharide units that can be combined into a large number of different saccharide sequences. This great structural variety can be explained by the mechanisms of the heparin biosynthesis [9].

As usual for proteoglycans, a short chain core protein rich in serine and glycine is first synthesized on the rER. The formation of the polysaccharide chains is initiated by xylosylation of serine units, and continues by transfer of D-galactose residues to yield the galactosyl-galactosyl-xylosyl trisaccharide linkage region that connects the polysaccharide chain proper with the core protein [10]. Subsequent chain elongation occurs by alternating transfer of N-acetyl D-glucosamine (GlcNAc) and D-glucuronic acid (GlcA) monosaccharide units, from the corresponding UDP sugar nucleotides, to the non-reducing termini of the nascent polymers.

The polysaccharides consisting of about 300 monosaccharide units are subsequently modified in a stepwise manner, through the following series of reactions: (a) N-deacetylation of GlcNAc units; (b) N-sulfation of the resulting free amino groups; (c) C-5-epimerization of GlcA to IdoA residues; (d) O-sulfation at C-2 of IdoA residues; and (e) O-sulfation at C-6 of GlcN units.

Due to the stepwise nature of the process, and the substrate specificities of the enzymes involved, the product of any given reaction will be the substrate for the subsequent reaction. The selection of target units is restricted by the substrate specificities of the enzymes catalyzing the various reactions [11]. Polymer modification is incomplete in the sense that each enzyme acts upon only a fraction of those monosaccharide units that satisfy its substrate specificty (Figure 1). Consequently, the structural complexity and heterogeneity of the polysaccharide chain under modification increase throughout the process. The extent of the initial N-deacetylation/N-sulfation reaction will determine the overall degree of polymer modification, and is therefore a factor of prime importance in the regulation of the process.

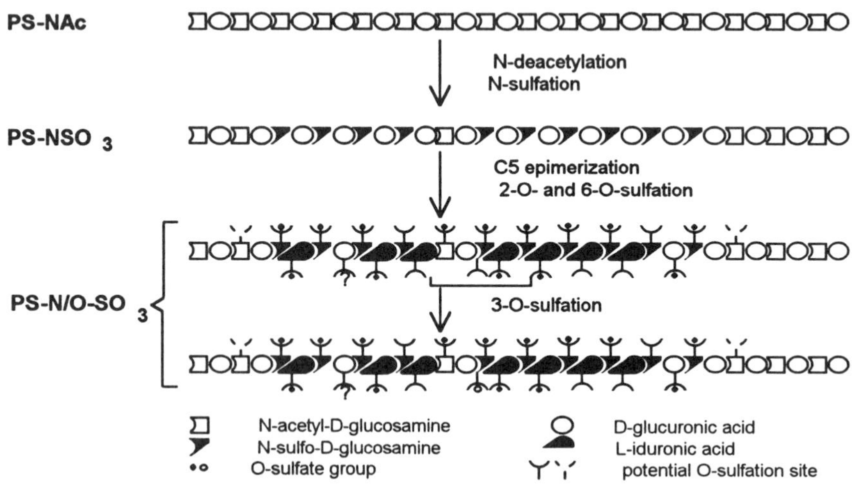

Figure 1 Scheme of polymer modification in the biosynthesis of heparin/heparan sulfate. The initial polymerization product (PS-NAc), consisting of GlcNAc and GlcA units in alternating sequences, traverses a series of modification reactions, each of which will affect a fraction of disaccharide units. The scheme illustrates the role of N-sulfate groups in the subsequent modification reactions. Both, O-sulfation, and GlcA C-5 epimerization thus are restricted to already N-sulfated regions of the polymer. The symbols have been chosen to illustrate the substrate specificity of the epimerase (the reducing terminus of the chain is to the right). A fraction of the potential GlcA target residues escapes epimerization and some O-sulfation sites remain unoccupied. The pentasaccharide sequence indicated by the bracket comprises a substrate recognition site in the GlcN-3-O-sulfotransferase reaction that concludes the formation of the "antithrombin III-binding site" as well as the overall modification process (the incorporated 3-O-sulfate residue is indicated by an open symbol). The GlcA unit shown to be sulfated (at C-2 or C-3) has been arbitrarily selected; current information implies only that it is located between two $GlcNSO_3$ units. (Redrawn from Ref. 9.)

The product with a high proportion (generally > 80 %) of N-sulfated GlcN residues is usually heavily O-sulfated, shows high IdoA/GlcA ratio, and is identified as "heparin". Conversely, the lesser modified product showing a lower proportion of N-sulfated GlcN units, a lower IdoA content, and a lower degree of O-sulfation is called "heparan sulfate".

After synthesis of the heparin proteoglycan (Figure 2), the GAG chains are cleaved by special peptidases in the endosomes [12]. Endoglycosidases degrade the chains and thus reduce the molecular weight (MW) to an average of 13 kd.

4. Chemistry and structure

There are three different levels of complexity of commercial heparin. First, heparin is a polydisperse mixture of GAGs with MW ranging from 2 to 50 kd. Even the heparin chain corresponding to the most prevalent degree of

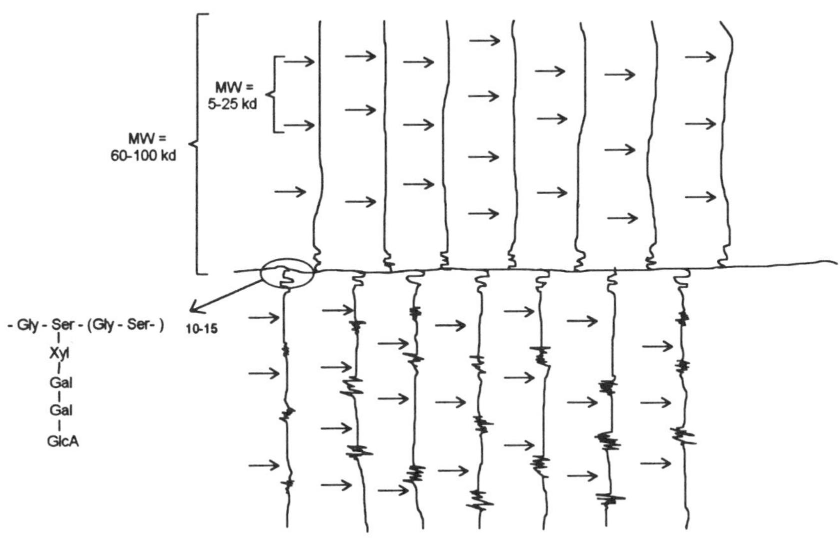

Figure 2 Schematic representation of the heparin proteoglycan. The polysaccharide chains, bound to a polypeptide matrix, the core protein, through a linkage region (circled), are cleaved by tissue endoglycosidases (arrows) after completion of postpolymer biosynthetic transformations. Irregular line region = AIII-binding site. (Redrawn from Ref. 21. with permission from S. Karger AG.)

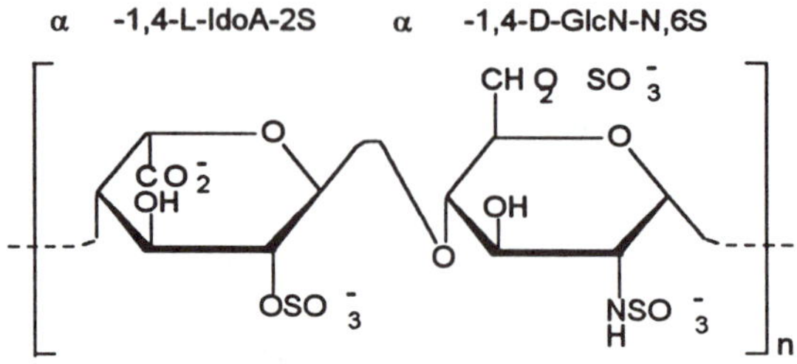

Figure 3 The disacchararide unit representing 60-90 % of the heparin chain.

polymerization represents a mere 5 % MW of a typical GAG heparin preparation [13]. The second level of complexity is associated with its primary structure or sequence [14]. Finally, there is variability of the secondary structure and the conformation in solution both of which have an important influence on the biological activities of heparin as well [15]. Consequently, by combination of several separation methods, more than 100 fractions of one heparin preparation may be obtained whose distribution pattern is different for each heparin resulting in a specific 'fingerprint' [16].

By identifying and quantifying the various monosaccharide units of heparin, the trisulfated disaccharide unit (1→4)-O-(α-L-idopyranosyluronic acid 2-sulfate)-(1→4)-O-(2-deoxy-2-sulfamino-α-D-glucosamine 6-disulfate) was shown to represent 60-90 % of the heparin chain (Figure 3). The remaining 10-40% consists of other disaccharide units having unsulfated GlcA instead of the IdoA-2S and/or GlcNAc instead of $GlcNSO_3$. In addition, the disaccharide units can differ in their degree and position of sulfation [17].

An important tool for establishing the structural features of GAGs is the so-called "oligosaccharide mapping technique" using chemical, enzymatic, and spectroscopic techniques originally developed to sequence nucleic acids and proteins [17]. It could be shown that the sequence in heparin is not accidental, but at least certain regions have defined sequences performing special functions. The best example is the "ATIII-binding site" (Figure 4), a specific pentasaccharide sequence which mediates the binding between heparin and certain lysyl residues on the plasma proteinase inhibitor antithrombin III (ATIII) [19,20].

As shown in Figure 2, the specific pentasaccharide sequence is present in only a fraction of chains of commercial heparins. Heparins obtained from porcine or bovine intestinal mucosa or from beef lung usually contain not more than one-third of chains with the active site for ATIII binding which account for most of the anticoagulant activity and make an important contribution to the antithrombotic properties [21]. About two thirds of the chains have a low affinity for ATIII. The relative proportion of low and high affinity chains in

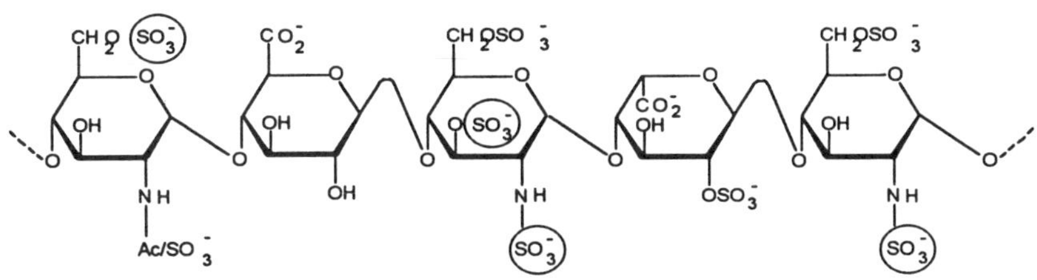

Figure 4 The ATIII-binding site in the heparin chain. The pentasaccharide sequence is composed of three GlcN units, a GlcA unit and an IdoA unit. The 3-O-sulfate group of the central GlcN (circled) is a marker component for the ATIII-binding site. The two further circled sulfate groups, as well as the GlcA unit, have been proven essential for the high affinity binding of ATIII.

unfractionated heparins (UH) varies according to the biological origin. Thus, clam heparin, which is much more heterogeneous than heparins present in mammalisan tissues, contains more of the pentasaccharide sequence than mammalian heparins [22].

5. Biological activities and medical applications

Although the biological role of endogenous heparin is not completely understood, this has not precluded its use for a variety of medical indications. Its use as an antithrombotic drug was the first and is still the most important therapeutic application. For six decades, heparin has been the drug of choice in the prevention and treatment of thromboembolic disorders. Given pre- and postoperatively, heparin reduces the frequency of deep vein thrombosis (DVT) and pulmonary embolism by up to 70% [23]. Further, heparin is used for the therapy of DVT and other venous as well as arterial thromboembolic disorders including pulmonary embolism, acute myocardial infarction, unstable angina pectoris, acute peripheral vascular occlusion, cerebral insult and disseminated intravascular coagulation (DIC Heparin is administered during most extracorporal therapy and hemodialysis [24]. Afterwards, it is neutralized by protamine sulfate or enzymatically eliminated by heparinase [25]. Due to its antiproliferative effect, heparin is presently under investigation for use after percutaneous transluminal coronary angioplasty (PTCA, restoration of the blood flow in obstructed arteries), in order to reduce the high rate of restenosis (30-40%) in which uncontrolled arterial smooth muscle cell proliferation is an important step [1].

Since orally administered heparin is inactive, it has to be injected intravenously. Despite its reduced bioavailability, it is also given by the subcutanous route.

Finally, sport injuries, bruises, varicose vein symptoms, surface thromboses and thrombophlebitis are indications for the local use of heparin.

Besides the effects on blood coagulation including fibrinolysis, heparin has a great variety of other activities. It releases several endogenous factors, such as lipoprotein lipase from the endothelium [26], forms complexes with lipoproteins [27] and inhibits smooth muscle cell proliferation [28], important components of atherogenesis [29]. The antiproliferative properties also are of interest with regard to the tumor therapy [30].

In various ways, heparin also influences angiogenesis playing an important role in tumor progression among other things. It stimulates endothelial cell migration and shows interactions with growth factors [31].

Moreover, heparin modulates the immune system and is able to regulate complement resulting in antiinflammatoric properties [32]. Finally, the antiviral effect, especially the inhibition of herpes-simplex virus type 1, has to be mentioned [33].

6. Relationship between heparin and other sulfated polysaccharides

Heparin's biological activities are primarily mediated through its binding and regulation of proteins. Up to now, the only heparin-protein interaction that has been demonstrated to be restricted to a specific saccharide sequence is the binding of heparin on ATIII mediated by the specific pentasaccharide sequence (34). Otherwise, the binding is non-specific and primarily based on electrostatic interactions. In fact like heparin, other sulfated polysaccharides bind to so-called heparin binding proteins such as serine proteinase inhibitors like heparin cofactor II (HCII), thrombin, platelet factor 4 (PF_4), histidine-rich glycoprotein, elastase, viral coat proteins, transcription factors, adhesive matrix proteins like laminin, vitronectin, fibronectin, thrombospondin and collagens, growth factors and proteins involved in lipid metabolism like lipoprotein and hepatic triglyceride lipase and apolipoproteins B and C [35]. Both, charge density and length of the charged segments along the polyanion chain or the total length of chains with homogeneously distributed charges affect the strength of these polyelectrolyte interactions. Thus, non-specific binding of polysulfates to basic proteins and to the endothelium increases with increasing degree of sulfation (DS) and MW of the polysaccharide [36].

Nevertheless, further cases where specific structures are part of the polysaccharide-protein interaction cannot be excluded as shown by the binding of the relatively low-sulfated heparan sulfate to thrombin and low-density lipoproteins [37, 38]. It has to be considered that only a few suitably spaced anionic groups in the carbohydrate chain could often achieve a strong binding to clusters of basic residues in certain regions of proteins [39].

Furthermore, molecular flexibility seems to play a role in modulating the binding and biological properties of GAG. An easy change in shape, favored by a high content of IdoA residues, provides a mechanism for a closer, more effective approach of the anionic groups of the GAG to appropriate basic groups of "receptors" than achievable with polysaccharide chains having the same charge density but containing a more rigid structure by the presence of GlcA [40].

These facts explain why heparin not only interferes with blood coagulation but influences many other physiological processes and why similar activities are also to be found with a great number of other sulfated polymers.

7. Mechanism of the antithrombotic activity

The antithrombotic action of heparin is due mainly to its ATIII-mediated blood anticoagulant activity.The α_2-globuline ATIII functions as the principal plasma inhibitor of most blood coagulation proteinases [20, 41]. This serine proteinase inhibitor forms tight, irreversible, equimolar complexes with its target enzymes by formation of an ester between an arginine residue of its active center and the serine residue of the active center of the enzyme. This slow, time-, temperature- and pH-dependent process is accelerated many thousandfold by heparin. An essential component of the mechanism of heparin's effect is the binding of a lysine-rich region of ATIII to the highly specific ATIII-binding site in the heparin molecule (Figure 5). This reversible, electrostatic interaction induces a conformational change in ATIII which considerably reinforces its anticoagulant activity. In agreement with this conformational change hypothesis, heparin oligosaccharides as small as the pentasaccharide were observed to be as effective as larger heparin chains in accelerating factor Xa (FXa) inhibition by ATIII, provided they contain the active site for ATIII.

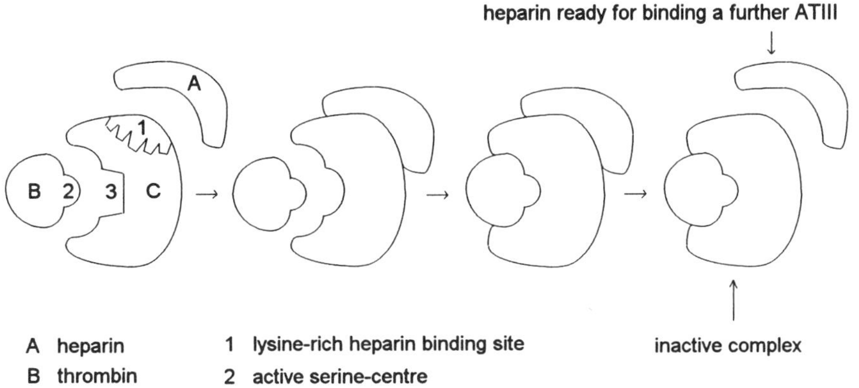

Figure 5 Mechanism of the rate enhancing effect of heparin on the enzyme inhibiting ATIII. (Redrawn from Ref. 42.)

This indicates that heparin chains just large enough to bind ATIII are sufficient to account for heparin's rate-enhancing effect on the ATIII-FXa reaction. In contrast, small heparin-derived oligosaccharides were essentially inactive in accelerating thrombin inhibition by ATIII. Instead, significant accelerating activity was not observed until the chain length reached 18 saccharides (Figure 6). To explain this chain length dependence of heparin, a second mechanism was postulated that heparin was acting as a surface or bridge to get ATIII and the enzyme closer together by the binding of both proteins to the same heparin chain, which requires a larger chain length.

By accelerating the thrombin inactivation rate of ATIII, heparin influences thrombin generation, the crucial process for thrombogenesis, in two ways [44]: it neutralizes thrombin traces required for feedback activation, that is the activation of FV, FVIII and of the platelets by thrombin, as well as the bulk amount of thrombin which has been generated due to activation of coagulation. Apart from its ATIII mediated action, heparin also stimulates the inactivation of thrombin by HCII. Finally, it directly interferes with some coagulation factors and modulates the enzymatic reactions by interaction with cofactors like Ca^{2+} and phospholipids [45, 30]. Here, displacement of the enzymes from the phospholipid matrix with consequential destruction of the tenase and prothrombinase and the interaction with the thrombin-catalyzed FVIII activation are discussed [46].

The antithrombotic activity of heparin is not limited to its anticoagulant effect; it also influences other factors significant for thrombogenesis.

Thus, heparin increases the electronegative potential and consequently the

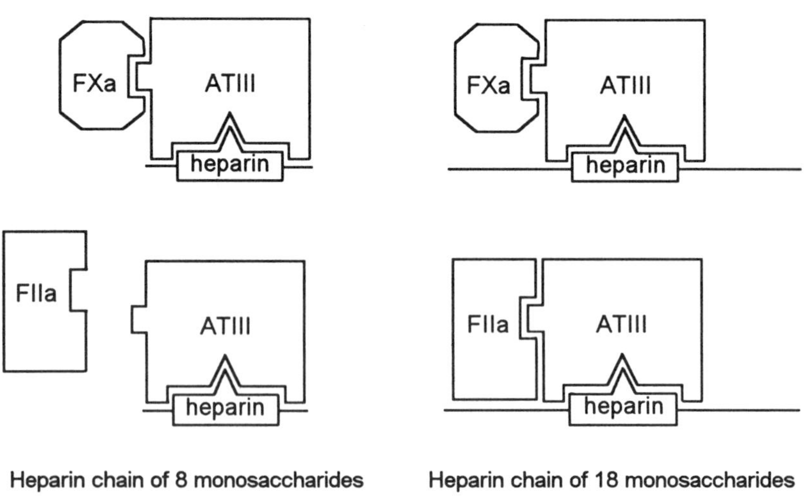

Figure 6 The influence of the heparin chain length on the inhibition of FXa and thrombin, respectively, by ATIII. (Redrawn from Ref. 43.)

athrombogenic character of the vessel wall [32]. Possibly, it increases the production and release of anticoagulantly active HS from the endothelium [47]. It affects platelet function and the release of substances from the platelets important to the hemostasis [48]. For instance, it stimulates the release of PF_4, a heparin neutralizing protein. From the endothelium, it releases the tissue factor pathway inhibitor (TFPI), which facilitates anticoagulant and antithrombotic actions [49]. By forming complexes with fibrinogen and plasmin, heparin also influences fibrinolysis [50].

B. Problems Related to Heparin

Although heparin drastically reduces the incidence of thrombosis and pulmonary embolism in hospitalized patients and also is used successfully in the treatment of thromboembolic disorders, physicians are not able to completely eliminate the occurence of thromboembolic events [51]. The major side effects of heparin are hemorrhages ranging from minor bleeding to severe, life threatening hemorrhages like intracranial, retroperitoneal or intraocular bleeding [52]. Hemorrhages occur even with low dose regimens of heparin (5000 units every 8 or 12 hours).

A further, serious side effect is heparin-induced thrombocytopenia (HIT) of Type II whose incidence is estimated to be at least 1-5 %. In contrast to the early onset of the mild Type I HIT which normally remains asymptomatic, this form appears later and is often associated with an increased risk of thrombotic complications [53, 54]. Type I HIT is caused by an immune mechanism, in which heparin bound to PF_4 or some other plasma proteins forms large immune complexes with antibodies [55]. These complexes bind to platelets resulting in platelet activation, reduced platelet survival, thrombocytopenia, and in severe cases, thrombosis [56]. In most Type I and all Type II patients, heparin treatment has to be terminated resulting in the need for an alternative antithrombotic drug which must not cross-react with the antibodies [57, 58].

It has to be considered that the action of heparin is dependent on ATIII so that it cannot be used in patients with hereditary or acquired ATIII deficiency. Since it is not effective either against clot-bound thrombin, there is a high risk of recurrent thrombosis especially after clot lysis and the release of the still active thrombin [59].

In addition, the prophylactic use of heparin in the prevention of DVT is not well accepted since two to three daily injections are required due to the the short half-life of heparin and its limited bioavailability when given subcutaneously [60]. Apart from these therapeutic aspects, there are some other disadvantages of heparin. This glycosaminoglycan is not homogeneous in MW, DS, and substitution pattern resulting in considerable variations in its physiological activities [61]. Since synthesis of the macromolecule is not economically viable, heparin has to be isolated from biological material with the risk of severe contaminations such as the bovine spongiforme encephalopathy (BSE) pathogen.

For all these reasons, despite its continuous importance in clinical medicine, the need for alternatives to heparin is obvious.

C. Low Molecular Weight Heparins

1. Introduction

In the past twenty years heparin itself has been the matter of intense research efforts. Important discoveries concerning its chemical and physico-chemical structure, structure-activity relationships, mechanism of anticoagulant action and pharmacological function have been the basis of attempts to develop new heparin derivatives with modified biological activities, improved pharmacological properties and facilitated modes of administration. Awareness of the influence of MW on biological properties and the original assumption that fragments with high anti-Factor Xa (anti-Xa) activity and high anti-Xa/ anti-thrombin activity ratio (aXa/aIIa ratio) would retain optimal antithrombotic activity with greatly reduced prohemorrhagic actions led to the development of the LMWH [62] whose efficacy is now well proven [24, 63-64].

2. Preparation and heterogeneity of LMWH

The typical LMWH consist of GAG chains of MW ranging from 2 to 8 kd with an average of 5 kd. However, the effort to acquire better defined, more homogeneous substances by reducing MW proved to be a misjudgement. Commercial LMWH are prepared by a variety of fragmentation methods [34, 39, 43], (Table 1): fractionation by gel chromatography or ultrafiltration, acid hydrolysis with sulfuric acid/chlorosulfonic acid, nitrous acid or isoamylnitrite, β-elimination by esterification and base treatment. Further, oxidative depolymerization methods are employed using periodate, hydrogen peroxide or hydroxyl and oxygen radicals, and finally, enzymatic treatment with polysaccharide lyases. Based on the method of preparation, different structural modifications are induced with the consequence that LMWH may differ considerably in their chemical and physical as well as in their biological and pharmacological properties, as for instance reflected by the aXa/aIIa ratios (Table 1). Consequently, despite many common properties, each LMWH has to be considered as an individual drug, which is critical from the point of view of standardization [43]. And still, like UH, each LMWH itself is a polydisperse, heterogeneous mixture of GAG.

Table 1 Various Commercial LMWH Preparations

Company	Code Numbers	INN (WHO)	Trade Name	Preparation method	MW [kd]	aXa/aIIa[a]
Choay/ Sanofi	CY216	Nardoparin calcium	Fraxiparine/ Seleparin	ethanol precipitation, nitrous acid depolymerization	4.5 (2.0-8.0)	3.2
Choay/ Sanofi	CY222	/	/	exhaustive nitrous acid depolymerization	2.5	?
KabiVitrum	Kabi 2165	Dalteparin sodium	Fragmin	nitrous acid depolymerization	5.0 (2.0-9.0)	2.0
Novo/ Leo	LHN1	Tinzaparin sodium	Logiparin/ Innohep	enzymatic depolymerization with heparinase	4.5 (3.0-6.0)	1.9
Pharmuka/ Rhône-Poulenc	PK10169	Enoxaparin sodium	Clexane/ Lovenox	β-elimination: benzylation followed by alkaline depolymerization	4.5 (3.0-8.0)	2.7
Opocrin	OP 2123	Parnaparin sodium	Fluxum	peroxidative depolymerization, ethanol extraction	3.5-5.0	?
Sandoz	/	Certoparin sodium	Sandoparin/ Mono-Embolex NM	isoamylnitrite depolymerization	6.0 (5.0-7.0)	2.0
Knoll	/	Reviparin sodium	Clivarin	nitrous acid depolymerization	3.9 (3.5-4.5)	5.0

[a]aXa/aIIa = ratio of the anti-Xa to the anti-thrombin activity

3. Comparison between UH and LMWH

Despite the differences between individual LMWHs, there are common differences from UH (Table 2). Many of these differences have favorable impact on the use of LMWH. Among other things, an important advantage of LMWH is that it may be administered subcutaneously only once a day due to its high bioavailability and longer half-life [43].

4. Mode of action

The original hypothesis for the connection between the anti-Xa activity and the antithrombotic effect as well as between the anti-IIa activity and the bleeding risk, formed the basis of the development of LMWH. Subsequently this was proven to be an oversimplification of the complex procedure of thrombogenesis [62]. Antithrombotic activity is not solely dependent on anti-Xa activity, or is a high

Table 2 Comparison between UH and LMWH

	UH	LMWH
Mean MW	13,000 d	5,000 d
MW range	3,000- 30,000 d	2,000-8,000 d
Activity in aPTT	strong	weak
Thrombin inhibition	strong	weak
Factor Xa inhibition	strong	strong
Anti-Xa/anti-IIa activity ratio	1	2-5
PF_4 release	existing	existing
Stimulation of ADP-induced platelet aggreg.	strong	weak
Inhibition of platelet adhesion to collagen	existing	none
Induction of thrombocytopenia (Type II)	1-5 %	probably less
Neutralization by PF_4 and other plasma proteins	strong	weak
Neutralization by protamin	yes	rest anti-Xa activity
Induction of lipolytic activity	strong	weak
Affinity for and binding to the endothelium	strong	weak
Half-life after intravenous injection	1 h	2 h
Bioavailability after subcutanous injection	10-25 %	90 %
Elimination	more cellular	more renal
Monitoring in prophylaxis	necessary	not necessary
Standardization	USP method	no uniform method
Product equivalence	yes	no

[a] aPTT = activated partial thromboplastin time

aXa/aIIa ratio necessarily an indicator of a stronger antithrombotic effect [65]. A certain degree of anti-IIa activity seems to be required for optimal antithrombotic action [66]. The prohemorrhagic effect is not only related to the anti-IIa activity, but is especially dependent upon interferences with the platelets and the vascular wall, the presence of impurities, etc.[67]. Consequently, the reduced bleeding risk of LMWH could be partly due to diminished interactions with platelets [43]. The reason for their well-documented antithrombotic activity is not their high aXa/aIIa ratio.

A special quality of LMWH is their antithrombotic effect *in vivo* in contrast to their low anticoagulant activity *in vitro* as well as ex vivo [60]. LMWH are supposed to produce their effects by influencing several hemostatic pathways, so that the degree of antithrombotic action is a cumulative result of these effects. Possibly, the mechanism of the plasmatic actions differs from that observed at cellular surfaces, such as the vasculature and the blood cells [60].

5. Use of LMWH in the prophylaxis and treatment of thrombosis

The introduction of LMWH has considerably ameliorated the initial treatment of DVT, and should now be regarded as the treatment of choice for most patients with DVTs [51]. A meta-analysis of the recent past showed that LMWH is at least as effective and safe as UH. A definite answer concerning the superiority of LMWH for the treatment of DVT awaits further studies. Investigations are currently underway evaluating the out-patient use of LMWH and their use in the treatment of acute pulmonary embolism [63].

Prophylaxis against postoperative venous thromboembolism with LMWH proved to be justified on the basis of efficacy, safety, better patient acceptability, decreased nursing time and cost efficacy [64]. Still, it has to be kept in mind that although the bleeding risk is reduced with the use of LMWH, it still remains [52, 63, 64]. Furthermore, LMWH is not a suitable alternative in HIT patients due to cross-reaction with the antibodies. Finally, just as UH, LMWH are heterogeneous mixtures of GAG molecules, their production is dependent on natural sources, and they still need ATIII for their action. Consequently, the search for an optimal antithrombotic drug is continuing.

D. Modifications of Heparin

To improve the properties of heparin like its bioavailability, the efficacy-safety ratio, or to modify its physiological activities, second and third generations of LMWH were created [60]. The second generation comprises ultra-low MWH (<2,500 d) and very-low MWH (<3,500 d) without anti-IIa activity, narrow cut LMWH (2,500 to 5,000 d) and wide range LMWH (2,500 to 10,000 d), medium molecular weight heparin (5,000 to 10,000 d), LMWH with varying degrees of ATIII affinity, and chemically modified such as oversulfated or, after deamination, acetylated LMWHs. The third generation represents chemically or affinity well-defined LMWH, oversulfated or in other ways chemically modified LMWH and heparin conjugates.

Chemical modification of heparin was under investigation more than twenty years ago in order to systematically elucidate the structure of the heparin components [68]. This studies included procedures like periodate oxidation, N-desulfation, O-desulfation, N-acetylation, oversulfation or carboxyl reduction.

An example of a chemically modified heparin is Suleparoide® which was introduced for the prophylaxis of thrombosis and the treatment of vascular disorders. It consists of a depolymerized, semisynthetic, N-desulfated and re-N-acetylated heparin derivative with a mean MW of about 9 kd thereby structurally resembling HS. *In vitro*, it is less active than heparin exhibiting its best effect on thrombin time (TT), which is primarily mediated by HCII, and a strong inhibition of intrinsic FXa generation [69]. However, significant *ex vivo* activity was only found in the Heptest®, which correlated well with the antithrombotic effect being much weaker than that of heparin in a modified Wessler model in rabbits. In addition to the antithrombotic effect, Suleparoide® exhibits a dose-dependent thrombolytic effect [70]. In contrast to the moderate effect in the thrombosis model, Suleparoide® was able to effectively maintain endothelial stability, an important response in thrombosis prevention [71].

In order to prolong the short half-life of heparin, several strategies have been explored. Apart from the previously described LMWH, another possibility is to ensure a permanent attachment of heparin or of heparin fragments to ATIII, by covalent coupling [72], or to increase the ATIII affinity of heparin by selective alteration of the pentasaccharide structure (see below). Another approach is the introduction of O-acyl residues into heparin, e.g. butyryl groups [73]. These heparin derivatives bind nonspecifically to plasma proteins resulting in a prolongation of the antithrombotic effect. Similar effects were found using other lipophilic substituents such as cholesterol hemisuccinate or L-ascorbic acid-6-palmitate [74].

III. MAMMALIAN GLYCOSAMINOGLYCANS

A. Naturally Occuring Mammalian Glycosaminoglycans

1. Introduction

Heparin belongs to the the group of mammalian GAG, which are found as polyanionic proteoglycans with important biological functions in both, intra- and extracellular locations. Relative to their chemical structure, seven groups are to be considered whose main disaccharide units are listed in Table 3. In contrast to other GAG, hyaluronic acid is not sulfated, and keratan sulfate contains no uronic acids.

Whereas the latter two do not play any role in the field of antithrombotic agents,

Table 3 The Main Disaccharide Units of Mammalian Glycosaminoglycans

Glycosaminoglycan	Disaccharide unit
Heparin	α-1,4-L-IdoA-2-S / α-1,4-D-GlcN-N,6-di-
Heparan sulfate	β-1,4-D-GlcA / α-1,4-D-GlcN-N-Ac-6-S
Chondroitin sulfate A	β-1,4-D-GlcA / β-1,3-D-GalN-N-Ac-4-S
Chondroitin sulfate C	β-1,4-D-GlcA / β-1,3-D-GalN-N-Ac-6-S
Dermatan sulfate (= chondroitin sulfate B)	α-1,4-L-IdoA / β-1,3-D-GalN-N-Ac-4-S
Keratan sulfate	β-1,3-D-Gal / β-1,4-D-GlcN-N-Ac-6-S
Hyaluronic acid	β-1,4-D-GlcA / β-1,3-D-GlcN-N-Ac

the other non-heparin GAG, originally byproducts of heparin, have caused great interest as antithrombotic drugs during the past few years, especially after the demonstration that the interaction with ATIII is not essential to produce antithrombotic effects [75, 76].

Thus, at present a number of mammalian GAG-derived drugs are used as antithrombotic agents. There are still some other indications for these natural products such as inflammations, hyperlipemia (e.g. Prisma®) and atherosclerosis, vascular disorders (e.g. Vessel®), AIDS and senile dementia (e.g. Arteroid®), wound healing, and finally the use as general cytoprotective agents in surgery or transplant procedures [77].

2. Heterogeneity of the glycosaminoglycan-derived drugs

However, quite a few products are empirically used without detailed information about their chemistry as well as pharmacology. They are mixtures of natural GAG obtained by extraction of different mammalian organs, whose compositions depend on the source of the material, the extraction method and the further treatment by fractionation and/or depolymerization. The most common sources are bovine and porcine tissue such as intestine mucosa, skin, spleen, kidney, pancreas, lung and trachea.

Therefore, each product represents an individual, polydisperse mixture of various GAG molecules resulting in a correspondingly complex pharmacological profile with many physiological functions, since each different GAG molecule shows a great number of interactions with endogenous humoral and cellular receptors.

Typically, many GAG produce their antithrombotic actions without extensively altering plasma clotting parameters [77]. In their action, indirect pathways may be involved such as the release of anticoagulant substance from the endothelium, i.e. TFPI, further the modulation of the endogenous GAG synthesis, the displacement of surface-bound GAG, or the influence on the blood rheology [77].

Due to their complexity, it is difficult to characterize these GAG-derived drugs, since common biochemical and pharmacological methods only reflect a few characteristics a time. Consequently, assay results strongly depend on the chosen test system and cannot be easily extrapolated to the clinical situation [77]. For lack of suitable alternatives, this class of antithrombotic agents is still characterized by classical methods. By combining as many as possible assays, this procedure may be valid for screening purpose.

Considering the development of glycosaminoglycan-derived drugs, the natural GAG present considerable therapeutic potential. However, much work has to be done to understand their mechanisms of action. For this, purification of the heterogeneous GAG mixtures, fractionation and isolation of the active components, chemical modification, as well as the synthesis of GAG analogues may be useful for the development of new drugs.

B. Heparan Sulfate

Heparan sulfate (HS), considered to be the biogenetic precursor of heparin, contains less sulfate groups and mainly GlcA as uronic acid. It occurs on the surfaces of virtually all animal cell types and in the extracellular matrix as a proteoglycan. It is a metabolically dynamic cell constituent which has a half-life on the cell surface of only 4 hours [78]. HS is generally a weaker anticoagulant *in vitro* than heparin and interacts neither with platelets nor with the antibodies in the serum of HIT patients [48, 79]. However, HS from different animals and tissues might have different structures and functions as well as antithrombotic activities. Anticoagulant activities *in vitro* are associated with the content of ATIII-binding material inidcating an ATIII dependent mechanism. They range from less than 5 U/mg to values similar to heparin with the best effect in the Heptest® [80]. According to Marcum et al. [81], such a specific population of HS having a high anticoagulant activity *in vitro* is responsible for the athrombogenic character of the vessel wall by binding and activating ATIII *in vivo*. In addition to the ATIII-mediated effect *in vitro*, HS binds directly to thrombin, probably through a certain oligosaccharide sequence [82]. Correspondingly, thrombin generation also is inhibited [83].

Despite its weak *in vitro* anticoagulant activity, HS is a potent antithrombotic agent *in vivo* [84]. As indicated by the low anticoagulant activity and the missing anti-Xa and anti-IIa activity ex vivo, the mode of action *in vivo* is different from that of heparin and, despite the *in vitro* results, independent of ATIII and HCII. Mechanisms, such as endogenous synthesis of GAG and displacement of surface-bound GAG may be primarily responsible for the antithrombotic activity [77]. Although the antithrombotic dosages of HS are higher than that of heparin, a better saftey index may be provided, since HS is 10-fold less hemorrhagic than heparin [85]. Prophylaxis with HS is in agreement with the concept to use substances which contribute to the biocompatibility of blood and vessel wall and predominate under normal conditions, without a risk of bleeding [86].

Clinically, HS has been studied in DVT, chronic venous insufficiency and intermittent claudication. Although only small numbers of patients were included in these trials, results obtained with this agent appear to be positive [77].

C. Dermatan Sulfate

Dermatan sulfate (DeS) which occurs mainly in mucosa and skin belongs to the chondroitin sulfate family which contains galactosamine instead of glucosamine. Its special quality is its high content of iduronic acid which confers high conformational flexibilitythereby favoring interactions with proteins [87]. Accordingly, the major difference between DeS from various species having low and high antithrombotic activity *in vivo* was shown to be due to their iduronic acid content.

Similar to HS, DeS is a relatively weak anticoagulant *in vitro* (70 times less than heparin), and apart from the inhibition of the thrombin-induced platelet aggregation [88], does not interact with platelets [48, 79]. DeS has neither ATIII-mediated anti-Xa nor anti-IIa activity. Its anticoagulant effect is mainly based on enhancement of HCII activity as shown by the lack of antithrombotic activity of a DeS hexadecasaccharide with low affinity to HCII [89]. To enhance HCII-mediated thrombin inhibition, fragments with at least 7-8 units are necessary and besides the major IdoA-GalNAc4SO_3 disaccharide unit, IdoA_2SO_3-conaining sequences are required [90]. Due to its anti-II activity, DeS is also able to inhibit thrombin generation and thrombin mediated amplification reactions, but less effectively than heparin [91, 92].

Recently, several DeS preparations have been developed for prophylaxis of venous thromboembolism. Compared to heparin, native DeS is a less active, but safer antithrombotic drug by intravenous administration due to the lack of bleeding potential, even at very high doses (up to 16 mg/kg). A clear profibrinolytic contribution to the antithrombotic properties of DeS could not be evidenced, but its prolonged action cannot be explained only by HCII potentiation [93]. For equivalent anti-IIa activities, DeS was a better inhibitor of fibrin accretion onto existing thrombi than heparin possibly by more effectively inhibiting thrombin within a thrombus [94].

Just as DeS was found as a contaminant of commercial heparin preparations [95], DeS contains heparin as a result of inadequate purification. However, the antithrombotic activity of DeS has been reported to be significantly enhanced by heparin. Since LMWH, low affinity heparin as well as the pentasaccharide did not show such an effect, the activity enhancement by UH is postulated to result mainly from its anti-IIa activity [96]. Heparin (0.25 %) potentiated the antithrombotic action of DS, whereas thrombin generation was not affected [97].

Similar to UH and HS, native DeS is poorly absorbed via the subcutanous route. To improve bioavailability, low molecular weight DeS have been developed such as Desmin 370 (OP 370). The latter reduced thrombus size via mechanisms other than inhibition of thrombus growth suggesting the involvement of a fibrinolysis-mediated mechanism [98]. Comparing a 4 kd fraction with an unfractionated DeS, the smaller molecules were two times less potent *in vitro*, but showed better pharmacokinetic properties including increased bioavailability and duration of action [99]. Testing DeSs in a MW range from 4 to 15 kd, it was established that the best compromise between biological activity, clearence, biological half-life, bioavailability and reproducibility after subcutanous injection would be a compound having a MW from 4-9 kd, since all parameters

Table 4 Comparison of the Effects of Heparin, Heparan Sulfate, and Dermatan Sulfate *in vitro* and of their Possible Sites of the Pharmacological Actions[a]

	Heparin	Heparan sulfate	Dermatan sulfate
SO_3^- / COO^-	2.-2.5	1-1.5	1
PT[b]	±	-	-
aPTT[c]	+++	±	±
Heptest®	+++	+	+
TT[d]	+++	-	+
ATIII-dependence	+++	±	±
HCII-dependence	++	±	+++
Direct proteinase inhibition	+	+	++
Thrombin generation inhibition	+++	++	++
Binding to lipoprotein lipase	+++	+	+
Binding to low density lipoprotein	+++	+	+
Endothelial release markers[e]	+++	+++	+++
Rheologic and flow measuremnts	+	+++	++
Action on endothelium	+++	++++	+++
Action on platelets	+++	±	±

[a] modified after sources: Ref. 76. and Ref. 86
[b] PT = prothrombin time
[c] aPTT = activated partial thromboplastin time
[d] TT = thrombin time
[e] tissue plasminogen activator, plasminogen activator inhibitor, 6-keto-prostaglandin $F_{1\alpha}$

increased with decreasing MW, but large inter-animal variations were seen over 9 kd [100].

Furthermore, the effect of oversulfation of DeS was examined [101]. Additional sulfate groups result in an increase of the anti-IIa activity up to a certain maximum DS. In contrast to their enhanced anticoagulant activity *in vitro* and *ex vivo*, the antithrombotic potency of oversulfated DeS was unchanged or even weaker above a certain DS [102].

A new succinyl DeS derivative was 2-3 times more active as an anticoagulant and antithrombotic than DeS. In contrast, the bleeding effect was comparable, so that this chemical alteration of natural DeS yields a new compound with an improved antithrombotic benefit/hemorrhagic risk ratio [103].

At present, several DeS preparations, e.g. MF 701 and OP 435 with high MW and OP 370 with low MW, are under preclinical or clinical investigation as prophylactic antithrombotics in surgery [69].

D. Chondroitin Sulfate

Chondroitin sulfates (ChS) are often found as contaminants of other GAG [17]. The chondroitin family comprises polydisperse, microheterogeneous sulfated copolymers of N-acetyl-D-galactosamine and uronic acid. The latter is glucuronic acid, with the exception of chondroitin sulfate B, i.e. DS (see above). The DS is generally not higher than 0.5. Depending on the predominant sulfate position, they are classified as chondroitin sulfate A (4-sulfate, Ch4S) and C (6-sulfate, Ch6S). As found by Volpi et al. [104], human plasma contains free ChS in a concentration of about 0.1 mg/100ml. As a component of thrombomodulin, chondroitin-4-sulfate is essentially involved in the inhibition of thrombin clotting activity [105].

The moderate antithrombotic action of chondroitin-4-sulfate *in vivo* is partly due to its anticoagulant effect which is only in part mediated by ATIII [106]. In order to improve the antithrombotic activity, a ChS from the respiratory tract of horned cattle was artifically sulfated resulting in the so-called semisynthetic heparin analogue (SSHA) with a mean MW of 7 kd [107]. With less than 5 U/mg, SSHA has low FXa inhibitory activity *in vitro* [108]. It seems independent of ATIII and may involve the extrinsic coagulation pathway [109]. Parenteral injection of SSHA releases anti-Xa clotting activity without a significant effect on overall clotting [108]. A second injection or a higher dose does not increase the level of induced anti-Xa clotting activity which is not neutralized by PF_4 or protamine in contrast to GAG.

In several clinical studies comparing the prophylactic effect of SSHA with standard low dose heparin, SSHA was shown to be as effective as heparin without an increased bleeding risk compared to heparin [107]. Thus, despite its weak anticoagulant activity, SSHA has a significant antithrombotic potential.

In contrast to the highly active heparin, sulfamino groups are not present in the sulfated galactosaminoglycans which exhibit weaker antithrombotic activity. In order to examine if the sulfamino groups are associated with the expression of the antithrombotic activty, the acetamido groups of the Ch4S, Ch6S and DeS were substituted by sulfamino residues [110]. However, this modification was shown to result in a decrease of the anticoagulant activity of DeS, and only a slight, insignificant increase of that of Ch4S and Ch6S. Accordingly, a loss of antithrombotic activity was observed *in vivo*. In contrast to this, the sulfamino group induced significant lipase-releasing activity.

E. Further Mammalian Glycosaminoglycan-Derived Drugs

1. Danaparoid (Orgaran®)

Danaparoid (earlier Org 10172 named, Orgaran®) is the most completely investigated example among the antithrombotic GAG-derived drugs. It is a depolymerized mixture of GAG extracted from animal intestinal mucosa with a mean MW of 6 kd. It consists of approximately 80 % low molecular weight HS, 10 % DeS, and 5 % ChS. Danaparoid is devoid of heparin or heparin fragments, but a small proportion of HS (4%) has high affinity to ATIII [111]. It catalyzes the inactivation of FXa by ATIII and of thrombin by ATIII and HCII exhibiting a much higher aXa/aIIa ratio compared to heparin and most LMWH. In contrast to heparin, danaparoid has only moderate inhibitory effects on activated partial thromboplastin time (aPTT) as well as thrombin generation, predominantly via formation of HCII-thrombin complexes [112]. But it suppresses factor IX (FIX) activation more effectively than heparin and delays the onset of FX activation nearly as effectively as heparin [113]. This coordinated inhibition of FIX and FX activation by danaparoid may contribute to its antithrombotic effectiveness.

In vivo, danaparoid inhibits the formation of various kinds of thrombi (clot-like and mixed, i.e. platelet containing, thrombi) with approximately the same potency as heparin [111]. In concert with its lack of effect on platelet function, it does not inhibit platelet deposition in experimental mixed thrombi unless very high doses are used and has only minimal effect on platelet degranulation during plug formation. On the other hand, it is more effective than heparin in preventing the extension of established venous thrombosis. At an equivalent antithrombotic effect, it is less hemorrhagic than heparin in various experimental bleeding models [112].

Although the antithrombotic mechanism of danaparoid is not yet well-established, it has been successfully used for prophylaxis of postsurgical thromboembolism. For general and hip surgery, danaparoid was shown to be better than dextran and at least as effective as low dose heparin or oral anticoagulation by warfarin without increased risk of bleeding complications [107]. Further, it is used in the prevention of ischemic complications associated with stroke and it is claimed to have a better efficacy-safety ratio than heparin [114].

In addition to its antithrombotic potency and its good efficacy-safety ratio, danaparoid is superior to UH as well as unfractioned HS and DeS in terms of its bioavailability and duration of action [115].

Recently, danaparoid has gained additional importance, since it was clinically demonstrated to be a valuable, alternative treatment for patients who suffer from HIT and require anticoagulation [116]. This use of danaparoid is based on the fact that this heparinoid causes platelet activitation with sera from HIT patients in the HIPA test only at 10 %, whereas LMWH show cross-reaction with the antibodies resulting in platelet aggregation at 98 % [117].

2. MPS, SMPS, Sulodexide®

Several other GAG mixtures are used clinically or are currently under development as antithrombotic agents [77]. For instance, MPS, a mixture of depolymerized oversulfated mucopolysaccharides, that is GAG, obtained from mammalian trachea, represents an antithrombotic drug for both human and equine usage in the treatment of joint diseases. SMPS or sulfated mucopolysaccharides represent a large class of drugs sold under various commercial names for various indications and consist of non-characterized GAG.

Sulodexide®, a medium MW GAG mixture consisting of 80 % heparin and 20 % DeS has been used clinically since 1974 [70]. Just as the case with heparin contaminated by DeS [95-97], the commercial mixture of heparin and DS acts synergistically catalyzing thrombin inhibition by ATIII and HCII simultaneously. Accordingly *in vivo*, a relatively small dose can inhibit thrombus formation as well as thrombus growth with less systemic anticoagulation than comparable heparin doses [118]. Besides preventing venous thrombus formation, Sulodexide® is able to promote thrombus dissolution by a mechanism that is partly correlated with local fibrinolysis stimulation [119]. A clinical study provided evidence that long-term therapy with Sulodexide® started early after an episode of myocardial infarction is associated with reductions in total mortality, rate of reinfarction and mural thrombus formation [120]. Furthermore, antithrombotic action was effective in the treatment of intermittent claudication without hemorrhagic risks [121]. Mauro et al. [122] were able to demonstrate that Sulodexide® administered as enteric-coated tablets, exhibited profibrinolytic and antithrombotic activity, whereas the absence of any influence on aPTT confirmed the lack of interference with the intrinsic pathway of coagulation after oral administration.

IV. SYNTHETIC LMW SULFATED CARBOHYDRATE DERIVATIVES

A. Background

In response to the awareness of various severe contaminants in natural products of animal origin such as the BSE pathogen and foot and mouth virus, there is increasing fear that many products from bovine or livestock origin may be contaminated. This has stimulated the development of new strategies within the pharmaceutical industry, which are independent from the use of natural resources, as described before. Many synthetic analogs of GAG are therefore now being developed such as heparin pentasaccharide, sulfated lactobionic acid and sulfonated polyphenolic compounds. They have been shown to produce dose dependent antithrombotic effects. Thus, each agent is being developed in a defined manner for a specific therapeutic indication in order to have a major impact on the management of thrombosis [77].

B. Pentasaccharide

One of the major breakthroughs in the development of antithrombotic agents was the identification of the well-defined pentasaccharide (PS) that specifically binds and activates ATIII. The experimental approach consisted of controlled cleavage of heparin by chemical or enzymatic means, separation of the fractions with high ATIII affinity by affinity chromatography, gel filtration, and finally determination of the structure of the biologically active oligosaccharides [123]. Final proof was obtained by the chemical synthesis of the corresponding PS which bound to ATIII with the same affinity as high-affinity heparin molecules [124]. By synthesis and testing of various PS analogues, the specificity of the interaction between the characteristic PS and ATIII has become more completely understood, and a number of structural features, such as the critical 3-O-sulfate group of the central glucosamine unit [125], have been identified as either essential for or only contributing to the ATIII accelerating effect.

Further development in this field was marked by several considerations. Originally, the synthesis comprised 75 steps and thus, industrial-scale production was out of the question for economical reasons unless the synthetic scheme was considerably simplified. Further, a better understanding of the heparin-mediated activation of ATIII was desirable. Finally, the hypothesis that anti-Xa activity is sufficient for an antithrombotic effect [43] was confirmed by various animal experiments with PS [126] and opened up the possibility of obtaining structurally defined antithrombotic derivatives by chemical synthesis.

This triggered the chemical synthesis of a large series of PS analogues ranging from closely related counterparts to highly modified variants [123]. With the aid of molecular modeling and through "tailored" molecular modifications of the PS, much knowledge has been gained concerning structure-activity relationships. On this basis not only more potent and simplified derivatives have been developed (see a selection in Figure 7), but also the molecular recognition between heparin and ATIII became better understood.

In the development of a PS-derived drug, the important discovery that the hydrogen atoms of the sulfated amino groups do not participate in the interaction process means that replacing the N-sulfate by O-sulfate does not reduce the activity, however considerably simplifying the synthesis [123]. The hydroxyl groups are not essential hydrogen bond donors. As shown by methylation such hydrophobic residues rather improve the interaction represented by the lower K_D value and increase the specific activity [123], (Figure 7).

In vitro anti-Xa activity ranges from 700 aXa-U/mg for the original PS to 1611 aXa-U/mg for the most active derivative which has no N-sulfate groups and contains the methylated hydroxyl groups. Ofosu et al. [91] showed that,

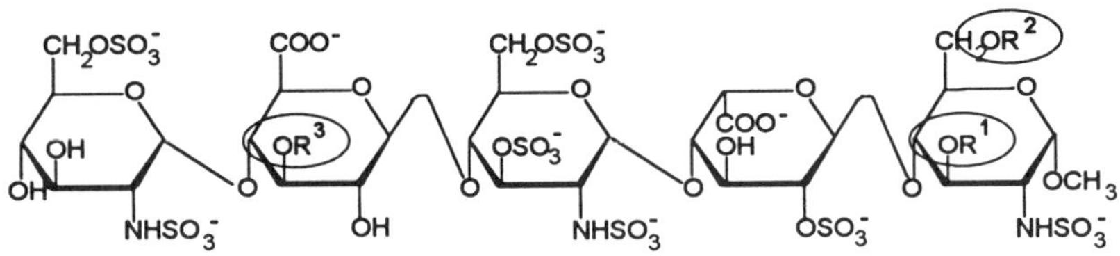

	aXa [U/mg]	K_D [nM]	
R1 = R2 = R3 = H	198	1560	
R1 = SO_3^-; R2 = R3 = H	***700***	***320***	***original pentasacccharide***
R1 = R2 = SO_3^-; R3 = H	1250	22	
R1 = R2 = R3 = SO_3^-	476	218	

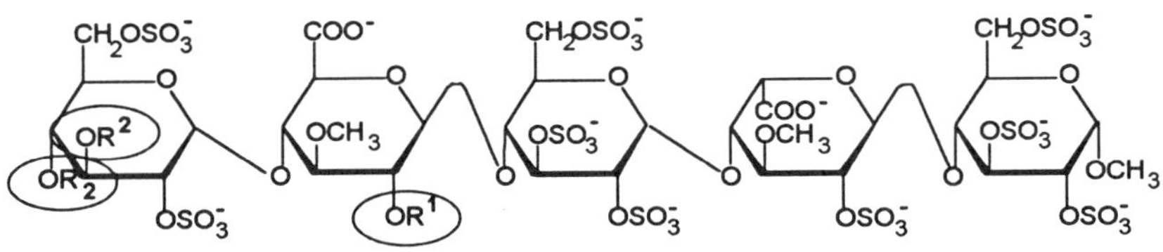

	aXa [U/mg]	K_D [nM]
R1 = R2 = CH_3	1323	16
R1 = CH_3; R2 = C_4H_9	1529	2
R1 = SO_3^-; R2 = CH_3	1159	139
R1 = CH_3; R2 = C_6H_{13}	910	9

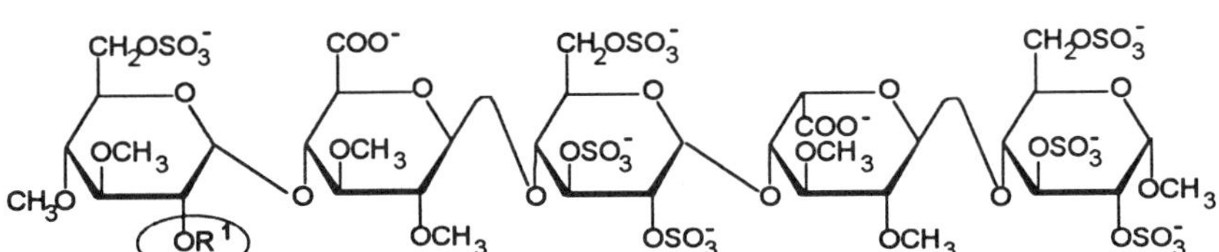

	aXa [U/mg]	K_D [nM]
R1 = SO_3^-	1318	28
R1 = CH_3	1611	13

Figure 7 A selection of synthetic pentasaccharides and their anti-Xa activities and dissociation constants K_D with ATIII. (Redrawn from Ref. 124.)

despite exclusively catalyzing the FXa inhibition, PS is able also to reduce thrombin generation. The two amplification reactions of coagulation resulting in an explosive thrombin generation are appararently catalyzed not only by endogenously generated thrombin but also by FXa.

In contrast to heparin, the PS do not interact with platelets [48]. Against the background of HIT becoming an increasingly common clinical problem, it is of a special interest that these oligosaccharides do not cause any platelet reaction in the presence of HIT-serum [127].

At present, several PS analogues are under preclinical investigation [128-129]. Just as the original PS, all the *in vitro* active analogues exert similar dose-dependent antithrombotic activity after intravenous as well as subcutaneous administration in animals. Compared to heparin, the half-life of the plasma anti-Xa activity as well as the duration of the antithrombotic effect is extremely prolonged ranging from 5 hours for the original PS to more than 5 days for some derivatives [130-132]. These oligosaccharides cause only minimal bleeding enhancement resulting in a good efficacy-safety ratio [129, 133]. Consequently, some of these new PS derivatives may be interesting candidates for future clinical development.

C. Aprosulate

Aprosulate (LW 10082) represents a further synthetically produced antithrombotic agent with low MW and a defined carbohydrate based structure just as the pentasaccharide analogues. It belongs to the class of polyanionic

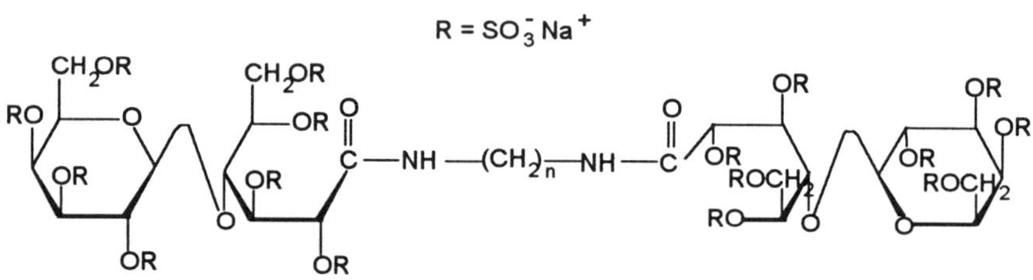

Substance	n	MW
LW 10082	3	2388
LW 10078	6	2430
LW 10086	9	2472
LW 10079	12	2514

Figure 8 Structure of sulfated bis-lactobionic acid amides. (Redrawn from Ref. 134 with permission from Thieme Medical Publishers, Inc.)

bis-lactobionic acid amides which are synthesized by condensation of the corresponding aldonic acids and α,ω-diamines followed by sulfation [134]. The compounds consist of two moieties of lactobionic acid connected by amide bonds to a diamine serving as a spacer group. The individual derivatives are completely sulfated containing 16 sulfate groups per molecule. Their MW depends on the the type of the diamine moiety and is in general not higher than 2600 d (Figure 8).

Depending on their structure, the derivatives exhibit moderate to good aPTT activity. Aprosulate, e.g. with the shortest diamine spacer has 42 U/mg. Activity decreased with increases in the number of methylene groups in the connecting alkyl chain. In the Heptest®, Aprosulate was 20 times less active than heparin, and in the amidolytic anti-Xa assay, the effect was still weaker. As demonstrated by the lack of activity in the corresponding amidolytic anti-IIa assay, these compounds act independently of ATIII, but catalyze the thrombin inhibition by HCII [135]. In addition, they inhibit prothrombin and FX activation and, in agreement with the aPTT-activity, they are particularly active as inhibitors of the intrinsic system of the FX activation.

The promising *in vitro* effects of the lactobionic acid amides, especially of LW 10082, could be confirmed *in vivo*. They demonstrated structure-activity relationships similar to the *in vitro* results [136]. Their bioavailability is better than that of DeS and HS. Bleeding time is prolonged at dosages 10 times higher than that of heparin resulting in a reduced bleeding risk on the basis of equally effective antithrombotic dosage. In sub subchronic toxicological studies in rats, LW 10082 was well tolerated up to a dosage of 100 mg/kg body weight. As a consequence of all these results, Aprosulate reached Phase I in clinical investigation. However, despite its efficacy in humans, these studies were not continued due to its chronic toxicity. Nevertheless, Aprosulate is a good mode for further developments in this field.

V. Genuine Polysaccharide sulfates

A. Carrageenans

Besides mammalian GAG, important natural sources for sulfated polysaccharides are the marine flora and fauna, that is algae and organisms like mussels and corals. Carrageenans, occuring in red algae, have been well-known for a long time and they are used therapeutically as antipeptic and immunosuppressive agents [137]. Their anticoagulant activity, which was first demonstrated by Elsner in 1937 [138], is considered to be a side effect.

In vitro, the carrageenans prolong aPTT and the PT, show only a small effect in the Heptest®, and inhibit the amidolytic thrombin but not FXa activity [137, 139]. Furthermore, fibrinolytic activity is found in the fibrin plate assay as well as in the euglobulin lysis time. The inhibition of Ristocetin®-induced platelet aggregation could be an indication of the hemorrhagic effect of the carrageenans [139].

Just as in their chemical structure carrageenans also vary widely in their anticoagulant activity. The three most important types of these β-1,3/1,4-D-galcatans are ι-, λ-, and κ-carrageenan, differing in their DS, their sulfation pattern, and their 3,6-anhydro galactose content. Generally, λ-carrageenan with the highest DS was most active followed by κ-carrageenan and then ι-carrageenan which had the lowest DS [137, 139].

B. Fucoidans

Whereas carrageenan-type sulfated polysaccharides are present in the red algae, brown algae contain sulfated fucans in their cell walls. These fucans comprise a wide, continuous spectrum ranging from high-uronic acid, low-sulfate containing polymers, that are glucuronofucoglycans and fucoglucuronans, to the relatively pure fucans, the so-called homofucans or fucoidans. The latter are primarily composed of α-(1,2)-linked units of L-fucose-4-sulfate with branching or a second sulfate unit in position 3 [140]. When extracted from brown algae, these sulfated polysaccharides with an original MW between 80 and 100 kd exhibit anticoagulant properties. The first one was isolated from *Fucus vesiculosus* by Springer et al. in 1957. Since then many other species, such as *Ascophyllum nodosum, Dictyota dichotoma*, *Ecklonia kurome*, *Eisenia bicyclis*, *Laminaria digitata* and *L. religiosa*, *Padina pavonia* and *P. tetrastomatica*, *Pelvetia canaliculata* and *P. wrightii*, *Sargassum linifolium* and *S. muticum*, *Undaria pinnatifida*, were shown to contain fucan sulfates with relatively high activities in comparison to heparin [141]. Due to differences in the sugar composition, MW, DS, sulfation pattern, and glycosidic branching, marked differences are found in their physiolgical activity. In the investigation of different fractions as well as chemically oversulfated derivatives of the fucan sulfate from *Ecklonia kurome*, antithrombin activity was shown to increase with growing MW as well as with DS until a plateau was reached [142, 143].

Fucan sulfates have a mode of action different from that of heparin. Several mechanisms contribute to their anticoagulant activity. They interfere with intrinsic coagulation and delay contact-triggered thrombin generation, but the most important point of attack is inhibition of fibrin polymerization by thrombin [144-145]. Besides a weak direct inhibiting effect, this is primarily due to the enhancement of thrombin inhibition by HCII more than 3500-fold [142, 144, 146]. In addition, by binding of fucan sulfate to fibrinogen, thrombin action and as a consequence, fibrin polymerization, is blocked for steric reasons [142,144, 147].

Apart from the anticoagulant acitivity, the fucan sulfates support fibrinolysis by stimulating the plasminogen activator-catalyzed plasminogen activation. In contrast to heparin, they do this even in the presence of high concentrations of plasminogen activator inhibitor. In addition, they promote clot lysis possibly by protecting plasmin activity from thé α_2-plasma inhibitor [148].

In agreement with the *in vitro* activity, *in vivo* fucoidans are able to prevent thrombosis [145, 146]. Compared to heparin, the duration of antithrombotic action is about twice as long as that of heparin [145].

In order the improve the activities of genuine fucan sulfates, chemical modifications of a fucoidan from *Fucus vesiculosus* were performed showing that oversulfation as well as the introduction of alkylamino residues increase the *in vitro* as well as the *in vivo* activity [146, 147].

In summary, the fucoidans, byproducts of alginate production in the food and cosmetic industries, are easily obtained. They may represent in the future a cheap and easy source of a new type of antithrombotics.

C. Further Natural Polysaccharide Sulfates

Apart from the carrageenans of the red algae and the fucan sulfates of the brown algae, the existence of other sulfated polysaccharides of marine origin with anticoagulant properties has been demonstrated [149, 150]. Here, again galactose and fucose are the main building components. For instance, from the red alga *Iridaea sp.* which is one of the sources for carrageenan, complex mixtures of sulfated D-galactans, different from the carrageenans, were isolated [151]. Sulfated polysaccharides containing L-galactose are abundant in the tunic of ascidians like *Ascidia*, *Styela*, or *Clavelina* [152]. From the latter, a sulfated α-L-galactofucan was isolated.

Marine mussels represent a further source of potential antithrombotic agents. Bergvquist et al. [153] analyzed a low molecular weight sulfated polysaccharide from mussel broth. It exhibited an antithrombotic effect, but showed only little influence on primary hemostasis indicating a good safety-efficacy ratio. Whereas this compound had no antithrombin effect and seemed to interact with the phospholipids, Nilsson et al. [154] reported on a mucopolysaccharide isolated from mussel broth which, after sulfation, acted as a pure antithrombin with a potency of about 40-50 U/mg.

Recently, the mode of the anticoagulant action of a depolymerized GAG derived from a *Holothurian sp.*, better known as sea-cucumber, was demonstrated [155]. This anticoagulant polysaccharide has two different inhibitory activities, one being a HCII-dependent thrombin inactivation and the other an ATIII- and HCII-independent inhibition of the interaction of FX with the intrinisc FXase complex. This is assumed to be due to binding to the FIXa-FVIIIa complex, which is a new dicovery in the field of anticoagulant mechanisms.

Sulfated polysaccharides from bacterial origin may be of special drug innovation interest since large amounts can be obtained by fermentation methods. Such an example is the semisynthetic galactan sulfate DH 6322, which is derived from a naturally occuring sulfated polysaccharide isolated from the culture medium of a strain of *Arthrobacter sp.* [156]. Its anticoagulant and fibrinolytic effects *in vitro* could be confirmed ex vivo [156, 157]. *In vivo*, DH 6322 effectively prevented thrombus formation in rats and also caused a significant reduction of thrombus size in progressive venous thrombosis indicating that its antithrombotic effects not only resulted from its anticoagulant action but also from its enhancement of the fibrinolytic potential.

Finally, it should be mentioned that the occurence of heparin is not only limited to vertebrates but extends to other organisms, i.e. in mollusks. Mollusk heparins are chemically indistinguishable from heparins present in mammalian tissues suggesting that heparin has maintained its structural features through evolution [158].

In summary, nature offers a great variety of sulfated polysaccharides of which only a small number of representatives are known. Based on the knowledge that sulfated polysaccharides generally exhibit physiological activities due to their polyanion character, consequently a potentially large number of substances might exist that could be used as antithrombotic drugs. So far however, little systematic work has been done, which is certainly due to the lack of appropriate methods for analysis as well as testing. It still remains possible that in the future by means of modern techniques these natural resources might be utilized for different medical purposes.

VI. (SEMI -)SYNTHETIC POLYSACHARIDE SULFATES

A. Sulfation of Polysaccharides

As an alternative to the naturally sulfated polysaccharides, a number of non-sulfated polysaccharides were used to synthesize molecules whose structures simulate that of heparin or which at least show comparable antithrombotic activity. Examples of readily available starting polymers that have no natural anticoagulant activity are agarose, alginic acid, starch, i.e. amylose and amylopectin respectively, arabinogalactan, cellulose, chitosan and chitin, curdlan, dextran, galactomannan, guaran, hyaluronic acid, inulin, lentinan, laminarin, locust bean gum, pectin, xanthan and xylan. The intent was to offer potential replacements for heparin, and further, to provide products for attachment to the surface of synthetic objects for implantation, with the objective of conferring a biologically acceptable, nonthrombogenic surface characteristic for such devices [68]. At present, a remarkable interest in sulfated polysaccharides like dextran sulfate (DxS) or pentosan polysulfate (PPS) can be noted. This is mainly due to their anti-human immunodeficiency virus (HIV) activity [159], where the anticoagulant activity still is an undesired side effect.

Based on the discovery that sulfate groups in the molecule are necessary for an anticoagulant effect, as long as fifty years ago sulfate esters of polysaccharides were produced that exhibited heparin-like activity *in vitro*. Introduction of the essential sulfate groups is not only the most important, but also the most difficult step in the synthesis. Since the first polysaccharide sulfation in 1819, several hundred methods have been described for the sulfation of carbohydrates, mainly in view of the production of commercial polysaccharide sulfates. However, most of these methods have either induced excessive polysaccharide degradation or led to a low, unevenly distributed sulfate content with concomitant low solubility of the derivative [160]. It took a long time and many efforts, until suitable sulfation methods were established [161].

Chemical characterization of the synthesized polysaccharide mixtures is difficult thereby complicating the evaluation of structure-activity relationships as the basis for the development of new drugs.

Regarding biological testing, it must be considered, that *in vitro* anticoagulant activity, which is generally tested first, does not absolutely correspond with antithrombotic activity. Furthermore, *in vitro* effects always depend on the selected test system as demonstrated by which were shown to be dependent on HCII in a purified system and independent of HCII in plasma [162].

For all of these reasons, contradictory results often occur clearly demonstrating that many questions still are to be answered in this promising area of research [163-167]. With DxS and PPS only two sulfated polysaccharides could be identified for clinical use, and the ultimate antithrombotic agent has not as yet been discovered. Still, research has revealed valuable information concerning structure-activity relationships which may be helpful in the search for the optimal antithrombotic drug.

A new limitation in the development of new antithrombotics is the prerequisite that they must not cross-react with heparin antibodies in HIT-patients. Since the activity as well as the cross-reactivity are improved with increasing DS [127], solving this problem may be a special challenge.

B. Dextran and Dextran Sulfates

1. Dextran

The use of unsulfated dextrans in the prophylaxis of postsurgical and medical thromboembolism has a long tradition. This bacterial polysaccharide is produced by *Leuconostoc sp.* and consists of α-1,6-linked glucose residues, highly branched in position 3 and 4. For medical use as plasma expander, the original high MW of more than 1000 kd is reduced by partial degradation. The antithrombotic effect of the various dextrans with defined MW ranging from 10 to 100 kd is exclusively due to physical effects, i.e. they do not directly influence blood coagulation. Because of their limited efficacy and their side effects such as hypervolemia, impairement of platelet function resulting in bleeding, and above all, serious anaphylactoid reactions, dextrans cannot be considered as ideal antithrombotic drugs.

2. Dextran sulfate

Since dextran was already used as an antithrombotic agent and pure dextrans of controlled MW were readily available, it was one of the first polysaccharides to be sulfated and to produce derivatives exhibiting heparin-like activity. However, in toxicity tests, some active preparations demonstrated fatal effects in doses only a few times greater than the minimum anticoagulant dose [166]. Therefore, it was of interest to establish structural requirements for active anticoagulants with low toxicity, an essential condition for potential clinical use. According to Ricketts [166], the toxicity increases with the MW and only molecules smaller than a critical size of about 10 kd are nontoxic. Maximum anticoagulant activity was attained at a DS higher than 1.3 as indicated by the sharp increase of activity at DS between 1.0 and 1.3. This relatively high DS requirement for anticoagulant effects suggests that structural features other than the DS are determining the efficacy [165]. The high toxicity could be drastically reduced by the introduction of sulfamide groups with retention of the anticoagulant activity [168].

Dextran sulfate (DxS) inhibits coagulation independent of ATIII. Like other sulfated polysaccharides, it activates HCII at a level dependent upon its DS and MW, and also exhibits a weak direct inhibitory effect on amidolytic thrombin activity [142]. The discovery that aPTT-activity in ATIII and HCII depleted plasma was similar to that in normal plasma suggests that DxS acts by directly preventing thrombin generation rather than enhancing thrombin inhibition [162]. In addition, DxS appears to interfere with fibrin polymerization just as found with fucan sulfates [162]. A special property of DxS is their activating effect on the contact phase of the intrinsic coagulation pathway. The formation of a complex with FXII, high molecular weight kininogen and prekallikrein initiates the conversion of prekallikrein to kallikrein and subsequently the FXII activation [169].

About thirty years ago, low molecular weight DxS were introduced as antithrombotic and antilipemic agents into clinical use. Nowadays, they have lost their importance for clinical practice due to their low safety-efficacy ratio and the availability of better alternatives. However, they still play an important role in the laboratory, e.g. as DxS cellulose columns for low-densitiy lipoprotein apheresis [169] or as additives in diagnostic assays [170]. During recent years, DxS again became important due to their inhibition of HIV replication [171].

C. Pentosan Polysulfate

1. Pentosan polysulfate (SP 54)

A second semisynthetic heparinoid, which has been well established in clinical practice for a long time, is pentosan polysulfate (PPS), also called SP 54, Hémoclar, PZ68 [172, 173]. It is obtained by sulfation of the pentosans extracted from the bark of the beech tree, *Fagus sylvatica*. The structure of the parent polysaccharide has been elucidated and consists of (1,4)-β-D-xylopyranose chains, branched in position 2 with 4-O-methyl-α-D-glucuronic acid, in an average proportion of 1 uronic acid to 9 xylose units. With a DS of 1.9, 95% of the free hydroxyl groups of PPS are sulfated. It has a mean MW of 4700 d ranging from 1800 to 9000 d.

In vitro, PPS is at 40 U/mg less active than heparin, but more active than HS, DeS or DxS. In contrast to earlier findings, it is practically devoid of anti-Xa activity [174, 175]. By interaction with FVIIIa, it significantly reduces intrinsic FXa generation and consequently also thrombin generation [174]. Its action was proved to be independent of ATIII [176]. In purified systems, or at relatively high concentrations in plasma, PPS potentiates the HCII thrombin reaction [177], but its anticoagulant effect can also be achieved independently of HCII catalysis [162]. In addition, PPS prolongs the clotting of purified fibrinogen by thrombin which is assumed to be due to interaction with the anionic binding site of thrombin [178]. These various effects are responsible for the relatively strong prolongation of the aPTT and the moderate effect in the TT. PPS enhances the fibrinolytic activity and activates FXII and prekallikrein similar to other sulfated polysaccharides but less than DxS [179- 180].

These anticoagulant and fibrinolytic *in vitro* effects could be confirmed *in vivo* [180]. In addition, ex vivo PPS elicits a slight dose-independent anti-Xa activity, which is thought to be connected with the lipolytic activity of PPS [181]. It has been demonstrated that the interaction of lipoproteins with the released hepatic triglyceride lipase generates this weak anti-Xa activity [182]. A further contribution to the anticoagulant action *in vivo* is the enhancement of the inhibition of coagulation by TFPI. This lipoprotein-associated coagulation inhibitor inhibits the tissue factor/FVIIa-induced coagulation in an FXa-dependent manner and is released by heparin as well as by sulfated polysaccharides. PPS is less potent than UH and LMWH, but more than DeS, DxS, and HS [183].

PPS shows good efficacy in the prevention of thromboembolisms, especially in patients with diminished fibrinolysis [184]. However, its bleeding risk is similar to heparin [185]. Further, it reduces the platelet count like heparin [181] and in the literature there are many reports of severe thrombocytopenia [58, 186]. *In vitro*, the plasma of these patients demonstrates platelet reactivity both by aggregometry and ^{14}C-serotonin release in the presence of PPS or heparin [186]. Conversly, PPS shows cross-reactivity with the serum of HIT-patients [57] supporting the same immune complex mechanism such as for heparin. Consequently, PPS can not be used for alternative anticoagulant treatment in HIT-patients.

2. Further pentosan polysulfates

Apart from the original PPS, other xylans were also isolated and sulfated resulting in potent anticoagulants as well. Thus, the xylans derived from corn cobs, larchwood and oatspelts provided heparinoids more active than PPS while oatspelt xylan sulfate was as active as heparin in aPTT and PT and more active in TT [187]. From a large series of semisynthetic sulfated polysaccharides, a larchwood xylan sulfate was most active and its effect in the aPTT was similar to heparin [188]. In addition, this xylan sulfate and heparin were the least toxic among the tested polysaccharide sulfates with LD_{50}s 5-100 fold higher than the effective dose.

D. Chitin and Chitosan Sulfates

Chitosan, an unbrached (1→4)-linked 2-amino-2-deoxy-β-D-glucan, is prepared by chemical N-deacetylation of chitin, which is the main structural element of the cuticles of crab, shrimp, and insects, and widespread in the cell walls of fungi. In comparison to other polysaccharides, chitin and chitosan appear to be especially suitable for producing heparin-like substances by sulfation because of their (1→4) linkages, linearity and the presence of amino and acetamido groups, which otherwise are difficult to introduce [163].

To produce heparin-like substances, sulfate and carboxyl groups have been introduced into chitosan. Sulfation of chitosan as well as chitin results in anticoagulant derivatives. The best activity was achieved with O-sulfated N-acetylchitosan with > 300 aPTT-U/mg, followed by the N,O-sulfated chitosan with > 200 U/mg, which was more active than heparin itself [189]. In contrast to these findings, O-sulfated chitosan was less active than heparin and N-sulfated chitosan was completely inactive. Conversion of position 6 into a carboxyl group in N-sulfated chitosan results in a product with 23 % of the activity of heparin [190], which can be increased by additional O-sulfation [189]. The carboxymethylation of position 6 [191] or the amino group [163, 192] increases the activity of sulfated chitin as well as chitosan. If the DS is high enough, they are as active as heparin. Generally, sulfation is essential for any anticoagulant activity, which can be improved by increasing DS [189, 191, 192]. A 6-sulfate group as the main active site should be accompanied by sulfation of other positions; the sulfamido group does not need to be present at a high DS; a carboxymethyl group adds to the biological activty [192]. In addition to the DS, MW is an important determinant of activity [163, 189, 192].

As reflected by the higher activity in the aPTT than in the PT and TT, chitosan sulfates seem to specifically interact with the intrinsic coagulation pathway [192]. Furthermore, they inhibit thrombin by a mechanism partly mediated by ATIII [193, 194]. Whether the in vitro effects correlate with in vivo activity must still be established.

E. Further Semisynthetic Polysaccharide Sulfates

Among the great variety of polysaccharide sulfates produced as potential anticoagulants, but never used as antithrombotic drugs, several examples have to be cited. Generally, it can be concluded, that all these heparinoids work as anticoagulants by mechanisms different from that of heparin.

The β-1,4-glucan cellulose sulfate is worth mentioning. Cellulose and many of its derivatives have found wide use in industrial applications. Sulfation of cellulose results in potent anticoagulants whose activities increase with increasing DS, especially in position 2 and 3 and, in contrast to many other observations, tend to decrease with increasing MW [167]. However, a high DS is also associated with a high toxicity, which is among other effects due to the formation of insoluble complexes with fibrinogen [166]. Similar to PPS, its anticoagulant activity is partly mediated by its action on FVIII [195]. As in the case of fucan sulfates and dextran sulfates [148, 168], toxicity is reduced by introduction of amino groups with retention of high anticoagulant activity which increases in the order 2-amino- < 6-amino- < aminodeoxycellulose sulfate [195].

Using the anomeric α-1,4 glucan amylose, Horton et al. [68] tried to simulate the structure of heparin by introducing an amino group and uronic acid in addition to sulfate esters. By a complicated synthetic procedure, various products with different activities as high as 45 U/mg, were obtained, but they were quite polydisperse due to degradation during the synthesis.

Also the attempt to transform alginic acid, the 1,4-linked copolymer of α-L-guluronic and β-D-mannuronic acid, into a molecule with a heparin-like structure did not succeed [196]. A fraction amounting to 7 % of the product mixture separated by ATIII affinity chromatography exhibited high anticoagulant activity, but its structure was completely different from that of heparin. Comparing the anticoagulant effect of a partially reduced sulfated alginic acid, a partially reduced aminated and sulfated alginic acid, and sulfated guaran, a β-1,4-mannan branched in position 6 with α-D-galactose, the aPTT-activities were 28, 39 and 70 U/mg [197]. All three derivatives produced an ATIII-dependent inhibition of thrombin but not of FXa, and guaran sulfate additionally interacted directly with the thrombin-fibrinogen reaction. As found for other heparinoids, the introduction of amino groups does not considerably influence the activity. Whereas the presence of uronic acids seems to reduce the anticoagulant effect as also shown by a large series of partially reduced and sulfated alginic acids differing in their uronic acid content [198]. In this study, the sulfate esters of several polysaccharides were investigated showing that for significant aPTT activity a DS of about 1 as well as a MW higher than 10 kd is necessary. In this case again, none of the derivatives had any anti-Xa activity. However the conclusion, that a high content of primary sulfate groups is essential, as well as the postulate of a DS of at least 1 could not be confirmed by later investigations (see part F).

The conclusion that activity was improved by glycosidic branching such as by guaran, locust bean gum, arabinogalcatan and galactomannan sulfate [198, 199], could also be confirmed by Uryu et al. [200] during the testing of sulfated polysaccharides as potential HIV inhibitors. Mannose-branched xylofuranan as well as ribofuranan sulfates exhibited higher anticoagulant activities than the corresponding unbranched derivaties. In agreement with many other findings, this study clearly demonstrates again that the anticoagulant activity improves with increasing DS as well as with increasing MW.

F. Total Synthesis of Polysaccharide Sulfates

A new approach to the synthesis of sulfated polysaccharides has been reported by Hatanaka et al. [168] producing heparinoids by total synthesis starting from ring-opening polymerization of anhydro sugar derivatives. In addition to the previous mentioned dextran type polysaccharide sulfates with sulfamide groups, sulfated α-1,5-D-pentofuranans and β-1,4-D-pentopyranans were synthesized using the pentoses xylose and ribose. Similar to the semisynthetic PPS, these pentosan sulfates produce high anticoagulant activities up to 69.1 U/mg exhibited by an α-1,5-D-xylofuranan sulfate. The furanan-type derivatives showed generally better effects than the pyranan-type ones. This is assumed to be due to a higher flexibility of the polymer backbone favoring the nonspecific interaction with blood coagulation factors.

With the means to produce pure, well defined polysaccharides by total synthesis, a new avenue is offered in the field of antithrombotic carbohydrate-derived drugs. In future, this could gain in importance as an alternative to the isolation of carbohydrate polymers from biological material.

G. Heparin-Like Compounds by Modification of a Bacterial Polysaccharide

The newest development in the field of heparinoids are heparin-like compounds prepared by chemical modification of the capsular polysaccharide from E. coli K5 [201-202]. Capsular polysaccharide K5 (PS K5), one of the major surface antigens of Escherichia coli K5, was previously found to have the same structure, i.e. $[-(\rightarrow 4)-\beta\text{-GlcA}(1\rightarrow 4)-\alpha\text{-GlcNAc}(1\rightarrow)-]_n$, as the non-sulfated precursor polysaccharide in heparin biosynthesis [203]. By N-deacetylation and

N-sulfation of PS K5, the sulfaminoheparosan [-(→4)-β-GlcA(1→4)-α-GlcNSO$_3$(1→)-]$_n$ was obtained. Sulfation under systematically different experimental conditions yielded products with various sulfation patterns. Derivatives with high affinity for ATIII and high anti-Xa activity contained the trisuflated aminosugar GlcNSO$_3$-3,6-SO$_3$, which was shown to be the marker component of the ATIII-binding pentasaccharide [201]. Thus, homogeneous heparin-like substances with special anticoagulant properties can be specifically produced by biotechnological methods. Maybe in the future the spectrum of structural variations can be expanded by the use of various enzyme induced modifications (204). For instance, D-glucuronyl C5 epimerase was purified from bovine liver and was able to convert approximately 30 % of the D-glucuronyl residues of the PS K5-derived sulfosaminoheparosan [205]. Recently, research has concentrated on the expression, regulation and the export of PS K5, in E. coli [206-209]. Based on this knowledge, it may be possible to produce PS K5 on an industrial scale by biotechnological methods. In this manner, real heparin substitutes could be created which chemically are more defined, homogeneous, not contaminated by any pathogen, and whose production is independent of the material delivered by the meat industry.

H. Structure-Activity Relationship of β-1,3-Glucan Sulfates

1. Introduction

In order to refine possible structure-activity relationships of anticoagulant sulfated polysaccharides, the influence of various structural parameters should be systematically evaluated by specific chemical modification of a model polysaccharide and followed by testing of the anticoagulant and antithrombotic properties of the produced derivatives. For this purpose, inexpensive, readily available and chemically well-defined natural polysaccharides are needed as basic polymers. We have chosen curdlan, a natural, linear β-1,3-glucan from the soil bacterium *Alcaligenes faecalis* produced by fermentation [210], further the β-1,3-glucan laminarin from brown algae (Figure 9). In contrast to the high molecular weight of the water-insoluble insoluble curdlan, laminarin has a DP of about 35 and is soluble in water. Pullulan, a bacterial α-1,4/1,6-glucan, was used for structural comparison.

Sulfation of these glucans results in potent anticoagulants. The dependence of anticoagulant activity on the DS and the MW and the influence of the substitution pattern of the glucose unit was investigated by careful selection of structural variations.

Finally, it should be established if glycosidic branching of the linear β-1,3-glucan can modify the physiological properties.

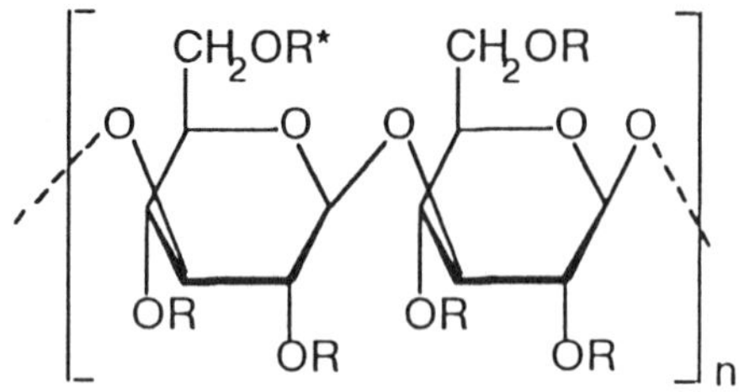

Figure 9 Repeating unit of β-1,3-glucan sulfate.

2. Synthesis of β-1,3-glucan sulfates

In order to synthesize substances with high and regular DS and controlled chain lengths, a sulfation method had to be established which is nondestructive and efficient. This includes the choice of a suitable sulfation reagent and the determination of optimal reaction conditions. To avoid acid-catalyzed degradation of the polysaccharide, an absolutely waterfree atmosphere during the sulfation is essential [161]. A high and regular DS is dependent upon activation of the polysaccharide prior to sulfation resulting in the disaggregation of highly-ordered cristalline regions making the hydroxyl groups more accessible to substitution [161, 211-212] SO_3-pyridine in DMF was proven to be a suitable sulfation reagent for the selected polysaccharides [212].

Since the procedure showed a direct relationship between the molar excess of the sulfation reagent and the resulting DS, laminarin sulfates with DS between 0.3 and 2.3 could be obtained by variation of the SO_3-pyridine amount (Figure 10).

Prior to sulfation the chain length of the high molecular weight curdlan was thermically reduced in order to obtain derivatives with different MW [213]. A second utilized method was the fractionation of polydisperse curdlan sulfates by GPC. The chromatography of a highly sulfated derivative with a DS of 1.2 and a less sulfated one with a DS of 0.6 resulted in distinct fractions representing MW ranging from 5 to 190 and from 12 to 163 kd, respectively [213].

Since the literature contained varied conclusions concerning the dependence of anticoagulant activity on the position of the sulfate groups, the investigation of this parameter was of special interest [125, 190-192, 214]. Kamide et al. [167] reported on an improvement of the anticoagulant activity of cellulose sulfates

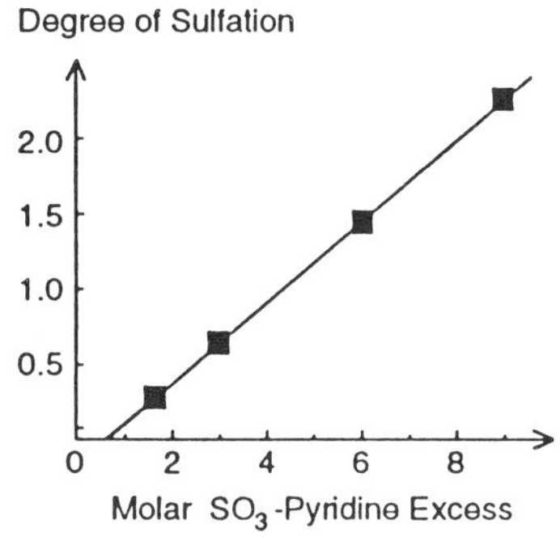

Figure 10 The influence of the molar SO_3-pyridine excess on the DS of laminarin.

by an increasing amount of C2- and C3-sulfate groups. However, it was not considered if this improvement was due as well to other factors such as the higher DS directly corresponding to the C2- and C3-sulfate increase. Therefore, curdlan sulfates with different kinds of substitution patterns had to be produced. The primary hydroxyl group on C6 of the glucose unit has more than ten-fold accessibility to sulfation relative to the secondary hydroxyl groups on C2 and C4. This preferred C6-substitution can be reduced in favor of that of C2 and C4 by protection of the primary hydroxyl group prior to sulfation [215]. The sulfation pattern was evaluated by a newly developed methylation procedure [216] based on that of Harris et al. [217]. The resulting methylated, ethylated alditol acetates were analyzed by combined gas-liquid chromatography-mass spectrometry (GLC/MS) providing qualitative and quantitative information about the polysaccharide structure and the distribution of sulfate groups. The large adamantoyl residue proved to be the most suitable of all candidate protection agents [218]. Whereas with the non-protected sulfated compounds up to a DS of 2 less than 50 % of the sulfate residues were bound constantly to C2 and C4, their substitution increased to more than half the sulfate groups after C6-protection. Such a sulfation pattern results in a more uniform charge distribution around the glucose monomers.

To investigate the influence of glycosidic branching of the linear main chain on physiological activity, curdlan was sugar-substituted in position 6 according to Demleitner et al. [219] prior to sulfation. Glucose, arabinose, rhamnose and the disaccharide gentiobiose were used as model side chains. These products differed in their sugar substitution as well as in their degree of branching.

In a manner similar to the C6-protected curdlan sulfates, more than 50 % of the sulfate esters were bound to secondary hydroxyl groups so that they were uniformly distributed around the glucose monomer. In terms of the complete polymer, the side chains are more substituted than the main chain and the molecule resembles a string of highly charged pearls [218].

3. Anticoagulant activity of β-1,3-glucan sulfates

The synthesized β-1,3-glucan sulfates (GluS) were tested in classical coagulation assays to evaluate structure activity relationships. These tests reflect the influence of tested compounds on the different stages of the blood coagulation cascade. The TT measures the inhibition of the thrombin-mediated fibrin-formation, that is the anti-IIa activity, the Heptes® determines the anti-Xa activity, and by means of the aPTT an intervention in the intrinsic pathway can be ascertained. In this test, most of the GluS showed the best activity which was as high as 46 U/mg determined by means of the IVth International Standard of Heparin.

A minimum population of sulfate groups was shown to be essential for any anticoagulant effect. The corresponding minimum DS is influenced by other structural parameters like the MW, the kind of glycosidic linkages of the main chain, the sulfation pattern and the presence of glycosidic side chains. Thus, a pullulan sulfate with a DS of 0.47 showed a remarkable anticoagulant effect, whereas a laminarin sulfate with the same MW and a higher DS of 0.64 had no significant anticoagulant activity. The two derivatives differed in the nature of the glycosidic linkages which influences the three-dimensional structure and the flexibility of the polymer. Therefore, the higher activity of the α-1,4/1,6-linked glucan sulfate compared to the β-1,3-linked one is assumed to be due to a higher flexibility allowing an improved interaction with the coagulation factors [200].

As presented by the series of laminarin sulfates differing only in their DS (Figure 11), above an average DS of 0.64 activity increases with increasing DS in all three coagulation tests until an optimum of 1.49. Similar to oversulfated fucans [143], a higher degree of substitution results in a decreased activity indicating that a high charge density above a certain limit prevents an optimal interaction with the coagulation enzymes.

The MW-dependent activity profile of two curdlan sulfate series is shown in Figure 12. Generally, the more highly charged derivatives are more active than the less sulfated compounds. Comparing fractions having the same DS, a decrease in MW results in a bell-shaped curve with the maximum effect at MW ranging from 60 to 40 kd. Activity in the TT continuously decreases from the most effective MW fraction to those with smaller MW, whereas activity in the Heptest® declines more slowly and rises again at MW smaller than 10 kd (data not shown).

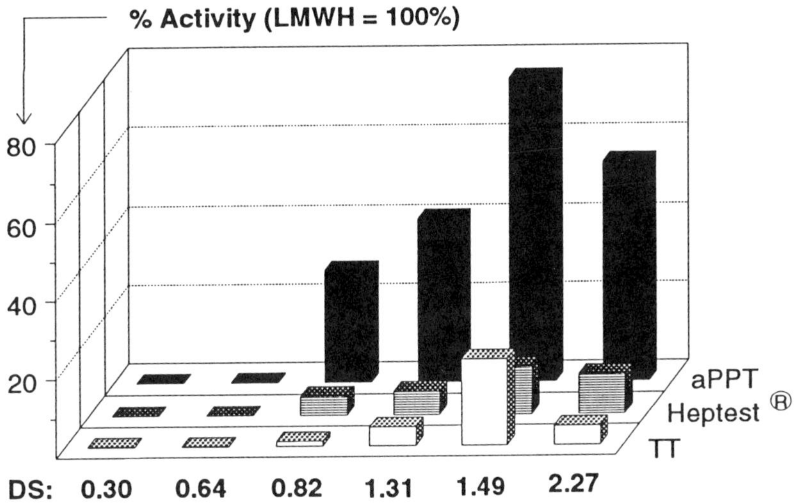

Figure 11 Anticoagulant activity of laminarin sulfates with different DS in aPTT, Heptest® and TT. The activities are given in % activity of the LMWH Sandoparin. (From Ref. 212.)

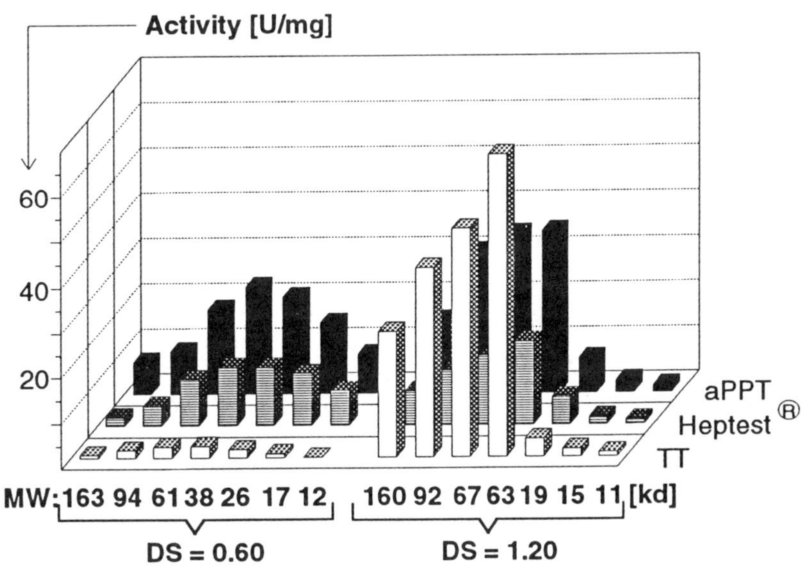

Figure 12 Anticoagulant activity (U/mg) of two series (DS = 0.6 and DS = 1.2) of curdlan sulfate fractions with different MW in aPTT, Heptest® and TT. (From Ref. 213.)

This results in an increase of the aXa/aIIa ratio with decreasing MW and demonstrates the direct influence of chain length on the activity profile [213]. The anticoagulation mechanism, which is responsible for prolonging clotting time in the Heptest® is favored by short chain molecules. In contrast to this, significant activity in the TT caused by an interference with thrombin-mediated fibrin formation requires a certain minimum chain length as well as a DS of at least 0.75 [218].

To demonstrate the influence of the sulfation pattern on anticoagulant activity, three representative C6-protected curdlan sulfates with more than 50 % of the sulfate residues bound to the secondary hydroxyl groups on C2 and C4 are compared with three so-called natural, mainly C6-substituted derivatives (Figure 13). The higher the portion of the C2/C4-sulfate groups the higher is the anticoagulant activity of derivatives with similar DS and MW. For instance, a curdlan sulfate with a DS of 0.85 and a MW of 50 kd and 74% C2/C6-sulfate groups exhibited an aPTT-activity of 46 U/mg, whereas a natural curdlan sulfate with the same MW, but an even higher DS of 1.16 had only 33 U/mg. Therefore, a more uniform charge distribution around the polymer chain obviouly results in anticoagulants with activities normally corresponding to higher MW and DS. Relative to the reported correlation between the MW of heparinoids and their toxicity [167] as well as between their DS and their bleeding risk [220], the question arises if such sulfated polysaccharides with relatively low DS and MW and yet good in vitro anticoagulant activities would operate in vivo as antithrombotics with a reduced risk of toxicity. It is further remarkable that activity increases to a different extent in the various tests depending on the DS and the MW of the respective C6-protected derivatives. This is clearly reflected by the changed aXa/aIIa ratios. Consequently, a higher C2/C4-sulfate content influences the quantity as well as the quality of the anticoagulant activity.

And finally, glycosidic branching of the main chain also modulates the anticoagulant activity (Table 5). To realize any effect, at least 20% of the glucose units had to be glycosidically branched at C6. In the TT, the branched curdlan sulfates, independent of their degree and their kind of branching, were generally more active than the respective linear derivatives with similar DS and MW. Since glycosidic substitution enhances the flexibility of the polymer chain, this effect is possibly based on an improved interaction with the coagulation enzymes. In contrast to the unspecific activity-increase in the TT, the effect in the aPTT is influenced by the quantity as well as the character of the side chain [218].

Consequently, these results support the thesis that not only charge density, polymer size and substitution pattern, but also the chain flexibility and the three-dimensional structure are determinants for interaction with coagulation enzymes.

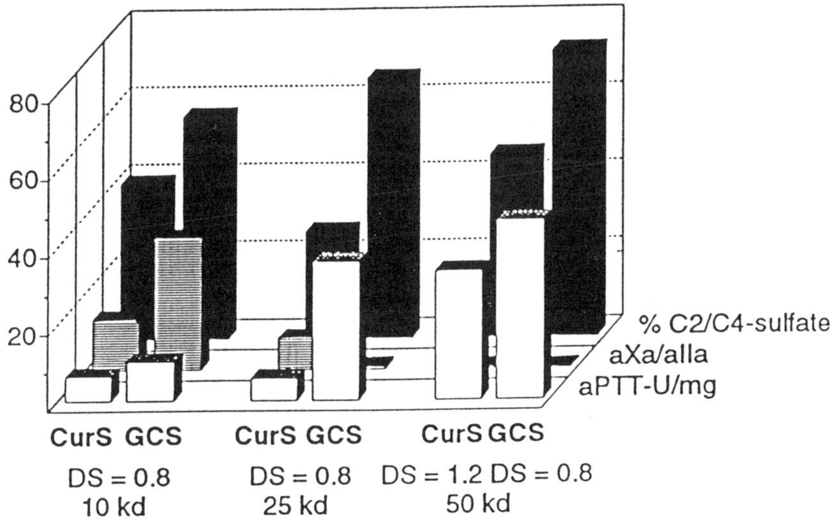

Figure 13 Anticoagulant activity (aPTT-U/mg), aXa/aIIa ratio and share of the sulfate residues bound to the secondary hydroxyl groups on C2 and C4 of three pairs of curdlan sulfates differing in their MW. Each pair consists of a genuine curdlan sulfate (CurS) with < 50 % C2/C4-sulfate groups and a C6-protected curdlan sulfate (GCS) with > 50 % C2/C4-sulfate groups. (From Ref. 216. With permission from Thieme Medical Publishers, Inc.)

Table 5 The Influence of the Degree (DB) and the Kind of Glycosidic Branching on the Anticoagulant Activity of Curdlan Sulfates in the aPTT and the TT

Curdlan sulfates[a]	aPTT	TT
Degree of branching:		
DB < 0.2	o	o
DB = 0.2-0.7	+	+
DB > 0.7	o	+
Branching by:		
Glucose	o/-	+
Gentiobiose	-	+
Rhamnose	+	+
Arabinose	+	+

o no influence on the activity
\+ increase in activity
\- decrease in activity
[a] the effect was assessed by comparison with the respective linear derivatives with similar DS and MW

4. Mechanism of the anticoagulant activity of β-1,3-glucan sulfates

The coagulation tests showed that structural variations modify the intensity of the anticoagulant effect as well as the activity profile of GluS. Therefore, it is supposed that these negatively charged heparinoids specifically interfere with several sites of the coagulation cascade and that their diverse effects are differently influenced by the various structural parameters.

Further coagulation assays and several test systems with chromogenic substrates supplied more detailed information about the mode of action of the GluS. Of special interest were the mechanisms of the inhibition of thrombin-mediated fibrin formation which required a certain minimum MW. Also of interest was the mode of action in the aPTT, where most derivatives showed the best activity, and finally the reason for shortening the PT, which measures the extrinsic pathway of the coagulation, by low GluS concentrations.

In Table 6, the anticoagulant properties of CS, a curdlan sulfate with a DS of 1.2, a MW of 40 kd and an aPTT-activity of 40 U/mg, are compared with those of a UH which is four times more active. As reflected by the high plasma con-entration required to double the baseline time in the PT, these heparinoids are like heparin poor inhibitors of the extrinsic system. The more potent clotting-inhibition in the aPTT indicates a possible interference with the intrinsic process of coagulation. In relation to heparin, the Heptest® activity was especially low so that a heparin-like anti-Xa activity can be excluded. In contrast to this, the effect in the thrombin time was much tronger. Consequently, apart from the interction with the intrinsic coagulation, represented by aPTT activity, the final step of the coagulation cascade, that is the conversion of fibrinogen to fibrin by

Table 6 Anticoagulant Properties of CS, a Curdlan Sulfate with a DS of 1.2, a MW of 40 kd and 40 aPTT-U/mg in Comparison to an UH, an Unfractionated Heparin with 155 aPTT-U/mg

	UH [μg/ml] [a]	CS [μg/ml] [a]	CS/UH [b]
PT	11.0	33.0	3.0
aPTT	2.5	10.0	4.0
Heptest®	1.0	20.0	20.0
TT	1.5	11.3	7.5
Fibrin polymerization (+ ATIII) [c]	0.5	0.5	1.0
Fibrin polymerization (- ATIII) [c]	11.0	0.5	0.05
Batroxobin time	- [d]	130.0	-

[a] plasma concentration [μg/ml] for doubling the baseline time
[b] ratio of the CS-concentration to the UH-concentration
[c] purified system: 0.125 % fibrinogen solution with/without 1 U/ml ATIII and 3 U/ml bovine thrombin
[d] no anticoagulant effect

thrombin, also seems to be affected. This was confirmed by the identical effects of CS and heparin in a fibrin polymerization assay in a purified system containing only fibrinogen, thrombin and ATIII [218, 221]. When the assay was performed in the absence of ATIII, the heparin effect was 22 times weaker as expected, whereas the action of the CS was unchanged. This indicates, that in contrast to heparin, the GluS operate independently of ATIII which would be a therapeutic benefit. Further, the low effect in the batroxobin time assay using batroxobin instead of thrombin for the proteolysis of fibrinogen shows that apart from the interaction with thrombin fibrinogen also is partly involved in

Table 7 Inhibition (IC_{50} [μg/ml]) of the Amidolytic Thrombin Activity by the Unfractionated Heparin UH and the Curdlan Sulfate CS in Different Chromogenic Substrate Assays

	UH [μg/ml] [a]	CS [μg/ml] [a]	CS/UH [b]
TRIS-buffer without ATIII [c]	-	-	-
human plasma [c]	0.25	15.0	100
TRIS-buffer with ATIII [c]	0.06	-	-
ATIII-depleted plasma [c]	-	25.0	-
TRIS-buffer with HCII [d]	0.9	0.9	1

[a] IC_{50} [μg/ml] = concentration for 50 % inhibition of the amidolytic thrombin activity
[b] ratio of the CS-concentration to the UH-concentration
[c] bovine thrombin
[d] different test system, human thrombin ⇒ also heparin showed an effect in contrast to the assay with ATIII-depleted plasma [135]

the anticoagulation by GluCS. A certain minimum DS and MW were shown to be necessary to produce an inhibition effect in this assay [218]. They possibly bind to fibrinogen themselves and prevent the fibrin polymerization for steric reasons in a manner similar to the fucan sulfates [142].

Tests with chromogenic substrates gave information on the interference of CS with thrombin (Table 7, 218, 221).The lack of effect of UH and CS in a purified buffer system containing only thrombin and the chromogenic substrate suggests that neither directly attack thrombin in such a way that the action against the small peptides is reduced. In plasma containing ATIII and other natural inhibitors, CS showed an effect. Due to the much greater difference between UH and CS in this test than in the thrombin time, its anti-IIa activity is supposed to be based on the acceleration of the thrombin-inactivation by a natural plasma inhibitor as well as by the inhibition of the specific reaction between thrombin and fibrinogen. The results of the fibrin polymerization assay, excluding ATIII as the inhibitor accelerated by CS, could be confirmed by the lack of effect in the amidolytic assay using a purified ATIII-system and further, by the very similar activity when plasma depleted from ATIII by monoclonal antibodies was used. The assumption that HCII is involved could be proved by a purified HCII system.

Table 8 shows the results of further amidolytic assays [221]. In the anti-Xa assay, CS showed only minor effects in comparison to UH. The inconsistency with Heptest® activity is due to the lack of specificity of this anti-Xa coagulation assay measuring not only the anti-Xa activity but also other effects like the anti-IIa activity [222]. Though the amidolytic activity of FXa is not significantly affected by the CS, its physiological activity in the form of the prothrombinase complex is impaired, as can be seen from the results of a prothrombin activation assay measuring the thrombin amount formed by the prothrombinase complex. However, this effect as well as the activity in the extrinsic FXa generation assay is possibly caused by an unspecific destruction of the phospholipid enzyme complexes due to the test compound's polyanion character. Since in the anti-Factor VIIa assay, which additionally measures the activation of FVII, the IC_{50} was nearly the same, a specific inhibition of this process can also be excluded. In contrast to this, the intrinsic FXa generation assay was inhibited more effectively. Here, CS produced the best activity in the amidolytic assays.This property is likely responsible for the high aPTT-activity of the GluS.

Table 8 Inhibition (IC_{50} [μg/ml]) of the Factor Xa and Factor VIIa Activity, the Thrombin and Factor Xa Generation by the Unfractionated Heparin UH and the Curdlan Sulfate CS Using Chromogenic Substrate Assays

	UH [μg/ml] [a]	CS [μg/ml] [a]	CS/UH [b]
Anti-Xa activity in TRIS-buffer	-	-	-
Anti-Xa activity in human plasma	0.10	50.0	500
Prothrombin activation	0.50	50.0	100
Extrinsic FXa generation [c]	0.35	30.0	85
Anti-Factor VIIa activity [d]	0.22	33.0	150
Intrinsic FXa generation [e]	0.48	7.0	15

[a] IC_{50} [μg/ml] = concentration for 50 % inhibition of the amidolytic FXa or thrombin activity, respectively
[b] ratio of the CS-concentration to the UH-concentration
[c] induced by the activated prothrombincomplex concentrate FEIBA®
[d] COA-SET FVII® (Kabi Diagnostica, Sweden)
[e] induced by kaolin

As seen in the prothrombin time, activating effects were found with CS at doses lower than the inhibiting concentrations also in the extrinsic FXa generation and the anti-FVIIa assay. As shown by an amidolytic prekallikrein activation assay, this observation was due to the stimulation of the kallikrein- and FXIIa-formation, respectively. Whereas 1 μg/ml UH or DxS, the classical FXII activator, did not show any effect, the same concentration of CS caused a 14-fold increase in the kallikrein formation [218, 221]. Since FXIIa not only activates coagulation, but also fibrinolysis, GluS could be of interest with respect to a fibrinolytic effect. This first indication of a fibrinolytic activity could be confirmed by the euglobulin lysis time [221]. Addition of CS solution to the clot resulted in lysis which began after 15 min. and was completed after 90 min.The results concerning the mode of action are summarized in Figure 14.

It could be established that, in contrast to heparin, the GluS act independently of ATIII. Besides unspecific effects on the phospholipid enzyme complexes due

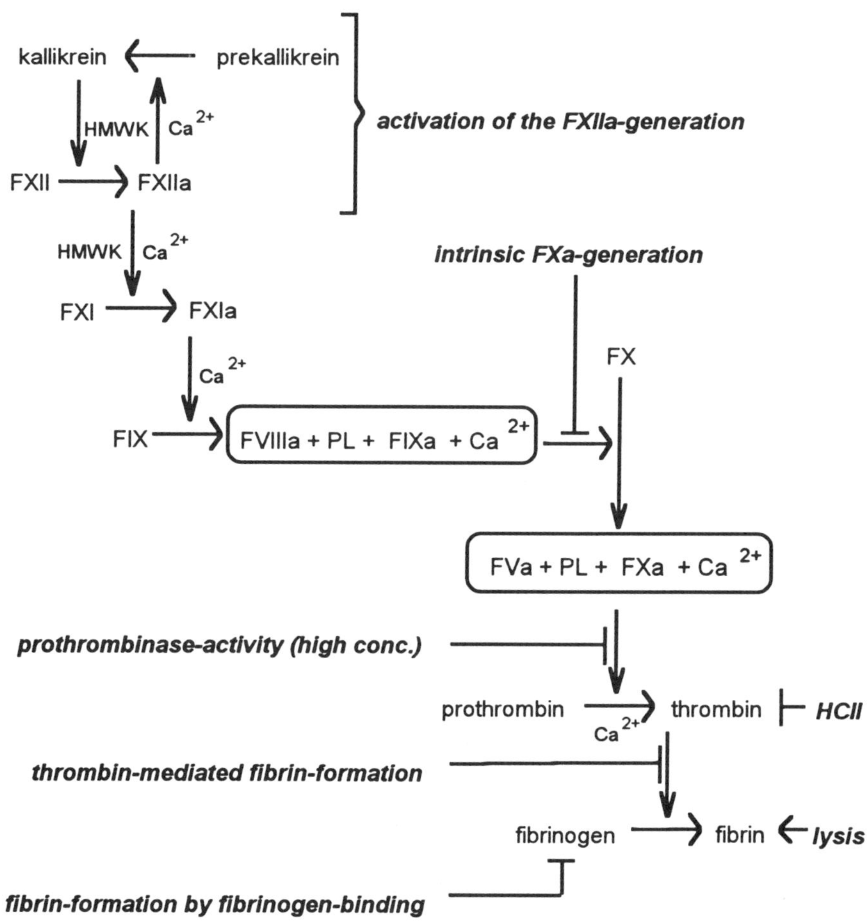

Figure 14 Scheme of the mode of action of anticoagulantly active β-1,3-glucan sulfates (explanation: see text).

to their polyanion character, already low concentrations inhibit the intrinsic FXa generation explaining the high aPTT-activities. In this way, these GluS, like heparin and many other heparinoids, ultimately reduce thrombin formation, an important process in thrombosis development [44]. Existing thrombin is neutralized with the participation of a plasma inhibitor, probably HCII. At the end of the coagulation cascade the derivatives directly inhibit thrombin mediated fibrin formation. Above a certain DS and MW they possibly bind to fibrinogen themselves due to steric reasons preventing fibrin polymerization. Finally, at the onset of the coagulation procedure FXIIa generation is stimulated.

5. Influence of β-1,3-glucan sulfates on the platelet aggregation

Apart from the coagulation, platelet aggregation is the most important event in thrombogenesis; on the other hand, it also plays an important role in the primary hemostasis [48]. The modulation, that is both activation and inhibition [223], of platelet function is one of the major complications of heparin therapy and can present clinical problems such as bleeding, HIT, and arterial thrombosis [79]. Due to the *in vitro* and *in vivo* results with heparinoids as well as LMWH, it is assumed, that a lesser interaction with platelets, especially inhibition of collagen induced aggregation, is connected with a smaller bleeding risk [58].

In contrast to heparin [224], the GluS themselves did not stimulate aggregation. In the presence of serum from HIT patients containing heparin antibodies the behavior of the GluS is different depending on their structure, as demonstrated by aggregometry as well as by ^{14}C-serotonin release assay or HIPA [127]. For instance, a predominantly C2/C4-substituted curdlan sulfate, which showed good antithrombotic effect *in vivo*, did not induce platelet aggregation in the presence of these antibodies so that such a heparinoid would be of a general interest as an alternative drug for HIT patients.

Studies on the influence on the agonist induced platelet aggregation demonstrated that, in contrast to heparin, neither a stimulation nor an inhibition was caused when aggregation was induced by arachidonic acid, ADP, collagen, epinephrine or Ristocetin® [215]. However, when human thrombin was used as the agonist, the GluS inhibited the aggregation similar to UH, LMWH, heparinoids, hirudin and other thrombin inhibitors [59, 225, 226]. This concentration dependent effect corresponded to the anti-II activity of the respective GluS. Since human thrombin is recognized not only to play a central role in the coagulation process but also is probably the most important platelet agonist *in vivo*, this effect could contribute to the inhibition of the formation of a platelet-rich thrombus [225, 226].

6. Antithrombotic activity of β-1,3-glucan sulfates

Two different animal models have been chosen to test the antithrombotic activity of the GluS *in vivo*: a FEIBA® induced stasis thrombosis model in rabbits (227), a modification of the widely used Wessler model [228], and a clamping induced occlusion model in rats [229].

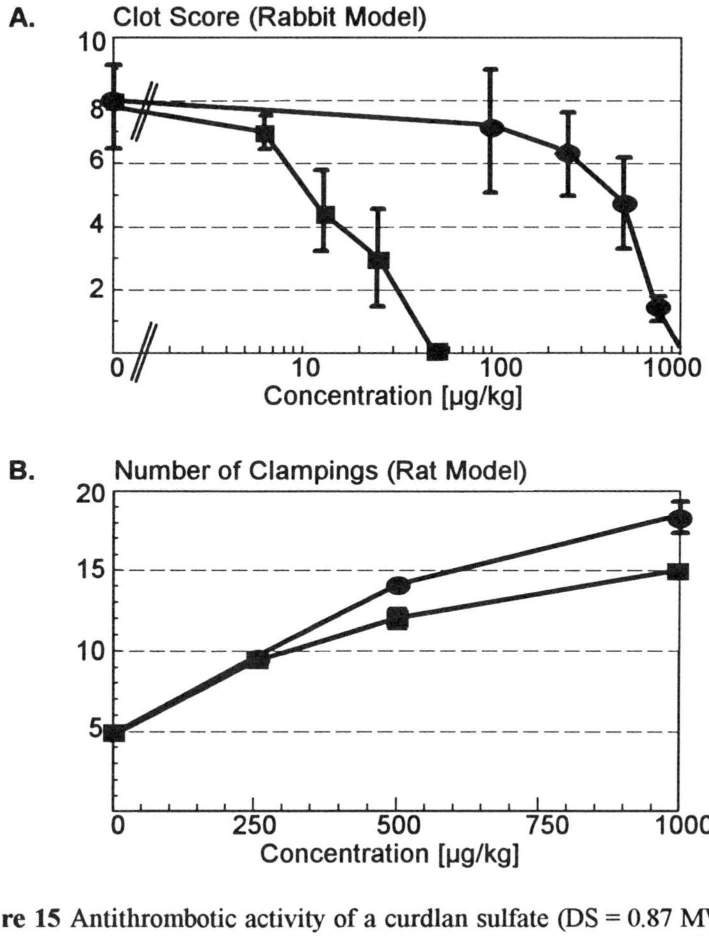

Figure 15 Antithrombotic activity of a curdlan sulfate (DS = 0.87 MW = 49 kd, 74 % of the sulfate groups on C2/C4) (-●-) in comparison to heparin (-■-) in: A: FEIBA® induced stasis thrombosis model in rabbits (antithrombotic effect reflected by the clot score). B: Clamping induced jugular vein occlusion model in rats (antithrombotic effect reflected by the number of clampings required for complete occlusion). (From Ref. 215. with kind permission from Elsevier Science Ltd., Kidlington, UK.)

Wessler-derived models mainly reflect the venous type of thrombosis in which the occlusion is due to fibrin formation. Since vessel wall properties are essential modulators in the process of thrombosis, the *in vivo* models should induce thrombosis via endothelial injury exclusively or combined with artificial hemostasis in animals [229]. This is the case, in contrast to the Wessler-derived models, in the clamping induced occlusion model, where repeated clamping of the jugular vein results in stasis as well as in a damage of the vessel wall with consequent adhesion and activation of platelets. In the following, typical deposition thrombi originate, which initially contain at most traces of fibrin and correspond more to the arterial type of thrombi [229].

As shown in Figure 15, in the rabbit stasis thrombosis model, the tested GluS produced concentration dependent antithrombotic actions, but the ED_{50} of 550 μg/kg was 32 times higher than the heparin ED_{50} of 17.2 μg/kg (230). This result agrees with the low activity of this heparinoid in the FEIBA® induced extrinsic FXa generation assay. In contrast to the weak effect in the rabbit model, heparin-like antithrombotic actions were produced in the rat model. At a 250 μg/kg dosage, both agents showed similar effects. At 500 and 1000 μg/kg, heparin was found to act only slightly better than the GluS.

Consequently, depending on their structure, the GluS are potent antithrombotic agents in the vascular injury model. Based on the *in vitro* results, thrombin seems to play an important role in their action. However, since the activities obtained in the rat model are higher than those found in the coagulation and aggregation assays, additional mechanisms of action such as modulation of the vascular wall may contribute to their antithrombotic action [215].

Therefore, as demonstrated by a large number of other heparinoids, global anticoagulant effects may not reflect the true antithrombotic potential of new sulfated polysaccharide derived heparinoids. To test the effects of such agents, different models must be considered to determine their differential antithrombotic actions.

VII. CONCLUSIONS

1. For about 60 years heparin has been the drug of choice in prophylaxis and treatment of thromboembolic disorders. However, there are some problems related to it, e.g. its lack of homogeneity and side effects such as bleeding and heparin-induced thrombocytopenia (HIT). Consequently, identification of alternatives to heparin is an important field of research. One of the various strategies is the development of substances structurally related to heparin, that is sulfated carbohydrates.

   In order to improve the efficacy-safety ratio of heparin, the class of Low Molecular Weight Heparins (LMWH) was developed. Their efficacy is recognized and they have better bioavailability and a prolonged duration of

action. However, the bleeding risk is not eliminated and they cannot be used as heparin substitutes in HIT patients. Like heparin, also the LMWH are heterogeneous mixtures and each of them has to be considered as an individual drug.
Furthermore, there are efforts to get better antithrombotics by chemical modification of heparin, for example Suleparoide® or heparins with lipophilic substituents.

2. Apart from heparin, other mammalian glycosaminoglycans also exhibit anticoagulant properties and present considerable therapeutic potential. But, just as heparin, they are heterogeneous mixtures and not much is known about their mode of action.
At present, especially heparan sulfate and dermatan sulfate preparations are under investigation. Simultaneously, there are efforts to optimize their activities by depolymerization and/or chemical modification. One such representative of the chondroitin sulfate family is the so-called semisynthetic heparin analogue (SSHA) obtained by fragmentation and sulfation of a natural chondroitin sulfate.
Despite incomplete knowledge about their chemistry and pharmacology, several glycosaminoglycan mixtures are now clinically used as antithrombotic drugs as well as for other indications. The most completely investigated example among the antithrombotic glycosaminoglycan-derived drugs is the well-established compound, danaparoid. Actually, it has gained in importance as the usual alternative for anticoagulant treatment of HIT patients.

3. In order to be independent of natural sources and to exclude the risk of biological contaminations, sulfated carbohydrate derivatives with low molecular weights are being developed by total synthesis. These chemically well defined substances represent the antithesis to the heterogeneous natural polymers.
An example is the naturally occuring heparin pentasaccharide, representing the ATIII-binding site of heparin. The original pentasaccharide sequence can be produced by a very complex chemical synthesis but this is not feasible used in clinical practice for economical reasons. In contrast to this, newer derivatives obtained by simplified synthetic procedures proved to be potent antithrombotics *in vivo*. In addition, these promising oligosaccharides can be used as heparin substitutes in HIT patients.
Aprosulate has to be mentioned as the most important representative of the group of bis-lactobionic acid amides. It exhibits good activity *in vitro* as well as *in vivo* by a mode of action different from that of heparin. However, due to chronic toxicity, the clinical studies have not been continued.

Nevertheless, Aprosulate is a good model for further developments in this area.

4. Besides the glycosaminoglycans in vertebrates, sulfated polysaccharides are widespread in the marine flora and fauna, such as the carrageenans in red algae or the fucoidans in brown algae which show anticoagulant activity dependent upon their structure. Research in this large field of potential heparinoids is still in its infancy and so far none of these has been clinically tested. It still remains open if in the future by means of modern techniques these natural compounds might be utilized for different medical purposes.

5. As an alternative to the naturally occuring sulfated carbohydrates, many other, non-sulfated polysaccharides were used to synthesize molecules whose structures simulate that of heparin or which at least show comparable antithrombotic effects. At this time, only two of these semisynthetic anticoagulants, dextran sulfate and pentosan polysulfate, have been accepted for clinical use and the optimal agent has not yet been discovered. However, as in the case of the systematic evaluation of structure-activity relationships of β-1,3-glucan sulfates described in this chapter, the research findings still may provide some valuable leads for the development of the optimal carbohydrate-based antithrombotic drug in the future.

   A recent strategy worth mentioning is the total synthesis of polysaccharide sulfates in order to produce well-defined polymers.
   And finally, there are programs seeking to produce heparin-like compounds by modification of bacterial polysaccharides. Through the use of modern biotechnological methods, it should be possible to obtain effective heparin substitutes in industrial scale quantities without the risk of contamination and independent of the material delivered by the meat industry.

In summary, in the field of anticoagulant and antithrombotic drugs, carbohydrates present a large potential beginning with the standard heparin up to the newest developments. Which of the various promising strategies to find carbohydrate-based heparin alternatives will prove to be the best, remains to be demonstrated.

REFERENCES

1. Petitou M. Drug based on carbohydrates - past and future. In: Kovác P, ed. Synthetic Oligosaccharides. Indispensable Probes for the Life Sciences. Washington, DC: American Chemical Society, 1994:19-35.
2. Fareed J, Hoppensteadt D, Walenga JM, Pifarré R. An overview of current anticoagulant and antithrombotic drugs. In: Pifarré R, ed. Anticoagulation,

Hemostasis, and Blood Preservation in Cardiovascular Surgery. Philadelphia: Hanley & Belfus, 1993:111-128.
3. Witczak ZJ. Carbohydrates as drugs and potential therapeutics. Curr Med Chem 1995; 1:392-405.
4. Rodén L. Highlights in the history of heparin. In: Lane DA, Lindahl U, eds. Heparin, Chemical and Biological Properties, Clinical Applications. London: Edward Arnold, 1989:1-23.
5. Jorpes E. The chemistry of heparin. Biochem J 1935; 29:1817-1830.
6. Conrad HE. Dissection of heparin - past and future. Pure & Appl Chem 1993; 65(4):787-791.
7. Nader HB, Dietrich CP. Natural occurence, and possible biological role of heparin. In: Lane DA, Lindahl U, eds. Heparin, Chemical and Biological Properties, Clinical Applications. London: Edward Arnold, 1989:81-96.
8. Hoving P, Piepkorn M, Linker A. Biological implications of the structural, antithrombin affinity and anticoagulant activity relationships among vertebrate heparin and heparan sulfates. Biochem J 1986; 237:573-581.
9. Lindahl U. Biosynthesis of heparin and related polysaccharides. In: Lane DA, Lindahl U, eds. Heparin, Chemical and Biological Properties, Clinical Applications. London: Edward Arnold, 1989:159-189.
10. Lindahl U, Kjellén L. Heparin or heparan sulfate - what is the difference? Thromb Haemost 1991; 66:44-48.
11. Lindahl U, Feingold DS, Rodén L. Biosynthesis of heparin. Trends Biochem Sci 1986; 11:221-225.
12. Hopwood JJ. Enzymes that degrade heparin and heparan sulphate. In: Lane DA, Lindahl U, eds. Heparin, Chemical and Biological Properties, Clinical Applications. London: Edward Arnold, 1989:191-227.
13. Linhardt RJ, Merchant ZM, Rice KG, Kim YS, Fitzgerald GL, Grant AC, Langer R. Evidence of random structural features in the heparin polymer. Biochemistry 1985; 24:7805-7810.
14. Linhardt RJ, Cohen DM, Rice KG. Nonrandom structural features in the heparin polymer. Biochemistry 1989; 28:2888-2894.
15. Linhardt RJ. Heparin: An important drug enters its seventh decade. Chemistry and Industry 1991; 1:45-50.
16. Nader HB, Takahashi HK, Guimarães JA, Dietrich CP, Bianchini P, Osima B. Heparin: A family of more than one hundred components revealed by the combination of electrofocusing, affinity chromatography and gel electrophoresis. Int J Biol Macromol 1981; 3:356-360.
17. Linhardt RJ, Rice KM, Kim YS, Lohse DL, Wang HM, Loganathan D. Mapping and quantification of the major oligosaccharide components of heparin. Biochem J 1988; 254:781-787.
18. Linhardt RJ, Loganathan D, Al-Hakim A. Oligosaccharide mapping of low molecular weight heparins: Structure and activity differences. J Med Chem 1990; 33:1639-1645.

19. Choay J: Structure and activity of heparin and its fragments: An overview. Semin Thromb Hemost 1989; 15:359-364.
20. Björk I, Olson ST, Shore JD. Molecular mechanisms of the accelerating effect of heparin on the reactions between antithrombin and clotting proteinases. In: Lane DA, Lindahl U, eds. Heparin, Chemical and Biological Properties, Clinical Applications. London: Edward Arnold, 1989:229-255.
21. Casu B. Heparin structure. Haemostasis 1990; 20(suppl 1):62-73.
22. Pejler G, Danielsson A, Björk I, Lindahl U, Nader HB, Dietrich CP. Structure and antithrombin-binding properties of heparin isolated from the clams *Anomalicardia brasiliana* and *Tivela mactroides*. J Bio Chem 1987; 262: 11413-11421.
23. Ofosu FA. Prophylactic doses of enoxaparin and heparin inhibit prothrombin activation. Semin Thromb Hemost 1993; 19(suppl 1):29-30.
24. Verstraete M. Pharmacotherapeutic aspects of high and low molecular weight heparin. Drugs 1990; 40:498-530.
25. Linhardt RJ, Langer R. New approaches for anticoagulation in extracorporal therapy. Biomater Artif Cells Artif Organs 1987; 15:91-100.
26. Engelberg H. Heparin and atherosclerotic process. Pharmacol Rev 1984; 36:91-110.
27. Cherchi GM, Formato M, Demuro P, Masserini M, Varani I, DeLuca G. Modifications of low density lipoprotein induced by the interaction with human plasma glycosaminoglycan protein complexes. Biochim Biophys Acta 1994; 1212:345-352.
28. Wright Jr TC, Castellot JJ, Diamond JR, Karnovsky MJ. Regulation of cellular proliferation by heparin and heparan suphate. In: Lane DA, Lindahl U, eds. Heparin, Chemical and Biological Properties, Clinical Applications. London: Edward Arnold, 1989:295-316.
29. Chan PAD. Cell biology of human vascular smooth muscle. Ann R Coll Surg Engl 1994; 76:298-303.
30. Bertolesi GE, Lauria de Cidre L, Eijan AM. Growth inhibition *in vitro* of murine mammary adenocarcinoma cells by heparin and chemically modified heparins. Tumour Biol 1994; 15:275-283.
31. Folkman J, Ingber DE. Angiogenesis: Regulatory role of heparin and related molecules. In: Lane DA, Lindahl U, eds. Heparin, Chemical and Biological Properties, Clinical Applications. London: Edward Arnold, 1989:317-333.
32. Jaques LB. Heparin: An old drug with a new paradigm (Part B). Hämostaseologie 1985; 5:121-126.
33. de Clercq E. New acquisitions in the development of anti-HIV agents. Antiviral Res 1989; 12:1-20.
34. Linhardt RJ, Wang H, Ampofo SA. New methodologies in heparin structure analysis and the generation of LMW heparins. In: Lane DA, Björk I, Lindahl U, eds. Heparin and Related Polysaccharides. New York: Plenum

Press, 1992:37-47.
35. Zhou F, Höök T, Thompson JA, Höök M. Heparin protein interactions. In: Lane DA, Björk I, Lindahl U, eds. Heparin and Related Polysaccharides. New York: Plenum Press, 1992:141-153.
36. Lindahl U, Höök M. Glycosaminoglycans and their binding to biological macromolecules. Ann Rev Biochem 1978; 47:385-417.
37. Fransson LA, Johannson BG. Interaction of heparin and heparan sulfate with thrombin. Int J Biol Macrom 1981; 3:25-30.
38. Fransson LA, Havsmark B. Interaction of heparin and heparan sulfate with low-density lipoproteins. Int J Biol Macrom 1982; 4:73-78.
39. Jaques LB. Heparin: An old drug with a new paradigm (Part A). Hämostaseologie 1985; 5:88-92.
40. Casu B, Petitou M, Provasoli M, Sinay P. Conformational flexibility: A new concept for explaining binding and biological properties of iduronic acid-containing glycosaminoglycans. Trends Biochem Sci 1988; 13:221-225.
41. Olson ST, Björk I. Role of protein conformational changes, surface approximation and protein cofactors in heparin-accelerated antithrombin-proteinase reactions. In: Lane DA, Björk I, Lindahl U, eds. Heparin and related polysaccharides. New York: Plenum Press, 1992:155-165.
42. Marcum JA, Rosenberg RD. The biochemistry, cell biology, and pathophysiology of anticoagulantly active heparin-like molecules of the vessel wall. In: Lane DA, Lindahl U, eds. Heparin, Chemical and Biological Properties, Clinical Applications. London: Edward Arnold, 1989: 275-294.
43. Holmer E. Low molecular weight heparin. In: Lane DA, Lindahl U, eds. Heparin, Chemical and Biological Properties, Clinical Applications. London: Edward Arnold, 1989: 575-595.
44. Hemker HC, Beguin S. Mode of action of heparin and related drugs. Semin Thromb Hemost 1991; 17(suppl 1): 27-34.
45. Baruch D, Lindhout T, Wagenvoord R, Hemker HC. Inhibition of thrombin-catalysed reactions in blood coagulation and platelet activation by heparin fractions in the absence of antithrombin III. Haemostasis 1986; 16:71-81.
46. Ofosu FA, Modi GJ, Cerskus AL, Hirsh J, Blajchman MA. Heparin with low affinity to antithrombin III inhibits the activation of prothrombin in normal plasma. Thromb Res 1982; 28:487-497.
47. Nader HB, Toma L, Pinhal MS, Buonassisi V, Colburn P, Dietrich CP. Effect of heparin and dextran sulfate on the synthesis and structure of heparan sulfate from cultured endothelial cells. Semin Thromb Hemost 1991; 17(suppl 1):47-56.
48. Messmore HL, Griffin B, Koza M, Seghatchian J, Fareed J, Coyne E. Interaction of heparinoids with platelets: comparison with heparin and low molecular weight heparin. Semin Thromb Hemost 1991; 17(suppl 1):57-64.
49. Abildgaard U. Tissue factor pathway inhibitor and heparin. In: Lane DA, Björk I, Lindahl U, eds. Heparin and related polysaccharides. New York:

Plenum Press, 1992:199-204.
50. Vairel EG. Quelques activities biologiques des heparines autres que leurs propriétés antithrombotiques. Ann Pharm Fr 1994; 52:81-88.
51. ten Cate JW, Koopman MMW, Prins MH, Büller HR. Treatment of venous thromboembolism. Thromb Haemost 1995; 74:197-203.
52. Ginsberg JS. Evaluation of the bleeding effects of antithrombotic drugs. In: Harenberg J, Heene DL, Stehle G, Schettler G, eds. New Trends in Haemostasis; Coagulation Proteins, Endothelium and Tissue Factors. Berlin, Heidelberg, New York: Springer-Verlag, 1990:263-269.
53. Chong BH. Heparin-induced thrombocytopenia. Aust N Z J Med 1992; 22: 145-152.
54. Warkentin TE, Kelton JG. Heparin and platelets. Hematol Oncol Clin North Am 1990; 4:243-264.
55. Greinacher A, Pötsch B, Amiral A, Dummel V, Eichner A, Mueller-Eckhardt C. Heparin-associated thrombocytopenia: Isolation of the antibody and characterization of a multimolecular PF_4-heparin complex as the major antigen. Thromb Haemost 1994; 71:247-251.
56. Greinacher A, Michels I, Liebenhoff U, Presek P, Mueller-Eckhardt C. Heparin-associated thrombocytopenia: Immune complexes are attached to platelet membrane by the negative charge of highly sulphated oligosaccharides. Br J Haematol 1993; 45:252-257.
57. Greinacher A, Michels I, Mueller-Eckhardt C. Heparin-associated thrombocytopenia: The antibody is not heparin specific. Thromb Haemost 1992; 67:545-549.
58. Tardy PB, Tardy B, Grelac F, Reynaud J, Mismetti P, Bertrand JC, Guyotat D. Pentosan polysulfate-induced thrombocytopenia and thrombosis. Am J Hematol 1994; 88:803-808.
59. Weitz J. New anticoagulant strategies; current status and future potential. Drugs 1994; 48:485-497.
60. Fareed J, Hoppensteadt DA, Walenga JM. Current perspectives on low molecular weight heparins. Semin Thromb Hemost 1993; 19(suppl 1): 1-11.
61. McGinnis DM, Outschoorn AS. Problems of component activities of heparin. Pharmacopeial Forum 1991; 5:2438-2441.
62. Coccheri S. Low molecular weight heparins: An introduction. Haemostasis 1990; 20(suppl 1):74-80.
63. Hirsh J, Siragusa S, Cosmi B, Ginsberg JF. Low molecular weight heparins (LMWH) in the treatment of patients with acute venous thromboembolism. Thromb Haemost 1995; 74:360-363.
64. Kakkar VV. Effectiveness and safety of low molecular weight heparins (LMWH) in the prevention of venous thromboembolism (VTE). Thromb Haemost 1995; 74:364-368.
65. Thomas DP, Merton RE, Gray E, Barrowcliffe TW. The relative antithrombotic effectiveness of heparin, a low molecular weight heparin, and a penta-

saccharide fragment in an animal model. Thromb Hemost 1989; 61:204-207.
66. Ofosu FA, Blajchman MA, Modi GJ, Smith LM, Buchanan MR, Hirsh J. The importance of thrombin inhibition for the expression of the anticoagulant activities of heparin, dermatan sulfate, low molecular weight heparin and pentosan polysulphate. Br J Haematol 1985; 60:695-704.
67. Levine MN, Hirsh J, Kelton JG. Heparin-induced bleeding. In: Lane DA, Lindahl U, eds. Heparin, Chemical and Biological Properties, Clinical Applications. London: Edward Arnold, 1989:517-531.
68. Horton D, Usui T. Sulfated carbohydrates obtained by chemical modification of polysaccharides. In: Schweiger RG, ed. Carbohydrate Sulfates. Washington DC; American Chemical Society, 1978:95-112.
69. Callas DD, Ahsan A, Iqbal O, Fareed J. Molecular weight and biochemical profile of a chemically modified heparin derivative, Suleparoide. Thromb Res 1993; 69:369-376.
70. Callas DD, Hoppensteadt DA, Jeske W, Iqbal O, Bacher P, Ahsan A, Fareed J. Comparative pharmacological profile of a glycosaminoglycan mixture, Sulodexide®, and a chemically modified heparin derivative, Suleparoide®. Semin Thromb Hemost 1993; 19(suppl 1):49-57.
71. Hladovec J, Kornalik F. Antithrombotic effects of heparin and related agents. Cor Vasa 1991; 33:68-74.
72. Mattson C, Hoylaert M, Holmer E, Uthne T, Collen D. Antithrombotic properties in rabbits of heparin and heparin fragments covalently coupled to human antithrombin III. J Clin Invest 1986; 43:1-5.
73. Saivin S, Petitou M, Lormeau JC, Dupouy D, Sié P, Caranobe C, Houin G, Boneu B. Pharmacologic properties of an unfractionated heparin butyryl derivative with long-lasting effects. J Lab Clin Med 1992; 119:189-196.
74. Piazolo L, Harenberg J, Malsch R, Heene DL. Improved pharmacodynamic of cholesterol-hemisuccinate- or L-ascorbic-acid-6-palmitate-LMM-heparin as lipophilically modified low molecular mass heparins. Thromb Haemost 1995; 73:968.
75. Bianchini P, Osima B, Parma B, Nader HB, Dietrich CP. Pharmacological activities of heparins obtained from different tissues: Enrichment of heparin fractions with high lipoprotein lipase, antihemolytic and anticoagulant activities by molecular sieving and antithrombin III affinity chromatography. J Pharmacol Exp Ther 1982; 220:406-410.
76. Marcum JA, Rosenberg RD. Anticoagulantly active heparan sulfate proteoglycan and the vascular endothelium. Semin Thromb Hemost 1987; 13:464-474.
77. Fareed J, Hoppensteadt D, Jeske W, Walenga JM. An overview of non-heparin glycosaminoglycans as antithrombotic agents. In: Poller L, ed. Recent Advances in Blood Coagulation. London: Churchill Livingstone, 1993:169-187.

78. Bienkowski MJ, Conrad HE. Kinetics of proteoheparan sulfate synthesis, secretion, endocytosis, and catabolism by a hepatocyte cell line. J Biol Chem 1984; 259:12989-12996.
79. Hoppensteadt D, Walenga J, Fareed J. Effect of dermatan sulfate and heparan sulfate on platelet activity compared to heparin. Semin Thromb Hemost 1991; 17(suppl 1):60-64.
80. Hubbard AR, Jennings CA, Barrowcliffe TW. Anticoagulant properties *in vitro* of heparan sulfates. Thromb Res 1984; 35:567-576.
81. Marcum JA, Rosenberg RD. Regulation of the blood coagulation mechanism by anticoagulantly active heparan sulfate proteoglycans. Biol Carbohydr 1991; 3:47-73.
82. Dawes J. Measurement of the affinities of heparins, naturally occuring GAGs, and other sulfated polymers for antithrombin III and thrombin. Anal Biochem 1988; 174:177-186.
83. Ofosu FA, Buchanan MR, Anvari N, Smith LM, Blajchman MA. Plasma anticoagulant mechanisms of heparin, heparan sulfate, and dermatan sulfate. Ann N Y Acad Sci 1989; 556:123-131.
84. Bianchini P, Osima B, Parma B, Nader HB, Dietrich CP. Lack of correlation between '*in vitro*' and '*in vivo*' antithrombotic activity of heparin fractions and related compounds. Heparan sulfate as an antithrombotic agent '*in vivo*'. Thromb Res 195; 39:21-28.
85. Hoppensteadt D, Walenga JM, Fareed J. Comparative antithrombotic and hemorrhagic effects of dermatan sulfate, heparan sulfate and heparin. Thromb Res 1990; 60:191-200.
86. Buchanan MR, Brister SJ, Ofosu F. Prevention and treatment of thrombosis: Novel strategies arising from our understanding the healthy endothelium. Wien Klin Wochenschr 1993; 105:309-313.
87. Casu B. Structural features and binding properties of chondroitin sulfates, dermatan sulfate, and heparan sulfate. Semin Thromb Hemost 1991, 17 (suppl 1):9-14.
88. Cofrancesco E, Colombi M, Gianese F, Cortellaro M. The effect of dermatan sulfate on *in vitro* human plasma coagulation, platelet aggregation and beta TG/PF_4 release. Thromb Res 1990; 57:405-414.
89. Sié P, Dupouy D, Caranobe C, Petitou M, Boneu B. Antithrombotic properties of a dermatan sulfate hexadecasaccharide fractionated by affinity for heparin cofactor II. Blood 1993; 81:1771-1777.
90. Mascellani G, Liverani L, Bianchini P, Parma B, Torri G, Bisio G, Guerrini M, Casu B. Structure and contribution of the heparin cofactor II-mediated inhibition of thrombin of naturally oversulphated sequences of dermatan sulphate. Biochem J 1993; 296:639-648.
91. Merton RE, Thomas DP. Experimental studies on the relative efficacy of dermatan sulfate and heparin as antithrombotic agents. Thromb Haemost 1987; 58:839-842.

92. Ofosu FA, Choay J, Anvari N, Smith LM, Blajchman MA. Inhibition of factor X and factor V activation by dermatan sulfate and a pentasaccharide with high affinity for antithrombin III in human plasma. Eur J Biochem 1990:193:485-493.
93. Santoro FM, Alvarez R, Fussi F. Pharmacological profile of a native dermatan sulfate. Thromb Res 1992; 67:201-211.
94. Van Ryn-McKenna J, Ofosu FA, Gray E, Hirsh J, Buchanan MR. Effects of dermatan sulfate and heparin on inhibition of thrombus growth *in vivo*. Ann N Y Acad Sci 1989; 556:304-312.
95. Perlin AS, Sauriol F, Cooper B, Folkman J. Dermatan sulfate in pharmaceutical heparins. Thromb Haemost 1987; 58:792.
96. Cadroy Y, Dol F, Caranobe C, Petitou M, Lormeau JC, Sié P, Choay J. Standard heparin enhances the antithrombotic activity of dermatan sulfate in the rabbit but CY216 does not. Thromb Haemost 1988; 59:295-298.
97. Thomas DP, Gray E, Merton RE. Potentiation of the antithrombotic action of dermatan sulphate by small amounts of heparin. Thromb Haemost 1990; 64:290-293.
98. Barbanti M, Calanni F, Milani MR, Marchi E, Semeraro N, Colucci M. Therapeutic effect of low molecular weight dermatan sulfate (Desmin 370) in rat venous thrombosis - evidence for an anticoagulant-independent mechanism. Thromb Haemost 1993; 69:147-151.
99. Dol F, Petitou M, Lormeau JC, Choay J, Caranobe C, Sié P, Saivin S, Houin G, Boneu B. Pharmacologic properties of a low molecular weight dermatan sulfate (4kd): Comparison with unfractionated dermatan sulfate (25kd). J Lab Clin Med 1990; 115:43-51.
100. Saivin S, Dol F, Caranobe C, Petitou M, Lormeau JC, Sié P, Houin G, Boneu B. Influence of molecular weight upon the anticoagulant and pharmacokinetic properties of dermatan sulfate in the rabbit. Thromb Res 1992; 66:527-535.
101. Maaroufi RM, Tapon-Bretaudière J, Mardiguian K, Sternberg C, Dautzenberg MD, Fischer AM. Influence of the oversulfation method and the degree of sulfation on the anticoagulant properties of dermatan sulfate derivative. Thromb Res 1990; 59:749-758.
102. Dol F, Petitou M, Choay J, Sié P, Houin G, Boneu B. Pharmacological properties of dermatan sulfate, of a low molecular weight dermatan sulfate (LMDS) and of two oversulfated derivatives. Folia Haematol Int Mag Klin Morphol Blutforsch 1989; 116:851-857.
103. Saivin S, Caranobe C, Petitou M, Sié P, Lormeau JC, Level M, Crepon B, Houin G, Boneu B. Antithrombotic activity, bleeding effect and pharmacodynamics of a succinyl derivative of dermatan sulphate in rabbits. Br J Haematol 1992; 80:508-513.
104. Volpi N, Cusmano M, Venturelli T. Qualitaitve and quantitative studies of heparin and chondroitin sulfates in normal human plasma. Biochim Biophys

Acta 1995; 1243:49-58.

105. Nawa K, Sakano K, Fujiwara H, Sato Y, Sugiyama N, Teruuchi T, Iwamoto T, Marumoto Y. Presence and function of chondroitin-4-sulfate on recombinant human soluble thrombomodulin. Biochem Biophys Res Commun 1990; 171:729-737.
106. Bjornsson TD, Nash PV, Schaten R. The anticoagulant effect of chondroitin-4-sulfate. Thromb Res 1982; 27:15-21.
107. Bergqvist D, Lindblad B, Mätzsch T. Glycosaminoglycans in prophylaxis against venous thromboembolism. In: Lane DA, Björk I, Lindahl U, eds. Heparin and Related Polysaccharides. New York: Plenum Press, 1992: 259-274.
108. Thomas DP, Barrowcliffe TW, Merton RE, Stocks DP, Dawes J, Pepper DS. *In vivo* release of anti-Xa clotting activity by a heparin analogue. Thromb Res 1980: 17:831-840.
109. Törngren S, Blombäck M, Almgård L-E, Norming U, Nyman CR. Conventional heparin and semisynthetic heparin analogue (SSHA) alteration of blood coagulation after embolic occlusion of human renal circulation. Thomb Res 1990; 59:237-246.
110. Pescador R, Porta R, Mantovani M, Prino G, Casu B, Naggi A, Torri G, Walenga JM, Hoppensteadt D, Fareed J. Pharmacological profile of sulf-amino-galactosaminoglycans. Semin Thromb Hemost 1991; 17(suppl 1): 74-79.
111. Meuleman DG. Orgaran (Org 10172): Its pharmacological profile in experimental models. Haemostasis 1992; 22:58-65.
112. Meuleman DG. Synopsis of the anticoagulant and antithrombotic profile of the low molecular weight heparinoid Org 10172 in experimental models. Semin Thromb Hemost 1989; 15:370-372.
113. Ofosu FA. Anticoagulant mechanisms of Orgaran (Org 10172) and its fractions with high affinity to antithrombin III (Org 10849). Haemostasis 1992; 22:66-72.
114. Gordon DL, Linhardt R, Adams HP. Low molecular weight heparins and heparinoids and their use in acute or progressing ischemic stroke. Clin Neuropharmacol 1990; 13:522-543.
115. Danhof M, de Boer A, Magnani HN, Stiekema JC. Pharmacokinetic considerations of Orgaran (Org 10172) therapy. Haemostasis 1992; 22:73-84.
116. Magnani HN. Heparin-induced thrombocytopenia (HIT): An overview of 230 patients treated with Orgaran (Org 10172). Thromb Haemost 1993; 70:554-561 and 1072 (erratum).
117. Greinacher A, Amiral J, Dummel V, Vissac A, Kiefel V, Mueller-Eckhardt C. Laboratory diagnosis of heparin-associated thrombocytopenia and comparison of platelet aggregation test, heparin-induced activation test, and platelet factor 4/heparin enzyme-linked immunosorbent assay. Transfusion 1994; 34:381-385.

118. Buchanan MR, Liao P, Smith LJ, Ofosu FA. Prevention of thrombus formation and growth by antithrombin III and heparin cofactor II-dependent thrombin inhibitors: Importance of heparin cofactor II. Thromb Res 1994; 74:463-475.
119. Barbanti M, Guizzardi S, Calanni F, Marchi E, Babbini M. Antithrombotic and thrombolytic activity of sulodexide in rats. Int J Clin Lab Res 1992; 22:179-184.
120. Condorelli M, Chiariello M, Dagianti A, Penco M, Dalla-Volta S, Pengo V, Schivazappa L, Mattioli G, Mattioli AV, Brusoni B, et al. IPO-V2: a prospective, multicenter, randomized, comparative clinical investigation of the effects of sulodexide in preventing cardiovascular accidents in the first year after acute myocardial infarction. J Am Coll Cardiol 1994; 23:27-34.
121. Catania G, Salanitri T. Pharmacological treatment of intermittent claudication: double blind study of Sulodexide vs placebo. Riv Eur Sci Med Farmacol 1992; 14:149-157.
122. Mauro M, Palmieri GC, Palazzini E, Barbanti M, Calanni-Rindina F, Milani MR. Pharmacodynamic effects of single and repeated doses of oral sulodexide in healthy volunteers. A placebo-controlled study with an enteric-coated formulation. Curr Med Res Opin 1993; 13:87-95.
123. van Boeckel CAA, Petitou M. The unique antithrombin III binding domain of heparin: A lead to new synthetic antithrombotics. Angew Chem Int Ed Engl 1993; 32:1671-1690.
124. Sinaÿ P, Jacquinet JC, Petitou M, Duchaussoy P, Lederman I, Choay J, Torri G. Total synthesis of a pentasaccharide fragment having high affinity for antithrombin III. Carbohydr Res 1984; 132:C5-C9.
125. Petitou M, Duchaussoy P, Lederman I, Choay J, Sinaÿ P. Binding of heparin to antithrombin III: A chemical proof of the critical role played by a 3-sulfated 2-amino-2-deoxy-D-glucose residue. Carbohydr Res 1988; 179:163-172.
126. Walenga JM, Petitou M, Lormeau JC, Samama M, Fareed J, Choay J. Antithrombotic activity of a synthetic heparin pentasaccharide in a rabbit stasis thrombosis model using different thrombogenic challenges. Thromb Res 1989; 54:399-410.
127. Greinacher A, Alban S, Dummel V, Franz G, Mueller-Eckhardt C. Characterization of the structural requirements for a carbohydrate based anticoagulant with a reduced risk of inducing the immunological type of heparin-associated thrombocytopenia. Thromb Haemost 1995; 74:886-892.
128. Hérault JP, Bârzu T, Crépon B, Bernat A, Lormeau JC, Herbert JM, Petitou M. Biochemical and pharmacological properties of O-sulfated, O-methylated analogues of the 'natural pentasaccharide'. Thromb Haemost 1995; 73:1321.
129. Herbert JM, H érault JP, Barzû T, Bernat A, Lormeau JC, Petitou M. Biochemical and pharmacological properties of SANORG 32701, a potent analogue of the 'natural pentasaccharide'. Thromb Haemost 1995; 73:1321.

130. Meuleman DG, Hobbelen PMJ, van Dinther TG, Vogel GMT, van Boeckel CAA, Moelker HCT. Antifactor Xa activity and antithrombotic activity in rats of structural analogues of the minimum antithrombin III binding sequence: Discovery of compounds with a longer duration of action than the natural pentasaccharide. Semin Thromb Hemost 1991; 17(suppl 1):112-117.
131. Hérault JP, Barzû T, Bernat A, Crépon B, Donat F, Petitou M, Lormeau JC, Herbert JM. Pharmacokinetics of three synthetic ATIII-binding pentasaccharides in various animal species - extrapolation to humans. Thromb Haemost 1995; 73:1320.
132. van Amsterdam RGM, Vogel GMT, Princen AWM, Meerbach CMJ, van As E, van Dinther TG, Meuleman DG. Residence times of antithrombotic pentasaccharides with very high binding affinity for antithrombin III in the pig. Thromb Haemost 1995; 73:969.
133. Hobbelen PM, van Dinther TG, Vogel GM, van Boeckel CA, Moelker HC, Meuleman DG. Pharmacological profile of the chemically synthesized antithrombin III binding fragment of heparin. Thromb Haemost 1990; 63:265-270.
134. Klauser RJ, Meinetsberger E, Raake W. Biochemical studies on lactobionic acid amides. Semin Thromb Hemost 1991; 17(suppl 1):118-125.
135. Klauser RJ. Interaction of the sulfated lactobionic acid amide LW 10082 with thrombin and its endogenous inhibitors. Thromb Res 1991; 62:557-565.
136. Raake W, Klauser RJ, Meinetsberger E, Zeller P, Elling H. Pharmacological profile of the antithrombotic and bleeding actions of sulfated lactobionic acid amides. Semin Thromb Hemost 1991; 17(suppl 1):129-135.
137. Kindness G, Long WF, Williamson FB. Enhancement of ATIII activity by carrageenans. Thromb Res 1979; 15:49-60.
138. Elsner H, Brosner W, Bürgel E. Über das Vorkommen von hochwirksamen, die Blutgerinnung hemmenden Stoffen in Rotalgen. Z Physiol Chem 1937; 246:244-247.
139. Güven KV, Özsoy Y, Ulutin ON. Anticoagulant, fibrinolytic and antiaggregant activity of carrageenans and alginic acid. Bot Mar 1991; 34:429-432.
140. Mabeau S, Kloareg B, Joseleau J-P. Fractionation analysis of fucans from brown algae. Phytochem 1990; 29:2441-2445.
141. Nishino T, Yokoyama G, Dobashi K, Fujihara M, Nagumo T. Isolation, purification, and characterization of fucose-containing sulfated polysaccharides from the brown seaweed *Ecklonia kurome* and their blood-anticoagulant activities. Carbohydr Res 1989; 186:119-129.
142. Nishino T, Aizu Y, Nagumo T. The influence of sulfate content and molecular weight of a fucan sulfate from the brown seaweed *Ecklonia kurome* on its antithrombin activity. Thromb Res 1991; 64:723-731.
143. Nishino T, Nagumo T. Anticoagulant and antithrombin activities of oversulfated fucans. Carbohydr Res 1992; 229:355-362.
144. Nishino T, Aizu Y, Nagumo T. Antithrombin activities of a fucan sulfate from the brown seaweed *Ecklonia kurome*. Thromb Res 1991; 62:765-773.

145. Mauray S, Sternberg C, Theveniaux J, Millet J, Sinquin C, Tapon-Bretaudière J, Fischer AM. *In vitro* and *in vivo* mode of action of fucoidan. Thromb Haemost 1995; 73:1319.
146. Church FC, Meade JB, Treanor RE, Whinna HC. Antithrombin activity of fucoidan. J Biol Chem 1989; 265:3618-3623.
147. Soeda S, Sakaguchi S, Shimeno H, Nagamatsu A. Fibrinolytic and anticoagulant activities of highly sulfated fucoidan. Biochem Pharmacol 1992; 43:1853-1858.
148. Soeda S, Ohmagari Y, Shimeno H, Nagamatsu A. Preparation of aminated fucoidan and its evaluation as an antithrombotic and anitlipemic agent. Biol Pharm Bull 1994; 17:784-788.
149. Efinmov VS, Usov AI, Ol'skaia TS, Baliunis AI, Rozkin MI. Comparative study of the anticoagulant activity of sulfated polysaccharids from red marine algae. Farmakol Toksikol 1983; 46:61-67.
150. Caporiccio B, Braun M, Vignaud M, Chalet M, Teste J, Codomier L, Catayee G. Characteristics of the action of an acidic sulfated polysaccharide on global blood coagulation *in vitro*. C R Seances Soc Biol Fil 1983; 177:412-420.
151. Ibañez CM, Matsuhiro B. Structural studies on the soluble polysaccharide from *Iridaea membranacea*. Carbohydr Res 1986; 146:327-334.
152. Pavão MSG, Mourão PAS, Mulloy B. Structure of a unique sulfated α-L-galactofucan from the tunicate *Clavelina*. Carbohyrdr Res 1990; 208:153-161.
153. Bergqvist D, Frohm B, Stenberg P, Nilsson IM. The influence of a new low molecular weight sulphated polysaccharide from mussels (Org 30016) on haemostasis. Thromb Res 1985; 38:233-242.
154. Nilsson IM, Rothman U, Stenberg P, Frohm B, Persson NH. A novel semi-synthetic sulphated polysaccharide with potent antithrombin activity. Br J Haematol 1982; 50:335-343.
155. Nagase H, Enjyoji K, Minamiguchi K, Kitazato KT, Kitazato K, Saito H, Kato H. Depolymerized holothurian glycosaminoglycan with novel anticoagulant actions: antithrombin III- and heparin cofactor II-independent inhibition of factor X activation by factor IX- factor VIIIa complex and heparin cofactor II-dependent inhibition of thrombin. Blood 1995; 85:1527-1534.
156. Kumada T, Inoue K, Korenaga H, Abiko Y. Antithrombotic and thrombolytic effects of galactan polysulfate (DH 6322) in rats. Thromb Res 1985; 39:21-28.
157. Kumada T, Abiko Y. Fibrinolytic action of a new semi-synthetic polysaccharide sulfate, galactan polysulfate (DH 6322), in the rat. Thromb Res 1985; 39:9-19.
158. Dietrich CP, Nader HB, de Pavia JF, Santos EA, Holme KR, Perlin AS. Heparin in molluscs: chemical, enzymatic degradation and ^{13}C and ^{1}H n.m.r. spectroscopical evidence for the maintenance of the structure through evolution. Int J Biol Macromol 1989; 11:361-366.

159. De Clercq E. Selective virus inhibitors. Microbiologica 1990; 13:165-178.
160. Whistler RL, Spencer WW. Preparation and properties of several polysaccharide sulfates. Arch Biochem Biophys 1961; 95:36-41.
161. Guiseley KB. Some novel methods and results in the sulfation of polysaccharides. In: Schweiger RG, ed. Carbohydrate Sulfates. Washington DC; American Chemical Society, 1978:148-162.
162. Sié P, Ofosu F, Fernandez F, Buchanan MR, Petitou M, Boneu B. Respective role of antithrombin III and heparin cofactor II in the *in vitro* anticoagulant effect of heparin and of various sulphated polysaccharides. Br J Haematol 1986; 64:707-714.
163. Muzzarelli RAA, Tanfani F, Emanuelli M, Pace DP, Chiurazzi E, Piani M. Sulfated N-(carboxymethyl)chitosan: novel blood anticoagulants. Carbohydr Res 1984; 126:225-231.
164. Nishimura S-I, Nishi N, Tokura S, Okiei W, Somorin O. Inhibition of the hydrolytic activity of thrombin by chitin heparinoids. Carbohydr Res 1986; 156:286-292.
165. Nagasawa K, Harada H, Hayashi S, Misawa T. Sulfation of dextran with piperidine-N-sulfonic acid. Carbohydr Res 1972; 21:420-426.
166. Ricketts CR. Dextran sulphate - a synthetic analogue of heparin. Biochem 1952; 51:129-133.
167. Kamide K, Saito M. Cellulose and cellulose derivatives: Recent advances in physical chemistry. Adv Polym Sci 1987; 83:1-56.
168. Hatanaka K, Yoshida T, Miyagara S, Sato T, Ono F, Uryo T, Kuzuhara H. Synthesis of new heparinoids with high anticoagulant activity. J Med Chem 1987; 30:810-814.
169. Kojima S, Harada-Shiba M, Nomura S, Kimura G, Tsushima M, Kuramochi M, Yamamoto A, Omae T. Effect of nafamostat mesilate on bradykinin generation during low-density apheresis using a dextran sulfate cellulose column. ASAIO-Trans 1991; 37:644-648.
170. Wagenvoord RJ, Hendrix HH, Kolde HJ, Hemker HC. Development of a rapid and sensitive chromogenic heparin assay for clinical use. Haemostasis. 1993; 23:26-37.
171. Mitsuya H, Looney DJ, Kuno S, Ueno R, Wong-Staal F, Broder S. Dextran sulfate suppression of viruses in the HIV family: inhibition of virion binding to CD4+ cells. Science 1988; 240:646-649.
172. Joffe S. Drug prevention of postoperative deep vein thrombosis. A comparative study of calcium heparinate and sodium pentosan polysulfate. Arch Surg 1976; 111:37-40.
173. Bergqvist D, Efsing HO, Hallbook T, Lindblad-B. Prevention of postoperative thromboembolic complications. A prospective comparison between dextran 70, dihydroergotamine heparin and a sulphated polysaccharide. Acta Chir Scand 1980; 146:559-568.
174. Wagenvoord R, Hendrix H, Soria C, Hemker HC. Localization of the inhibi-

tory site(s) of pentosan polysulphate in blood coagulation. Thromb Haemost 1988; 60:220-225.
175. Czapek EE, Kwaan HC, Szczecinski M. The effect of a sulfated polysaccharide on antithrombin III. J Lab Clin Med 1980; 95:783-790.
176. Fischer AM, Dautzenberg MD, Aurousseau MH, Beguin S, Goudemand J, Hemker HC. Comparison between the effect of pentosan polysulphate, heparin and antithrombin III injections in antithrombin III. Thromb Res 1985; 37:295-307.
177. Scully MF, Kakkar VV. Identification of heparin cofactor II as the principal plasma cofactor for the antithrombin activity of pentosan polysulfate (SP 54). Thromb Res 1984; 36:187-194.
178. Ofosu FA, Fernandez F, Anvari N, Caranobe C, Dol F, Cadroy Y, Petitou M, Mardiguian J, Sié P, Boneu B. Further studies on the mechanisms for the antithrombotic effects of sulfated polysaccharides in rabbits. Thromb Haemost 1988; 60:188-192.
179. Vinazzer H. Effect of pentosan polysulfate on fibrinolysis: basic tests and clinical application. Semin Thromb Hemost 1991; 17:375-378.
180. Vinazzer H, Stemberger A, Haas S, Blümel G. Influence of heparin of different heparin fractions and of a low molecular weight heparin-like substance on the mechanism of fibrinolysis. Thromb Res 1982; 27:341-352.
181. Sié P, Albarede JL, Robert M, Bouloux C, Lansen J, Chigot C, Correll S, Thouvenot JP, Boneu B. Tolerance and biological activity of pentosan polysulfate after intramuscular or subcutanous administration for ten days in human volunteers. Thromb Haemost 1986; 55:86-89.
182. Williams SP, Barrowcliffe TW. The effects of post-heparin plasma lipases on anti-Xa activity. Thromb Res 1985; 37:371-377.
183. Wun TC. Lipoprotein-associated coagulation inhibitor (LACI) is a cofactor for heparin: synergistic anticoagulant action between LACI and sulfated polysaccharides. Blood 1992; 79:430-438.
184. Vinazzer H. Antithrombotic therapy in patients with known risk factors for thromboembolism. Folia Haematol Int Mag Klin Morphol Blutforsch 1989; 116:927-934.
185. Buchanan MR, Ofosu FA, Fernandez F, van Ryn J. Lack of relationship between enhancing bleeding induced by heparin and other sulfated polysaccharides and enhanced catalysis of thrombin inhibition. Semin Thromb Hemost 1986; 12:324-327.
186. Goad HE, Horne MK 3rd, Gralnick HR. Pentosan-induced thrombocytopenia: Support for an immune complex mechanism. Br J Haematol 1994; 88:803-808.
187. Campbell A, Nesheim ME, Doctor VM. Mechanism of potentiation of antithrombin III [ATIII] inhibition by sulfated xylans. Thromb Res 1987; 47:341-352.
188. Doctor VM, Lewis D, Coleman M, Kemp MT, Marbley E, Sauls V. Anti-

coagulant properties of semisynthetic polysaccharide sulfates. Thromb Res 1991; 64:413-425.
189. Hirano S, Tanaka Y, Hasegawa M, Tobetto K, Nishioka A. Effect of sulfated derivatives of chitosan on some blood coagulant factors. Carbohydr Res 1985; 137:205-215.
190. Horton D, Just EK. Preparation from chitin of (1-4)-2-amino-2-deoxy-β-D-glucopyranuronan and 2-sulfamino analog having blood-anticoagulant properties. Carbohydr Res 1973; 29:173-179.
191. Nichimura SI, Nishi N, Tokura S, Okiei W, Somorin O. Inhibition of the hydrolytic activity by chitin heparinoids. Carbohydr Res 1986; 156:286-292.
192. Muzzarelli RAA, Giacomelli G. The blood anticoagulant activity of N-carboxymethylchitosan trisulfate. Carbohydr Polym 1887; 7:87-96.
193. Drozd NN, Bashkov GV, Makarov VA, Kheilomskii AB, Gorbacheva IN. Mechanisms of anticoagulant effect if chitosan sulfuric acid ester. Vopr Med Khim 1992; 38:12-14.
194. Okiei W, Nishimura S, Somorin O, Nishi N, Tokura S. Inhibitory action of sulphated chitin derivatives on the hydrolytic activity of thrombin. In: Muzzarelli RAA, Jeuiaux Ch, Gooday GW, eds. Chitin in Nature and Technology. New York: Plenum Press, 1986:453-460.
195. Ishizu A, Kamide K. Chemical modification of cellulose. In: Hon DN-S, Shiraishi N, eds. Wood and Cellulose Chemistry. New York: Marcel Dekker, 1990:757-800.
196. Larm O, Larsson K. The preparation of a heparin analogue from alginic acid. Carbohydr Res 1979; 73:332-336.
197. Andersson L-O, Hoffman J, Holmer E, Larm O, Larsson K, Söderström G. Mechanisms of anticoagulant effects of some sulfated polysaccharides. Thromb Res 1982; 28:741-747.
198. Hoffman J, Larm O, Larsson K, Andersson LO, Holmer E, Söderström G. Studies on the blood-anticoagulant activity of sulphated polysaccharides with different uronic acid content. Carbohydr Polym 1982; 2:115-121.
199. Bode V, Franz G. Physiological activity of new heparinoids derived from plant polysaccharides. Arch Pharm 1991; 324:363-365.
200. Uryu T, Yoshida T, Ikushima N, Hatanaka K, Kaneko Y, Mimura T, Nakashima H, Yamamoto N. Sulfated polysaccharides as a potent and selective inhibitor of AIDS virus infection. Polym Sci I 1991; 2:989-996.
201. Casu B, Grazioli G, Razi N, Guerrini M, Naggi A, Torri G, Oreste P, Tursi F, Zoppetti G, Lindahl U. Heparin-like compounds prepared by chemical modification of capsular polysaccharide from E. coli K5. Carbohydr Res 1994; 263:271-284.
202. Lormeau JC, Chevallier B, Salome MLV, T Tenaille d'Estais GEM. Héparosanes-N,O-sulfatés, leur procédé de préparation et les compositions pharmaceutiques qui les contiennent EP 0 489 647 A3 1991.
203. Vann WF, Schmidt MA, Jann B, Jann K. The structure of the capsular

polysaccharide (K5 antigen) of urinary-tract-infective Escherichia coli 010:K5:H4. A polymer similar to desulfo-heparin. Eur J Biochem 1981; 116:359-364.
204. Kusche M, Hannesson HH, Lindahl U. Biosynthesis of heparin. Use of Escherichia coli K5 capsular polysaccharide as a model substrate in enzymic polymer-modification reactions. Biochem J 1991; 275:151-157.
205. Campbell P, Hannesson HH, Sandback D, Rodèn L, Lindahl U, Li JP. Biosynthesis of heparin/heparan sulfate. Purification of the D-glucuronyl C-5 epimerase from bovine liver. J Biol Chem 1994; 269:26953-26958.
206. Jann K, Jann B. Capsules of Escherichia coli, expression and biological significance. Can J Microbiol 1992; 38:705-710.
207. Stevens MP, Hanfling P, Jann B, Jann K, Roberts IS. Regulation of Escherichia coli K5 capsular polysaccharide expression: Evidence for involvement of RfaH in the expression of group II capsules. FEMS Microbiol Lett 1994; 124:93-98.
208. Rosenow C, Esumeh F, Roberts IS, Jann K. Characterization and localization of the KpsE protein of Escherichia coli K5, wich is involved in polysaccharide export. J Bacteriol 1995; 177:1137-1143.
209. Rosenow C, Roberts IS, Jann K. Isolation from recombinant Escherichia coli and characterization of CMP-Kdo synthetase, involved in the expression of the capsular K5 polysaccharide (K-CKS). FEMS Microbiol Lett 1995; 125:159-164.
210. Harada T. The story of research into curdlan and the bacteria producing it. Trends in Glycoscience and Glycotechnology 1992; 4:309-317.
211. Yalpani M. Studies in Organic Chemistry: Polysaccharides - Synthesis, Modification and Structure/Property Relations. 1st ed. Amsterdam, Oxford, New York, Tokyo: Elsevier, 1988:211-216.
212. Alban S, Kraus J, Franz G. Synthesis of laminarin sulfates with anticoagulant activity. Drug Res 1992; 42:1005-1008.
213. Alban S, Franz G. Anticoagulant activity of β-1,3-glucan sulfates in dependence on their molecular weight. Pure & Appl Chem 1994; 66:2403-2406.
214. Uchiayma H, Metori A, Ogamo A, Nagasawa K. Contribution of chemical 6-O-sulfation of the aminodeoxyhexose residues in whale heparin with high affinity for antithrombin III to its anticoagulant properties. J Biochem (Tokyo) 1990; 107:377-380.
215. Alban S, Jeske W, Hoppensteadt D, Fareed J, Franz G. Synthesis and testing of a defined curdlan sulfate with anticoagulant, antiaggregation and antithrombotic activity. Pharm Pharmacol Lett 1995; 5:70-73.
216. Alban S, Franz G. Gas-liquid chromatography-mass spectrometry analysis of anticoagulant active curdlan sulfates. Semin Thromb Hemost 1994; 20: 152-158.
217. Harris JP, Henry RJ, Blakeney AB, Stone BA. An important procedure of the methylation analysis of oligosaccharides and polysaccharides. Carbohydr

Res 1984; 127:59-73.
218. Alban S. Synthese und physiologische Testung neuartiger Heparinoide. Ph.D. disseration, University of Regensburg, Regensburg, Germany, 1993.
219. Demleitner S, Kraus J, Franz G. Synthesis and antitumour activity of derivatives of curdlan and lichenan branched at C-6. Carbohydr Res 1992; 226:239-246.
220. Van Ryn-McKenna J, Ofosu FA, Hirsh J, Buchanan MR. Antithrombotic and bleeding effects of GAGs with different DS. Br J Haematol 1989; 71: 265-269.
221. Alban S, Franz G. Mode of action of anticoagulantly active β-1,3-glucan sulfates. Thromb Haemost 1995; 73:1317.
222. Bara L, Mardiguian J, Samama M. *In vitro* effect on Heptest of low molecular weight heparin fractions and preparations with various anti-IIa and anti-Xa activities. Thromb Res 1990; 57:585-592.
223. Fernandez F, N'guyen P, van Ryn J, Ofosu FA, Hirsh J, Buchanan MR. Hemorrhagic doses of heparin and other glycosaminoglycans induce a platelet defect. Thromb Res 1986; 43:49-495.
224. Godal HC. Heparin-induced thrombocytopenia. In: Lane DA, Lindahl U, eds. Heparin, Chemical and Biological Properties, Clinical Applications. London: Edward Arnold, 1989:533-548.
225. Messmore HL, Griffin J, Fareed J, Coyne E, Seghatchian J. In-vitro studies of the interaction of heparin, low molecular weight heparin and heparinoids with platelets. Ann NY Acad Sci 1989; 556:217-231.
226. Vermylen JG. Effect of heparin and low molecular weight heparin on platelets. Semin Thromb Hemost 1993; 19(suppl 1):20-21.
227. Fareed J, Walenga JM, Kumar A, Rock A. A modified stasis thrombosis model to study the antithrombotic actions of heparin and its fractions. Semin Thromb Hemost 1985; 11:155-175.
228. Wessler S, Reimer SM, Sheps MC. Biologic assay of a thrombosis inducing activity in human serum. J Appl Physiol 1959; 14:943-946.
229. Raake W, Elling H. A new model for testing antithrombotic agents. Thromb Res 1992; 53:73-77.
230. Alban S, Walter J, Welzel D, Franz G, Fareed J. Anticoagulant, antithrombotic actions of a semisynthetic β-1,3-glucan sulfate. Thromb Res 1995; 78:201-210.

7
Heparin Oligosaccharides: New Analogues Development and Applications

Robert J. Linhardt
University of Iowa
Iowa City, Iowa

Toshihiko Toida
Chiba University
Chiba, Japan

I. HEPARIN'S HISTORY

The history of heparin (1-4) began in Baltimore with Jay McLean, a second year medical student at Johns Hopkins University (5,6). McLean undertook a research project in 1916 with the physiologist William Howell with the aim of discovering substances within the blood that caused it to clot. To his surprise, McLean isolated fractions from mammalian tissues that instead of clotting blood prevented its coagulation (7). Howell quickly recognized the importance of the discovery of his student suggesting its therapeutic use to treat coagulation disorders. This discovery was not without its controversies. First, McLean misnamed the extract "heparophosphatide" (1). Heparin, as it was later renamed by Howell, was ultimately established to be a sulfated polysaccharide leaving doubts in the minds of some whether McLean had ever prepared heparin.

An understanding of heparin's structure developed slowly. It didn't take Howell long to realize that heparin had no phosphate and was a carbohydrate, correctly identifying one of its sugars as a uronic acid in 1928 (8). In 1935-6, Sune Bergstrom a student working with Eric Jorpes in Sweden, correctly identified glucosamine as the second sugar component in heparin (9). Jorpes and later Arthur Charles established that heparin contained a high level of covalently linked sulfate making it one of the strongest acids in nature (10). It was not until 1950 that Jorpes was able to show that the glucosamine in heparin was primarily *N*-sulfated (5,6). Heparin's uronic acid was initially identified as D-glucuronic acid by Melville L. Wolfrom in 1946 (5,6). Tony Cinfonelli and Al Dorfman reported in 1962 that they had found L-iduronic acid in heparin (5,6). Finally, Arthur Perlin correctly identified L-iduronic acid as the major uronic acid residue in heparin using a newly developed method called NMR spectroscopy (11).

By the 1920s several groups were manufacturing heparin. At Hynson, Wescott & Dunning Co. in Baltimore, another of Howell's students, Dr. Dunning, had designed the first commercial process for manufacturing heparin from dog liver, a particularly rich source (2). In Toronto, Charles Best, began a research program aimed at the commercial production of beef lung heparin (3). Eric Jorpes, then a graduate student, transferred this technology from Toronto to Stockholm (4). By 1935, Jorpes in Stockholm along with Arthur Charles and David Scott of Connaught Laboratories at the University of Toronto had prepared sufficient heparin for human trials. In 1935, Clarence Crafoord successfully began using heparin postoperatively in surgery patients (4), a medical practice that continues to this day.

II. HEPARIN: STRUCTURE AND SIGNIFICANCE

Heparin is a polydisperse, highly sulfated, linear polysaccharide comprised of repeating 1→4 linked uronic acid and glucosamine residues (12,13). Although heparin has been used clinically as an anticoagulant for the past 60 years, its precise structure remains unknown (5). The failure to completely understand heparin's structure is not the result of a lack of effort but rather is due to its extremely complex structure. Heparin is biosynthesized as a proteoglycan (PG, M_r ~x10^6) consisting of a central core protein from which

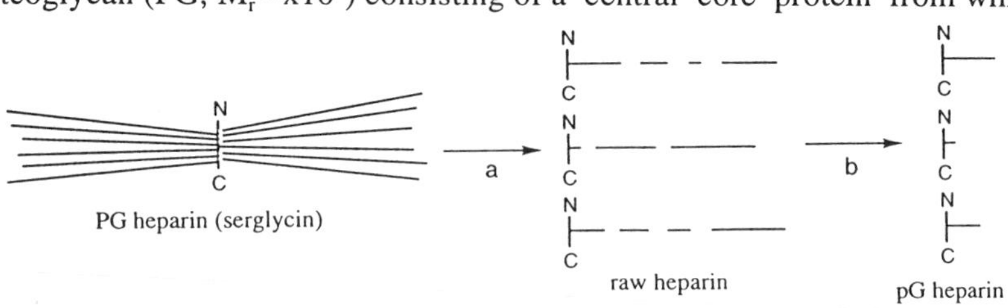

Figure 1 Structure of heparin proteoglycan. The heparin proteoglycan (PG-heparin or serglycin) is released from mast cells on degranulation as raw heparin formed through the action of β-endoglucuronidase and proteinases (a). Raw heparin can be separated (b) into peptidoglycan (pG) heparin and glycosaminoglycan (GAG) heparin. Pharmaceutical heparin is prepared by bleaching raw heparin to remove peptide.

approximately eleven long linear polysaccharide chains extend (Figure 1) (14). The drug heparin is recovered from porcine intestinal mucosa or bovine lung as a glycosaminoglycan (GAG) or simply a linear polysaccharide chain without any associated protein (12,13). This processing includes treatment with alkali, proteases, bleaching agents (15,16). While it was originally believed that this harsh commercial processing of tissues from which the drug heparin was prepared resulted in the characteristic complexity of GAG heparin (15), proteases and β-glucuronidases present in the tissue itself are actually responsible for most of the differences between PG heparin and GAG heparin (17).

The structural complexity of heparin can be considered at several levels. At the PG level, different numbers of the polysaccharide chains (possibly having different saccharide sequences) can be attached to the various serine residues present in the protein core. Once freed from core protein, through the action of tissue proteases, peptidoglycan heparin is formed (*i.e.*, a small peptide to which a single long polysaccharide chain (M_r 100,000) is attached) (14). This peptidoglycan is short-lived as it is immediately processed by a β-endoglucuronidase to a number of smaller polysaccharide chains (only one, corresponding to the original site of attachment to the core protein, would contain peptide material) called GAG heparin. At the GAG level, some of heparin's structural complexity results from its polydispersity. GAG heparin contains polysaccharide chains having from 15 to 100 saccharide residues. Even the heparin chain corresponding to the most prevalent degree of polymerization (dp) represents < 5 mol% of a typical GAG heparin preparation (18). GAG heparin has yet another level of structural complexity, associated with its primary structure or sequence (19). Heparin has as many as ten different saccharide residues (13) that comprise its sequence. While not all sequences are biosynthetically favored, a large number of sequences are certainly possible (20).

The structural features of PG heparin have been primarily established by studying its biosynthesis (20-22). The structural features of GAG heparin have relied on biosynthetic, chemical, enzymatic, and spectroscopic techniques. Recent efforts in our laboratory (23,24) as well as others (25) have sought to use techniques originally developed to sequence nucleic acids and proteins to establish the sequence of heparin.

Heparan sulfate is structurally similar to heparin (Figure 2). Both of these molecules contain polysaccharide side chains comprised of alternating 1→4 linked sulfated and or acetylated glucosamine and sulfated or nonsulfated iduronic or glucuronic acid residues (13). The core protein of PG heparin and PG heparan sulfate are different. Heparan sulfate PG is primarily found in the extracellular matrix and in cell membranes (25,26) while heparin is only intracellular (27). Although structurally similar, GAG heparin and GAG heparan sulfates have different ratios of *N*-acetylation to *O*-sulfation (28) and can often be distinguished by their different sensitivity towards heparin lyases (29).

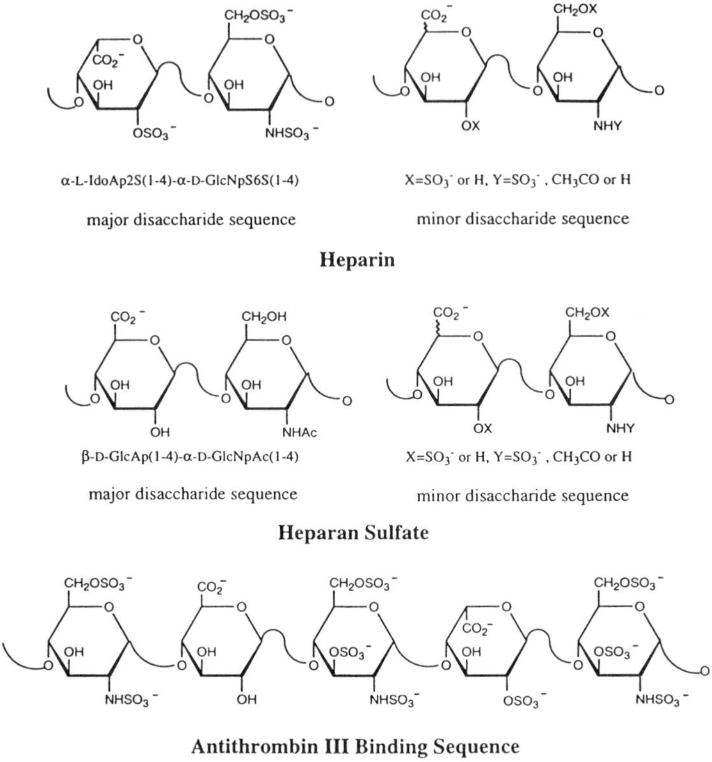

Figure 2 Structure of heparin and heparan sulfate glycosaminoglycans and the ATIII binding site. The major and minor disaccharide sequences of heparin and heparan sulfate and the pentasaccharide sequence that binds antithrombin III are shown.

Heparin's primary therapeutic application is as an anticoagulant (5). It may be more appropriate to consider heparin a polyelectrolytic drug having a multiplicity of biological activities (30,31). Heparin is the strongest acid present in the body and thus is present under physiologic conditions as a highly charged polyanion (31). Virtually any cationic protein (pI>7) is capable of binding to heparin under physiological conditions and protein activity is often altered by heparin. Some heparin-binding proteins strongly interact with heparin because of the presence of basic amino acid residues clustered on the protein's surface. Arginine and lysine residues, in the interacting protein, form strong ion-pairs with heparin's sulfate groups (32,33). Hydrophobic amino acid residues are capable of interacting with the *N*-acetyl groups present on certain glucosamine residues (34). Over the past two decades, a growing number of biological activities (30,35) have been shown to be regulated by heparin, including the interaction with proteins important in angiogenesis (35,36) as well as those that regulate the inflammatory response (37). The discovery of these new activities have brought about an urgency to understand heparin's structure and to elucidate its structure-activity relationship (SAR).

A new class of drugs called low molecular weight heparins have proven to be effective antithrombotic agents (38,39) for use in a wide variety of disease states ranging from deep vein thrombosis (40,41) to non-hemmorrhagic stroke (42). These agents are prepared chemically or enzymatically from heparin and offer certain therapeutic advantages. Finally, small, structurally defined heparin oligosaccharides and semi-synthetic or synthetic heparin analogs are under extensive investigation in an effort to exploit the newer biological activities associated with heparin.

III. PROTEOGLYCAN HEPARIN

A. Biosynthesis

The biosynthesis of heparin has been well studied in a mastocytoma cell culture system (20,21,22). The core protein, which contains a high number of serine-glycine repeats, is first synthesized in the rough endoplasmic recticulum (43). To this core a linkage region consisting of three neutral sugars is attached to serine through the action of a xylosyl transferase (43). Onto this neutral sugar linkage region a repeating copolymer of 1→4 linked glucuronic acid and *N*-acetylglucosamine is assembled through the stepwise addition of UDP-sugars (44). The linear polysaccharide chain is extended by approximately 300 sugar units before its synthesis terminates (21). During chain elongation the polysaccharide is the partially de-*N*-acetylated and sequentially *N*- and *O*-sulfated through the action of a multienzyme complex (45). A unique ATIII binding site is introduced through the action of a 3-*O*-sulfotransferase (20-22). The structural variability in the heparin polymer is primarily the result of the incomplete nature of these modifications (20-22,45). The sequence heterogeneity in the polysaccharides chains results in a family of PG heparin macromolecules.

B. Biological Functions

PG heparin is primarily found in the granules of mast cells (27). When mast cells degranulate heparin is released but as GAG heparin, the result of processing by proteases and endo-β-glucuronidases (17,27). Although the GAG heparin released on mast cell degranulation demonstrates anticoagulant activity, the role of this activity is unclear. There is no evidence that endogenous mast cell heparin plays a role in maintaining blood flow through the vasculature even though this is the primary application for exogenously administered GAG heparin (30,46). The true biological function of heparin still remains hotly contested.

Syndecans are a family of heparan sulfate proteoglycans, localized to the cell surface of most tissues that have a wide variety of biological roles (25,26). Structurally related to heparin, heparan sulfate PG, has been isolated from the endothelial surface responsible for the natural blood compatibility of vessels (46,47). The biological activities of heparan sulfate PGs primarily result from the specific interaction of proteins with their GAG chains. Among Syndecan's functions are: 1. organization of extracellular matrix, through binding to collagens, fibronectin, thrombospondin, tenascin, etc.; 2. organization of the epithelia; 3. affording non-thrombogenic vascular endothelial surfaces, through binding to antithrombin III, protein C, protease nexins; 4. a co-receptor for fibroblast growth factor (FGF) (26,48,49); and 5. regulation of development in early embryogenesis and in cancer. Syndecans are expressed in tissue-specific patterns, their expression is highly regulated and this regulation may be affected by their cell surface residence time. Syndecans are shed from the cell surface as the result of their cleavage at a protease-susceptible site near the plasma membrane. The extracellular matrix (ECM) at the basal cell surface might slow Syndecan release.

Although the biological roles of endogenous PG heparin and PG heparan sulfate are not completely understood this has not precluded the use of GAGs derived from these natural products as well as GAG fractions, oligosaccharides and synthetic analogs, for a variety of medical applications.

IV. GLYCOSAMINOGLYCAN HEPARIN

A. Preparation

GAG heparin is prepared from animal tissues that are rich in mast cells such as porcine intestinal mucosal and bovine lung (50). Other species of mammals as well as birds, fish and even invertebrates such as lobster and clams, which do not have a blood coagulation system, also contain heparin (51). The method of commercial processing of heparin varies between manufacturers and is generally regarded as trade secrets (15). The basic approach involves the collection of the appropriate tissue followed by proteolytic treatment or salt extraction followed by alcohol precipitation of the sodium salt or the cetylpyridinium salt. Using this approach, our laboratory prepared, purified, and characterized a human heparin from the mast cells contained in a large vascular tumor (16).

Treatment with base to remove residual protein by β-eliminative cleavage of xylose *O*-serine glycoside and/or bleaching with oxidizing agents is commonly used to prepare the pure white drug form of heparin. The major criteria for purity is a high specific activity expressed as USP units per milligram (15). The extensive processing of commercial heparin preparations leads to little if any modification of the primary structure of GAG heparin (15,16). Studies on conditions required for desulfation of heparin have confirmed heparin's stability under conditions used in its processing (52).

Oligosaccharide mapping, involves the partial enzymatic depolymerization of heparin followed by the analysis of the oligosaccharide products by either strong anion exchange high pressure liquid chromatography (23,53-55) or gradient polyacrylamide gel (23,56,57) or capillary elctrophoresis (58,59), and demonstrates remarkable similarities between heparins prepared from a single type of tissue by different manufacturers. Heparins prepared under mild conditions (*i.e.* without bleaching) are nearly indistinguishable from highly processed samples.

B. Properties

GAG heparin is a polydisperse preparation with a M_r ranging from 5,000-40,000 having an M_r (avg) of 13,000 (60,61). Its general physical properties (62) including pKa (63) and molecular weight (64) have been studied using a variety of analytical methods. Heparin and heparan sulfate are composed of alternating D-glucosamine residues linked 1→4 to either L-iduronic acid or D-glucuronic acid (12). The principle repeating unit in heparin is →4)-α-D-glucosamine-2,6-disulfate (1→4)-α-L-iduronic acid-2-sulfate (1→, corresponding to 75-90% of its sequence (16). Disaccharide analysis is performed by complete enzymatic or chemical depolymerization of heparin followed by disaccharide analysis using HPLC or CE (16,58,59,65). Typical disaccharide analyses are shown in Table 1. Heparan sulfate typically consists

Table 1 Structure and Frequency of Disaccharide Sequences Found in Heparin and Heparan Sulfate

Disaccharide	Sequence[a]	Heparin mole%[b]	Heparan sulfate mole%
a	→4)-α-L-IdoAp2S(1→4)-α-D-GlcNp2S,6S(1→	50.3	0.5
b	→4)-α-L-IdoAp2S(1→4)-α-D-GlcNp2S(1→	3.9	1.0
c	→4)-α-L-IdoAp2S(1→4)-α-D-GlcNpAc,6S(1→	2.1	n.d.[c]
d	→4)-α-L-IdoAp(1→4)-α-D-GlcNp2S,6S(1→	10.3	1.4
e	→4)-β-D-GlcAp2S(1→4)-α-D-GlcNp2S,6S(1→	n.d.	n.d.
f	→4)-β-D-GlcAp2S(1→4)-α-D-GlcNp2S(1→	n.d.	n.d.
g	→4)-β-D-GlcAp(1→4)-α-D-GlcNp2S(1→	4.1	4.4
h	→4)-β-D-GlcAp(1→4)-α-D-GlcNp2S(1→	2.9	14.1
i	→4)-β-D-GlcAp(1→4)-α-D-GlcNpAc(1→	11.1	75.7

[a]S, sulfate. Disaccharide sequences were determined using heparin lyase I and III.
[b]mole% found in a typical commercial porcine mucosal heparin and heparan sulfate, and calculated from the results based on CE and ^{1}H-NMR analyses (65).

of 50-75% →4)-α-D-N-acetylglucosamine(1→4)-β-D-glucuronic acid together with lesser amounts of →4)-α-D-*N*-acetyl glucosamine-6-sulfate (1→4)-β-D-glucuronic acid (1→ and →4)-α-D-glucosamine-2-sulfate (1→4)-β-D-glucuronic acid (1→. All the disaccharide units found within heparin also occur in heparan sulfate and vice versa. Larger oligosaccharide sequences correspond to protein binding sites associated with many of the biological activities of heparin and heparan sulfate. For example, both heparin and heparan sulfate can contain a specific pentasaccharide sequence having a central 2,3,6 trisulfated D-glucosamine residue that facilitates a specific high-affinity interaction with antithrombin III (ATIII) (Figure 2). The disaccharide sequences in Table 1 do not account for all the mass of heparin and heparan sulfate, suggesting other minor disaccharide sequences may be present. But these await independent confirmation.

Disaccharide sequences are the building blocks of heparin and heparan sulfate and represent an alphabet through which information can be stored in the heparin polymer. We will see how this is the case for the ATIII binding site, which is defined by a particularly rare saccharide sequence (Figure 2), later in this chapter. Until recently, there was little information on the sequence of the heparin polymer or indeed whether or not there was even a sequence. Early researchers viewed heparin as a polymer having a random sequence where the only restrictions were the type of saccharide units present, the 1→4 linkage configuration, and the required alternating substituted uronic acid and glucosamine residues. Both biosynthetic studies and the frequent occurence of what should have been rare sequences, associated with the ATIII binding site, have changed this perception.

One approach to understanding order (sequence or information storage) in the heparin polymer has used computer or mathematical simulation (18,19,66). For example, the major repeating unit in the heparin polymer, a trisulfated disaccharide (Figure 2, Table 1), could be: 1. clustered in one region of the heparin polymer; 2. distributed uniformly through the polymer; 3. distributed randomly throughout heparin; or 4. occur in unique well defined positions within the polymer. The results of computer and mathematical simulations demonstrated that while the major disaccharide repeating sequence might be randomly distributed through the heparin polymer (18), the sugar units found in the ATIII binding site of heparin could not be randomly arranged (19,53).

Classical sequencing methods have also been used to study the primary structure of heparin. GAG heparin can be partially depolymerized (either enzymatically or chemically) to prepare transient oligosaccharides (*i.e.*, oligosaccharides not found in the final products because they still contain cleavable sites). These large oligosaccharides are then be purified to homogeneity (67). If a transient oligosaccharide is then exhaustively depolymerized to smaller, previously characterized final oligosaccharides their composition can be determined (68,69). By introducing a tag such as a radiolabel at one end of the transient oligosaccharide a sequence can be determined (69,70).

Sophisticated mass spectrometry and high field (500-600 MHz) NMR spectroscopy has made it possible to determine the structure of very large oligosaccharides. Fast atom bombardment MS analysis has been successfully applied to determine structures as large as decasaccharides (71). This method, however, both fails to position sulfate groups at a specific site on a saccharide residue and fails to distinguish glucuronic from iduronic acid residues. MS/MS analysis has recently been successfully applied to small oligosaccharides to localize sulfate groups (72). Using both ^{1}H and ^{13}C NMR spectroscopy it is possible to definitively assign structures as large as tetradecasaccharides (67). Such studies are primarily limited by the lack of pure, homogeneous oligosaccharides larger than a tetradecasaccharide. Particularly interesting are oligosaccharides, isolated from peptidoglycan heparin, that contain a xylose residue at their reducing end. Sugahara and coworkers (73) have isolated such oligosaccharides (Figure 3) from both heparin and heparan sulfate using enzymatic means and have characterized their structures by NMR spectroscopy. These oligosaccharides extend slightly beyond the linkage region and thus provide an interesting insight into the sequence variability of these chains immediately adjacent to their core attachment sites.

Oligosaccharide mapping (23), originally introduced as a technique to compare heparins prepared from different tissues and species has also been used to examine heparin (16,23) and heparan sulfate for the presence of structural motifs (25,65,74). Gallagher and coworkers have extensively used this approach to study heparan sulfate (25,74) and have shown it to be a well organized molecule with structural domains that are presumably important in the binding and presentation of growth factors (25,26).

New methods of sequencing heparin are now being proposed which involve sequencing an intact GAG chain. This represents the polysaccharide equivalent of Maxim-Gilbert sequencing of nucleic acids (75). These methods depend on the introduction of a reading frame from which the sequence can be determined. One such reading frame is the reducing terminus of the polysaccharide chain where it attaches to the core protein. Our laboratory labeled peptidoglycan heparin with an amine-reactive hydrophobic fluorescent tag (24). The labeled heparin molecules were enriched using hydrophobic chromatography after which the tag was removed by alkali β-elimination exposing the xylose residue in the heparin chain for labeling with a hydrophilic fluorescent tag. Treatment with specific heparin lyases and analysis by PAGE afforded the predominant sequence of the first 13 saccharide residues of heparin (24).

C. Medical Applications of Heparin

Heparin is the most commonly used clinical anticoagulant. Over 33 metric tons of heparin is manufactured worldwide each year representing over 500 million doses (5). Heparin is administered *intravenously* during most extracorporeal procedures (where blood is removed from the body and passed through a device) such as kidney dialysis and membrane oxygenation used in heart bypass procedures (76). Systemic heparinization is also widely used in treating deep vein thrombosis (77) and a variety of other surgical procedures

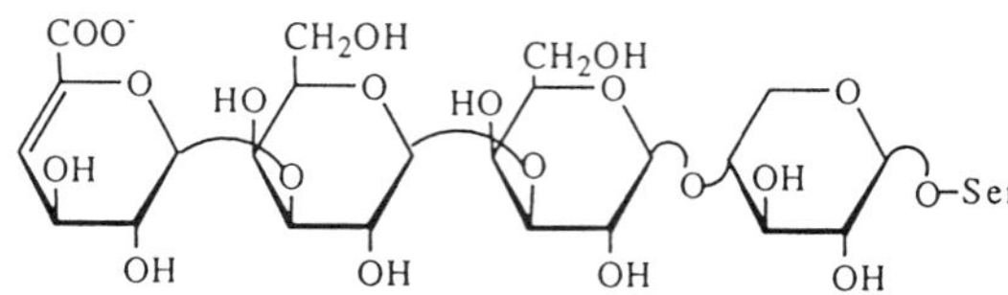

ΔUAp(1-3)-β-D-Galp(1-3)-β-D-Galp(1-4)-β-D-Xylp(1-O-Ser)

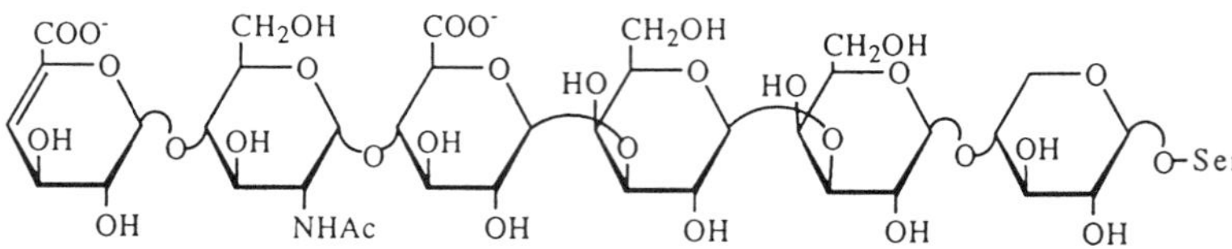

ΔUAp(1-4)-α-D-GlcNpAc(1-4)-α-D-GlcAp(1-3)-β-D-Galp(1-3)-β-D-Galp(1-4)-β-D-Xylp(1-O-Ser)

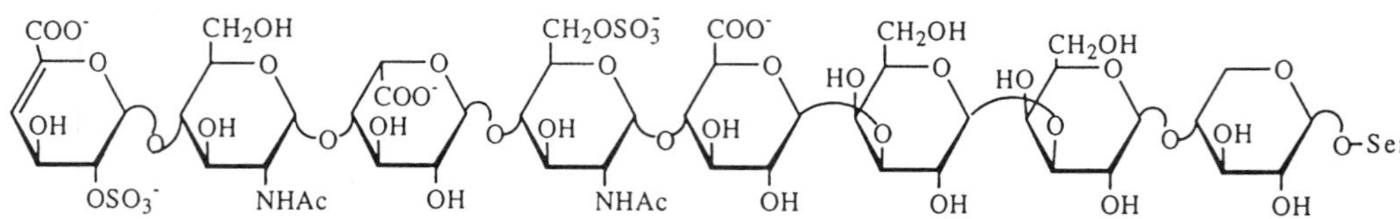

ΔUAp2S(1-4)-α-D-GlcNpAc(1-4)-α-L-IdoAp(1-4)-α-D-GlcNpAc6S(1-4)-α-D-GlcAp(1-3)-β-D-Galp(1-3)-β-D-Galp(1-4)-β-D-Xylp(1-O-Ser)

Figure 3 Structure of heparin and heparan sulfate linkage regions (24,73). These linkage region oligosaccharides were obtained by exhaustively treating heparin and heparan sulfate with heparin lyases. The unsaturated sugar (ΔUAp) at the nonreducing end is an artifact of this enzymatic treatment. These are only representative linkage region structures as more structures are currently being determined.

(77). Heparin is also given by the *subcutaneous* route and, despite its reduced bioavailability, a low level of heparinization can often be maintained this way for prolonged periods of time (77). While heparin is not orally active, the preparation of orally active formulations, derivatives, and analogues represents an active research area (see below). Heparin's major sideeffect, hemorrhagic complications (77), is closely linked to its anticoagulant activity. After administering heparin, it is often necessary to neutralize its anticoagulant activity and this is usually done by administering a cationic polypeptide-based drug called protamine (77). Protamine is not without its own set of sideeffects and the resulting protamine heparin complex has been associated with an immune related loss of platelets (77-79).

V. HEPARIN AND ITS ANTITHROMBOTIC ACTIVITY

A. Blood Coagulation

Hemostasis, the spontaneous cessation of bleeding from ruptured blood vessels, is a broad physiological process of which the blood coagulation system is just one part. *In vivo* hemostasis involves plasma coagulation factors, platelets, monocytes, and endothelial cells which line the blood vessels.

The coagulation cascade is divided into the extrinsic and intrinsic pathway (77). The extrinsic pathway is activated by tissue damage releasing thromboplastin, calcium and phospholipid from the cell membrane. These factors activate factor VII, transforming factor X into its proteolytically active form, factor Xa. The intrinsic pathway is activated by blood contacting non-endothelial surfaces followed by the activation of circulating coagulation factors leading to the formation of factor Xa. The remaining steps, referred to as the final common pathway (80), results from factor Xa catalyzing the conversion of prothrombin to thrombin which cleaves fibrinogen to form a fibrin monomer producing a clot.

B. Heparin's Effect on Serine Proteases

The blood coagulation enzymes are serine proteases (81) with trypsin-like specificity for arginyl linkages. Unlike trypsin they have a higher degree of specificity for the linkages they cleave (82,83) and often require cofactors to accelerate their rate of reaction (84). The entire coagulation process is under the control of a group of glycoprotein serine protease inhibitors (Serpins) the most important of which is antithrombin III (85).

Antithrombin III (ATIII) is a single chain, anionic (pI 4.9-5.3) (86) glycoprotein of molecular weight 58,000 (87). All the coagulation proteases (except for VIIa) are inhibited by ATIII which forms an equimolar covalent complex with these enzymes. ATIII has a high affinity for these enzymes (low Km) but a slow turnover rate (low Kcat). The reaction affords an intact protease covalently linked to the amino terminal portion of ATIII. The ATIII that is released is proteolytically modified at a single site near the carboxyl terminus (88). Heparin binds to thrombin and ATIII in a ternary complex accelerating the rate of thrombin inhibition by ATIII by 2000-fold (89).

Heparin cofactor II (HCII) is structurally similar to ATIII with a molecular weight of 65,000 (90) and a pI of 4.9-5.3 (86) having a similar carboxyl terminal sequence but a distinctly different amino terminal sequence (91). The physiological role of HCII might be as a reserve of thrombin inhibitor when the plasma concentration of ATIII becomes abnormally low (92). Unlike ATIII, HCII can inhibit thrombin but no other coagulation proteases (89,93). In addition to this unusual specificity, HCII also can be potentiated by GAGs, other than heparin, including dermatan sulfate (94) and heparan sulfate both of which are found lining the luminal surface of the endothelium (95). The mechanism by which HCII inhibits thrombin is similar to that proposed for ATIII (89).

C. Other Antithrombotic Actions of Heparin

Platelets entrapped in the fibrin clot are an essential component of the hemostatic plug sealing a wounded vessel. Heparin is known to activate platelets, an important component of thrombosis (96), resulting in their aggregation and loss. Thrombocytopenea is a severe side-effect associated with heparinization (77). Heparin analogs are being examined to reduce this undesirable effect on platelets. Heparin also exerts its antithrombotic effects at the vessel wall (97). Heparin releases tissue factor pathway inhibitor (TFPI), a potent anticoagulant, from the endothelium (97). Once a clot is formed, heparin can act to accelerate fibrinolysis causing the dissolution of the clot (98). Fibrinolytic drugs, including streptokinase, urokinase, and tissue plasminogen activator, particularly promising for the treatment of coronary clots formed in heart attacks (99-100), have been tested in conjunction with heparin to prevent reocclusion (101).

VI. HEPARIN'S OTHER BIOLOGICAL ACTIVITIES

A. Antiatherosclerotic Activity

1. Effect on lipoprotein lipase

When heparin is administered intravenously, it causes the release of lipoprotein lipase (LPL) from the endothelium (102-105). This may result in increased triglyceride lipolysis occurring in the bloodstream, thereby lowering the concentration of cholesterol-rich remnant particles in contact with the arterial wall (102-105). The effect of heparin on the release and activation of LPL has been studied in a number of different animal models including humans (102,103). Two major problems stand in the way of the application of heparin as an antiatherosclerotic agent. These are its primary activity as an anticoagulant (31) and heparin's lack of oral bioavailability (106). Heparin analogs might be used to circumvent these problems.

2. Effect on Smooth Muscle Proliferation

The proliferation of smooth muscle cells following damage to endothelium is important part of atherogenesis (107-109). Both anticoagulant and non-anticoagulant heparin have demonstrated the ability to inhibit the proliferation of smooth muscle cells (107,108,110). This activity results from heparins interaction with growth factors including fibroblast growth factor (FGF) and endothelial cell growth factor (ECGF) (109). Heparan sulfate (111) and dermatan sulfate (94), present on the luminal surface of the endothelium, probably play a physiological role in the regulation of smooth muscle proliferation during atherogenesis (102). Heparin's anticoagulant and antiproliferative activities are separable (108). While preliminary animal studies demonstrated that periodate oxidized heparin (Astenose®) with little or no anticoagulant activity was useful in treating restenosis injury (110), the manufacturer of this agent decided against taking this drug into clinical studies possibly because of its low potency and complex chemical structure (110). More extensive structure-activity studies will certainly be required to exploit heparin or a heparin analog as an antiproliferative agent. The effect of heparin and heparin analogs on smooth muscle proliferation remains an active area of research interest.

B. Ability to Inhibit Complement Activation

Heparin's principle location is the granules of tissue mast cells. Because heparin's primary location is so closely linked to the immune response, its ability to regulate complement has become an active area of interest. The major activities of the complement system are opsonization/phagocytosis and cell lysis together with roles in vascular permeability and neutrophil chemotaxis. The major function of the complement system is to recognize non-self material. It performs this function through both antibody-dependent and antibody-independent mechanisms. Carbohydrate containing substances, such as heparin, are most often involved in these antibody-independent mechanisms (112,113).

Recognition of a target results in assembly of C3 convertases that cleave and activate C3 into C3b that binds covalently through exposed thioester groups in its structure principally to the hydroxyl groups of carbohydrates on the complement activating surface. Once large amounts of C3b are immobilized on the target surface, phagocytosis and other biological responses occur through the interaction of C3b with receptor. When the surface is a cell then the membrane attack complex may assemble and cause lysis.

Complement can be activated through the classical and alternative pathways, which converge on the activation of C3. The classical pathway is initiated by the interaction of two IgG or one IgM with C1q, causing a conformational change in C1q that activates C1r and then C1s giving rise to serine proteases that in turn activate C4 and C2 that form a C4bC2a complex.

In this complex, called the classical pathway, C3 convertase converts C3 to C3b. The C3b and its homologue C4b localize at the surface where they are formed and either react with that surface or are rapidly hydrolyzed or inactivated. Surface bound C3b binds to form C4b2a3b and this complex then interacts with C5 to begin the assembly of a membrane attack complex leading to lysis of the target cell. Heparin has two important roles in regulating the classical pathway. Heparin inhibits the activity of $\overline{\text{C1s}}$ and also binds to and augments the activity of C1INH, an inhibitor of $\overline{\text{C1s}}$. Antibody-independent activation of the alternative pathway mainly involves the neutral carbohydrates of glycoproteins forming an activating surface. Factor D, a serine protease, cleaves B when it is bound to C3b to form C3bBb complex that can then serve to convert C3 to C3b. The original C3b in this complex may be formed through the thermal unfolding of C3 exposing its reactive thioester group. If C3b is deposited on a non-activating surface it complexes with factor H and is rapidly inactivated by factor I. Heparin inhibits the interaction of factor B with C3b. Heparin's structure-activity relationship on the complement system, however, is still poorly understood. Although studies indicate the importance of *O*-sulfation (114,115) and a minimum molecular size (60), a specific binding site for a complement factor such as C3b (similar to the ATIII binding site) has not yet been implicated in heparin's ability to inhibit complement activity (68). Heparin analogs might prove to be useful new agents to inhibit complement activation in the treatment of diseases such as hereditary angioedema, systemic lupus and rheumatoid arthritis (116). Heparin also inhibits human leukocyte elastase which has a well established role in rheumatoid arthritis as well as in emphysema (117-119).

C. Antiinflammatory Activity

In addition to its effect on complement, heparin also may act as an antiinflammatory agent through its interaction with selectins and chemokines (120).

Selectins are a family of transmembrane glycoproteins found on endothelium, platelets and leukocytes (121). L-selectin, found on the surface of lymphocytes as well as neutrophils and monocytes, is constituitively expressed. When cells are activated, L-selectin activity on the cell surface may increase. In doing so, selectins appear to play a role in lymphocyte migration and the homing of lymphocytes to the lymph nodes. In acute inflammation neutrophils move through the vascular wall. Initially neutrophils anchor to the endothelium through a carbohydrate-protein interaction (121). Although the putative ligand on the endothelium responsible for leukocyte interaction with selectins is sialyl Lewis X, recently heparan sulfate has also been shown to play a role in this interaction (122). Our laboratory has demonstrated a heparin tetrasaccharide can interact tightly with both L- and P-selectin (37). This tetrasaccharide can competitively displace these selectins from immobilized sialyl Lewis X at two orders of magnitude lower concentration than can soluble sialyl Lewis X, the putative natural ligand (37). This result suggests that such a tetrasaccharide sequence in heparan sulfate may play an important role *in vivo* in the natural function of the selectins.

Chemokines are soluble GAG-binding proteins that are involved in the recruitment and activation of leukocytes (123). There are at least 15 known chemokines and *in vitro* they seem to exhibit overlapping activities. Some selectivity is observed *in vivo* that may correlate to differential localization of chemokines in tissues. For example IL-8 binds preferentially to the endothelium of post capillary venules while MIP-1β is found at the HEV of lymph nodes. The GAG chains of cell surface proteoglycans have been proposed to play a role in this differential localization. The heparan sulfate family of GAGs are most responsible for chemokine binding. The interactions of chemokines PF4, IL8, Groα and NAP-2 for heparin give K_{diss} values ranging from 10^{-6} to 10^{-8}M (123). Selectivity in heparin and heparan sulfate binding was also demonstrated by carrying out affinity electrophoresis using heparan sulfate (123). This selectivity might be exploited by developing heparin analogs that interact specifically with certain chemokines. Such agents might be useful in treating acute inflammation.

D. Angiogenic and Antiangiogenic Activities

Angiogenesis or neovascularization is defined as the formation of new blood vessels. This process, originally identified with vascularization in the placenta (124) has been extended to other capillary growth including tumor angiogenesis (125). After a tumor takes hold it grows slowly remaining quite small (1-2mm^3) until new capillaries come to within 150-200 μm (the distance which oxygen can diffuse) (126). Once vascularized, the tumor cells multiply at a rapid rate resulting in serious damage to the tissue surrounding the growing tumor. The process of angiogenesis, or the ingrowth of blood vessels, requires the enzymatic degradation of extracellular matrix underlying tissue (basement membrane), the movement of endothelial cells into this space and their replication (127). Heparin may play a variety of roles in angiogenesis. Immediately before capillary ingrowth, mast cells containing heparin congregate. The heparin from these mast cells can stimulate endothelial cell migration and this activity can be blocked by the addition of protamine (128). Heparin can localize activate and stabilize growth factors that stimulate angiogenesis such as fibroblast growth factor (FGF), endothelial cell growth factor (ECGF) (129-131).

The FGFs are a family of nine, highly homogeneous protein mitogens that have a profound effect on angiogenesis. These growth factors often exhibit overlapping biological activities (132,133). FGFs transduce their signals by binding to a group of cell surface tyrosine kinase receptors, FGFR1-FGFR4.

FGFs also interact with low affinity (K_{diss} 10^{-9}-10^{-7}) heparan sulfate PG receptors (Syndecans (26)). Cells that make FGFR but do not make HSPG fail to respond mitogenically to acidic FGF (FGF1) or basic FGF (FGF2) without the addition of heparin.

The FGFs are a multi-ligand, multi-receptor family in which one receptor can bind several ligands with high affinity. Various interactions are coordinated through heparan sulfate PG which either promotes or restricts growth factor binding to a particular receptor (133). Heparin and heparin oligosaccharides (in the presence of angiostatic steroids) (36,134) and even heparinase (135) can inhibit angiogenesis. Specific sequences in the heparin polymer (136) are capable of very tight chelation of copper, long implicated as an angiogenesis modulator (124). Folkman and Klagsbrun proposed that in the presence of other soluble factors such as steroids, heparan sulfate PG lining the endothelium restrain capillary growth (129). This quiescent microvasculature (125) can rapidly respond to heparin modulated growth factors produced during ovulation, by wounds (126) or inflammation (as occurs in stroke where the damaged blood brain barrier requires repair (127)), in immune reactions, and in tumors (128).

To fully understand the role of heparin in the process of angiogenesis, specific saccharide sequences within the heparin polymer that bind growth factors (137-139) and growth modulators such as copper need to be elucidated. Once the necessary sequences have been determined, heparin-oligosaccharides or small molecule analogues might be used as important new drugs to regulate angiogenesis.

E. Heparin and Cancer

Heparin and other proteoglycans have a role in neoplasia (140). In addition to heparin's role in regulating the growth of endothelial cells, heparin is also capable of controlling the proliferation of other cells through its interaction with growth factors. Basic FGF might be closely related to the malignant growth of prostate cancer in addition to tumor angiogenesis (141). Epidermal growth factors (EGF) and their receptors (EGFR) are also important in human cancers (142). This family of growth factors includes heparin-binding EGF, making the study of heparin or heparin analogs important in the design of new anticancer agents (142).

F. Antiviral Activity

Heparan sulfate proteoglycans line the surface of most mammalian tissues (25,26). Therefore it is not surprising that viruses make use of these molecules as receptors to gain entry into cells. Herpes simplex virus (HSV) uses heparan sulfate PG to infect cells (143,144). It does so by anchoring on heparan sulfate and localizing at the cell surface prior to its entry into the cell. Human immunodeficiency virus (HIV) binds heparan sulfate PG on T-lymphocytes through an extended loop in its GP120 coat protein (145). Heparin can block this interaction (145) as can other highly sulfated glycosaminoglycans (146). Heparin has been clinically tested in the treatment of AIDS and has shown itself effective (as has dextran sulfate) in blocking HIV replication *in vitro* (145). Human milk contains a highly sulfated chondroitin sulfate that is capable of inhibiting GP120 binding to its host cell CD4 receptor (146). This glycosaminoglycan probably acts as a general antiviral agent, protecting newborn infants during the first months of life against a wide range of viral challenges.

G. Alzheimer's Disease

In Alzheimer's disease heparan sulfate PGs are found in deposited in the senile plaques and neurofibrillary tangles (147). Amyloid peptide contains a specific amino acid sequence capable of binding heparin only at the low pH present within these plaques (148). Heparin also inhibits a protease nexin found predominantly in the brain and having an activity associated with Alzheimer's disease (149). Basic FGF, found bound to heparan sulfate PGs in Alzheimer's plaques as well as in inclusions in Parkinson disease, suggests a possible role of heparan sulfate PG in the formation of intraneuronal inclusions (150). While the precise role of heparan sulfate in Alzheimer's disease pathology is still unknown, it is possible that heparin or heparin analogues may be useful agents in slowing or reversing the formation of senile plaque with the brain (151).

H. Heparins Regulation of Nuclear Events

The presence of heparan sulfate in the nucleus was first reported by Conrad and coworkers (152). Heparin fractions, low molecular weight heparin, and heparin oligosaccharides having different molecular weight, net charge, and primary structures show differential activity towards the regulation of gene transcription (153). It is unclear how heparin exhibits this activity, *i.e.*, through growth factor signaling or by directly entering the nucleus and interacting with nuclear proteins such as topoisomerase (154,155), histones, (156,157) and transcription factors (157). Heparin and its analogues might be capable of regulating cell growth and replication at the gene level making such agents exquisitely potent new drugs.

I. Interaction of Heparin with Proteins

Heparin has been found to bind to a large number of proteins (Table 2). These proteins can be broken into classes including: 1. enzymes; 2. protease inhibitors; 3. lipoproteins; 4. growth factors; 5. chemokines; 6. selectins; 7. extracellular matrix proteins; 8. receptor proteins; 9. viral coat proteins; 10. nuclear proteins; and 11. other proteins (157). The specificity and avidity of these interactions

Table 2. Heparin Binding Enzymes and Proteins

I. Enzymes

1. Lipolytic enzymes
   - Diamine oxidase[a] (259)
   - Hepatic triglyceride lipase[a] (260)
   - Human skin esterase[b] (261)
   - Lipoprotein lipase[a] (104,105,260)
   - Phospholipase A2[a] (262,263)
   - Phospholipase C[a] (264)

2. Kinase (265)
   - Casein kinase I, II[b] (266,267)
   - Dictyostelium myosin II heavy chain kinase A[a] (268)
   - Diacylglycerol kinase (269)
   - Enterokinase (265)
   - Glycogen synthase kinase[b] (270)
   - 1,4,5-triphosphate 3-kinase (271)
   - Phosphoglycerate kinase (265)
   - Posphotidylinositol 4-kinase (269)
   - Protein kinase C[a] (272)
   - Pyruvate kinase (265)
   - Rhodopsin kinase (265)
   - RNA-dependent protein kinase[a] (273,274)
   - Streptokinase (275)
   - Urokinase (276)

3. Phosphatases (277)
   - Acid phosphatase[b] (277)
   - Alkaline phosphatase[b] (277)
   - Protein phosphatase 1[b] (277)

4. Enzymes acting on carbohydrates
   - Chondroitinase (278)
   - Glucosamine-6-sulfatase (279)
   - Heparan sulfate lyase (278,280)
   - Heparan sulfate 6-sulfotransferase (281)
   - Hyaluronidase[b] (282)
   - Lysozyme[b] (283)
   - N-Acetyl-α-glucosaminidase[b] (282)
   - Neuraminyl transferase[b] (284)

5. Proteases/Esterases
   - Acetylcholine esterase (285)
   - Cathepsin G[b] (286)
   - Coagulation factors (Xa, XII, XI, IX,VIII) (77,80)
   - Complement convertases (112)
   - Neutrophil elastase[b] (117-119)

Table 2 (continued)

Plasminogen[a] (287)
Protein C[b] (288)
Trypsin[b] (291)
Tryptase (a mast cell trypsin-like protease) (292,293)

6. Enzymes acting on or associated with nucleic acids
   Angiogenin (294)
   Deoxyribonuclease I, II[b] (295)
   DNA polymerase[b] (296)
   Endonuclease (297)
   Ribonuclease[b] (298)
   RNA polymerase[b] (299,300)
   Topoisomerase I, II[b] (154,155)

7. Other enzymes
   Adenylate cyclase[b] (301)
   Myosin ATPase[b] (302)
   Progelatinase A (gelatinase A)[a] (303)
   Superoxide dismutase (EC SOD) (304)
   THF synthetase (305)
   Xanthine oxidase (306)

II. Protease inhibitors (Serpins) (307,308)
Antistasin (309)
Antithrombin III (ATIII) (91,310)
C1 Inhibitor proteins (112)
Heparin cofactor II (HCII) (91,92)
Histidine-rich glycoprotein (157)
Leuserpin (311)
Mucus proteinase inhibitor (312)
Plasminogen activator inhibitor type 1 (PAI-1) (313)
Protease nexin (314)
Thrombomodulin (315)
Tissue factor pathway factor (TFPI) (316)
Tissue-type plasminogen activator (t-PA) (317)

III. Lipoproteins
VLDV and LDL (102)
Apolipoprotein B-100 (318)
Apolipoprotein E (319)

IV. Growth factors (320)
Acidic fibroblast growth factor (aFGF, FGF1) (141,321)
Basic fibroblast growth factor (bFGF, FGF2) (141,321)
Epidermal growth factor (HB-EGF) (142)
Hepatocyte growth factor (HGF) (322)
Keratinocyte growth factor (KGF, FGF4) (323)
Platelet-derived growth factor (PDGF) (321,324)

Table 2 (continued)

Schwannoma-derived growth factor (325)
Smooth muscle cell growth factor (326)
Transforming growth factor β-1 (TGF-β1) (327)
Vascular endothelial growth factor (VEGF) (328)

V. Chemokines
Gro-1-α (329)
γ-Interferon (330)
Interleukin 8 (IL-8) (329,331)
Neutrophil activating peptide 2 (NAP-2)[d] (332)
Platelet factor IV (331,333)

VI. Selectins
L-selectin (37,122)
P-selectin (37,122)

VII. Extracellular matrix proteins
Collagens I-VI (334,335)
Fibronectin (336)
Hexabrachions (337)
Laminin (338)
Tenascin (339)
Thrombospondins 1 and 2 (340)
Vitronectin (S-protein) (341)

VIII. Receptor proteins (also see kinases)
Androgen receptor[c] (157)
CD4 receptor (342)
Estrogen receptor[c] (157)
FGF receptor (FGFR1-4) (343,344)
Glucocorticoid receptor[c] (157)
Glycoprotein 330 (LDL receptor) (345)
Insulin like growth factor binding proteins (346)
L-type Ca^{2+} channel (347)
Mineral corticoid receptor (157)
Progesterone receptor[c] (157)
Thyroid receptor (157)
Vitamin D receptor[c] (157)

IX. Viral coat proteins
gC-II (348)
gp120 of HIV-1 (145,349)
gp140 and gp160 of HIV-2 (349)
Herpes simplex virus -1 (HSV-1) (350)
Pseudorabies glycoprotein g III (351)

X. Nuclear proteins[c] (157)
c-fos (352)
c-jun (352)

Table 2 (continued)

CREB[c] (157)
C/ESB[c] (157)
fra 1[c] (157)
GCN4[c] (157)
jun-B[c] (157)
jun-D[c] (157)
Histones (156)
Transcription factors[c] (153,157)
YAP-1[c] (157)

XI. Other proteins

90 kDa stress protein (353)
Alzheimer β-amyloid precursor protein (APP) (354)
Factor J (FJ) (355)
Fibrin (356)
Immunoglobulin G (209,357)
N-CAM (358)
Placenta Heparin Binding Lectin (156)
Platelet/endothelial cell adhesion molecule-1 (GMP-140) (359)
Protein C inhibitor (276)

a. Activated by heparin
b. Inhibited by heparin
c. Predicted heparin binding protein based on sequence motif
d. Mammalian heparanase

probably result from ionic interactions between the sulfate groups of heparin and basic amino acid residues present in the interacting protein (158). The half sulfate esters and sulfamide groups in heparin are chemically unique anions. Sulfate groups differ in both pKa and distribution of negative charge from carboxylate groups and phosphate groups. In addition, phosphate monoesters have a net charge of -2 compared to the -1 charge present in sulfate monoesters. Protonated monosulfate esters are very strong acids that give rise to a very weak (soft) base on ionization because three equivalent oxygen atoms carry the -1 charge. Thus, it is not surprising that the guanidinium group of arginine forms the strongest interaction of all amino acids with the sulfate groups in heparin (32,33). While lysine also can interact with heparin's sulfate groups, histidine, when localized on the surface of a protein, is insufficiently basic to ion pair with sulfate at physiologic pH. In low pH environments, such as those present in the senile plaque of Alzheimer's patients, histidine can be actively involved in binding heparin (148). In addition to basic amino acids, hydrogen bond forming amino acids such as glutamine and asparagine also appear to be involved in promoting protein-heparin interactions (32). Hydrophobic amino acids such as leucine, isoleucine, phenylalanine and valine might contribute to binding by interacting with *N*-acetyl groups present in heparin (34).

Cardin and Weintraub (158) proposed a specific linear sequence motif of basic amino acids found in heparin binding proteins. This motif requires all the basic residues to segregate to a single side of the peptide when it adopts an α-helical conformation. While many heparin-binding proteins display such a motif, others do not presumably because these binding sites either do not lie in an α-helical domain or do not reside in a linear domain (as is the case for the heparin binding domain in ATIII (159)).

The study of heparin-protein/peptide interactions has become a very active area of research involving a multiplicity of modern chemical and biochemical techniques including: site directed mutagenesis (160); the study of synthetic peptides (32,33,34); microtitration calorimetry (33,160); NMR spectroscopy (34,161); affinity electrophoresis (123,162); affinity chromatography (34); and plasmon resonance spectroscopy (137); and the use of combinatorial libraries (32). A better understanding of heparin-protein interactions will be essential in developing heparin and heparin analogues as drugs to both agonize and antagonize the activities based on these interactions.

VII. RATIONALE FOR DEVELOPING ANALOGUES TO REPLACE HEPARIN

A. Structural Requirements for Active Analogues

Heparin's biological activities are primarily mediated through its binding to proteins and its regulation of their activities. To substitute for heparin, an analog must bind to these proteins and regulate the same activities. Heparin's binding is primarily through electrostatic interactions and depends on its high charge density. Thus, these analogs are often highly charged polyanions, prepared enzymatically or chemically, and can substitute for heparin by the appropriate positioning of these charged groups. Ideally, these heparin substitutes should exhibit tighter binding to these proteins in order to have higher potency. Binding can be enhanced by adding hydrophobic sites to these analogs that lead to enhanced binding through their interaction with hydrophobic residues present at or near the heparin-binding domain of the target protein.

Only the heparin ATIII binding site has been studied sufficiently to lead to a well defined structure activity relationship (SAR). The synthetic heparin pentasaccharide binding site and heparin have comparable ATIII binding avidity (heparin-ATIII, ~ synthetic pentasaccharide-ATIII, $K_{diss}= 10^{-7}$ (163,164)) as well as comparable activity towards the inhibition of factor Xa. Structure modification studies have even been successfully undertaken to prepare analogs with even higher binding affinities (165-167). Some information is available on the structure of heparin's binding site for other proteins including: basic FGF (138,139); acidic FGF (137) and L-selectin (37,168). In none of these cases have the structural details of the binding site been fully established. The first step in drug discovery typically begins with the identification of an active, lead structure, from a natural product. Thus, additional efforts are required to prepare and characterize heparin fractions and heparin oligosaccharides that interact tightly with target protein(s).

B. Combating Heparin's Side Effects

Although heparin's major activity, as an anticoagulant is widely exploited, it also results in heparin's most common side-effect (77). Bleeding complications can range from mild mucosal oozing to intracranial hemorrhaging. Low molecular weight (LMW) heparins were developed as a new class of therapeutic agents called antithrombotics and offer several advantages over the anticoagulant heparin (169). These advantages include a reduced affect on platelets and increased specificity of action. LMW heparins catalyze the ATIII mediated inhibition of Factor Xa to a greater extent than the inhibition of Factor IIa. This greater specificity towards an early step in the coagulation cascade may be partially responsible for the reduced hemorrhagic sideeffects of LMW heparins. The decreased effect of LMW heparin on platelets also reduces hemorrhagic complications (77). Thus, heparin's hemorrhagic side-effects have been reduced by simply modifying its molecular weight. In the future more subtle structural modifications may be used to prepare antithrombotic heparin analogs that are entirely free from hemorrhagic side-effects. Heparin-induced osteoporosis (decalcification of the bones) is a secondary side-effect that might be greatly reduced or eliminated by using a heparin analog with increased anticoagulant potency thereby reducing the dose of the agent (77). Many of heparin's other secondary sideeffects can also be reduced by this approach.

C. Exploiting Heparin's Side Effects

Not all of heparin's sideeffects are undesirable. Some of these sideeffects could be exploited and new pharmacologically active agents prepared if only these activities could be enhanced. For example, heparin releases and activates lipoprotein lipase (LPL) and inhibits smooth muscle cell proliferation but does so only at doses that fully anticoagulate (103,110). Heparin that is devoid of anticoagulant activity and has high LPL releasing and smooth muscle cell inhibitory activities (such as Astenose®) might represent a useful agents in the treatment of atherosclerosis (103). Heparin inhibits complement activation but only at doses ten times those required for full anticoagulation (68). Recent results in our laboratory have demonstrated that it is possible to prepare heparin-oligosaccharides equipotent with heparin (on a weight basis) towards complement but without anticoagulant activity (68,116,170). Such a drug or one that contained equal potency as an anticoagulant and as a complement activation inhibitor might be useful in preventing both coagulation and complement activation in extracorporeal therapy. There are many other heparin sideeffects which might be similarly exploited in new classes of drugs.

D. Controlling Pharmacokinetics

A major problem with heparin is its poor bioavailability when administered by certain routes (171). Heparin is primarily administered *intravenously* and acts systemically as an anticoagulant (77). It has also has been administered by the *subcutaneous* route; however, only a fraction of heparin ever makes it to the circulation and then only after a long delay (169,171). This delayed delivery of low levels of heparin into the circulation is often desirable and has been effectively used in the treatment of deep vein thrombosis (40,41).

Both *intravenous* and *subcutaneous* routes require hospitalization of the patient being treated. If an orally active heparin could be prepared which was safe it would permit the treatment of outpatients (106,172). This type of therapy might be useful for disorders which do not generally call for hospitalization, such as the prophylactic use to reduce VLDLs and LDLs in the treatment of atherosclerosis (103).

There is a continuing controversy regarding heparin's oral bioavailability. Heparin is a large molecule (M_r (avg) 12,000) that is highly charged (~ -50). Such polymers are not expected to be orally absorbed by passive mechanisms and no mechanism for active uptake for heparin by the digestive tract has been reported. While claims have been made that heparin is gastrointestinally or sublingually active (173,174,175) these claims have not survived critical evaluation (176,177,178). No definitive studies, however, have ever been performed that answer the question of heparin's oral bioavailability. There are several reasons why these studies have not been done. First, no biosynthetically labeled (*i.e.*, ^{14}C or ^{3}H) heparin has been prepared in sufficient quantities and chemical labeling is either unstable or significantly alters the structure of heparin. Second, bioassays for circulating heparin (or heparin in urine) following oral administration are susceptible to interference from components of the biological fluids, including endogenous glycosaminoglycans. Third, since heparin is a complex mixture it is difficult to chemically monitor its presence in biological fluids following oral administration and even if it is chemically measured it is difficult to distinguish exogenous heparin from endogenous heparin-like substances.

Heparin administered orally in high doses has been found in low levels in the urine suggesting that it has limited bioavailability (179). There is a recent report that heparin is completely bioavailable (180). In this study, heparin was orally administered to rabbits. Neither control or experimental animals showed any circulating heparin as measured by coagulation assays. When endothelium (both capillary and wide vessel) was removed from both groups of animals and glycosaminoglycans were extracted, heparin (or a closely related heparin-like substance) was observed only in the experimental group. The amount of heparin measured on (in) the endothelium of the experimental group suggested that heparin was completely bioavailable by the oral route but that it all localized on (in) the endothelium and none circulated. An alternative explanation is that the heparin was not absorbed but instead somehow stimulated the production of endogenous heparin-like substances in the endothelium.

There have been a number of reports of biological effects when heparin is administered orally. In the past most of these reports involved the relatively uncontrolled use of oral heparin by physicians on small groups of patients. Of particular interest is the work of Dr. Hyman Engelberg, a Beverly Hills cardiologist. Dr. Engelberg's studies span a decade and uniformly suggest that relatively large doses of orally administered heparin display several biological activities, particularly relating to the treatment of atherosclerosis (104). Recent animal studies that are well controlled also suggest that oral heparin can exert biological affects. One such study has demonstrated that oral heparin can lower blood pressure and treat hypertension in rats (181,182).

It is possible that heparin (or a heparin oligosaccharide) might exert biological affects without being orally adsorbed. The mechanism for these activities might involve the interaction of heparin at receptors in the gut or their action as biological response modifiers by their uptake into cells within the digestive tract. Mast cells, the natural source of heparin, are found in the intestinal mucosa and it might be that these cells or other nearby cells respond to the presence of orally administered exogenous heparin. These explanations are of course highly speculative but warrant further investigation.

Several approaches have been used for improving heparin's oral bioavailability. The reduction of heparin's molecular weight has resulted in some improvement in its bioavailability. Lasker has studied low molecular weight heparins (M_r (avg) 3,260 - 5,300) in both rats (183) and mice (184) and has demonstrated that an oral dose of 10 mg/kg of low molecular weight heparin (M_r avg 4,000) resulted in a plasma heparin concentration of 1.5 μg/ml 30 min following administration to fasted rats as measured by factor Xa assay. Heparin (M_r (avg) 11,600) showed <0.5 μg/ml in an identical experiment. Studies in monkeys (185) showed similar results giving a curvilinear dose response curve between 7 and 16 mg/kg. Doutremepuich (186,187) randomly administered 60 mg/kg of low molecular weight heparin (M_r (avg) 6,000) to anesthetized rabbits in gelatin capsules resulting in a maximum plasma level of 0.5 μg/ml, 120 min following administration as measured using a factor Xa assay. Heparin (M_r (avg) 14,000) showed no activity in an identical experiment.

These studies suggest that the oral bioavailability of heparin can be enhanced by reducing its molecular weight. The use of defined heparin oligosaccharides might further enhance this oral bioavailability.

Low molecular weight heparin (M_r avg 6,000) has been prepared in proteinoid microspheres and administered to human volunteers in orange drink. At a dose of 37,600 IU (~3.5 mg/kg) a peak plasma concentration of 0.7 μg/ml was observed after 1 h (188). These microspheres do not appear to represent a significant improvement over the bioavailability of low molecular weight heparin administered either directly or in gelatin capsules. The oral bioavailability of heparin can be reportedly increased through the use of adjuvants. Reber and Struder (189) showed enhanced gastrointestinal absorption of calcium heparin in the presence of sulfoxides while Tidball and Lipman (190) found sodium EDTA was capable of enhancing jejunal absorption of heparin. Acidic buffers (64), sulfonated surfactants (191), olive oil dispersions (192) bile salt (193) and monoolein (194) formulations all enhance heparins oral bioavailability. Adjuvants represent an additional level of formulation complexity and the ones used to stimulate heparin uptake have toxicity problems. While some additional studies are warranted, there are probably better approaches to enhance oral bioavailability. Liposomes have been used for heparin formulations to enhance bioavailability and to alter heparin pharmacodynamics. Liposomally entrapped heparin prolongs its anticoagulant effect when administered *intravenously* (194). Liposomes have been shown to enhance the *percutaneous* delivery of heparin (195). When liposomally entrapped heparin is administered orally 100 mg/kg was required for anticoagulant (~5 μg/ml plasma) of beagle dogs (196,197). Because of the successful application of liposomes in the oral delivery of other large molecules this approach warrants further investigation. Maximization of heparin loading in liposome formulations may give substantial enhancement of oral bioavailability. Heparin shows enhanced oral bioavailability when administered as the amino acid salt forms (198,199). The glycine salt of low molecular weight heparin shows slightly improved oral bioavailability when compared with the sodium salt form (186,187). A low molecular weight heparin formulated as the tetra-alkyl ammonium salt form was considerably more bioavailable than the sodium salt form (200,201). These tetra-alkyl ammonium salts of heparin and low molecular weight heparin are actually very hydrophobic and are soluble in both aqueous and in organic solvents. This approach appears to have promise since it represents a change in heparin's fundamental physical properties, from highly hydrophilic to hydrophobic. Tetra-alkyl ammonium salts are, however, often toxic. The coupling of large lipophilic alkyl chains to the reducing ends of heparin had little effect on increasing heparin's hydrophobic character (202).

In summary, while orally activity of heparin is highly desirable and has been vigorously pursued, no significant progress has been made in this area. Heparin analogs of low molecular weight and reduced or masked charge may represent the best approach towards achieving oral bioavailability.

A major second problem associated with the use of heparin has been the lack of understanding of its metabolism and its erratic and sometimes unpredictable rate of clearance (77). Many of these problems are due to the absence of a sufficiently sensitive assay for the measurement of chemical heparin in the plasma (203). Currently the only methods having the required sensitivity are bioassays which are also effected by subtle changes in the levels of coagulation proteins in the plasma and are subject to numerous additional interferences. The preparation of homogeneous heparin analogs that are pure single entities would go a long way towards solving these problems in facilitating the development of sensitive chemical assays.

VIII. FRACTIONATED HEPARINS

Heparin can be fractionated, on the basis of size, using gel permeation chromatography (GPC) (204). Both low pressure and high pressure GPC has been used to obtain heparin fractions of average molecular weight between 5,000 and 40,000 (204). Because of its high charge density, heparin exists as an extended helical rod and thus has a greater molecular size than does a globular protein of similar molecular mass (205). Some activities of heparin that vary according to chain size are shown in Table 3.

Being a sulfated polysaccharide (with an average of 2.7 sulfate groups per repeating disaccharide unit), heparin can be fractionated on the basis of its charge using anion exchange chromatography. Strong anion exchange (SAX) chromatography of heparin typically involves the use of a salt gradient (53,55). Heparin fractions can be prepared with a degree of sulfation ranging from 2 to 3 sulfate groups per disaccharide repeating unit (206). Some of heparin's biological activities vary with degree of sulfation making this a useful technique to separate or enrich specific heparin based activities (Table 4).

A major breakthrough in heparin research came with the observations that there were two populations of heparin chains, having different affinities to ATIII, which could be separated by affinity chromatography (207).

Pharmaceutical heparin having an anticoagulant activity of 150 U/mg can be fractionated into 33 wt% high-affinity heparin (500 U/mg) and 66 wt% low-affinity heparin (50 U/mg) (208). Affinity electrophoresis fractionates heparin on a similar basis (123,162). The affinity separation is dependent on tight-binding of the high affinity fraction ($K_{diss}=10^{-8}$) and weak-binding (through non-specific interactions) of the low-affinity fraction ($K_{diss}=10^{-5}$). Heparin has been fractionated by its affinity towards other proteins (Table 5). While immunoaffinity chromatography and electrophoresis have been successfully used to fractionate a wide variety of proteins, this approach has not been useful in fractionating heparin. Heparin is remarkably non-immunogenic. No true (IgE-mediated) allergic reactions to heparin have ever been demonstrated in

Table 3 Heparin Activities as a Function of Polymer Chain Length

	Approx. Mr	Anti IIa Activity (U/mg) mediated by(206) ATIII	HCII	Relative Inhibition of Complement Activation(68)	Relative Lipase Releasing Activity(103)
Tetrasaccharides[a]	1200	<1	<1	4	8
Hexasaccharides	1900	<1	<1	5	10
Octasaccharides	2400	<1	<1	10	9
Decasaccharides	2900	<1	<1	25	30
Dodecasaccharides	3500	5	2	35	--[b]
Tetradecasaccharides	4100	13	16	45	---
Hexadecasaccharides	4700	17	16	55	---
Octadecasaccharides	5300	26	18	--	---
Eicosasaccharides	5900	24	14	–	---
Heparin fraction 1	4000	6	23	100	32
Heparin fraction 2	5700	26	44	100	---
Heparin fraction 3	10,700	38	115	100	---
Heparin fraction 4	14,500	196	271	---	---
Heparin fraction 5	16,900	130	331	---	---
Heparin fraction 6	25,300	92	356	100	---
Unfract. heparin	14,000	171	225	100	100

a. All heparin and heparin oligosaccharides are from porcine intestinal mucosa.
b. Not determined.

Table 4 Heparin Activity as a Function of Degree of Sulfation or Charge

	Anti IIa Activity (U/mg) mediated by(206) ATIII	HCII
Heparin fraction 1[a]	17	90
Heparin fraction 2	108	261
Heparin fraction 3	125	336
Heparin fraction 4	227	468
Heparin fraction 5	540	448
Unfractionated heparin	171	225

a. Heparin fractionated on the basis of charge using DEAE Sephadex with salt gradient elution. The degree of sulfation of heparin fractions 1 to 5 ranged from 2.73 - 2.83 sulfates/disaccharide repeating unit.

Table 5 Proteins That Have Been Used to Affinity Fractionate Heparin

Acidic fibroblast growth factor (aFGF) (137,360)
Antithrombin III (ATIII) (361)
Basic fibroblast growth factor (bFGF) (360,362)
Chemokine Gro-1-α (329)
Chemokine (IL-8) (329)
Chemokine NAP-2 (329)
Collagens (I-V1) (335)
Complement C3b (282)
Concanavalin A (363)
Fibronectin (360)
Heparin cofactor II (HCII) (206)
Lipoprotein lipase (103)
Nerve growth factor (360)
Platelet factor IV (PF4) (329,364)
Poly(lysine) (365)
Protamine (366)
Thrombin (FIIa) (367)
Tissue plasminogen activator (t-PA) (368)

patients (only one unconfirmed report of a human antibody to heparin is presented in the scientific literature (209)) despite heparins widespread clinical use for nearly 60 years (5,6). Attempts to generate either polyclonal or monoclonal antibodies to unmodified heparin chains in animals using a variety of approaches have failed.

IX. LOW MOLECULAR WEIGHT HEPARINS

Low molecular weight (LMW) heparins are generally prepared through the controlled, partial, chemical or enzymatic depolymerization of commercial GAG heparin (Figure 4, Table 6). A convenient method to cleave heparin involves its

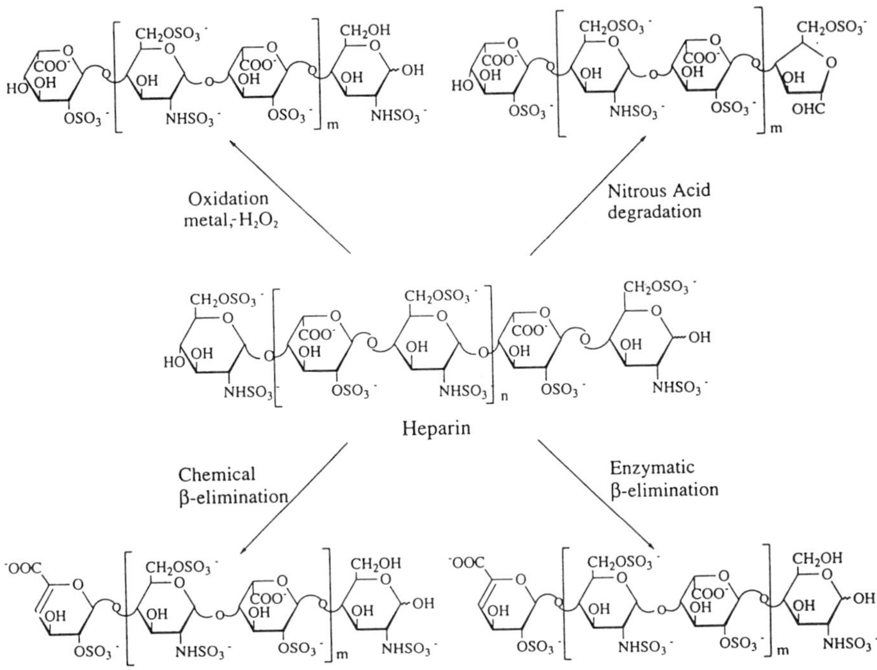

Figure 4 Preparation of low molecular weight heparins. Heparin (center) contains chains with the structure shown where n≈20. Partial depolymerization of this heparin chain through radical oxidation, nitrous acid degradation, chemical β-elimination and enzymatic β-elimination affords low molecular weight heparin m≈7 with the structures that are shown.

Table 6 Properties of Commercial LMW Heparin

Agent	Trade name	Source	Origin	Preparation	Mr Wt Avg[b]	Potency[c]
CY216	Nadroparin calcium[a] Fraxiparine Fraxiparin Fraxiparina Seleparina	Sanofi-Choay (France)	PM	EtOH fractionation	5,500	95
CY22		Sanofi-Choay (France)	PM	Nitrous acid degradation	3,400	60
Kabi 2165	Fragmine Dalteparin sodium[a] Tedelparin Fragmin Low Liquemine	KABI AB (Sweden)	PM	Nitrous acid degradation	6,400	142
LMWH/DHE	Certiparin sodium[a] Sandoparin	Sandoz AG (Germany)	PM	Isoamylnitrite degradation	6,300	
RD11885	Ardeparin sodium[a] Normiflo	Hepar/Wyeth (U.S.A.)	PM	Peroxidative degradation	6,200	62
OP2123	Parnaparin sodium[a] Fluxum Minidalton	Opocrin (Italy)	BM	Peroxidative degradation	6,500	83

Table 6 (Continued)

Agent	Trade name	Source	Origin	Preparation	Mr Wt Avg[b]	Potency[c]
PK10169	Enoxaparin sodium[a] Clexane Lovenox Klexane	Rhone Poulenc (France)	PM	Benzylation/β-elimination	3,800	96
LHN-1	Tinzaparine sodium[a] Logiparin	Novo (Denmark)	PM	Enzymatic β-elimination	4,900	87
Reveparin	Reveparin sodium[a] Clivarin Clivaparin	Knoll AG (Germany)	PM	Nitrous acid degradation	4,000	
Std LMWH85/600		NIBSC (U.K.)	PM	Nitrous acid/GPC		100
Heparin		Hepar (U.S.A.)	PM		14,000	150

[a]International and the United States nomenclature
[b]By gel filtration chromatography (GPC) (369)
[c]Relative anti-Xa potency against NIBSC standard low-molecular-weight heparin (370)

treatment under relatively mild conditions with nitrous acid to cleave heparin at some of the glucosamine residues containing an *N*-sulfate group (52,210). This method of preparing LMW-heparins introduces an artifact into the oligosaccharide's reducing end, a ring-contracted anhydromannose residue (57,210). Depolymerization by nitrous acid treatment is seldom run to completion to preserve *N*-sulfated glucosamine residues essential for ATIII binding (52,163,164).

Oxidative cleavage of sugars by Smith degradation permits the cleavage of heparin at susceptible unsulfated glucuronic and iduronic acid residues having vicinal diol functionality (52) resulting in an acyclic dialdehydic residue, which is sensitive to hydrolysis under relatively mild conditions. (Reduction of this residue affords Astenose®, a drug having approximately the same MW as heparin.) Controlled, partial oxidation, followed by hydrolysis, is required to protect unsulfated uronic acid residues required for ATIII binding. Heparin can be oxidatively cleaved using a variety of reagents including: hydrogen peroxide; Cu^{+}/O_2; Fe^{++}/O_2; peracetic acid; amyl nitrite; and hypochlorous acid (52). These reactions rely on oxygen or hydroxyl radical initiated oxidation of primarily the unsulfated uronic acid residues and again must be performed in a controlled fashion to maintain biological activity.

Heparin can be eliminatively cleaved at its iduronic acid residue (the $^4C_1$ conformer of this residue contains the appropriate anti-relationship between the 5-position proton and the 4-position glycoside anion) using base. Eliminative cleavage can be further facilitated by forming the benzyl ester of uronic acid prior to base treatment. Only the glucuronic acid residues survive such eliminative cleavage and these can then be debenzylated (52,211). The resulting oligosaccharides have an unsaturated uronic acid residue at their non-reducing end. This residue may prolong the *in vivo* half-life of these oligosaccharides by blocking their biotransformation by exoglucuronidases (212). This unsaturated sugar, however can be easily removed by treatment with ozone followed by acidic work-up.

More selective eliminative depolymerization of heparin and heparan sulfate can be accomplished enzymatically. Heparin lyase I cleaves heparin's most common linkage 2,6 disulfated glucosamine (1→4) 2-sulfated iduronic acid (29,213). This enzyme will cleave this linkage even if the 6-sulfate is missing or in the presence of a 3-*O*-sulfate in the glucosamine residue, found in heparin's ATIII binding site (29,214). Controlled enzymatic depolymerization is required to retain heparin's ATIII binding site (215). It now appears that heparin lyase I preferentially cleaves heparin's ATIII binding site possibly because of its high degree of sulfation (216). This makes heparin lyase a more useful tool in preparing oligosaccharides with low ATIII mediated anticoagulant activity (for applications such as inhibition of complement activation (68,116,170)). Heparin lyase II has a different specificity and can also be used to prepare low molecular weight heparin (29,214). As with the LMW heparin prepared by chemical eliminative cleavage, the unsaturated sugar residue formed can either be left on each chain to block biotransformation (212), or can be removed by ozonolysis (52,69,217).

Although each LMW heparin is different (*i.e.*, each possesses structural artifacts generated in their preparation) they all have several features in common (57). LMW heparins are polydisperse with M_r (avg) ~ 5000 (Table 6) and having M_r ranging from 2,000-8,000 (218,219). These LMW heparins have been mapped using SAX-HPLC, gradient PAGE and capillary electrophoresis (57,58). LMW heparins are primarily comprised of the same small number of major oligosaccharides that make up standard commercial heparins (57). In addition to these major oligosaccharides these oligosaccharide maps also show uncharacterized oligosaccharides that may correspond to artifacts introduced in the depolymerization step (52,57,58,220,221). More structural studies on LMW heparin preparations are required to more fully understand and to rationalize the differences in their biological activities (57).

The anticoagulant activities of many of the LMW heparins are presented in Table 6. A ratio of ATIII mediated anti-Xa to ATIII mediated anti-IIa activities of 5-25 is observed in LMW heparins resulting in their inhibition of coagulation early in the coagulation cascade (208), presumably reducing their hemorrhagic side-effects (38). These LMW heparins also appear to exhibit a decreased effect on platelets reducing both their activation and aggregation (38). They show increased bioavailability following *subcutaneous* administration and increased half-life (169). The results of clinical trials are promising suggesting that they are safer than heparin, generally resulting in less hemorrhagic complications and have greater bioavailability and longer half-lives resulting in better dose control (42). These improved biological activities have paved the way for clinical trials and approval of several of these low molecular weight heparins.

X. HEPARIN ANALOGS

A. Other Glycosaminoglycans

Heparin is only one member of the family of GAGs. Heparan sulfate, as discussed above, has a similar structure (Figure 2) and exhibits many of the same biological activities as heparin. One factor that has held back the development of drugs based on heparan sulfate is its high price and very limited availability (65). Recently, our group has described the properties of a new commercial heparan sulfate that is prepared from a sidestream obtained in the manufacturing of heparin (65). With increased availability of heparan sulfate it is likely that drug development based on its structure will be accelerated.

Dermatan sulfate PG resides in the same locations as heparin sulfate PG and might be responsible for many of the same biological activities (94). Dermatan sulfate (Figure 5) has been used clinically as an antithrombotic agent (94,222,223) and its structure activity relationship has been studied using many of the same techniques developed for heparin (222,223). Chondroitin sulfates (Figure 5) have been studied to a lesser extent. Both dermatan and chondroitin sulfates are commercially available in quantities required for drug development.

Mixtures of several glycosaminoglycans have been studied as antithrombotic agents and have been used in clinical studies for the treatment of non-hemorrhagic stroke (224).

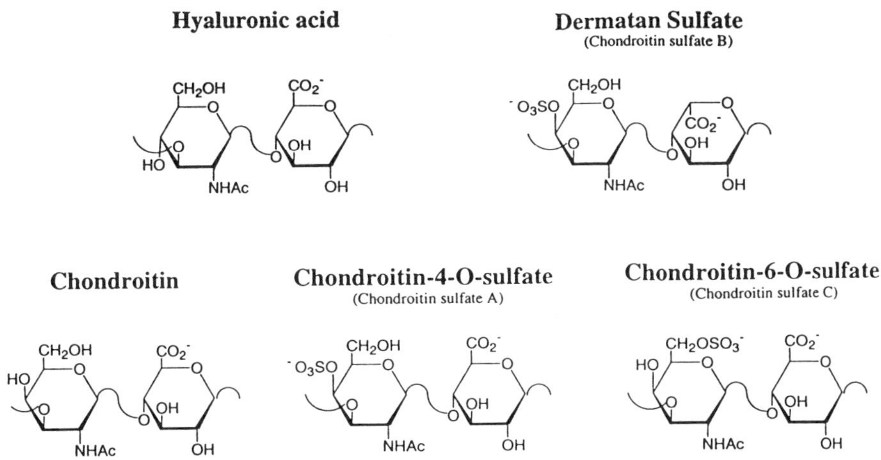

Figure 5 Structure of chondroitin sulfate, dermatan sulfate and hyaluronic acid polymers. The structures of the major disaccharide units contained in these glycosaminoglycans are shown where Ac is acetate.

B. Sulfation of Natural Polysaccharides

Chitin, poly(2-acetamido-2-deoxy-D-glucopyranose) is the major organic component of the exoskeleton found in insects (225). Chitin can be de-*N*-acetylated to prepare chitosan, which on chemical sulfation (Figure 6) and/or carboxymethylation results in a polymer having certain structural similarities to heparin (226). These chitosan derivatives show anticoagulant activity related to their degree of sulfation. Sulfated, carboxymethylated chitosan inhibits thrombin activity through ATIII to almost the same degree as heparin.

Pentosan, a (1→4)-β-linked xylopyranose with a single branded 4-*O*-methyl-α-D-glucopyranosyl acid, is extracted from the bark of the beech tree, *Fagus sylvantica* (227). When fully sulfated by chemical methods (Figure 6) it has been demonstrated to be an anticoagulant with one-tenth of heparin's activity on a weight basis (228). Its primary anti-IIa activity has been proposed to be HCII mediated (229). This heparin analogue also has a demonstrated anti- heparin

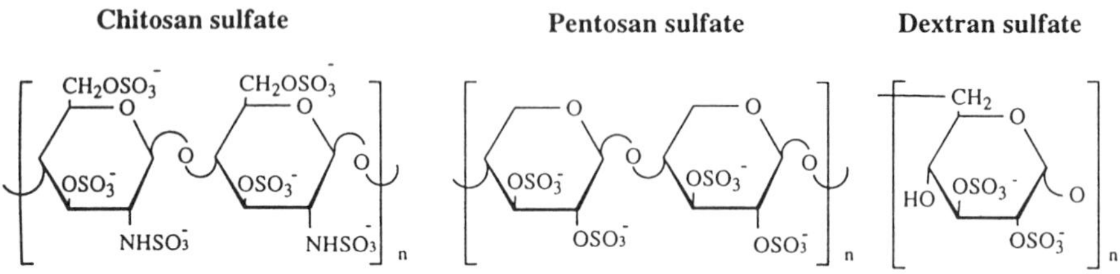

Figure 6. Structure of Chitin, Pentosan and Dextran Sulfates. Representative saccharide units comprising each of these polymers are shown where n is the degree of polymerization.

activity and can compete with heparin for antithrombin III binding, reducing the rate of heparin catalyzed ATIII mediated inhibition of thrombin (230).

Dextran, a (1→4)-β-D-,(1→3)-α-D-branched glucan polymer (225), can be chemically sulfated to prepare dextran sulfate (Figure 6) (231). Both dextran (a plasma extender) and dextran sulfate have been used as pharmaceuticals (232). Dextran sulfate has low anticoagulant activity with high LPL releasing activity (233). This has permitted the exploitation of this agent as an antiatherosclerotic in Japan (233). Dextran sulfate has been used as a heparin replacement in anticoagulation and has recently been immobilized on plastic tubes to prepare non-thrombogenic surfaces (234). Dextran sulfate is an inhibitor of HIV binding to T-lymphocytes but its low oral bioavailability has precluded its use in the treatment of AIDS (180).

C. Fully Synthetic Sulfated Polymers

Synthetic polymers such as poly(vinyl sulfate) and poly (anethole sulfonate) are highly charged heparin-like polyanions (Figure 7) that exhibit anticoagulant activity (235). This activity has only been exploited *in vitro* in preventing coagulation of collected blood or plasma samples during storage for assay. The use of these agents *in vivo* has largely been precluded by their high toxicity. These synthetic polymers are resistant to metabolism and thus remain in the body for extended periods during which they can display severe sideeffects.

D. Small Sulfated and Sulfonated Molecules

Small sulfated and sulfonated molecules are currently being explored as heparin analogs (Figure 8). These small molecules might have improved bioavailability because of their reduced size and charge. They may also include

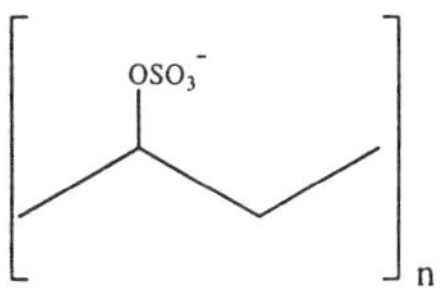

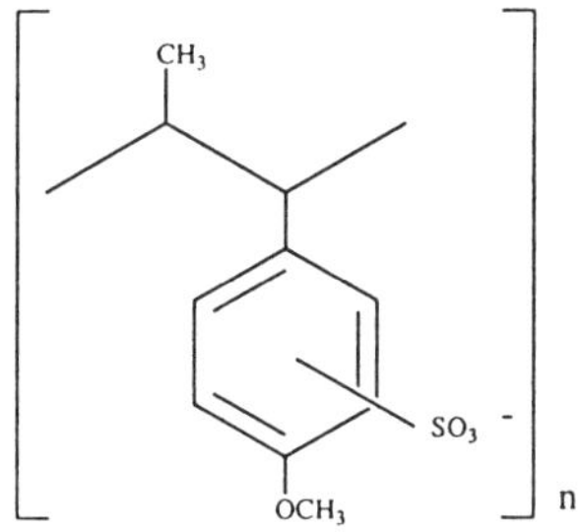

Polyvinyl sulfate **Polyanethole sulfate**

Figure 7 Structure of poly(vinylsulfate) and poly(anetholesulfonate) where n is the degree of polymerization.

hydrophobic sites in their structures that can enhance protein binding. While sulfate monoesters more closely resemble functional groups in the natural product, sulfonates are structurally similar giving a nearly identical delocalization of negative charges on the three oxygen atoms bound to sulfur (236).

Suramin, synthesized in 1917, was the first heparin analog and has been used clinically in a variety of applications (237). These applications include its activity as an anthelmintic, an antiprotozoal, an antineoplastic (238,239) and an antiviral agent (240,241). Despite its potent activities suramin has a very long half-life in the body and exhibits a wide range of toxic side-effects (237).

Naphthalene sulfonates have been investigated as investigated as antiviral agents (242). A wide range of these compounds have been synthesized and show potent anti-HIV activity. Limited toxicological data is currently available on these compounds.

A series of simple aliphatic disulfates and disulfonates have recently been prepared and tested for their ability to arrest amyloidosis *in vivo* as potential agents for the treatment of Alzheimer's disease (151). When administered orally these simple anionic heparin analogs, ranging in molecular weight between 135 and 1,000, reduced murine splenic AA amyloid progression.

Sucrose octasulfate is used to prepare an insoluble aluminum salt called sucralfate (Carafate®) which is used in the treatment of ulcers. Sucralfate coats the ulcer bed but does not reduce gastric acid and until recently its mechanism of action was not well understood. Folkman and coworkers (243) discovered that sucralfate bound basic FGF with greater avidity than did heparin. Basic FGF bound to sucralfate was protected against inactivation even under the strongly acidic conditions present in the stomach. Folkman hypothesized that the anti-ulcer activity of sucralfate was the result of its localization of stabilized basic FGF at the ulcer wound, promoting its healing. Recently, sucrose octasulfate was co-crystallized with basic FGF (244). This is the first structure of sulfated carbohydrate-protein complex to be solved using x-ray crystallography and

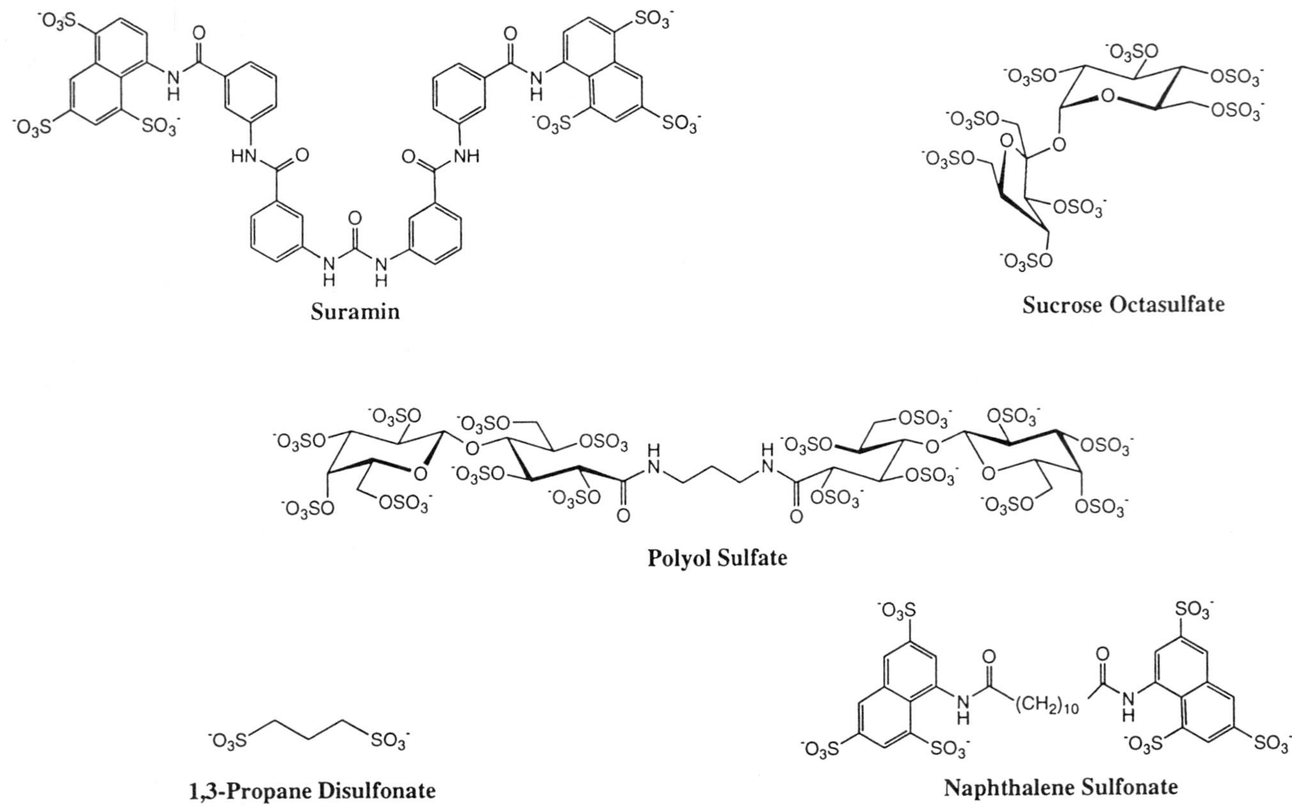

Figure 8 Structures of suramin, sucrose octasulfate and representative structures of naphthalene sulfonates, aliphatic disulfates and disulfonates.

represents an important step in understanding the biological activity of heparin and heparin analogs (244). The solution structure of sucrose octasulfate has also recently been studied in detail using NMR spectroscopy (245).

Synthetic, highly sulfated, lactobionic acids are very active in potentiating the of HCII mediated inhibition of thrombin (246). The regulation of the coagulation cascade at this step resembles dermatan sulfate's activity (94) and might represent an interesting new class of antithrombotic agents. There have also been some reports of some oral bioavailability of certain sulfated lactobionic acids (247) potentially opening up many new applications for these agents. Similar compounds have also been developed and tested for their activity to inhibit smooth muscle cell proliferation (248).

Sulfated cyclodextrins, such as β-cyclodextrin tetradecasulfate, have demonstrated anticancer activity presumably through an antiangiogenic mechanism (239). β-cyclodextrin tetradecasulfate has a high binding affinity for FGF. Interestingly, sulfated cyclodextrins also inhibit complement activation (249). There have been reports of some oral bioavailability of β-cyclodextrin derivatives (250).

XI. HEPARIN OLIGOSACCHARIDES

Chemically defined, homogeneous heparin oligosaccharides (Figure 9) have clear advantages over the currently used polydisperse micoheterogeneous preparations (including LMW heparins) but disadvantages over the simpler synthetic analogs described above. It should be possible to prepare heparin oligosaccharides that are extremely selective agents, *i.e.*, one that regulates complement activation but not coagulation or one that targets a specific step in the coagulation cascade. The concentration homogeneous drug entities are easier to monitor *in vivo* and hence, better dose control may be possible reducing heparin's sideeffects (77). Homogeneous heparin oligosaccharides should also make metabolism SAR studies easier to perform resulting in the faster development of second and third generation agents.

The first total synthesis of a heparin-oligosaccharide was accomplished in 1984 resulting in the preparation of a pentasaccharide ATIII binding site (Figure 2) (251). This synthesis both confirmed the reported structure and activity described for the natural product and also permitted the full elucidation of the SAR of ATIII mediated activity (252). The initial synthetic scheme affording ATIII binding pentasaccharide required multiple synthetic and separation steps resulting in a very low overall yield (251). Subsequent synthesis reported by a number of groups have reduced the required steps and increased the yield (253,254,255). It is still unclear whether such a total synthesis approach can result in a cost-competitive pharmaceutical. Chemical synthesis of the ATIII binding pentasaccharide was a major scientific achievement because it resulted in the independent confirmation of structure and it has permitted the synthesis of unnatural derivatives (167).

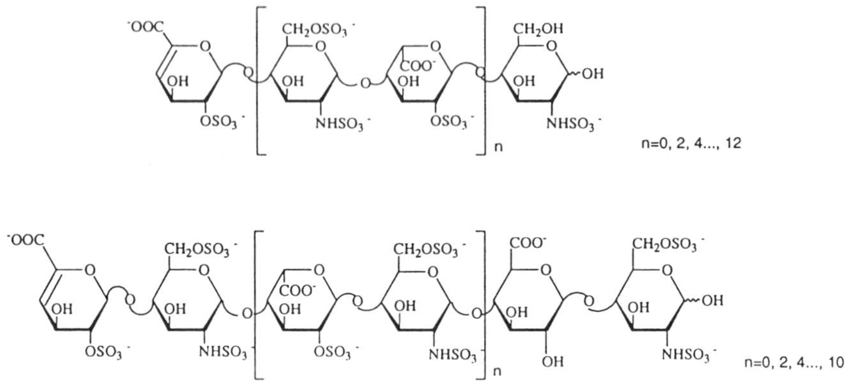

Figure 9 Structure of heparin oligosaccharides formed on partial depolymerization of heparin with heparin lyase I. The unsaturated uronic acid at the nonreducing end is an artifact of enzymatic treatment and n is the degree of polymerization of the internal disaccharide unit.

Heparin oligosaccharides, ranging from disaccharide to tetradecasaccharide, can be prepared from the natural product in a single enzymatic step using a heparin lyase (67,256). This approach has afforded oligosaccharides having important biological activities (37,116,137,170,257). The simplicity of this approach and the low cost of the heparin starting material certainly makes it competitive with classical chemical synthesis. One major drawback of this approach is that often the desired sequence may either be absent from (or present in very small amounts) the starting heparin or it may not be possible to enzymatically recover the desired sequence from the polymer. One approach to these problems is to first use a heparin lyase to prepare a collection of small oligosaccharide building blocks. These building blocks might then be combined either chemically or enzymatically to prepare a wide variety of heparin sequences. Initial studies on the use of heparin lyase to synthesize large oligosaccharides from smaller ones by reverse enzymatic catalysis have been unsuccessful. Efforts are now being directed at chemically ligating two small oligosaccharides to prepare a single larger oligosaccharide (258).

XII. HEPARIN ANALOGUES: POTENTIAL AS NEW THERAPEUTIC AGENTS

Whether structurally defined, homogeneous heparin-analogues will ever be widely used as therapeutic agents depends on two factors: first, improved anticoagulant properties including better bioavailability, pharmacokinetics, dose control and dose monitoring, reduced side-effects, low toxicity, easier reversal and higher specificity; second, the therapeutic exploitation of other activities such as complement regulatory activity, angiogenesis regulatory activity, smooth muscle proliferative regulation, lipoprotein lipase release and activation activities, antiviral, and other activities. Before these activities can be exploited their SAR must be better understood and these activities must be separated. The absence of sideeffects (particularly hemorrhagic complications) as well as high potency (a high therapeutic index) will be essential in the development of these agents. An understanding of how heparin acts and what its natural physiological roles are may lead to unforeseen applications for new heparin analogs. Ultimately, only by understanding heparin's SAR and physiological roles can heparin-analogue with greatly improved properties and potency, the "next generation" of heparin-based drugs, be developed.

XIII. REFERENCES

1. Foster AB, Huggard AJ. The chemistry of heparin. Adv Carbohydr Chem Biochem 1955; 10:335-368.
2. Coyne E. Title. In: Lundblahd RL, Brown WV, Mann KG, Roberts HR, eds. Chemistry and Biology of Heparin. Elsevier North Holland Inc., 1981:9-17.
3. Caen JP. Heparin: a 70 year history from 1914 to 1984. Nouv Rev Fr Hematol 1984; 26:209-210.
4. Roden L, Feingold DS. A vintage year for Jorpes, Crafood and heparin. TIBS 1985; 10:407-409.
5. Linhardt RJ. Heparin: An important drug enters its seventh decade. Chem & Indust 1991; 2:45-50.
6. Roden L. Highlights in the history of heparin. In: Lane DA, Lindahl U (eds). Heparin Chemical and Biological Properties Clinical Applications, Boca Raton: CRC Press, 1989:1-24.
7. McLean J. The thromboplastic action of cephalin. Am J Physiol 1916; 41:250-257.
8. Howell WH. The purification of heparin and its chemical and physiological reactions. Bull Johns Hopkins Hosp 1928; 42:199-206.
9. Jorpes E, Bergstrom S. The aminosugars of heparins. Z Physiol Chem 1936; 244:253-256.
10. Charles AF, Todd AR. Observations on the structure of the barium salt of heparin. Biochem J 1940; 34:112-118.
11. Perlin AS, Mazurek M, Jaques LB, Kavanaugh LW. A proton magnetic resonance spectral study of heparin. L-iduronic acid residues in commercial heparins. Carbohydr Res 1968; 7:369-379.
12. Casu B. Structure and biological activity of heparin. Adv Carbohydr Chem Biochem 1985; 43:51-134.
13. Comper WD. Heparin (and related polysaccharides). Vol. 7. New York:

Gordan and Breach Science Publ., 1981.
14. Horner AA, Young E. Asymmetric distribution of sites with high affinity for antithrombin III in rat skin heparin proteoglycans. J Biol Chem 1982; 257:8749-8754.
15. Coyne E. Heparin manufacturing processes. Fed Proc 1977; 36:32.
16. Linhardt RJ, Ampofo SA, Fareed J, Hoppensteadt D, Mulliken JB, Folkman J. Isolation and characterization of a human heparin. Biochemistry 1992; 31:12441-12445.
17. Jacobsson KG, Lindahl U. Degradation of heparin proteoglycan in cultured mouse mastocytoma cells. Biochem J 1987; 246:409-415.
18. Linhardt RJ, Merchant ZM, Rice KG, Kim YS, Fitzgerald GL, Grant AC, Langer R. Evidence of random structural features in the heparin polymer. Biochemistry 1985; 24:7805-7810.
19. Linhardt RJ, Cohen DM, Rice KG. Non-random structural features in the heparin polymer. Biochemistry 1989; 28:2888-2894.
20. Kjellen L, Pettersson I, Unger E, Lindahl U. Two enzymes in one: N-deacetylation and N-sulfation in heparin biosynthesis are catalyzed by the same protein. Adv Exp Med Biol 1992; 313:107-111.
21. Lindahl U, Feingold DS, Roden L. Biosynthesis of heparin. TIBS 1986; 11:221-225.
22. Lindahl U, Kjellen L. The Biology of the Extracellular Matrix: Proteoglycans. New York: Academic Press, 1987.
23. Linhardt RJ, Liu J, Han X-J. Mapping and sequencing of oligosaccharides by electrophoresis. Trends in Glycosci Glycotechnol 1993; 5:181-192.
24. Liu J. Approach for Sequencing Glycosaminoglycan Chains of Proteoglycans. Ph.D. Thesis, June 1993.
25. Gallagher JT, Turnbull JE, Lyon M. Heparan sulphate proteoglycans: molecular organisation of membrane--associated species and an approach to polysaccharide sequence analysis. Adv Exp Med & Biol 1992; 313:49-57.
26. Bernfield M, Kokeyesi R, Kato M, Hinkes MT, Spring J, Gallo RL, Lose EJ. Biology of Syndecans. Ann Rev Cell Biol 1992; 8:365-393.
27. Stevens RL, Fox CC, Lichtenstein LM, Austen KF. Identification of chondroitin sulfate E proteoglycans and heparin proteoglycans in the secretary granules of human lung mast cells. Proc Natl Acad Sci USA 1988; 85:2284-2287.
28. Gallagher JT, Walker M. Molecular distinctions between heparan sulfate and heparin. Biochem J 1985; 230:665-674.
29. Linhardt RJ. Polysaccharide lyases for glycosaminoglycan analysis. In: Varki A, ed. Current Protocols in Molecular Biology, Analysis of Glycoconjugates. V 2, 17.13.17-17.13.32, Wiley Interscience, 1994.
30. Linhardt RJ, Loganathan D. Heparin, Heparinoids and Heparin Oligosaccharides: Structure and Biological Activities. In: Gebelein G, ed.

Biomimetic Polymers. New York: Plenum Press, 1990:135-175.
31. Jaques LB. Heparin an old drug with a new paradigm. Science 1979; 206:528-533.
32. Caldwell EEO, Nadkarni VD, Fromm JR, Fromm SJ, Linhardt RJ, Weiler JM. Importance of basic amino acids in heparin and heparan sulfate protein binding sites. Eur J Biochem 1995, submitted.
33. Fromm JR, Hileman RE, Caldwell EEO, Weiler, JM, Linhardt RJ. Differences in the interaction of heparin with arginine and lysine and the importance of basic amino acids in the binding of heparin to aFGF. Biochemistry 1995, submitted.
34. Bae J, Desai UR, Pervin A, Caldwell EO, Weiler JM, Linhardt RJ. Interaction of heparin with synthetic antithrombin III peptide analogues. Biochem J 1994; 301:121-129.
35. Folkman J. Regulation of angiogenesis: a new function of heparin. Biochem Pharmacol 1985; 34:905-909.
36. Folkman J, Langer R, Linhardt RJ, Haudenschild C, Taylor S. Angiogenesis inhibition and tumor regression caused by heparin or a heparin fragment in the presence of cortisone. Science 1983; 221:719-725.
37. Nelson RM, Cecconi, Roberts WG, Aruffo A, Linhardt RJ, Bevilacqua MP. Heparin Oligosaccharides bind P- and L-selectin and inhibit acute inflammation. Blood 1993; 82:3253-3258.
38. Fareed J. Development of heparin fractions. Sem Thromb Hemostas 1985; 11:227-236.
39. Green D, Hirsh J, Heit J, Prins M, Davidson B, Lensing AW. Low molecular heparin: a critical analysis of clinical trials. Pharmacol Rev 1994; 46:89-109.
40. Kakkar VV, Djazaeri B, Fok J, Fletcher M, Scully MF, Westwick J. Low-molecular weight heparin and prevention of deep vein thrombosis. British Medical J 1982; 284:375-379.
41. Bratt G, Tornebohm E, Granquist S, Aberg W, Lockner D. A comparison between low molecular weight heparin (KABI2165) and standard heparin in the intravenous treatment of deep venous thrombosis. Thromb Haemostas 1985; 54:813-817.
42. Biller J, Massey EW, Marler JR, Adams HP, Davis JN, Bruno A, Henriksen RA, Linhardt RJ, Goldstein LB, Alberts M, Kisker CT, Toffol GJ, Greenberg CS, Banwart KG, Bertels C, Beck DW, Walker M, Maganani HN. A dose escalation study or org 10172 (Low molecular weight heparinoid) in stroke. Neurology 1989; 39:262-265.
43. Hoffman HP, Schwartz NB, Roden L, Prockop DJ. Location of xylosyltransferase in the cisternae of the rough endoplasmic reticulum of embryonic cartilage cells. Connect Tissue Res 1984; 12:151-163.
44. Lidholt K, Riesenfeld J, Jacobsson K-G, Feingold DS, Lindahl U. Biosynthesis of heparin. Biochem J 1988; 254:571-578.

45. Campbell P, Hannesson HH, Sandback D, Roden L, Lindahl U, Li JP. Biosynthesis of heparin/heparan sulfate. Purification of the D-glucuronyl C-5 epimerase from bovine liver. J Biol Chem 1994; 269(43):26953-26958.
46. Marcum JA, Rosenberg RD. Anticoagulantly active heparan sulfate proteoglycan and the vascular endothelium. Semin Thromb Haemostasis 1987;13:464-474.
47. Pejler G, Backström G, Lindahl U, Paulsson M, Dziadek M, Fujiwara S, Timps R. Structure and affinity for antithrombin of heparan sulfate chains derived from basement membrane proteoglycans. J Biol Chem 1987; 262:5036-5043.
48. Turnbull JE, Gallagher JT. Heparan sulphate: functional role as a modulator of fibroblast growth factor activity. Biochem Soc Trans 1993; 21(2):477-482.
49. Gallagher JT, Turnbull JE. Heparan sulphate in the binding and activation of basic fibroblast growth factor. Glycobiology 1992; 2(6):523-528.
50. Hovingh P, Piepkorn M, Linker A. Biological implications of the structural, antithrombin affinity and anticoagulant activity relationship among vertebrate heparins and heparan sulfates. Biochem J 1986; 237:573-581.
51. Nader HB, Dietrich CP. Natural occurrence, and possible biological role of heparin. In: Lane DA, Lindahl U (eds). Heparin Chemical and Biological Properties Clinical Applications. Boca Raton:CRC Press, 1989:81-96.
52. Linhardt RJ. Chemical and enzymatic methods for the depolymerization and modification of heparin. In: Ogura H, ed. Carbohydrates-Synthetic Methods and Applications in Medicinal chemistry. Tokyo/Weinheim: Kodansha/VCH Publishers, 1992:387-403.
53. Linhardt RJ, Rice KG, Kim YS, Lohse DL, Wang HM, Loganathan D. Mapping and quantification of the major oligosaccharide components of heparin. Biochem J 1988; 254:781-787.
54. Loganathan D, Wang HM, Mallis LM, Linhardt RJ. Structural variation of the antithrombin III binding site and its occurrence in heparin from different sources. Biochemistry 1990; 29:2611-2617.
55. Rice KG, Kim YS, Grant AC, Merchant ZM, Linhardt RJ. High-performance liquid chromatographic separation of heparin derived oligosaccharides. Anal Biochem 1985; 150:325-331.
56. Rice KG, Rottink MK, Linhardt RJ. Fractionation of heparin-derived oligosaccharides by gradient polyacrylamide-gel electrophoresis. Biochem J 1987; 244:515-522.
57. Linhardt RJ, Loganathan D, Al-Hakim A, Wang HM, Walenga JM, Hoppensteadt D, Fareed J. Oligosaccharide mapping of low molecular weight heparins: Structural differences and their relationship to activity. J

Med Chem 1990; 33:1639-1645.

58. Desai UR, Wang HM, Ampofo SA, Linhardt RJ. Oligosaccharide compositional analysis of heparin and low molecular weight heparins by capillary electrophoresis. Anal Biochem 1993; 213:120-127.
59. Pervin A, Al-Hakim A, Linhardt RJ. Separation of glycosaminoglycan derived oligosaccharides by capillary electrophoresis using reverse polarity. Anal Biochem 1994; 221:182-188.
60. Sharath MD, Weiler JM, Merchant ZM, Kim YS, Rice KG, Linhardt RJ. Small heparin fragments regulate the amplification pathway of complement. Immunopharm 1985; 9:73-80.
61. Grant AC, Linhardt RJ, Fitzgerald G, Park JJ, Langer R. Metachromatic activity of heparin and heparin fragments. Anal Biochem 1984; 137:25-32.
62. Nieduszynski I. General physical properties of heparin. In: Lane DA, Lindahl U (eds) Heparin Chemical and Biological Properties Clinical Applications. Boca Raton:CRC Press, 1989:51-63.
63. Wang HM, Loganathan D, Linhardt RJ. Determination of the pKa of glucuronic acid and heparin's carboxyl groups by ^{13}C NMR. Biochem J 1991; 278:689-695.
64. Desai UR, Linhardt RJ. Molecular weight of heparin using ^{13}C nuclear magnetic resonance spectroscopy. J Pharm Sci 1995; 84:212-215.
65. Griffin CC, Linhardt RJ, Van Gorp CL, Toida T, Hileman RE, Schubert RL, Brown SE. Isolation and characterization of heparan sulfate from crude porcine intestinal mucosa peptidoglycan heparin. Carbohydr Res 1995, in press.
66. Cohen DM, Linhardt RJ. Randomness in the Heparin Polymer: Computer simulations of alternative action patterns of heparin lyase. Biopolymers 1990; 30:733-741.
67. Pervin A, Gallo C, Jandik K, Han X-J, Linhardt RJ. Preparation and structural characterization of large heparin-derived oligosaccharides. Glycobiology 1995; 5:83-95.
68. Linhardt RJ, Rice KG, Kim YS, Engelken J, Weiler J. Homogeneous, structurally defined heparin-oligosaccharides with low anticoagulant activity inhibit the generation of the amplification pathway C3 convertase *in vitro*. J Biol Chem 1988; 263:13090-13096.
69. Mallis LM, Wang HM, Loganathan D, Linhardt RJ. Sequence analysis of highly sulfated heparin-derived oligosaccharides using fast atom bombardment mass spectroscopy. Anal Chem 1989; 61:1453-1458.
70. Rice KG. A Sequencing Strategy for Heparin. Ph.D. Thesis, University of Iowa, Iowa City, USA, 1987.
71. Lamb DJ, Wang HM, Mallis LM, Linhardt RJ. Analysis of glycosaminoglycan-derived oligosaccharides using fast atom bombardment mass spectrometry. Carbohydr Res 1992; 225:137-145.
72. Lamb DJ, Wang HM, Mallis LM, Linhardt RJ. Negative-ion fast atom

bombardment tandem mass spectrometry to determine sulfate and linkage position in glycosaminoglycan-derived disaccharides. J Amer Soc Mass Spectr 1992; 3:797-803.

73. Sugahara K, Tohno-oka R, Yamada S, Khoo K-H, Morris HP, Dell A. Structural studies on the oligosaccharides isolated from bovine kidney heparan sulphate and characterization of bacterial heparitinases used as substrates. Glycobiology 1994; 4:535-544.
74. Gallagher JT, Turnbull JE, Lyon M. Patterns of sulphation in heparan sulphate: polymorphism based on a common structural theme. Int J Biochem 1992; 24(4):553-560.
75. Maxim AM, Gilbert W. A new method for sequencing DNA. Proc Natl Acad Sci USA 1977; 74:560-564.
76. Langer R, Linhardt RJ, Cooney CL, Klein M, Tapper D, Hoffberg SM, Larsen A. An enzymatic system for removing heparin in extracorporeal therapy. Science 1982; 217:261-263.
77. Freedman MD. Pharmacodynamics, Clinical Indications and Adverse Effects of Heparin. J Clin Pharmacol 1992; 32:584-596.
78. Weiler JM, Freiman P, Sharath MD, Metzger WJ, Smith JM, Richerson HB, Ballas ZK, Halverson PC, Shulan DJ, Matsuo S, Wilson RL. Serious adverse reactions to protamine sulfate: are alternatives needed? J Allergy Clin Immunol 1985; 75:297-303.
79. Sharath MD, Metzger WJ, Richerson HB, Scupham RK, Meng RL, Ginsberg BH, Weiler JM. Protamine-induced fatal anaphylaxis. J Thorac Cardiovasc Surg 1985; 90:86-90.
80. Bouma BN, Griffin JH. Initiation mechanisms: the contact activation on system in plasma. New Compre Biochem 1986; 13:103-128.
81. Davie EW, Fujikawa K, Kurachi K, Kisiel W. The role of serine proteases in the blood coagulation cascade. Adv Enzymol Related Areas Mol Biol 1979; 48:277-318.
82. Davie EW, Fujikawa K. Basic mechanisms in blood coagulation. Ann Rev Biochem 1975; 44:799-829.
83. Jackson CM, Nemerson Y. Blood coagulation. Ann Rev Biochem 1980; 49:765-811.
84. Tans G, Rosing J. Multicomponents enzyme complexes of blood coagulation. New Compre Biochem 1986; 13:59-85.
85. Guillin M, Bezeaud A. The blood coagulation process and its regulation. Path Biol 1985; 33:917-932.
86. Kim YS, Lee KB, Linhardt RJ. Microheterogeneity of plasma glycoproteins antithrombin III and heparin cofactor II and their carbohydrate analysis. Thromb Res 1988; 51:97-104.
87. Kurachi K. Schmer G, Hermodson MA, Teller DC, Davie EW. Characterization of human, bovine, and horse antithrombin III. Biochem 1976; 15:368-372.
88. Longas MO, Finlay TH. The covalent nature of the human antithrombin

III-thrombin bond. Biochem J 1980; 189:481-489.
89. Rosenberg RD, Damus PS. The purification and mechanism of action of human antithrombin-heparin cofactor. J Biol Chem 1973; 248:6490-6505.
90. Tollefsen DM, Blank MK. Heparin cofactor II. J Biol Chem 1982; 257:2162-2169.
91. Griffith MJ, Noyes CM, Church FC. Reactive site peptide structural similarity between heparin cofactor II and antithrombin III. J Biol Chem 1985; 260:2218-2225.
92. Tran TH, Zbinden B, Lammle B, Duckert F. Methodology and clinical significance of heparin cofactor II. Semin Thromb Haemost 1985; 11:342-346.
93. Tollefsen DM. Activation of heparin cofactor II by heparin and dermatan sulfate. Nouv, Rev Fr Hematol 1984; 26:233-237.
94. Linhardt RJ, Hileman RE. Dermatan sulfate as a potential therapeutic agent. Gen Pharmacol 1995; 26:443-451.
95. Buonassisi V. Sulfated mucopolysaccharide synthesis and secretion in endothelial cell cultures. Exp Cell Res 1973; 76:363-368.
96. Harker LA. New antithrombotic strategies for resistant thrombotic processes. J Clin Pharmacol 1994; 34:3-16.
97. Gerlach E, Becker BF. Interaction of blood and the vascular wall: Hemostatic aspects. Zeitschrift fur Kardiologie 1993; 5:13-21.
98. Verstraete M. Recent developments in fibrinolysis. Klin Wochenschr 1988; 66:5-14.
99. Verstraete M, Bory M, Collen D, Erbel R, Lennane RJ, Mathey D, Michels HR, Schartl M, Uebis R, Bernard R, et al. Comparative randomized study of the effectiveness of intravenous recombinant tissue-type plasminogen activator and intravenous streptokinase in patients with acute myocardial infarct. Klin Wochenschr 1988; 66:77-85.
100. Crabbe SJ, Cloninger CC. Tissue plasminogen activator: A new thrombolytic agent. Clin Pharm 1987; 6:373-386.
101. Stassen JM, Juhan-Vague I, Alessi MC, DeCock F, Collen D. Potentiation by heparin fragments of thrombolysis induced with human tissue-type plasminogen activator or human single-chain urokinase-type plasminogen activator. Thromb Haemostasis 1987; 58:947-950.
102. Engelberg H. Heparin and the atherosclerotic process. Pharmacol Rev 1984; 36:91-110
103. Merchant ZM, Erbe EE, Eddy WP, Patel RJ, Linhardt RJ. Effect of very low molecular weight heparin-derived oligosaccharides on lipoprotein lipase release in rabbits. Atherosclerosis 1986; 62:151-158.
104. Santamarina-Fiojo S, Brewer HB Jr. Lipoprotein lipase: structure, function and mechanism of action. Int J Clin Lab Res 1994; 24:143-147.
105. Sivaram P, Choi SY, Curtis LK, Goldberg IJ. An amino-terminal

fragment of apolipoprotein B binds to lipoprotein lipase and may facilitate its binding to endothelial cells. J Biol Chem 1994; 269:9409-9412.

106. Sue TK, Jaques LB, Yuen E. Effects of acidity, cations and alcoholic fractionation on absorption of heparin from gastrointestinal tract. Can J Physiol Pharmacol 1976; 54:613-617.

107. Castellot JJ Jr., Beeler DL, Rosenberg RD, Karnovsky MJ. Structural determinants of the capacity of heparin to inhibit the proliferation of vascular smooth muscle cells. J Cell Physiol 1984; 120:315-320.

108. Castellot JJ Jr., Choay J, Lormeau J-C, Petitou M, Sache E, Karnovsky MJ. Structural determinants of the capacity of heparin to inhibit the proliferation of vascular smooth muscle cells II. J Cell Biol 1986; 102:1979-1984.

109. Wright TC, Castellot JJ, Diamond JR, Karnovsky MJ. Regulation of cellular proliferation by heparin and heparan sulfate. In: Lane DA, Lindahl U, eds. Chemical and Biological Properties Clinical Applications. Boca Raton:CRC Press, 1989:295-316.

110. Sternbergh WC, Sobel M, Mackoul RG. Heparinoids with low anticoagulant potency attenuate postischemic endothelial cell dysfunction. J Vascular Surg 1995; 21:477-483.

111. Parthasarathy N, Goldberg IJ, Sivaram P, Mulloy B, Flory DM, Wagner WD. Oligosaccharide sequences of endothelial cell surface heparan sulfate proteoglycan with affinity for lipoprotein lipase. J Biol Chem 1994; 269:22391-22396.

112. Edens RE, Weiler JM, Linhardt RJ. Heparin, not just an anticoagulant anymore: Six and one half decades of studies on the ability of heparin to regulate complement activation. Compl Prof 1993; 1:96-120.

113. Sim RB, Malhotra R. Interactions of carbohydrates and lectins with complement. Biochem Soc Trans 1994; 22:106-111.

114. Loos M, Volanakis JE, Stroud RM. Mode of interaction of different polyanions with the first (C1, C1), the second (C2) and the fourth (C4) component of complement. Immunochem 1976; 13:257-261.

115. Caughman GB, Boackle RJ, Vesely J. A postulated mechanism of heparin's potentiation of C1 inhibitor function. Mol Immunol 1982; 19:287-295.

116. Linhardt RJ, Weiler JM. Oligosaccharide heparin fragments as inhibitors of complement cascade. US Patent 4,916,219, 1990.

117. Jordan RE, Kilpatrick J, Nelson RM. Heparin promotes the inactivation of antithrombin by neutrophil elastase. Science 1987; 237:777-779.

118. Redini F, Tixier J-M, Petitou M, Choay J, Robert L, Hornerbeck W. Inhibition of leukocyte elastase by heparin and its derivatives. Biochem J 1988; 252:515-519.

119. Faller B, Cadene M, Bieth JG. Demonstration of a two-step reaction mechanism for the inhibition of heparin-bound neutrophil elastase by

alpha 1-proteinase inhibitor. Biochemistry 1993; 32:9230-9235.
120. Bevilacqua MP, Nelson RM, Venot A, Linhardt RJ, Stamenkovic, I. Carbohydrate-protein interactions in vascular biology. Ann Rev Cell Biol 1995; in press.
121. McEver RP. Selectins. Curr Opin Immunol 1994; 6:75-84.
122. Norgard-Sumnicht KE, Varki NM, Varki A. Calcium dependent heparin-like ligands for L-selectin in non-lymphoid endothelial cells. Science 1993; 261:480-483.
123. Witt DP, Lander AD. Differential binding of chemokines to glycosaminoglycan subpopulations. Curr Biol 1994; 4:394-400.
124. Long WF, Williamson FB. Heparan structure and the modulation of angiogenesis. Med Hypoth 1984; 13:385-394.
125. Denekamp J. In: Hammerson F, Hudlicka O, eds. Progress in Applied Microcirculation, Karger, Basel, 1984.
126. Knighton DR, Hunt TK, Scheuenstuhl H, Halliday BJ, Werb Z, Banda MJ. Oxygen tension regulates the expression of angiogenesis factor by macrophages. Science 1983; 221:1283-1285.
127. Beck DW, Olson JJ, Linhardt RJ. Effect of heparin, heparin fragments, and cortisone on cerebral endothelial cell growth *in vitro* and *in vivo*. J Neuropath Exper Neurol 1986; 45:503-512.
128. Folkman J. Angiogenesis and its inhibitors. Important Adv Oncol 1985; 42:42-62.
129. Folkman J, Ingber DE. Angiogenesis: regulatory role of heparin and related molecules. In: Lane DA, Lindahl U, eds. Chemical and Biological Properties Clinical Applications. Boca Raton:CRC Press, 1989:317-333.
130. Mason JJ. The Ins and outs of fibroblast growth factors. Cell 1994; 78:547-552.
131. Shiokawa K. Asano M, Shiozaki C. Function, molecular structure and gene expression of fibroblast growth factor (FGF/HBGF), Jap J Clin Med 1992; 50:1893-1901.
132. McKeehan WL, Kan M. Heparin sulfate fibroblast growth factor receptor complex: Structure function relationships. Molec Reprod and Develop 1994; 39:69-81.
133. Reich-Slotky R, Bonneh-Barkay D, Shaoul E, Bluma B, Srahn CM, Ron D. Differential effect of cell-associated heparan sulfates on the binding of keratinocyte growth factor (KGF) and acidic fibroblast growth factor to the KGF receptor. J Biol Chem 1994; 269:32279-32285.
134. Crum R, Szabo S, Folkman J. A new class of steroids inhibits angiogenesis in the presence of heparin or a heparin fragment. Science 1985; 230:1375-1378.
135. Sasisekharan R, Moses MA, Nugent MA, Cooney CL, Langer R. Heparinase inhibits neovascularaization. Proc Nat Acad Sci (USA) 1994; 91:1524-1528.

136. Grushka E, Cohen AS. The binding of Cu(II) and Zu(II) ions by heparin. Analyt Lett 1982; 15:1277-1288.
137. Mach H, Volkin DB, Burke CJ, Middaugh CR, Linhardt RJ, Fromm J, Loganathan D, Mattsson L. Nature of the interaction of heparin with acidic fibroblast growth factor. Biochemistry 1993; 32:5480-5489.
138. Walker A, Turnbull JE, Gallagher JT. Specific heparan sulfate saccharides mediate the activity of basic fibroblast growth factor. J Biol Chem 1994; 269(2):931-935.
139. Ishihara M. Structural requirements in heparin for binding and activation of FGF-1 and FGF-4 are different from that for FGF-2. Glycobiology 1994; 4:817-824.
140. Iozzo RV. Proteoglycans and neoplasia. Cancer Metastasis Rev 1988; 7:39-50.
141. Matuo Y, McKeehan WL, Yan GC, Nikolaropoulos S, Adams PS, Fukabori Y, Yamanaka H, Gaudreau J. Potential role of HBGF (FGF) and TGF-beta on prostate growth. Adv Exp Med & Biol 1992; 324:107-114.
142. Prigent SA, Lemoine NR. The type 1 (EGFR-related) family of growth factor receptors and their ligands. Prog Growth Factor Res 1992; 4:1-24.
143. Herold BC, Visallai RJ, Susmarski N, Brandt CR, Spear PG. Glycoprotein C-independent binding of herpes simplex virus to cells requires cell surface heparan sulfate and glycoprotein B. J Gen Virol 1994; 75:1211-1222.
144. Herold BC, Gerber SI, Polonsky T, Belval BJ, Shaklee PN, Home K. Identification of structural features of heparin required for inhibition of herpes simplex virus type 1 binding. Virology 1995, in press.
145. Rider CC, Coombe CR, Harrop HA, Hounsell EF, Bauer C, Feeny J, Mulloy B, Mahmood N, Hay A, Parish CR. Anti-HIV-1 activity of chemically modified heparins: correlation between binding to the V3 loop of gp120 and inhibition of cellular HIV-1 interaction in vivo. Biochemistry 1994; 33:6974-6980.
146. Newburg DS, Linhardt RJ, Ampofo SA, Yolken RH. Human milk glycosaminoglycan inhibits binding to the HIV glycoprotein, GP120, to its host cell receptor. J Nutrition 1995; 125:419-424.
147. Buee L, Ding W, Delacourte A, Fillit H. Binding of secreted human neuroblastoma proteoglycans to the Alzheimer's amyloid A4 peptide. Brain Research 1993; 601:154-163.
148. Brunden KR, Richter-Cook NJ, Chaturvedi N, Frederickson RC. pH-dependent binding of synthetic beta-amyloid peptides to glycosaminoglycans. J Neurochem 1993; 61:2147-2154.
149. Scott RW, Bergmann BL, Bajpai A, Hersh RT, Rodriguez H, Jones BN, Barreda C, Watts S, Baker JB. Protease nexin. J Biol Chem 1985; 260:7029-7034.
150. Perry G, Richey P, Siedlak SL, Galloway P, Kawai M, Cras P. Basic

fibroblast growth factor binds to filamentous inclusions of neurodegenerative disease. Brain Research 1992; 579:350-352.

151. Kisilevsky R, Lemieux LJ, Fraser PE, Kong X, Hultin PG, Szarek WA. Arresting amyloidosis in vivo using small-molecule anionic sulphonates or sulfates implications for Alzheimer's disease. Nature Medicine 1995; 1:143-148.

152. Ishihara M, Fedarko NS, Conrad HE. Transport of heparan sulfate into the nuclei of hepatocytes. J Biol Chem 1986; 261:13575-13580.

153. Zvibel I, Halay E, Reid LM. Heparin and hormonal regulation of mRNA synthesis and abundance of autocrine growth factors: relevance to clonal growth of tumors. Molec and Cell Biol 1991; 11:108-116.

154. Ishii K, Futaki S, Uchiyama H, Nagasawa K, Andoh T. Mechanism of inhibition of mammalian DNA topoisomerase I by heparin. Biochem J 1987; 241:111-119.

155. Ackerman P, Gover CV, Osheroff N. Phosphorylation of DNA topoisomerase II in vivo and in total homogenates of Drosophila Kc cells. J Biol Chem 1988; 263:12653-12660.

156. Kohneke-Godt B, Gabius H-J. Heparin-binding lectin from human placenta: Further characterization of ligand binding and structural properties and its relationship to histones and heparin-binding growth factors. Biochemistry 1991; 30:55-65.

157. Jackson RL, Busch SJ, Cardin AD. Glycosaminoglycans: molecular properties, protein interactions, and role in physiological processes. Physiol Rev 1991; 71:481-537.

158. Cardin AD, Weintraub HJR. Molecular modeling of protein-glycosaminoglycan interactions. Arteriosclerosis 1989; 9:21-32.

159. Griffith MJ. Inhibitors: antithrombin III and heparin. New Compr Biochem 1986; 13:259-283.

160. Thompson LD, Pantoliano MW, Springer BA. Energetic characterization of basic fibroblast growth factor-heparin interaction: identification of the heparin binding domain. Biochemistry 1994; 33:3831-3840.

161. Mikhailov D, Mayo KH, Pervin A, Linhardt RJ. ^{13}C-NMR relaxation study of heparin-disaccharide interactions with tripeptides GRG and GKG. Biochemistry 1995, submitted.

162. Edens RE, Fromm JR, Fromm SJ, Linhardt RJ, Weiler JM. Two-dimensional affinity resolution electrophoresis demonstrates that three distinct heparin populations interact with antithrombin III. Biochemistry 1995; 34:2400-2407.

163. Atha DH, Stephens AW, Rimon A, Rosenberg RD. Sequence variation in heparin octasaccharides with high affinity for antithrombin III. Biochem 1984; 23:5801-5812

164. Lindahl U, Thurnberg L, Backström G, Reisenfeld J, Nordling K, Bjork I. Extension and structural variability of the antithrombin-binding sequence of heparin. J Biol Chem 1984; 259:12368-12376.

165. van Boeckel CAA, van Aelst SF, Beetz T, Lucas H. XIVth Intl Carbohydr Sympos, Stockholm, Sweden, August, 1988, B-88.
166. van Boeckel CAA, Basten JEM, Lucas H, Van Aelst SF, Intl Carbohydr Sympos, Stockholm, Sweden, August, 1988. B-89.
167. van Boeckel CAA, Beetz T, van Aelst SF. Synthesis of a potent antithrombin activating pentasaccharide: a new heparin-like fragment containing two 3-*O*-sulfated glucosamines. Tetrahedron Lett 1988; 29:803-806.
168. Varki A. Selectin ligands. Proc Natl Acad Sci USA, 1994; 91:7390-7397.
169. Breddin HK, Fareed J, Bender N, eds. Low-Molecular-Weight Heparins, 4th Congress on Thrombosis and Haemostasis Research. In: Haemostasis, Vol 18. 1988:1-87.
170. Linhardt RJ, Merchant ZM, Weiler JM, Sharath MD. Low molecular weight heparin fragments as inhibitors of complement activation. US Patent 4,847,338, 1989.
171. Schran HF, Bitz DW, DiSerio FJ, Hirsh J. The pharmacokinetics and bioavailability of subcutaneously administered dihydroergotamine, heparin and the dihydroergotamine-heparin combination. Thromb Res 1983; 31:51-67.
172. Lasker SE, Chium ML. Orally active anticoagulant. U.S. Patent No. 3,766,167, 1973.
173. Litwins J, Vorzimer JJ, Sussman LN, Applezweig N, Etess D. Sublingual administration of heparin. Proc Soc Exp Biol Med 1951; 77:325-326.
174. Loomis TA. Absorption of heparin from the intestine. Proc Soc Exp Biol Med 1959; 101:447-449.
175. Windsor E, Cronheim GE. Gastrointestinal absorption of heparin and synthetic heparinoids. Nature 1961; 190:263-264.
176. McDevitt E, Heubner RD, Wright IS. Ineffectiveness of heparin by sublingual administration. J Am Med Assoc 1952; 148:1123-1124.
177. Windsor E, Freeman L. An investigation of routes of administration of heparin other than injection. Am J Med 1964; 37:408-416.
178. Wright IS. An evaluation of anticoagulant therapy. Am J Med 1953; 14:720-730.
179. Larsen AK, Hetelekidis S, Langer R. Disposition and anticoagulant activity of biologically active heparin fragments in the rat. J Pharma Experimental Ther 1984; 231:373-380.
180. Jaques LB, Hiebert LM, Wice SM. Evidence from endothelium of gastric absorption of heparin and of dextran sulfates 8000. J Lab Clin Med 1991; 117:122-130.
181. Vasev S, Prabharakan V, Sampson CA. Heparin lowers blood pressure and vascular calcium uptake in hypertensive rats. Scand J Clin Lab Invest 1991; 51:321-327.

182. Vasev S, Sampson CA, Longerich L, Parai S. Oral heparin prevents hypertension and elevated cytosolic calcium in salt-sensitive rats. Artery 1992; 19:225-245.
183. Lasker SE. Low molecular weight heparin-like preparations with oral activity. Sem Thromb Hemostas 1985; 11:37-39.
184. Lasker SE. Low molecular weight derivative of heparin that is orally active in mice. Adv Exp Med Biol 1975; 59:119-127.
185. Lasker SE, Lee BY, Madden RE. Low molecular weight heparins: oral absorption in monkeys. Thrombos Haemostas 1987; 58:555.
186. Doutremepuich CH, Toulemonde F, Muary MO, Kuttler MC, Masse A, Oca C, Quilichini R. Oral administration of low molecular weight heparin fractions in rabbits. Thromb Hemostas 1983; 50:74.
187. Doutremepuich C, Toulemonde F. Oral administration of low molecular weight heparin fractions. Sem in Thromb Hemostas 1985; 11:247
188. Santiago N, Rivera T, Mayer E, Milstein S. Proteinoid microspheres for the oral delivery of heparin. Proc Intern Symp Control Rel Bioact Mater 1991; 19:514-515.
189. Reber K, Studer A. Forderung der gastrointestinalen resorption von heparin durch calcium bindungsmittle. Experimentia 1963; 19:141-142.
190. Tidball CS, Lipman RI. Enhancement of jejunal absorption of heparinoid by sodium ethylene diaminetetra acetate in the dog. Proc Soc Exp Biol Med 1962; 111:713-715.
191. Engel RH, Riggi SJ. Intestinal absorption of heparin facilitated by sulfated or sulfonated surfactant. J Pharm Sci 1969; 58:706-710.
192. Guarini S, Ferrari W. Olive oil provoked bile-dependent absorption of heparin from gastrointestinal-tract in rats. Pharmacol R 1985; 17:685-697.
193. Guarini S, Ferrari W. Sodium deoxycholate promotes the absorption of heparin administered orally, probably by acting on gastrointestinal mucosa in rats. Experientia 1985; 41:350-352.
194. Kim TD, Sakon M, Kawasaki T, Kambayashi J, Ohshiro T, Mori T. Studies on liposome-encapsulated heparin. Thromb Res 1986; 43:603-612.
195. Artmann C, Roding J, Ghyczy M, Pratzel HG. Liposomes from soya phospholipids as percutaneous drug carriers. Arzneim - Forsch/Drug Res 1990; 40:1365-1369.
196. Ueno M, Nakasaki T, Horikoshi I, Sakuragawa N. Oral administration of liposomally-entrapped heparin in beagle dogs. Chem Pharm Bull 1982; 30:2245-2247.
197. Horikoshi I, Ueno M, Adachi A, Nakasaki T. Oral administration of liposomally-entrapped heparin in beagle dogs - effect of electrostatically charged liposomes on the gastrointestinal absorption. Thromb Haemostas 1983; 50:66
198. Koh TY. Intestinal absorption of heparin. Can J Biochem 1969;

47:951-954.
199. Koh TY, Bharucha KR. Intestinal absorption of stable heparinic acid complexes. J Lab Clin Med 1972; 80:47-55.
200. Zoppetti G. Heparin by alternative administration routes. In Heparin in the 90's: Present and Future, 105-107, Florence, 1990.
201. Dal Pozzo A, Acquasaliente M, Geron MR, Andriuoli G. New heparin complexes active by intestinal absorption: I-multiple ion pairs with basic organic compounds. Thromb Res 1989; 56:119-124.
202. Liu J, Pervin A, Gallo CM, Desai UR, VanGorp CL, Linhardt RJ. New approaches for the preparation of hydrophobic heparin derivatives. J Pharm Sci 1994; 83:1034-1039.
203. Klein M, Drongowski RA, Linhardt RJ, Langer R. A colorimetric assay for chemical heparin in plasma. Anal Biochem 1982; 124:59-64.
204. Laurent T, Tengbland A, Thunberg L, Höök M, Lindahl U. The molecular-weight-dependence of the anti-coagulant activity of heparin. Biochem J 1978; 175:691-701.
205. Khorramian BA, Stivala SS. Small-angle X-ray scattering of high- and low-affinity heparin. Arch Biochem Biophys 1986; 247:384-392.
206. Kim YS, Linhardt RJ. Structural features of heparin and their effect on heparin cofactor II mediated inhibition of thrombin. Thromb Res 1989; 53:55-71.
207. Höök M, Bjork I, Hopwood J, Lindahl U. Anticoagulant activity of heparin: separation of high-activity and low-activity heparin species by affinity chromatography on immobilized antithrombin. FEBS Lett 1976; 66:90-93.
208. Kim YS. Interaction of Heparin with Antithrombin III and Heparin Cofactor II. Ph.D. Thesis, University of Iowa, 1988.
209. Szondy E, Beaumont JL. Purification of a monoclonal IgA with anti-heparin activity. Immunobiol 1980; 157:407-413.
210. Shivley JE, Conrad HE. Formation of a hydrosugars in the chemical depolymerization of heparin. Biochemistry 1976; 15:3932-3942.
211. Mardiguian J. Sulfated oligosaccharides as antithrombotics and hypolipemics. US Patent 4,440,926, April 3, 1984.
212. Larsen AK, Linhardt RJ, Rice KG, Wogan G, Langer R. Resistance of heparinase-derived heparin fragments to biotransformation. J Biol Chem 1989; 264:1570-1577.
213. Linhardt RJ, Galliher PM, Cooney CL. Polysaccharide lyases. Appl Biochem Biotech 1986; 12:135-177.
214. Linhardt RJ, Rice KG, Merchant ZM, Kim YS, Lohse, DL. Structure and activity of a unique heparin-derived hexasaccharide. J Biol Chem 1986; 261:14448-14454.
215. Langer RS, Linhardt RJ, Cooney CL, Fitzgerald G, Grant A. Heparinase derived anticoagulants. US Patent 4,396,762, 1983.
216. Linhardt RJ, Wang HM, Loganathan D, Bae JH. Search for the heparin

antithrombin III binding site precursor. J Biol Chem 1992; 267:2380-2387.

217. Linker A, Meyer K, Hoffman P. The production of unsaturated uronides by bacterial hyaluronidase. J Biol Chem. 1955; 219:13-25.
218. Edens RE, Al-Hakim A, Weiler JM, Rethwisch DG, Fareed J, Linhardt RJ. Gradient polyacrylamide gel electrophoresis for determination of the molecular weight of heparin and low molecular weight heparin derivatives. J Pharm Sci 1992; 81:823-827.
219. Desai UR, Linhardt RJ. Molecular weight of low molecular heparins by ^{13}C NMR spectroscopy. Carbohydr Res 1994; 255:193-212.
220. Bienkowski MJ, Conrad HE. Structural characterization of the oligosaccharides formed by depolymerization of heparin with nitrous acid. J Biol Chem 1985; 260:356-365.
221. Fussi F. Low-molecular-weight heparins endowed with elevate pharmacological properties. US Patent 4,281,108, July 28, 1981.
222. Linhardt RJ, Desai UR, Liu J, Pervin A, Hoppensteadt D, Fareed J. Low molecular weight dermatan sulfate as an antithrombotic agent: Structure activity relationship studies. Biochem Pharmacol 1994; 47:1241-1252.
223. Linhardt RJ, Al-Hakim A, Liu J, Hoppensteadt D, Fareed J, Mascellani G, Bianchini P. Structural features of dermatan sulfates and their relationship to anticoagulant and antithrombotic activities. Biochem Pharm 1991; 42:1609-1619.
224. Gordon DL, Linhardt R, Adams HP. Low molecular weight heparin and heparinoids in cerebrovascular disease. Clin Neuropharm 1990; 13:522-543.
225. Kennedy JF, White CA. Bioactive Carbohydrates. New York: Ellis Horwood Limited, 1983.
226. Nishimura S-I, Nishi N, Tokura S. Inhibition of the hydrolytic activity of thrombin by chitin heparinoids. Carbohydr Res 1986; 156:286-292.
227. Aspinall GO. Structural chemistry of the hemicellulose. Adv Carbohydr Chem Biochem. 1959; 14:429-468
228. Scully MF, Kumudini M, Weerasinghe KM, Ellis V, Djazaeri B, Kakkar VV. Anticoagulant and antiheparin activities of a pentosan polysulfate. Thrombosis Res 1983; 31:87-97.
229. Dol F, Sie P, Dupouy D, Boneu B. Effect of pentosan polysulfate administration in man on the formation of covalent complexes between heparin cofactor II and thrombin generated in plasma by contact activation. Thromb Haemostas 1986; 56:295-301.
230. Scully MF, Kakkar VV. The antiheparin effect of a heparinoid, pentosan polysulfate. Biochem J 1984; 218:657-665.
231. Rickets CR, Walton KWWH. Anticoagulants. US Patent No 2715091, 1955.
232. Osol A, ed. In: 16th Remington's Pharmaceutical Sciences. Easton, PA: Mack Publishing Co., 1980:760-761.

233. Windholz M., ed. The Merck Index, 9th Ed., Rahway: Merck Publishing Co., p. 2911, 1976.
234. Oshima G. Trigger action of dextran sulfate on inactivation of thrombin. Thromb Res 1988; 49:353-361.
235. Langer R, Linhardt RJ, Klein M, Galliher PM, Cooney CL, Flanagan MM. A system for heparin removal. In: Cooper S, Hoffman, A, Pepes N, Ratner B (eds), Biomaterials: Interfacial Phenomenon and Applications, ACS Press, Washington D.C., 1982, pp 493-509.
236. Personal communication from Dr. William Gleason, University of Minnesota, St. Paul, MN.
237. Hawking F. Suramin: with special reference to onchocerciasis. Adv Pharmacol Chemother 1978; 15:289-322.
238. Cole DJ, Ettinghausen SE, Pass HI, Danforth DN, Linhan MW, Myers CW, Cooper MR, Sindelar WF. Postoperative complications in patients receiving suramin therapy. Surgery 1994; 116:90-95
239. Mitchell MA, Wilks JW. Inhibitors of angiogenesis. Eds. Ann Reports in Medicinal Chemistry V 27, Academic Press Inc., San Diego, 1992.
240. De Clercq E. Suramin: a potent inhibitor of the reverse transcriptase of RNA tumor viruses. Cancer Letters 1979; 8:9-22.
241. Mitsuya H, Popovic M, Yarchoan R, Matsushita S, Gallo RC, Broder S. Suramin protection of T cells in vitro against infectivity and cytopathic effect of HTLV-III. Science 1984; 226:172-174.
242. Tan GT, Wickramasinghe A, Verma S, Singh R, Hughes SH, Pezzuto JM, Baba M, Mohan P. Potential anti-AIDS naphthalenesulfonic acid derivatives. J Med Chem 1992; 35:4846-4853.
243. Folkman J, Shing Y. Control of angiogenesis by heparin and other sulfated polysaccharides. In: Lane DA, Björk I, Lindahl U, eds. Heparin and Related Polysaccharides. New York: Plenum Press, 1992:355-364.
244. Zhu X, Hsu BT, Rees DC. Structural studies of the binding of the anti-ulcer drug sucrose octasulfate to acidic fibroblast growth factor. Structure 1993; 1:27-34.
245. Desai UR, Vlahov IR, Pervin A, Linhardt RJ. Conformational analysis of sucrose octasulfate by high resolution nuclear magnetic resonance spectroscopy. Carbohydr Res 1994, in press.
246. Walenga J, Fareed J, Schumacher H. The 30th Annual Meeting of the American Society of Hemotology, August 1988.
247. Personal communication from Dr. Jawed Fareed, Loyola University, Maywood, IL.
248. Fügedi P, Petõ C, Ishihara M, Musser J. Synthesis of sulfated derivatives of tethered oligosaccharides and their bFGF-binding activity. XVIIth International Carbohydrate Symposium, Ottawa Canada, 1994.
249. Lewis AJ, Bernstein S. C6-Modified cyclodextrin sulfate salts as compliment inhibitors. US Patent 4,258,180, 1981 and Modified cyclodextrin sulfate salts as complement inhibitors. US Patent

4,247,535, 1981.
250. Gerloczy A, Hoshino T, Pitha J. Safety of oral cyclodextrins: effects of hydroxy-propyl cyclodextrins, cyclodextrin sulfates and cationic cyclodextrins on steroid balance in rats. J Pharm Sci 1994; 83:193-196.
251. Sinay P, Jacquinet J-C, Petitou M, Duchaussoy P, Lederman I, Choay J, Torri G. Total synthesis of a heparin pentasaccharide fragment having high affinity for antithrombin III. Carbohydr Res 1984; 132:c5-c9.
252. Petitou M, Duchaussoy P, Lederman I, Choay J, Sinay P. Binding of heparin to antithrombin III: a chemical proof of the critical role played by a 3-sulfated 2-amino-2-deoxy-D-glucose residue. Carbohydr Res 1988; 179:163-172.
253. Petitou M, Duchaussoy P, Lederman I, Choay J, Jacquinet J-C, Sinay P, Torri G. Synthesis of heparin fragments: a methyl alpha-pentaoside with high affinity for antithrombin III. Carbohydr Res 1987; 167:67-75.
254. van Boeckel CAA, Beetz T, Vos JN, De Jong AJM, Van Aelst SF, Van den Bosch RH, Mertens JMR, Van der Vlught FA. Synthesis of a pentasaccharide corresponding to the antithrombin III binding fragment of heparin. J Carbohydr Chem 1985:4:293-321.
255. Ichikawa Y, Monden R, Kuzuhara H. Synthesis of methyl glucoside derivatives of tri- and penta-saccharides related to the antithrombin III-binding sequence of heparin, employing allobiose as a key starting-material. Carbohydr Res 1988; 172:37-64.
256. Rice KG, Linhardt RJ. Study of defined oligosaccharides substrates of heparin and heparan monosulfate lyases. Carbohydr Res 1989; 190:219-233.
257. Vokin DB, Tsai PK, Dabora JM, Gress JO, Burke CJ, Linhardt RJ, Middaugh CR. Physical stabilization of acidic fibroblast growth factor by polyanions. Arch Biochem Biophys 1993; 300:30-41.
258. Kerns RJ, Vlahov IR, Linhardt RJ. Capillary electrophoresis for monitoring chemical reactions: Sulfation and synthetic manipulation of sulfated carbohydrates. Carbohydr Res 1995; 267:143-152.
259. Baenziger NL, Mack P, Jong YJ, Dalemar LR, Perez N, Lindberg C, Wilhelm B, Haddock RC. An environmental regulated receptor for diamine oxidase modulates human endothelial cell/fibroblast histamine degradative uptake. J Biol Chem 1994; 269:14892-14898.
260. Persson E. Lipoprotein lipase, hepatic lipase and plasma lypolytic activity. Acta Med Scand Suppl. 1988; 724:1-56
261. Posner I, Desanctis J. The kinetics of heparin inhibition of the esterase and basal lipase activities of lipoprotein lipase. Arch Biochem Biophys 1987; 253:475-485.
262. Rehfeldt W, Resch K, Goppelt-Struebe M. Cytosolic phospholipase A2 from human monocytic cells: characterization of substrate specificity and Ca(2+)-dependent membrane association. Biochem J 1993; 293:225-261.
263. Manjunath P, Soubeyrand S, Chandonnet L, Roberts KD. Major

proteins of bovine seminal plasma inhibit phospholipase A2. Biochem J 1994; 303:121-128.
264. Wichelhaus DP, Khuja A, Jones CT. GTP gamma S effects on phosphatidylinositol phospholipase C alpha isoenzyme activity isolated from guinea pig uterine smooth muscle at different stages of pregnancy. J Develop Physiol 1992; 18:179-186.
265. Pfaff M, Anderer FA. Casein kinase II accumulation in the nucleolus and its role in nucleolar phosphorylation. Biochim Biophys Acta 1988; 969:100-109.
266. Leiva L, Carrasco D, Taymor A, V'eliz M, Gonzalez C, Allende CC, Allende JE. Casein kinase II is a major protein phosphorylating activity in the nuclei of Xenopus laevis oocytes. Biochem Int 1987; 14:707-717.
267. Sutherland C, Renaux BS, McKay DJ, Walsh MP. Phosphorylation of caldesmon by smooth-muscle casein kinase II. J Muscle Res Cell Motil 1994; 15:440-456.
268. Medley QG, Bagshaw WL, Truong T, Cote GP. Dictyostelium myosin II heavy-chain kinase A is activated by heparin, DNA and acidic phospholipids and inhibited by polylysine, polyarginine and histones. Biochim Biophys Acta 1992; 1175:7-12.
269. Farooqui AA, Yang HC, Horrocks LA. Purification of lipases, phospholipases and kinases by heparin-sepharose chromatography. J Chromatogr 1994; 673:149-158.
270. Singh TJ. Inhibition of glycogen synthase (casein) kinase-1 by heparin. Arch Biochem Biophys 1988; 260:661-666.
271. Conigrave A, Patwardhan A, Broomhead L, Roufogalis B. A purification strategy for inositol 1,4,5-trisphosphate 3-kinase from rat liver based upon heparin interaction chromatography. Cell Signal 1992; 4:303-312.
272. Castellot JJ Jr, Pukac LA, Caleb BL, Wright TC Jr, Karnovsky MJ. Heparin selectivity inhibits a protein kinase C-dependent mechanism of cell cycle progression of calf aortic smooth muscle cells. J Cell Biol 1989 109:3147-3155.
273. Sihag RK, Jeng AY, Nixon RA. Phosphorylation of neurofilament proteins by protein kinase C. FEBS Lett 1988; 233:181-185.
274. Hovanessian AG, Galabru J. The double-stranded RNA-dependent protein kinase is also activated by heparin. Eur J Biochem 1987; 167:467-473.
275. Nikandrov VN, Voitiakov VI, Naumovich SA, Vorob'eva GV, Tsymanovich SG. Formation of streptokinase-heparin complexes and their properties. Vopr Med Khim 1987; 33:54-58.
276. Geiger M, Priglinger U, Griffin JH, Binder BR. Urinary protein C inhibitor. Glycosaminoglycans synthesized by the epithelial kidney cell line TCL-598 enhance its interaction with urokinase. J Biol Chem 1991; 266:11851-11857.
277. Deana AD, Pinna LA. Identification of pseudo 'phosphothreonyl-specific' protein phosphatase T with a fraction of polycation-stimulated

protein phosphatase 2A. Biochim Biophys Acta 1988; 968:179-185.
278. Ototani N, Yosizawa Z. Interaction of mucopolysaccharidases with glycosaminoglycans on glycosaminoglycan-bound AH-sepharose 4B. J Biochem 1978; 84:1005-1008.
279. Freeman C, Hopwood JJ. Human glucosamine-6-sulphatase deficiency. Diagnostic enzymology towards heparin-derived trisaccharide substrates. Biochem J 1992; 282:605-614.
280. Bar-Ner, M, Eldor A, Wasserman L, Matzner Y, Cohen IR, Fuks Z, Vlodavsky I. Inhibition of heparanase-mediated degradation of extracellular matrix heparan sulfate by non-anticoagulant heparin species. Blood 1987; 70:551-557.
281. Habuchi H, Habuchi O, Kimata K. Purification and characterization of heparan sulfate 6-sulfotransferase from the culture medium of chinese hamster ovary cells. J Biol Chem 1995; 270:4172-4179.
282. Unpublished data from our laboratory.
283. van Damme MP, Moss JM, Murphy WH, Preston BN. Binding properties of glycosaminoglycans to lysozyme--effect of salt and molecular weight. Arch Biochem Biophys 1994; 310:16-24.
284. Schwarting GA, Gajewski A, Carroll P, DeWolf WC. Inhibition of ganglioside sialyltransferase activity and stimulation of neutral glycolipid exocytosis by heparin. Arch Biochem Biophys 1987; 256:69-77.
285. Torres JC, Behrens MI, Inestrosa NC. Neural 16S acetylcholinesterase is solubilized by heparin. Biochem J 1983; 215:201-204.
286. Evangelista V, Piccardoni P, Maugeri N, De Gaetano G, Cerletti C. Inhibition by heparin of platelet activation induced by neutrophil-derived cathepsin G. Eur J Pharmacol 1992; 216:401-405.
287. Soeda S, Ohki H, Shinemo H, Nagamatsu A. Further characterization of the binding of plasminogen to heparin: evidence for the involvement glycine residues. Biochim Biophys Acta 1989; 999:29-35.
288. Geiger M, Heeb MJ, Binder BR, Griffin JH. Competition of activated protein C and urokinase fore a heparin-dependent inhibitor. FASEB J 1988; 2:2263-2267.
289. Griffith MJ. Kinetics of the heparin-enhanced antithrombin III/thrombin reaction. J Biol Chem 1982; 257:7360-7365.
290. Olson ST, Halvorson HR, Björk I. Quantitative characterization of the thrombin-heparin interaction. Discrimination between specific and nonspecific binding models. J Biol Chem 1991; 266:6342-6352.
291. Finotti P, Manente S. Heparin-induced structural and functional alterations of bovine trypsin. Biochim Biophys Acta 1994; 1207:80-87.
292. Alter SC, Metcalf DD, Bradford TR, Schwartz LB. Regulation of human mast cell tryptase. Biochem J 1987; 248:821-827.
293. Fiorucci L, Erba F, Ascoli F. Bovine tryptase: purification and characterization. Biol Chem Hoppe-Seyl 1992; 373:483-490.
294. Soncin F, Shapiro R, Fett JW. A cell-surface proteoglycan mediates

human adenocarcinoma HT-29 cell adhesion to human angiogenin. J Biol Chem 1994; 269:8999-9005.
295. Harosh I, Binninger DM, Harris PV, Mezzina M, Boyd JB. Mechanism of action of deoxyribonuclease II from human lymphoblasts. Eur J Biochem 1991; 202:479-484.
296. Sellmann E, Schroder KL, Knoblich IM, Westermann P. Purification and characterization of DNA polymerases from Bacillus species. J Bacteriol 1992; 174:4350-4355.
297. Harosh I, Mezzina M, Harris PV, Boyd JB. Purification and characterization of a mitochondrial endonuclease from Drosophila melanogaster embryos. Eur J Biochem 1992; 210:455-460.
298. Gauthier D., Ven Murthy MR. Efficacy of kinase inhibitors during brain polysome isolation. Neurochem Res 1987; 12:335-339.
299. Johnson RS. Nonspecific interaction of Escherichia coli pyrenyl RNA polymerase holoenzyme with synthetic polynucleotides as monitored by fluorescence spectroscopy. Biochemistry 1991; 30:198-206.
300. Zhang-Keck ZY, Stallcup MR. Optimized reaction conditions and specific inhibitors for initiation of transcription by RNA polymerase II in nuclei from cultured mammalian cells. J Biol Chem 1988; 263:3513-3520.
301. Mioh H, Chen JK. Acidic heparin binding growth factor transiently activates adenylate cyclase activity in human adult arterial smooth muscle cells. Biochem Biophys Res Commun 1987; 146:771-776.
302. Tersariol IL, Dietrich CP, Nader HB. Interaction of heparin with myosin ATPase: possible involvement with the hemorrhagic activity and a correlation with antithrombin III high affinity-heparin molecules. Thromb Res 1992; 68:247-258.
303. Crabbe T, Ioannou C, Docherty AJ. Human progelatinase A can be activated by autolysis at a rate that is concentration-dependent and enhanced by heparin bound to the C-terminal domain. Eur J Biochem 1993; 218:431-438.
304. Sandström J, Carlsson L, Marklund SL, Edlund T. The heparin-binding domain of extracellular superoxide dismutase C and formation of variants with reduced heparin affinity. J Biol Chem 1992; 267:18205-18209
305. Staben C, Whitehead TR, Rabinowitz JC. Heparin-agarose chromatography for the purification of tetrahydrofolate utilizing enzymes. Anal Biochem 1987; 162:257-264.
306. Adachi T, Fukushima T, Usami Y, Hirano K. Binding of human xanthine oxidase to sulphated glycosaminoglycans on the endothelial-cell surface. Biochem J 1993; 289:523-527.
307. Carrell RW, Christey PB, Boswell DR. Serpins: antithrombin and other inhibitors of coagulation and fibrinolysis evidence from amino acid sequences. In: Verstraete M, Vermylen J, Lijenen HR, Arnout J (eds). Thrombosis, Haemostasis, Leuven: Leuven Univ. Press, 1987:1-15.

308. Hermans JM, Stone SR. Interaction of activated protein C with serpins. Biochem J 1993; 295:239-245.
309. Manely GD, Owen TJ, Krstenansky JL, BranKamp RG, Cardin AD. Heparin binding properties of the carboyxl terminal domain antistansin 93-119. Adv Exp Med Biol 1992; 313:135-140.
310. Wu YI, Sheffield WP, Blajchman MA. Defining the heparin-binding domain of antithrombin. Blood Coag Fibrin 1994; 5:83-95.
311. Ragg HT, Ulshöfer, Gerewitz J. On the activation of human leuserpin-2, a thrombin inhibitor, by glycosaminoglycans. J Biol Chem 1990; 265:5211-5218.
312. Mellet, P, Ermolieft J, Bieth, JG. Mapping the heparin-binding site of mucus proteinase inhibitor. Biochemistry 1995; 34:2645-2652.
313. Ehrlich HJ, Keijer J, Preissner KT, Gebbink RK, Pannekoek H. Functional interaction of plasminogen activator inhibitor type 1 (PAI-1) and heparin. Biochemistry 1991; 30:1021-1028.
314. Stone SR, Hermans JM. Inhibitor mechanism of serpins. Interaction of thrombin with antithormbin and protease nexin 1. Biochemistry 1995; 34:5164-5172.
315. Bourin MC. Thrombomodulin: a new proteoglycan. Structure-function relation. Ann Biol Clin 1991; 49:199-207.
316. Enjyoji K-I, Miyata T, Kamikubo Y-I, Kato H. Effect of heparin on the inhibition of factor Xa by tissue factor pathway inhibitor. Biochemistry 1995; 34:5725-5735.
317. Fischer B. Binding of one-chain tissue-type plasminogen activator to fibrin, partially plasmin-degraded fibrin, lysine and heparin. Biomed Biochem Acta 1991 50:47-54.
318. Camejo G, Olofsson S-O, López, Carlsson P, Bondjers G. Identification of apo B-100 segments mediating the interaction of low density lipoproteins with arterial proteoglycans. Arteriosclerosis 1988; 8:368-377.
319. Cardin AD, Hirose N, Blankenship DT, Jackson RL, Harmony JAK. Binding of a high reactive heparin to human apolipoprotein E: identification of two heparin-binding domains. Biochem Biophys Res Commun 1986; 134:783-789.
320. Burgess WH, Maciag T. The heparin-binding (fibroblast) growth factor family of proteins. Annu Rev Biochem 1989; 58:575-606.
321. Remmers EF, Sano H, Wilder RL. Platelet-derived growth factors and heparin-binding (fibroblast) growth factors in the synovial tissue pathology of rheumatoid arthritis. Sem Arth Rheum 1991; 21:191-199.
322. Naka D, Ishii T, Shimomura T, Hishida T, Hara H. Heparin modulates the receptor-binding and mitogenic activity of hepatocyte growth factor on hepatocytes. Exp Cell Res 1993; 209:317-324.
323. Pillai S, Gilliam L, Conrad HE, Holleran WM. Heparin and its non-anticoagulant analogues inhibit human keratinocyte growth without

inducing differentiation. J Invest Dermatol 1994; 103:647-650.
324. Khachigian LM, Chesterman CN. Platelet-derived growth factor and alternative splicing: a review. Pathol 1992; 24:280-290.
325. Kimura H, Fischer WH, Schubert D. Structure, expression and function of a schwannoma-derived growth factor. Nature 1990; 348:257-260.
326. Casscells W. Smooth muscle cell growth factors. Prog Growth Factor Res 1991; 3:177-206.
327. McCaffrey TA, Falcone DJ, Du B. Transforming growth factor-beta 1 is a heparin-binding protein: identification of putative heparin-binding regions and isolation of heparins with varying affinity for TGF-beta 1. J Cell Physiol 1992; 152:430-440.
328. Higashiyama S, Abraham JA, Miller J, Fiddes JC, Klagsbrun, M. A heparin-binding growth factor secreted by macrophage-like cells that is related to EGF. Science 1991; 251:936-939.
329. Witt DP, Lander AD. Differential binding of chemokines to glycosaminoglycan subpopulations. Current Biol 1994; 4:394-400.
330. Lortat-Jacob H, Grimaud J.-A. Binding of interferon-gamma to heparan sulfate is restricted to heparin-like domains and involves carboxylic-but not N-sulfated-groups. Biochim Biophys Acta 1992; 1117:126-130.
331. Oppenheim JJ, Zachariae CO, Mukaida N, Matsushima K. Properties of the novel proinflammatory supergene "intercrine" cytokine family. Ann Rev Immun 1991; 9:617-648.
332. Hoogwert AJ, Leone JW, Reardon IM, Howe WJ, Asa D, Heinrikson RL, Ledbetter SR. CKC chemokines connective tissue activating peptide-III and neutrophil activating peptide-2 one heparin/heparan sulfate-degrading enzymes. J Biol Chem 1995; 270:3268-3277.
333. Maccarana M, Lindahl U. Mode of interaction between platelet factor 4 and heparin. Glycobiol 1993; 3:271-277.
334. San Antonio JD, Lander AD, Karnovsky MJ, Slayter HS. Mapping the heparin-binding sites on type I collagen monomers and fibrils. Cell Biol 1994; 125:1179-1188.
335. San Antonio JD, Karnovsky M, Gay S, Sanderson RD, Lander AD. Interaction of syndecan-1 and heparin with human collagens. Glycobiology 1994; 4:327-332.
336. Calaycay J, Pande H, Lee T, Borsi L, Siri A, Shively JE, Zardi L. Primary structure of a DNA- and heparin-binding domain (Domain III) in human plasma fibronectin. J Biol Chem 1985; 260:12136-12141.
337. Marton LS, Gulcher JR, Stefansson. Binding of hexabrachions to heparin and DNA. J Biol Chem 1989; 264:13145-13149.
338. Frenette GP, Ruddon RW, Krzesicki RF, Naser JA, Peters BP. Biosynthesis and deposition of a noncovalent laminin-heparan sulfate proteoglycan complex and other basal lamina components by a human malignant cell line. J Biol Chem 1989; 264:3078-3088.
339. Aukhil J, Paritosh J, Yingzhuo Y, Erickson HP. Cell- and heparin-

binding domains of hexabrachion arm identified by tenascin expression proteins. J Biol Chem 1993; 268:2542-2553.
340. Legrand C, Morandi V, Mendelovitz S, Shaked H, Hartman JR, Panet A. Selective inhibition of platelet macroaggregate formation by a recombinant heparin-binding domain of human thrombospondin. Arter Thromb 1994; 14:1784-1791.
341. Su HR, Boackle RJ. Heparin mediates binding of S-protein/vitronectin to the envelope glycoprotein of the human immunodeficiency virus and CD4. Int Arch Aller Immun 1994; 105:238-244.
342. Schols D, Baba M, Pauwels R, Desmyter J, De Clercq E. Specific interaction of aurintricarboxylic acid with the human immunodeficiency virus/CD4 cell receptor. Proc Natl Acad Sci USA 1989; 86:3322-3326.
343. Shi E, Kan M, Xu J, Wang F, Hou J, McKeehan WL. Control of fibroblast growth factor receptor kinase signal transduction by heterodimerization of combinatorial splice variants. Mol Cell Biol 1993; 13:3907-3918.
344. Khazaie K, Schirrmacher V, Lichtner RB. EFG receptor in neoplasia and metastasis. Cancer & Metastasis Rev 1993; 12:255-274.
345. Kounnas MZ, Chappell DA, Strickland DK, Argraves WS. Glycoprotein 330, a member of the low density lipoprotein receptor family, lipoprotein lipase in vitro. J Biol Chem 1993; 268:14176-14181.
346. Booth BA, Boes M, Andress DL, Dake BL, Kiefer MC, Maack C, Linhardt RJ, Bar K, Caldwell E, Weiler J, Bar RS. IGFBP-3 and IGFBP-5 association with endothelial cells: Role of C-terminal heparin binding domain. Growth Regulations 1995; 5:1-17.
347. Kraus H-G, Scheffauer, F, Romainin C, Schindler HG, Glossmann, H. Heparin binds with high affinity to voltage-dependent L-type Ca^{2+} channels. J Biol Chem 1990; 265:11156-11166.
348. Kari B, Gehrz R. A human cytomegalovirus glycoprotein complex designated gC-11 is a major heparin-binding component of the envelope. J Virol 1992; 66:1761-1764.
349. Mbemba E, Gluckman JC, Gattegno L. Interactions of HIV-1 and HIV-2 envelope glycoproteins with sulphated polysaccharides and mannose-6-phosphate. Glycobiology 1994; 4:13-21.
350. WuDunn D, Spear PG. Initial interaction of herpes simplex virus with cells is binding to heparan sulfate. J Virol 1989; 63:52-58.
351. Mettenleiter TC, Zsak L, Zuckermann F, Sugg N, Kern H, Ben-Parat T. Interaction of glycoprotein gIII with a cellular heparin-like substance mediates adsorption of pseudorobies virus. J Virology 1990; 64:278-286.
352. Busch SJ, Jackson RL, Sassone-Corsi P. Heparin inhibits transcriptional activation by Fos and Jun oncoproteins. Trans Assoc Am Physic 1990; 103:21-27.
353. Itoh H, Tashima Y. Domain structure of the 90-kDa stress protein: heparin and antibody-binding domain. Int J Biochem 1993; 25:157-161.

354. Bush AI, Pettignell WH Jr., de Paradis M, Tanzi RE, Wasco W. The amyloid beta-protein precursor and its mammalian homologues. Evidence for a zinc-modulated heparin-binding superfamily. J Biol Chem 1994; 269:26618-26621.
355. Gonzalez-Rubio C, Jimenez-Clavero MA, Fontan G, Lopez-Trascase M. The inhibitory effect of factor J on the alternative complement pathway. J Biol Chem 1994; 269:26017-26024.
356. Hotchkiss KA, Chesterman CN, Hott PJ. Inhibition of heparin activity in plasma by soluble fibrin: evidence for ternary thrombin-fibrin-heaprin complex formation. Blood 1994; 84:498-503.
357. Levy DE, Horner AA, Soloman A. Immunoglobulin-sulfated polysaccharide interactions. J Exp Med 1981; 153:883-896.
358. Cole GJ, Akeson R. Identification of a heparin binding domain of the neural cell adhesion molecule N-cam using synthetic peptides. Neuron 1989; 2:1157-1165.
359. Skinner MP, Fournier DJ, Andrews RK, Gorman JJ, Chesterman CN, Berndt MC. Characterization of human platelet GMP-140 as a heparin-binding protein. Biochem Biophys Res Comm 1989; 164:1373-1379.
360. Lee MK, Lander AD. Analysis of affinity and structural selectivity in the binding of proteins to glycosaminoglycans: Development of a sensitive electrophoretic approach. Proc Nat Acad Sci (USA) 1991; 88:2768-2772.
361. Denton J, Lewis WE, Nieduszynski IA, Phelps CF. Fractionation of heparin using antithrombin III reversibly bound to concanavalin A-sepharose. Anal Biochem 1981; 118:388-391.
362. Walter A, Turnbull JE, Gallagher JT. Specific heparan sulfate saccharides mediate the activity of basic fibroblast growth factor. J Biol Chem 1994; 269:931-935.
363. Fernandez deRecondo ME, Legorburu C, Monge JC, Recondo EF. Heparin and concanavalin A interaction: Isolation of fraction with higher anticoagulant activity. Biochem Biophys Res Commun 1988; 155:216-223.
364. Denton J, Lane DA, Thunberg L, Slater AM, Lindahl U. Binding of platelet factor 4 to heparin oligosaccharides. Biochem J 1983; 209:455-460.
365. Mohammad SF, Chuang HYK, Mason RG. Quantitative removal of heparin from plasma and other aqueous solutions by affinity adsorption on poly (L-lysine)-Sepharose 4B. Thromb Res 1980; 20:599-609.
366. Teng CL, Kim JS, Port FK, Wakefield TW, Till GO, Yang VC. A protamine filter for extracorporeal blood heparin removal. ASAIO-Trans 1988; 34:743-746.
367. Nordenman B, Bjork I. Function of heparin by chromatography on immobilized thrombin. Thromb Res 1980; 19:711-718.
368. Andrade-Gordon P, Strickland S. Fractionation of heparin by

chromatography on a tissue plasminogen activator-Sepharose column. Proc Nat Acad Sci (USA) 1990; 87:1865-1869.

369. Fareed J, Walenga JM, Racanelli A, Hoppensteadt D, Huan X. Comparative study on the in vitro and in vivo activities of seven low-molecular-weight heparins Haemostasis 1988; 18:3-15.
370. Fareed J, Walenga JM, Racanelli A, Hoppensteadt D, Huan X, Messmore HL. Validity of the newly established low-molecular-weight heparin standard in cross-referencing low-molecular-weight heparins. Haemostasis 1988; 18:33-47.

8
Chemoenzymatic Synthesis of Optically Active *myo*-Inositol Polyphosphate

Shoichiro Ozaki and Ling Lei
Ehime University
Matsuyama, Japan

I. INTRODUCTION

Communication between cells is essential for the maintenance of life processes in complex organisms. Nature uses a wide variety of chemical signals to pass information from one cell to another. These chemical messengers may be classified by their function, as neurotransmitters, hormones, growth factors, etc. The chemical messengers released by cells are received by specific receptors located on the surface of the receiving cells. These receptors process the incoming signal from the outer surface of the cell membrane to produce a signal inside the target cell. Receptors use a variety of mechanisms to achieve this signal transduction [1]. Once the message has been passed across the cell membrane, the target cell then responds to this intracellular signal, and then the stimulation of one cell leads, via extracellular chemical messengers, receptors, signal transduction, and the production of an intracellular signal, to response in a second physically distant cell.

In 1975 Michell suggested that the receptor-controlled hydrolysis of inositol phospholipids could be directly linked to changes in calcium levels within cells [2]. It is now known that phosphatidylinositol 4,5-bisphosphate (PIP_2) is hydrolyzed by a G protein-controlled enzyme, phospholipase C (PLC), releasing two second messengers, diacylglycerol (DAG) and D-inositol 1,4,5-triphosphate (1,4,5-IP_3). 1,4,5-IP_3 binds to specific receptors on the endoplasmic reticulum and stimulates the release of calcium from intracellular storage sites. Then it is metabolized to other inositol phosphate derivatives, IP_4, IP_5, IP_6, IP_2, and IP_1, in a complex metabolic cycle [3-4].

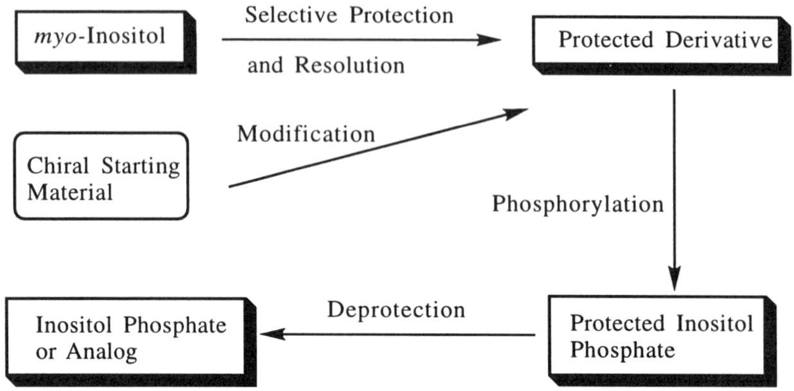

Figure 1

In order to clarify the individual function of the inositol phosphates involved as metabolites in the metabolic cycle, coupled with the fact that only minute amounts of these metabolites were available from natural sources, worldwide interest in the chemical synthesis of metabolites was stimulated in

Scheme 1

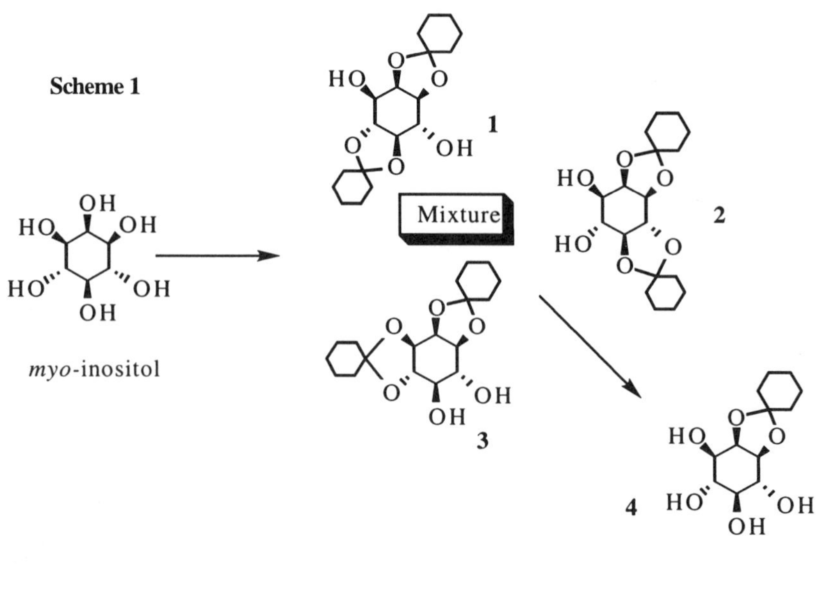

recent years. Many laboratories have subsequently contributed methodology and techniques for the synthesis. Due to the commercial availablility of pure myo-inositol, most syntheses use the parent cyclitol as starting material. One of the main problems in the synthesis of a given inositol phosphate is the preparation of a suitably protected inositol derivative, with free hydroxyl groups at the desired positions, and subsequest optical resolution.

Scheme 2

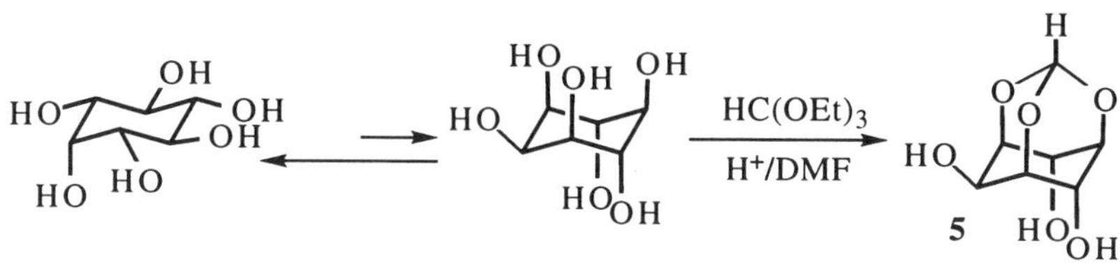

5

Generally, the syntheses start with either the protection of inositol by acetonization with cyclohexanone or the formation of inositol orthoformate by reaction of inositol with triethyl orthoformate (Scheme 1 and Scheme 2), leaving the possibility for further selective protection and deprotection of the products.

Reaction of inositol with cyclohexanone, or more efficiently with a cyclohexanoneacetal under acidic conditions leads to the formation of the three isomeric bisacetals (1, 2, and 3) (Scheme 1). Each of the three isomeric bisacetals may be obtained in pure (Scheme 1) form by a combination of crystallization and chromatography. All of the bisacetals give the monoacetal 4 on mild acid hydrolysis due to selective cleavage of the less stable trans acetal group. Each of the bisacetals 1-3 and 4 have proved useful for the synthesis of inositol phosphates, due to the relative stability of the acetal protecting group, and the differential reactivities displayed by their free hydroxyl groups.

In addition to the cyclohexanone acetals, other acetals including the isopropylidene and the cyclopentylidene acetals have been studied and used in syntheses. The overall reactivity patterns of three acetals are very similar to the cyclohexylidene compounds.

Treatment of inositol with triethylorthoformate at high temperature, in the presence of an acid catalyst results in the formation of the crystalline orthoformate derivative 5 (Scheme 2). This provides an interesting protected inositol, in which the 1, 3, and 5- hydroxyl groups are protected simultaneously. In addition, the normal axial/equatorial relationship of the remaining free hydroxyl groups has been reversed. The spatial juxtaposition of the axial hydroxyl group in 5 allows highly selective monoalkylations and monophosphorylations to be performed.

Thus, most of the reported synthetic methods were based on further selective protection, deprotection, and optical resolution from 1-5. This chapter focuses on developing new efficient chemoenzymatic methods for preparation of optically active inositol derivatives, further selective protection of the hydroxyl groups, and effective syntheses of biologically interesting inositol phosphates from inositol intermediates 2-5.

II. ENZYMATIC RESOLUTION OF 1,2 : 5,6-DI-O-CYCLOHEXYLIDENE-MYO-INOSITOL [5]

A. Enzymatic Esterification

In addition to the preparation of optically active myo-inositol derivatives from other chiral materials [6], many methods reported for this purpose use racemic or meso myo-inositol derivatives. The products are resolved by the use of different optically pure reagents [7], in the cumbersome procedures of conversion of the racemic inositol derivatives into a pair of diastereomeric isomers, followed by separation of the diastereomers. Resolution may also be accomplished by the use of chiral HPLC columns to separate individual enantiomers in small amounts [8].

Enzymatic reactions are generally conducted in aqueous media because of the prevalent notion that an aqueous environment is optimal for maintaining the catalytically active conformations of the enzymatic protein for binding and catalysis. Enzyme catalyzed selective hydrolysis in aqueous media was used to resolve some racemic inositol derivatives [9], or to yield optically active intermediates from meso inositol carboxylates.[10] However, as we know, most compounds of interest to organic chemists are insoluble in water. Also, water often promotes unwanted side reactions and decomposes a number of common organic reagents. Finally, the thermodynamic equilibria of many processes are unfavorable in water.

It has been known for a few years that a variety of enzymes are catalytically active in non-polar organic media, and some of these enzymes have been used for the transformation of water-insoluble substrates and for peptide synthesis via reversal of the hydrolytic process. The enantioselective properties of the lipases, in particular, have attracted the attention of synthetic organic chemists because of their synthetic utility. Lipases are uniquely stable in non-polar organic solvents and have the remarkable ability of assuming a variety of conformations to accommodate substrates of varying sizes and stereochemical complexities. Moreover, compounds that are unstable in water could be used as substrates in nonaqueous media. More importantly, lipase-catalyzed esterification reactions in organic solvents are often more enantioselective than the corresponding hydrolytic reactions in water. For these reasons, non-aqueous biocatalytic technology is becoming increasingly important as a method for the preparation of homochiral (optically active) compounds [11].

However there are no reports concerning the application of enzymatic esterification in organic solvent for the resolution of inositol derivatives. The need for various optically active inositol derivatives in our research prompted us to investigate the applicability of this method. First an important intermediate for synthesis of inositol phosphate, 1,2:5,6-di-O-cyclohexylidene myo-inositol (**2**), which is readily available by acetalization of myo-inositol with cyclohexanone in a single step [12], was chosen as a model substrate for our examination. This racemic material was subjected to esterification in the presence of lipases. Fortunately only one enantiomer was acylated based . Another enantiomer remained in the form of alcohol, so that the two enantiomers could be easily separated by common methods (e.g. chromotography). Factors such as different enzymes, solvents and those substrates have been found to affect the resolution.

Scheme 3

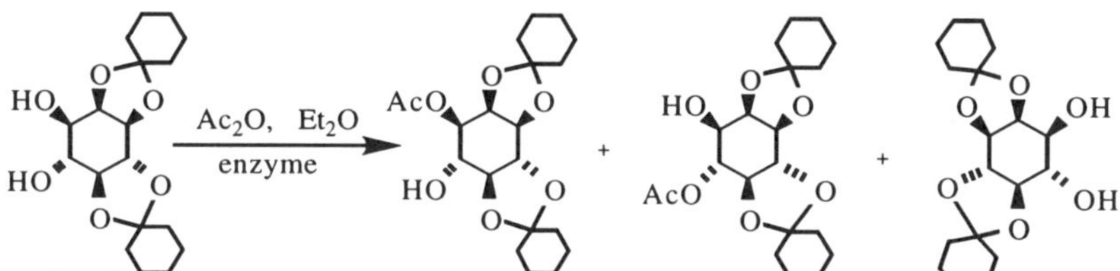

DL-**2** D-**6** D-**7** L-**2**

	D-6	D-7	L-2
Lipase P (23 h)	25% (100% e.e.)		73% (36% e.e.)
Lipase AY (16 h)		48% (100% e.e)	51% (98% e.e.)

The resolution of the racemic **2** is shown in Scheme 3. Among several commercially available hydrolytic enzymes, the lipase from Pseudomonas sp. [Amano Lipase P] and the lipase from Candida cydindracea [Amano Lipase AY] exhibited high regio- and enantio- selectivity in acetylation of the racemic 2. Lipase AY exclusively acetylates the hydroxyl group at the C-4 position of the D-enantiomer almost completely (48%) to give two optically active inositol intermediates D-**7** and L-**2**. The enantiomeric excess of D-**7** and L-**2** was 100% and 98% respectively. Lipase P preferably acetylates the hydroxyl group at the C-3 position of the same enantiomer. Optically pure D-3-O-acetyl-1,2:5,6-di-O-cyclohexylidene myo-inositol (D-**6**} was obtained in the yield of 25% and diol L-2 in the yield of 73% (36% e.e).

The general mechanism of enzyme catalyzed esterification is depicted in Scheme 4. [11d] in which **E** is the hydrolytic enzyme, RCOOR' is the acylating

Scheme 4

reagent, ERCOOR' is a noncovalent enzyme-substrate complex (the Michaelis complex) and RCOOE is the covalent acyl-enzyme intermediate. The reactivity and selectivity depends on the structure of RCOOR', R"OH and the type of the enzyme.

Some different active esters such as trichloroethyl acetate, vinyl acetate, and phenyl acetate have been used as acylating reagents. The results are listed in Table 1. It has been found that all these reagents are active for our resolution. Phenyl and nitrophenyl acetate, which have not been commonly applied as acyl donors in enzymatic reaction, exhibited high activities in the enzymatic reaction. All of the above compounds were, however, less active than acetic anhydride.

Table 1. Enzymatic Esterification of the Racemic **2**

Lipase	Acyl Donor (eq)	reaction time	yield (ee)		
			D-**6**	D-**7**	L-**2**
P	Ac_2O(4)	23 h	25% (100%)	trace	73% (36%)
P	$AcOCH_2CCl_3$(4)	72h	trace		
AY	$AcOCH_2CCl_3$(4)	18 h		29% (100%)	63% (58%)
AY	$AcOCH_2CCl_3$(6.5)	19 h		38% (100%)	48% (84%)
AY	$AcOCH_2CCl_3$(19)	18 h		37% (100%)	61% (80%)
AY	$AcOCH{=}CH_2$(4)	17 h		23% (100%)	70% (52%)
AY	$AcOC(CH_3){=}CH_2$(4)	19 h		15% (100%)	67% (44%)
AY	AcOPh(4)	19 h		38% (100%)	53% (92%)
AY	$AcOC_6H_4NO_2$(4)	17 h		40% (100%)	55% (84%)
AY	Ac_2O(4)	16 h		48% (100%)	51% (98%)

As shown in Table 1, only half of the D-enantiomer (25%) was acylated in the case of Lipase P. It was considered that increasing the mass equivalent of enzyme might improve the yield, so the effect of concentration of enzyme was investigated. The results are listed in Table 2. The yield and selectivity has not changed with the change in the concentration of Lipase P. But in the case of

Lipase AY, D- and L- enantiomer could be resolved completely with the use of two enzyme mass equvalents. Prolongation of reaction time in the case of lipase P did not give significant improvement as was expected (Table 3).

Table 2. Effect of Mass Equivalent of Enzyme on the Resolution of DL-**2**

Lipase	mass eq of enzyme	reaction time	yield (ee)		
			D-**6**	D-**7**	L-**2**
P	1	21 h	25% (98%)		73% (34%)
P	2	23 h	25% (100%)		73% (36%)
P	4	21 h	28% (100%)		65% (54%)
AY	2	16 h		48% (100%)	51% (98%)
AY	1	24 h		26% (100%)	73% (44%)

Table 3. Effect of Reaction Time on the Resolution of DL-**2**

reaction time	yield (ee)		
	D-**6**	D-**7**	L-**2**
21 h	25% (98%)	0	73% (34%)
33 h	32% (100%)	trace	65% (58%)
48 h	31% (100%)	trace	67% (54%)

Conditions: Lipase P (1 eq) was used.

The analysis of optical purity of the products has been performed by HPLC with the use of a chiral column. It has been found that the D- and L-enantiomer of racemic **2** could be baseline separated by a commercial Chiralcel OD column (isopropanol/hexane=1:25). The retention time of D- and L-enantiomers were 15 minites and 19 minites respectively (flow rate 0.8 ml/min). The optical purity of the products D-**6** and D-**7** was conveniently analyzed after the acetates were hydrolyzed (Scheme 5). The absolute configuration was confirmed by comparing the specific rotation of L-2 ($[a]_D^{21}$ +19.3°, c 0.75, $CHCl_3$) and hydrolyzed products of **6** and **7** with that of the known L-1,2:5,6-di-O-cyclohexylidene-myo-inositol [9a)] ($[a]_D^{20}$ +18.4°, c 1.0, $CHCl_3$) and the retention times in the HPLC chromatogram.

Scheme 5

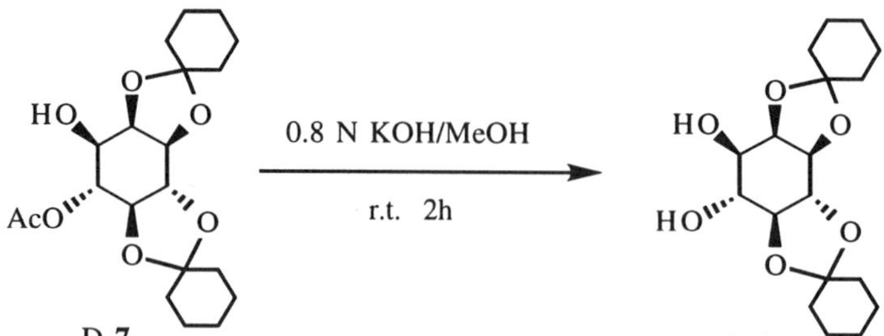

D-7 D-2

B. Enzymatic Hydrolysis

Enzymatic hydrolysis has been used to prepare optically active materials for years [13]. Various kinds of optically pure materials have been obtained by this method. Recently it has been reported that optically active inositol derivatives could be obtained by this method [9-10]. As an alternative method, the resolution of inositol derivatives by enzymatic hydrolysis was investigated in our laboratory and compared with the results of esterification.

Scheme 6

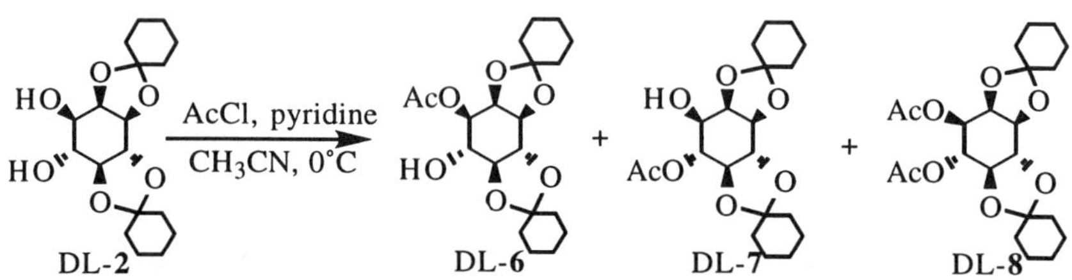

DL-2 DL-6 DL-7 DL-8

In order to resolve the hydroxyl group containing material by enzymatic hydrolysis, the hydroxyl group has to be first acetylated chemically. The preparation of material, 3-O-acetyl-1,2:5,6-di-O-cyclohexylidene-myo-inositol (DL-**6**) is outlined in Scheme 6. The hydroxyl group at the C-3 position was predominantly acetylated to give DL-6 with a 51% yield with a 31% recovery of starting material. Only a trace amount of 4-acetylated and diacetylated products DL-**7** and DL-**8**, respectively, were detected by TLC.The mono-acetate DL-**6** was subjected to enzyme-catalyzed hydrolysis. The results are shown in Scheme 7.

Lipase AY and Lipase P catalyzed the hydrolysis of the D-enantiomer of racemic 6 which was the same enantiomer as it selected in esterification to yield the complementary products [11d]. It is noteworthy that some amount of product **7** was obtained. It was assumed that the migration of the acetyl group occurred in such a phosphate buffer [14]. The reaction rate of the hydrolysis of D-**7** by Lipase AY might be much faster than D-**6**, therefore the optically pure L-**7** was obtained. Lipase A hydrolyzed the acetate of the L-enantiomer **7**, but the optical purities of the products were modest.

Scheme 7

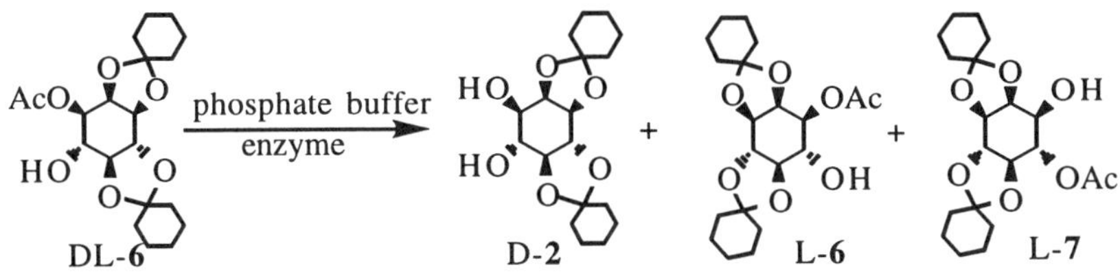

	D-2	L-6	L-7
Lipase AY 2 days	19% (100% e.e.)	78% (36% e.e.)	trace
Lipase AY 6 days	44% (100% e.e.)	35% (74% e.e)	16% (100% e.e.)
Lipase P 6 days	20% (100% e.e.)	54% (44% e.e)	22% (46% e.e.)

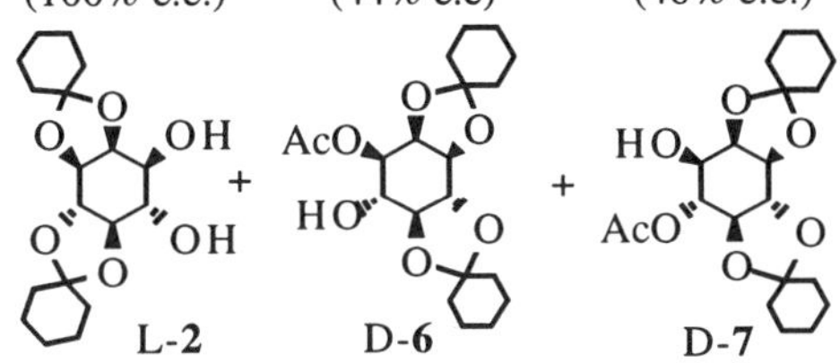

	L-2	D-6	D-7
Lipase A 1 day	52% (50% e.e.)	39% (60%e.e.)	trace

One of the problems in enzymatic hydrolysis is the solubility of material in waters. In principle, some amount of water miscible organic solvent must be used to dissolve the material, but addition of 50% acetone inactivated both Lipase A and Lipase P. The question of selection of a proper solvent system is discussed in more detail in the next section.

The hydrolysis of racemic **6** has been examined by changing the amount of enzyme used. It has been found that reaction rate of hydrolysis is enhanced with increasing amounts of enzyme, but the enantiomeric excess of product was about the same (Table 4).

Table 4. Effect of the Amount of Enzyme on the Resolution of DL-**6**

DL-**6**	Lipase A	Reaction Time	Yield (e.e.)	
			L-2	D-**6**
40.1 mg	8.4 mg	1 day	27% (48%)	35%
49.7 mg	30.2 mg	1 day	52% (50%)	39%
43.2 mg	87.5 mg	5 h	38% (54%)	46%

III. ENZYMATIC RESOLUTION OF 1,2:3,4- DI-O-CYCLOHEXYLIDENE-MYO-INOSITOL [5]

A. Enzymatic Esterification

Scheme 8

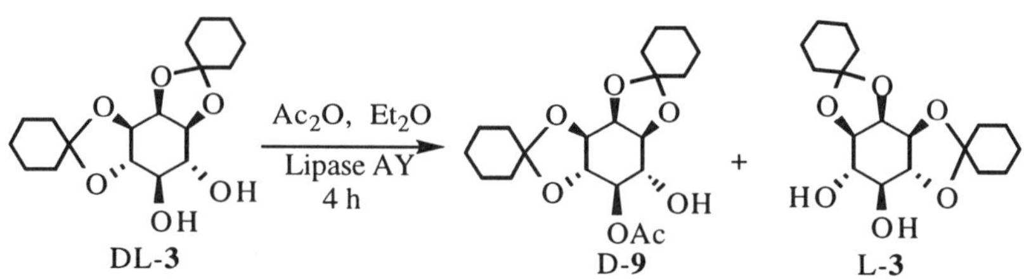

50% (96% e.e.) 48% (100% e.e.)

The resolution of racemic 1,2:3,4-di-O-cyclohexylidene-myo-inositol (DL-**3**), which is another useful intermediate for synthesis of inositol phosphate derivatives, was carried out in the same way as that of DL-**2** (Scheme 8). Some commercially available hydrolytic enzymes have been screened. It has been found that only Lipase AY was active in this reaction. It acetylated the hydroxyl group at the C-5 position of the D-enantiomer of the racemic 3 exclusively to give two optically active products D-**9** and L-**3**. The enantiomeric excess was 96% and 100%, respectively enabling complete resolution of the racemic material. The optical purity of the products was analyzed by using the same HPLC column as above after hydrolysis of the acetyl group (isopropanol/hexane 1:9). The retention times of the L- and D-enantiomers of **3** were 9.0 and 13.3 min., respectively (flow rate 0.6 ml/min.). The absolute configuration of the products was determined by comparing the corresponding specific rotation of the obtained **3** with that of the known D- and L-1,2:3,4-di-O-cyclohexylidene-myo-inositol [15].

Resolution of the diol DL-**3** with lipase AY was investigated using acid anhydride with different chain lengths. Reaction proceeded smoothly. Racemic 3 could be effectively resolved (Table 5). Considering the convenience of separation, acetic anhydride is recommended.

Table 5. Enzymatic Resolution of DL-**3** Using Different Acid Anhydride

Acid Anhydride	Reaction Time	Yield (ee)	
		D-**9**	L-**3**
acetic anhydride	4 h	50% (96%)	46% (100%)
propionic anhydride	24 h	54% (84%)	37% (100%)
hexanoic anhydride	24 h	50% (92%)	43% (100%)
lauric anhydride	13 h	not determined	52% (88%)

Table 6. The Effect of Solvent on the Resolution of DL-**3**

Solvent	Reaction Time	Yield (ee)	
		D-**9**	L-**3**
Et_2O	4 h	50% (96%)	48% (100%)
Benzene	10.5 h	44% (100%)	54% (88%)
AcOEt	30 h	13% (94%)	86% (18%)
Acetone	48 h	8%*	85% (0)
THF	24 h	No Reaction	
Dioxane	30 h	No Reaction	

* Acetylated product was mixture of 5- and 6-acetylated products, optical purity not determined

Solvent effect on the resolution was studied in the case of the reaction of DL-**3** with acetic anhydride in the presence of Lipase AY (Table 6). It can be seen that the solvent affected resolution significanttly. Diethyl ether and benzene were fairly effective solvents for this resolution, but Lipase AY lost its activity when water miscible solvents such as acetone, dioxan, and THF were used.

In the case of diethyl ether as solvent, the results were almost the same when the reaction was carried out for 4 hours and 10 hours. This means that no more reaction took place after all of the D-enantiomer was acetylated.

B. Enzymatic Hydrolysis

The enzymatic resolution of racemic 1,2:3,4-di-O-cyclohexylidene-myo-inositol by hydrolysis was carried out in the same way mentioned in the earlier section (Scheme 9). Racemic 1,2:3,4-di-O-cyclohexylidene-myo-inositol was acetylated with acetic anhydride in the presence of pyridine to give the racemic **9** (24%) and di-acetylated product **10** (73%). The mono-acetylated product (DL-**9**) was subjected to enzymatic hydrolysis.

Scheme 9

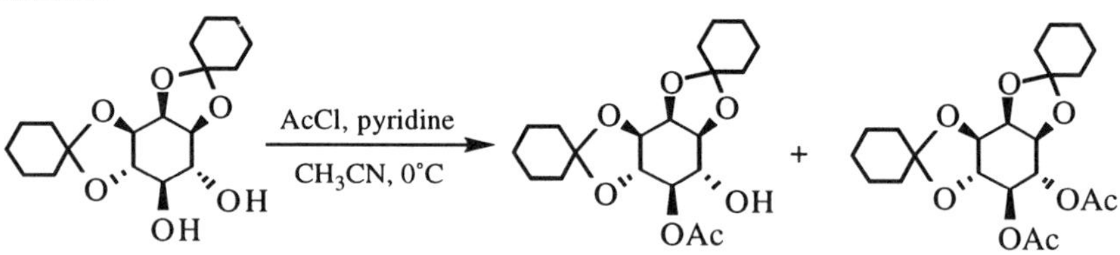

DL-**3** DL-**9** DL-**10**

Lipase AY catalyzed the hydrolysis of D-enantiomer of racemic **9** which was the same enantiomer as it selected in esterification to yield the complementary product L-**9** (Scheme 10) [11d]. However,the optical purities of the obtained products were not so good as that of esterification precess. Lipase A and PLE-A both showed activity in hydrolysis of the L-enantiomer.

The kinetic resolution of racemic 1,2:5,6-di-O-cyclohexylidene-myo-inositol (DL-**2**) and racemic 1,2:3,4-di-O-cyclohexylidene-myo-inositol (DL-**3**) was performed by enzyme-catalyzed esterification and hydrolysis in organic solvents. Along with the resolution, selectively protected products D-**6**, L-**6**, D-**7**, L-**7**, D-**9**, and L-**9** were obtained. All of these optically active materials are useful intermediates in synthesis of naturally occurring products. This provides a new and efficient method for resolution and selective protection of inositol derivatives. The application of these selectively protected materials in synthesis will be described later.

Scheme 10

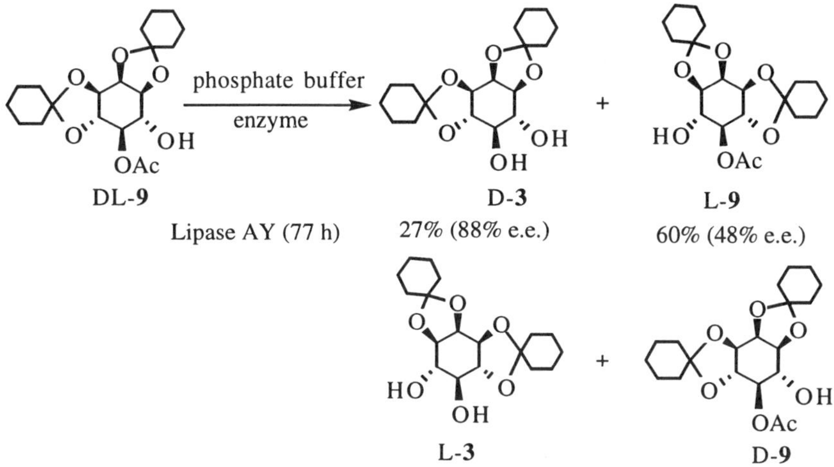

DL-**9** → D-**3** + L-**9**

	D-**3**	L-**9**
Lipase AY (77 h)	27% (88% e.e.)	60% (48% e.e.)

	L-**3**	D-**9**
Lipase A (9 h)	54% (56% e.e.)	46% (58% e.e.)
PLE-A (48 h)	36% (46% e.e.)	63% (10% e.e.)

IV. ENZYMATIC RESOLUTION OF 2,3-O-CYCLOHEXYLIDENE -MYO-INOSITOL [16].

A. Enzymatic Resolution

Because 2,3-O-cyclohexylidene-myo-inositol (4), which can be prepared in a single step from inositol in a very high yield, is an important precursor for the synthesis of various naturally occurring myo-inositol derivatives [17-18] development of an efficient procedure for optical resolution of **4** is imperative.

Scheme 11

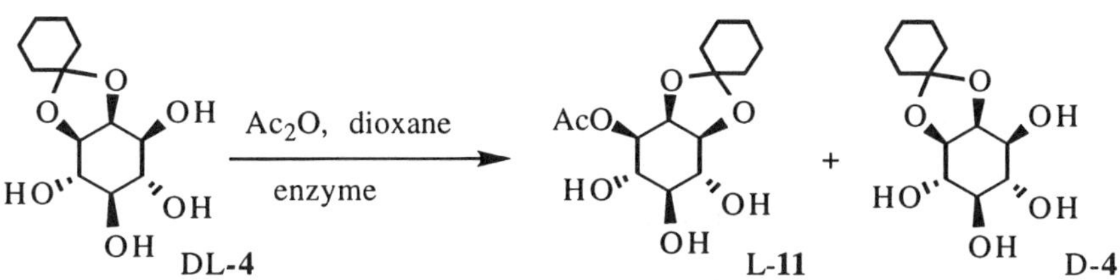

DL-**4** → L-**11** + D-**4**

There are, however, few method available for the resolution of **4**. Bruzik and his co-workers have developed a technique for this purpose by preparing an optically active analog D-2,3-O-(1'R,2'R,4'R-1', 7', 7'-trimethyl [2.2.1] bicyclohept-2'-ylidene)-myo-inositol which is then treated in the same way as 2,3-O-cyclohexylidene-myo-inositol for the synthesis of various optically active myo-inositol derivatives [18b]. A problem arises from the difficulty in the preparation of this analog, because the acetalization of inositol with D-camphor dimethyl acetal gives four diastereomers and the separation of these diastereomers is difficult.

We have investigated the resolution of racemic 4 by the enzyme-catalyzed enantioselective esterification in organic solvents (Scheme 11). Among commercially available hydrolytic enzymes, we found two lipases from Pseudomonas sp. (Amano Lipase P and Lipase CES) which were effective in the resolution in dioxanea as solvent. The representative results are listed in Table 7. Lipase P and Lipase CES selectively acetylated the hydroxyl group at the C-1 position of the L-enantiomer. Increasing the mass equivalent of the Lipase P and the reaction time improved the optical purity of the product D-**4**. Lipase CES acetylated the L-enantiomer exclusively to give the optically active 1-acetylated product L-**11** (49%, 98% ee) and D-**4**(49%, 100% ee).

Table 7. Enzymatic Resolution of Racemic 2,3-Mono-*O*-cyclohexylidene-*myo*-inositol (**4**)

Enzyme	Mass Equiv	Ac_2O Equiv	Reaction Time (h)	Yield (%) (e.e.)	
				L-**11**	D-**4**
Lipase P	5	6	13	32 (84%)	67 (30%)
Lipase P	8	8	22	46 (90%)	53 (68%)
Lipase P	11	8	46	49 (80%)	43 (100%)
Lipase CES	10	6	46	49 (98%)	49 (100%)

Table 8. Recovery and Reuse of Lipase CES

Run	Treament of Enzyme Before Reuse	Reaction Time (h)	Yield (ee)	
			L-**11**	D-**4**
1	dried under vacuum for 15 min	3 days	trace reaction	
2	dried under vacuum for 3 h	2 days	trace reaction	
3	water added, soaked 20 min, dried under vacuum for 3 h	68 h	44% (88%)	51%(46%)

The results of the study on the recovery and reuse of enzyme are listed in Table 8. The enzyme was recovered from the reaction system by filtration. After washing with MeOH, the enzyme was recycled. It has been found that the recovered enzyme was almost inactive (Run 1 and 2). Since we considered that the hydration water of enzyme, which is essential for enzyme activity [11c,19] was extracted by the dioxane, water was added to soak the enzyme powder to hydrate the enzyme protein. After excess water was evaporated and the enzyme was dried under vacuum, it was reused under the same condition as it was used for the first time (Run 3). The fact that the enzyme was reactivated by hydration indicates the importance of hydration water for enzyme activity in an anhydrous organic solvent.

B. Determination of Optical Purity and Absolute Configuration

The absolute configuration of D-4 was determined by comparing its specific rotation ($[\alpha]_D^{21}$ +36.9°, c 1.30, MeOH) with that of the known D-2,3-cyclohexylidene-myo-inositol ($[\alpha]_D^{24}$ +42.4°, c 0.33, EtOH) [15].

Scheme 12

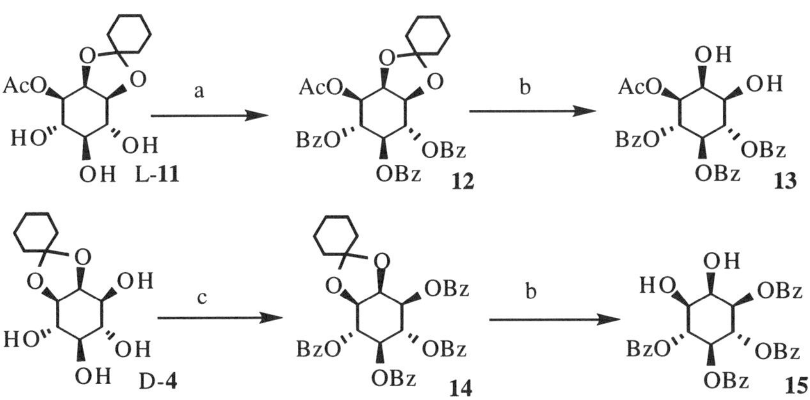

a) BzCl, 6 eq; NEt_3, 6 eq; DMAP, CH_2Cl_2, r. t., overnight; b) 80% AcOH, reflux, 2 h; c) BzCl, 8 eq; NEt_3, 8 eq; other conditons were the same as in a).

The configuration was further confirmed by transforming D-4 to D-1,4,5,6-0-tetrabenzoyl-1-2,3-0-cyclohexylidene-myo-inositol and comparing specific rotation value with that of L-1,4,5,6-O-tetrakisbenzoyl-2,3-O-cyclohexylidene-myo-inositol derived from L-quebrachitol [20]. The optical purity was determined by HPLC analysis with the use of a chiral column. D- and L-Enantiomers of racemic 1-O-acetyl-4,5,6-tri-O-benzoyl-myo-inositol (DL-**13**) and 1,4,5,6-tetra-O-benzoyl-myo-inositol (DL-**15**) were readily separated by a Chiralcel OD column (Daicel Chemical Industries, isopropanol/hexane=1/5). Retention times of the L- and D-enantiomers were 7.8 and 10.5 min for 13, and 8.1, and 12.5 min for 15, (flow rate 1.0 ml/min). The optical purities of D-**4** and L-**11** were therefore conveniently determined as tetrabenzoylated derivative

15 and 1-O-acetyl-4,5,6-tri-O-benzoyl-L-myo-inositol (**13**) respectively (Scheme 12).

V. ENZYMATIC ESTERIFICATION OF 1,3,5-O-MYO-INOSITOL ORTHOFORMATE

A. Enzymatic Esterification

In addition to a resolution of a racemic mixture, another method for preparation of optically active material is enantioselective acylation of meso material. myo-Inositol as well as inositol 1,3,5-O-orthoformate (**5**), readily available from inositol in one step [21] are both meso compounds. Enzymatic asymmetrization of the latter compound has been achieved by PLE-catalyzed hydrolysis of 4,6-di-O-acetyl-inositol 1,3,5-O-orthoformate [10]. We have investigated whether the asymmetrization of inositol 1,3,5-O-orthoformate can be performed by enzymatic esterification, which is more convenient than hydrolysis. The results are as follows.

Scheme 13

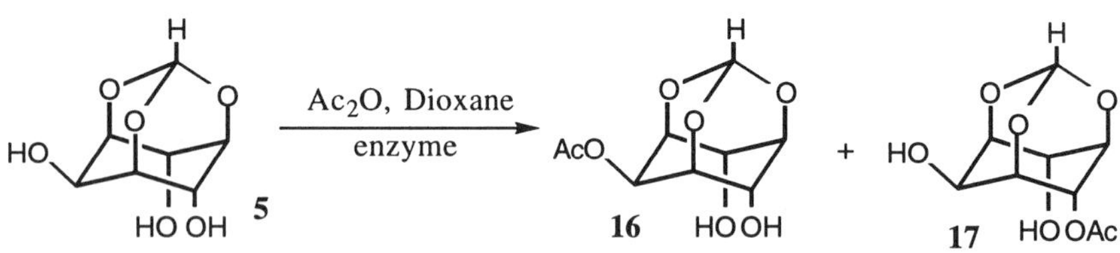

5 16 17

Table 9. Enzymatic Esterification of *myo*-Inositol 1,3,5-*O*-Orthoformate (**5**)

Enzyme	Solvent	Reaction Time	Yield		
			16	**17**	**5** (S.M.)
Lipase P	dioxane	24 h	50%		40%
Lipase AK	dioxane	24 h	57%		42%
Lipase CES	dioxane	24 h	45%	trace	42%
PLE	dioxane	36 h		40%	56%
Lipase M	dioxane	72 h	trace	22%	73%
phospho-lipase C	dioxane	1 week	trace	56%	35%
PLE	acetone	72 h	trace	51%	35%
Lipase M	acetone	72 h		52%	45%

First a few of some commercially available enzymes were screened (Scheme 13). Considering solubility of the material, all experiments were carried out in dioxane. The results are listed in Table 9. High regioselectivity was observed. PLE, Lipase M and phospholipase C selectively acylate the hydroxyl group of inositol 1,3,5-O-orthoformate at C-4 position to give 4-O-acetyl-inositol 1,3,5-O-orthoformate (**17**). Lipase AK, CES, and P acylated the hydroxyl group at C-2 position to give a meso derivative, 2-O-acetyl-inositol 1,3,5-O-orthoformate (**16**). Other enzymes showed no activity toward the acylation.

Scheme 14

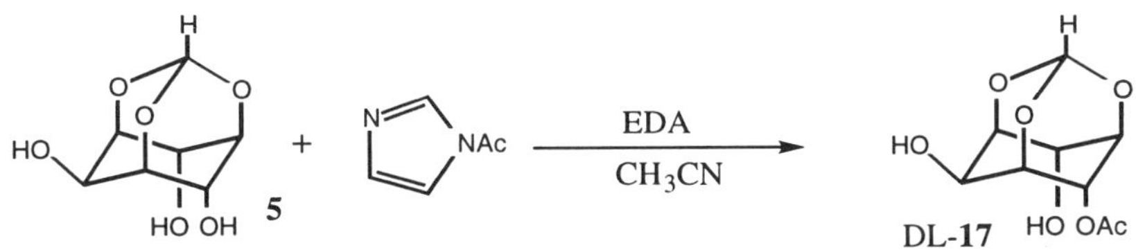

Scheme 15

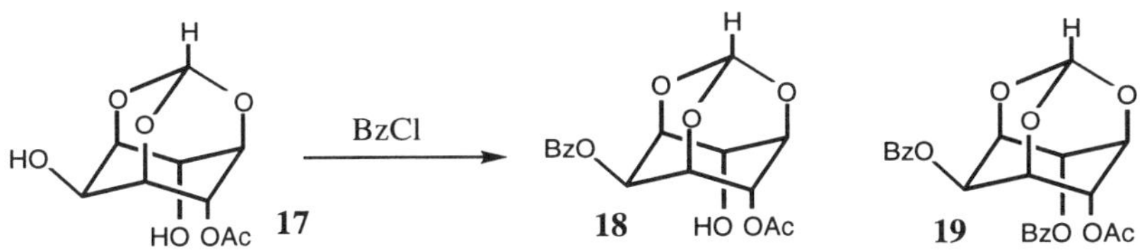

	18	19
pyridine	28%	57%
NEt_3/CH_2Cl_2	78%	0

Scheme 16

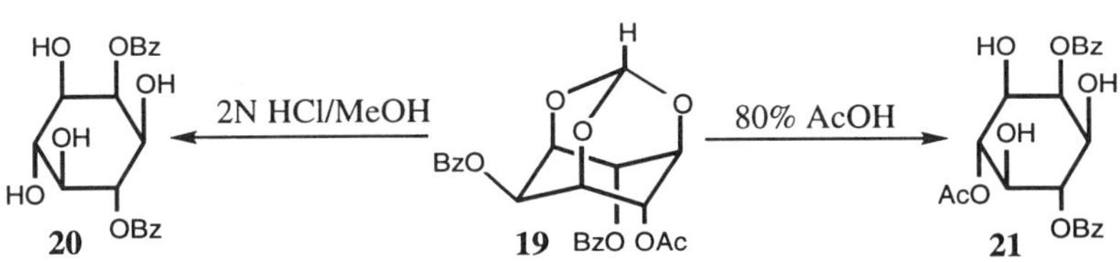

B. Determination of Optical Purity and Absolute Configuration

In order to determine the optical purity of 4-O- acylated product **17** obtained from enzymatic esterification, an authentic racemate was necessary. This authentic sample was synthesized chemically, as shown in Scheme 14 using N-acetyl imidazole as an acyl donor.

The reaction of 5 with N-acetyl-imidazole carried out in acetonitrile using EDA(ethyl diisopropyl amine) as a base proceeded in a highly selective manner to give 4-O-Acetyl-inositol 1,3,5-orthoformate (**17**) in 87% yield (Scheme 14). The acetate **17** was used to synthesize 4-O-acetyl-2-O-benzoyl-inositol 1,3,5-orthoformate (**18**), 4-O-acetyl-2,6-di-O-benzoyl-inositol 1,3,5-orthoformate (**19**), 4-O-acetyl-2,6-di-O-benzoyl-inositol (**21**) and 2,6-di-O-benzoyl-inositol (**20**) (Scheme 15 and Scheme 16). The determination of optical purity was attempted by chiral HPLC analysis. (Sumichiral OA and Chiralcel OD column). The above synthetic racemic authentic samples (**17-21**) were used to determine the final analytical condition. After tedious testing of various conditions, we found that the D and L-enantiomers of compound **17-21** could not be separated by the two columns. The determination of optical purity was therefore performed by ^{1}H NMR analysis after the compounds were transformed into Mosher's esters (Scheme 17) [22]. Unfortunately, all of the products of enzymatic esterification turned out to be racemic mixture. Thus, only regioselectivity but not enantioselectitvity was observed in enzymatic esterification of 5. Recently, similar enzymatic esterification of 5 was reported by M. P. Schneider [23]. Even though only the 2-acetylated product (**16**) was obtained from the enzymatic esterification of **5**, optically active compounds were obtained in successive procedures starting from **16**.

Scheme 17

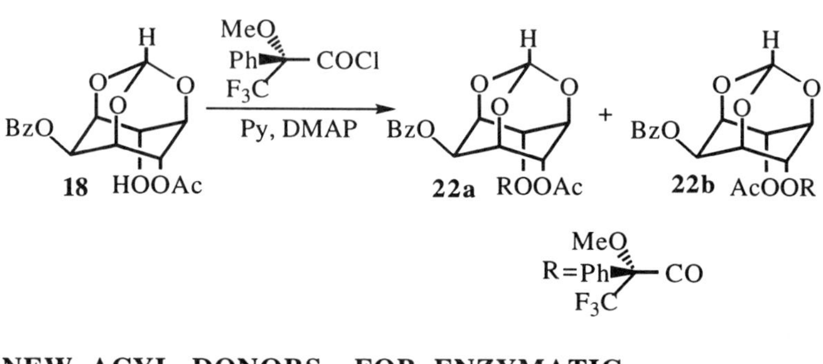

VI. NEW ACYL DONORS FOR ENZYMATIC ESTERIFICATION

A. Introduction

As shown in the previous enzymatic resolution of 1,2:5,6-di-O-cyclohexylidene-myo-inositol (**2** (Table 1), the results of resolution were greatly affected by the activity of the acyl donor. In general the initially discouraging results could be greatly improved by changing the acyl donor. Acetic anhydride gave the best results in the previous resolution. In some cases however, acid anhydrides could not be used as acyl donors for enzymatic esterifications. For example, if basic solvents like pyridine and triethylamine were used, the esterification proceeds well with acid anhydride even in the absence of enzyme. It is desirable to develop some highly active acyl donors for enzymatic reactions.

Scheme 18. Synthesis of Acyl Donor

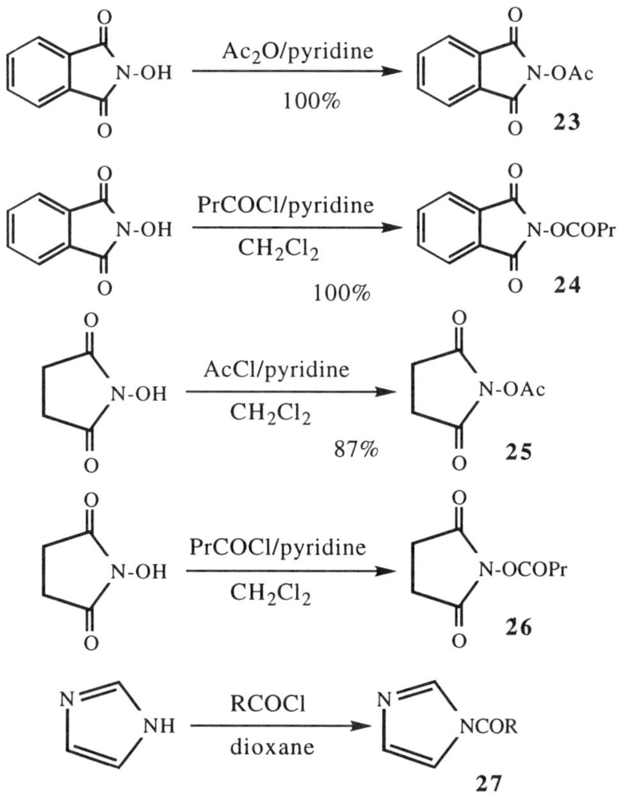

The reagents with activated carboxyl group such as N-acetoxy-, and butyroxy-succinimide and phthalimide were often used as acyl donors for acylation of amino groups of amino acide in peptide syntheses. We thus considered compounds with active carboxylic group as useful acyl donors for enzymatic transesterifications. Thus, N-acetoxy and N-butyroxy-succinimide and N-acetoxy- and butyroxy-phthalimide, acylated imidazole derivatives were synthesized and their utility in enzymatic esterification was investigated.

B. Synthesis

The synthesis of N-acetoxyphthalimide (**23**) proceeded well by direct acylation of N-hydroxyphthalimide with acetic anhydride in pyridine. Compound

23 was obtained quantitatively after general work-up.

The reaction of butyrylchloride with N-butyroxyphthalimide gave the corresponding ether **24** in quantitative yield. In contrast, acylation of N-hydroxysuccinimide with acetic anhydride in pyridine did not proceed at all, and a combination of acetyl chloride-triethylamine in CH_2Cl_2 did not give good results either.

The combination of AcOH-DCC in CH_2Cl_2 resulted in products which were difficult to separate. Both N-acetoxy and N-butyroxy succinimide (**25** and **26**) were obtained by using acid chloride in the presence of pyridine in CH_2Cl_2. The synthesis of acylated immidazole (**27**) was performed by using the corresponding acid chloride and two equivalents of imidazole in dioxane. All of the synthesized acyl donors were subsequenlty used for enzymatic transesterifictions.

C. Activity of the Synthesized Acyl Donors for Enzymatic Acylation

The activities of the synthesized acyl donors as acyl donors in enzymatic esterifications were tested with phenethyl alcohol (**28**) as an acyl accepter (Scheme 19). The results were compared with those obtained for trichloroethyl and vinyl esters. All reactions were carried out in the presence of Lipase AY. Due to solubility of the synthesized donors, benzene was used as the solvent in all enzymatic reactions.

Scheme 19

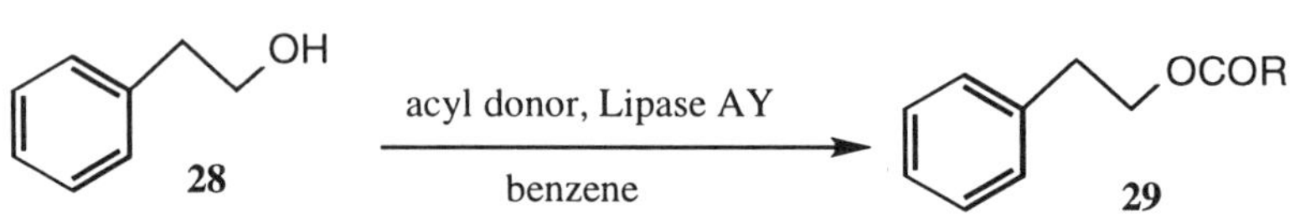

As shown in Figure 2 and Figure 3, all of the synthesized reagents were active for Lipase AY catalyzed esterification. Butyryl deirvatives showed higher activity than the coresponding acetyl derivatives, and phthalimide derivatives were more active than the corresponding succinimide derivatives; however, all compounds **23-27** were less active than the commonly used acyl donors. The activity decreased in the following order; trichloethyl acetate, vinyl acetate, N-acetoxy phthalimide,N-butyryloxy phtalimide, N-acetoxysuccinimide (Fig. 2) and trichloroethyl butyrate, N-butyroxyphthalimide, N-butyroxy succinimide (Fig. 3). Reaction rate increased with the increasing mass equivalent of enzyme, as shown in Figs. 2 and 3. The synthetic reagents could be used in other solvents such as ethyl ether, pyridine and dioxane. In pyridine, N-butyroxyphthalimide even showed higher activity than trichloroethyl butyrate.

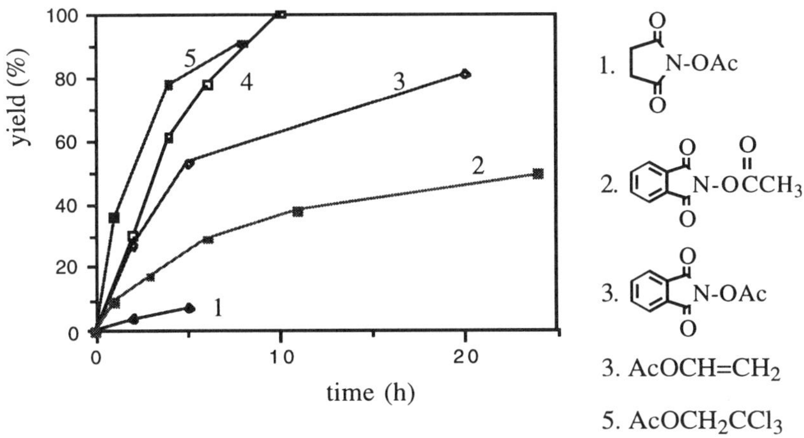

Figure 2. Enzymatic Esterification of **28**

2 Mass equivalent Lipase AY was used in these experiments.

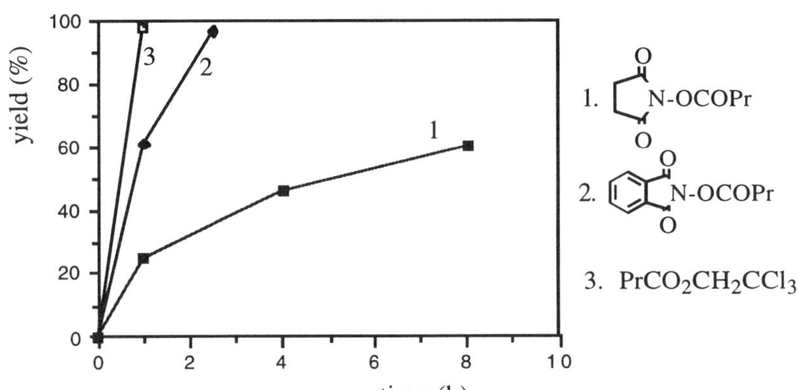

Figure 3. Enzymatic Esterification of **28**

4 Mass equivalent of Lipase AY was used in these experiments. Therefore, reaction proceeded faster than that shown in Fig. 2.

D. Practical Application of the Synthetic Reagents

N-Acetoxyphthalimide was used as an acyl donor in the resolution of racemic α-phenethyl alcohol and 1,2:3,4-di-*O*-cyclohexylidene-*myo*-inositol(**3**). Reactions proceeded well in both cases. In the case of 1,2:3,4-di-*O*-cyclohexylidene-*myo*-inositol (Scheme 20). the reaction was quenched by filtering the enzyme after 48 h. Separation of the reaction mixture gave D-5-*O*-acetyl-1,2:3,4-di-*O*-cyclohexylidene-*myo*-inositol (37%) in an almost optically pure form and L-1,2:3,4-di-*O*-cyclohexylidene-*myo*-inositol (50%) with 64% e.e. optical purity.

Scheme 20

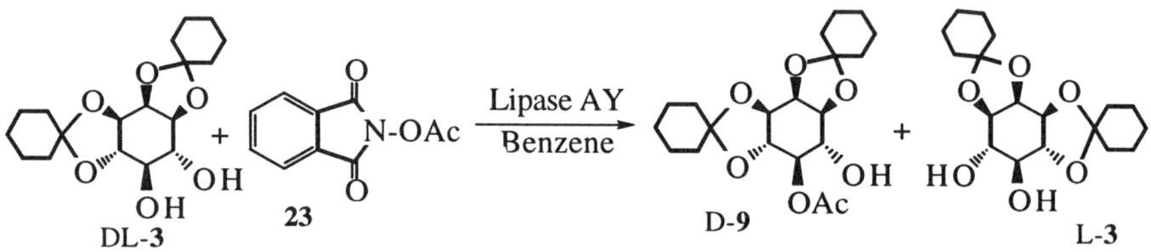

37% (100% e.e.) 50% (64% e.e.)

VII. Enzymatic Resolution of Hindered *myo*-Inositol derivative [24]

A. Introduction

Lipase-catalyzed kinetic resolution of racemates has been developed as a common and efficient method for the preparation of optically active materials in organic synthesis. However this methodology encounters problems with sterically hindered starting materials [11c]. Recently, several reports appeared concerning resolution of hindered molecules. For example, N. J. Turner and coworkers studied resolution of the mixed oxalates of tertiary alcohol (**30**) by hydrolysis of the carboxyl group distal from chiral center. Although PPL (porcine pancreatic lipase) did hydrolyze the less hindered carboxyl group, better results were obtained when the carboxyl group nearer to the chiral center was hydrolyzed [25]. In the preparation of optically active 4-aryl-1,4-dihydro-2,6-dimethyl-pyridine derivatives from 4-aryl-1,4-dihydro-2,6-dimethyl-3,5-pyridine-dicarboxylate (**31**), K. Achiwa and coworkers found that the alkyl ester **31** was inert to enzymatic hydrolysis because of the effect of steric hindrance. They have successfully solved this problem by introduction of an acyloxymethyl group to the molecule (**32**) and hydrolysis of the distal carboxyl group [26].

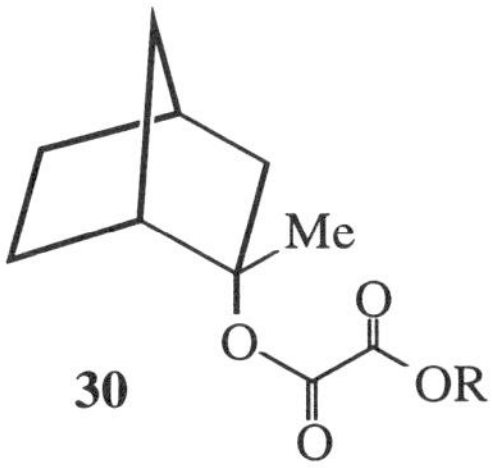

30

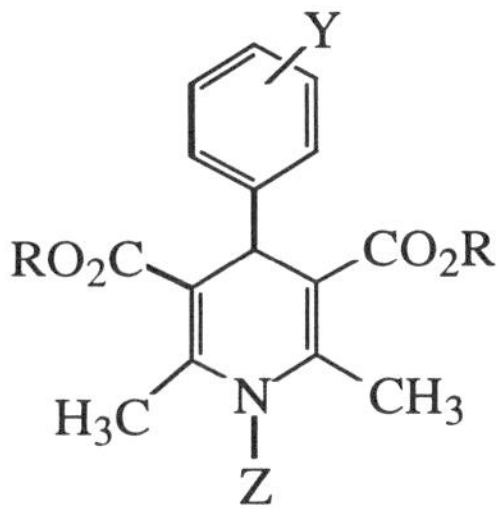

31 R = alkyl
32 R = acyloxymethyl

In our application of enzymes for preparation of optically active *myo*-inositol derivatives, we found that some *myo*-inositol derivatives are very sterically hindered and resistant to enzymatic reaction. To minimize the effect of sterical hindrance, we have derived a sterically hindered hydroxyl group with a side chain, which contains a terminal hydroxyl group, and used the terminal hydroxyl group for resolution.

B. Enzymatic Esterification

Scheme 21

$HOCH_2CO_2H \xrightarrow[\text{pyridine}]{\text{TBDMS-Cl}} TBDMSOCH_2CO_2TBDMS$ (**33**) $\xrightarrow[CH_2Cl_2,\ DMF]{ClCOCOCl} TBDMSOCH_2COCl$ (**34**)

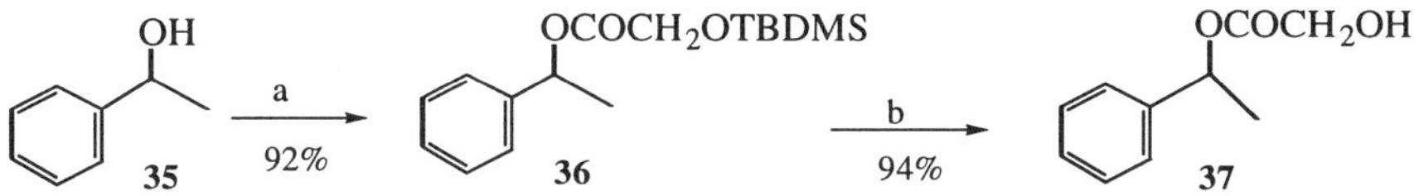

35 36 37

(a) $TBDMSOCH_2COCl$, pyridine, ether, 0°C to room temperature; (b) Bu_4NF, AcOH, THF

sec-Phenethyl alcohol (**35**) was selected as a material for model tests. Thus, a glycolic acid side chain with a terminal hydroxyl group on it was attached to the chiral hydroxyl group of **35** by an ester bond. The preparation of material is shown in Scheme 21. Both the carboxyl and the hydroxyl group of glycolic acid were protected by silylation to give **33**. The acid chloride **34** with a protective group on the hydroxyl group, was then formed under neutral conditions from **33** [27]. This acid chloride was used for the acylation of **35** without purification to give phenethyl 2-*O*-t-butyldimethylsilyloxyacetate (**36**) in 92% yield. After the silyl group had been removed [28], phenethyl 2-hydroxyacetate (**37**) was used as material for enzymatic esterification. The results are outlined in Scheme 22. Compared with the enzymatic esterification of *sec*-phenethyl alcohol itself, (Scheme 23) the esterification of the primary hydroxyl group proceeded much faster it has been reported [29], but enantioselectivity decreased drastically.

Scheme 22

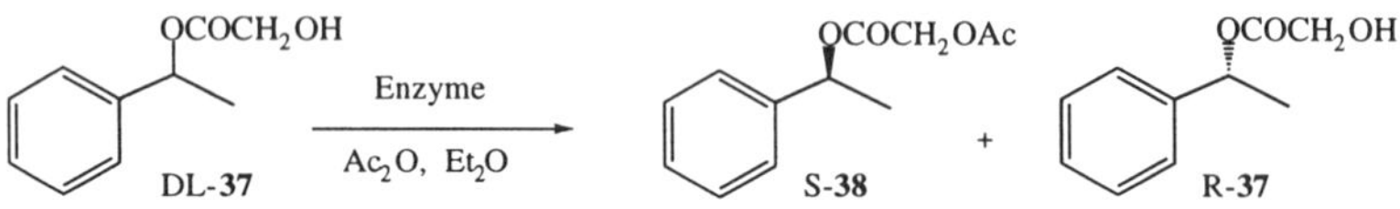

enzyme	Ac_2O	time	Conversion	Yield (ee)		E
Lipase P 1 eq.	2	3h	59%	46% (9% ee)	42% (13% ee)	1.3
"	2	5h	71%	55% (8% ee)	31% (20% ee)	1.4
Lipase PS 1 eq.	2	5h	85%	55% (6% ee)	12% (35% ee)	1.5

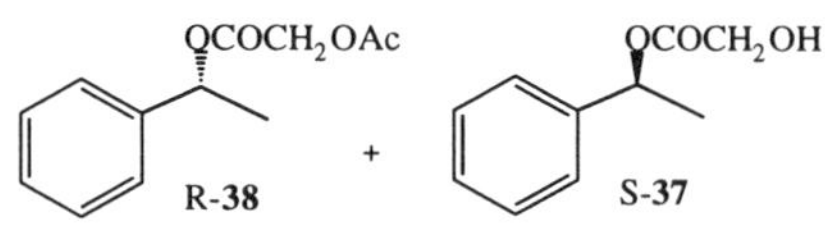

SP 435 1 eq.	2	15 min	85%	68% (4% ee)	11% (57% ee)	1.6

Scheme 23

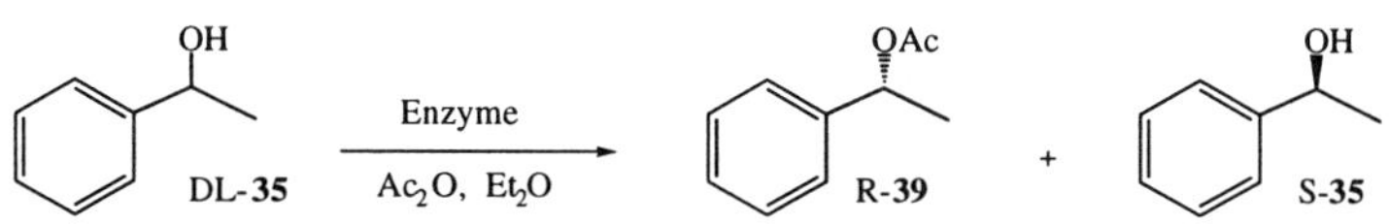

enzyme	Ac_2O	time	conversion	Yield (ee)		E
Lipase PS 1 eq.	2	4 h	41%	36% (80% ee)	62% (56% ee)	16
SP 435 1 eq	2	30 min	43%	38% (87% ee)	51% (66% ee)	30
Lipase P 1 eq.	2	13 h	51%	45% (93% ee)	47% (96% ee)	97

The E value decreased to 1.3 for the primary hydroxyl group from 97 for the secondary one in the case of Lipase P. One interesting finding is the selectivity of Lipase P and Lipase PS for the primary hydroxyl group was opposite to the

secondary one.
In order to determine optical purities and absolute configurations of the products from the enzymatic reactions, some corresponding optically pure authentic samples have been synthesized as shown in Scheme 24.

$$E = \frac{\ln [(1-c)(1-ee_s)]}{\ln [(1-c)(1+ee_s)]} \quad \text{or} \quad E = \frac{\ln [(1-c(1+ee_p)]}{\ln [(1-c(1-ee_p)]}$$

$$c = \frac{ee_s}{ee_s + ee_p}$$

ee_s=enantiomer eccess of starting material : ee_p=ee of product

The optical purities and absolute configurations of **39** and **35** (Scheme 23) were determined by comparing the specific rotation value with the ones of optically pure S-**35** and its acetate (S-**39**) [30]. The optical purity and absolute configuration of **37** and **38** (Scheme 22) were determined by comparing the specific rotations with the ones of optically pure S-**37** and S-**38** derived from S-**35**.

Scheme 24

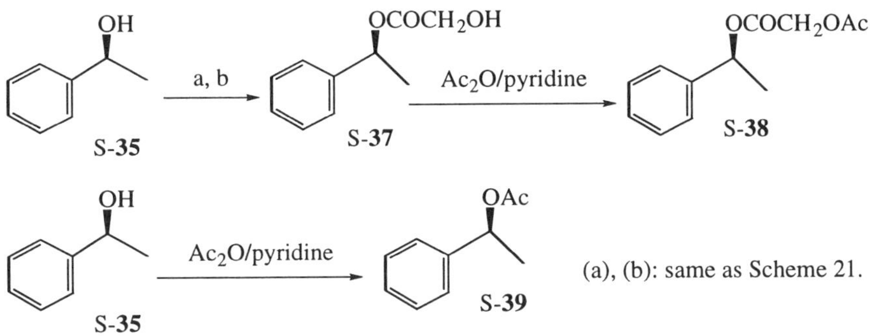

(a), (b): same as Scheme 21.

Scheme 25

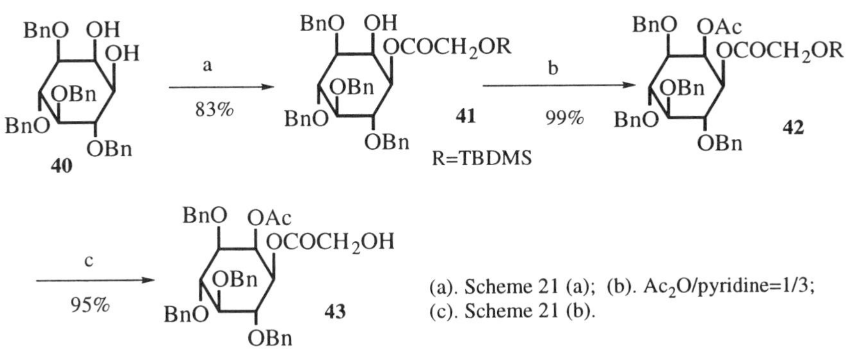

(a). Scheme 21 (a); (b). Ac_2O/pyridine=1/3;
(c). Scheme 21 (b).

3,4,5,6-*O*-Tetrabenzyl-*myo*-inositol (**40**), easily available from *myo*-inositol in high yield [17], is an important precursor for the preparation of naturally occurring inositol derivatives [15,17-18,31]. The optical resolution of this material is usually accomplished by formation of its diastereomeric derivatives and their separation [15,17-18,31-32].

Scheme 26

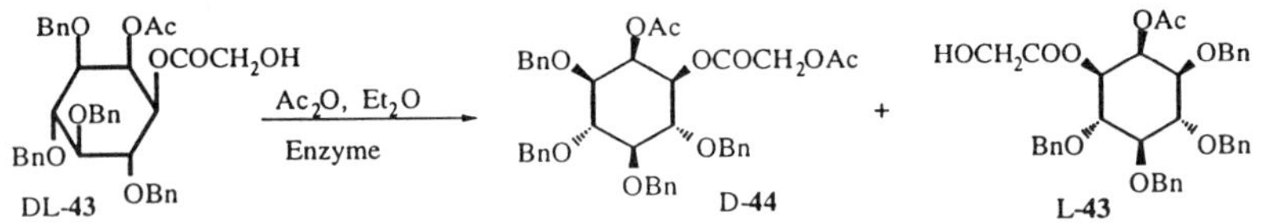

Enzyme (eq)	Ac_2O (eq)	Reaction Time	Conversion	Yield (e.e.)		E
KWI-56 (3.2 eq)	3	45 min	49%	39% (66% ee)	49% (64% ee)	10
KWI-56 (1.5 eq.)	2	2 h	55%	53% (62% ee)	44% (76% ee)	10
Lipase PS (2 eq)	2	12 h	30%	25% (32% ee)	68% (14% ee)	2
Lipase CES (2.6 eq)	3	6 h	38%	37% (30% ee)	51% (18% ee)	2
Lipase P (3 eq.)	2	36 h	25%	16% (36% ee)	62% (12% ee)	2

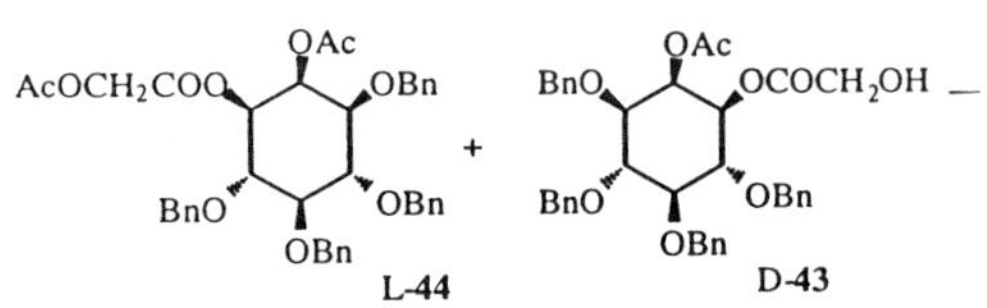

SP 435 (4 eq)	3	1 h	67%	64% (46% ee)	25% (94% ee)	9
SP 435 (2 eq.)	2	45 min	58%	56% (64% ee)	38% (88% ee)	13
Lipase AY (4.8 eq)	3	3 h	38%	39% (78% ee)	49% (48% ee)	13
Lipase AY 2.4 eq	2	18 h	24%	27% (74% ee)	61% (24% ee)	9

In our research on the preparation of optically active inositol intermediates and synthesis of inositol phosphates, we have attempted to resolve this material by enzymatic esterification of the hydroxyl groups and enzymatic hydrolysis of the corresponding acetate. The unsuccesful results prompted us to develop a modification for enzymatic resolution of such hindered hydroxyl compounds. The preparation of starting materials is shown in Scheme 25. The hydroxyl group at the C-1 position of **40** was acylated with 2-(*O*-t-butyldimethylsilyloxy) acetyl chloride to give **41** in 83% yield. In order to prevent the migration of the acyl group, the hydroxyl group at the C-2 position was protected by acetylation. After cleavage of the silyl group, the compound **43** was obtained in 95% yield.

Scheme 27

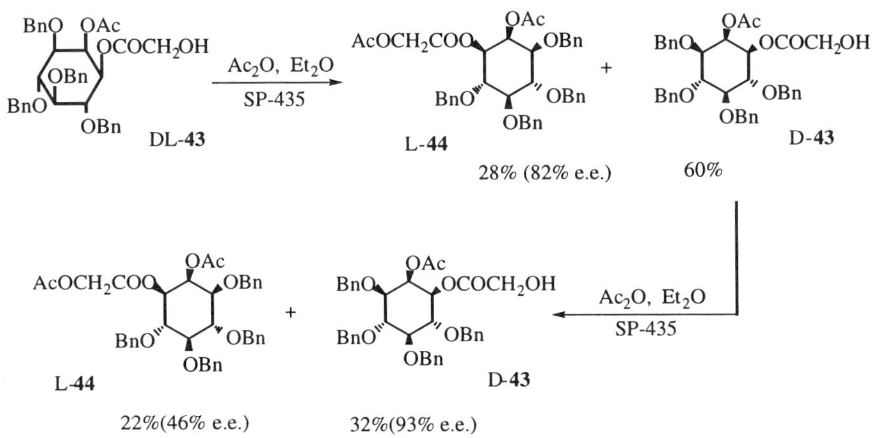

Compound **43** was then subjected to enzymatic esterification. The results are outlined in Scheme 26 and 27. Optical purities were analyzed by HPLC analysis using a chiral column (Chiralcel OD, isopropanol/hexane=1/5) after products (D or L-**43** and **44**) were hydrolyzed to tetrabenzyl inositol (**40**) (Scheme 28). The absolute configuration was determined by comparing the specific rotation of the hydrolyzed product **40** with known data [31c].

Scheme 28

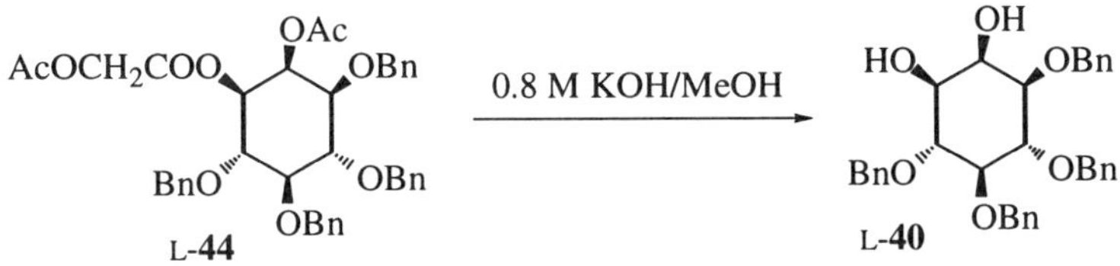

Among the enzymes screened, Lipase P, PS, CES (Amano, lipases from *Pseudomunas sp.*), and KWI-56 (Kurita, lipase from *Pseudomonas sp.*) recognized the D-enantiomer through the distal hydroxyl group. Low selectivities were observed with Lipase P, PS, and CES, while KWI-56 showed higher selectivity (E=10).

Scheme 29

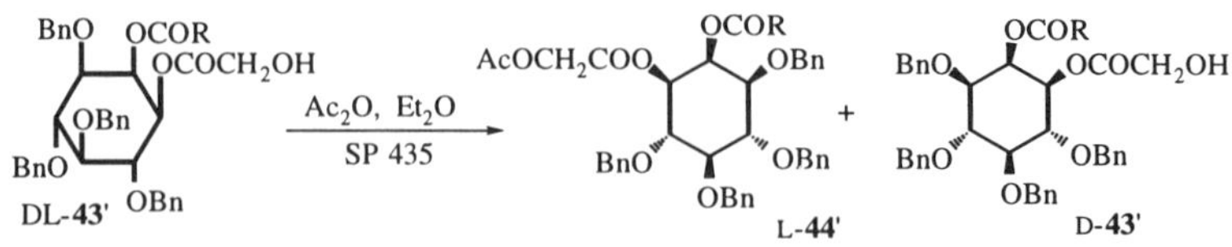

R	Reaction Time	Conversion	Yield (e.e.)		E
Me	45 min	58%	56% (64% ee)	38% (88% ee)	13
Et	30 min	56%	47% (56% ee)	39% (72% ee)	7.5
$(CH_2)_4Me$	1 h	59%	51% (52% ee)	37% (74% ee)	6.6
Ph	1 h	51%	52% (42% ee)	44% (44% ee)	3.7

Scheme 30

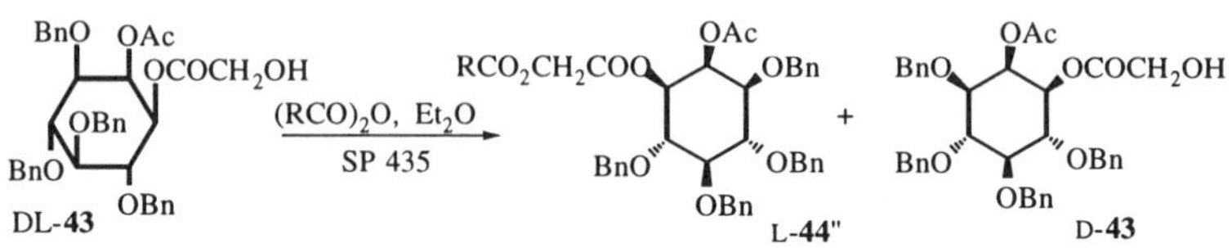

R	Reaction Time	Conversion	Yield (e.e.)		E
Me	45 min	58%	56% (64% ee)	38% (88% ee)	13
Me_3C	2.5 h	43%	41% (32% ee)	49% (24% ee)	2.4
$(CH_2)_4Me$	45 min	50%	53% (2% ee)	45% (2% ee)	1.1

Lipase AY (Amano, lipase from *Candida rugosa*) and immobilized enzyme SP-435 (Novo, lipase from *Candida antarctica*) acetylated the hydroxyl group of the L-enantiomer with modest selectivity. (E values were 9 and 13 respectively).

The D-enantiomer could be obtained in high optical purity (94% e.e.) at the expense of chemical yield. When the esterification was stopped at low conversion, and the recovered unesterified material was recycled, both enantiomers were obtained at relatively high purity (28% yield, 82% e.e. for L-**44** and 32% yield, 93% e.e. for D-**43**) (Scheme 27).

The effect of substituents at the C-2 position and the chain length of the acyl donor has been investigated. The results are shown in Scheme 29 and Scheme 30. The selectivity (E value) was strongly affected by the substituents and acyl donors decreasing with the increase of the chain length of acyl donors and the size of the substituents.

C. Enzymatic Hydrolysis

The effect of steric hindrance on the rates and selectivity of enzymatic hydrolysis was studied with the use of of the racemic substrate **44** because there are three ester bonds which could be hydrolyzed by the enzyme.

Scheme 31

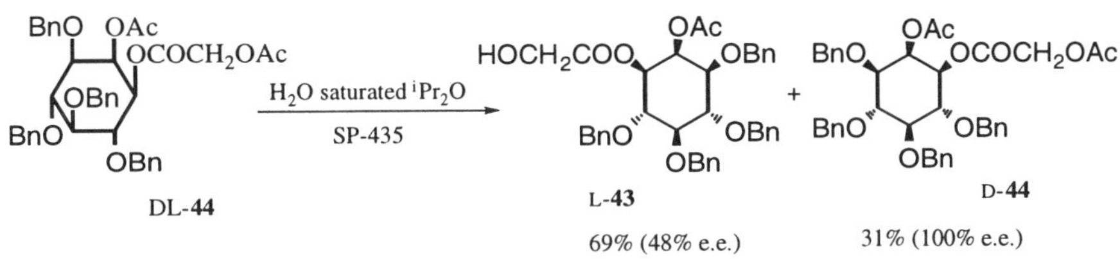

The effect of steric hindrance should be manifested by the regioselectivity of the reaction, i.e. hydrolysis of **44** should occur at the less hindered acetate. Considering the different conformations the enzyme may adopt in organic and aqueous media and the solubility of the material, the hydrolysis was carried out in water-saturated isopropyl ether (Scheme 31). As it had been expected, SP435 hydrolyzed the less hindered acetate exclusively to give **43**.

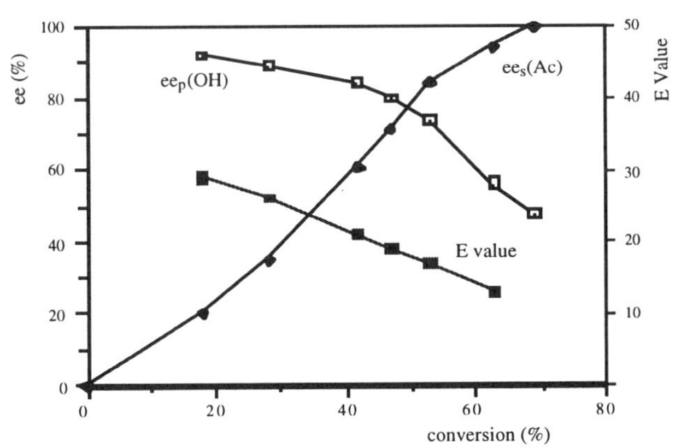

Figure 4. Kinetics of Enzymatic Hydrolysis of the ester **44**.

The reaction was followed by HPLC analysis (Chiralcel OD column, iPrOH/Hexane=1/20, 0.9 ml/min, flow rate). The progress of conversion of DL-**44** into alcohol **43** and the optical purity of both products were determined at the same time. The results are shown in Figure 4. The change of the optical purities of both products during the hydrolysis were clearly observed. The optical purity of L-**43** decreased with the proceeding of hydrolysis while the optical purity of D-**44** increased. So it is easy to determine when the reaction should be stopped based on the results shown in Figure 4. After 5 days at room temperature, about 69% of the material was hydrolyzed to give L-**43** with enantiomeric excess of 48% and D-**44**, approximately 31% in optically pure form. Isolation and removal of the acyl group at D-**44** afforded the optically pure D-**40**. The calculated E value decreased during the hydrolysis indicating the reversibility of the reaction.

In summary, 3,4,5,6-*O*-tetrabenzyl-*myo*-inositol (**40**), which is resistant to direct enzymatic esterification, was resolved by a modified approach involving esterification of the more remote and less hindered hydroxyl group. The application of this method for resolution of various sterically hindered alcohols and effect of the length of side chain on the recognition of asymmetric center by enzyme are under investigation.

VIII. SYNTHESIS OF D-MYO-INOSITOL 1,4,5-AND 1,4,6-TRISPHOSPHATE

A. Synthesis of D-*myo*-Inositol 1,4,5- and 1,4,6-Trisphosphate from 1,2:5,6-Di-*O*-cyclohexylidene-and 1,2:3,4-Di-*O*-Cyclohexylidene-*myo*-Inositol

It has been reported that *myo*-inositol 1,4,5-trisphosphate 1,4,5-IP_3 could be easily synthesized from 1,2:5,6-di-*O*-cyclohexylidene-*myo*-inositol [12].

The key steps of the synthesis are the regioselective protection of the hydroxyl group at C-4 position by benzylation and the selective cleavage of the transacetal group to produce a 1,4,5-triol suitable for further phospholylation. A combination of this procedure with the enzymatic resolution method described in the preceeding section provides a convenient route to optically active 1,4,5-IP_3. Thus, selective benzylation of L-1,2:5,6-di-*O*-cyclohexylidene -*myo*-inositol by reported methods gave 4-*O*-benzyl-1,2:5,6-di-*O*-cyclohexylidene-*myo*-inositol (**45**) in 83% yield [9a]. D-*myo*-Inositol 1,4,5-trisphosphate can be synthesized from it. L-*myo*-Inositol 1,4,5-trisphosphate can be synthesized from D-1,2:5,6-*O*-cyclohexylidene-*myo*-inositol in the analogous way (Scheme 32).

Scheme 32

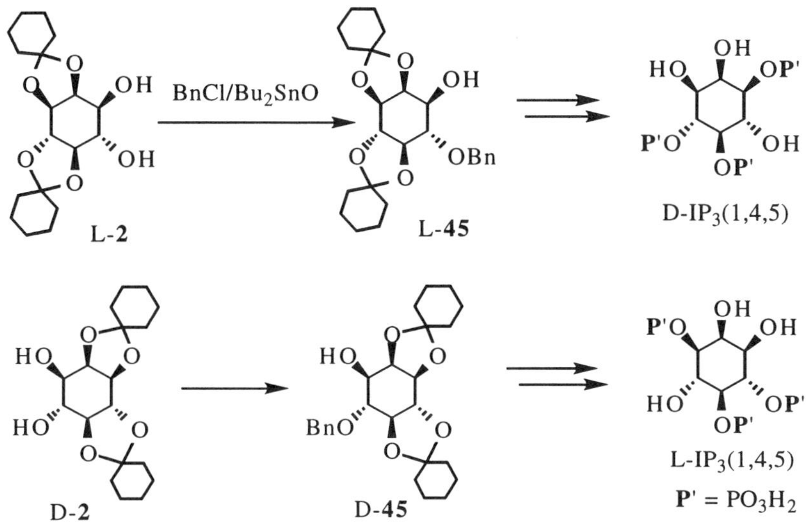

Both D- and L-*myo*-inositol 1,4,5-trisphosphate can be also effectively synthesized from the early described L-**7** and D-**7** respectively.

Scheme 33

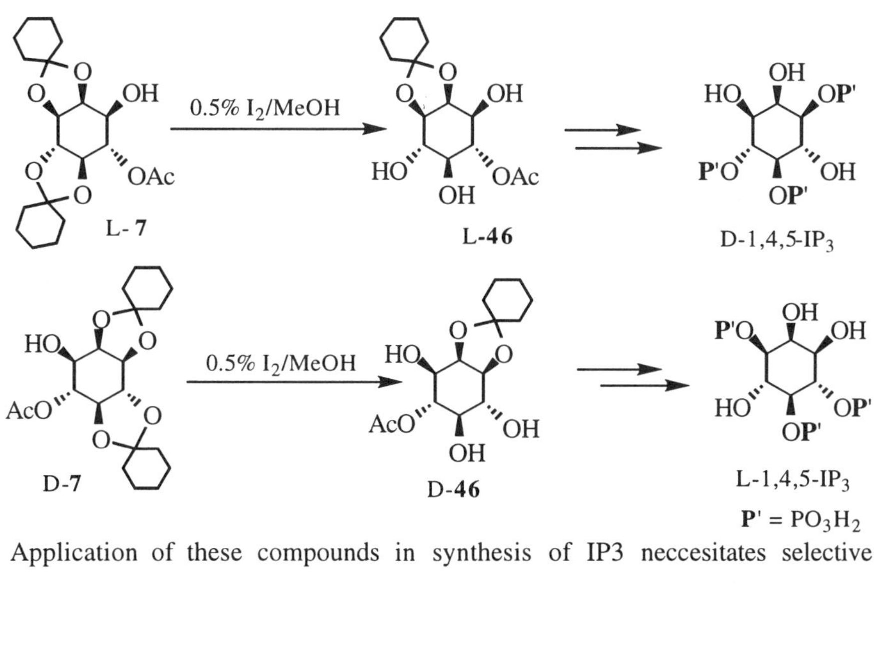

Application of these compounds in synthesis of IP3 neccesitates selective removal of the trans-acetal moiety. Deprotection under acidic conditions resulted in the migration of the cis-acetal to give a mixture of products. It has been found that the selective cleavage of trans-acetal in L-or D-7 proceeded smoothly with I_2 in MeOH to give 6-*O*-Acetyl-2,3-*O*-cyclohexylidene *myo*-inositol (**46**) in 60% yield (Scheme 33). Synthesis of 1,4,5-IP_3 from 46 will be described elsewhere.

The derivatives D- and L-**9** described earlier in this work are useful in the synthesis of D- and L-(1,4,6)IP_3 as shown in Scheme 34. The selective deprotection of trans-acetal could be performed either under acidic condition (aqueous acetic acid) or using iodine in MeOH to give 5-*O*-acetyl-2,3-*O*-cyclohexylidene-*myo*-inositol (**47**) in the 57% and 72% yield respectively. Synthesis of the D- and L-(1,4,6)IP_3 from L- and D-**47**, respectively, will be described later.

Scheme 34

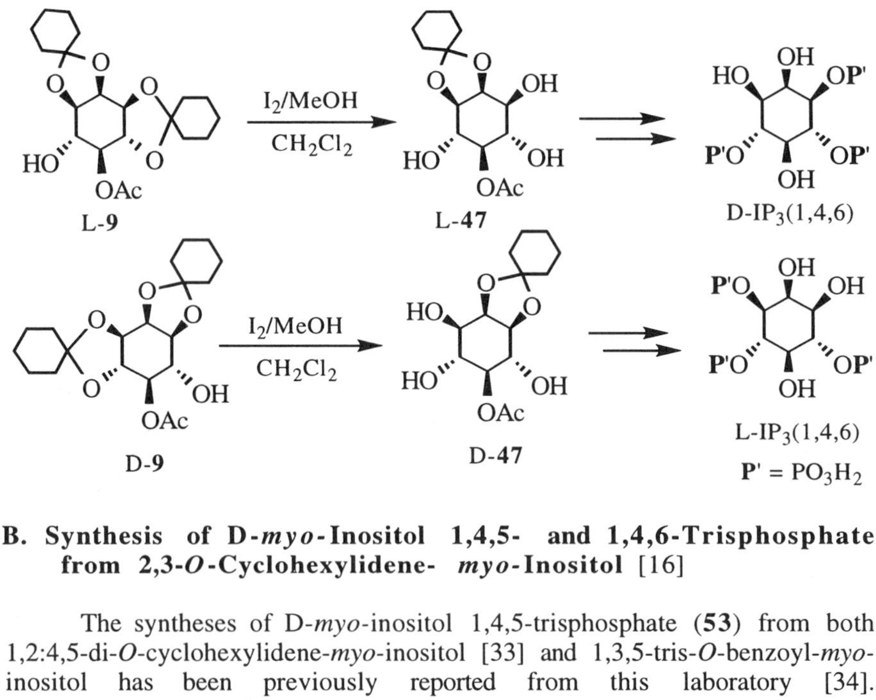

B. Synthesis of D-*myo*-Inositol 1,4,5- and 1,4,6-Trisphosphate from 2,3-*O*-Cyclohexylidene- *myo*-Inositol [16]

The syntheses of D-*myo*-inositol 1,4,5-trisphosphate (**53**) from both 1,2:4,5-di-*O*-cyclohexylidene-*myo*-inositol [33] and 1,3,5-tris-*O*-benzoyl-*myo*-inositol has been previously reported from this laboratory [34].

Scheme 35

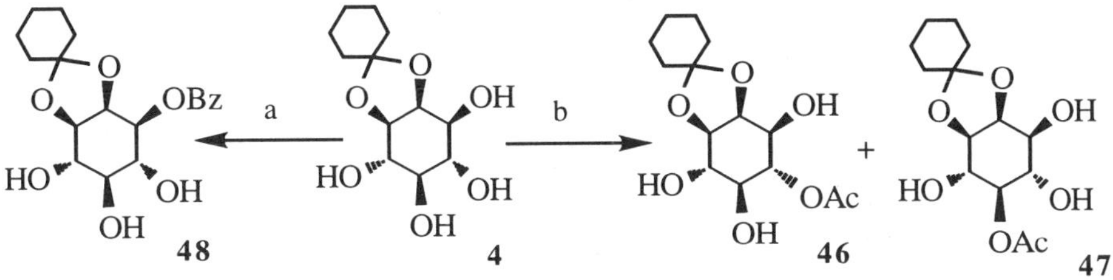

a) BzCl, 2 eq; pyridine; b) Ac_2O, 6 eq; 4A MS, DMA

Other research groups used 1,2:4,5-di-*O*-isopropylidene-*myo*-inositol (Gigg's group) [35] and 1,2:5,6-di-*O*-cyclohexylidene-*myo*-inositol (Vacca's [12] and Chen's [9a] group) and some other inositol derivatives [36] as starting materials for the synthesis.

Scheme 36

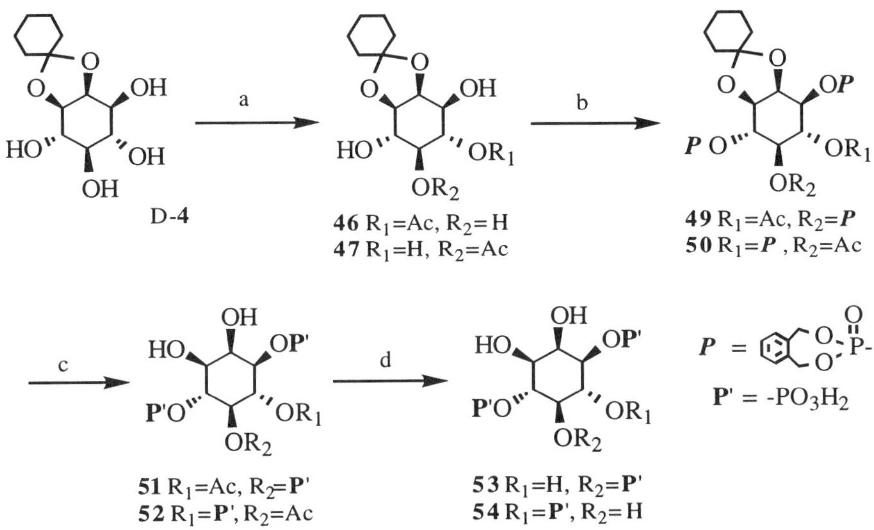

a) Ac_2O, 4A MS, DMA; b) ref. 34; c) H_2 Pd/C, H_2O/MeOH;
d) concd. NH_4OH

The common drawbacks of these methods are low yields of starting materials and multiple protection steps resulting in the overall low yield of **53**.

A new possibility was explored using the readily available acetal **4**. 1,2-Acetal **4** could be selectively protected at the hydroxyl group at the C-1 position with bulky acylating or silylating reagents [18]. Independently, we have studied the selective protection of this material by direct acylation with benzoyl chloride, benzoic anhydride, acetyl chloride, and acetic anhydride in such solvents as pyridine, DMF, and DMA at different temperatures (Scheme 35). Treatment of **4** with benzoyl chloride (2 eq.) in pyridine at room temperature gave predominantly 1-*O*-benzoylated product (**48**, 47%). In contrast, when acetic anhydride (6 eq.) in DMA was used as an acylating reagent at room temperature in the presence of powdered molecular sieve 4A, the 6- and 5-mono-*O*-acetylated products (**46** and **47**) were obtained in approximately 1:1 ratio. The 6-acetylated product, features a proper protection pattern for the synthesis of inositol 1,4,5-trisphosphate (**53**). The 5-acetylated product, is properly protected for the synthesis of inositol 1,4,6-trisphosphate (**54**). Thus, the protection was accomplished in one step. No satisfactory results were obtained in the cases of acylation with benzoic anhydride and acetyl chloride.

The synthesis of **53** and **54** starting from optically pure D-**4** is shown in Scheme 36. The 6- and 5-acetylated products (**46** and **47**) were isolated as a mixture in 74% total yield. The mixture was subjected to phosphorylation using the method developed in this laboratory [37] to give the mixture of phosphorylated products **49** and **50**, which was separated by silicagel chromatography. The yield of **49** and **50** was 41% and 48%, respectively, from the mixture of **46** and **47**. The removal of *O*-xylylene group by catalytic hydrogenolysis and removal of cyclohexylidene group by methanolysis catalyzed by the formed phosphate acids in **49** and **50** and, followed by hydrolysis of the acetyl group in concentrated ammonia solution at room temperature, and purification by cellulose chromatography afforded **53** and **54** from **49** and **50** respectively. Thus **53** was synthesized in the overall yield of 30% from D-**4**. Based on the yield of the starting material [17], the overall yield of **53** from *myo*-inositol was 13%.

In summary, we have described three methods for synthesis of D-*myo*-inositol 1,4,5-trisphosphate from the optically pure materials obtained by enzymatic reactions. Small number of steps in the last procedure combined with the ready availability of the starting material provide a new practical route for the synthesis of our target product. The application of other optically active inositol derivatives in the synthesis of naturally occurring inositol phosphatesis is described in the following sections.

IX. A CONVENIENT CHEMOENZYMATIC SYNTHESIS OF D- AND L-MYO-INOSITOL 1,4,5,6-TETRAKISPHOSPHATE [38]

We have previously reported a practical procedure for kinetic resolution of 2,3-*O*-cyclohexylidene-*myo*-inositol (**4**) by enzymatic esterification in anhydrous organic solvent (Scheme 11). We now report the use of both enantiomers of 4 in the synthesis of D- and L-*myo*-inositol 1,4,5,6-tetrakisphosphate (D-**56** and L-**56**) [39]. The former was recently prepared by Bruzik and coworkers [40].

As previously reported, Lipase CES (a lipase from *Pseudomonas sp.*)

catalyzed esterification of racemic 2,3-*O*-cyclohexylidene-*myo*-inositol (**4**) gave L-1-*O*-acetyl-2,3-*O*-cyclohexylidene-*myo*-inositol (L-**11**) and D-2,3-*O*-cyclohexylidene-*myo*-inositol (D-**4**).

Scheme 37

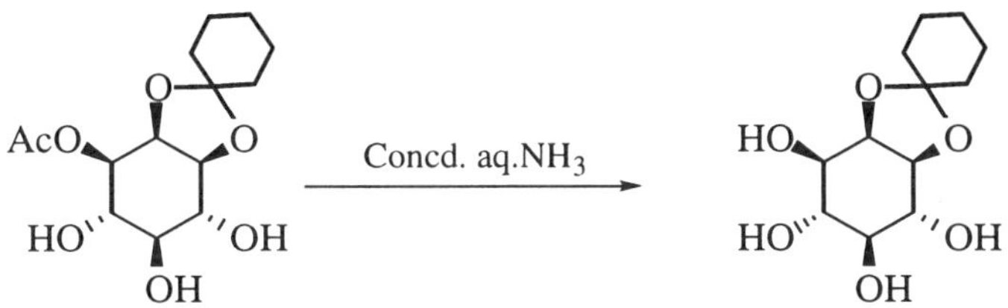

L-**11** L-**4**

The acetyl group of L-**11** was hydrolyzed in concentrated aqueous ammonia solution to give L-**4** (Scheme 37). D-**4** and L-**4** were used for the synthesis of D-**56** and L-**56** respectively.

Scheme 38

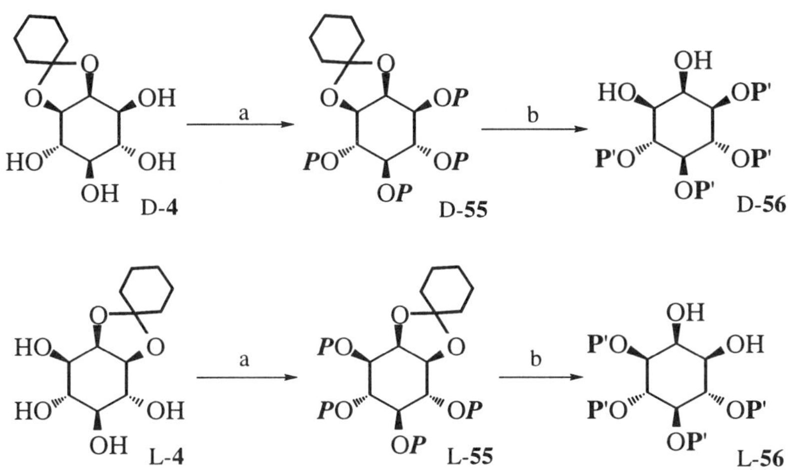

a. $(BnO)_2PN^iPr_2$, MCPBA; b. H_2 , 5%Pd/C. *P*=$(BnO)_2PO$, **P'**=PO_3HNa

The synthesis of D-**56** from optically pure D-**4** is shown in Scheme 38. The phosphitylation of D-**4** proceeded smoothly by reaction with dibenzyl *N,N*-diisopropyl-phosphoramidite in the presence of 1*H*-tetrazole followed by oxidation with m-chloroperbezoic acid to give fully protected D-*myo*-inositol 1,4,5,6-tetrakisphosphate (D-**55**). The removal of the benzyl group on the phosphoric ester was performed by catalytic hydrogenolysis. The hydrolysis of the cyclohexylidene group was catalyzed by acidic phosphate groups formed during hydrogenolysis. After purification by means of cellulose chromatography (PrOH/aq. NH_3/H_2O=5/4/1) and ion exchange column, the final product D-**56** was obtained as a sodium salt. The product gave satisfactory NMR spectra compared with reported ones [39]. L-Enantiomer was obtained from L-**4** in the same way.

The procedure described here provides a short practical route to both enantiomers of *myo*-inositol 1,4,5,6-tetrakisphosphate. Synthesis of L-enantiomer has not been described before.

X. SYNTHESIS OF D-MYO-INOSITOL 1,2,4,5,6-PENTAKISPHOSPHATE AND MYO-INOSITOL 1,3,4,5,6-PENTAKISPHOSPHATE

A. Synthesis of D-*myo*-Inositol 1,2,4,5,6-Pentakisphosphate

The synthesis of D-*myo*-inositol 1,2,4,5,6-pentakisphosphate is shown in Scheme 39. Triol L-**11** was prepared by Lipase CES catalyzed acetylation of 2,3-*O*-cyclohexylidene-*myo*-inositol as described earlier in this work. Depending on the enzyme used, sometimes this material contains other regio-isomers such as 4-, 5-, or 6-acetylated products; however, the triol **11** can be purified by recrystallization from AcOEt-hexane. The cyclohexylidene group was hydrolyzed in trifluoroacetic acid-methanol (1:1, v/v). After the solvent was evaporated to dryness, the residue was dried under vacuum to give **57** quantitatively. No acetyl group migration was observed during hydrolysis of the cyclohexylidene moiety. Compound **57** was used for phosphorylation without further purification.

Scheme 39

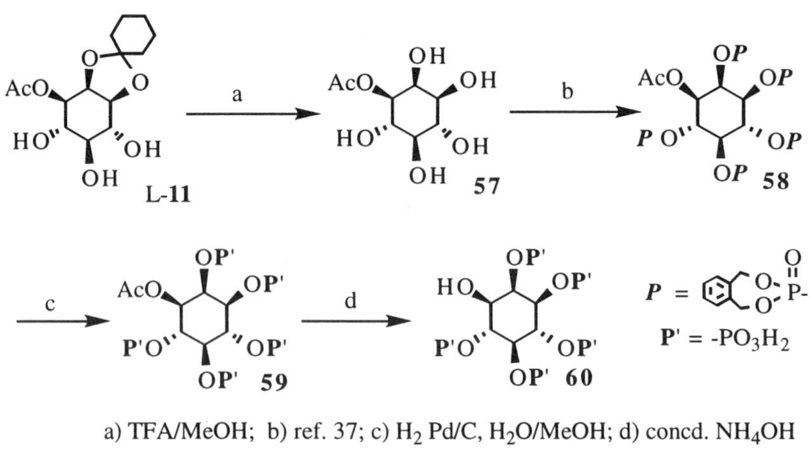

a) TFA/MeOH; b) ref. 37; c) H_2 Pd/C, H_2O/MeOH; d) concd. NH_4OH

The phosphorylation of the acetate **57** proceeded well to afford the fully protected D-*myo*-inositol 1,2,4,5,6-pentakisphosphate D-(1,2,4,5,6)IP_5 (**58**) in a 89% yield. Deprotection of **58** was performed by hydrogenolysis followed by hydrolysis of the acetyl group in the concentrated aqueous ammonia. The final product D-(1,2,4,5,6)IP_5 (**60**) was obtained as a mono-sodium salt. Thus, a new optically active inositol phosphate derivative was effectively synthesized. The biological activity of this product is under investigation.

B. Synthesis of *myo*-Inositol 1,3,4,5,6-Pentakisphosphate [41]

Finally, another biologically interesting metabolite, [42] inositol 1,3,4,5,6-pentakis-phosphate and its 2-*O*-substituted analogs were synthesized as shown in Scheme 40.

Scheme 40

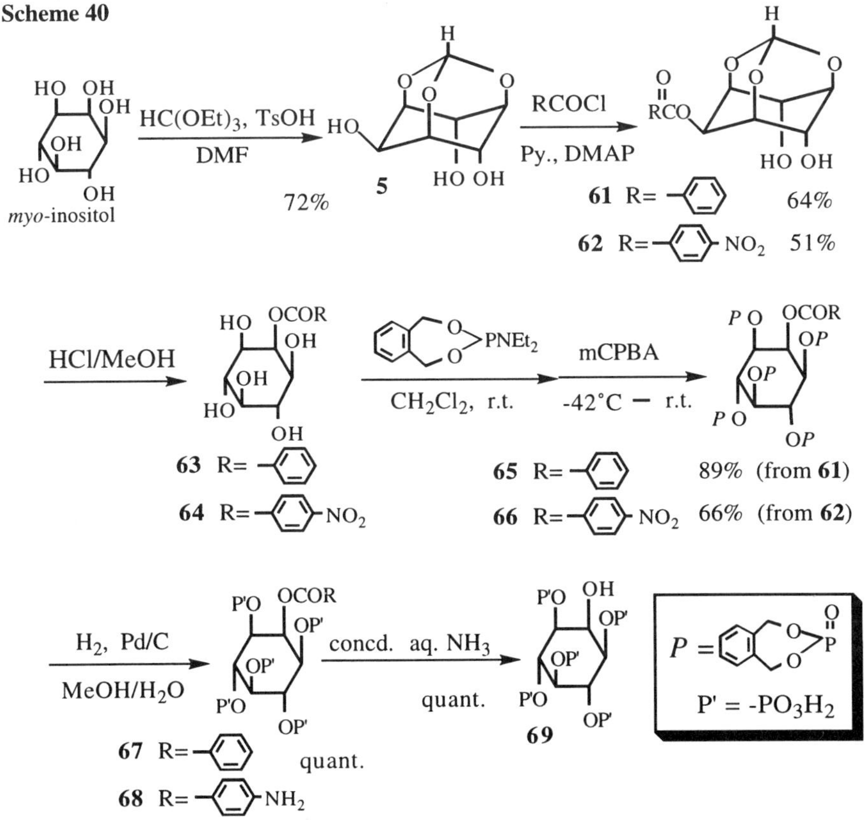

myo-Inositol 1,3,5-orthoformate (**5**), [21] was selectively protected at the C-2 position when it was treated with benzoyl chloride or *p*-nitro-benzoyl chloride in pyridine to give **61** and **62**, respectively. The hydrolysis of the orthoformate **61** proceeded smoothly in 4 M HCl/MeOH; however, HCl treatment of **62**, caused migration of the *p*-nitrobenzoyl group. Higher lability of the p-nitrophenyl group is probably due to the effect of a strong electron-withdrawing group (-NO_2). The successful removal of orthoformate could be accomplished with 1 M HCl/MeOH to give the pentol **64**. Intermediates **63** and **64** were subjected to phosphorylation and deprotection of the ester group on the phosphate function without purification to give the 2-*O*-substituted inositol 1,3,4,5,6-pentakisphosphate analogs **67** and **68**. Finally, the benzoyl group of **67** was hydrolyzed in concentrated aqueous ammonia to afford inositol 1,3,4,5,6-pentakisphosphate (**69**). Because the benzoates at the C-2 position of **67** and **68** are sensitive to alkaline hydrolysis, cellulose chromatography with an eluent containing 40% of concentrated aqueous ammonia (n-PrOH/aq.NH_3/H_2O= 5/4/1), gave a mixture containing 25% hydrolyzed product **69** and an unhydrolyzed benzoates **67** and **68**. An eluent containing less than 3% of concentrated aqueous ammonia is therefore recommended. A biological study of inositol 1,3,4,5,6-pentakisphophate is under investigation.

Conclusion

Many myo-inositol acetals and esters are D-L-mixtures. These racemic compounds can be separated by enzymatic acylations or hydrolysis to optically active compounds. Using these optically active compounds, many inositol polyphosphates such as D- and L-(1,4,5)IP3 , D- and L-(1,4,6)IP3, D- and L-(1,4,5,6)IP4, D- and L-(1,2,4,5,6)IP5 were obtained easily.

References

1. Billington DC. Recent developments in the synthesis of *myo*- inositol phosphates. Chem Soc Rev. 1989; 18: 83- 122.
2. Michell RH. Inositolphospholipids and cell surface receptor function. Biochim Biophys Acta, 1975; 415: 81-147.
3. a) Shears SB. Metabolism of the inositol phosphates produced upon receptor activation. Biochem J. 1989; 260: 313-324.; b) Potter BVL, Recent advances in the chemistry and biochemistry of Inositol Phisphates of biological interest. Natural Products Reports. 1990; 7: 1-24.
4. Billington DC. The Inositol Phosphates, VCH, 1993.
5. a) Ling L, Watanabe Y, Akiyama T, Ozaki S. A new efficient method for resolution of *myo*- inositol derivatives by enzyme catalyzed regio- and enantiop selective esterification in organic solvents. Tetrahedron Lett. 1992; 33: 1911-1914.; b) Ling L, Li X,Watanabe Y, Akiyama T, Ozaki S. Bioorg & Med Chem. 1991; 1, 155-158.
6. a) Tegge W, Ballou CE, Proc Natl Acad Sci USA. 1989; 86: 94.; b) Falck JR, Yadagiri P. Enantiospecific synthesis of D-myo-Inositol 1,4,5-trisphosphate from (-) Quinic acid. J Org Chem. 1989; 54: 5851-5852.;

(c) Watanabe Y, Mitani M, Ozaki S. Synthesis of optically active derivatives starting from D-glucurono-6,3,-lactone. Chem Lett. 1987; 123-126.; d) Akiyama T, Takechi N, Ozaki S. Chiral synthesis of D-myo-1-phosphate starting from L-quebrachitol. Tetrahedron Lett. 1990; 31: 1433-1434.

7. a) Ozaki S, Kondo Y, Shiotani N, Ogasawara T, Watanabe Y. J Chem Soc Perkin Trans. 1, 1992, 729-737.; b) Ozaki S, Watanabe Y, Ogasawara T, Kondo K, Shiotani N, Nishii H, Matsuki T. Total synthesis of optically active myo-inositol 1,4,5-trisphosphate. Tetrahedron Lett. 1986; 27: 3157-3161; c) Vacca JP, Desolms SJ, Huff JR. Total synthesis of D- and L-myo-Inositol 1,4,5-trisphosphate. J Am Chem Soc. 1987; 109: 3478-3479.; d) Stepanov AE, Runova OB, Schlewer G, Spiess B, Shvets VI. Total synthesis of chiral inositol 1,4,5-triphosphate. Tetrahedron Lett. 1989; 30: 5125-5128.; e) Baker R, Kulagowski JJ, Billington DC, Leeson PD, Lennon IC, Liverton N.Synthesis of 2- and 6-deoxyinositol 1-phosphate and the role of the adjacent hydroxy groups in the mechanism of inositol monophosphatase. J Chem Soc Chem Commun. 1989; 1383-1385.; f) Young RC, Downes CP, Eggleston DC, Jones M, Macphee CH, Rana KK, Ward JG. Total synthesis of the four stereoisomers of dihexadecanoyl phosphatidyl inositols. J Med Chem. 1990; 33: 641-646.; g) Tagliaferri F, Wang SN, Berlin WK, Outten RA, Shen TY. Glycosyl inositols I. Synthesis of 1-substituted *chiro*-inositol derivatives Tetrahedron Lett. 1990; 31: 1105-1108.; h) Westerduin P, Willems HAM, van Boeckel CAA. Tetrahedron Lett. 1990; 31: 6915.
8. Ozaki S, Kohno M, Nakahira N, Bunya M, Watanabe Y. Synthesis of optically active myo-inositol 1,3,4-triphosphate. Chem Lett. 1988, 77-80.
9. a) Liu Y-C, Chen C-S. An efficient Synthesis of optically active D-myo-inositol 1,4,5,-triphosphate. Tetrahedron Lett. 1989; 30: 1617-1620.; b) Honing H, Seufer-Wasserthal P, Stutz AE, Zenz E. Facile chemo-enzymatic synthesis of selectively protected derivatives of deoxy-inositols.Tetrahedron Lett. 1989; 30: 811-814.
10. Baudin G, Glanzer BI, Swaminathan KS, Vasella A. Helv Chim Acta. 1988; 71: 1367.
11. a) Roberts SM, Turner JN. J Biotech. 1992; 22: 227.; b) John VT, Abraham G. Biocatal Ind. 1991; 193.; c) Chen C-S, Sih CS. General aspects and optimization of enantioselective biocatalysis in organic solvents. The use of lipases. Angew Chem Int Ed Engl. 1989; 28: 695.; d) Klibanov AM. Lipase catalyzed transformations using enolesters as acylating reagents.Acc Chem Res. 1990; 23: 114.; e) Wang, Y-F, Lalonde JJ, Momongan, M, Bergbreiter DE, Wong C-H. Lipase catalyzed irreversible transesterification using enol, esters as reactive acylating agents. J Am. Chem. Soc. 1989; 110: 7200-7204.
12. a) Vacca JP, deSolms ST, Huff JR, Billington DC, Baker R, Kulagowski

JJ, Mawer IM. Synthesis of 1D and 1L-inositol 1,3,4,5-tetraphosphate. Tetrahedron. 1989; 45: 5679.; b) Plourde R, d'Alarcao M, Synthesis of potentially insulin mimetic phosphodisaccharide.Tetrahedron Lett. 1990; 31, 2693-2697.

13. a) Zhu LM, Tedford M. Application of pig liver testerase. Tetrahedron. 1990; 46: 6587.; b) Whitesides GM, Wong C-H. enzymes as catalysts in synthetic organic chemistry.Angew Chem Int Ed Engl. 1985; 24: 617.; c) Chen CS, Fujimoto Y, Girdaukas G, Sih CJ. Quantitative analysis of biochemical kinetics resolutions of enantiomers. J Am Chem Soc. 1982; 104: 7294-7297.
14. Gopalsamy A, Anderson WK. Tetrahedron Lett. 1991; 32: 4669-4672.
15. Sadovnikova MS, Kuznetsova ZP, Shvets VI, Evstigneev RP. Zh Org Khim. 1975; 11: 1211-1218.
16. a) Ling L, Ozaki S. Enzym aided synthesis of D-*myo*-inositol 1,4,5-trisphosphate. Tetrahedron Lett. 1993; 34: 2501-2504.; b) Ling L, Ozaki S. A chemoenzymatic synthesis of D-*myo*-inositol 1,4,5-trisphosphate.Carbohydr Res. 1994; 256: 49-58.
17. Massy DJR, Wyss P. The synthesis of DL-1-(hexadecanoyloxy)methyl-1O-hexadecanoyl-inositols as inhibitor of phospholipase C. Helv Chim Acta. 1990; 73, 1037-1043.
18. a) Bruzik KS, Myers J, Tsai MD. Practical synthesis of enantiomerically pure *myo*- inositol derivatives. Tetrahedron Lett. 1992; 33: 1009-1012.; b) Bruzik KS, Tsai MD. Efficient synthesis of enantiomerically pure and regiospecifically protected myo-inositols. J Am Chem Soc. 1992; 114: 6361-6369.
19. Gupta MN. Enzyme function in organic solvents. Eur J Biochem. 1992; 203: 25-32.
20. Akiyama T, Takechi N, Ozaki S. Achiral synthesis of D-*myo*-inositol 1-phosphate starting from l-quebrachitol. Bull Chem Soc Jpn. 1992; 65: 366-372.
21. a) Baudin G,Glanzer BI, Swaminathan KS, Vasella A. A synthesis of lD- and lL-*myo* -inoisitol 1,3,4,5-tetraphosphate. Helv Chim Acta, 1988; 71: 1367-1378.; b) Lee WH, Kishi Y. Synthesis of mono and unsaturated bis ortho esters of *scyllo*-Inositol. J Org Chem. 1985; 50: 4402- 4404.
22. a) Dale JA, Mosher HS. Nuclear magnetic resonance enantiomer reagents. Configurational correlations via nuclear magnetic reseonance chemical shifts of diastereomeric mandelate, O-methylmandelate and α - methoxy-α trimethyulphenylacetate (MPTA) esters. J Am Chem Soc. 1973; 95: 512-519. b) Dale JA, Dull DL, Mosher, HS. Methoxytrifluoromethyl-phenylacetic acid. Aversatile reagent for the determination of enantiomeric composition of alcohol. J Org Chem. 1969; 34: 2543-2446.
23. Andersch P, Schneider MP. Enzyme assisted synthesis of enantiomerically pure *myo*- inositol derivatives- chiral building blocks fro inositol polyphosphates. Tetrahedron:Asymmetry. 1993; 4: 2135-2138.
24. Ling L, Ozaki S. Enantiomeric resolution of streically hindered inositol

derivatives. Bull Chem Soc Jpn. 1995; 68: 1200-1205.
25. Brackenridge I, McCague R, Roberts SM, Turner NJ. Enzymatic resolution of oxalate of a tertiary alcohol using porcine pancreatic lipase. J Chem Soc Perkin Trans.1, 1993; 1093-1094.
26. a) Hirose Y, Kariya K, Sasaki I, Kurono Y, Achiwa K. Protease-catalyzed enantioselective synthesis of optically active 1,4-dihydroxy pyridines. Tetrahedron Lett. 1993; 34: 3441-3444.; b) Y. Hirose, K. Kariya, I. Sasaki, Y. Kurono K, Achiwa K. A novel synthesis of valnidipine. Tetrahedron Lett. 1993; 34: 5915-5918.; c) Ebiike H,Terao Y, Achiwa K. Acyloxymethylas an activating group in lipase-catalyzed hydrolysis. Tetrahedron Lett. 1991; 32: 5805-58; Ebiike H,. Maruyama K, Achiwa K. Lipase catalyzed hydrolysis of 4-alkyl1,4-dihydropyridine derivatives. Chem Pharm Bull. 1992; 40: 1083-1088.
27. Wissner A, Grudzinskas CV.Reaction of t-butylmethylsilylester with oxalylchloride-dimethylformamide. J Org Chem. 1978; 43: 3972-39.
28. Greene TW, Wuts PGM. Protective Groups in Organic Synthesis; John Wiley & Sons, Inc. New York. 1991.
29. a) Jouglet B, Rousseau G. Enzymatic resolution of N-hydroxymethyl-γ-butyrolactams. An access to optically active γ-butyrolactams. Tetrahedron Lett. 1993; 34: 2307-2310.; b) Khushi T, O'Toole KJ, Sime JT Biotransformation of phosphonate esters. Tetrahedron Lett. 1993; 34: 2375-2378.; c) Barth S, Effenberger F. Lipase-catalyzed resolution of racemic 2-alkyl substituted 1-alkanols. Tetrahedron: Asymmetry. 1993; 4: 823-833.
30. Catalog Handbook of Fine Chemicals, Aldrich.
31. a) Watanabe Y, Inada E, Jinno M, Ozaki S. Phosphonium salt methodology for the synthesis of phosphonic monoesters and diesters its application to selective phosphorylation. Tetrahedron Lett. 1993; 34: 497-5009.; b) Garegg PJ, Lindberg B. Carbohydr Res. 1988; 173: 205-210.; c) Shvets VI, Klyashchitskii BA, Stepanov AE, Evstigneeva RP. Tetrahedron 1973; 29: 331-340.; d) Baker R, Kulagowski JJ, Billington DC, Leeson PD, Lennon IC, Liverton N. Synthesis of 2- and 6-deoxy inositol-1-phosphate. J Chem Soc Chem Commun. 1989, 1383-1384.
32. Aneja R, Parra A. Facile optical resolution of DL-1,4,5,6-tetra-O-benzyl-*myo*- inositol: key synthesis for the phosphoinositides. Tetrahedron Lett. 1994; 35: 525-526.
33. a) Ozaki S, WatanabeY, Ogasawara T, Kondo Y, Shiotani N, Matsuki T. Tetrahedron Lett. 1986; 27: 3157-3161.;
34. Watanabe Y, Fujimoto T. Shinohara T, Ozaki S. A shotr step synthesis of optically active *myo*--inositol 1,3,4,5-tetrakis(phosphate) and *myo*-inositol 1,4,5,tris(phosphate) from 1,3,4,-triO-benzyl-*myo*-inositol. J Chem Soc Chem Commun. 1991; 428-429.
35. Gigg J,Gigg R. Payne, S, Conant R. The allyl group for protection in carbohydrate chemistry Part 18. Allyl and benzyl ethers of myo-Inositol. Intermediates for the synthesis of myo-inositol triphosphates. J Chem

Soc Perkin Trans. 1, 1987; 423-4.

36. a) Reese CB, Ward JG. Synthesis of D-*myo*-inositol 1,4,5-triphosphate. Tetrahedron Lett. 28, 2309-2312.; b) Aguilo A, Lomas MM. Penades S.Efficient synthesis of *myo*-inositol 1,4,5-trisphosphate. Tetrahedron Lett. 1992; 33: 401-40 ; c) Salamonczyk GM, Pietrusiewicz KM. Expedient synthesis of D-*myo*- inositol 1,4,5,-trisphosphate and 1,4-bisphosphate. Tetrahedron Lett. 1991; 32: 6167-616.
37. Watanabe Y, Komota K, Ebisuya K, Ozaki S. An efficient phosphorylation method using a new phospoitylating agent 2-diethylamino-1,3,2-benzodiaphosphoephane. Tetrahedron Lett. 1990; 31: 255- 256.
38. Ozaki, L. Ling, T. Ogasawara, Y. Watanabe Y, Hirata M.A convenient chemoenzymatic synthesis of D- and L-*myo*-inositol 1,4,5,6-tetrakispfosphate. Carbohydr Res. 1944; 259: 307-310.
39. Mayr GW, Dietrich W. The only inositol tetrakisphosphate detectable in avian erythocytes. Is the isomer lack phosphate at position 3: a NMR study. FEBS Lett. 1987; 213: 278.
40. Pietrusiewicz KM, Salamonczyk GM, Bruzik KS. The synthesis of Homochiral inositol phosphates from *myo*- inositol. Tetrahedron. 1992; 48, 5523-5242.
41. Ozaki S, Koga Y, LingL, Kimura Y, Hirata M, Watanabe Y. Synthesis of 2-substituted *myo*-inositol 1,3,4,5-tetrakisphosphate and 1,3,4,5,6-pentakisphosphate analogs. Bull Chem Soc Jpn. 1994; 67: 1058-1068.
42. Hughes PJ, Shears SB. Inositol 1,3,4,5,6-pentakisphosphate and inositol hexakisphosphate inhibit inositol-1,3,4,5-tetrakisphosphate 3-phosphatase in rat parotoid glands. J Biol. Chem. 1990; 265: 9869-9875.

9
Phosphatidylinositol-Specific Phospholipase C as a Target of Rational Drug Design

Karol S. Bruzik
University of Illinois at Chicago
Chicago, Illinois

I. INTRODUCTION

Receptor-mediated metabolism of inositol phospholipids has been a subject of intense research during the last decade, starting with the discovery of the turnover of inositol phospholipids in response to multiple extracellular signals, such as growth factors, neurotransmitters, hormones and even physical factors (1). At the center of this response is receptor-mediated activation of phosphatidylinositol-specific phospholipase C (PI-PLC) and ensuing hydrolysis of inositol phospholipids to produce water soluble, small molecular weight inositol phosphate second messengers (Figure 1) (2). While originally the proposed mechanism was relatively simple, starting with phosphatidylinositol 4,5-bisphosphate to give a single inositol 1,4,5-trisphosphate (IP_3) second messenger, further development in this area showed enormous complexity of inositol phospholipid and phosphate metabolism, and involvement of many other protein factors such as G-proteins, protein tyrosine kinases, protein kinases C, cytoskeletal proteins, etc. Many new structures of inositol phosphates have been discovered recently, including phospholipase C resistant 3-phosphorylated phospholipids (3), inositol pyrophosphates (4) and prostaglandylinositol 1,2-cyclic phosphate (5). Some of these metabolites function as second messengers themselves. Numerous reviews concerning regulation of PI-PLC have appeared recently (6-11), therefore this aspect of PI-PLC will be omitted here. In addition, several reviews on the strategies of development of anti-signaling compounds as potential drugs have also been published (12-16). This work is focused on strategies to develop inhibitors of phospholipase C as potential drug candidates, based on its chemical mechanism (17) and structure of its active site.

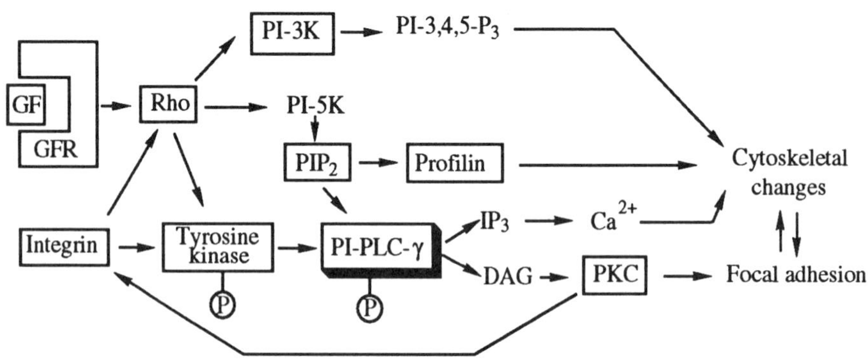

Figure 1 General scheme of signal transduction involving phosphoinositides. There are four major second messengers derived from natural phosphatidyl--inositols: IP_3, DAG, PI-3,4,5-P_3 and phosphoinositol glycans (PIG, not shown) produced by activation of two major types of receptors: membrane spanning G-protein-associated receptors activated by small peptides, and tyrosine kinase receptors activated by growth factors. Formation of PIG second messengers occurs most likely through G-linked receptors (redrawn from ref. 2).

Due to its key role in cellular signal transduction PI-PLC is a ubiquitous enzyme in most prokaryotic and eukaryotic organisms. To date multiple forms of this enzyme have been found, many of which have been characterized, sequenced, cloned and expressed (6-11, 17). The structure-function studies utilizing site-directed mutagenesis (SDM) and X-ray crystallography have just begun, and will undoubtedly intensify in the future to provide further details of the enzyme active site structure, catalytic mechanism, interfacial activation and binding, and modes of regulation. The first crystal structure of bacterial PI-PLC has been solved (18a), and it is anticipated that analogous data for mammalian enzymes will be forthcoming in the near future.

A. Function of PI-PLC

PI-PLC comprises a broad family of proteins of various sizes, sequences and functions (6-11, 17). All known activities can be roughly separated into five categories: mammalian PI-PLC-β, -γ, -δ, bacterial PI-PLC, and glycosylphosphatidylinositol-specific phospholipase C from unicellular parasites (GPI-PLC) (6-11, 17, 19). The earlier identified PI-PLC-α (20) is currently suspected to be derived from one of the major mammalian types. Likewise, a broad spectrum of mammalian PI-PLC activities reported earlier is now believed to have resulted from the above mentioned isozymes by proteolytic cleavage or aggregation. Still, reports of mammalian PI-PLC activities requiring metals other than calcium (21), and showing specificity for unphosphorylated substrates (22), suggest that many more members of the PI-PLC family are yet to be discovered. Of the types listed in Table 1 only the functions of PI-PLC-β and -γ isozymes are clearly understood. PI-PLC-β responds to the signals of small molecular weight mitogenic peptides such as bombesin. These oligopeptides bind to seven-pass membrane spanning receptors associated with G-proteins (8). PI-PLC-γ responds to signals of a variety of growth factors, such as EGF, PDGF, TGF, FGF, CSF, etc. (6,7).

Table 1 Summary of the major known PI-PLC types

PI-PLC type, source	MW [kDa]	Substrate	Metal Ion	Regulation
α, guinea pig	60	PIP_2>PIP>PI	Ca^{2+}	none
β, bovine brain	154	PIP_2>PIP>PI	Ca^{2+}	G-protein
γ, bovine brain	148	PIP_2>PIP>PI	Ca^{2+}	phosphorylation
δ, bovine brain	85	PIP_2>PIP>PI	Ca^{2+}	unknown
bacterial	34	PI ~ GPI	none	none
GPI-PLC	39	GPI>>PI	none	unknown

Binding of the growth factor to the specific tyrosine kinase receptor causes its conformational change and dimerization followed by mutual phosphorylations of each monomer at tyrosine residues. Phosphorylated receptor sequesters from the cytosol a variety of proteins bearing sequence homology domains (SH2) such as PI-PLC-γ and PI 3-kinase (23). In the case of the PDGF receptor it has been proven that binding of PLC to the receptor causes activation of these proteins and stimulates hydrolysis of phosphatidylinositol 4,5-bisphosphate, and further sends the mitogenic signal down to the nucleus. In some instances, however, such as with bFGF, the PI-PLC limb of the signal can be shut off without apparent effect on cell proliferation, attesting to the degeneracy of signal transduction pathways. Some growth factors such as insulin (24) and TGF (25) activate hydrolysis of glycosylated phosphatidylinositols; the former is presumed to occur through G-protein dependent phospholipase C (24).

II. PI-PLC AS A POSSIBLE TARGET OF DRUG DESIGN

The process of structure-based drug design based on enzyme inhibition starts with identification of a target enzyme, followed by determination of important topological features of its binding cavity, and synthesis of drug candidates based on the complementarity of structures of the enzyme and the synthetic ligands. The knowledge of the three-dimensional structure of the enzyme is a primary source of leads. However, in the absence of such data certain topological details of the enzyme can be obtained using conventional active-site mapping. The second source of leads is the catalytic mechanism of the enzyme, whereby a latent chemical functionality can be included in the substrate molecule (suicide substrate). Such a substrate can release a reactive intermediate upon undergoing catalytic turnover, thereby covalently modifying enzymic residues leading to a loss of enzyme activity. A third approach has been to identify natural products which narrowly target the selected enzyme, and to synthesize their synthetic analogs to establish important binding determinants. All such approaches have been pursued with respect to PI-PLC inhibition, as will be discussed later. A combination of all of these strategies will be possible, after the 3D-structure of the enzyme has been determined. Since the three-dimensional structure is only available thus far for bacterial PI-PLC (18a), the current design of inhibitors for mammalian PI-PLC has to be based on (i) mechanistic considerations, (ii) the structure of bacterial PI-PLC and assumption of similarity between mammalian and bacterial enzymes, (iii) the results of mapping the active site using phosphatidylinositol (PI) analogs, and (iv) leads obtained from inhibition studies with natural products.

The evaluation of the enzyme as a possible target of rational drug design is based, among other factors, on the assessment of physiological consequences of its inhibition. Since PI-PLC holds a central role in signal transduction, its inhibition should have a far-reaching impact on cell physiology. Inhibition of PI-PLC should enable growth control of cells with deranged signaling pathways, such as cancer cells, and of inflammation processes involving metabolites of arachidonic acid. Selectivity of inhibition is an important condition of successful development of an enzyme inhibitor as a drug. This issue is especially important for PI-PLC since various types of this enzymes are responsible for different physiological functions. Therefore broad inhibition would result in high toxicity of an inhibitor. Selective inhibition of PI-PLC species presents a challenge, since one has to distinguish between related enzymes with analogous mechanisms and very close substrate specificities. It is therefore unlikely that inhibitor design purely on the basis of an enzyme mechanism and a substrate structure will attain the needed selectivity towards major PI-PLC subtypes (α, β, γ, δ). It could be, however, expected that once good general leads are discovered, specificity can be achieved by adding on various chemical residues and exploring diversity among PI-PLC subtypes. Still, since major PI-PLC types do differ in their substrate specificity, e.g. melanoma PI-PLC (PI substrate), GPI-PLC (GPI) and major types of PI-PLC (PIP_2), the distinction between these classes of enzyme should be realized more easily. It is worth mentioning that in addition to PI-PLC, other enzymes involved in phosphoinositide metabolism such as phosphatidylinositol 3-kinase (15a) and inositol 1-phosphate phosphatase (15b) are also attractive targets of drug design. Furthermore, blocking of the IP_3 signal could achieved using synthetic IP_3-receptor antagonists (16).

A. Cancer-PI-PLC Connection

Both products of PI-PLC reaction, inositol trisphosphate and diacylglycerol, mediate cell growth, differentiation and proliferation (1). Conversely, inhibition of formation of these metabolites by antibodies to PI-PLC substrate, PIP_2, inhibits cell growth (26). Likewise, antibodies to PI-PLC-γ block serum and *ras*-stimulated synthesis of DNA (27). This is a consequence of the fact that PI-PLC is a transducer of the signal of growth factors, many of which are oncogene products such as *sis* (PDGF), *erbB* (EGF) and *fms* (CSF-1 receptor). Since cell proliferation and differentiation are strictly controlled by the balance of oncogenes and tumor suppressor genes, it is logical that overexpression of oncogenes and deletion of tumor suppressor genes results in oncogenesis. Tumor-related changes in phosphoinositide metabolism can occur at several levels: increased expression of growth factor receptors, increased expression of PI-PLC, and increased production of PI-PLC substrate. Microinjection of both PI-PLC-β and -γ into the quiescent NIH 3T3 cells causes induction of DNA synthesis and change of cell morphology, reminiscent of malignant transformation (28). Expectedly, recent reports indicate multifold increase of expression of PI-PLC-γ and EGF receptor in tumor tissues and malignant cell lines, such as primary human breast carcinoma (29), colorectal cancer tissues (30), familial adenomatous polyposis (31), and the surface of human colorectal cancer cells (32). The marked stimulation of overexpressed normal and tumor colon PI-PLC by bile acids can further enhance colon carcinogenesis (33). The elevated PI-PLC activity in colorectal cancer tissue correlated with enhanced phosphoinositide metabolism (34). The *in vitro* experiments with cell cultures confirmed the role of PI-PLC-γ or -β in induction of DNA synthesis (35), induction of an S-phase and transformation (28). In addition to increased expression of PI-PLC in tumors, the steady-state activities of PI 4-kinase and PIP 5-kinase are also elevated in ovarian and breast carcinomas (36). All above results strongly support the role of PI-PLC in regulation of the growth and differentiation pathways, the defects of which are a likely cause of neoplastic transformation (37). It is worth pointing out, however, that several signals are generated in response to stimulation of the growth factor receptor, including signals generated from cleavage of inositol phospholipids. Hydrolysis of phosphoinositides is not the only, and may be not the most important factor in mitogenesis. For example, site directed mutagenesis of FGF receptor removing the ability of the receptor to activate PLC, has no effect on differentiation of PC12 cells (38, 39), and overexpression of PI-PLC by stable transfection of cells and the resulting enhanced turnover of PIP_2 does not bring about a mitogenic response (40).

B. Possible Role of PI-PLC in Angiogenesis

Compelling evidence shows that angiogenesis is stimulated by basic fibroblast growth factor (bFGF) (41) and by vascular endothelial growth factor (VEGF) (42), both of which are strongly expressed in tumor cells, and increase inositol trisphosphate and diacyl glycerol levels through the mechanism involving tyrosine-kinase stimulation of PI-PLC (43). Furthermore, binding of FGF and VEGF to their receptors is heparin-dependent (44), consistent with the known role of heparin as an angiogenesis factor. The recent finding that about 20% of heparan sulfate is GPI-linked, and can be released by bacterial PI-PLC (45) further suggests the involvement of PI-PLC in angiogenesis. The overall mechanism of angiogenesis could be quite complex and could require the intervention of other factors such as angiogenin, angiostatin and others, some of which may also be PI-PLC-dependent.

III. CHEMICAL MECHANISM OF PI-PLC

From the mechanistic standpoint it is important to distinguish between two PI-PLC groups of different origin: (i) mammalian and (ii) bacterial. These two broad classes of PI-PLC differ in their molecular size (mammalian enzymes have higher molecular weights), metal ion requirements (mammalian, but not bacterial enzymes, require calcium), and regulation (mammalian, but not bacterial enzymes, are regulated) (Table 1).

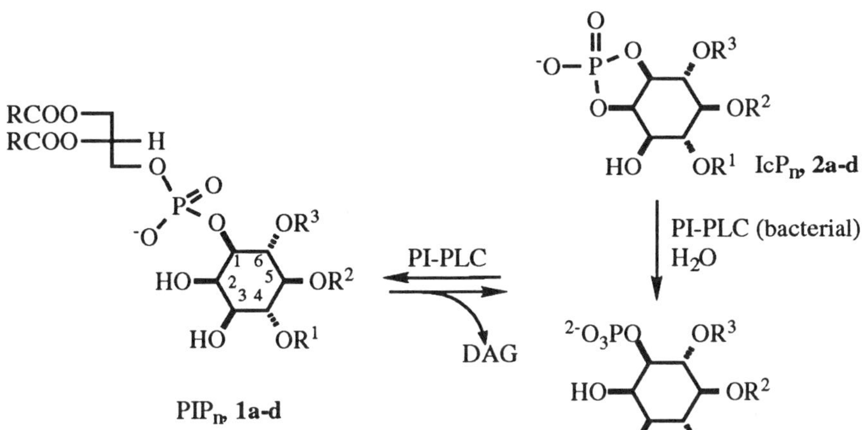

$R = C_nH_{2n+1}$, DAG = diacylglycerol

1-3a (PI, IcP, I-1-P): $R^1 = R^2 = R^3 = H$

1-3b (PIP, IcP-4-P, I-1,4-P_2): $R^1 = PO_3^{2-}$, $R^2 = R^3 = H$

1-3c (PIP_2, IcP-4,5-P_2, I-1,4,5-P_3): $R^1 = R^2 = PO_3^{2-}$, $R^3 = H$

1-3d (GPI, cPIG, PIG): $R^1 = R^2 = H$, R^3 = glucosaminoglycan

Scheme 1 Formation of inositol 1,2-cyclic and acyclic 1-phosphates by mammalian and bacterial enzymes. The conversion of IcP to IP is realized only by bacterial enzymes.

Another apparent difference is that the action of mammalian enzymes on phosphatidylinositols (**1**) produces simultaneously 1,2-cyclic phosphate (**2**) and acyclic inositol 1-phosphate (**3**), while the bacterial enzymes produce exclusively the 1,2-cyclic phosphate as an initial product (Scheme 1). The bacterial enzymes, however, can further hydrolyze the cyclic phosphate into the 1-phosphate at a low rate, while the cyclic phosphate is not a kinetically competent substrate with mammalian enzymes (17, 46, 72). This difference suggested large disparities in the catalytic mechanism of the two classes of enzymes. Clarifying this difference was an important research goal in view of the fact that bacterial PI-PLC are much more readily available and more stable; therefore, if the catalytic mechanisms were analogous, the initial phase of inhibitor design could be performed with bacterial PI-PLC as a surrogate enzyme.

A. Steric Course of PI-PLC Catalyzed Cleavage of Phosphatidylinositol

The mechanism of both types of enzymes was investigated using the stereochemical course of nucleophilic displacement at phosphorus as a criterion (Scheme 2) (17,46). To this end we have synthesized phosphatidylinositol analogs stereospecifically labeled in the phosphate group with a sulfur atom or oxygen isotopes, and furthermore have determined their absolute configurations at phosphorus (46, 47). The reaction of the *R*p-phosphorothioate analog of PI (DPPsI, **4a**) with both *B.cereus* and guinea pig enzymes (PI-PLC-α) indicated that both enzymes produced the phosphorothioate analog of IcP of the *trans*-configuration (IcPs, **5a**), consistent with inversion of configuration at phosphorus (47). Furthermore, only one of the isomers, Rp-DPPsI (**4a**), was a reactive substrate (47). Hydrolysis of *S*p-[^{17}O]DPPI (**4b**) by both *B.thuringiensis* and PI-PLC-β afforded [^{16}O,^{17}O,^{18}O]IP (**6**) with overall retention of configuration (46), meaning that hydrolysis of IcP to IP by the bacterial enzyme proceeded with a net inversion. The inversion for individual DPPI $\rightarrow$ IcP steps, and an overall retention for the DPPI $\rightarrow$ IP conversion with the mammalian enzyme, indicated that the hydrolysis of PI to IP catalyzed by this enzyme consists in fact of two sequential elementary nucleophilic displacement steps, each proceeding with inversion (46).

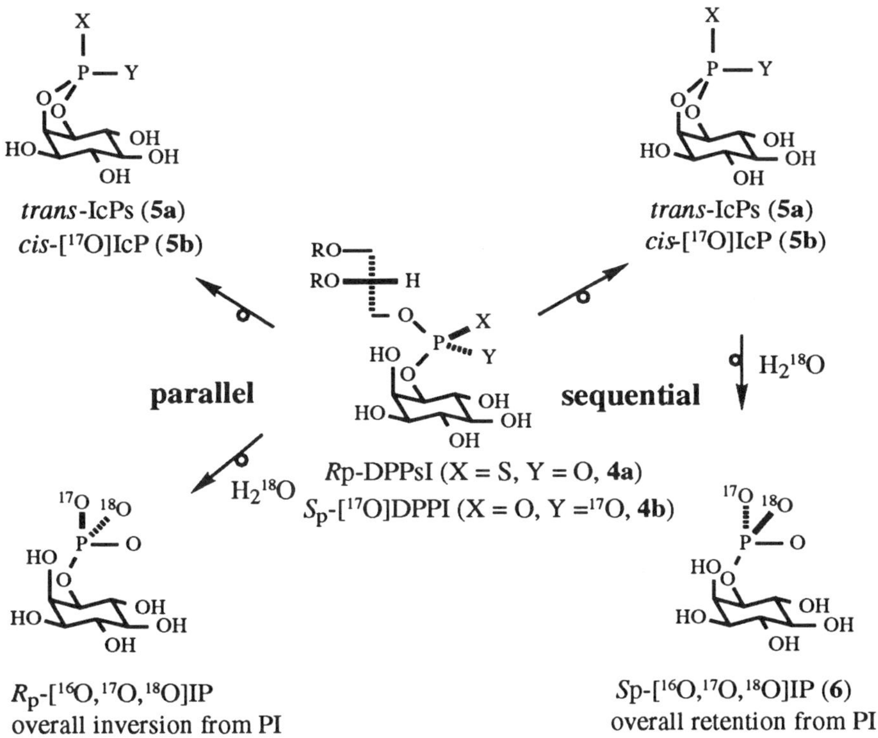

Scheme 2 Possible stereochemical courses of PI-PLC. Since a single displacement at the phosphorus atom proceeds with inversion the parallel reaction should yield both IcP and IP with inversion, while sequential reaction should give IcP with inversion, and IP with retention.

The consequences of this result are abundant: (A) First, it indicates that despite the apparent difference in chronology of product formation, the mechanisms of both enzymes are quite analogous. The reaction starts in each case by the attack of the inositol 2-hydroxyl group at the phosphorus atom leading to 1,2-cyclic phosphate (Scheme 3). The observed difference in the time of appearance of IcP and IP for both enzymes can be explained by interfacial activation of PI-PLC by the aggregated substrate, and by assuming that binding of the cyclic intermediate is affected by the number of phosphate groups.

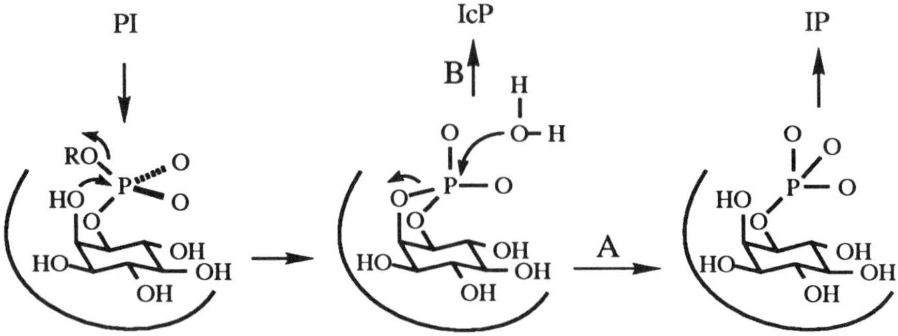

Scheme 3 Explanation of the disparity of the formation of IcP and IP by mammalian and bacterial enzymes. It is postulated that mammalian enzymes display lower diffusion-off rates allowing a certain amount of intermediate IcP to be converted into IP before diffusion takes place (path A). Bacterial enzymes follow path B in which all IcP formed diffuses from the active site prior to hydrolysis.

Phospholipase C from *B. cereus* is activated by binding to the aggregated substrate interface of PI (48,49). In contrast, such allosteric activation could not result from enzyme interaction with monomeric IcP, leading to binding, but poor catalysis. Removal of the hydrophobic residues from the substrate results in a tremendous decrease of the cleavage rates. For example, the deacylated PI and simple short chain alkylphosphoinositols are cleaved 100-500 times slower by the *B.thuringiensis* enzyme compared to the natural PI (50, 51). Likewise, the removal of fatty acid chains from phosphatidylinositols 4-phosphate renders the resultant phosphodiester (Gro-PIP) totally resistant to hydrolysis by PI-PLC-α (52). The complete resistance of Gro-PIP could be due to a greater interfacial activation effect in mammalian PI-PLC, or simply due to the fact that mammalian enzymes are in general about 100-fold less efficient catalysts. In the absence of interfacial activation their activity could be too low to measure.

A different phosphorylation status of preferred substrates of both enzymes could further influence the off rates of the cyclic intermediates. In mammalian enzymes, binding energy provided by the presence of phosphate groups at the 4,5-positions could reduce the off rate as compared to the bacterial enzymes, resulting in significant progress of the second step before the diffusion takes place. Consistent with this hypothesis is the strong dependence of the percentage of the cyclic product on the phosphorylation status of the substrate:

PIP_2>PIP>PI (53). We conclude, that the intermediate cyclic has to be hydrolyzed by the mammalian enzyme during the short time prior to its release (Scheme 3), and when the enzyme is still bound to the interface (46). After the intermediate leaves the active site, the initial acid-base equilibrium is reestablished, to make the enzyme unsuitable for catalysis of the second step.

(B) Inversion of configuration in both steps is consistent with an in-line displacement mechanism, whereby the 2-hydroxyl group attacks the phosphorus atom with a simultaneous departure of the diacylgycerol leaving group (Scheme 4). The transition state (**TS**) for such reaction consists of the three key enzymic elements: (i) a base B_1 abstracting the proton from the 2-hydroxyl group of inositol, (ii) the proton acid interacting with the phosphate group, and (iii) a proton acid HB_2 assisting the leaving group. The reduction in the rate of cleavage upon sulfur substitution by two-orders of magnitude (47), and high enzyme stereoselectivity with regard to configuration at phosphorus in the analog **4** indicate that protonation or hydrogen bonding to the phosphate group takes place in PI-PLC active site. The pentacoordinated structure of the transition state and interaction with the general acid HB_2 imposes further constraints on possible torsional angles about P-O and O-C bonds of the scissile ester group. The proposed mechanism is highly reminiscent of the mechanism of ribonucleases (54).

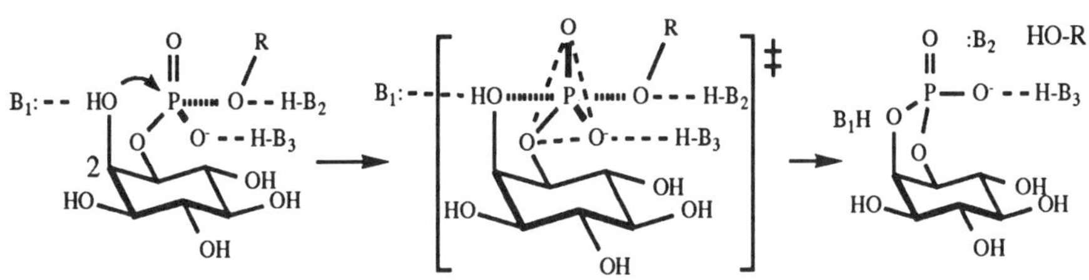

TS

Scheme 4 Catalytic mechanism of PI-PLC active site employing the pentacoordinated transition state **TS**. Note that the apical-apical arrangement of the leaving group and of the incoming 2-OH group oxygen atom, and the hydrogen bonding to the leaving group imposes significant constraints on the flexibility of the bound conformation in the ground state.

(C) Observation that both PI-PLC types are completely non-stereospecific with respect to the configuration at the C-2 carbon atom of diacylglycerol suggests that the active site interactions do not extend past the phosphodiester oxygen. This information is corroborated by the fact that PI-PLC efficiently hydrolyze the lyso-PI and single chain analogs of PI, and that the bacterial PI-PLC catalyzes the reverse transesterification process of IcP with a wide variety of primary alcohols (17, 51). Chemical residues attached to the phosphodiester oxygen are therefore only important for the enzyme as far as they define aggregation properties of substrates, and their partitioning into membrane.

B. Three Dimensional Structure of the Bacterial PI-PLC

The above mechanistic results were fully confirmed by the first molecular structure of the complex of the *B.cereus* PI-PLC with *myo*-inositol reported recently by Heinz et al. (Figure 2) (18a).

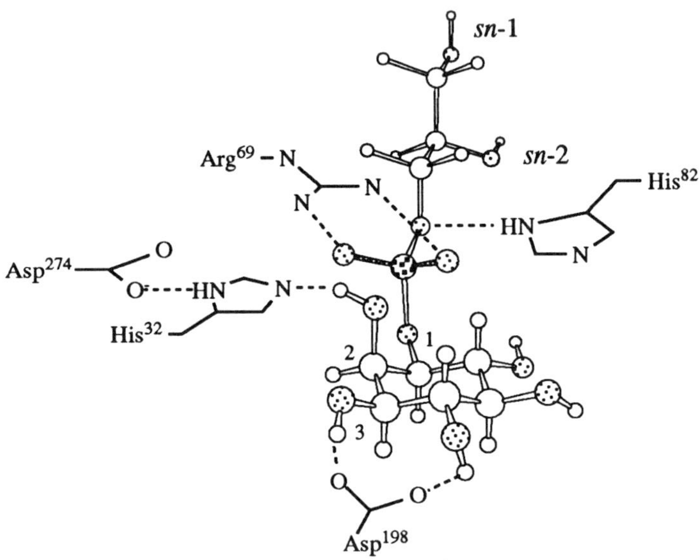

Figure 2 Sketch of the *B.cereus* PI-PLC active site (adapted from Heinz et al. 1995, ref. 18a). Note the presence of the catalytic triad including Asp-274, His-32 and inositol 2-hydroxyl group, and similarity of this active site to those of ribonucleases.

The catalytic machinery is composed by two histidine residues, His-32 abstracting the proton from OH-2 of inositol and His-82 conveniently positioned for protonation of the leaving group. The phosphate moiety is most likely interacting with Arg-69. His-32 is a part of the catalytic triad consisting also of the aspartate Asp-274 and the inositol 2-hydroxyl group. Thus, the function of inositol 2-hydroxyl group is analogous to that of the 2'-hydroxyl group of ribonucleotide, and to the role fulfilled typically by serine (such as in serine proteases). The roles of His-82 and Arg-69 are only putative since no structure of PI-PLC complex with a more elaborate substrate analog is available yet. The binding cavity appears to be quite shallow; therefore as stated earlier, it should not include interactions with diacylglycerol sites. It is worth pointing out here that the structure of the active site of PI-PLC is unrelated to that of the unspecific PLC (18b).

C. Similarity of PI-PLC to RNases

Comparison of the X-ray structures of PI-PLC and RNases shows far reaching analogies. The analogy stems from the fact that the function of both enzymes is to cleave a phosphodiester bond by a transesterification mechanism. Earlier stereochemical studies showed that the steric course of RNase is also inversion of configuration in each displacement step (55), indicating a trigonal bipyramidal structure of the transition state. First, as might be expected the catalytic histidines are virtually superimposable between the two enzyme types (18a), consistent with the same separation of the two apical oxygen atoms in the transition states of both reactions. Furthermore, the hydrogen bonding to phosphate in RNase A is realized by the protonated Lys-41 (54b). Interestingly, the most recent addition to the RNase family, human and bovine angiogenin (56), have the active site which is obscured by the Gln-117 residue (57), which makes the enzyme highly specific towards certain RNA structures and regulated by interaction with receptor proteins. It is unknown, however, whether this regulation model is related to regulation of some PI-PLC species.

IV. STRUCTURE-ACTIVITY RELATIONSHIP FOR PI-PLC INHIBITORS

Detailed investigations of the SAR for PI-PLC presents several problems. (A) First, due to the presence of many different types of PI-PLC in mammalian tissues, there is a strong need for obtaining inhibition data for a single purified species of the enzyme. Unfortunately, since purification of mammalian PI-PLC is difficult, much of the inhibition data reported have been obtained using impure or only slightly purified phospholipases. (B) Since phospholipase C is an interfacially activated enzyme, the observed rate effects could be due to the active site binding, as well as changes in the interfacial binding. (C) Positively charged molecules tend to interact with the negatively charged substrate (PIP_2), rather than with the enzyme. (D) On the whole, cell assays inhibition of IP_3 and diacylglycerol formation could be due to effects other than the direct PI-PLC inhibition. These problems have to be resolved by using pure enzyme preparations, and employing a clean kinetic model of scooting mode kinetics, or nonaggregated substrates. The first issue discussed below is to provide a list of important structure determinants used by PI-PLC for substrate recognition.

A. Diacylglycerol Residue

Both bacterial and mammalian PI-PLC display a complete lack of stereospecificity towards hydrolysis of PI diastereomers (**7**) with *R* or *S* configurations at the glycerol C-2 carbon (Figure 3) (46). Both enzyme types also hydrolyze lyso-PI (PI deacylated at the 2-position). Alkyl inositol phosphates (**8**), the single chain analogs of PI, and short double chain esters are also substrates of bacterial enzyme, although the cleavage rates depend strongly on the alkyl chain length (48). Since PI-PLC has a strong preference for substrates aggregated in micelles, the length of the alkyl chain is important inasmuch as it affects the critical micelle concentration (cmc) of the substrate (48, 59). With intermediate chain lengths a large rate increase is observed upon passing the cmc concentration (48, 59). The removal of both acyl chains results in very slow turnover in the case of bacterial enzymes (50, 51), and complete cessation of the cleavage by the guinea pig enzyme (52). Interestingly, this effect is not caused by the diminished binding since the deacylated phosphatidylinositol 4-phosphate (**9**) is a potent inhibitor of PI-PLC-α (52). The origin of such slow turnover could be due to: (i) loss of interfacial activation of the enzyme; (ii) ligand binding in a nonproductive orientation in the active site; (iii) lack of induction of conformational change in the enzyme by a truncated substrate.

B. Primary vs. Secondary Alkyl Inositol Phosphate Esters

Although the precise structure of diacylglycerol is unimportant, the replacement of the primary glycerol C-3 by a secondary carbon such as in the

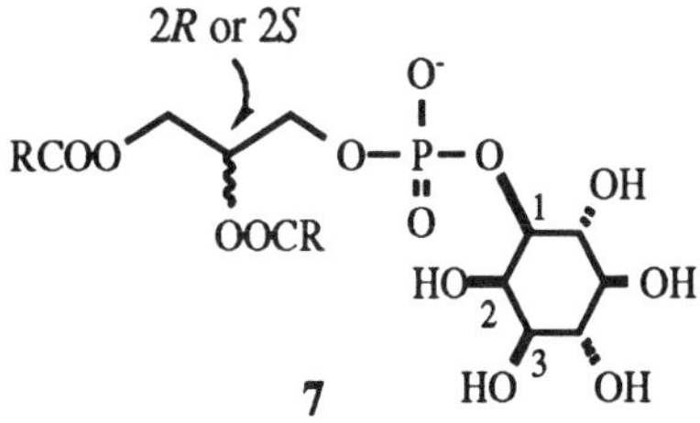

7

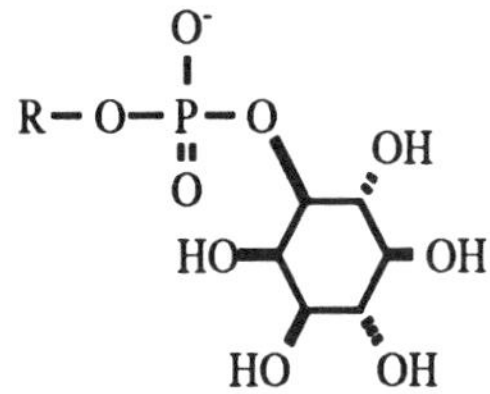

8, R= 1° alkanols
10, R = 4-NO_2-Ph; **11**, 1-Naphth
12, R = cholesteryl

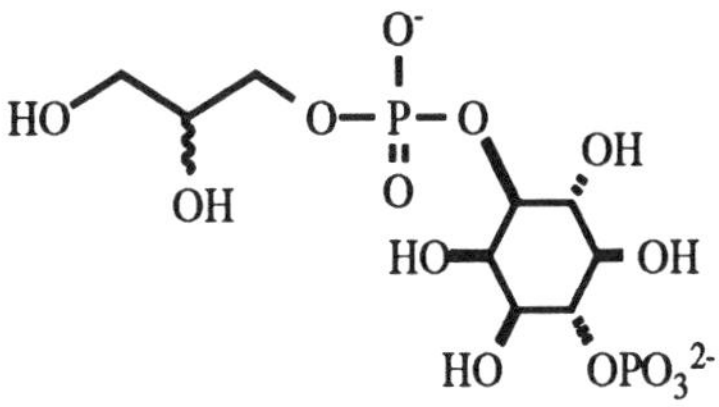

9, PI-PLC-α, K_i 40 μM

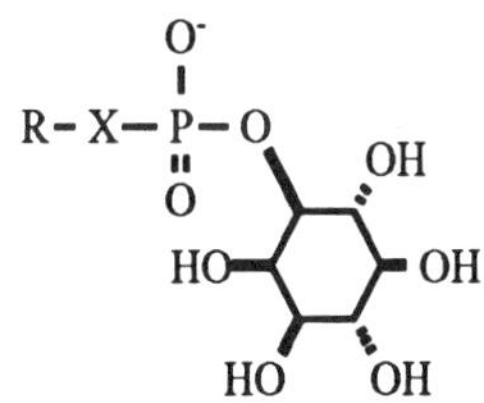

Bacterial PI-PLC
13, X = S, R = hexadecyl, K_m 22 μM
14, X =CH_2, R = alkyl, K_m ca. 4 mM

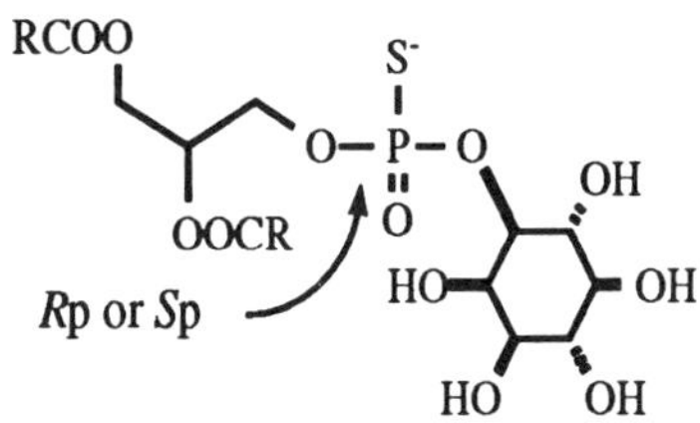

15, *S*p; **16**, *R*p

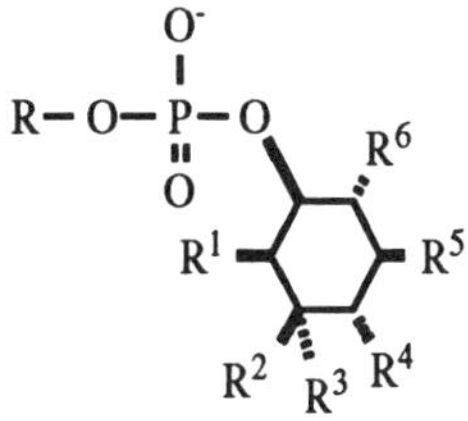

17, $R^2 = R^4 = R^5 = R^6$ = OH, $R^1 = R^3$ = H
18, $R^2 = R^4 = R^5 = R^6$ = OH, R^1 = OMe, OAc; R^3 = H
19, $R^2 = R^4 = R^5 = R^6$ = OH, R^2 = H, $R^3 = PO_3^{2-}$
20, $R^1 = R^3 = R^4 = R^5 = R^6$ = OH, R^2 = H
21, $R^1 = R^2 = R^4 = R^5$ = OH, R^3 = H, R^6 = glucosaminoglycan
R = diacylglycerol

Figure 3 Structures of the phosphatidylinositol analogs **7-21**.

p-nitrophenyl- (**10**) and naphthyl- (**11**) inositol phosphates causes a large increase in K_m values, while the k_{cat} values are less affected (60). Similarly, the cleavage of cholesterol inositol phosphate (**12**) is extremely slow (61). Likewise, the secondary alcohols do not participate in the reverse transesterification reaction catalyzed by *B.cereus* PI-PLC, although substitution at the alcohol C-2 with a wide variety of residues is universally accepted by the enzyme (17, 51). These results can be explained if the antiperiplanar conformation about both phosphodiester bonds is assumed in the transition state (Figure 2). The presence of the extra substituent at alkyl C-1 carbon would interfere with stabilization of the phosphate group by Arg-69 or His-82. An antiperiplanar conformation about P-O bond is, however, not found in any of the known crystal structures of phospholipids (62) and its adoption may cause loss of binding energy.

C. Bridging Oxygen Atom

Since the catalytic mechanism requires electrophilic assistance of the His-82 to the negative charge being developed on the glycerol O-3 oxygen atom in the transition state, the presence of oxygen is essential at this position. Consequently, replacement of oxygen by sulfur, a poor electron donor in hydrogen bonds, as in thiolester **13**, results in ca. 10^2-fold reduction of V_{max}, largely due to the fall in k_{cat}, despite the fact that thiols have lower pKa than the corresponding alcohols (63). As could be expected, the phosphonate analogs are completely resistant to bacterial enzyme (64), although they also exhibit lower binding affinity than that of PI.

D. Phosphoryl Group

In most phosphodiesterases protonation of the phosphodiester or metal chelation is required to lower the repulsive electrostatic interaction of the incoming nucleophile and the phosphate group in the transition state. In mechanistically closely related RNases hydrogen bonding to the negatively charged nonbridging oxygen atom in the ground state by the His-119 or Lys-41 is postulated (54b). The replacement of the nonbridging phosphoryl oxygen with a sulfur atom reduces such stabilizing interactions and results in lowering the cleavage rates with most phosphodiesterases. The large sulfur substitution effect indicates protonation of the nonbridging oxygens in the transition state. Interestingly, replacement of the *pro*-S oxygen with sulfur affords a completely resistant analog **15**, while the analogous replacement of the *pro*-R oxygen in **16** results in approximately 10^2-fold reduction of the cleavage rate (47). This result suggests that protonation or hydrogen bonding occurs mainly at the *pro*-S oxygen, consistent with the position of the His-82 being closer to this atom.

E. Structure of Inositol Core

1. 2-Hydroxyl group

The 2-deoxy-PI (**17**) has been shown to be hydrolytically resistant to melanoma PI-PLC, but its binding is several fold weaker than that of PI (65). Likewise, the 2-alkyl analogs of PI (**18**) (66, 48) and 2-palmitoylated GPI (67) are resistant to cleavage by *B.cereus* PI-PLC. The importance of the 2-OH group for catalysis is understood since it constitutes a part of the catalytic triad in *B.cereus* enzyme together with His-32 and Asp-274 (18a).

2. 3-Hydroxyl group

The presence of the free equatorial hydroxyl group at the 3-position is important since the 3-phosphorylated PIs (**19**) are not substrates of mammalian PI-PLC (3). Furthermore, the *chiro*-inositol analog of PI (**20**) with an inverted hydroxyl group at the 3-position is hydrolyzed about 10^3-fold slower than the natural PI (68). The nature of the influence of this hydroxyl group on the reaction rate is not known. It is possible that the lower rate of formation of *chiro*-inositol 1,2-cyclic phosphate (*chiro*-IcP) is due to its higher energy. In a model study *chiro*-IcP was found to be about 10^2 times more susceptible to base-catalyzed hydrolysis than the corresponding *myo*-IcP (61). An important factor in enzymic hydrolysis could also be interaction of the 3-hydroxyl group with Asp-198 in *B.cereus* PI-PLC (Figure 2) (18a).

3. Hydroxyl groups at the 4- and 5-positions

The 4- and 5-phosphate groups are important for catalysis by mammalian PI-PLC since 4,5-bisphosphate is a preferred substrate. The K_m values for PI and PIP_2 by PI-PLC-β and γ- are frequently in the same range (17), whereas the catalytic rates are frequently tenfold or more higher with PIP_2 than with PI. In contrast, binding of PI to PI-PLC-δ is stronger than that of PIP_2 (52,69). In the latter cases weaker binding affinity is overcome by higher catalytic rates, such that PIP_2 is eventually a preferred substrate for PI-PLC-δ.

4. 6-Hydroxyl group

The importance of the 6-position depends on the PLC type. The bacterial PI-PLC binds GPI (**21**) more strongly than the nonglycosylated PI; however, the cleavage is faster with the latter (17). In fact it is possible that the functional substrate of bacterial PI-PLC is GPI rather than PI (70). The crystal structure of *B.cereus* PI-PLC shows that this position is open to the aqueous phase, therefore it should accommodate a variety of substitution (18a). In contrast to PI-PLC, GPI-PLC is very specific and accepts only 6-glycosylated PI as a substrate (71).

5. Inositol configuration

Bacterial PI-PLC have been shown to be very specific towards 1D-enantiomer of PIs (60, 63). The further hydrolysis of IcP also occurs only with 1D-IcP. (72). While no analogous data exist on the cleavage by mammalian enzymes, PI-PLC-δ has been shown to bind specifically 1D-inositol 1,4,5-trisphosphate, but not its enantiomer. An interesting controversy arises from the proposal that 1D-*chiro*-inositol could be a component of inositol phospholipids (73). As will be discussed later, since 1D-*myo*-inositol 1-phosphate is stereochemically related to 1L-*chiro*-inositol-2-phosphate, it is highly unlikely that phospholipids with 1D-*chiro*-inositol could be cleaved by any of the known PI-PLC isoforms.

F. Inhibitory Analogs of PI

1. Modification of phosphatidylinositol

Structural requirements for PI analogs to be competitive inhibitors of PI-PLC are far more relaxed than those for substrates. For example, the inhibition study using a number of PI analogs with the phosphatidate structure replaced by nonionic alkyl or acyl groups (compounds **22-27**, Figure 4) indicated that the phosphodiester group is unimportant for binding by both platelet (74) and guinea pig enzyme (PI-PLC-α, 75). The presence of the negatively charged sulfonate or phosphate groups at the 4- and 5-positions (compounds **22-24**) was, however, absolutely critical for the platelet (74), but not for the guinea pig enzyme (compound **25**) (75). Furthermore, the removal of the hydroxyl groups from the 2,3,5, and 6-positions (compound **29**) did not lower binding to guinea pig enzyme significantly, while the presence of the 4-hydroxyl group (**28**), but not the 2-hydroxyl group (**29**) was critical (75). It appears, therefore, that some selectivity among the PI-PLC isozymes could be attained using inhibitor design on the substrate structure level.

2. Transition state analogs

The design of transition state analogs for phosphotransferases has so far not met with success due to the paucity of stable pentacoordinated structures mimicking transition states of phosphoryl transfer reactions. The only example of such an approach to PI-PLC inhibition the involved assumption that the fluorinated IcP analog (**30**) should more easily form the pentacoordinated intermediate (**31**) due to a higher propensity of the fluoromethylene group to occupy the apical position. The IcP analog **30** has, unfortunately, displayed rather low inhibitory potency (76). Another approach employing complexes of inositol and tetravanadate was also relatively unsuccessful (IC_{50} >2 mM) (77). A large variety of PI analogs including phosphonate (**32**), sulfonate (**33**) and acetal (**34**) surrogates of the phosphoryl group shown in Figure 4 (also reviewed in ref. 17) has been synthesized in recent years. However, most were not tested for PI-PLC inhibition.

G. Irreversible Inhibitors

1. Mechanism based inhibitors of PI-PLC

The postulated mechanism involving participation of the acid-base pair in the active site suggests that substitution of the 2-hydroxyl group of PI with an epoxy residue (compound **35**) could provide an affinity reagent. The removal of the phosphate group should direct hydrogen bonding of the Arg-69 or His-82 to the axial oxygen of epoxide resulting in its activation (Figure 5). The epoxide would then be opened by the histidine nucleophile to produce irreversibly an inactive enzyme. Similarly, the phosphatidate moiety could be substituted with an epoxide residue (structure **36**). The combination of such two modifications

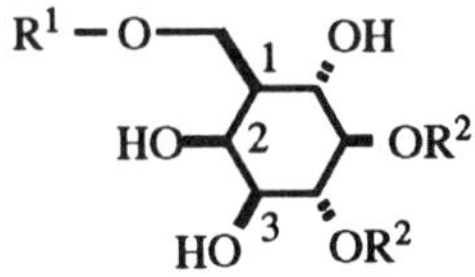

platelet PI-PLC
22, R^1 = Pal, R^2 = SO_3^-, IC_{50} 22 μM
23, R^1 = Pal, R^2 = PO_4^{2-}, IC_{50} 14 μM

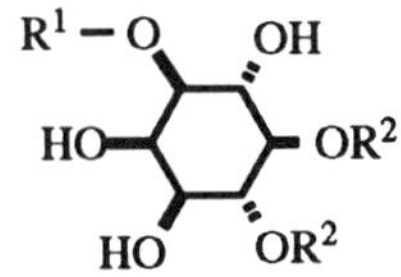

platelet PI-PLC
24, R^1 = Pal, R^2 = SO_3^-, IC_{50} 18 μM
25, guinea pig, R^2 = H, IC_{50} = 28 μM

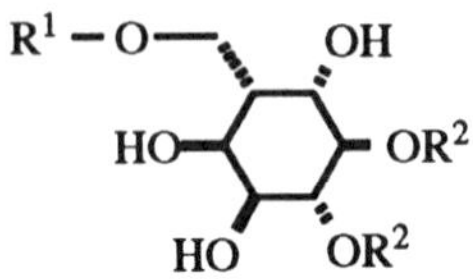

platelet PI-PLC
26, R^1 = Pal, R^2 = SO_3^-, IC_{50} 30 μM
27, R^1 = Pal, R^2 = PO_4^{2-}, IC_{50} 14.5 μM

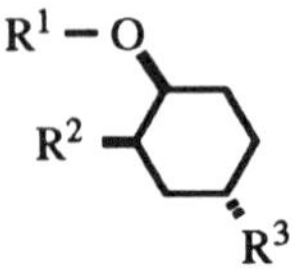

guinea pig PI-PLC, R^1 = $C_{16}H_{33}PO_2$-
28, R^2 = OH, R^1 = H, IC_{50} = >300 μM
29, R^2 = H, R^3 = OH, IC_{50} = 25 μM

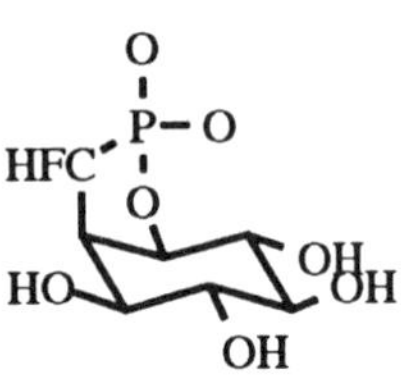

30

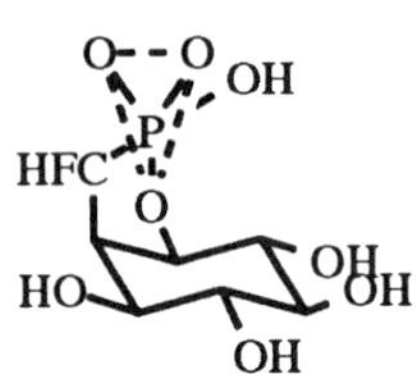

31

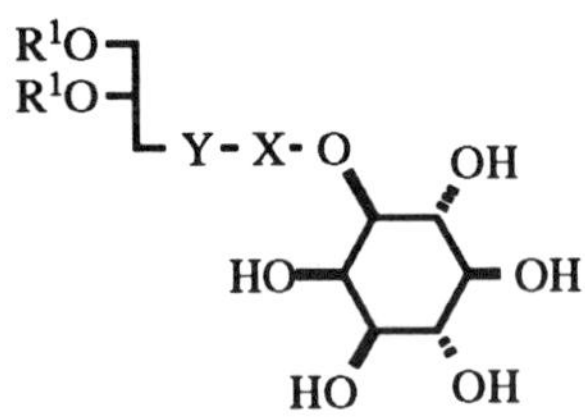

R^1 =palmitoyl
32, X =PO_2^-, Y=CH_2
33, X = SO_2, Y=CH_2
34, X=CH_2, Y=O

Figure 4 Structure of the phosphatidylinositol analogs **22-34**.

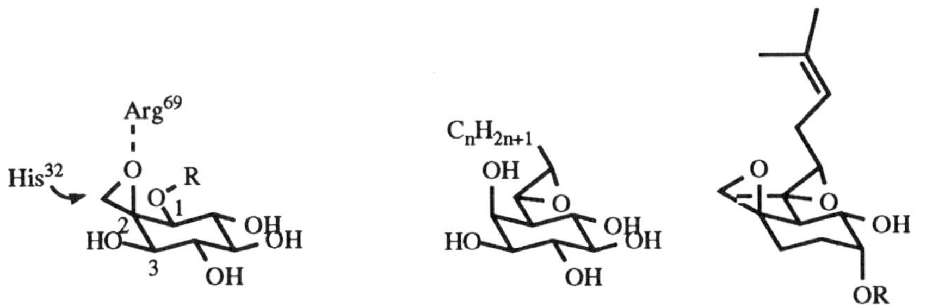

35, R = phosphatidyl; palmitoyl; octyl

36

37a, Fumagillin
R = OC(CH=CH)COOH
37b, AGM-1470
R = $CONHCOCH_2Cl$

Figure 5 Possible mechanism of inactivation of PI-PLC by the spiro-epoxide analogs of phosphatidylinositol (**35**), and structure of the antiangiogenic drug candidate AGM-1470 (**37**).

in one molecule would result in a bisepoxide, which should be an efficient protein crosslinking agent. The natural product which has a similar structure, fumagillin (**37a**), was discovered in the 1950s, and analogs of this compound have been recently shown to have strong antitumor properties. The most potent analog of fumagillin, AGM-1470 (**37b**), has been shown to be a potent suppressor of angiogenesis (78). Coupling this structure with the ribonucleolytic activity of angiogenin, or with the known role of PI-PLC in transduction of the signals of angiogenic factors, bFGF and VEGF, suggests that the biological target of AGM-1470 could be one of these two enzymes. Work is underway in our laboratory aimed at determining the biological target of this agent.

The synthesis of several epoxy analogs of PI is shown in Schemes 5 and 6(61).

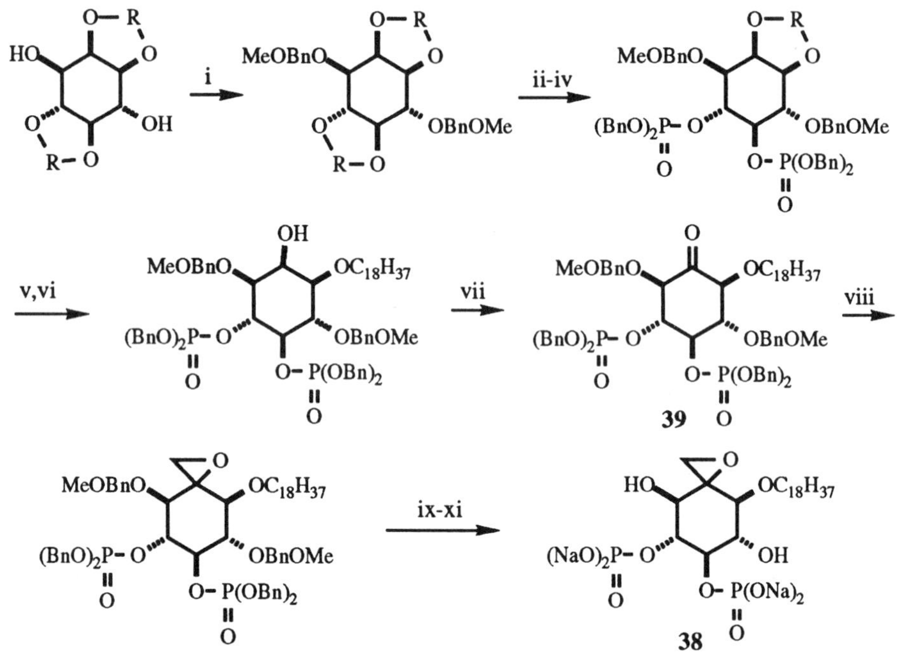

axial and equatorial epoxides

R = cyclohexylidene; i: MeOBnCl/NaH; ii: PPTS/MeOH; iii: $(BnO)_2PNiPr_2$, tetrazole; iv: MCPBA; v: PPTS/MeOH; vi: (a) $Bu_2Sn{=}O$, (b) CsF, $I\text{-}C_{18}H_{37}$; vii: Ac_2O, DMSO; viii: CH_2N_2 or $TMSCHN_2$; ix: isomer separation; x: DDQ; xi: H_2/Pd-C

Scheme 5 Synthesis of the diastereomeric epoxy analogs of PIP_2.

In brief, the syntheses consist of a number of regioselective protective steps, required to obtain differential protection of the hydroxyl groups to be converted into phosphate, long chain ether or ester, and an epoxide group. Synthesis of the PIP_2 analog (**38**, Scheme 5) was especially challenging due to the difficulty in achieving selective deprotection of inositol and phosphate groups in the presence of the fragile epoxide functionality (61). The key step (viii) of forming a *spiro*-epoxide residue is carried out by addition of diazomethane to the carbonyl group of the respective ketone (**39**, Scheme 5; **40**, Scheme 6). Interestingly, addition of diazomethane to the ketone **40** (Scheme 6) is a highly stereoselective process giving the axial epoxide **35** as a final main product (20:1 ratio), as determined by the NOESY spectra of the epoxide-reduced product **41**.

In contrast, the analogous reaction with the ketone **39** (Scheme 5) leads to formation of a 1:1 mixture of the axial and equatorial epoxides. Incubation of the *spiro*-epoxide analog **35** with *B.thuringiensis* PI-PLC led to its time dependent inactivation (Figure 6). The analogs of PIP_2 are currently undergoing screening with mammalian PI-PLC

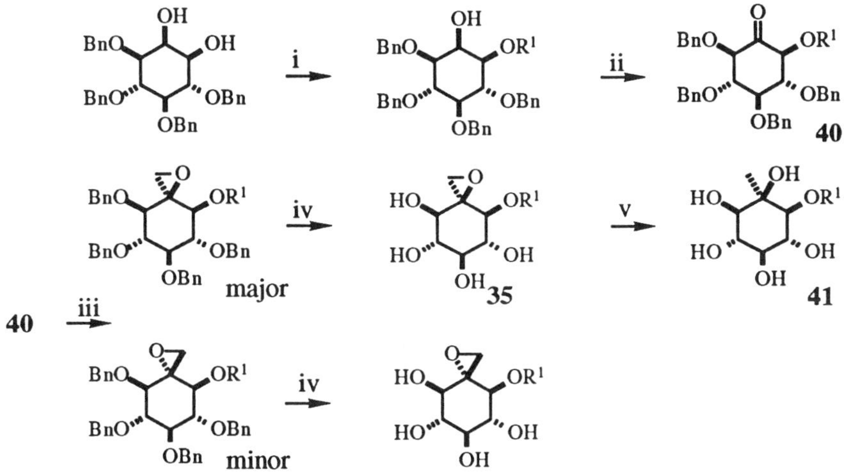

$R^1 = C_8H_{17}$ or $C_{15}H_{31}C{=}O$; i: (a) $Bu_2Sn{=}O$, (b) CsF, $I\text{-}C_8H_{17}$ or $C_{15}H_{31}COCl$; ii: $(COCl)_2$, DMSO, $EtiPr_2N$; iii: CH_2N_2; iv: H_2/Pd; v: H_2/Pd

Scheme 6 Synthesis of the epoxy analog of PI.

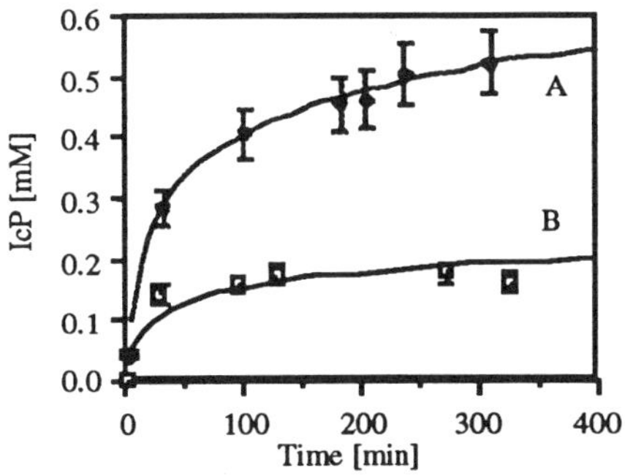

Figure 6 Inactivation of *B.thuringiensis* PI-PLC by epoxy analog **35** ($R = C_8H_{17}$). The time course of PI (1 mM) hydrolysis was followed in the absence (curve A) and presence (curve B) of **35** (1 mM) by HPAEC-PAD.†

2. Other irreversible inactivators of PI-PLC

A series of nitrochromenes and nitrocoumarins has been synthesized with the assumption that enzymic protonation of the nitrogen atom of the R group in compounds **42** and **43** should create a powerful acylating agent, which should irreversibly modify the enzyme (79). The most efficient representative of the series are shown in Figure 7 (compounds **42-46**). These compounds showed time dependent inactivation of melanoma PI-PLC-γ with IC_{50} in the range 3-6 μg/mL. PI-PLC inhibitory potency correlated with inhibition of inositol metabolism and cell growth inhibition (79). An anti-inflammatory sesquiterpenoid, manoalide (**47**), was also shown to inhibit irreversibly PI-PLC-α in the time dependent manner with the IC_{50} in the range 3-6 μM (80). Inhibition of PI-PLC correlated with inhibition of calcium transients, and could be related to the known anti-inflammatory activity of manoalide (80).

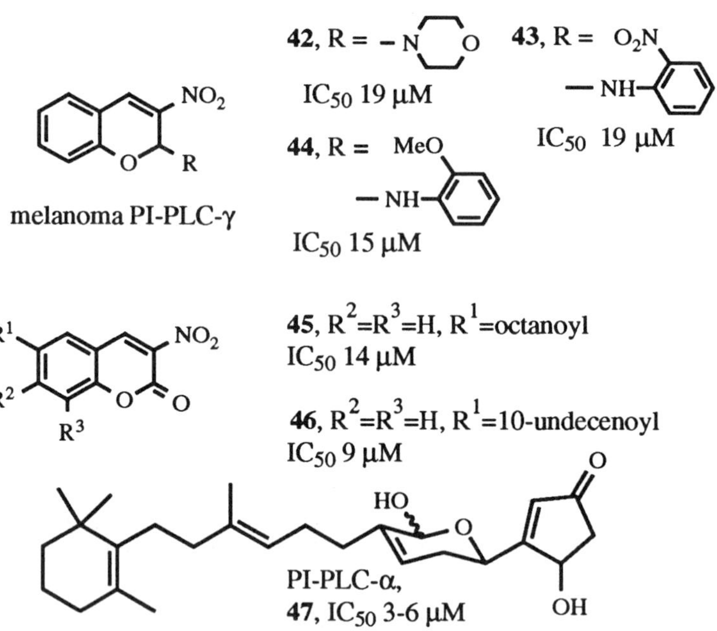

Figure 7 Structures of the irreversible inhibitors of PI-PLC **42-47**.

H. Inhibitors Structurally Unrelated to Phosphatidylinositols

Structures of PI-unrelated compounds with strong PI-PLC inhibition potency and the corresponding available IC_{50} values are shown in Figures 8 and 9. The most potent inhibitor discovered to date is the antitumor lysolipid analog, ET18-OCH_3 (edelfosine, **48**) which inhibits PI-PLC-β at a concentration lower than 100 nM, and PI-PLC-γ at 400 nM (12b, 81). The even simpler lysolipid analog, and a recently introduced antitumor agent, hexadecylphosphocholine (miltefosine, **49**), is also a modest inhibitor of PI-PLC-δ_1 (82). It is not clear whether the cytostatic activity of these compounds is due to PI-PLC inhibition, or a broader mechanism (81b). The PI-PLC activity is also inhibited by a variety of aminocyclitol antibiotics, such as tobramycin (**50**) (83), neomycin (12b), gentamycin, streptomycin and others (reviewed in ref. 17). The inhibitory power of these agents correlates with their basicity. It is likely that the inhibitory effect arises from ionic interactions of the protonated amino groups with the negatively charged head group of phosphatidylinositol 4,5-bisphosphate, rather than by a direct enzyme inhibition. Consistent with this hypothesis is the fact that inhibition of PIP_2 cleavage is stronger than that of PI. Based on this approach a series of positively charged replicas of phosphatidylinositol has been synthesized featuring an amino group in positions spatially corresponding to phosphate groups in PIP_2 (84). The most effective agents of the series (**51-53**) displayed modest inhibition of platelet PI-PLC (IC_{50} values in the range 20 μM). A number of natural products, potent inhibitors of mammalian PI-PLC, shown in Figures 8 and 9 such as caloporoside (**54**) (85), vinaxanthone (**55**) (86), hispidospermidine (**56**, Figure 9) (87), fluvirucin (**57**) (88), aminosteroid U73,122 (**58**) (89), compound 48/80 (**59**) (90), edein (**60**) (83) and streptothricin (**61**)(83) have been identified through routine screening procedures. These compounds displayed PI-PLC inhibition at 3-16 μM concentration level. Finally, certain oligopeptides derived from the natural sequence adjacent to the PI-PLC-γ SH2 domain showed inhibitory potency, giving rise to speculation about PI-PLC self-regulation (91). For example, the octapeptide Tyr-Arg-Lys-Met-Arg-Leu-Arg-Tyr derived from the PI-PLC-γ_1 sequence was modestly inhibitory at 200 μM (91); the unrelated Myroridin K, a natural peptide with the sequence Arg-Arg(OH)-Lys-Thr-Lys-Orn-Tyr, was inhibitory with IC_{50} at 8.3 μM (83), and the cyclic peptide Q12713, of an undisclosed sequence, was an inhibitor with IC_{50} at 8.5 μM (92). The origin of inhibition remains unknown, beyond the fact that the peptides are noncompetitive inhibitors.

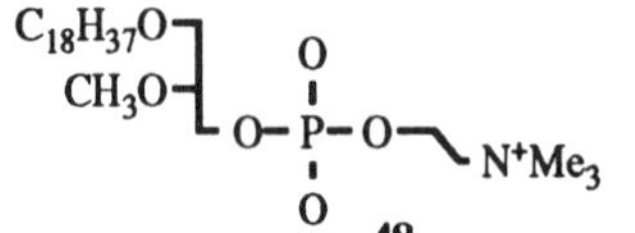

48

PI-PLC-γ, IC_{50} 40 μM
PI-PLC-β, IC_{50} 0.1 μM

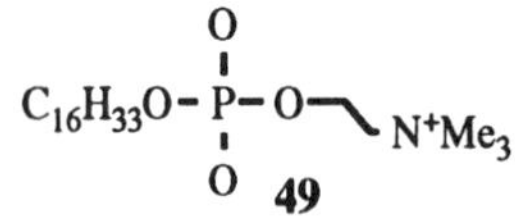

49

PI-PLC-δ_1, IC_{50} 30 μM

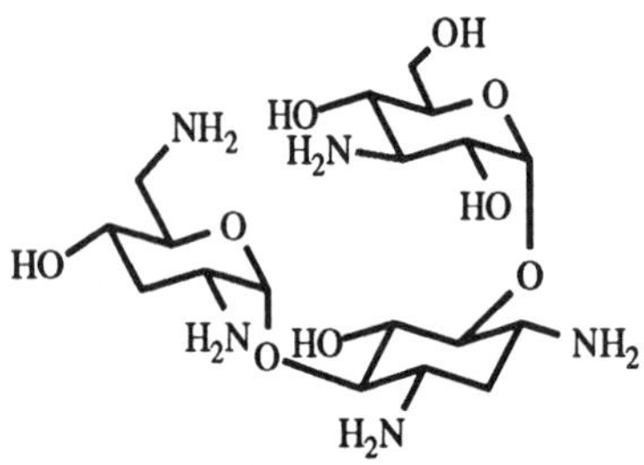

50, Tobramycin
PI-PLC-γ_1, IC_{50} 9.7 μM

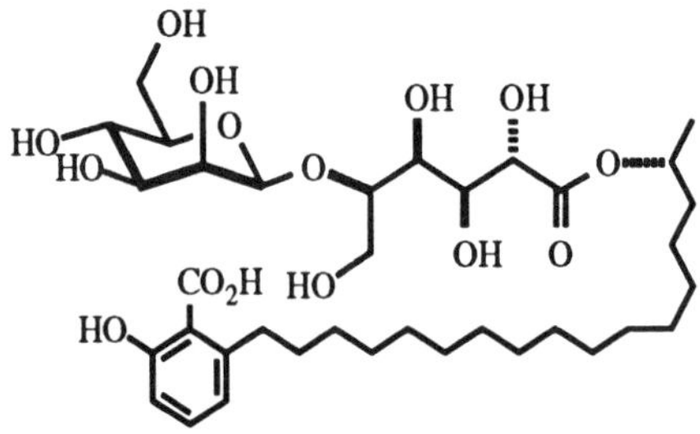

54, Caloporoside
pig brain, Ki 12.3 μM

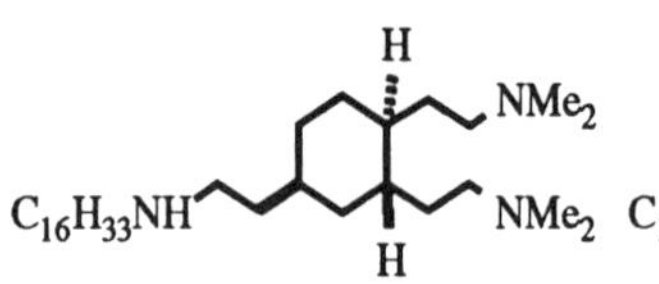

51, IC_{50} 19 μM

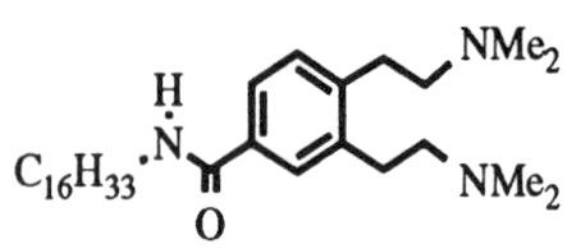

52, IC_{50} 22 μM

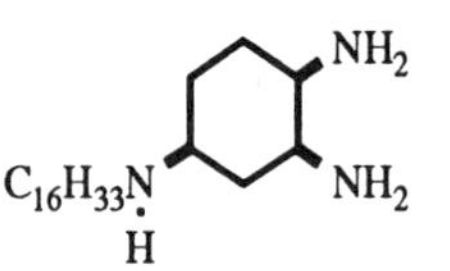

53, IC_{50} 21 μM

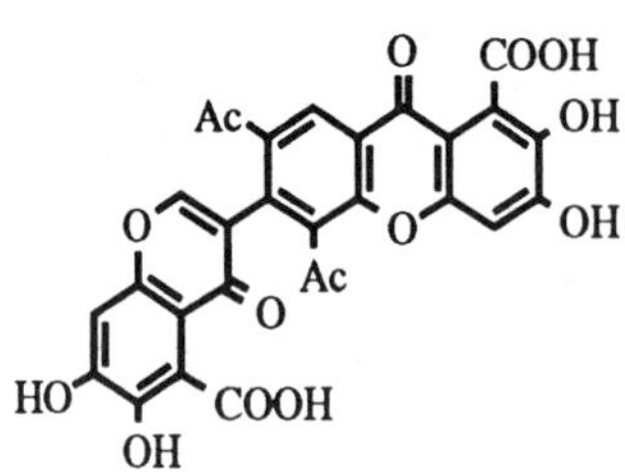

55, Vinaxanthone
various PI-PLC, IC_{50} 5.4-44 μM

Figure 8 Structures of the miscellaneous inhibitors of PI-PLC **48-55** with indicated IC_{50}.

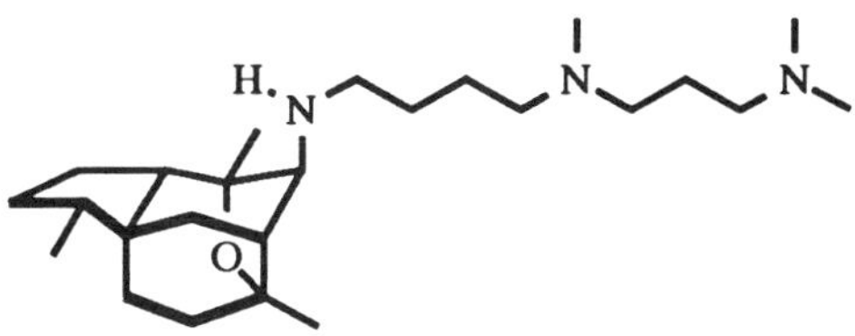

56, Hispidospermidin
rat brain PI-PLC, IC_{50} 16 μM

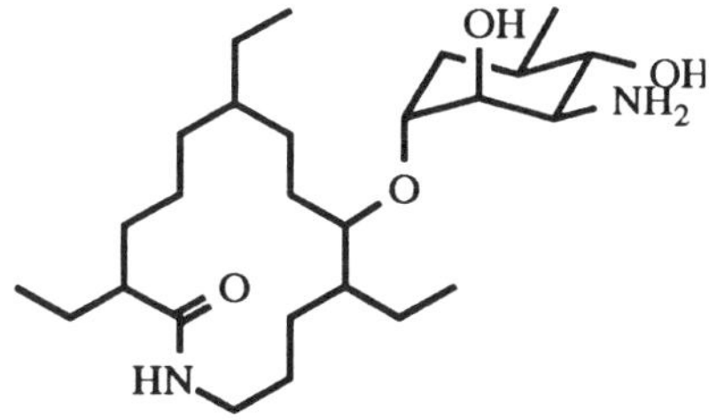

57, Fluvirucin B_2
PI-PLC from A431 cells
IC_{50} 3.5 μM/mL

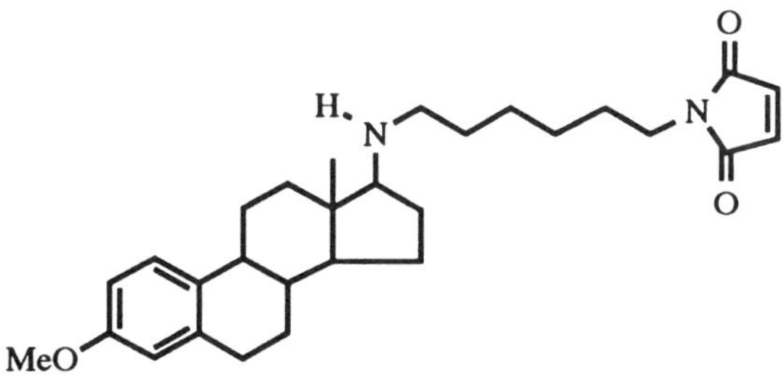

58, U-73,122
PI-PLC from 3T3 cells
IC_{50} 11 μM

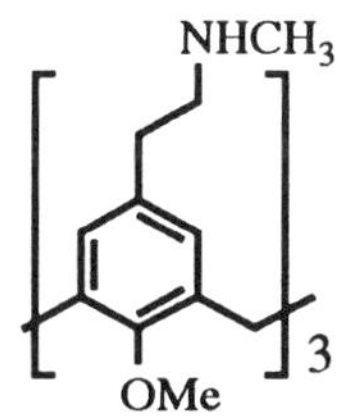

59, Compound 48/80
platelet PI-PLC, IC_{50} 3μM

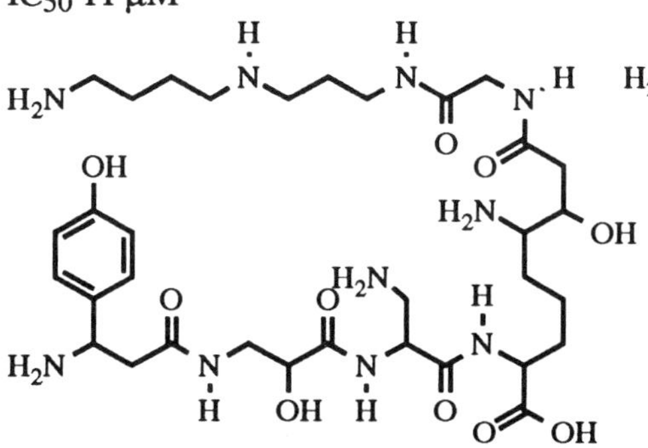

60, Edein A_1
PI-PLC-γ, IC_{50} 16.1 μM

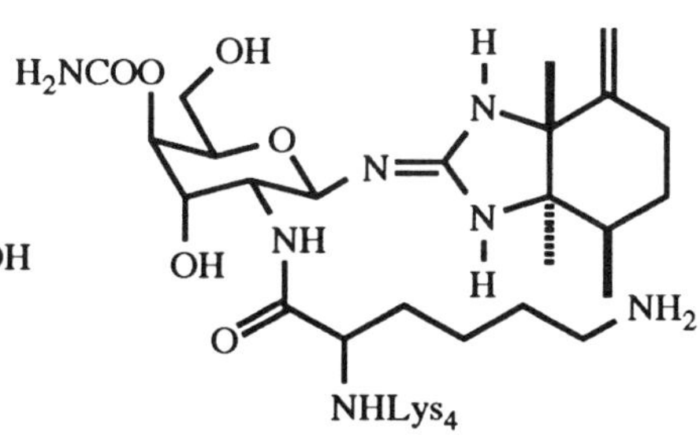

61, Streptothricin B
PI-PLC-γ_1, IC_{50} 6.7 μM

Figure 9 Structures of the miscellaneous inhibitors of PI-PLC **56-61** with indicated IC_{50}.

V. THE ROLE OF PI-PLC IN TRANSDUCTION OF INSULIN SIGNAL

A. The Second Messenger Hypothesis

The mechanism of the transduction of the insulin signal is a key to solving a problem of the ethiology of non-insulin dependent diabetes mellitus (NIDDM). Two major mechanisms, possibly operating simultaneously, are being currently considered. The first one involves tyrosine phosphorylation of the recently discovered insulin receptor substrate (IRS-1) and binding of the src homologous proteins by the stimulated insulin receptor (93, 94). Another possibility involves a signaling mechanism using second messengers derived from glycosylated phosphatidylinositol (24). The scenario combining both mechanisms is outlined in Figure 10.

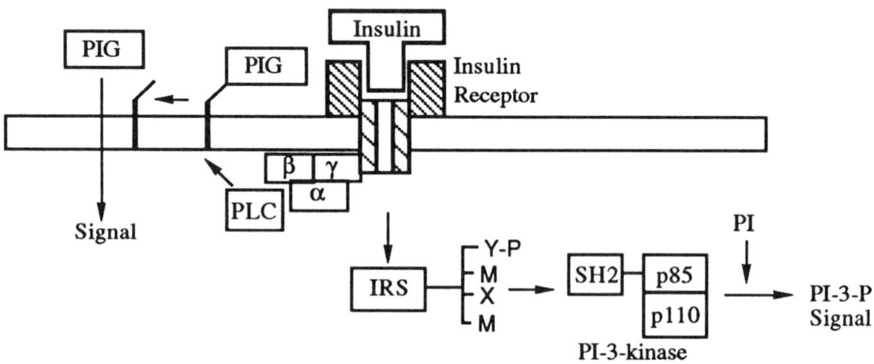

Figure 10 Signaling mechanism *via* PIG insulin mediators and 3-phosphorylated phosphatidylinositols. Binding of insulin to its receptor stimulates a putative GPI-PLC which cleaves GPI to form PIG. PIG are the second messengers involved in regulation of various enzymatic activities.

Binding of an agonist to the insulin receptor is communicated to the glycosylphosphatidylinositol-specific phospholipase C (GPI-PLC) by a G-protein. GPI-PLC cleaves the glycan head groups of membrane-bound glycosylphosphatidylinositols (GPI) to form the water-soluble phosphoinositol glycans (PIG) (24, 95-98). PIGs are then transported inside the cell, where they regulate activity of insulin-sensitive enzymes. Alternatively, PIG can be derived from biosynthetic precursors of mature GPI. Since mammalian tissues contain a broad family of closely related GPI molecules (99-102), the action of GPI-PLC should also provide a range of analogous PIGs. Each of the PIGs may be specific in the regulation of a single enzyme or a narrow group of insulin-sensitive enzymes. The critical element for efficient transduction of the insulin signal by this mechanism is an adequate supply of all types of the precursor GPI. Any metabolic disorder resulting in a decreased level or an absence of a certain molecular species of GPI would result in a limited response of a tissue to the insulin stimulus (24, 95). If deficiency of the insulin mediator precursor could be proven quantitatively and its chemical structure determined, supplementing it could become a viable therapy of NIDDM.

B. Structure of the Putative Second Messengers

Glycosylphosphatidylinositols serve as general anchoring devices for many different membrane proteins (19, 103). Recently determined structures of GPI used for protein anchoring are shown in Figure 11. All known GPI-anchors incorporate Manα1$\rightarrow$2Manα1$\rightarrow$6Manα1$\rightarrow$4GlcNα1$\rightarrow$6PI as a main glycan chain. Molecular species of GPI differ in the attachment of branched glycans at the Man-1, the presence/number of phosphoethanolamine (PE) groups attached to the Man-1, and a structure of the hydrophobic residue (diacyl glycerol, 1-alkyl-2-acylglycerol or ceramide) (19, 103). Biosynthetic intermediates with a varying length of the glycan and a different number of PE groups are also present in cell membranes in addition to mature GPI (99). It was generally understood that all determined GPI structures, and also all other known inositol phosphates and phospholipids, incorporate *myo*-inositol, one of the nine geometrical isomers of cyclohexane hexaol. However, several papers reported the presence of *chiro*-inositol (101, 104-106). In bovine liver *chiro*-inositol accounts for up to 9% of all inositol phospholipids (68). The PIG-insulin mediators generated by the treatment of hepatoma cells with *Staphylococcus aureus* PI-PLC (104), and from rat liver in response to insulin stimulation (105) have been reported to contain *chiro*-inositol exclusively or predominantly. In *chiro*-inositol-containing PIG (*chiro*-PIG) glucosamine is replaced by galactosamine (73, 105). The structure of *chiro*-inositol-containing GPI has been determined only in terms of the monosaccharide composition (73, 101, 104, 105), and there is no general consensus on their precise structure. Recently, Larner's group reported the presence of a family of at least eleven structurally distinct species of glycosylphosphatidylinositol in bovine liver, five of which contained exclusively *chiro*-inositol, and two of which did not contain hexosamine (101). Both enantiomers of *chiro*-inositol were found recently in urine and plasma of healthy human subjects (107), with the 1D-enantiomer predominating.

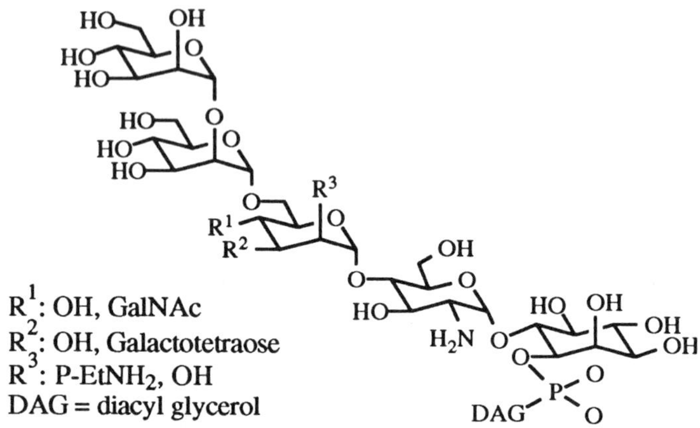

Figure 11 Structure of GPI involved in protein anchoring. In some GPI *chiro*-inositol and galactosamine may be present instead of *myo*-inositol and glucosamine.

C. Metabolism of *chiro*-Inositol

It is unknown at present how *chiro*-inositol is biosynthesized in higher animals. The likely pathways may include inosose dehydrogenase, such as the one identified in *Chlorella fusca* (108). Involvement of other pathways is also likely since both 1L- and 1D-enantiomers have been identified in human blood and urine (107). The uptake of inositol into the phospholipid pool is, however, a stereospecific process since all known *myo*-inositol phospholipids have the 1D-configuration (Figure 12). Assuming that *chiro*-inositol phospholipids are produced by the same pathway as are their *myo*-inositol counterparts it would ultimately lead to 1L-*chiro*-phosphatidylinositol (Figure 12), from which *chiro*-GPI could be assembled by sequential glycosylation analogous to *myo*-GPI. The kinetics of labeling indicated that oxidation-reduction of *myo*-PI or *myo*-GPI could also be involved (106) in the biosynthesis of *chiro*-GPI. The above considerations suggest that it is more likely that *chiro*-inositol phospholipids are of 1L- than 1D-configuration, unless different sets of enzymes are involved in assembling the *chiro*- and *myo*-GPI. Alternatively, the *chiro*-inositol insulin mediators could be generated by phospholipase D (109) (GPI-PLD), in which case the configuration of *chiro*-inositol could be less important for the enzyme. The question concerning the biosynthetic origin of *chiro*-GPI would, however, still remain.

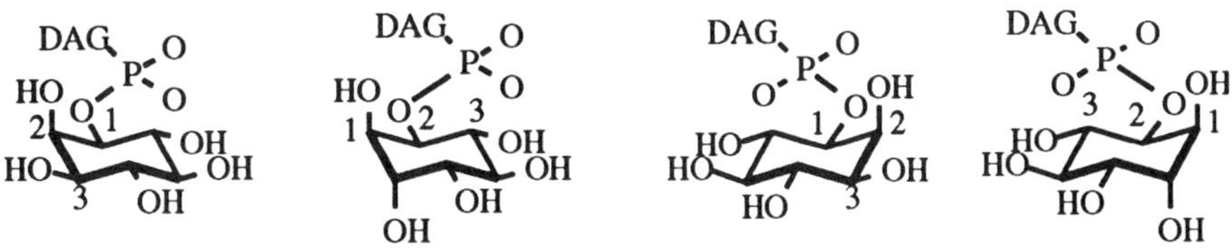

1D-*myo*-PI 1L-*chiro*-PI 1L-*myo*-PI 1D-*chiro*-PI

DAG = diacylglycerol

Figure 12 Stereochemical relationship between *myo*- and *chiro*-PI. Inversion at the 3-carbon atom converts phosphatidyl-1D-*myo*-inositol into phosphatidyl-1L-2-*chiro*-inositol. This is due to different numbering and stereochemical reference in *myo*- and *chiro*-inositols. Thus, 1D-*chiro*-inositol-containing PI or GPI is related to phosphatidyl-1L-*myo*-inositol, and is therefore not expected to be cleaved by PI-PLC.

D. *chiro*-Inositol in Diabetes

According to the most recent work the 1D-enantiomer is significantly elevated in the urine and plasma of NIDDM and IDDM patients (107). In contrast, Larner's group reported earlier the deficiency of 1D-*chiro*-inositol in NIDDM patients (110) and in diabetic animals (111). The latter findings along with the previous determination of the 1D-configuration in the insulin mediator precipitated a series of studies in which (i) the concentration of 1D-*chiro*-inositol in the urine and plasma has been proposed as a diagnostic parameter for NIDDM (112), (ii) dietary 1D-*chiro*-inositol has been used to lower blood sugar levels in human subjects and experimental animals (113, 114), and (iii) a synthetic method for 1D-*chiro*-inositol has been developed (115). The hypothesis has been put forward that a deficiency of D-*chiro*-inositol-containing GPI in NIDDM patients would result in the lack of the corresponding insulin mediators (*chiro*-PIG), causing a limited response to insulin (24). The levels of *chiro*-inositol in diabetes are controversial, since actually higher levels have been found by another research group (107, 116).

E. Can *chiro*-GPI be Cleaved by Enzymes Related to Known PI-PLC?

GPI-specific PLC has been found in higher organisms (117, 118), and GPI-PLC from *Trypanosoma brucei* is homologous (119) to bacterial PI-PLC (120). GPI-PLC should accept substrates with stereochemistry related to natural phosphatidylinositol, i.e. having 1L-configuration of *chiro*-inositol rather than 1D (Figure 12). To resolve the problem of inconsistency of the determined configuration of the *chiro*-inositol component of the insulin mediator, and preference of PI-PLC for 1D-configuration of *myo*-inositol, we have carried out the synthesis of both 1D- and 1L-diastereomers of phosphatidyl-2-*chiro*-inositol (**20**, Figure 3), and subjected both isomers to reaction with *B.thuringiensis* PI-PLC. Only the 1L-*chiro*-isomer was a substrate of this enzyme, and the 1D-isomer was completely resistant (68). It is worthwile to point out that the *B.thuringiensis* PI-PLC is closely related to the staphylococcal enzyme used by Mato et al. (104) to generate *chiro*-inositol insulin mediators from rat liver. We therefore conclude that if the *chiro*-GPI was to function as a substrate of insulin-activated GPI-PLC, it would have to incorporate 1L-*chiro*-inositol. Indeed, degradation of the total bovine lipid followed by purification of the inositol fraction, exhaustive benzoylation and chiral HPLC showed that only 1L-*chiro*-inositol is present in the phospholipid pool (Figure 13).

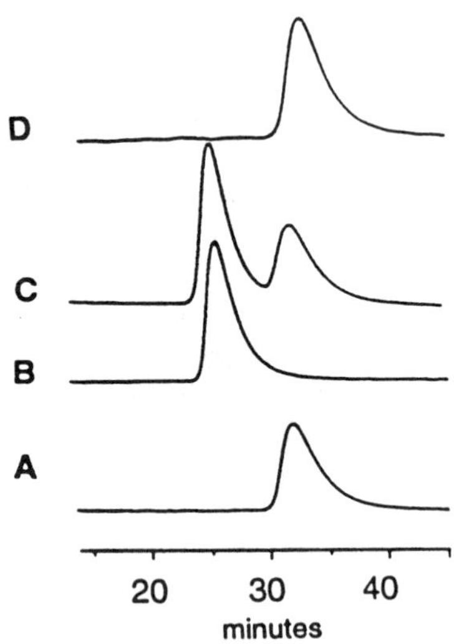

Figure 13 Determination of the absolute configuration of *chiro*-inositol from bovine liver lipid by chiral HPLC (68). Hexabenzoyl-*chiro*-inositols were separated on the polymethacrylate CHIRALPAK OT (+) column, using isocratic elution with methanol-hexane-isopropanol (50:25:25, v/v), 10°C, flow rate 0.6 mL/min, UV detection at 230 nm. (A) 1L-Bz_6-*chiro*-inositol, t_R 31.8 min.; (B) 1D-Bz_6-*chiro*-inositol, t_R 25.4 min.; (C) the mixture of 1D- (60%) and 1L- (40%) enantiomers, t_R 25.3 and 32.0 min.; (D) Bz_6-*chiro*-inositol from bovine liver, t_R 32.8min.

This result is consistent with the known stereochemical preference of PI-PLC and PI synthase. The original report of 1D-configuration gave rise to a number of misguided synthetic studies aimed at synthesis of *chiro*-inositol containing insulin mediators (121, 122).

F. Structure of PI-PLC-Sensitive Glycosylphosphatidylinositols

The first step of the simplest biosynthetic scheme of *chiro*-GPI could involve synthesis of *chiro*-PI employing PI-synthase. This enzyme displays tolerance to modifications at *myo*-inositol 3-position, suggesting that it could take *chiro*-inositol as a substrate (123). Furthermore, the presence of the unknown isomer of glycerophosphoinositol, possibly containing *chiro*-inositol, was observed in the product of deacylation of phospholipid from barley seeds (124). We have therefore embarked on a search for the putative *chiro*-phosphatidylinositol as a first biosynthetic intermediate. The methylamine digests of the total bovine and rat liver lipid were analyzed by high performance anion exchange chromatography coupled with pulsed amperometric detection (HPAEC-PAD) against standards of glycerophospho-*chiro*-inositol and *chiro*-inositol phosphate. The typical chromatographic run is shown in Figure 14 (125).

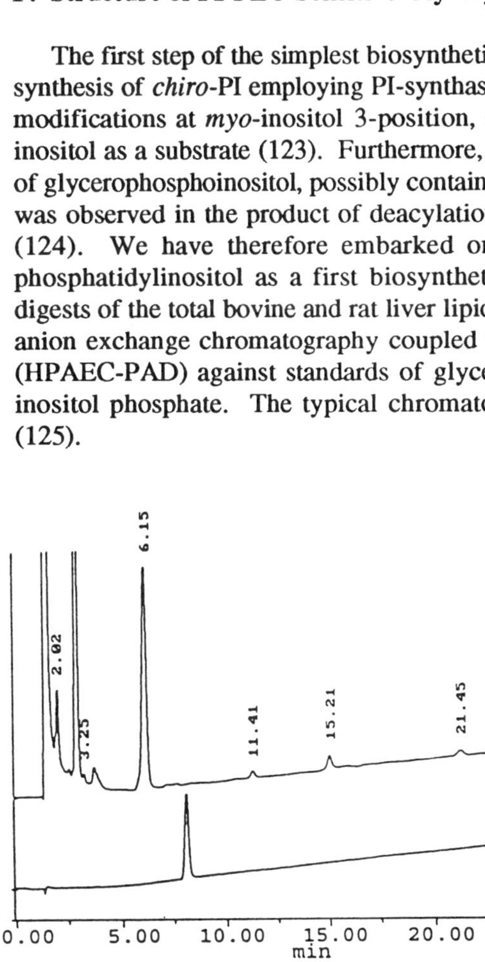

Figure 14 HPAEC-PAD analysis of the methylamine digest of bovine liver lipid (top). The identities of peaks are: glycerophospho-*myo*-inositol (Gro-PI, 6.1 min), IcP (11.4 min), IP+glycerol phosphate (15.2 min), and glycerophosphoserine (21.4 min), respectively. The minor component eluting shortly after Gro-PI does not match the glycerophospho-*chiro*-inositol standard (bottom). Chromatographic conditions: Carbopack PA1 column eluted with isocratic 0.1 M NaOH and sodium acetate gradient (40-200 mM in 20 min.) at 1.0 mL/min flow rate.

Repetitive analyses showed several minor peaks eluting closely to the glycerophospho-*chiro*-inositol (Gro-PcI) standard, none however, matched the chemically synthesized standard. The detection limit of the HPAEC method was estimated as 2% of the content of glycerophospho-*myo*-inositol.

Biosynthetic pathways of GPI in mammalian tissues produce a variety of intermediates, each of which could be a substrate for PI-PLC (19, 99). Unlike lower organisms, however, most of the mammalian GPI precursors are palmitoylated at the 2-hydroxyl group, therefore they are not susceptible to PI-PLC cleavage (99, 103). Nevertheless, the presence of a small pool of PI-PLC sensitive GPI has been recently established (100). One way to look at the population of PI-PLC-sensitive GPI is to employ ^{31}P NMR. The action of PI-PLC on the total lipid produces cyclic inositol phosphates which have characteristic chemical shifts approximately 16 ppm downfield from H_3PO_4, clearly removing it from the chemical shift range of other biophosphates. ^{31}P NMR spectra of cyclic inositol phosphates derived from various sources are shown in Figure 15 (125).

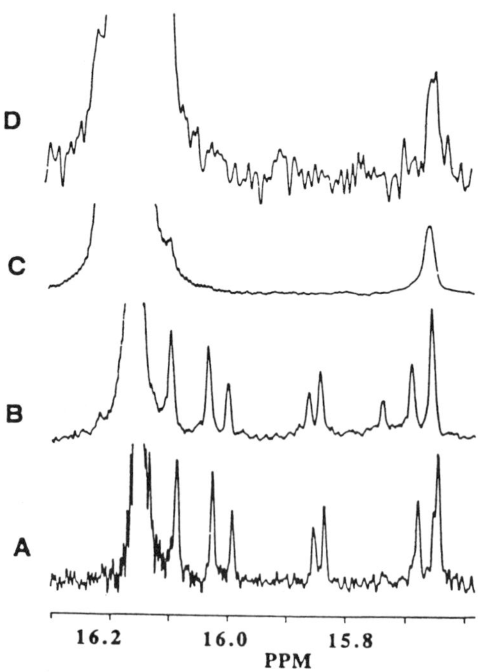

Figure 15 Partial ^{31}P NMR spectra of inositol cyclic phosphates produced by *B.thuringiensis* PI-PLC from the total phospholipid of (A) soybean, (B) soybean with *chiro*-IcP added; (C) bovine liver, and (D) rat liver. Phospholipids samples (≈ 1 g) were dispersed in sodium deoxycholate solution (2 g, pH 7.0) and treated with PI-PLC (50 µg) for 1-24 h. The product was acidified to pH 5.0, extracted with chloroform and loaded on the $HCOO^-$-Dowex anion exchange column. Spectra show the fraction collected between 50-200 mM formate.

Soybean lecithin, known to contain *chiro*-inositol phospholipids, produced at least 9 different minor cyclic inositol phosphates with chemical shifts in the range 15.6-16.1 ppm, in addition to the main signal of IcP at 16.15 ppm. Consistent with HPAEC results shown in Figure 14 *chiro*-IcP was not one of these products. Interestingly, both bovine and rat liver lipid gave rise to only one single minor signal (1.5% of IcP) in addition to that of IcP. Subsequent HPAEC-PAD analysis (Figure 16) showed that the minor signal at 15.6 ppm is likely to be a composite of several components eluting shortly after IcP.

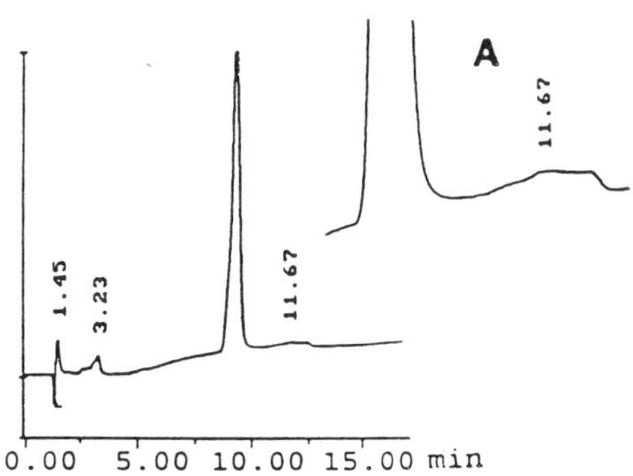

Figure 16 HPAEC-PAD analysis of the inositol cyclic phosphate fraction derived from bovine liver. The sample shown in Figure 15C has been passed through the C-8 column and the Biogel P-2 prior to analysis. The inset shows the presence of several closely eluting peaks between IcP and I-1-P. Chromatographic conditions are analogous to those in Figure 14, except 40-500 mM NaOAc gradient over 30 min was used.

We have thus elaborated methods for quantitation of PI-PLC sensitive glycolipids and their chromatographic separation. The current work is aimed at preparative separation of these minor components and determination of their structure by combination of physicochemical and enzymatic methods.

VI. FUTURE OUTLOOK

The current rapid progress in understanding the role of PI-PLC in growth factor signaling, the mechanism of its action and recent determination of the three-dimensional structure of mammalian PI-PLC active site will shift the current substrate-based approach of inhibitor design into active site structure-based development of inhibitors with high affinity and selectivity. It is especially important to develop inhibitors with specificity towards major types of enzymes (β, γ, δ), and further achieve selectivity among PI-PLC subspecies (i.e. γ_1, γ_2, etc.).

This will narrow the physiological effects of inhibitors and result in their lower toxicity.

Further progress in unraveling the role of phosphoinositol glycans in insulin signaling requires determination of their precise chemical structure. Since quantities of available insulin mediators are insufficient to accomplish this task, it is particularly important to establish structures of their precursor glycosylphosphatidylinositols, and to develop sensitive methods for their detection and quantitation in animal tissues. Having such analytical methods at hand would be especially beneficial in studying the differences in the content of these precursors in pathological conditions, and their changes in response to physiological factors. Establishing correlations between levels of these precursors in healthy and NIDDM subjects would prove their importance in NIDDM.

Note Added in Proof: The recent X-ray structure of the catalytically active deletion variant of PI-PLC-δ shows a mechanism somewhat different from that of bacterial enzyme involving binding of the phosphate group to Ca^{2+} and activation of the 2-hydroxyl group by the Ca^{2+} ion and two glutamate residues (126).

VII. ACKNOWLEDGMENT

This work was partially supported by Research Grant GM30327 from the National Institutes of Health. The author thanks Dr. Ming-Daw Tsai for his support and advice.

Abbreviations: Bn, benzyl, *chiro*-IcP, *chiro*-inositol 1,2-cyclic phosphate; CSF, colony stimulating factor; DAG, diacylglycerol; EGF, epidermal growth factor; FGF, fibroblast growth factor; GPI, glycosylphosphatidylinositol; Gro-PcI, glycerophospho-chiro-inositol; HPAEC, high performance anion exchange chromatography; IcP, inositol 1,2-cyclic phosphate; IP, I-1-P, inositol 1-phosphate; IP_3, inositol 1,4,5-trisphosphate; PAD, pulsed amperometric detection; Pal, palmitoyl; PDGF, platelet-derived growth factor; PI, phosphatidylinositol; PI-PLC, phosphatidylinositol-specific phospholipase C; PIP, phosphatidylinositol 4-phosphate; PIP_2, phosphatidylinositol 4,5-bisphosphate; TGF, transforming growth factor.

VIII. REFERENCES

1. Berridge MJ. Inositol trisphosphate and calcium signaling. Nature 1993; 361:315-325.
2. Clark EA, Brugge JS. Integrin and signal transduction pathways: The road taken. Science 1995: 268: 233-239.
3. (a) Lips DL, Majerus PW, Gorga FR, Young AT, Benjamin TL. Phosphatidylinositol 3-phosphate is present in normal and transformed fibroblasts and is resistant to hydrolysis by bovine brain phospholipase C. II. J Biol Chem 1989; 264: 8759-8763.
4. Menniti FS, Miller RN, Putney JW, Shears, SB. Turnover of inositol pyrophosphates in pancreatoma cells. J Biol Chem 1993; 268: 3850-3856.
5. Wasner HK, Salge U, Gebel M. The endogeneous cyclic AMP antagonist, cyclic PIP: its ubiquity, hormone stimulated synthesis and identification as prostaglandylinositol cyclic phosphate. Acta Diabetol 1993; 30: 220-232.
6. Rhee SG, Choi KD. Regulation of inositol phospholipid-specific phospholipase C isozymes. J Biol Chem 1992; 267:12393-12396.
7. Rhee SG, Choi, KD. Multiple forms of phospholipase C isozymes and their activation mechanisms. Adv Second Messenger Phosphoprotein Res 1992; 28:35-61.
8. Harden, TK. G-Protein regulated phospholipase C. Identification of component proteins. Adv Second Messenger Phosphoprotein Res 1992; 26:11-34.
9. Jones G, Carpenter G. Regulation of phospholipase C isozymes. Prog Growth Factor Res 1992; 4:97-106.
10. Meldrum E, Parker PJ, and Carozzi A. The PtdIns-PLC superfamily and signal transduction. Biochim Biophys Acta 1991; 1092:49-71.
11. Dennis EA, Rhee SG, Billah MM, Hannun YA. Role of phospholipases in generating lipid second messengers in signal transduction. FASEB J 1991; 5:2068-2077.
12. (a) Powis G, Alberts DS. Inhibiting intracellular signaling as a strategy for cancer chemoprevention. Eur J Cancer 1994; 30A:1138-1144. (b) Powis G, Phil D. Inhibitors of phosphatidylinositol signaling as antiproliferative agents. Cancer Metast Rev 1994; 13:91-103. (c) Powis G, Workman P, Signaling targets for the development of cancer drugs. Anticancer Drug Design 1994; 9:263-277.

13. Kozikowski AP, Fauq AH, Malaska MJ, Tueckmantel W, Ognyanov VI, Powis G. Inositol analogs as potential anticancer therapeutics. Curr Med Chem 1994; 1:1-12.
14. Brunton VG, Workman P, Cell-signaling targets of antitumor drug development. Cancer Chemother Pharmacol 1993; 32:1-19.
15. (a) Vlahos CJ. Phosphatidylinositol 3-kinase Inhibitors and Thei r Effects on Cell Signaling Pathways, Drugs Future 1995; 20:165-171. (b) Atack JR, Fletcher SR. Inhibitors of Inositol Monophosphatase. Drugs Future 1994; 19: 857-866.
16. Potter BVL, Lampe D. Chemistry of Inositol Lipid Mediated Cellular Signalling. Angew Chem Int Ed Engl 1995; 34: 1933-1972.
17. Bruzik KS, Tsai MD. Toward the mechanism of phosphoinositide-specific phospholipase C. BioMed Chem 1994; 2:49-72.
18. (a) Heinz DW, Ryan M, Bullock TL, Griffith OH. Crystal structure of the phosphatidylinositol-specific phospholipase C from *Bacillus cereus* in complex with *myo*-inositol. EMBO J 1995; 15:3855-3863. (b) Hansen S, Hough E, Svensson LA, Wong YL, Martin SF. Crystal Structure Of Phospholipase C from *Bacillus Cereus* Complexed with a Substrate Analog. J Mol Biol 1993, 234: 179-187.
19. Englund PT. The structure and biosynthesis of glycosyl phosphatidylinositol protein anchors. Ann Rev Biochem 1993; 62:121-138.
20. Bennett CF, Angioli MP, Crooke ST. Purification of guinea pig uterus phosphoinositide-specific phospholipase C. Meth Enzymol 1991; 197:526-535.
21. Melin PM, Pical C, Jergil B, Sommarin M. Polyphosphoinositide phospholipase C in wheat rot plasma membranes. Partial purification and characterization. Biochim Biophys Acta 1992; 1123:163-169.
22. Perrella FW, Jankewicz R, Dandrow EA. Phospholipase C from human melanoma: Purification and characterization of phosphatidylinositol-selective enzyme. Biochim Biophys Acta 1991; 1076:209-214.
23. Kapeller R, Cantley LC. Phosphatidylinositol 3-kinase. Bioassays 1994; 16:565-576.
24. Larner J, Romero G. Insulin mediators and the mechanism of insulin action. Adv Pharmacol 1993; 24:21-50.
25. Vivien D, Petitfrere E, Martiny L, Sartelet H, Galera P, Haye B, Pujol JP. IPG (inositolphosphate glycan) as a cellular signal for TGF-beta-1 modulation of chondrocyte cell cycle. J Cell Phys 1993; 155:437-444.

26. Matuoka K, Fukami K, Nakanishi O, Kawai S. Takenawa T. Mitogenesis in response to PDGF and bombesin abolished by microinjection of antibody to PIP_2. Science 1989; 239:640-643.
27. Smith MR, Liu YL, Kim H, Rhee SG, Kung HF. Inhibition of serum and *ras*-Stimulated DNA synthesis by antibodies to phospholipase C. Science 1990; 247:1074-1077.
28. Smith MR, Ryu SH, Suh PG, Rhee SG, Kung HF. S-Phase induction and transformation of quiescent NIH 3T3 cells by microinjection of phospholipase C. Proc Natl Acad Sci USA 1989; 86:3659-3663.
29. Artega CL, Johnson MD, Todderud G, King LE, Carpenter, G. Elevated content of tyrosine kinase substrate phospholipase C-γ_1 in primary human breast carcinomas. Proc Natl Acad Sci USA 1991; 88:10435-10439.
30. Noh DY, Lee YH, Kim SS, Kim YI, Ryu SH, Suh PG, Park JG. Elevated content of phospholipase C-γ_1 in colorectal cancer tissues. Cancer 1994; 73:36-41.
31. Park JG, Lee YH, Kim SS, Park KJ, Noh DY, Ryu SH, Suh PG. Overexpression of phospholipase C-γ in familial polyposis. Cancer Res 1994; 54:2240-2244.
32. Yeatman TJ, Mao WG, Karl RC, Djeu JY. Phospholipase C-γ_1 is present on the surface of human colorectal cancer cells. Biochem.Biophys Res Commun 1994; 201:1043-1049.
33. Nomoto K, Morotomi M, Miyake M, Xu DB, Logerfo PP, Weinstein IB. The effect of bile acids on phospholipase C activity in extracts of normal human colon mucosa and primary colon tumors. Mol Carcinogen 1994; 9: 87-94.
34. Homma MK, Homma Y, Namba M, Yuasa Y. Enhanced phosphoinositide metabolism in colorectal cells derived from familial adenomatus polyposis patients. J Cell Biochem 1994; 55:477-485.
35. Smith MR, Liu YL, Matthews NT, Rhee SG, Sung WK, Kung HF. Phospholipase C-γ_1 can induce DNA synthesis by a mechanism independent of its lipase activity. Proc Natl Acad Sci USA 1994; 91:6554-6558.
36. Singhal RL, Yeh YA, Look KY, Sledge GW, Weber G. Coordinated increase in activities of the signal transduction enzymes PI kinase and PIP kinase in human cancer cells. Life Sci 1994; 55:1487-1492.
37. Rillema JA. Possible role of phospholipase C in the regulation of cell division in normal and neoplastic cells. Med Hypo 1989; 29:1-4.
38. Spivakkroizman T, Mohammadi M, Jaye M, Schlessinger J. Point mutation in the fibroblast growth factor receptor eliminates

phosphatidylinositol hydrolysis without affecting neuronal differentiation of PC12 cells. J Biol Chem 1994; 269:14419-14423.
39. Peters KG, Marie J, Wilson E, Ives HE, Escobedo J, Del Rosario M, Mirda D, Williams LT. Point mutation of an FGF receptor abolishes phosphatidylinositol turnover and Ca^{2+} flux, but not mitogenesis. Nature 1992; 358:678-680.
40. Margolis B, Zilberstein A, Franks C, Felder S, Kremer S, Ulrich A, Rhee SG, Skorecki K, Schlesinger J. Effect of phospholipase C-γ overexpression on PDGF-induced second messengers and mitogenesis. Science 1990; 248:607-616.
41. Rifkin DB, Moscatelli D. Recent developments in cell biology of basic fibroblast growth factor. J Cell Biol 1989; 109:1-6.
42. Senger DR, van De Water L, Brown LF, Nagy JA, Yeo KT, Berse B, Jackman RW, Dvorak AM, Dvorak HF. Vascular permeability factor (VPF, VEGF) in tumor biology, Cancer Metastasis Rev. 1993; 12:303-324.
43. Brock TA, Dvorak HF, Senger DR. Tumor secreted vascular permeability factor increases cytosolic Ca^{2+} and von Hilldebrand factor release in human endothelial cells. Am J Pathol 1991; 138:213-221.
44. Tessler S, Rockwell P, Hicklin D, Cohen T, Levi BZ, Witte L, Lemishka IR, Neufeld G. Heparin modulates the interaction of VEGF (165) with soluble and cell associated flk-1 receptors. J Biol Chem 1994; 269:12456-12461.
45. Misra KB, Kim KC, Cho SY, Low MG, Bensadoun, A. Purification and characterization of adipocyte heparan sulfate proteoglycans with affinity for lipoprotein lipase. J Biol Chem 1994; 269:23838-23844.
46. Bruzik KS, Morocho AM, Jhon DY, Rhee SG, Tsai MD. Phospholipids chiral at phosphorus. Stereochemical mechanism for the formation of inositol 1-phosphate catalyzed by phosphatidylinositol-specific phospholipase C. Biochemistry 1992; 31:5183-5193.
47. (a) Lin G, Bennett CF, Tsai MD. Phospholipids chiral at phosphorus. Stereochemical mechanism of reactions catalyzed by phosphoinositide-specific phospholipase C from *Bacillus cereus* and guinea pig uterus. Biochemistry 1990; 29:2747-2757. (b) Bruzik KS, Lin G, Tsai, MD. Phosphorothioate Analogues of Phosphatidyinositol and Inositol 1,2-Cyclic Phosphate: Application to the Mechanism of Phospholipase C. ACS Symp. Ser. 1991; 463:172-85.

48. Lewis KA, Garigapati VR, Zhou C, Roberts MF. Substrate requirements of bacterial phosphatidylinositol-specific phospholipase C. Biochemistry 1993; 32:8836-8841.
49. Volwerk JJ, Filhuth E, Griffith OH, Jain MK. Phosphatidylinositol-specific phospholipase C from B*acillus cereus* at the lipid water interface: Interfacial binding, catalysis and activation. Biochemistry 1994; 33:3364-3474.
50. Shashidhar MS, Volwerk JJ, Griffith OH, Keana JFW. A chromogenic substrate for phosphatidylinositol-specific phospholipase C: 4-Nitrophenyl *myo*-inositol 1-phosphate. Chem Phys Lipids 1991; 60:101-110.
51. Bruzik KS, Guan Z. to be published
52. Cruz-Rivera M, Bennett CF, Crooke, ST. Glycero-3-phospho-D-myo-inositol-4-phosphate (GroPIP) is an inhibitor of phosphoinositide-specific phospholipase C. Biochim Biophys Acta 1990; 1042:113-118.
53. Kim JW, Ryu SH, Rhee SG. Cyclic and noncyclic inositol phosphates are formed at different ratios by phospholipase C isozymes. Biochem Biophys Res Commun 1989; 163: 177-182.
54. (a) Hershlag D. Ribonuclease revisited: Catalysis via the classical general acid-base mechanism or triester-like mechanism. J Am Chem Soc 1994; 116:11631-11635. (b) Messmore JM, Fuchs DN, Raines RT. Ribonuclease A: Revealing Structure-Function Relationship with Semisynthesis. J Am Chem Soc 1995; 117:8057-8060.
55. Eckstein F. Phosphorothioate analogues of nucleotides. Tools for the investigation of biological processes. Angew Chem Int Ed Engl 1983; 22:423-506.
56. Shapiro R, Riordan JF, Vallee BL. Characteristic ribonucleolytic activity of human angiogenin. Biochemistry 1986; 26:5141-5146.
57. Acharya KR, Shapiro R, Allen SC, Riordan JF, Vallee BL. Crystal structure of human angiogenin reveals the structural basis for its functional divergence from ribonuclease. Proc Natl Acad Sci USA 1994; 91:2915-2919.
58. Kume T, Taguchi R, Tomita M, Tokuyama S, Morizawa K, Nakachi O, Hirano J, Ikezawa H. The study of phosphatidylinositol-specific phospholipase C from *Bacillus thuringiensis*: Synthesis of homogeneous substrates, substrate specificity and other properties. Chem Pharm Bull 1922; 40:2133-2137.
59. Rebecchi MJ, Eberhardt R, Delaney T, Ali S, Bittman R. Hydrolysis of short acyl chain inositol lipids by phospholipase C-δ_1. J Biol Chem 1993; 268:1735-1741.

60. Leigh JA, Volwerk JJ, Griffith OH, Keana JF. Substrate specificity of pho sphatidylinositol-specific phospholipase C from *Bacillus cereus* examined using the resolved enantiomers of synthetic *myo*-inositol 1-(4nitrophenyl phosphate). Biochemistry 1992; 31:8978-8993.
61. Bhamare N, Wang Y, Tsai M-D, Bruzik KS. Design, Synthesis and Biochemical Applications of Analogs of Phosphatidylinositol. Phosphorus Sulfur Rel Elements, in press.
62. Pascher I, Lundmark M, Nyholm PG, Sundell S. Crystal structures of membrane lipids. Biochim Biophy Acta 1992; 1113:339-373.
63. Hendrickson EK, Johnson JL, Hendrickson HS, Johnson JL, Khan TH, Chial HJ. Kinetics of *Bacillus cereus* phosphatidylinositol-specific phospholipase C with thiophosphate and fluorescent analogs of phosphatidylinositol. Biochemistry 1992; 31:12169-12172.
64. Shashidhar MS, Volwerk JJ, Keana JFW, Griffith OH. Inhibition of phosphatidylinositol-specific phospholipase C by phosphonate substrate analogs. Biochim Biophys Acta 1990; 1042: 410-412.
65. Seitz SP, Kaltenbach RF, Vreekamp RH, Calabreses JC, Perrella FW. Synthesis and enzymatic properties of a deoxy analog of phosphatidylinositol. BioMed Chem Lett 1992; 2:171-174.
66. Garigapati VR, Roberts MF. Synthesis of phosphatidyl-2-O-alkylinositols as potential inhibitors for PI-specific PLC. Tetrahedron Lett 1993; 34:5579-5582.
67. Roberts WL, Santikarn S, Reinhold VN, Rosenberry TL. Structural characterization of the glycoinositol phospholipid membrane anchor of human erythrocyte acetylcholinesterase by fast atom bombardment spectrometry. J Biol Chem 1988; 263:18776-18784.
68. Bruzik KS, Hakeem AA, Tsai MD. Are D- and L-*chiro*-Phosphoinositides substrates of phosphatidylinositol specific phospholipase C? Biochemistry 1994; 33:8367-8374.
69. McDonald LJ, Mamarack MD. Purification and characterization of bovine heart phosphoinositide-specific phospholipase C. Kinetic analysis of the Ca^{2+} requirement and La^{3+} inhibition. Biochemistry 1989; 28:9926-9932.
70. Ikezawa H. Bacterial PI-PLCs unique properties and usefulness in studies of GPI anchors. Cell Biol Int Rep 1991; 15:1115-1131.
71. Mensa-Wilmot K, Englund PT. Glycosylphosphatidylinositol-specific phospholipase C of *Trypanosoma brucei*: expression in *E.coli*. Mol Biochem Parasitol 1992; 56:311-322.
72. Volwerk JJ, Shashidhar MS, Kuppe A, Griffith OH. Phosphatidylinositol-specific phospholipase C from *Bacillus cereus* combines intrinsic

phosphotransferase and cyclic phosphodiesterase activities: a phosphorus-31 NMR study. Biochemistry 1990; 29:8056-8062.

73. Larner J, Huang LC, Schwartz CFW, Oswald AS, Shen TY, Kinter M, Tang G, Zeller K. Rat liver insulin mediator which stimulates pyruvate dehydrogenase contains galactosamine and D-*chiro*-inositol. Biochem Biophys Res Commun 1988; 151:1416-1426.
74. Kaufmann F, Massy DJR, Pirson W, Wyss P. Synthesis and biological evaluation of inositol derivatives as inhibitors of phospholipase C. In: Reitz AB, ed. Inositol Phosphates and Derivatives. Synthesis, Biochemistry and Therapeutic Potential. ACS Symp Ser 1991; 463:202-213.
75. Ward JG, Young RC. Synthesis and biological evaluation of inositol phospholipid analogues. In: Reitz AB ed. Inositol Phosphates and Derivatives. Synthesis, Biochemistry and Therapeutic Potential. ACS Symp Ser 1991; 463:214-228.
76. Campbell AS, Thatcher GRJ. Synthesis of an analogue of D,L-*myo*-inositol-1,2-cyclic phosphate: Inhibition of phosphatidylinositol-specific phospholipase C. Tetrahedron Lett 1991; 32:2207-2210.
77. Campbell AS, Thatcher GRJ. Tetravanadate is an inhibitor of phosphatidylinositol-specific phospholipase C. BioMed Chem Lett 1992; 2:655-658.
78. Ingber D, Fujita T, Kishimoto S, Sudo K, Kanamaru T, Brem H, Folkman J. Synthetic analogues of fumagillin that inhibit angiogenesis and suppress tumor growth. Nature 1990; 348:555-557.
79. Perrella FW, Chen SF, Behrens DL, Kaltenbach RF, Seitz SP. Phospholipase C inhibitors: A new class of cytotoxic agents. J Med Chem 1994; 37: 2232-2237.
80. Bennett CF, Mong S, Wu HLW, Clark MA, Wheeler L, Crooke ST. Inhibition of phosphoinositide-specific phospholipase C by manoalide. Mol Pharm 1987; 32:587, 593.
81. (a) Powis G, Seewald MJ, Gratas C, Melder D, Riebow J, Modest EJ. Selective inhibition of phosphatidylinositol phospholipase C by cytotoxic ether lipid analogues. Cancer Res 1992; 52: 2835-2840. (b) For overview of lysolipid analogs as potential drugs see: Lohmeyer M, Bittman R. Antitumor Ether Lipids and Alkylphosphocholines. Drugs Future 1994: 19: 1021-1037.
82. Pawelczyk T, Lowenstein JM. Inhibition of phospholipase C-δ by hexadecylphosphorylcholine and lysophospholipids with antitumor activity. Biochem Pharm 1993; 45:493-497.

83. Hill SR, Bonjouklian R, Powis G, Abraham RT, Ashenedl CL, Zalkow LH. A multisample assay for inhibitors of phosphatidylinositol phospholipase C: Identification of naturally occurring peptide inhibitors with antiproliferative activity. Anticancer Drug Design 1994; 9:353-361.
84. Ji YH, Pantze M, Pirson W, Wyss P. Inhibitors of phospholipase C as potential antiinflamatory agents. Abstract Int Symp Med Chem Basel, Sept. 13-17, 1992.
85. Weber W, Schu P, Anke T, Velten R, Steglich W. Caloporoside, a new inhibitor of phospholipase C from *Caloporus dichrous* (fr.) *Ryv.* J Antibiot 1994; 47:1188-1194.
86. Aoki M, Itezono Y, Shirai H, Nakayama N, Sakai A, Tanaka Y, Yamaguchi A, Shimma N, Yokose K, Seto H. Structure of a novel phospholipase C inhibitor, Vinaxanthone (Ro 09-1450), produced by *penicillium vinaceum.* Tetrahedron Lett 1991; 32:4737-4740.
87. Yanagisawa M, Sakai A, Adachi K, Sano T, Watanabe K, Tanaka Y, Okuda T. Hispidospermidin, a novel phospholipase C inhibitor produced by *Chaetosphaeronema hispidium* (Cda) Moesz NR 7127. J Antibiot 1994; 47: 1-5.
88. Ut H, Imoto M, Umezawa K. Inhibition of phosphatidylinositol-specific phospholipase C activity by fluvirucin B2. J Antibiot. 1995; 48: 387-390.
89. Powis G, Lowry S, Forrai L, Secrist P, Abraham R. Inhibition of phosphoinositide phospholipase C by compounds U-73122 and D-609. J Cell Pharmacol 1991; 2: 257-262.
90. Bronner C, Wiggins C, Monte D, Mårki F, Capron A, Landry T, Franson RC. Compound 48/80 is a potent inhibitor of phospholipase C and a dual modulator of phospholipase A2 from human platelet. Biochim Biophys Acta 1987; 920: 301-305.
91. Homma Y, Takenawa T. Inhibitory effect of src homology (SH2/SH3) fragments of phospholipase C-γ on the catalytic activity of phospholipase C isoforms. J Biol Chem 1992; 267:21844-21849.
92. Ogawara H, Higashi K, Manita S, Hidaka M, Kato H, Takenawa T. An inhibitor of inositol phospholipid specific phospholipase C. Biochim Biophys Acta 1993; 1175:289-192.
93. Kahn CR. Insulin action, diabetogenes, and the cause of type II diabetes. Diabetes 1994; 43:1066-1084.
94. Myers MG, White MF. The new elements of insulin signaling. Insulin receptor substrate-1 and proteins with SH2 domains. Diabetes 1993; 42: 643-50

95. Romero G. Inositolglycans and cellular signaling. Cell Biol Int Rep 1991; 15:827-852.
96. Romero G, Luttrell L, Rogol A, Zeller K, Hewlett E, Larner J. Phosphatidylinositol-glycan anchors of membrane proteins: potential precursors of insulin mediators. Science 1988; 240:509-511.
97. Saltiel AR, Cuatrecasas P. In a search of second messenger for insulin. Am J Physiol 1988; 255: C1-C11.
98. Saltiel AR, Cuatrecasas P. Insulin stimulates the generation from hepatic plasma membranes of modulators derived from an inositol glycolipid. Proc Natl Acad Sci USA 1986; 83:5793-5797.
99. Hirose S, Prince GM, Sevlever D, Lakshmeswari R, Rosenberry TL, Yeda E, Medof ME. Characterization of putative glycoinositol phospholipid anchor precursors in mammalian cells. J Biol Chem 1992; 267:16968-16974.
100. Deeg MA, Murray NR, Rosenberry TL. Glycan components in the glycoinositol phospholipid anchor of human erythrocyte acetylcholinesterase. J Biol Chem 1992; 267:18581-18588.
101. Pak Y, Larner J. Identification and characterization of *chiro*-inositol containing phospholipids from bovine liver. Biochem Biophys Res Commun 1992; 184: 1042-1047.
102. Gaulton GN. Differential regulation of glycosylated phosphatidylinositol subtypes by insulin. Diabetes 1991; 40:1297-1304.
103. McConville MJ, Ferguson, MAJ. The structure, biosynthesis and function of glycosylated phosphatidylinositols in the parasitic protozoa and higher eukaryotes. Biochem J 1993; 294:305-324.
104. Mato JM, Kelly KL, Abler A, Jarrett L, Corkey BE, Cashel JA, Zopf D. Partial structure of an insulin-sensitive glycophospholipid. Biochem Biophys Res Commun 1987; 146:764-770.
105. Larner J, Huang LC, Tang G, Suzuki S, Schwartz CFW, Romero G, Roulidis Z, Zeller K, Shen TY, Oswald AS, Luttrell L. Insulin mediators: Structure and function. Cold Spring Harbor Symp Quant Biol 1988; LIII:965-971.
106. Pak Y, Paule CR, Bao CR, Huang LC, Larner J. In vivo conversion of [^{3}H]myo-inositol to [^{3}H]chiro-inositol in rat tissues. Proc Natl Acad Sci USA 1992; 90:7759-7763.
107. Ostlund RE, McGill JB, Herskowitz I, Kipnis DM, Santiago JV, Sherman WR. D-*chiro*-Inositol metabolism in diabetes mellitus. Proc Natl Acad Sci USA 1993; 90:9988-9992.

108. Woeber G, Ruis H, Hoffman-Ostenhof O. Studies on the biosynthesis of cyclitols. XXVII. A dehydrogenase system that can catalyze the epimerization of *myo*-inositol to D-*chiro*-inositol. Monatsh Chem 1971; 102:459-464
109. Rademacher TW, Caro H, Kunjara S, Wang DY, Greenbaum AL, McLean P. Inositolphosphoglycan second messengers. Brazilian J Med Biol Res 1994; 27:327-341.
110. Kennington AS, Hill CR, Craig J, Bogardus C, Raz I, Ortmeyer HK, Hansen BC, Romero G, Larner J. Low urinary *chiro*-inositol excretion in non-insulin dependent diabetes mellitus. N Engl J Med 1990; 323: 373-378.
111. Huang LC, Fonteles MC, Houston DB, Zhang C, Larner J. *chiro*-Inositol deficiency and insulin resistance. III. Acute glycogenic and hypoglycemic effects of two inositol phosphoglycan insulin mediators in normal and streptozocin-diabetic rats in vivo. Endocrinology (Baltimore) 1993; 132:652-657.
112. Kennington AS, Larner J. Quantitative analysis of *chiro*-inositol as diabetic condition predictor. PCT Int Appl 1991; WO 91 12,335; C. A. 115: 180781e.
113. Larner J, Kennington A, Huang LC. Method of reducing elevated blood sugar levels in humans using *chiro*-inositols. PCT Intl Appl 1992; WO 92 12,706; C. A. 117:163880k.
114. Larner J, Kennington A. D-*chiro*-Inositol as a dietary supplement for insulin resistant diabetics. PCT Int Appl 1991; WO 90 10,439; C. A. 114:157211q.
115. Kennington A, Larner J, Hill C, Stringfield B, Carta G, Kirwan DJ. Method for producing D-*chiro*-inositol. U.S. US 1992; 5,091,596; C. A. 116: 236091v.
116. Ostlund, RE, Sherman, WR. Measurement of D-*chiro*-inositol in clinical studies. Diabetes Care 1995; 18:1074-1075.
117. Fox JA, Soliz NM, Saltiel AR. Purification of a phosphatidylinositol-glycan-specific phospholipase C from liver plasma membranes: A possible target of insulin action. Proc Natl Acad Sci USA 1987; 84:2663-2667.
118. Stieger S, Diem S, Jakob A, Brodbeck, U. Enzymatic properties of phosphatidylinositol-glycan-specificphospholipase C and phosphatidyl-inositol -glycan-specific phospholipase D from rat serum. Eur J Biochem 1991; 197:67-73.

119. Hereld D, Hart GW, Englund PT. cDNA encoding the glycosyl-phosphatidylinositol-specific phospholipase C from *Trypanosoma brucei*. Proc Natl Acad Sci USA 1988; 85:8914-8918.
120. Kuppe A, Evans LM, McMillen DA, Griffith OH. Phosphatidylinositol-specific phospholipase C of *Bacillus cereus*: Cloning, sequencing, and relationship to other phospholipases. J Bacteriol 1989; 171:6077-6083.
121. Reddy KK, Falck JR, Capdevila J. Insulin second messengers.Synthesis of 6-O-(2-amino-2-deoxy-α-D-glucopyranosyl)-D-*chiro*-inositol phosphate. Tetrahedron Lett 1993;7869-7872.
122. Berlin WK, Zhang WS, Shen TY. Glycosyl-inositol Derivatives. Synthesis of hexosamine-inositol-phosphates related to putative insulin mediators. Tetrahedron 1991; 47:1-20.
123. Johnson SC, Dahl J, Shih TL, Schedler DJA, Anderson L, Benjamin TL, Baker DC. Synthesis and evaluation of 3-modified 1D-*myo*-inositols as inhibitors and substrates of phosphatidylinositols synthase and inhibitors of *myo*-inositol uptake by cells. J Med Chem 1993; 36:3628-3635.
124. Murthy, PPN, Matyshak GP, Keranen LM, Lam P, Mueller HH, Bhuvarahamurthy N. Evidence of two isomers of phosphatidylinositol in plant tissue. Plant Physiol 1992; 98:1498-1501.
125. Bruzik KS, Hakeem AA. unpublished.
126. Essen L-O, Perisic O, Cheung, R, Katan M, Williams, RL. Crystal Structure of a Mammalian Phosphoinositide-Specific Phospholipase C δ. Nature 1996; 380:595-601.

10

Carba-Sugars as Indispensable Units of Aminocyclitol Antibiotics and Related Biologically Active Compounds

Seiichiro Ogawa
Keio University
Yokohama, Japan

I. INTRODUCTION

5a-Carba-hexopyranose **1** having *α-talo* configuration was first chemically synthesized [1] as a racemate by McCasland *et al.* in 1966. Their carbocyclic (acyclic) analogue of hexopyranose was initially called pseudo-sugar. They succeeded in a synthesis of a total of three isomers [2-3] with α-*galacto* **2** and *β-gulo* configurations **3** (Fig. 1). He anticipated that, owing to its close resemblance to true sugars, it might be accepted by enzymes and incorporated into biological systems in place of sugars, exhibiting some biological activity, e.g. an inhibition of growth of malignant cells. In fact, his prediction has come true by

Fig. 1 Three 5a-carba-hexopyranoses synthesized by McCasland *et al.* and nomenclature of carba-sugars.

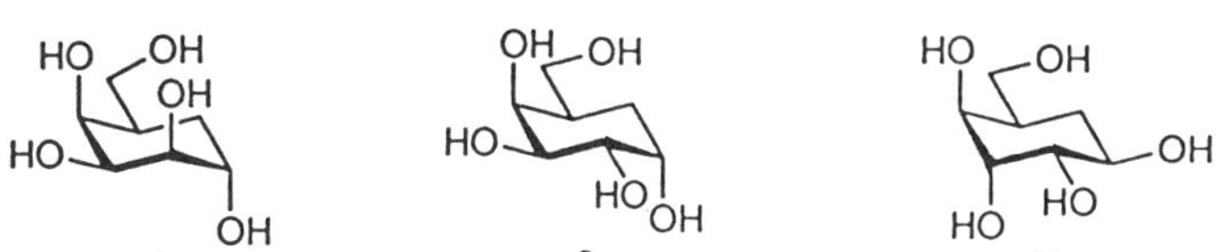

1 α-talo

2 α-galacto

3 β-gulo

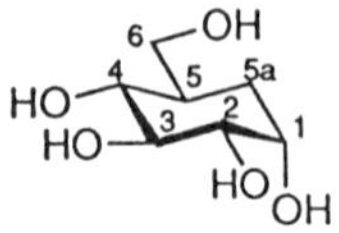

5a-Carba-α-D-glucopyranose

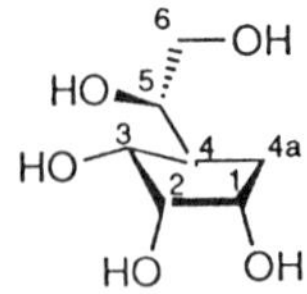

4a-Carba-α-D-glucofuranose

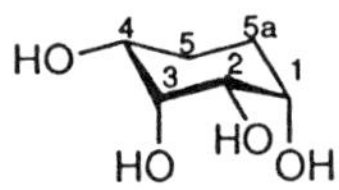

5a-Carba-α-D-ribopyranose

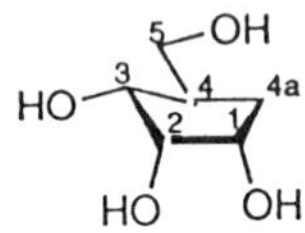

4a-Carba-α-D-ribofuranose

Fig. 2 Naturally occurring 5a-carba-hexose and biologically active compounds containing derivatives thereof.

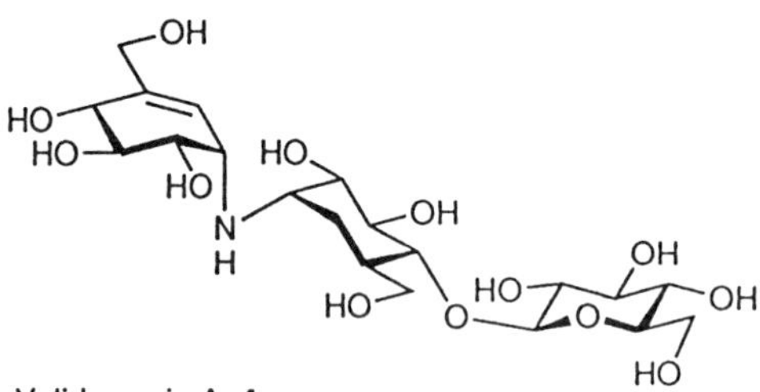

Validamycin A: **4**
Antibiotic (Iwasa *et al.* 1971)

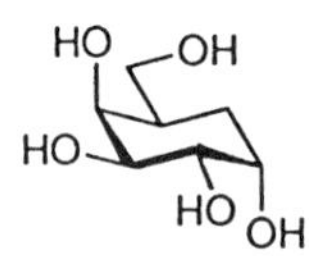

5a-Carba-α-D-galactopyranose: **5**
Antibiotic (Miller *et al.* 1973)

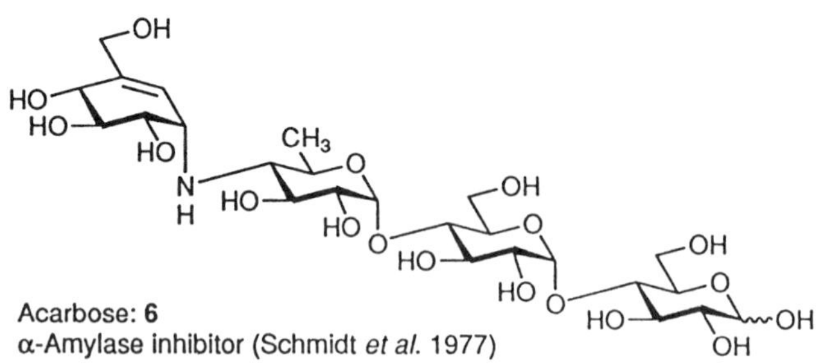

Acarbose: **6**
α-Amylase inhibitor (Schmidt *et al.* 1977)

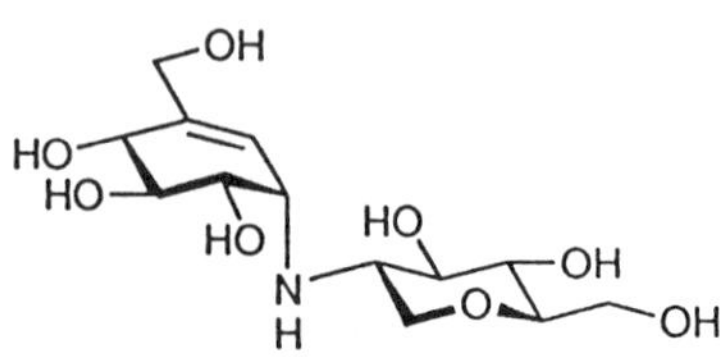

Salbostatin: **7**
Trehalase inhibitor (Vertesy *et al.* 1994)

consecutive discoveries of important antibiotic validamycins [4] **4** in 1970, and α-glucosidase inhibitor acarbose **6** and its homologs [5] in 1977 (Fig. 2). Interestingly, 5a-carba-α-D-galactopyranose **5** was isolated [6] as a weak antibiotic from the fermentation broth of some *Streptomyces* species. Very recently, isolation of a new trehalase inhibitor salbostatin **7** was reported [7].

From 1960–1980, clinically useful aminocyclitol (aminoglycoside) antibiotics such as streptomycins **8**, neomycins, kanamycins **9**, spectinomycin **10**, gentamicins, fortimicins **11** (Fig. 3) etc. have attracted the

Fig. 3 Aminocyclitol antibiotics and aminocyclitols.

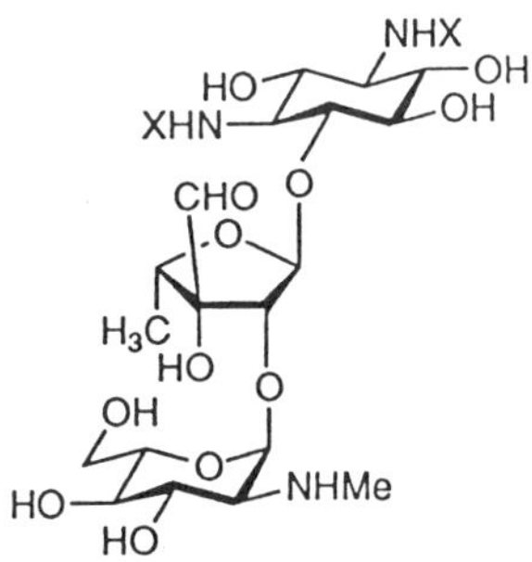

X = -C(=NH)NH_2

Streptomycin: **8**

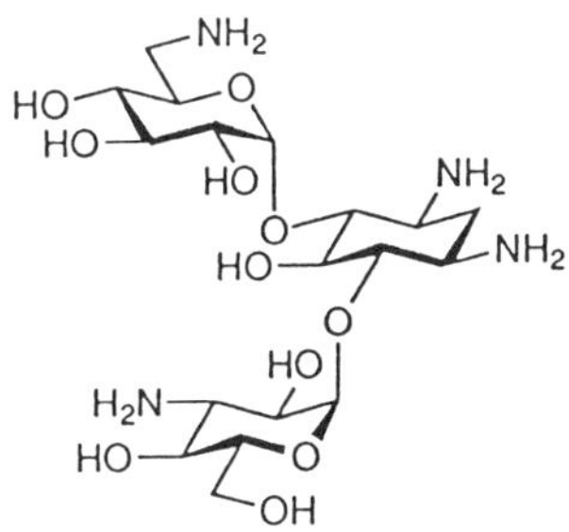

Kanamycin A: **9**

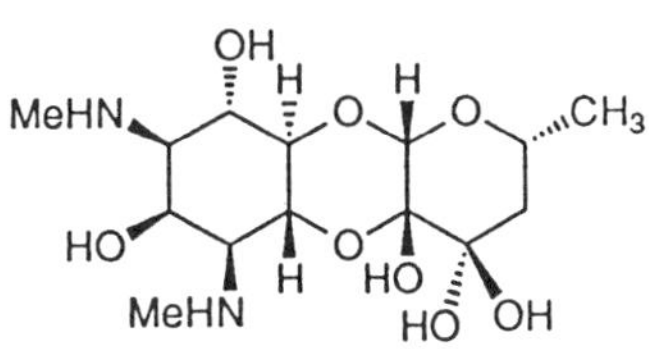

Spectinomycin: **10**

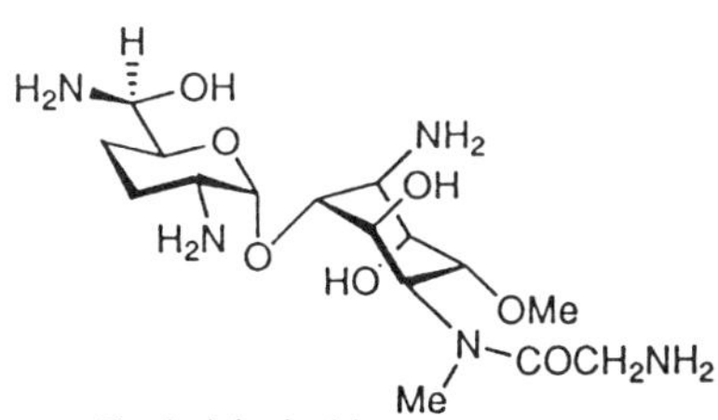

Fortimicin A: **11**

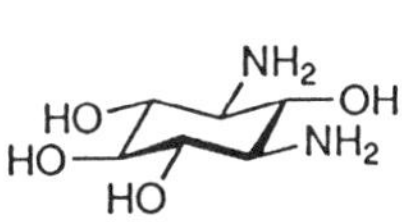

Streptamine: **12**

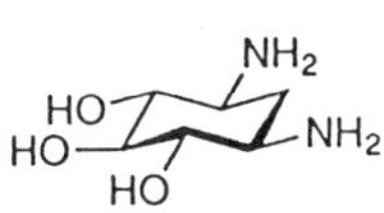

Deoxystreptamine: **13**

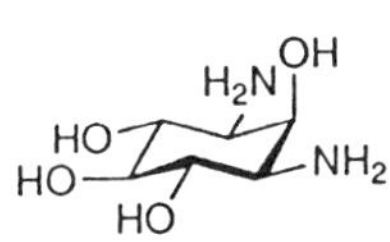

Epistreptamine: **14**

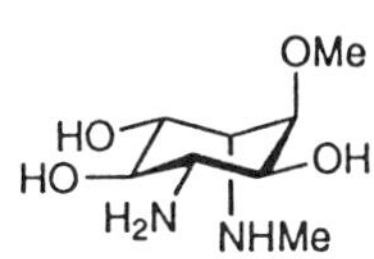

Fortamine: **15**

Fig. 4 Nucleoside antibiotics composed of 4a-carba-pentofuranoses.

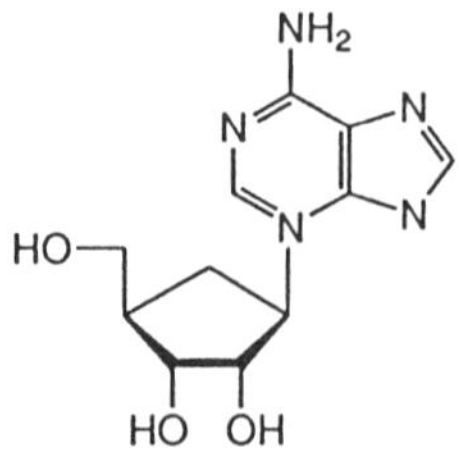

Aristeromycin: **16**

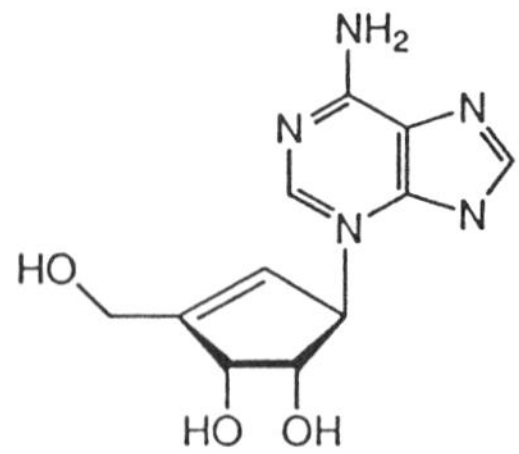

Neplanosin: **17**

attention of a great number of biochemists and synthetic cabohydrate chemists, and efforts have been made towards structural modification and biosynthetic studies in order to obtain safer and more potent agents, broadening the spectra of activity, especially against resistant bacteria [8]. Therefore, some rare aminocyclitols, the constituents of such antibiotcs: streptamine **12**, deoxystreptamine **13**, epistreptamine **14** , fortamine **15**, etc. have been targeted (Fig. 3) for both total synthesis, as well as, muta- and semisynthesis. The author had once been engaged in such an investigation on neomycins to prepare some aminocyclitols of biological interest [9]. Many excellent review articles and books [10-12] have appeared.

The cyclohexane containing class of antibiotic compounds, especially cyclohexanetetraols (dideoxyinositols) would belong to the 5a-carba-pentopyranoses according to McCasland's proposal (Fig. 1). Furthermore, the carbocyclic nucleoside antibiotics, such as aristeromycin **16** and neplanosins **17** (Fig. 4), are branched-chain cyclopentanepolyols possibly called 4a-carba-pentofuranoses. However, naturally occurring pentoses which play a role in biological systems, adopt the furanose rather than the pyranose form, and the carbocyclic nucleosides are usually referred to as the nucleoside analogues. Therefore, the author prefers to describe the hexopyranose analogues, the (hydroxymethyl)branched-chain deoxyinositol derivatives, as 5a-carba-hexopyranoses and closely related bioactive compounds thereof. The tentative carba-sugar-nomenclature was first introduced and recommended [13] by Professor Horton in 1989. Since then, the chemistry of carba-sugars has received recognition as that of one of the members of carbohydrate mimics. This has stimulated the development of carba-sugars as model compounds for biochemical and biological studies or as lead compounds for the preparation of biologically active substances [14].

This chapter comes, from a synthetic standpoint, examining agriculturally or clinically useful drugs of this kind, and the development of new types of enzyme-inhibitors in the hope of obtaining therapeutic drugs, based on the structure-activity relationship of naturally occurring inhibitors.

II. CURRENTLY USED DRUGS COMPOSED OF CARBA-SUGARS, SOME ANTIBIOTICS AND GLYCOHYDROLASE INHIBITORS

A. Antibiotic Validamycins

In 1971, Iwasa and his coworkers [15] isolated validamycin complex from the fermentation broth of *Streptomyces hygroscopicus* subsp. *limoneus*. The complex has been shown to exhibit a strong effect in the control of sheath blight of rice plants, and is now widely used in Japan and other rice-producing countries in Asia. Structural elucidation of the validamycin complex provided,

Fig. 5 Naturally occurring carba amino sugars and carba-disaccharides

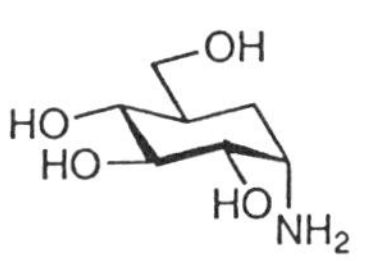

Validamine: **18**

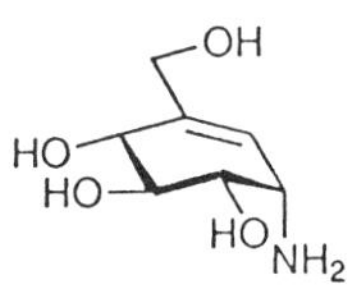

Valienamine: **19**

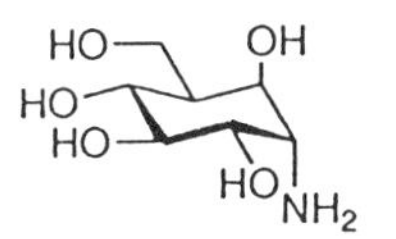

Hydroxyvalidamine : **20**

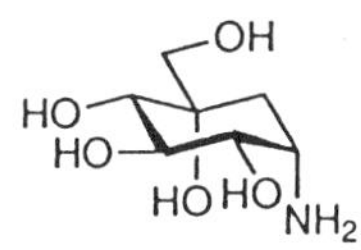

Valiolamine: **21**

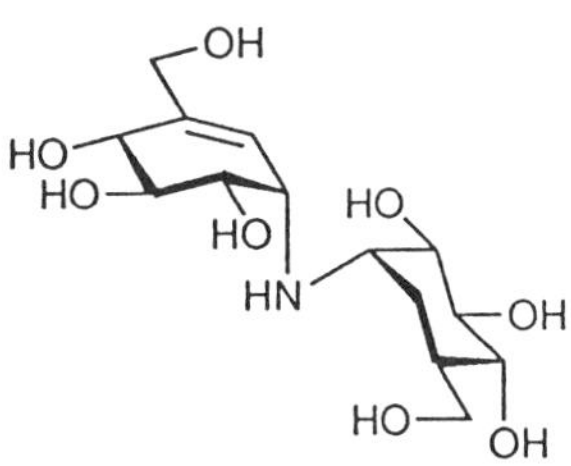

Validoxylamine A: **22**

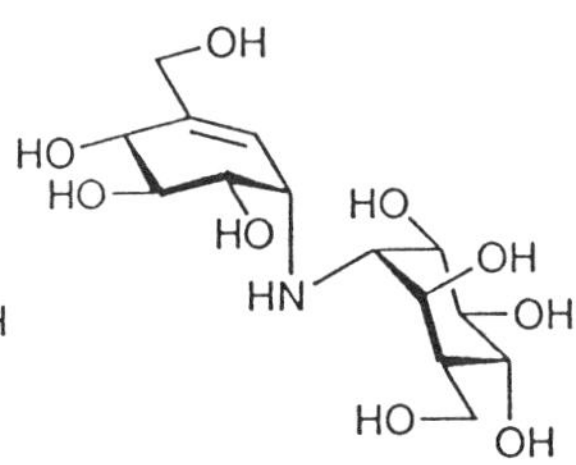

Validoxylamine B: **23**

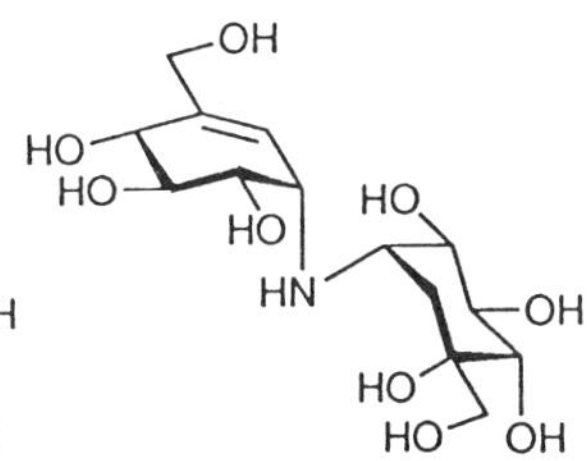

Validoxylamine G: **24**

through chemical and microbial degradation studies [16], an important class of α-glucopyranosylamine type carba-sugars: validamine **18**, valienamine **19**, hydroxyvalidamine **20**, and valiolamine **21**, and α,α-trehalose type carba-disaccharides linked by way of an imino bridge: validoxylamines A **22**, B **23**, and G **24** (Fig. 5).

Eight validamycins (A–H) and three validoxylamines have so far been isolated from the fermentation broth and fully characterized [17]. Validamycins A **4**, C, D, E, F, and H are D-glucopyranosides of the common carba-disaccharide validoxylamine A, while validamycins C and G are of validoxylamines B and G, containing hydroxyvalidamine and valiolamine, respectively, as carba-amino sugar residues. A considerable difference in activity would depend surely on their isomeric structures. Validamycin A **4** is the major component of the complex and most active against sheath-blight desease caused by *Rhizoctonia solani*. Validamycin A is neither antimicrobial nor fungicidal, but controls disease on host plants by retarding hyphal elongation of the fungs. On the other hand, for all carba-glycosylamines the components of the validamycins were found substantially as minor components in the fermentation broth.

We have been interested in the synthesis of biologically active carba-sugar derivatives, since the first synthesis of validamine **18** was readily achieved [18] in 1977, starting from the conveniently obtained (±)-*endo*-7-oxabicyclo-[2.2.1]hept-5-ene-2-carboxylic acid **25**, *endo*-adduct of furan and acrylic acid.

Fig. 6 *endo*-Adduct of furan and acrylic acid: starting material for carba-sugar synthesis. Optical resolution *via* (+)- and (-)-methylbenzylamine salts

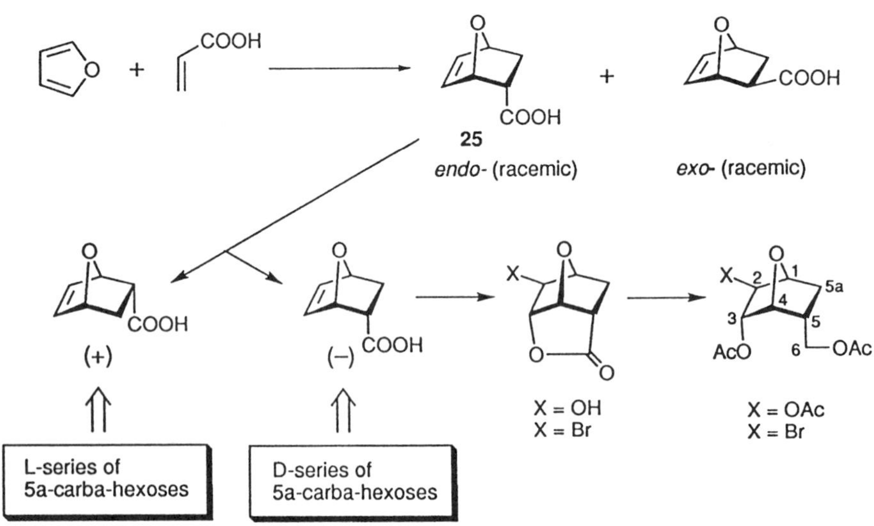

Fig. 7 Preparation of carba-sugars and some versatile synthetic intermediates.

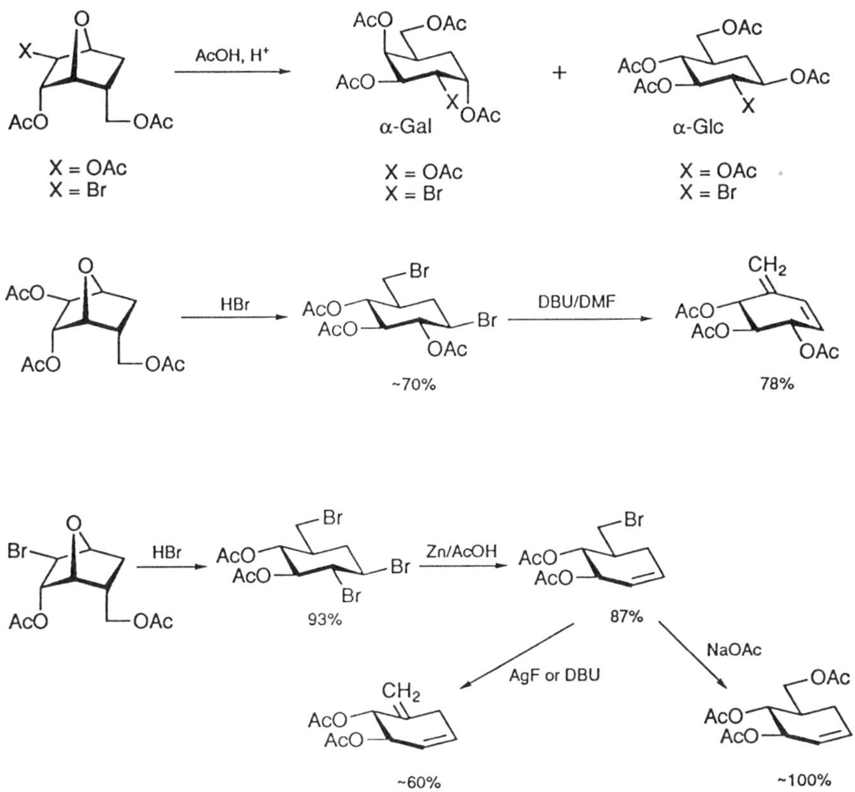

The racemic acid could easily be optically resolved [19] through fractional crystallization of the diastereoisomeric salts with α-methylbenzylamines, giving each enantiomer. The acids have been shown to be transformed into very practical and versatile synthetic intermediates (Fig. 6) for the total synthesis of various kinds of the carba-sugar class of compounds, including validamycins and acrabose derivatives.

Thus, the acid **25** was transformed, through hydroxylactonization, reduction by LAH, and acetylation, into the 1,4-anhydro-5a-carba-α-D-glucopyranose triacetate which on acetolysis produced two carba-hexoses with α-*galacto* and β-*gluco* configurations [20]. Starting from bromolactonization of the acid, the corresponding bomodeoxy compounds were obtained. On the other hand, the 1,4-anhydro carba-sugars could be easily converted into the versatile intermediates [21] (Fig. 7).

First total syntheses of validamine [21] **18** and valienamine **19** and its β-isomer [22] **26** from the bromo derivatives are shown in Fig. 8.

Fig. 8 Synthesis of validamine, and α- and β-valienamines.

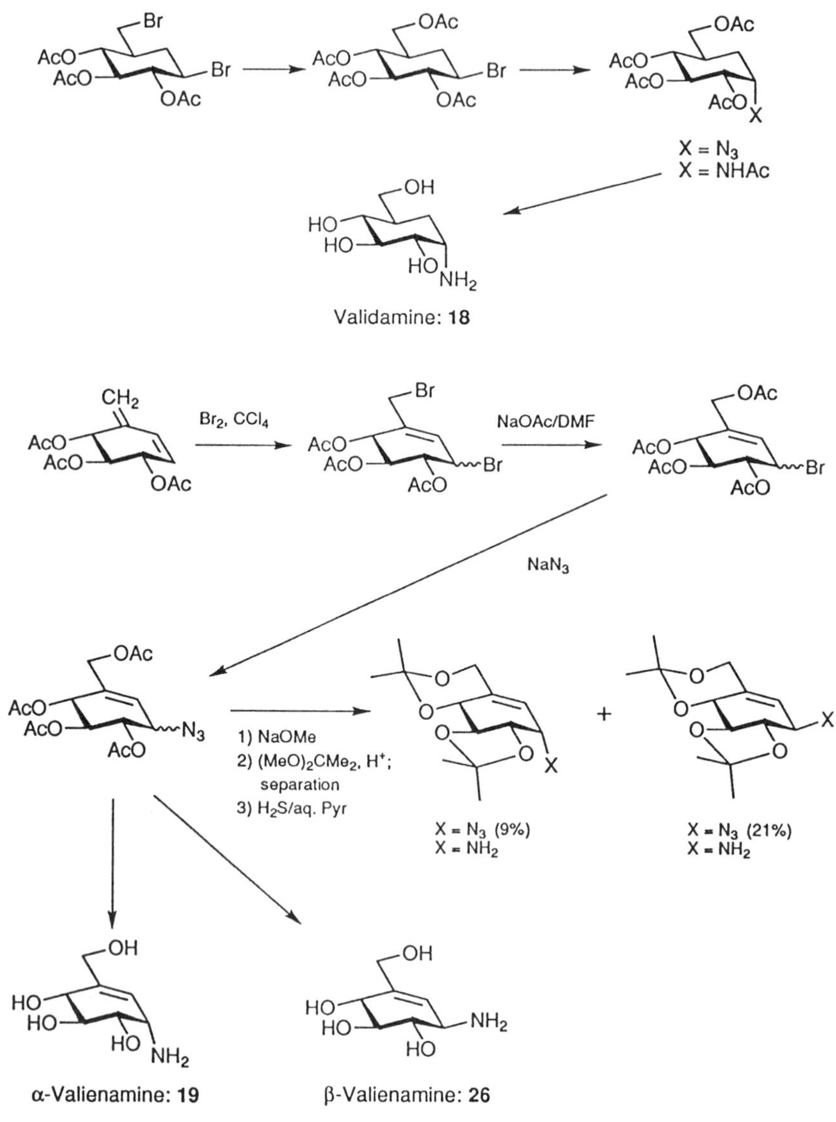

The outline of the total synthesis [23] of validamycin A is shown in Fig. 9. The major coupling product obtained from the donor of valienamine residue and the carba sugar epoxide was successively deoxygenated through themethylthiothiocarbonyl derivative, and deprotected under Birch conditions to afford the antibiotic. Thus, coupling between the appropriately protected amino and the epoxide derivatives has been demonstrated to be generally applicable for construction of an imino bridge between the carba-sugar residues.

Fig. 9 Total synthesis of validoxylamine A and validamycin A.

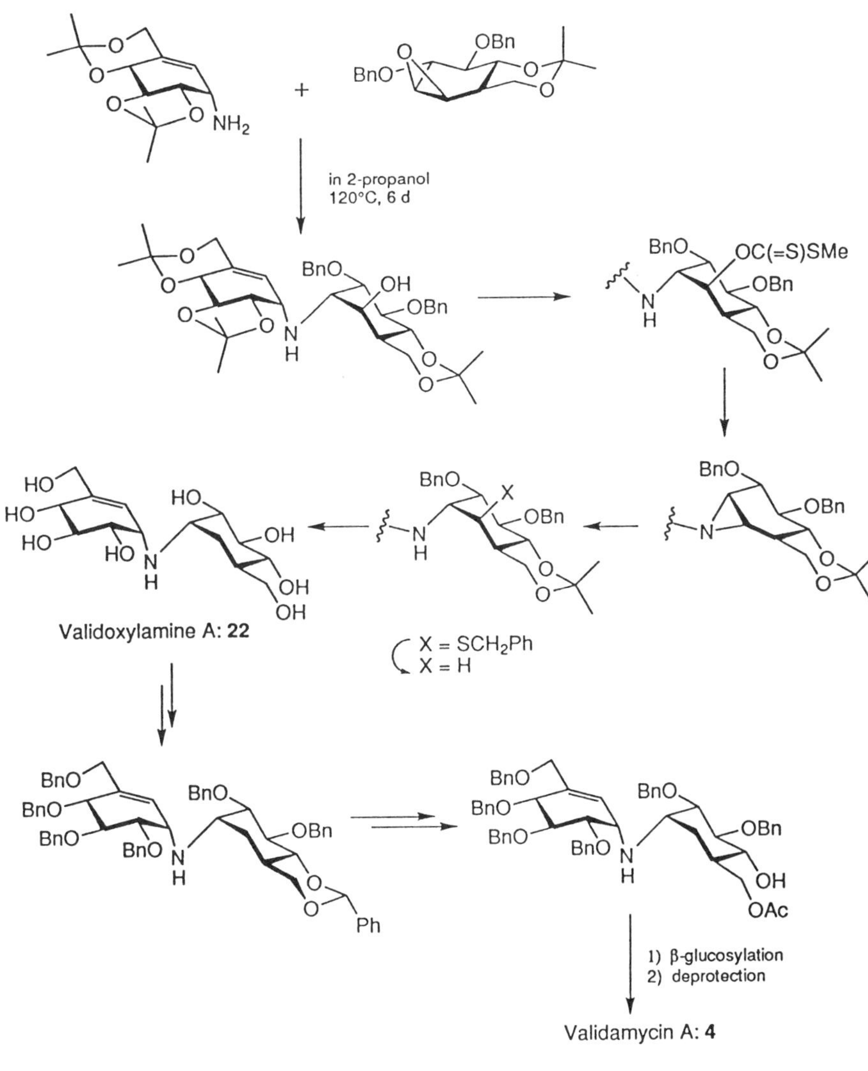

B. Acarbose and Its Homologues, and Related Pseudo-Oligosaccharidic α-Glucosidase Inhibitors

Sulfonylureas and biguanides are the orally active pharmacological agents widely utilized presently to control some types of diabetes mellitus. The recent development of glucosidase inhibitors has established a new therapeutic principle [24] concerning retardation and/or prevention of the digestion of carbohydrates by competitive enzyme inhibition at the brush border of the small intestine.

A number of α-glucosidase inhbitors have been isolated from the culture broths of microorganisms and fully characterized. They inhibit one or several enzymes present in the digestive tract. These enzymes are involved in the degradation of oligosaccharides and starch. D-Glucose generated by enzymatic splitting of starch can be effectively absorbed by the intestinal mucosal epihelium. Therefore, the action of α-glucosidase inhibitors takes place mostly in the intestinal lumen.

1. Acarbose and its homologues

In 1975, new pseudo-oligosaccharides exhibiting potent inhibition against intestinal α-glucosidase of mammals were found in the culture filtrate of *Actinoplanes* strains. These inhibitors are members of a homologous series of pseudo-oligosaccharides [25] of the general formula **28**. The essential core structure pseudo-disaccharide moiety consists of valienamine **19** and 4-amino-4,6-dideoxy-D-glucopyranose was obtained as α-methyl acarviosin **27** by methanolysis of acarbose (Fig. 10). Compound **27** was shown to be a strong α-glucosidase inhibitor.

Fig. 10 Acarbose and several related carba-oligosaccharidic α-amylase inhibitors

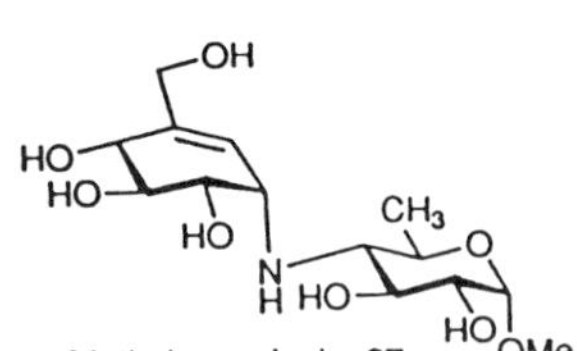

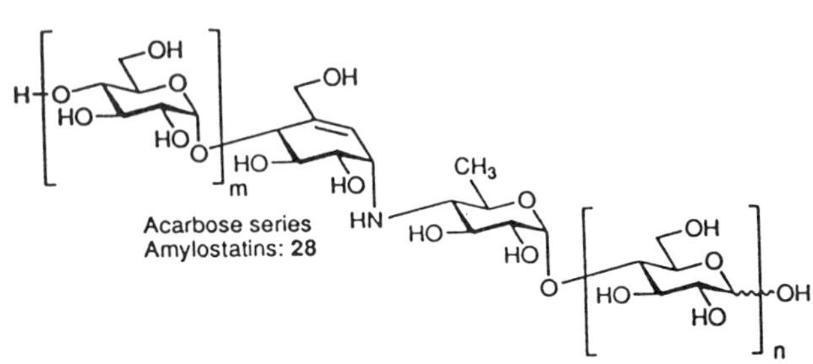

Fig. 10 Continued

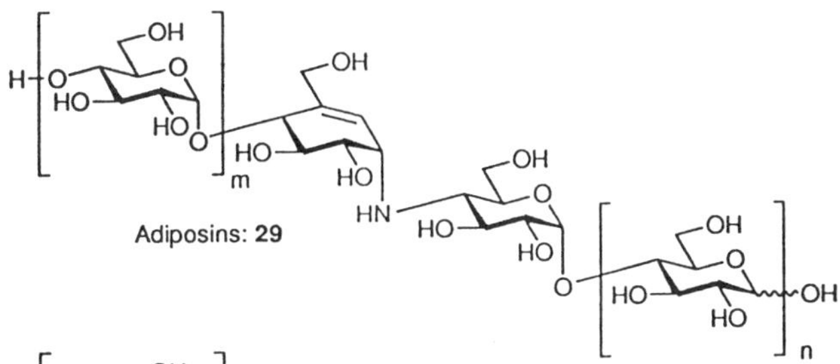

Adiposins: 29

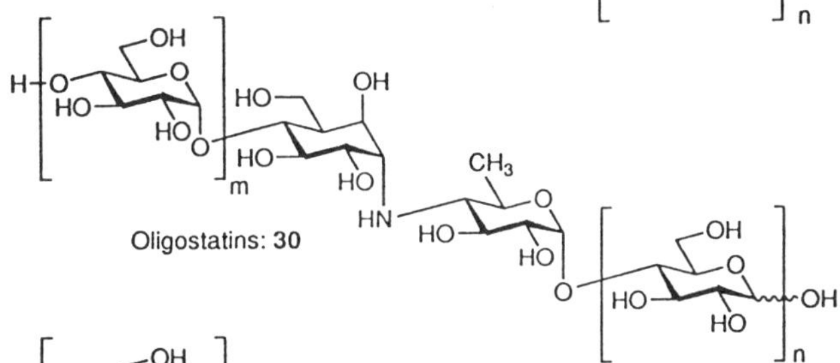

Oligostatins: 30

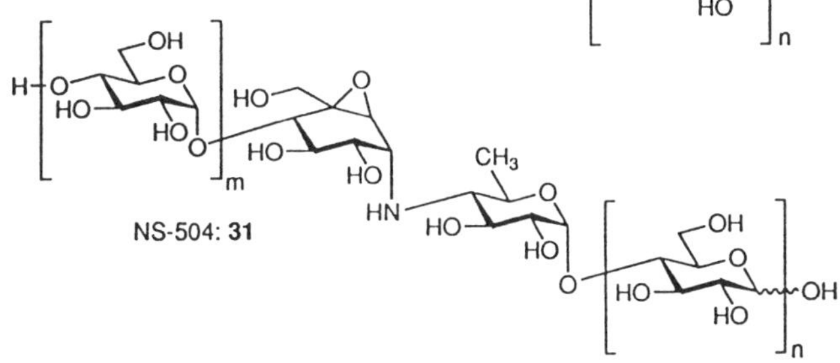

NS-504: 31

Among the homologues, the most important one is acarbose **6** in which a maltose unit is attached to the reducing end of acarviosin.

Interestingly, the individual members of this series of pseudo-oligosaccharides inactivate α-amylase and sucrase quite differently. Thus, amylase inhibition appears to be maximum with homologues with four and five glucose units. On the other hand, the sucrase inhibition has been shown to peak with the homologues containing two glucose residues, that is, acarbose. At present, acarbose is being used as an effective therapeutic drug to control diabetes.

2. Related pseudo-oligosaccharidic α-glucosidase inhibitors

Discovery of acarbose and its homologues has stimulated much interest in the isolation of α-glucosidase inhibitors of this kind in culture broths of *Streptomycetes*. As a result a closely related homologous series of α-glucosidase inhibitors amylostatins **28**, adiposins **29**, oligostatins **30**, trestatins, and amino-oligosaccharides NS-504 **31** have been found and characterized consecutively by a number of research groups in Japan. The characteristic feature of these inhibitors is that they are essentially analogues of acarbose and its homologues, possessing one or several glucose units linked to each core pseudo-disaccharide composed of the carba-sugar residue, valienamine or its oxygenated derivatives, and 4-amino-4-deoxy- or 4,6-dideoxy-D-glucopyranose bonded by way of an imino bridge.

Total syntheses of acarbose and adiposin-2 by our group [27] are shown in Fig. 11. Reaction of the protected valienamine and the anhydro-trisaccharide derivatives proceeded effectively to afford the coupling products in appropriate yields. The free pseudo-tetrasaccharides were readily obtained by deprotection of the products.

Fig. 11 Total synthesis of acarbose and adiposin 2.

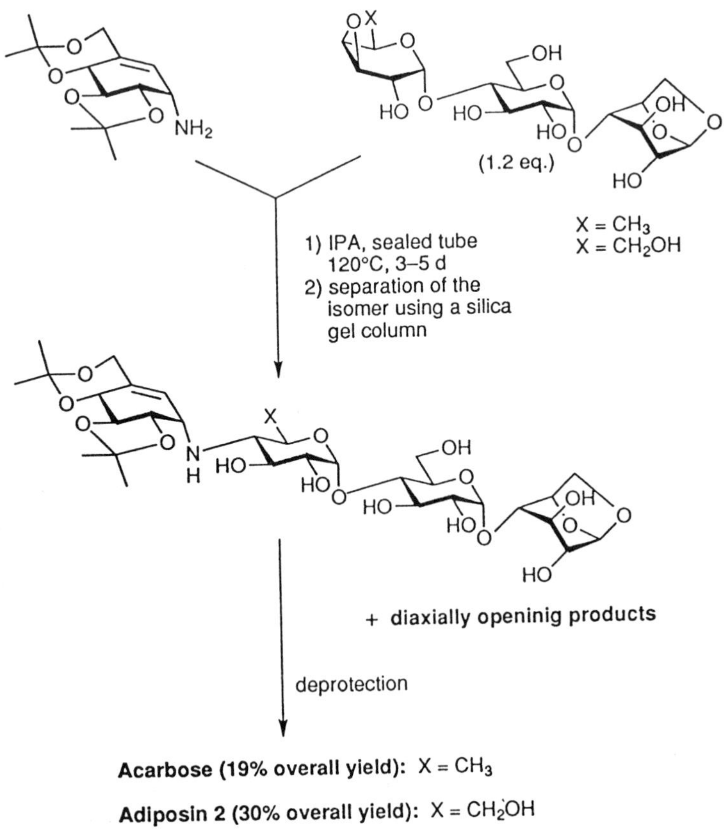

C. Semisynthetic α-Glucosidase Inhibitors

During studies [28] on the biologiocal effects on enzymes of carba-amino sugars isolated from validamycin complex, these compounds have been shown to possess more or less inhibitory potency against α-glucosidases such as maltase and sucrase. Valiolamine **21** possessed the strongest inhibitory activity against porcine maltase (IC_{50} 2.2 x 10^{-6} M) and sucrase (IC_{50} 4.9 x 10^{-8} M) among the related pseudo amino sugars tested.

In recent years, deoxynojirimycin (DNJ) **32** and especially its *N*-butyl derivative **33** have attracted much attention as a glucosidase I inhibitor useful a against the HIV-virus [29] (Fig. 12). Based on its topological resemblance to the postulated glucosyl cation (half-chair) transition state in hydrolysis of the α-glucoside, it was surely reasonable to assume that unsaturated 5a-carba-α-D-glucopy-ranosylamine valienamine type compounds would be good inhibitors of α-glu-cosidase [30] (Table 1).

Fig. 12 Anti-diabetic agents or anti-human immunodeficiency virus agent.

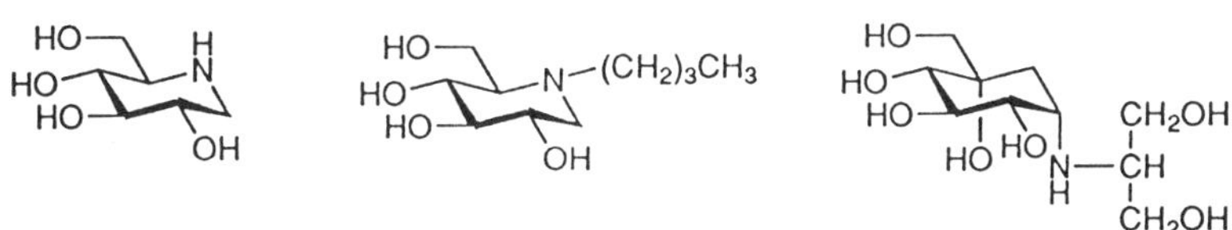

DNJ: **32** Moranoline | *N*-Butyl DNJ: **33** | AO-128: **34** *N*-(1,3-dihydroxyprop-2-yl)valiolamine

Table 1 Inhibitory Activity against α-Glucosidases I and II, and α-Glucosidase obtained from Rat Liver Microsomes

	IC_{50} (M)		
Compound	α-Glucosidase I	α-Glucosidase II	α-Glucosidase
17	1.6 x 10^{-4}	2.7 x 10^{-4}	
18	7.8 x 10^{-4}	1.0 x 10^{-3}	
19	1.2 x 10^{-5}	1.2 x 10^{-5}	1.4 x 10^{-5}
32	4.4 x 10^{-6}	4.6 x 10^{-6}	3.0 x 10^{-7}

Therefore, extensive investigation has been carried out in order to develop drugs, especially more effective derivatives of valiolamine. A number of pseudo-disaccharide derivatives containing valiolamine or valienamine were synthesized by reductive amination of the valiolamine and the keto sugar derivatives with sodium cyano borohydride [31], and tested as inhibitors of α-glucosidases. Very recently, the *N*-(1,3-dihydroxyprop-2-yl) derivative [31] **34** of valiolamine has been introduced as an effective therapeutic agent for diabetes.

III. STUDIES ON STRUCTURE-ENZYME INHIBITORY ACTIVITY RELATIONSHIPS OF CARBA-OLIGOSACCHARIDE DERIVATIVES

A. Glycosidase Inhibitor Validoxylamine A and Related Carba-disaccharides

Extensive studies on the biological properties of validamycins led to the discovery that their essential components, α,α-trehalose type pseudo-disaccharides possess strong *in vitro* inhibitory potency against trehalase [32]. However, they do not have *in vivo* activity. Trehalases are known to be very specific glycosidases which can hydrolyze only the α-glycosidic linkage of substrate α,α-trehalose **35**. Based on structural feature, validoxylamine A **22** is a mimic of the substrate and is likely to bind nicely to an active site of enzymes. Therefore, we synthesized [33] three symmetrical trehalose type carba-disaccharides **36**, **37**, and **38**, composed of validamine, valienamine, and valiolamine, respectively (Fig. 13). They were shown to possess strong activity against trehalase from muscidae, being roughly compatible to the parent validoxylamine A. Interestingly, the valienamine part is not essential for a trehalase inhibitor. These results suggested that symmetrical structures mimicking the substrate play a key role as trehalase inhibitors.

Fig. 13 Inhibitory activity (IC_{50} mg/ml) of validoxylamine A analogues against trehalase from muscidae.

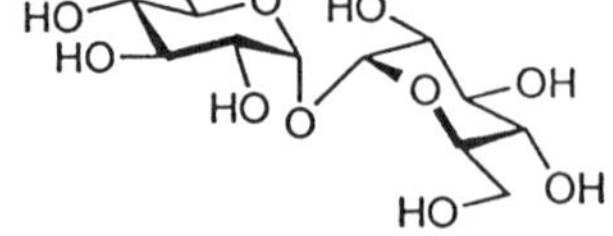

α,α-Trehalose **35**

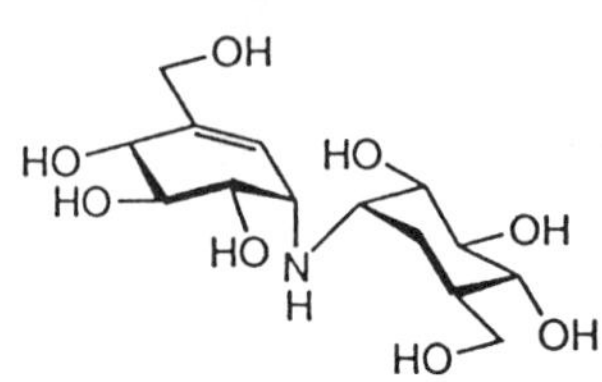

Validoxylamine A: **22**
IC_{50} 6.7 μg/ml

Fig. 13 Continued

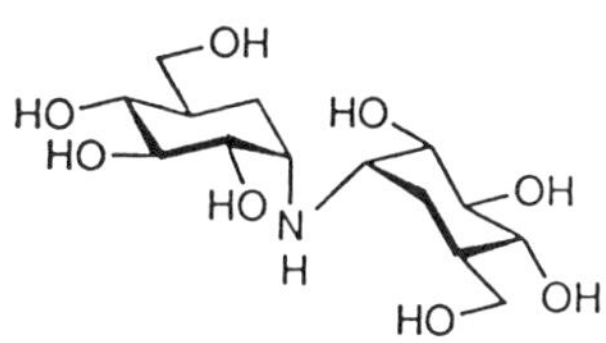

(Validamine)$_2$: **36**
IC_{50} 7.3 μg/ml

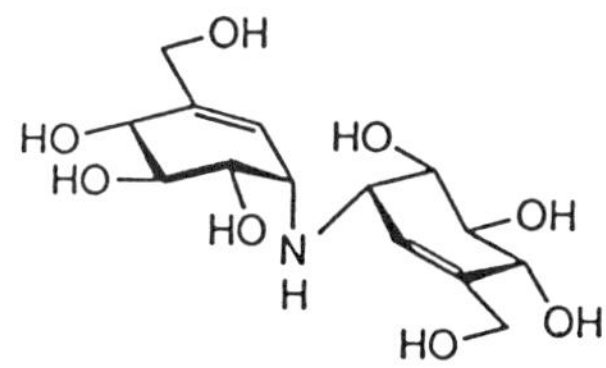

(Valienamine)$_2$: **37**
IC_{50} 12.7 μg/ml

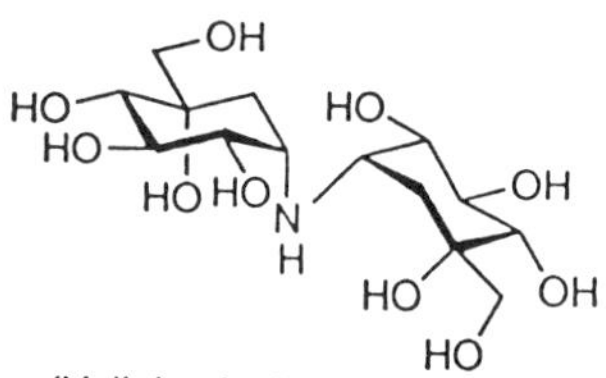

(Valiolamine)$_2$: **38**
IC_{50}15.1 μg/ml

B. α-Glucosidase Inhibitor α-Methyl Acarviosin and Related Carba-maltose Derivatives

Methyl acarviosin **27** has been recognized as an essential structure of acarbose, exhibiting potent inhibitory activity against α-glucosidase. During the course of study on chemical modification of **27**, it was found that the sugar moiety could be replaced by simple structures without affecting its inhibitory potency. The inhibitory activity of the 6-hydroxyl analogue **39** rather increases 10-fold when the sugar part is replaced by the 1,6-anhydro form, by the removal of two hydroxyl groups [34]. Apparently, the α-valienamine part mimics the postulated flattened half-chair transition state in hydrolysis of α-glucoside. Therefore, by simple analogy based on topological resemblance, three acarviosin analogues **40** [35], **41** [36], and **42** [37] were designed by modifying the valienamine part to mimic each substrate structure, with expectation that they would possess inhibitory activity against β-glucosidase, α-mannosidase, and chitinase, respectively (Fig. 14). However, our efforts were not well rewarded. Only the α-mannose-type analogue **40** was found to be a weak α-mannosidase inhibitor.

Fig.14 Inhibitory activity of methyl acarviosin analogues against α-glucosidase (Baker's yeast).

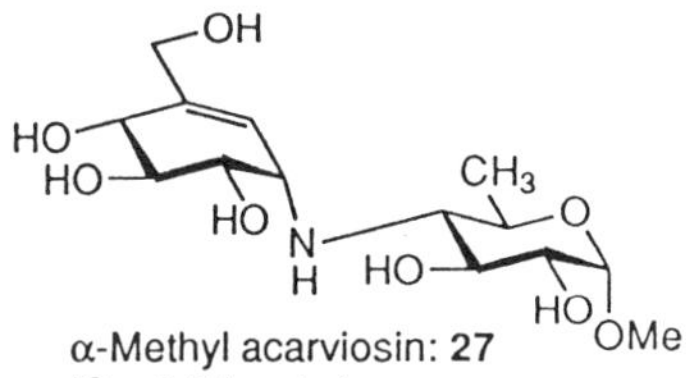

α-Methyl acarviosin: **27**
IC_{50} 0.34 μg/ml

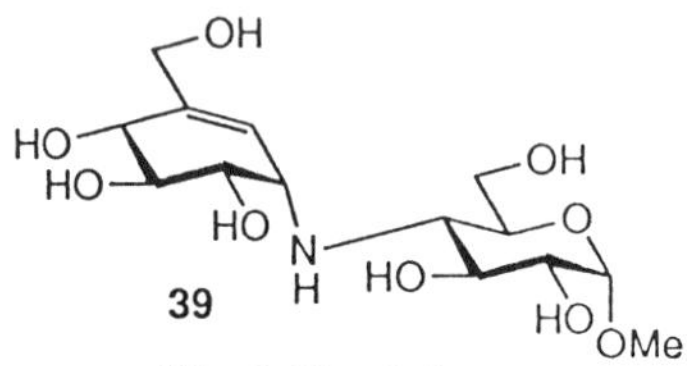

39
IC_{50} 0.98 μg/ml

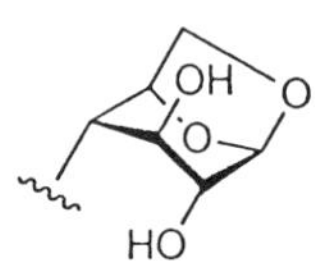

IC_{50} 0.18 μg/ml

IC_{50} 0.03 μg/ml

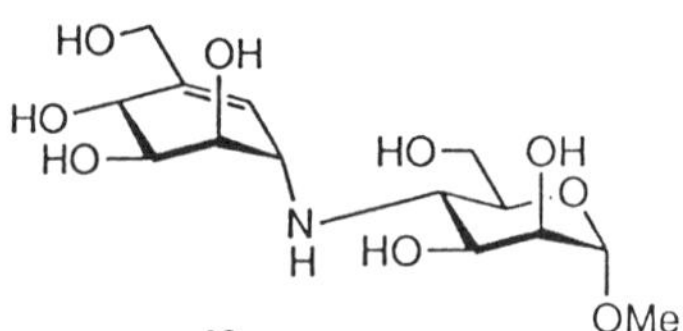

40
Weak inhibitor of α-mannosidase

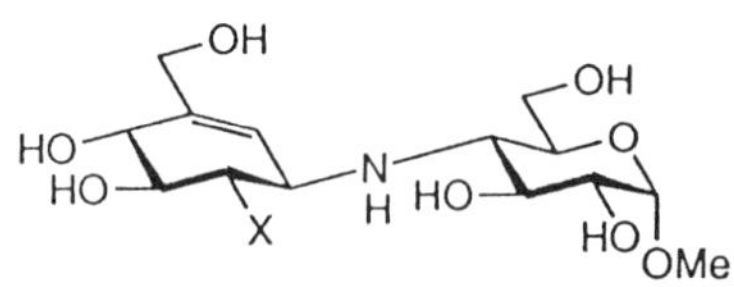

41: X = OH
No inhibition against β-glucosidase

42: X = NHAc
No inhibition agaisnt β-glucosidase and chitinase

C. Structure-Activity Relationship of Trehalase Inhibitors: Trehazolin

In 1990, trehalostatin **43**, a potent and specific inhibitor against blowfly trehalase, was isolated by Murao *et al.*. [38] from the culture broth of *Amycolatopsis trehalostatica*. Shortly after the discovery, Ando *et al.*. [39] disclosed the finding of the trehalase inhibitor trehazolin **44** from the culture broth of *Micromono-spora*, strain SANK 62390, and showed it to be identical to **43** by comparison of physical and spectroscopic data (Fig. 15). They, however, assigned a different structure to it. Later, we established their identity by synthesis of each aminocyclitol [40], and determined the absolute structure to be **44** by a total synthesis [41] of both structures proposed by two research groups.

Fig. 15 Trehazolin and its 7-epimer formally assigned to trehalostatin.

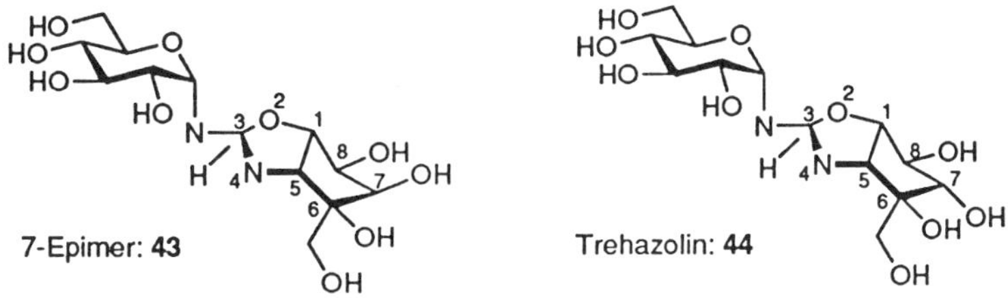

Outlined in Figs. 16, and 17 is the preparation [41-43] of some stereoisomers of the aminocyclopentanepolyol part. The synthesis was carried out unambiguous-ly starting from 5-aminocyclopentane-1,2,3,4-tetraol derivatives, which were readily obtained by base-catalyzed nitromethane cyclization [44] of the dialde-hyde derived from *myo*-inositol.

Fig. 16 Synthesis of 5-aminocyclopentanepolyols from *myo*-inositol (Angyal *et al.*)

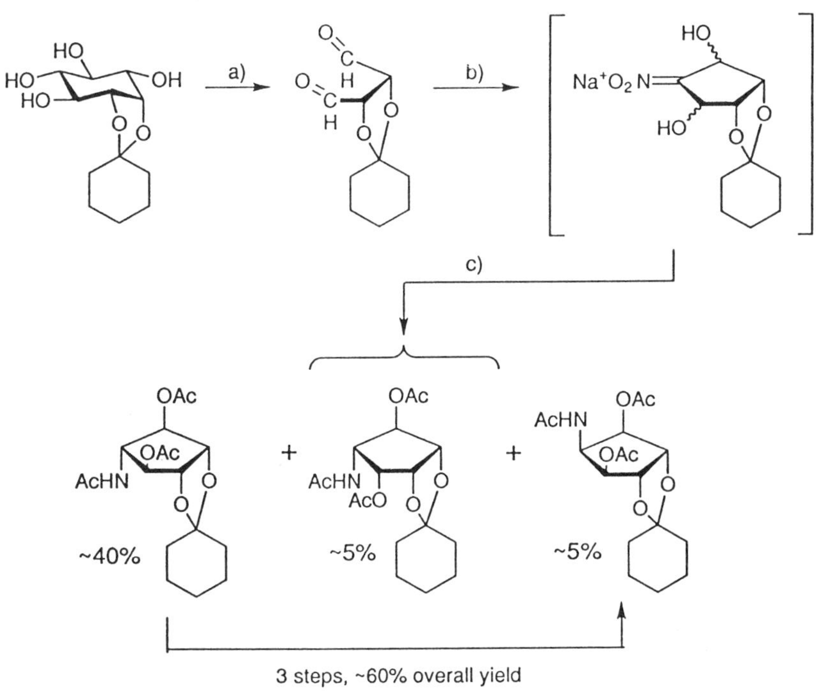

a) $NaIO_4$, H_2O; b) $MeNO_2$, MeOH, MeONa; c) H_2, Raney Ni; Ac_2O, Pyrd

Fig. 17 Synthesis of some stereoisomers of 5-amino-1-*C*-(hydroxymethyl)cyclopentane-1,2,3,4-tetraol

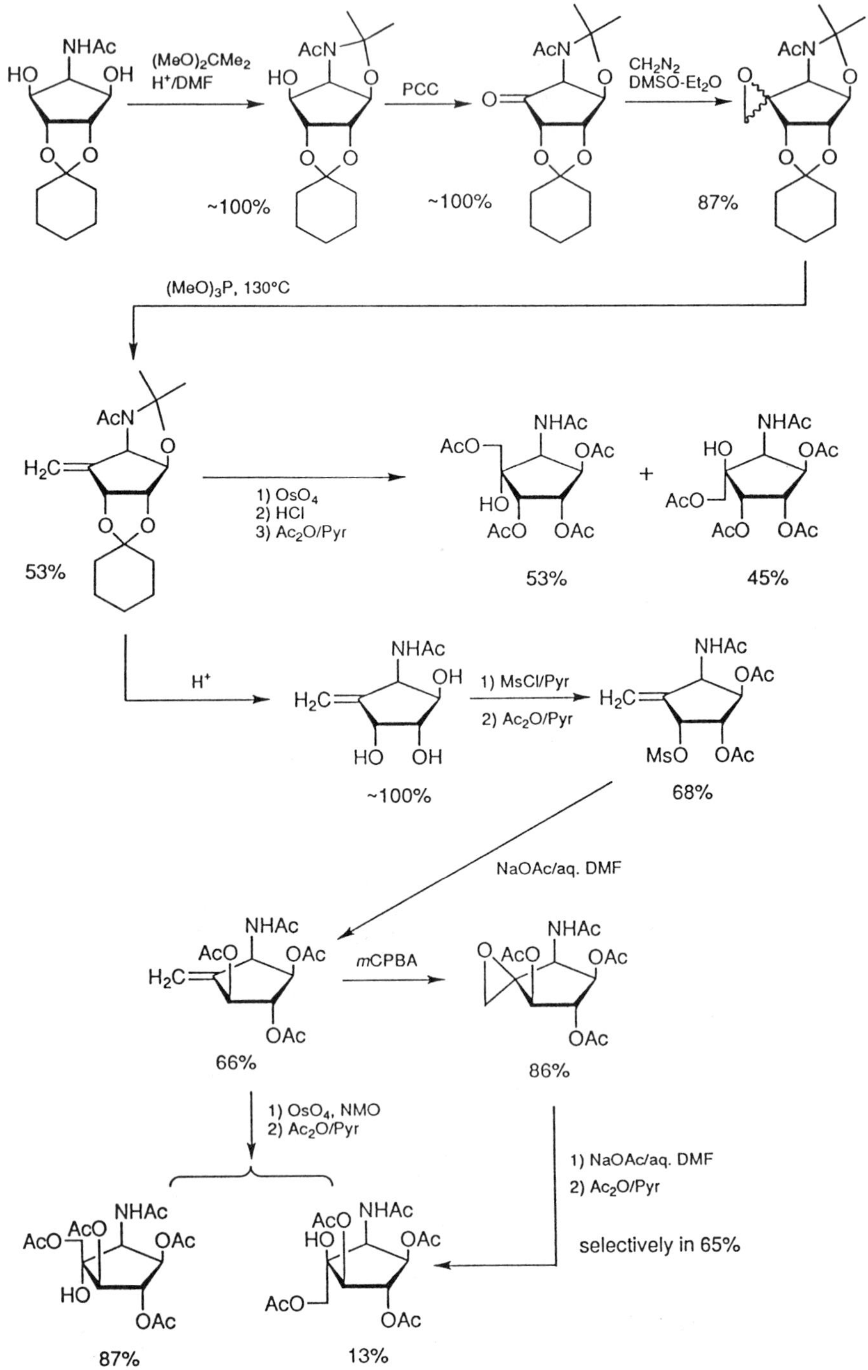

Fig. 18 Total synthesis of trehazolin and its 7-epimer.

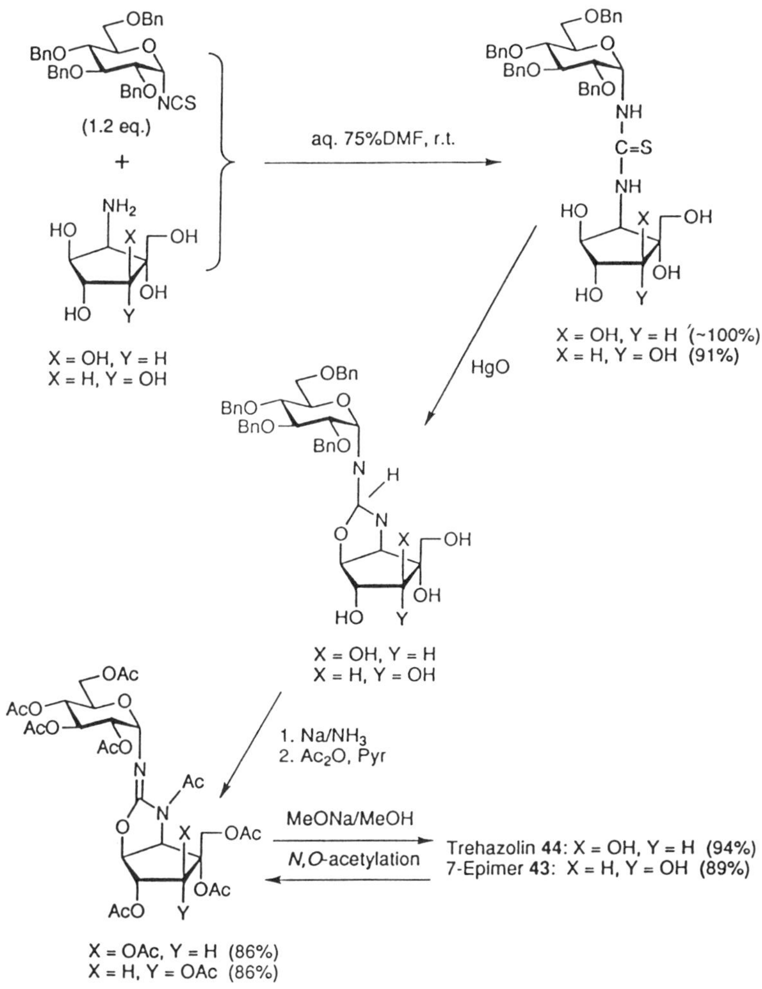

The total synthesis [41] of trehazolin **44** and its 7-epimer **43** was achieved readily by cyclization of the thioureas followed by deprotection under Birch conditions (Fig. 18). This procedure was generally applicable for preparation of trehazolin analogues.

In the continuation of our synthetic studies on glycoside inhibitors [45], especially trehalase inhibitors, we became interested in the structure-activity relationship of unique trehazolin structures.

Table 2 Inhibitory Activity of Trehazolin and Its Seventeen Analogues Against Silkworm Trehalase

Compound	Inhibitory activity IC_{50} (M)
Trehazolin **44**	4.9×10^{-8}
Trehazolin (diastereoisomer)	1.2×10^{-6}
6-Deoxy	2.1×10^{-5}
6-De(hydroxymethyl)	7.8×10^{-6}
6-De(hydroxymethyl) (diastereoisomer)	1.6×10^{-7}
7-Deoxy	6.3×10^{-5}
8-Deoxy	$>2.7 \times 10^{-4}$
9-Deoxy	$>2.7 \times 10^{-4}$
3'-Deoxy	1.2×10^{-6}
6'-Deoxy	7.9×10^{-6}
6-Epi	3.0×10^{-7}
7-Epi	$>2.7 \times 10^{-4}$
6,7-Diepi	9.2×10^{-7}
6,7-Diepi (diastereoisomer)	2.6×10^{-5}
2'-Epi	3.7×10^{-5}
4'-Epi	1.1×10^{-5}
5a'-Carba **45**	4.9×10^{-8}
5a'-Carba-5'(5a')-eno **46**	3.1×10^{-7}

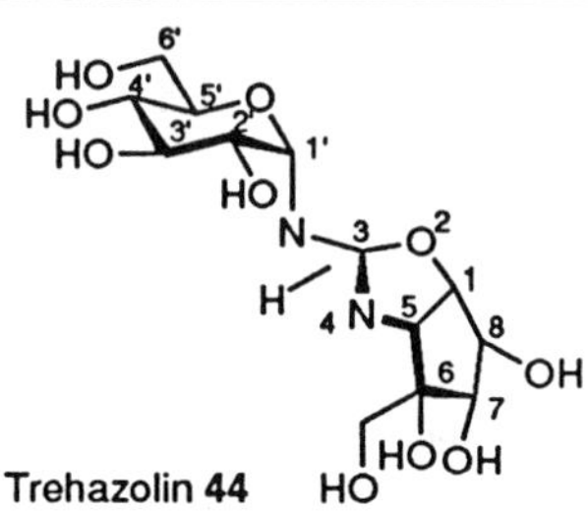

Trehazolin **44**

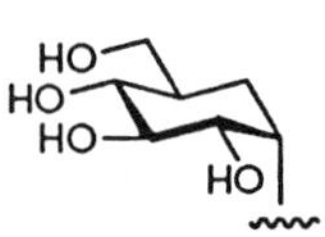

5a'-Carba **45**

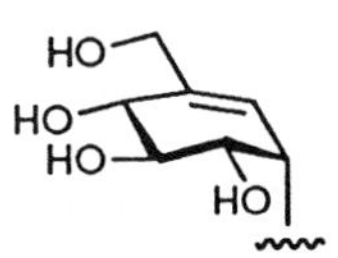

5a'-Carba-5'(5a')-eno **46**

More than twenty trehazolin analogues [41-44] were prepared and subjected to bioassay of their inhibitory activity against silkworm trehalase, of which a part of the data was listed in Table 2. Both the epimers and the deoxy derivatives had decreased potency compared with trehazolin. However, 5a'-carbatrehazolin **45** was shown to be a potent inhibitor almost comparable to trehazolin. These results suggested that the four hydroxyl groups of the D-glucopyranose residue of trehazolin **44** were topologically essential for its binding to the active site of the enzyme through hydrogen bonding. The pseudo-disaccharide structures of **44** as well as **45** may be close mimics of the substrate α,α-trehalose or its intermediate which is involved in the substrate-recognition and/or hydrolytic step. In fact, the analogue **46** containing an α-valienamine moiety therefore shows rather low activity owing to a conformational deformation of the 5a-carbahexopyranose ring. In contrast, in the case of α-methyl acarviosin the α-valienamine moiety is conversely thought to be a mimic of a flattened half-chair conformation of a glucosyl cation probably involved during hydrolysis of α-glucoside. Therefore, the aminocyclopentanepolyol moiety including the cyclic isourea function would constitute charge-distribution parts for binding the active site of the enzymes. All three 6-, 7-, and 8-hydroxyl groups on the cyclopentane ring are thought to be topologically very important as mimics of those of the α-glucopyranose residue of the substrate.

According to the above results, some stereoisomers of the simple cyclic isoureas of this kind would be expected to possess inhibitory activity against some glycosidases.

IV. DEVELOPMENT OF NEW GLYCOHYDROLASE INHIBITORS COMPOSED OF CARBA-SUGARS AND THE LIKE

A. Carba-glycoceramides, Potent and Specific Glycocerebrosidase Inhibitors

We first prepared [45] some 5a-carba-sugar analogues of the glycosylamides [46] **47** to know whether the glycolipid analogues composed of carba-sugars would be useful model compounds (Fig. 19). The carba-sugar analogues were shown to possess a comparable potency as an immunomodulator to the corresponding true sugar analogues, suggesting that the sugar residues would possibly be replaced by the carba-sugars without affecting its biological properties.

Glycosphingolipids (GSLs), for example glucosylceramide (GlcCer) **48** plays important roles in biological systems as essential constituents of membranes and cell walls. Biological functions of these compounds have therefore been extensively studied. In order to understand their functions, it would be necessary to provide structural analogues as well as naturally occurring compounds in pure form.

Fig. 19 Biological activity* (*in vivo*) of some 5a-carba-glycosylamides

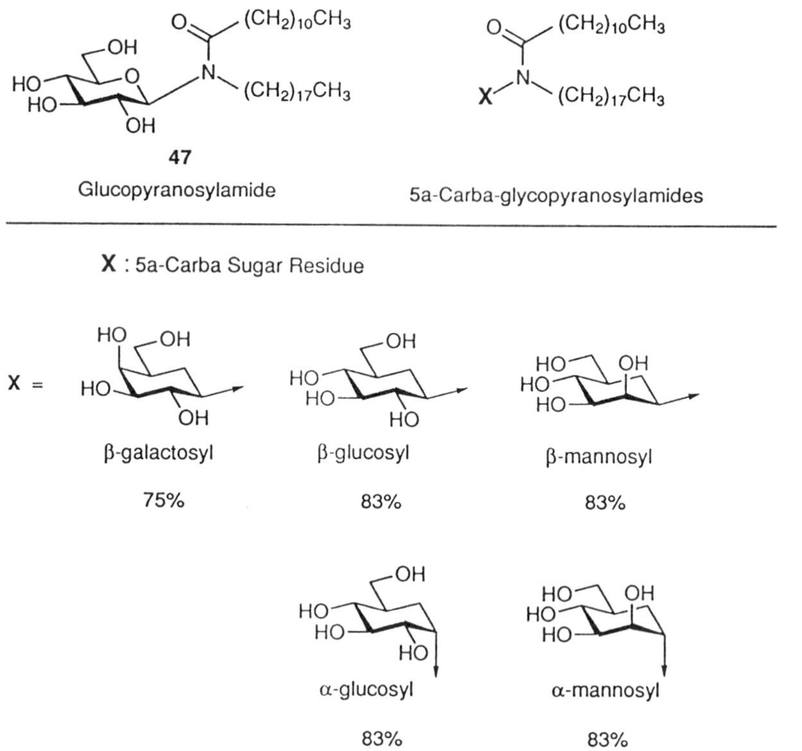

* Surviving mice (difference vs control) 7 days after pre-treatment with 10 mg/kg body weight with the samples, followed by infection with a 10-fold LD_{50} of *Escherichia coli* C14

A number of glycosphingolipids are derived from GlcCer, which is enzymatically generated from ceramide and UDP-glucose. The transfer of D-galactose to GlcCer forms lactosylceramide, which is the precursor for GSLs such as ganglio-, globo-, neolacto-, and lacto series. Among ceramide analogues, a mixture of 1-phenyl-2-decanoylamino-3-morpholino-1-propanol (PDMP), the D-*threo*-PDMP **49** had been shown [47,48] to strongly inhibit UDP-glucose: *N*-acylsphingosine glucosyltransferase (GlcCer synthase) (Fig. 20).

Fig. 20 5a-Carba-sugar analogues of glycosylceramides.

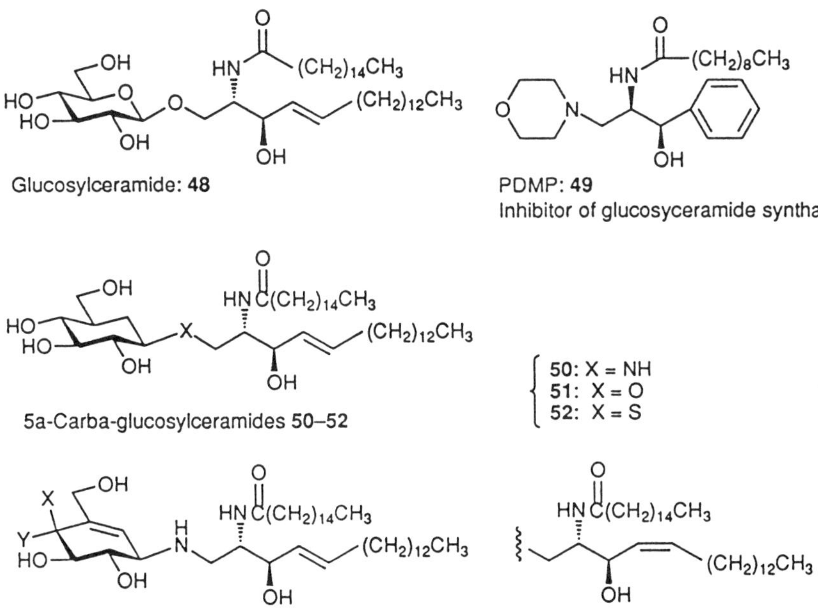

Glucosylceramide: **48**

PDMP: **49**
Inhibitor of glucosyceramide synthase

5a-Carba-glucosylceramides **50–52**

50: X = NH
51: X = O
52: X = S

E-53: X = H, Y = OH
E-54: X = OH, Y = H

Z-53: X = H, Y = OH
Z-54: X = OH, Y = H

Since all GSLs normally undergo catabolic degradation, the blockage of GlcCer biosynthesis should lead to reduction in the cellular content of all the GSLs biosynthesized from GlcCer. Therefore, potent and specific enzyme-inhibitors would control the biosynthetic pathways effectively, being possibly useful for diagnosis or as drugs to control diseases caused by a genetic disorder of complex GSLs metabolism due to lack of some particular enzymes.

Our 5a-carba sugar analogues have initially been designed to be active enzyme-inhibitors for glycosyltransferases and/or glycosylhydrolases in the processes of GSLs biosynthesis. We have recently elaborated [49] a general method for the synthesis of carbocyclic analogues **50**, **51**, and **52** of glycosylceramides, the sugar residues of which being comprised of 5a-carba-β-D-glucopyranosyl residue connected with each other by imino, ether, and sulfide linkages, respectively. The synthetic sequence of **50** is shown in Fig. 21. Two aziridines newly prepared were shown to be versatile sphingosine-precursors useful for introduction of various nucleophiles to the C-1 position.

Fig. 21 Synthesis of 5a-carba-sugar analogue of glucosylceramide.

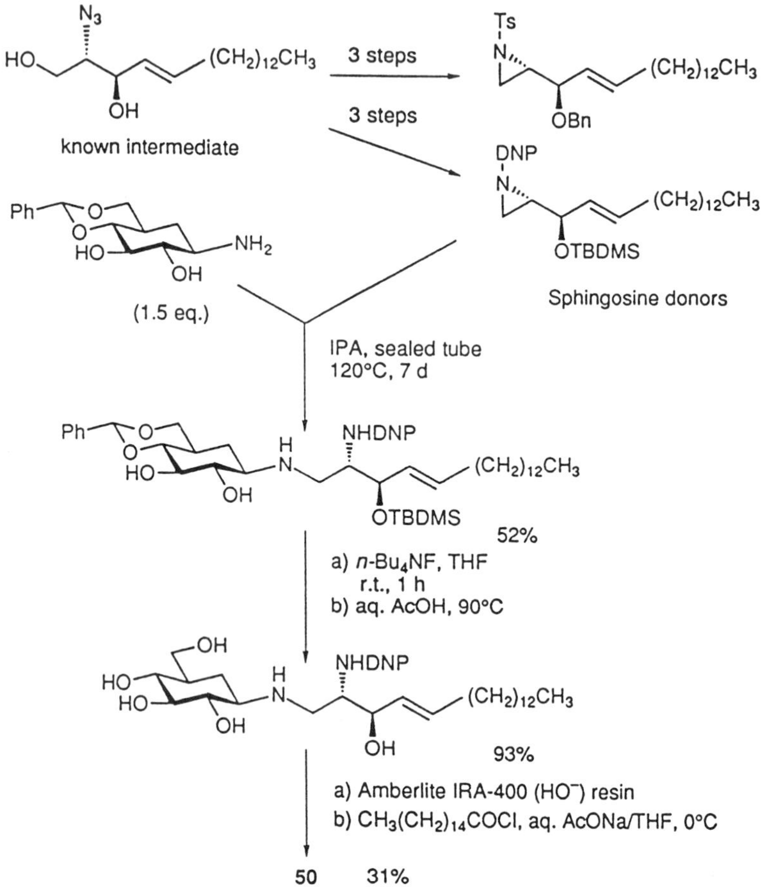

The inhibitory activity of the glycosylceramide analogues against β-glucocerebrosidase and β-galactocerebrosidase was assayed. The imino-linked analogue **50** has been shown [49] to inhibit moderately both β-glyco-cerebrosidases. In contrast, the corresponding ether **51** and sulfide-linked analogues **52** show no marked inhibitory potencies against both enzymes.

On the other hand, similar types of analogues **E,Z-53** [50] in which the 5a-carba-β-D-glucose residue has an unsaturation between C-5 and C-5a (β-valienamine residue) almost completely inhibited β-glucocerebrosidase activity at a concentration of 10 μM, and the IC_{50} values of the *E*- and *Z*-isomers are 0.3 and 0.1 μM, respectively (Table 3).

Table 3 Inhibitory Activities of Four Unsaturated 5a-Carba-glycocerebroside Analogues against β-Glucocerebrosidase and β-Galactocerebrosidase

	Inhibition at 10μM (%); [IC_{50}, μM]	
Compound	β-Glucocerebrosidase[a]	β-Galactocerebrosidase[b]
E-isomer **E-53**	95.2 [0.3]	19.4
Z-isomer **Z-53**	97.7 [0.1]	28.6
E-isomer **E-54**	6.7	90.3 [2.7]
Z-isomer **Z-54**	9.6	78.4 [4.5]

[a] Glucocerebrosidase was assayed with the fluorogenic substrate, NBD-glucosylceramide, with microsomal fraction of mouse liver in a total volume of 0.2 ml contain-ing glucocerebroside from Gaucher spleen, Trition X-100, sodium taurochlorate and phosphate–citrate (pH 5.5). — [b] Galactocerebrosi-dase was assayed with NBD-galactocerebroside, with microsomal fraction of mouse liver in a total volume of 0.2 ml containing stearoyl galactocerebroside, oleic acid, sodium taurocholate and phosphate-citrate (pH 4.5).

Furthermore, very interestingly, the *β-galacto* type analogues **E,Z-54** showed potent and specific inhibitory activity against β-galactocerebrosidase, and the IC_{50} values of the *E* and *Z*-isomers are 2.7 and 4.5 μM, respectively. The topological features of the unsaturated 5a-carba-sugar residues are likely to be recognized by the enzymes. However, both analogues did not inhibit GlcCer synthase at all.

Since PDMP was demonstrated to inhibit specifically GlcCer synthase but not β-glucocerebrosidase, it would be quite interesting to elucidate the structure-activity relationships of these glycolipids analogues, especially focusing on their selectivity of inhibition against glycosidase and glycotransferases in glycolipid biosynthesis. Attempts have now been made to prepare 1-phenyl-2-decanoylamino-3-(β-valienaminyl)-1-propanol **55** and its 4'-O-β-D-galactopyranosyl derivative **56** (Fig. 22). Preliminary bioassay demonstrated that the former was also a strong glucocerebrosidase inhibitor [51].

Fig. 22 Synthesis of potent β-glucosyl and β-galactosylcerebrosidase inhibitors composed of carba-sugars.

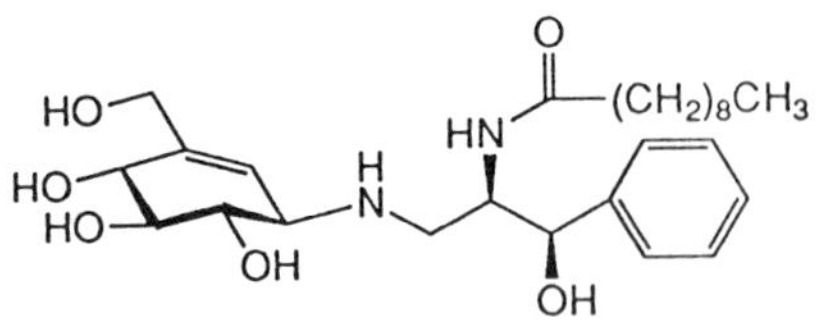

55

β-Valienaminyl derivative (*R,R*)-isomer

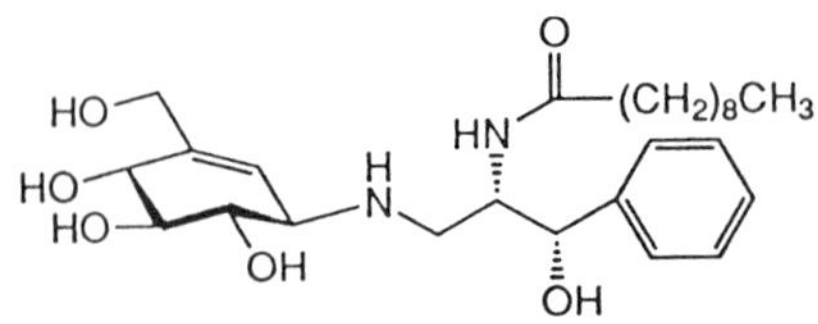

(*S,S*)-isomer

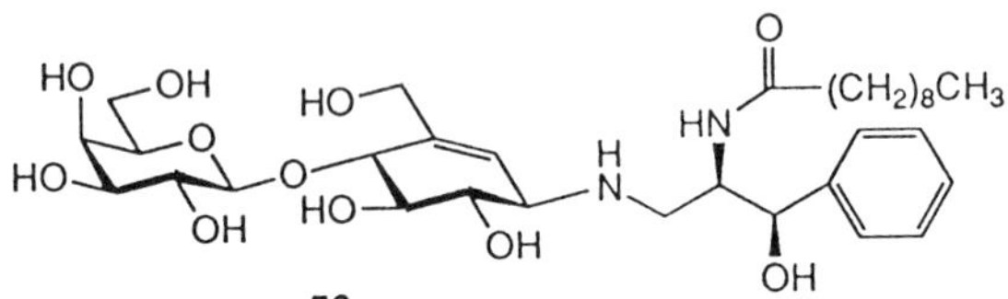

56

4-O-(β-D-Galactopyranosyl)-β-valienamine derivative

B. Future Development of Potent Glycohydrolase Inhibitors Composed of 5a-Carba-Sugars and Branched-Chain Aminocyclopentanepolyols

1. Glucocerebrosidase inhibitors composed of β-valienamine

The above findings prompted us to develop new carba-sugar-type glycocerebrosidase inhibitors with simple structures. The *N*-alkyl derivatives of D-glucopyranosylamine were known to be strong β-glucosidase inhibitors [52]. Thus, a homologous series of *N*-alkyl-β-valienamines **57–59** were synthesized and subjected to biological assay on inhibitory potential against β-glucocerebrosidase [53] and other enzymes (Table 4). The results demonstrated that the complex ceramide portion could be replaced by a simple aliphatic chain without affecting the activity.

Moreover, taking advantage of an easy access to *N*-alkyl-β-valienamine derivatives, several modifications of the carba-sugar parts have been carried out in order to obtain novel glycotransferase inhibitors of this kind. *N*-Alkyl 5a-carba-β-lactosylamine **60** and its derivative **61** are our next targeted compounds for this purpose (Fig. 23).

Fig. 23 *N*-Alkyl β-valienamines and derivatives, new glucocerebrosidase inhibitors.

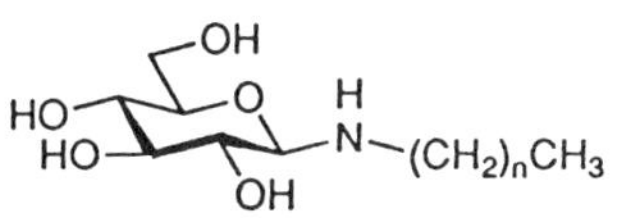

n = 5 ~ 17

Very potent but not so stable

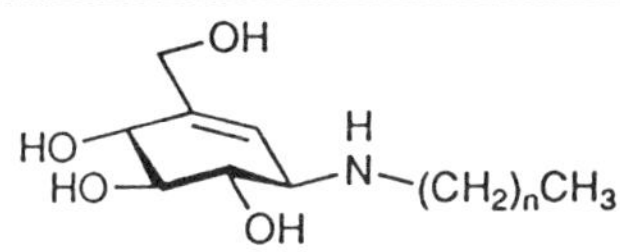

57: n = 3
58: n = 5
59: n = 7

Very potent and stable

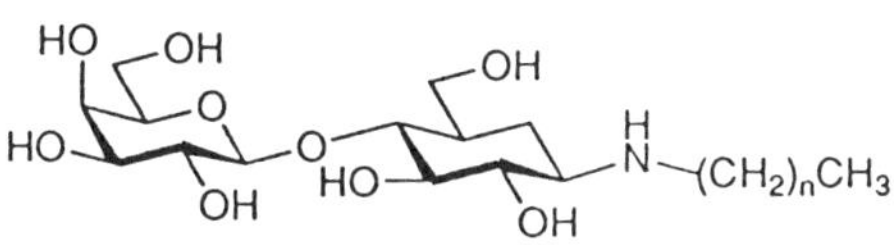

60

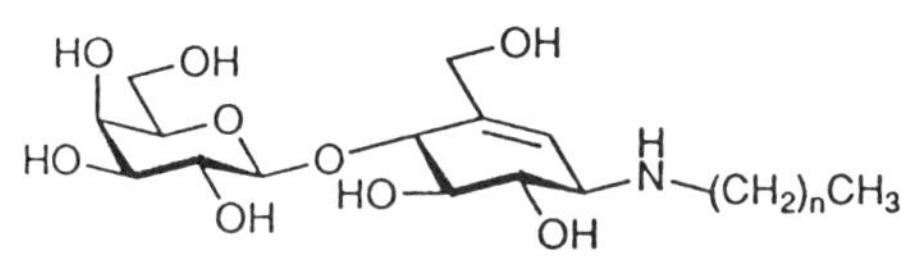

61

Table 4 Inhibitory Activity of Some *N*-Alkyl β-valienamines against β-Glucocerebrosidase and GlcCer Synthase (Mouse Liver Microsome), and α-Glucosidase (Baker's Yeast)

	IC_{50} (M)		
Compound	β-Glucocerebrosidase	GlcCer synthase	α-Glucosidase
57	1.1×10^{-5}	NI	
58	3.0×10^{-7}	NI	5.0×10^{-5}
59	3.0×10^{-8}	NI	1.7×10^{-5}

NI: No notable inhibition

2. Inhibitory activity of 5-aminocyclopentane-1,2,3,4-tetraol derivatives against some glycohydrolases

The chemical modifications of trehazolin described above suggested that, on the basis of its structural features, the aminocyclitol (trehazolamine) moiety **D-62**, 1D-(1,3/2,4,5)-5-amino-1-*C*-(hydroxymethyl)cyclopentane-1,2,3,4-tetraol, with a cyclic isourea may be assumed to correspond with one of the two α-D-glucopyranose residues as the mimic of symmetric α,α-trehalose (Fig. 24). On the other hand, the cyclic isourea moiety would constitute the charge distribution part for binding to the active site of the enzyme. Therefore, by analogy, it may be possible to design a new sugar hydrolase inhibitors composed of modified aminocyclitols. Initially, we became interested in the biological activity of the aminocyclitols, and tested all available samples in our laboratory for several glycohydrolases (Fig. 24).

Fig. 24 Inhibitory activity of 5-amino-1-*C*-(hydroxymethyl)cyclopentane-1,2,3,4-tetraols against α-glucosidase (Baker's yeast).

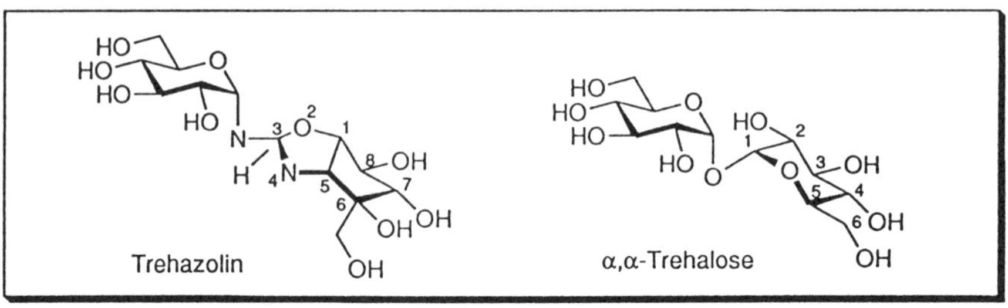

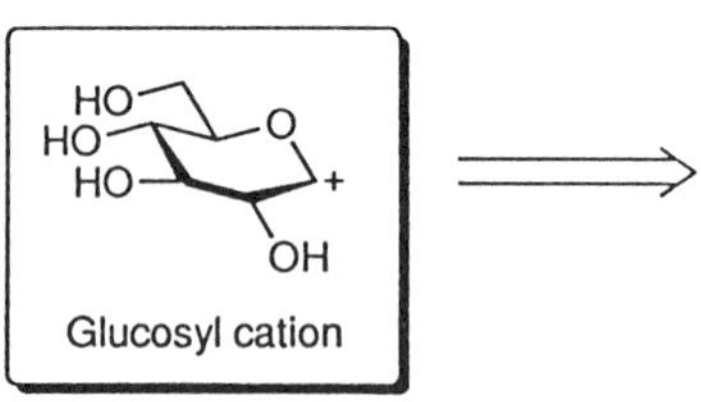

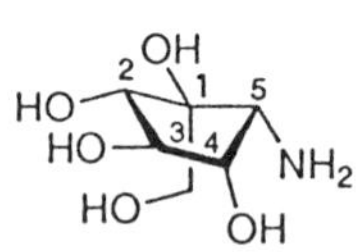

Trehazolamine **D-62**
(2.6 x 10⁻⁴)

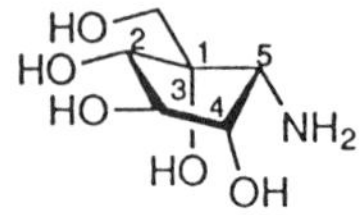

Epitrehazolamine: **L-63**
(4.0×10^{-7})

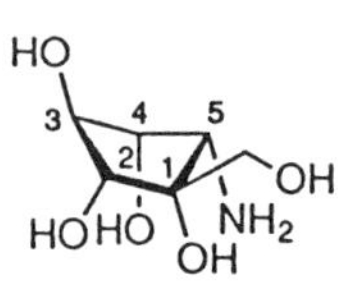

D-63
(1.6×10^{-4})

Since compound **D-62** had been described [54] as a mild inhibitor against β-glucosidase, we reinvestigated the inhibitory activity of **D-62**, as well as its 1-epimers [42] 1D,1L-(1,2,4,5/3)-5-amino-1-*C*-(hydroxymethyl)-cyclopentane-1,2,3,4-tetraols **D,L-63** against α- and β-glucosidases. When the structural features of **D-62** and **D,L-63** were compared, **L-63** was considered to have more similarity to the flattened half-chair conformation of the glucosyl cation probably formed during hydrolysis [55]. As had been expected, it was demonstrated that **L-63** possessed a strong potency against Baker's yeast α-glucosidase and a mild activity against almonds β-glucosidase, while **D-63** had a weak potency against α-glucosidase. On the other hand, **D-62** was shown to be a weak α-glucosidase inhibitor (IC_{50} 2.62×10^{-4} M) and a mild β-glucosidase inhibitor (IC_{50} 2.12×10^{-5} M). These results would seemingly support the feasibility of designing new glycosidase-inhibitors based on the above structural features.

Next, readily available two 5-aminocyclopentane-1,2,3,4-tetraols **64** and **65**, which mimic the half-chair conformation of galactosyl cation, were chosen to test the inhibitory activity against galactosidases (Fig. 25). Among them, as had been expected on the basis of the structural model, only **D-64** was shown to be a mild inhibitor against *E. coli* β-galactosidase. Interestingly, **D-64** and **65** also possessed a weak potency against β-glucosidase.

Fig. 25 Inhibitory activity (IC_{50}, M) of 5-amino-1-*C*-(hydroxymethyl)cyclopentane-1,2,3,4-tetraols against β-galactosidase (*E. coli*).

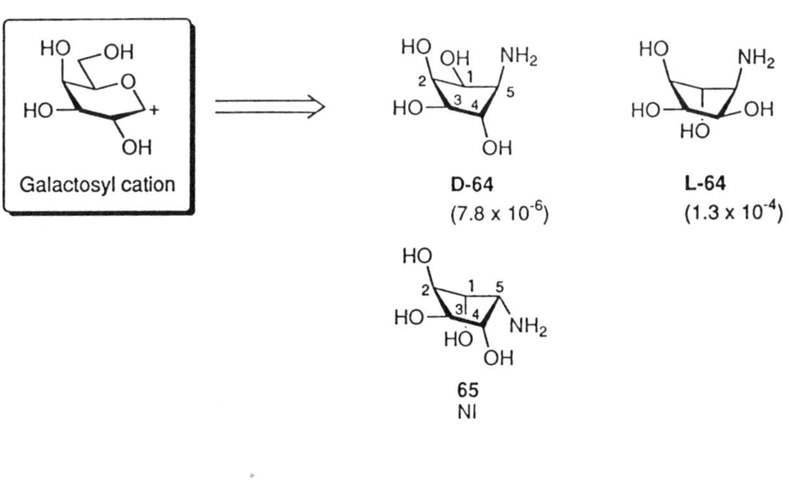

Fig. 26 Potent glycohydrolase inhibitors of *N*-phenyl cyclic isoureas of 5-amino-cyclopentane-1,2,3,4-tetraol derivatives.

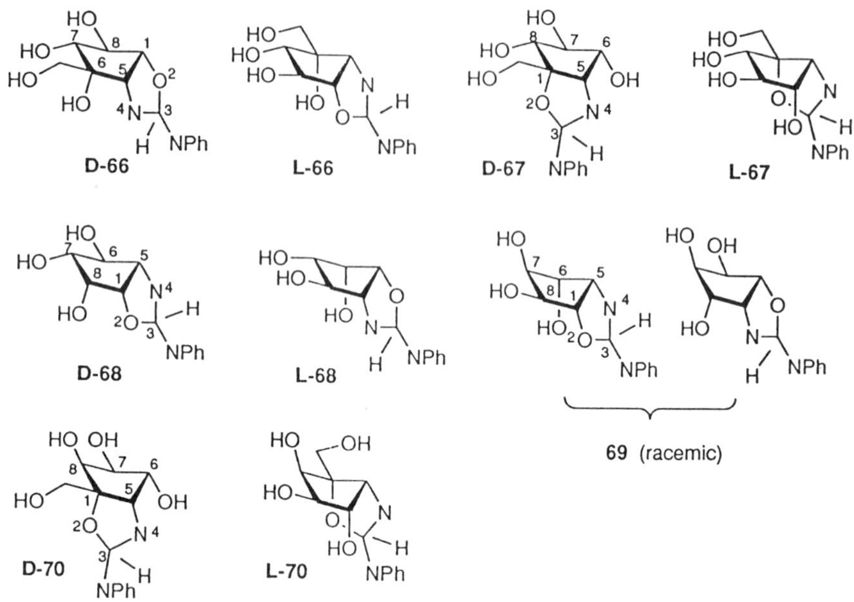

3. *N*-Alkyl and *N*-phenyl cyclic isourea derivatives of 5-aminocyclopentanetetraol derivatives: new type of α-glucosidase inhibitors

The above observation has stimulated us to study in detail the biological activity of the derivatives readily accessible by modification of the aminocyclitols of this kind. Therefore, in order to increase the inhibitory activities by changing somewhat charge distributions and their conformations, *N*-phenyl cyclic isourea groups with hydrophobic functions were first introduced. Thus, nine *N*-phenyl isoureas **D,L-66, D,L-67, D,L-68, 69**, and **D,L-70** were synthesized conventionally [39] from the corresponding aminocyclitols by coupling with phenylisothiocyanate, followed by treatment with yellow mercury(II) oxide (Fig. 26).

Inhibitory activities of the *N*-phenyl cyclic isourea derivatives [56] were listed in Table 5. Significant improvements in rather specific inhibitory activities against α-glucosidase were observed for derivatives **L-66** and **L-67** (diastereomeric pairs). Also, the derivatives **D,L-68** were shown to act strongly

Table 5 Inhibitory Activity of Stereopisomeric 5-Aminocyclopentanepolyols against Two Glycosidases

	Inhibitory activity IC_{50} (M)	
Compound	α-Glucosidase (Baker's yeast)	β-Galactosidase (Bovine liver)
D-66	2.3×10^{-6}	3.0×10^{-6}
L-66	2.9×10^{-8}	1.5×10^{-4}
D-67	1.0×10^{-6}	1.9×10^{-6}
L-67	1.0×10^{-8}	5.2×10^{-5}
D-68	NI	NI
L-68	NI	NI
69 (racemic)	NI	2.5×10^{-7}
D-70	NI	5.7×10^{-7}
L-70	NI	2.4×10^{-5}

NI: No notable inhibition, $IC_{50} > 4.2 \times 10^{-4}$

on both β-glucosidase and β-galactosidase. The configurations of the hydroxyl groups, including the oxygen atoms of the isoureas, seemed to be important for obtaining specificity on action. However, the topology of hydroxyl groups on the positions (C-8 of **D-68**, and C-7 of **69** and **D-70**) possibly corresponding to the C-4 of glycosyl cations were not as important for its binding to the enzymes.

In fact, *N*-phenyl isourea derivative **69** (racemic) of **65** was a potent β-galactosidase inhibitor as well as a mild β-glucosidase inhibitor. It is interesting to note that the respective enantiomers **D-66**, **D-67** and **L-68** in all cases still possess considerable activity toward β-galactosidase. Therefore, these new inhibitors would belong both to the mannostatin [57] type inhibitor that has an exocyclic nitrogen and to the deoxynojirimycin type inhibitors that have a nitrogen in the ring.

Alternatively, in order to improve the potency, *N*-butyl and *N*-butyl cyclic isourea functions were first introduced [61] into **L-63**, giving **L-71** and **L-72**, respectively (Fig. 27). These derivatives were shown to possess very strong activity against α-glucosidase (Baker's yeast) (Table 6). The latter had a strong potency agaisnt glucosidase I (yeast) almost comparable to *N*-butyl DNJ **33** (Table 7).

Fig. 27 Development of strong α-glucosidase inhibitors by chemical modification of 5-amino-1-C-(hydroxymethyl)cyclopentane-1,2,3,4-tetraol

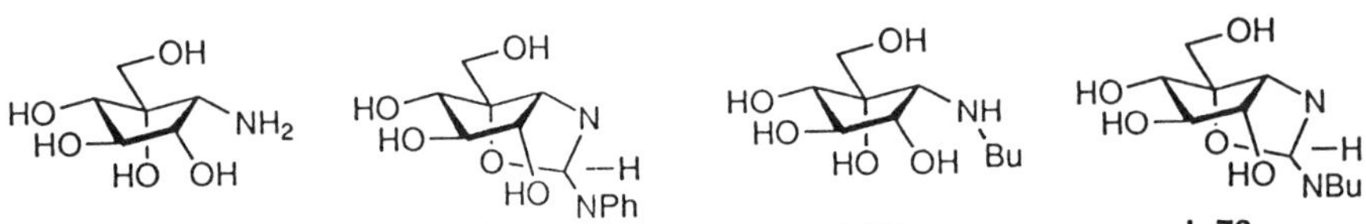

Table 6 Inhibitory Activity of Aminocyclopentanepolyol and Derivatives against Three Enzymes

Compound	IC_{50} (M) β-Glucocerebrosidase (Mouse liver microsome)	β-Glucosidase (Almonds)	α-Glucosidase (Baker's yeast)
DNJ **32**	9.0×10^{-5}	1.0×10^{-4}	1.6×10^{-4} [Ki; 6×10^{-5}]
N-Bu-DNJ **33**	9.4×10^{-5}	5.0×10^{-4}	$>3 \times 10^{-4}$
L-63	6.2×10^{-7}	3.0×10^{-5}	9.5×10^{-7} [Ki; 5×10^{-7}]
L-67	$>3 \times 10^{-4}$	3.2×10^{-4}	9.0×10^{-12} [Ki; 4×10^{-12}]
L-72	1.0×10^{-4}	4.6×10^{-4}	9.6×10^{-14} [Ki; 3×10^{-14}]

Table 7 Inhibitory Activity of 5-Aminocyclopentanepolyol Derivatives against α-Glucosidase I (Yeast)

Compound	Ki (μM)
N-Bu-DNJ **33**	3.6
L-63	300
L-67	51
L-71	180
L-72	4.2

In summary, the present results have demonstrated that the *N*-substituted 3-amino-2-oxa-4-azabicyclo[3.3.0]oct-3-ene-6,7,8-triols and the 6-(hydroxymethyl) derivatives thereof would be model compounds suitable for designing new potential sugar-hydrolase inhibitors.

REFERENCES

1. McCasland GE, Furuta S, Durham LJ. Alicyclic carbohydrates. XXIX. the syntesis of a pseudo-hexose (2,3,4,5-tetrahydroxycyclohexanemetha-nol). J. Org. Chem. 1966; 31:1516-1521.
2. McCasland GE, Furuta S, Durham LJ. Alicyclic carbohydrates. XXXII. synthesis of pseudo-β-DL-gulopyranose from a diacetoxybutadiene. proton magnetic resoance studies. J. Org. Chem. 1968; 33:2835-2841.
3. McCasland, Furuta S, Durham LJ. Alicyclic carbohydrates. XXXIII. epimerization of pseudo-α-DL-talopyranose to pseudo-α-DL-galactopyranose. proton magnetic resonance studies. J. Org. Chem. 1968; 33: 2841-2844.
4. Iwasa T, Yamamoto H, Shibata M. Studies on validamycins, new antibiotics. I. J. Antibiot. 1970; 32:595-602.
5. Schmidt DD, Frommer W, Junge B, Müller L, Wingender W, Truscheit E. Schafer D. α-Glucosidase inhibitors. new complex oligosaccharides of microbial origin. Naturwissenschafter, 1977; 64:535-536.

6. Miller TW, Arison BH, Albers-Schonberg G. Isolation of a cyclic antibiotic: 2,3,4,5-tetrahydroxycyclohexanemethanol. Biotech. Bioeng. 1973; 15:1075-1080.
7. Vertésy L, Fehlhaber H-W, Schulz A. The trehalase inhibitor salbostatin, a novel metabolite from streptomyces albus, ATCC 21838. Angew. Chem. Int. Ed. Engl. 1994; 33:1844-1845.
8. Umezawa H. Biochemical mechanism of resistance to aminoglycosidic antibiotics. Advans. Carbohydr. Chem. Biochem. 1974; 30:183-225.
9. Shier WT, Ogawa S, Hichens M, Rinehart KL Jr. Chemistry and biochemistry of the neomycins. XVII. bioconversion of aminocyclitols to aminocyclitol antibiotics. J. Antibiot. 1973; 26:551-55.
10. Umezawa S. Structures and syntheses of aminoglycoside antibiotics. Advan. Carbohydr. Chem. Biochem. 1974; 30:111-182.
11. Cox DA, Richardson K, Ross BC. The aminoglycosides" in Ed. Smmes PG. "Topics in antibiotic chemistry, vol. 1, 1977; 1-90. Ellis Horwood Ltd., New York.
12. Rinehart KL Jr, Suami T. Aminocyclitol antibiotics. ACS Symposium Series 125. American Chgemical Society, 1980.
13. Suami T, Ogawa S. Chemistry of carba-sugars (pseudo-sugars) and their derivatives. Advan. Carbohydr. Chem. Biochem. 1990; 48:21-90.
14. Ogawa S, Furuya T, Tsunoda H, Hindsgaul O, Stangier K, Palcic MM. Synthesis of β-D-GlcpNAc-(1→2)-5a-carba-α-D-Manp-(1→6)-β-D-Glcp-$O(CH_2)_7CH_3$: a reactive acceptor analog for N-acetylglucosaminyltransferase-V. Carbohydr. Res. 1995; 271:197-205 and references cited therein.
15. Suami T, Ogawa S, Chida N. The revised structure of validamycin A. J. Antibiot. 1980; 33:98-99.
16. Horii S, Iwasa T, Mizuta E, Kameda Y. Studies on validamycins, new antibiotics. VI. validamine, hydroxyvalidamine and validatol, new cyclitol. J. Antibiot. 1971; 24:59-63. Kameda Y, Asano N, Yoshikawa M, Takeuchi M, Yamaguchi T, Matsui K, Horii S, Fukase H. Valiolamine, a new α-glucosidase inhibiting aminocyclitol poduced by streptomyces hygroscopicus. J. Anitibiot. 1984; 37:1301-1307.
17. Asano N, Kameda Y, Matsui K. Validamycin H, a new pseudo-tetrasaccharide antibiotic. J. Antibiot. 1990; 43:1039-1041 and references therein.
18. Suami T, Ogawa S, Nakamoto K, Kasahara I. Synthesis of penta-N,O-acetyl-DL-validamine. Carbohydr. Res. 1977; 58:240-244.
19. Ogawa S, Iwasawa Y, Nose T, Suami T, Ohba S, Ito M, Saito Y. Total synthesis of (+)-(1,2,3/4,5)-2,3,4,5-tetrahydroxycyclohexane-1-methanol and (+)-(1,3/2,4,5)-5-amino-2,3,4-trihydroxycyclohexane-1-methanol [(+)-validamine]. X-ray crystal structure of (3S)-(+)-2-exo-bromo-4,8-dioxatricyclo[4.2.1.$0^{3,7}$]nonan-5-one. J. Chem. Soc., Perkin Trans. 1. 1985:903-906.
20. Ogawa S, Ara M, Kondoh T, Saitoh M, Masuda R, Toyokuni T, Suami T. Pseudo-sugars. VI. synthesis of six isomers of 5-hydroxymethyl-1,2,3,4-

cyclohexanetetrol (pseudo-hexopyranose) and their derivatives. Bull. Chem. Soc. Jpn. 1980; 53:1121-1126.

21. Ogawa S, Toyokuni T, Omata M, Chida N, Suami T. Pseudo-sugars. 5. synthesis of DL-validatiol and DL-deoxyvalidatol, and their epimers. Bull. Chem. Soc. Jpn. 1980; 53:1121-1126.
22. Toyokuni T, Ogawa S, Suami T. Synthetic studies on the validamycins. IV. synthesis of DL-valienamine and related branched-chain unsaturated aminocyclitols. Bull. Chem. Sopc. Jpn. 1983; 56:1161-1170.
23. Miyamoto Y, Ogawa S. Synthetic studies on antibiotic validamycins. part 13. total synthesis of (+)-validamycins A and E, and related compounds. J. Chem. Soc., Perkin Trans. 1, 1989:1013-1018.
24. Schmidt DD, Frommer W, Junge B, Müller L, Wingender W, Truscheit E. α-Glucosidase inhibitors of microbial origin. First International Symposium on Acarbose 1982; 5-49.
25. Junge B, Heiker F-R, Kurz J, Müller L, Schmidt DD, Wunsche C. Untersuchungen zur struktur des α-D-glucosidaseinhibitors acarbose. Carbohydr. Res. 1984; 128:235-268.
26 Truscheit E. Microbial α-glucosidase inhibitors: chemistry, biochemistry, and therapeutic potential. Progress in clinical biochemistry and medicine. 1988; 7:17-99.
27. Shibata Y, Ogawa S. Total synthesis of acarbose and adiposin-2. Carbohydr. Res. 1989; 189:309-322.
28. Kameda Y, Asano N, Yamaguchi T, Matsui K. Validoxylamine as trehalase inhibitors. J. Antibiot. 1987; 40:563-56 .
29. Gruters RA, Neefjes JJ, Tersmette M, de Goede RE, Tulp A, Huisman HG, Miedemaf, Ploegh HL. Interference with HIV-induced syncytium formation and viral infectivity by inhibitors of trimming glucosidase. Nature 1987; 330:74-77.
30. Takeuchi M, Kamata K, Yoshida M, Kameda Y, Matsui K. Inhibitory effect of pseudo-aminosugars on oligosaccharide glucosidases I and II and on lysosomal α-glucosidase from rat liver. J. Biochem. 1990; 108:42-46.
31. Horii S, Fukase H, Matsuo T, Kameda Y, Asano N, Matsui K. Synthesis and α-D-glucosidase inhibitory activity of N-substituted valiolamine derivatives as potential oral antidiabetic agents. J. Med. Chem. 1986; 29:1038-1046.
32. Asano N, Takeuchi M, Kameda Y, Matsui K, Kono Y. Trehalase inhibitors, validoxylamine A and related compounds as insecticides. J. Antibiot. 1990; 43:722-726.
33. Ogawa S, Sato K, Miyamoto Y. Synthesis and trehalase-inhibitory activity of an imino-linked dicarba-a,a-trehalose and analogues thereof. J. Chem. Soc., Perkin Trans. 1. 1993:691-696.
34. Shibata Y, Kosuge Y, Mizukoshi T, Ogawa S. Chemical modification of the sugar part of methyl acarviosin: syntheis and inhibitory activities of nine analogues. Carbohydr. Res. 1992; 228:377-398. Ogawa S, Aso D. Chemical modification of the sugar moiety of methyl acarviosin: synthesis

and inhibitory activity of eight analogues containing a 1,6-anhydro bridge. Carbohydr. Res. 1993; 250:177-184.
35. Ogawa S, Nakamura Y. Synthesis and enzyme-inhibitory activity of methyl acarviosin analogues having α-manno configuration. Carbohydr. Res. 1992; 226:79-89.
36. Ogawa S, Uchida C, Shibata Y. Alternative synthesis and enzyme-inhibitory activity of methyl 1'-epiacarviosin and its 6-hydroxy analog. Carbohydr. Res. 1992; 223:279-286.
37. Ogawa S, Tsunoda H. Synthesis of 1,4-imino-linked carbadisaccharides composed of a 2-amino-5a-carba-2-deoxyglucopyranosylamine residue and its congeners. Liebigs Ann. Chem. 1993:755-769.
38. Murao S, Sakai T, Gibo H, Nakayama T, Shin T. A novel trehalase inhibitor, torehalostatin, produced by amycolatopsis trehalostatica HG-25. Agric. Biol. Chem. 1991; 55:895-897.
39. Ando O, Satake H, Itoi K, Sato A, Nakajima M, Takahashi S, Haruyama H, Okuma Y, Kinoshita T, Enokita R. Trehazolin, a new trehalase inhibitor. J. Antibiot. 1991; 44:1165-1168.
40. Ogawa S, Uchida C. Synthesis of aminocyclitol moieties of trehalase inhibitors, trehalostatin and trehazolin. confirmation of the correct strucutre of the inhibitor. J. Chem. Soc., Perkin Trans. 1. 1992:1939-1942.
41. Uchida C, Yamagishi T, Ogawa S. Total synthesis of the trehalase inhibitors trehalostatin and trehazolin, and of their diastereoisomers. final structural confirmation of the inhibitors. J. Chem. Soc., Perkin Trans. 1, 1994:589-602.
42. Uchida C, Kitahashi H, Yamagishi T, Iwaisaki Y, Ogawa S. Synthesis of trehazolin analogues containing modified aminocyclitol moieties. J. Chem. Soc., Perkin Trans. 1, 1994:2775-2785.
43. Uchida C, Kitahashi H, Watanabe S, Ogawa S. Synthesis of trehazolin analogues containing modified sugar moieties. J. Chem. Soc., Perkin Trans. 1, 1995:1707-1717.
44. Uchida C, Yamagishi T, Kitahashi H, Iwaisaki Y, Ogawa S. Further chemical modification of trehalase ionhibitor trehazolin: structure and inhibitory activity relationshipof the inhibitor. Bioorg. Med. Chem. in press.
45. Tsunoda H, Ogawa S. Synthesis of some 5a-carbaglycosylamides, glycolipid analogs of biological interests. Liebigs Ann. Chem. 1994:103-107.
46. Lockhoff O. Glycolipids as immunomodulators: synthesis and properties. Angew. Chem. Int. Ed. Engl. 1991; 30:1611-1620; Angew. Chem. 1991; 103:1639-1648.
47. Inokuchi J-i, Radin NS. J. Lipid Res. 1987; 28:565-571.
48. Radin NS, Inokuchi J-i. Use of PDMP for the study of glycosphingolipid functions. TIGG. 1991; 3:200-213.

49. Tsunoda H, Ogawa S. Synthesis of 5a-carba-b-D-glycosylceramide analogs linked by imino, ether and sulfide bridges. Liebigs Ann. Chem. 1995:267-277.
50. Tsunoda H, Inokuchi J-i, Yamagishi K, Ogawa S. Synthesis of glycosyl-ceramide analogs composed of imino-linked unsaturated 5a-carbaglycosyl residues: potent and specific gluco- and galactocerebrosidase inhibitors. Liebigs Ann. Chem. 1995:279-284
51. Ogawa S, Mito T, Taiji E, Inokuchi J-i. unpublished results.
52. Legler G. Glycoside hydrolases: mechanistic information from studies with reversible and irreversible inhibitors. Advan. Carbohydr. Chem. Biochem. 1990; 48:319-384.
53. Ogawa S, Ashiura M, Uchida C, Inokuchi J-i. unpublished results.
54. Uchida C, Ogawa S. unpublished results.
54. Ando O, Nakajima M, Hamano K, Itoi K, Takahashi S, Takamatsu Y, Sato A, Enokita R, Okazaki T, Haruyama H, Kinoshita T. J. Antibiot. 1993; 46:1116-1125.
55 Look GC, Fotsch CH, Wong CH. Enzyme-catalyzed organic synthesis: practical routes to aza sugars and their analogs for use as glycoprocessing inhibitors. Acc. Chem. Res. 1993; 26:182-190.
56. Uchida C, Kimura H, Ogawa S. Potent glycosidase inhibitors, N-phenyl cyclic isourea derivatives of 5-amino- and 5-amino-1-C-(hydroxymethyl)-cyclopentane-1,2,3,4-tetraols. Bioorg. Med. Chem. Lett. 1994; 4:2643-2648.
57. Aoyagi T, Yamamoto T, Kojiri K, Morishima H, Nagai M, Hamada M, Takeuchi, T, Umezawa H. Mnnostatin A and B: new inhibitors of α-D-mannosidase, produced by streptoverticillium verticillus var. quintum ME3-AG3: taxonomy, production, isolation, phisico-chemical properties and biological activities. J. Antibiot. 1989; 42:883-889.

11
Azasugars: Chemistry and Their Biological Activity as Potential Anti-HIV Drugs

Leon A.G.M. van den Broek
N.V. Organon
Oss, The Netherlands

I. INTRODUCTION

The α-glucosidase inhibitors 1-deoxynojirimycin (**1**, dNM or DNJ, Figure 1) and castanospermine (**3**, CAS), and many other monocyclic and bicyclic polyhydroxylated alkaloids (also referred to as iminosugars or azasugars) are valuable tools for the study of the enzymatic steps involved in asparagine-linked N-glycoprotein biosynthesis. These compounds exert their biological effect by competitively inhibiting specific glycoside processing enzymes. This is achieved by mimicking the enzymes' normal carbohydrate substrates such as D-glucose- and D-mannose-terminated oligosaccharides. Both dNM and CAS closely resemble D-glucose; in fact CAS can be considered as a bicyclic derivative of dNM.

Considerable effort has also been made to develop therapeutic agents on the basis of the α-glucosidase I inhibition properties of these molecules. N-Butyl-dNM (**2**, Figure 1) and 6-O-butyryl-CAS (**4**) are two examples of more potent lipophilic derivatives of these natural products that show antiviral activity both in vitro and in animal models. Azasugars inhibit HIV replication by a mechanism different from that of the nucleoside analogues and therefore have attracted considerable attention. By altering the glycosylation of the HIV-1 envelope glycoproteins gp120 and gp41, the infectivity of the virus is reduced. Notably, N-butyl-dNM (**2**) is effective against human immunodeficiency virus (HIV) infection in Rhesus primates [1]. Preliminary positive results from a Phase I/II clinical trial in HIV-infected patients with N-butyl-dNM in combination with zidovudine (AZT) have been reported recently [2].

In addition to antiviral activity, CAS also shows anti-inflammatory (i.e. immunosuppressive) activity both in vitro and in animal models. Importantly, CAS not only inhibited the development of adjuvant arthritis in rats but also inhibited the progression of the disease [3].

The ester prodrug 6-O-butyryl-CAS (**4**) is almost 40 times more active in vitro against HIV-1 infection than CAS (**3**) itself [4]. However, in an animal model it showed no advantage in terms of antiviral efficacy or toxicity [5]. Recently, the perbutyrylated ester of N-butyl-dNM was described [6-7]. This prodrug was developed to eliminate the gastrointestinal side-effects associated with oral administration of N-butyl-dNM in animals. This prodrug, however, unlike compound **4**, is not antiviral in vitro but showed efficacy in animal studies [1].

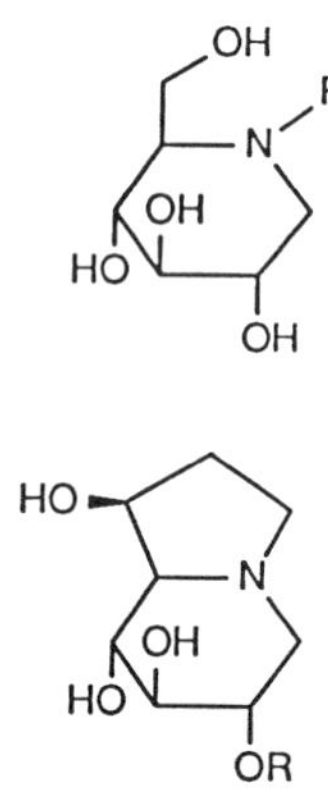

1 R=H, dNM
2 R=$(CH_2)_3CH_3$

3 R=H, CAS
4 R=$C(O)C_3H_7$

Figure 1 The structures of the α-glucosidase I inhibitors 1-deoxynojirimycin (dNM, **1**), N-butyl-dNM (**2**), castanospermine (CAS, **3**) and 6-O-butyryl-CAS (**4**).

From these and earlier studies the following conclusions regarding the use of the α-glucosidase I inhibitors dNM and CAS as potential therapeutics can be drawn:

1. Ki or IC_{50} values for dNM and CAS analogues are in the range of 0.1-1.0 μM, e.g. (8).
2. IC_{50} values of enzyme activity in intact cells are 1,000-10,000-fold higher and this huge difference is ill-understood, e.g. [9]; this may be due to poor membrane transport.
3. Analogues with a flattened conformation at the carbon atom next to the nitrogen atom frequently are shown to inhibit β- instead of the therapeutically relevant processing and intestinal α-glucosidases, e.g. [10]; these compounds are relevant, however, for studying the mechanism of action of such enzymes.
4. Several stereoisomers of dNM and CAS have been synthesized "accidentally" leading to (multi) epimers showing no glucosidase inhibitory activity, e.g. [11]; the gluco configuration is a prerequisite.
5. From in vivo experiments it is clear that of the present glycosidase inhibitors high dosages are necessary for antiviral or immunomodulatory effect, e.g. [2-3].
6. Fortunately, azasugars show low cytotoxicity in vivo, even at relatively high concentrations.
7. Administering high dosages in vivo is a disadvantage, in terms of dosage format and amount of material, thus compounds with higher enzyme inhibitory activity should be found, preferably with activity in nM range.
8. Membrane penetration of compounds should be improved for efficient therapeutic effect.
9. Therefore a need for structural analogues of dNM and CAS still remains.

There has been a number of reviews of various aspects of the chemistry, biochemistry and biological activities of azasugars in general and of dNM and analogues in particular [12-16]. The aim of this article is to review the literature of analogues of dNM published up to August 1995 with some emphasis on our own contributions to this field, to draw together the key biological observations with regard to α-glucosidase I inhibition and antiviral activity and to give conclusions of the present status of azasugar antiviral therapeutics.

II. NOVEL DNM ANALOGUES

Since our 1993 review [12], a considerable number of new structures have been prepared which are derived from dNM. These compounds are described below under A, B, C and D.

A. Oxygen-Substituted N-Alkyl-DNM Derivatives

One series of novel analogues of dNM was derived from N-decyl-dNM (**5**, Figure 2). This compound showed potent glucosidase inhibitory activity in human hepatoma HepG2 cells at 0.1 mM but was cell toxic at higher concentrations (Table 1) [9,17]. It was postulated that this was due to the amphiphilic character of the molecule. A series of oxygen-substituted side-chain analogues of **5**, i.e. compounds **6-11** (Figure 2, Table 1), was synthesized [9]. From these derivatives only N-(7-oxadecyl)-dNM (**8**) showed complete inhibition of α-glucosidase I at 3 mM in HepG2 cells as well as potent inhibition of HIV-1 infection. No toxic effects were observed for this compound. The position of the oxygen atom in the N-decyl side-chain is of importance, since compound 7 (N-(3-oxadecyl)-dNM) is less active than **8** and, moreover, is toxic to HepG2 cells at 3 mM. In compound **8** the amphiphilic properties of **5** are eliminated completely, while in **7** some amphiphilic character is still present.

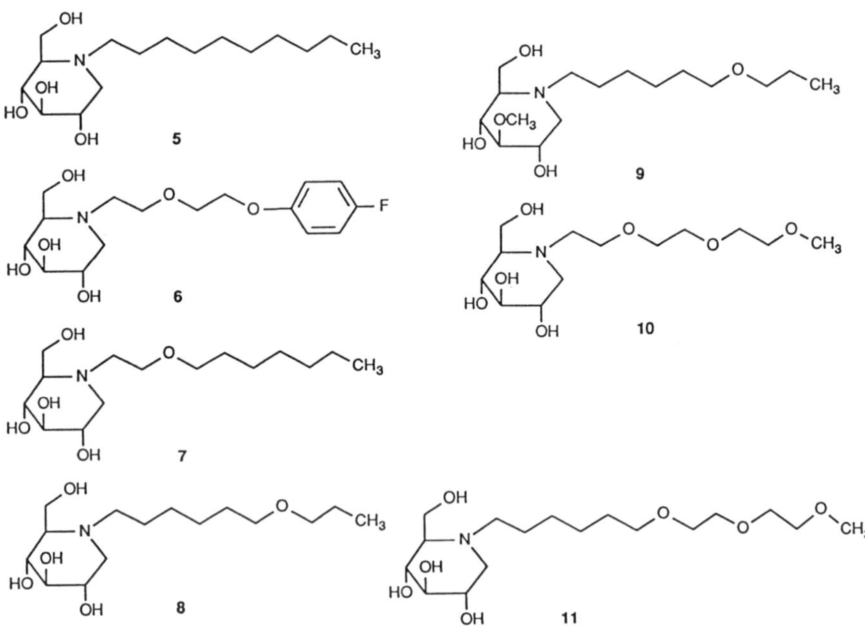

Figure 2 The structures of N-decyl-dNM (**5**) and six oxygen-substituted side-chain analogues of **5**, including the potent glucosidase inhibitor N-(7-oxadecyl)-dNM (**8**).

Compound **8** inhibits purified porcine liver α-glucosidase I with an IC_{50} value of 0.28 μM as compared to 0.57 μM for N-butyl-dNM [9]. N-(7-Oxadecyl)-dNM was also investigated in vivo. CAS was previously shown to inhibit adjuvant-induced arthritis when given orally and the rats ingested on average 60-65 mg/kg per day [3]. Compound **8** also reduced adjuvant-induced arthritis in rats at a dose of 150 mg/kg per day [12]. Both compounds showed no significant toxic effects while high doses were necessary to show the biological effect. From these studies it is clear that a need for more active dNM and CAS analogues exists.

Table 1. Estimated number of glucose residues (Glc_n) retained on α_1-antitrypsin synthesized in HepG2 cells at 3 mM drug concentration is taken as a measure of α-glucosidase inhibitory activity. The toxicity, as shown by reduced glycoprotein amounts detected on SDS-PAGE gels observed in the HepG2 assay is also shown at the lowest concentration found. Syncytia formation at 1 mM drug concentration is taken as a measure of antiviral (HIV-1) activity: ++: no syncytia (= total inhibition), +: clear inhibition, +/-: inhibition noticeable, -: no inhibition.

No.	Glc_n	Toxicity HepG2	HIV-I
1	Glc_{1-2}	none	+/-
2	Glc_{2-3}	none	++
3	Glc_{2-3}	none	++
5	Glc_{1-2}	0.1 mM	++
6	Glc_{1-2}	none	+
7	Glc_2	3 mM	++
8	Glc_3	none	++
9	Inactive	none	-
10	Inactive	none	+/-
11	Glc_{1-2}	none	+/-

B. DNM Nitrone Cycloadducts

In Figure 3, the structures of four natural products with a bicyclic sugar-like structure are shown. Nagstatin **(12)** is an N-acetyl-β-D-glucosaminidase inhibitor (ID_{50} 1.5 ng/ml for de-branched nagstatin lacking the acetic acid fragment) although the pyranose sugar moiety possesses the N-acetyl-β-D-galactoseamine configuration [18]. Compounds with a flattened conformation due to an sp^2 C-1 atom, such as nagstatin, frequently are found to be β-glycosidase inhibitors. Kifunensine (**13**) is a potent α-mannosidase inhibitor (IC_{50} 20-50 nM) with the pyranose sugar fragment in the mannose configuration [19]. However, the annulated oxamide function forces the sugar ring from the normal ${}_1C^4$ into the ${}_4C^1$ conformation, thereby changing the stereochemical orientation of the hydroxyl groups [20]. Allosamizoline (**14**) is the unusual sugar mimic of the trisaccharide allosamidin which is a chitinase inhibitor (IC_{50} 0.35 μM) (21,22). The heterobicyclic structure is a mimic for the sugar N-acetyl-β-D-glucosamine coupled at the C4-OH position. In trehazolin (**15**) and its C2-epimer trehalostatin a similar heterobicyclic structure to allosamizoline, but now mimicking an α-D-glucose fragment, is present but coupled at the other end of the molecule. Please, notice the branching at C1 which pushes the hydroxymethyl group in a distinct conformation. Trehazolin and trehalostatin (IC_{50} 1 μM) are trehalase inhibitors [22-23]. Studies looking into structure-activity relationships for these products hopefully will be published in the future, since we could learn from these how to substitute sugar rings with other, heterocyclic, structures.

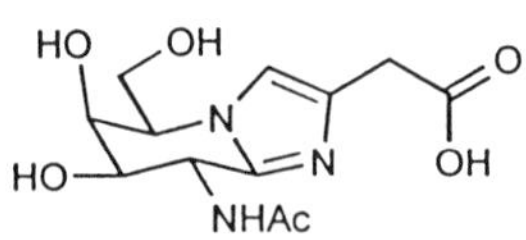

12, Nagstatin
N-acetyl-glucosaminidase inhibitor

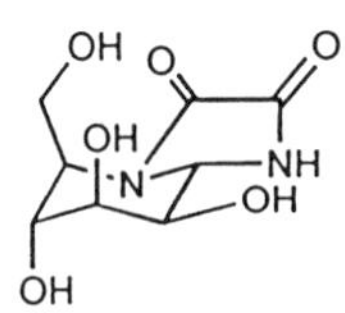

13, Kifunensine
mannosidase inhibitor

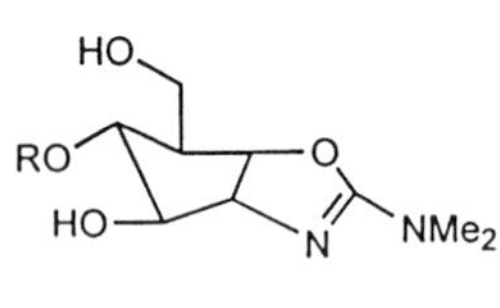

14, Allosamizoline, R=H
chitinase inhibitor

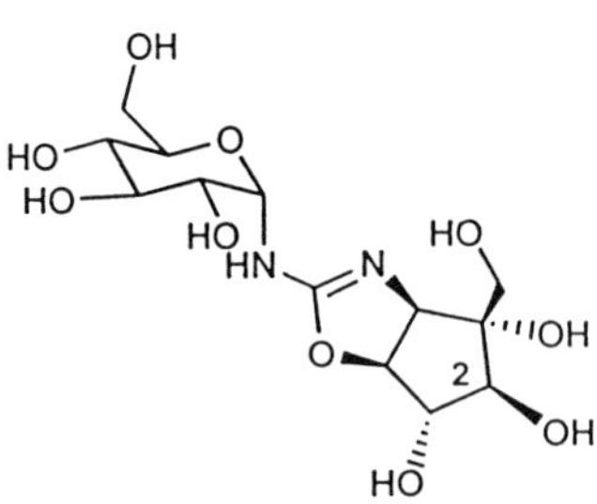

15, Trehazolin
trehalase inhibitor

Figure 3 The structures of four natural product glycosidase inhibitors with a heterobicyclic structure.

In the heterocyclic structures shown in Figure 3 the exact sugar configuration of the enzyme substrates which are mimicked may not be present, but still potent enzyme inhibition is achieved. In some, the attached heterocycle steers the conformation of the hydroxyl functions and interacts itself with the enzyme.

In a novel route to synthesize dNM derivatives we wanted to use the 1,3-dipolar cycloaddition reaction of nitrones with substituted alkenes. The resulting hexahydro-isoxazolo-pyridines are fairly rigid bicyclic systems which still possess a basic nitrogen atom for interaction with the enzyme α-glucosidase I. Moreover, cycloadducts derived from a dNM nitrone possess structural similarity to the natural products shown in Figure 3.

Recently, we started out to synthesize a dNM nitrone derivative (Figure 4). The secondary amine **16** could be converted into two regioisomeric nitrones in excellent yield, i.e. **17** and **18**, using 2,2-dimethyldioxirane [24].

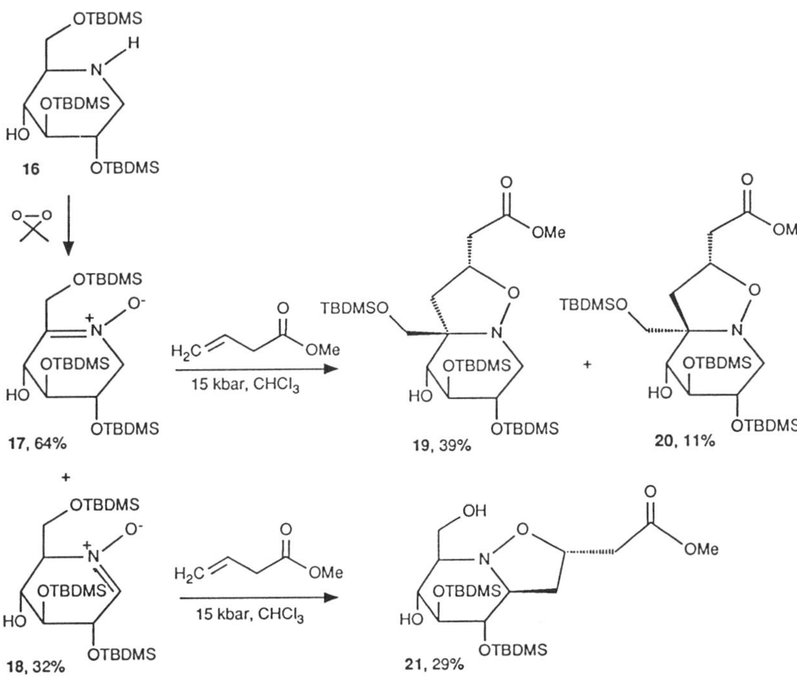

Figure 4 Synthesis of dNM nitrones **17** and **18** using 2,2-dimethyldioxirane and 1,3-dipolar cycloaddition reaction with methyl vinylacetate under high pressure.

Reaction of nitrone **17** with methyl vinylacetate resulted in the formation of two cycloadducts, i.e. **19** and **20**. Nitrone **18** was also allowed to react with methyl vinylacetate and gave exclusively the exo cycloadduct **21**. In this cycloadduct the C6-OTBDMS group was lost. These cycloaddition reactions had to be carried out at 15 kbar pressure because nitrones **17** and **18** are thermolabile.

No further reactions have been carried out with these cycloadducts so far. But clearly one can see various possibilities in going further to synthesize interesting dNM as well as CAS analogues, see for instance [25] and [26].

Of these cycloadducts, compound **19** is of particular interest since substitution at C5 forces the C6-OH group in the "active" gauche-gauche conformation [12,27]. In the natural product valiolamine (**52**, Figure 6), an aminoglucose α-amylase inhibitor, an α-C5-OH group is present which may also direct the C6-OH group in the gauche-gauche conformation [12]. After reduction of the N-O function in compound **20** the liberated hydroxyl function possibly could take the position of the C6-OH group in the gluco configuration in dNM. Note the structural resemblance between nitrone cycloadduct **21** and the natural product nagstatin **12** (Figure 3).

The dNM nitrone cycloaddition route, as well as substitution reactions on the nitrone function, may provide a whole new series of dNM and CAS analogues with different groups and rings attached at different positions to the dNM piperidine ring using a variety of available dipolarophiles and nucleophiles.

C. DNM Derivatives From the Literature

Several new dNM analogues have been described in the literature recently since our last review article on dNM analogues [12] and these 23 structures, i.e. **22-44** [18,28-47] are shown in Figure 5. Various groups have shown that interesting chemistry can lead to polyhydroxylated piperidine derivatives, some of which can be modified even further. These analogues belong to either one of two groups: a group of compounds in which all glucose hydroxyl groups are still in place and a group of compounds in which inversions, deletions and/or substitutions have been introduced. Interestingly, compounds **22**, **25**, **34** and **35** are expected to have the C6-OH group in the preferable "active" gauche-gauche conformation [12] (see also under II.B with regard to valiolamine (**52**, Figure 6)).

Unfortunately, very few biological data have been described for these analogues. Compound **23** (R=(2-ethyl)butyl) showed an ID_{50} of 6 μg/ml against HIV-1 [29], while compounds **32** [38] and **38** [44] were reported to possess no

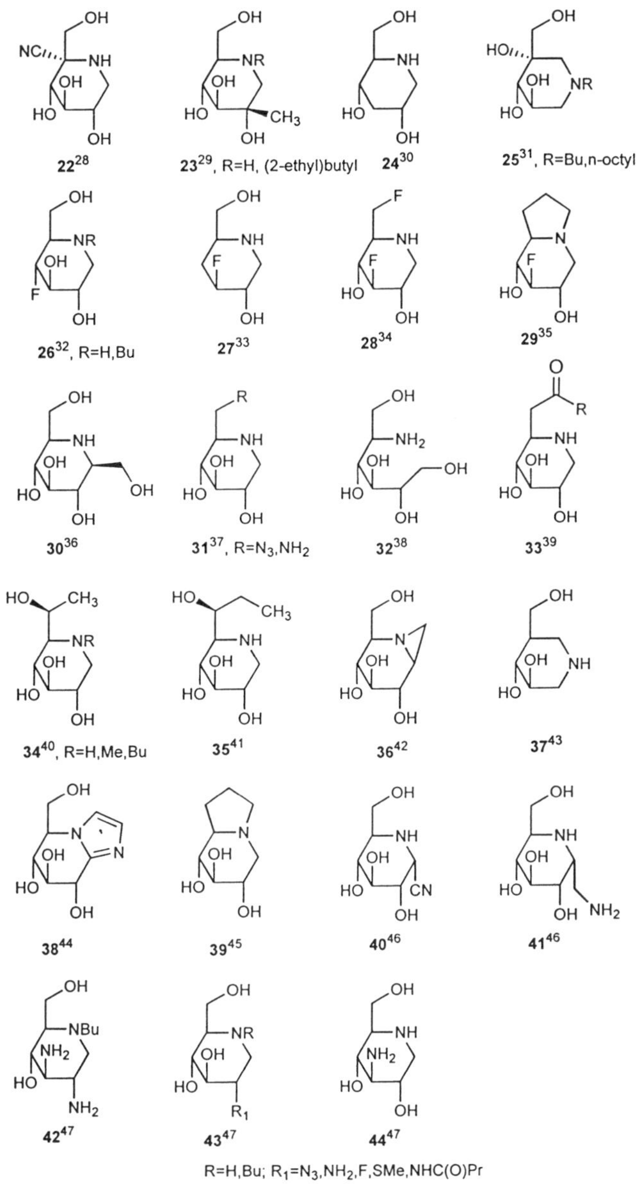

Figure 5 The structures of twenty-three novel dNM derivatives.

anti-HIV activity. Compounds **25** [31], **30** [36] and **37** [43] showed a preference for inhibiting β-glucosidase. Compound **35** (41), structurally closely related to CAS, is a potent inhibitor of yeast α-glucosidase with Ki of 3 μM. This is supportive of the proposed "active" gauche-gauche conformation of the C6-OH group [12]. It would be of interest to N-alkylate this compound; this side-chain then could interact with the hydrophobic site of the enzyme [12] (see below under II.D, Figure 8). Compound **41** is an inhibitor of α-glucosidase with Ki of 31 μM [46].

A general conclusion on the basis of all biological data on glucosidase inhibitors described so far is that all four hydroxyl functions in the gluco configuration must be present in the azasugar for activity. Also, most compounds are structurally closely related to dNM or CAS and no analogues have been reported that are more active against the processing α-glucosidase I enzyme than e.g. N-(7-oxadecyl)-dNM (IC_{50} 0.28 μM) or CAS (IC_{50} 0.1 μM) [12]. All compounds with inversions, deletions and/or substitutions show less or no activity or inhibit enzymes different from α-glucosidase I. Therefore, in order to arrive at compounds with activity in the nM range we probably have to look for structures diverging from dNM and CAS, as for example derivatives of the natural products depicted in Figure 3.

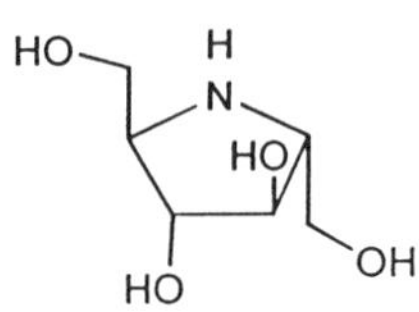

45, DMDP

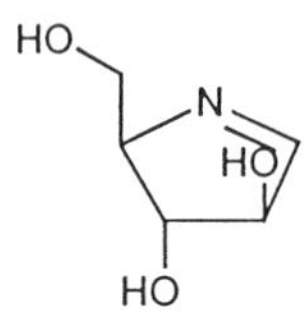

46, Nectrisine

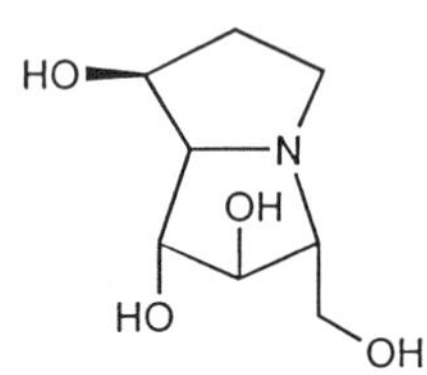

47, Australine

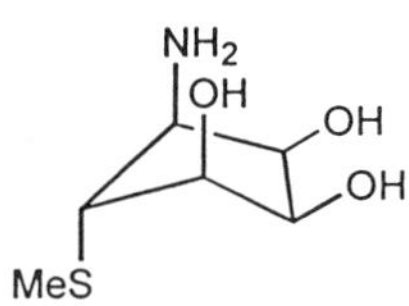

48, Mannostatin A

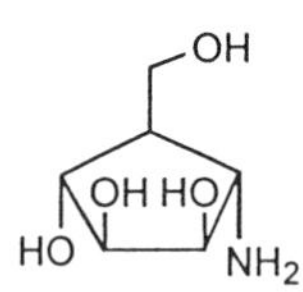

49

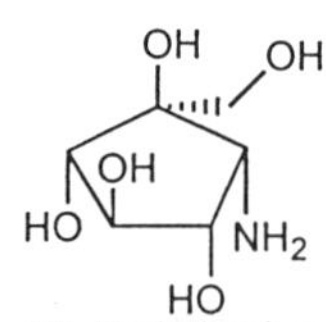

50, Trehalamine

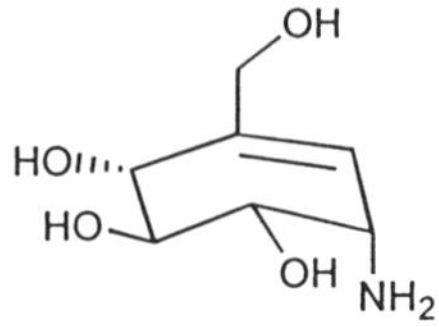

51, Valienamine

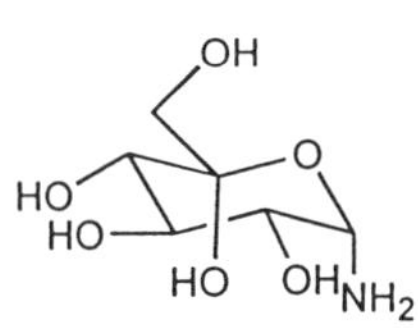

52, Valiolamine

Figure 6 The structures of three pyrrolidine azasugar analogues of dNM and CAS and of five non-azasugar glycosidase inhibitors.

Few structures have been described in which the azaglucopyranose ring has been replaced, except for the heterobicyclic structure in trehazolin (**15**) (Figure 3) and several polyhydroxylated pyrrolidine derivatives of dNM and CAS: ring-contracted azasugars like DMDP (**45**) [48], nectrisine (**46**) [49] and australine (**47**) [48] (Figure 6). In Figure 6, the structures of five non-azasugar compounds are also shown. The aminocyclopentane derivative mannostatin A (**48**) is a potent α-mannosidase inhibitor (IC_{50} 10-15 nM) (50); mannostatin B is the corresponding sulfoxide derivative. In these compounds one cannot recognize the mannose configuration. The aminocyclopentane **49** with the mannose configuration, is also an α-mannosidase inhibitor (IC_{50} 1 μM) [51]. The trehalamine **50**, the aminocyclopentitol part of trehazolin **15**, has recently been described [52]. Valienamine (**51**), the aminopseudosugar moiety of the α-amylase inhibitor acarbose [53] as well as of the trehalase inhibitor salbostatin (54) and valiolamine (**52**), the aminoglucose part of the α-amylase inhibitor AO-128 [55], are structures is which the glucose configuration is present.

Hopefully, the discovery of novel natural products other than those shown in Figures 3 and 6 and the screening of libraries of (heterocyclic) compounds from different origin could yield new α-glucosidase inhibitors with nM activity. So far, according to the literature, screening has not been employed for finding novel glycosidase inhibitors in general and α-glucosidase I inhibitors in particular.

D. Disaccharides with aglycon selectivity

One interesting way of increasing the activity of existing monosaccharide inhibitors against the processing enzymes is by taking into account the sugar linkage using interactions of the inhibitor with the aglycon binding site of the enzyme and prepare sequence-specific glycosidase inhibitors. Recently, several groups have reported the synthesis of dNM derivatives with aglycon specificity. Many glycosidases which recognize the same glycon moiety may differ only with the substrate aglycon portion and the type of glycosidic linkage.

DNM is thought to mimic the terminal glucose residue of the N-glycoprotein processing oligosaccharide precursor $Glc_3Man_9GlcNAc_2$-Asn. It is thought that in α-glucosidases a β-oriented carboxylate binds either the protonated substrate, the transition state or the glycopyranosyl cation itself (Figure 7). In its protonated form, N-alkyl-dNM mimics the oxocarbenium ion intermediate of the hydrolase and inhibits the enzyme competitively by binding to the enzyme active site carboxylate as exemplified for N-butyl-dNM in Figure 8 (the N-alkyl substituent interacts with a hydrophobic pocket in the active site) [56]. The same active site also interacts with the aglycon portion of the natural

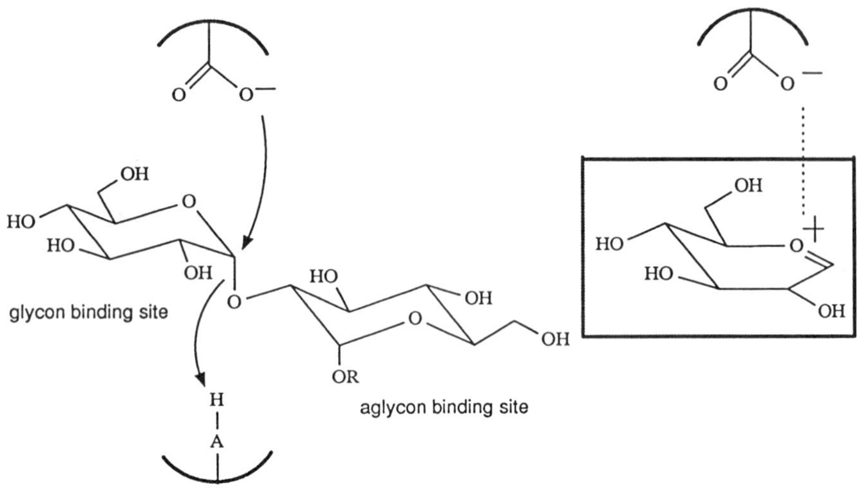

Figure 7 Mechanism of hydrolysis of α-glucosidase I showing the terminal Glc(α1,2)Glc oligosaccharide, the β-oriented catalytic carboxylate, glycon and aglycon binding sites and the proposed oxocarbenium ion intermediate.

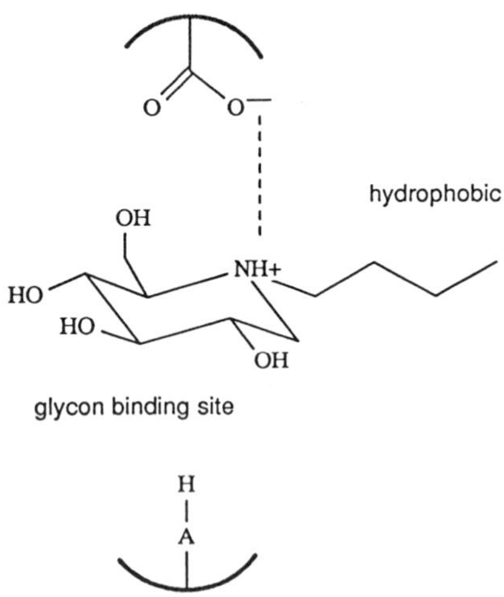

Figure 8 The binding of N-butyl-dNM (**2**) to α-glucosidase I showing the glycon binding site, interaction of the ammonium nitrogen atom with the active site carboxylate and interaction of the N-butyl side-chain with the enzyme hydrophobic pocket.

substrate, as evidenced by the linkage specificity shown by many glycosidases, e.g. α-1,2 for α-glucosidase I versus α-1,3 for α-glucosidase II.

A few naturally-occurring glycosidase inhibitors, such as the amino pseudosugar acarbose (a valienamine (**51**, Figure 6) containing tetrasaccharide) [53] and trehazolin (**15**, Figure 3) [23], are multi-subunit pseudo-oligosaccharides; they possess both a "glucopyranosyl" and an "aglycon" mimic, and both portions are important for efficacy and specificity. Several designed, linkage-spanning, pseudo-disaccharides that incorporate this multi-subunit attribute of the natural inhibitors have been prepared, and improvements in efficacy and selectivity over the simpler corresponding "glycopyranosyl" mimics have been recorded [57] and some of these are described below.

In Figure 9, the structures of some thirteen disaccharide mimics, i.e. **53-65**, are shown. Much of the work is done on setting up the methodology for synthesizing such compounds. Compound **53**, mimicking Glc(α1,3)Man, is a potent endomannosidase inhibitor (IC_{50} 1.7 μM) [58]. It does not span the linkage which is recognized by this enzyme, but its structure is of interest. This is because the second sugar on the glycon side is also recognized by the enzyme, and being present in the inhibitor clearly contributes considerably to the enzyme inhibitory activity. Recently, several analogues of **53**, α- and β-homomannojirimycin linked to D-glucose and D-mannose, have been reported [59].

The disaccharide **54** was described as a potent inhibitor of intestinal sucrase [60], where the β-glucose unit mimics the β-fructofuranosyl moiety of sucrose. In structure **55** the heteroaromatic ring was proposed to serve as a glycopyranosyl surrogate and this compound showed the best inhibition of β-glucosidase (Ki 0.8 mM) [57].

The azapyranosyl thioglycoside **56** was synthesized as a potential endo-glycosidase inhibitor [61]. The aza-C-disaccharide **57** mimics the disaccharide Man(β1,6)Gal and was proposed to be inert towards acidic or enzymatic hydrolysis while presenting the strong binding properties of the parent azasugar dNM [62]. The synthesis of oligosaccharide analogues containing various glycoside-linked dNMs and related moieties at the reducing end have also been reported (references cited in 62).

Compound **58** showed a remarkably improved inhibition (Ki 2.3 μM) of β-glucosidases as compared to dNM itself (Ki 58.8 μM) [63]. The structure of **59** is derived from that of trehazolin (**15**) and was found to be a potent inhibitor of α-glucosidase (Ki 9.3 μM); trehazolin and trehalamine (**50**), the pseudo-glucopyranosyl amino portion of **15** bound about an order of magnitude more weakly than **59** [64]. The amidine derivative **60** was developed as a potential β-glucosidase inhibitor (65). The corresponding α-anomer and N-Me α-anomer

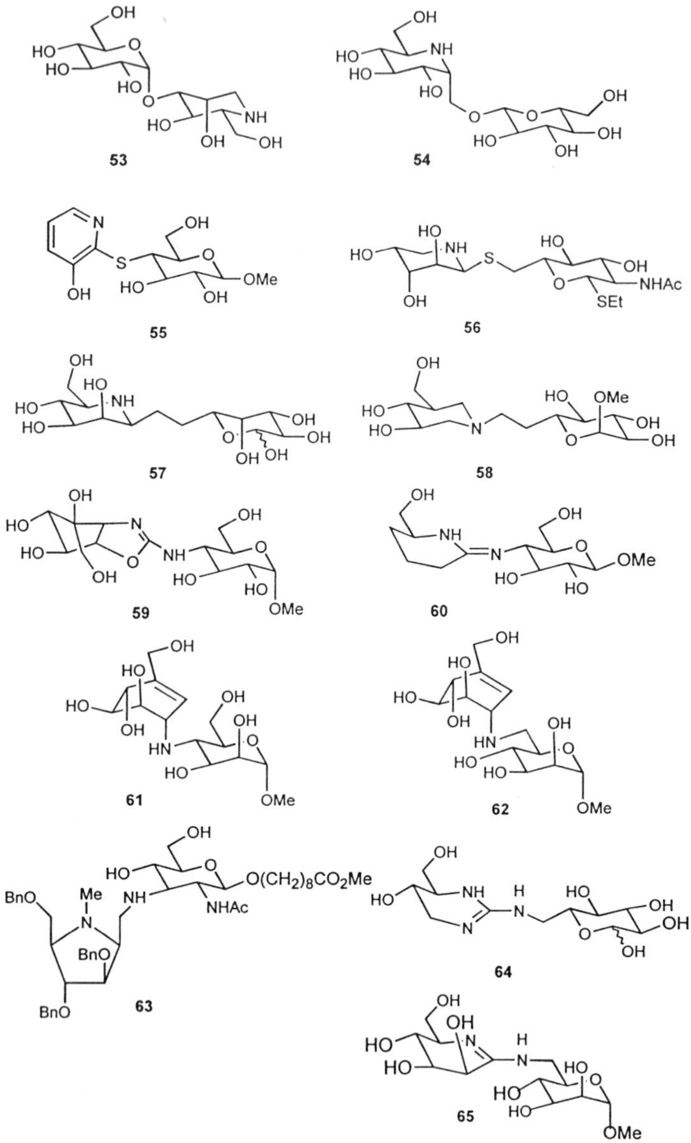

Figure 9 The structures of thirteen disaccharide mimics which are proposed to interact with the enzyme aglycon binding site.

were also prepared. Compounds **61** and **62** are valienamine (**51**, Figure 6)-derived structures and **62** showed inhibition of α-mannosidase (Ki 30 μM) whereas **61** was inactive [66]. Interestingly, upon N-acetylation of **62** the inhibitory activity was removed completely, pointing to the importance of the basic NH group in the inhibitory process.

The disaccharide mimic **63** was described recently as a new homoaminoazasugar type of glycosidase inhibitors (after removal of benzyl protecting groups) with potential aglycon specificity [46]. The guanidinium derivative **64** and its galactoguanidinium analogue were shown to be inhibitors of the corresponding β- and α-glycosidases [67]. Corresponding N-(substituted)phenyl-guanidinium derivatives were also prepared and shown to inhibit the corresponding glycosidases. Finally, the amidine **65** has just been published [68]. This pseudo-dimannoside inhibited α-mannosidase with Ki of 2.6 μM. In a program on the synthesis of linkage-spanning disaccharides with a β-oriented basic nitrogen function we have prepared several compounds. We tested the activity of the mannosamine disaccharide **66** (Figure 10) against α-glucosidase I [56].

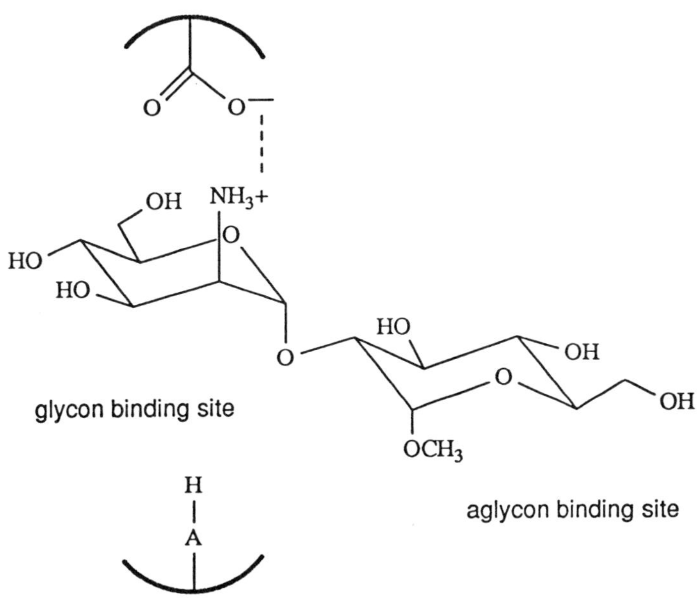

Figure 10 Interaction of aminodisaccharide **66** with the enzyme glycon and aglycon binding site showing binding of the ammonium nitrogen atom to the enzyme active site carboxylate.

Aminodisaccharide **66** shows a remarkably high affinity for α-glucosidase I isolated from porcine liver with an IC_{50} of 15.7 μM (the monosaccharide α-methyl-mannosamine shows no activity) as compared to 0.57 μM for N-butyl-dNM. This aminodisaccharide has the α-1,2 linkage present by which the terminal glucose in the oligosaccharide processing precursor is attached. We believe that the aminodisaccharide binds with the mannosamine fragment to the glycon binding site, where the β-amino group binds to the β-positioned carboxylate in the enzyme active site (Figure 10). In doing so, hydrolysis of the interglycosidic, as well as of the reducing glycosidic linkage is (minimized or) even prevented. The α1,2-linked glucose residue binds to the aglycon binding site of the same active site of the enzyme, thereby fulfilling the α1,2-linkage specificity of α-glucosidase I.

Even better at linkage-spanning, glycosidase inhibitors with effective selectivity of the aglycon portion and which are derived from **66** could be synthesized in the following ways by:

1. Eliminating the labile glycosidic linkages (i.e. via pseudo-disaccharides or C-disaccharides or via aza-C-disaccharides containing N-alkyl-dNM, as described above for the structures shown in Figure 9).
2. Incorporating the missing C2-OH group via branching at C2 to obtain the gluco configuration (as present in e.g. compounds **50** and **59**).
3. Incorporating N-alkyl groups for favorable hydrophobic interaction with he enzyme (e.g. N-7-oxadecyl substituent [9,17]).

III. CONCLUSIONS

The following conclusions regarding the present status on azasugar antiviral activity can be drawn:

1. A large number of publications are devoted to the preparation of dNM, CAS and analogues, which is due to the stereochemical challenge of these polyhydroxylated alkaloids.
2. Often enzyme activity was measured looking into the mechanism of enzyme action, but as often the important processing enzymes were not studied; for antiviral (HIV-1) activity only α-glucosidase I seems to be of interest.
3. Antiviral activity data often were not included.
4. Few derivatives show higher activity than N-butyl- or N-7-oxadecyl-dNM and CAS; no inhibitor with low nM activity against α-glucosidase I has been reported so far.

5. Compounds with aglycon specificity plus additional (hydrophobic) interaction sites with the processing enzyme(s) may possess higher affinity and as such offer a good perspective for future compounds with high activity.
6. No new in vivo data on azasugar antiviral activity have been reported.
7. Due to the mode of action of azasugars being N-glycoprotein processing inhibitors, combination therapy with nucleoside antivirals such as AZT may offer the greatest utility for this class of antiviral drugs [15].

Since our 1993 review [12] considerable attention has been paid to the synthesis of new analogues, but the amount of relevant biological information, such as anti-HIV activity, is limited. Hopefully, the synthesis of novel dNM and CAS analogues derived from the structures shown in Figures 3-6, 9 and 10, will lead to more potent α-glucosidase I inhibitors with antiviral activity. Also, from isolation from nature and screening, interesting and novel lead compounds could be obtained. Lastly, there is a clear need for structure-activity relationship studies, such as described in [69] of the novel lead compounds in order to be able to arrive at new and potent antiviral (HIV-1) drugs.

IV. ACKNOWLEDGMENTS

Stan van Boeckel is gratefully acknowledged for critically reading of the manuscript and for his support to the azasugar project. I would like to thank Dirk Jan Vermaas and Merian Kat-van den Nieuwenhof for their contribution to the synthesis of azasugar inhibitors and the analytical department of N.V. Organon for analysis of compounds. The high pressure cycloaddition reactions were carried out at the department of organic chemistry of the University of Nijmegen; Hans Scheeren and Rene Aben are acknowledged for their assistance in these reactions. Also, I would like to thank Agnes Tan for determining α-glucosidase inhibitory activity in the HepG2 assay, Terry Butters for data on the purified processing enzyme and Folkert van Kemenade for determining the antiviral activity of compounds.

REFERENCES

1. N-Butyl-1-deoxynojirimycin, SC-48334. Aidsdrug Citations Text, AIDS Compact Library CD-ROM. February 1, 1993. Macmillan New Media, Cambridge, MA, USA.
2. Fischl M, Resnick L, Coombs R, Pottage J, Smith S, Wallemark C et al. The preliminary efficacy and safety of N-butyl-deoxynojirimycin (SC-48334), an α-glucosidase I inhibitor, in combination with Zidovudine

(ZDV). Abstract at IXth International Conference on AIDS Berlin, June 6-11, 1993, abstract PO-B26-2039, pag. 475.

3. Willenborg DO, Parish CR, Cowden WB. Inhibition of adjuvant arthritis in the rat by phosphosugars and the α-glucosidase inhibitor castanospermine. Immunol Cell Biol 1992; 70:369-377.
4. Sunkara PS, Taylor DL, Kang MS, Bowlin TL, Liu PS, Tyms AS, Sjoerdsma A. Anti-HIV activity of castanospermine analogues. Lancet 1989; 1206.
5. Ruprecht RM, Bernard LD, Bronson R, Gama Sosa MA, Mullaney S. Castanospermine vs. its 6-O-butanoyl analog: a comparison of toxicity and antiviral activity in vitro and in vivo. J Acquired Immune Defic Synd 1991; 4:48-55.
6. Bryant M, Mueller R, Partis R, Stolzenbach J, Wyand M, Tiemeier D, Marr J. Efficacy and safety of α-glucosidase inhibitor prodrug SC-49483 in Rhesus monkeys infected with SIV. Abstract at IXth International Conference on AIDS Berlin, June 6-11, 1993, abstract WS-A11-2, pag. 27.
7. Bryant M, Mueller R, Schmidt M, Tiemeier D, Jacobs G, Platt F, Butters T, Karlsson N, Houseman K, Marr J. Anti-HIV properties of α-glucosidase inhibitor SC-48334, the active component of prodrug SC-49483. Abstract at IXth International Conference on AIDS Berlin, June 6-11, 1993, abstract WS-A17-6, pag. 33.
8. Schweden J, Borgmann C, Legler G, Bause E. Characterization of calf liver glucosidase I and its inhibition by basic sugar analogs. Arch Biochem Biophys 1986; 248:335-340.
9. Van den Broek LAGM, Vermaas DJ, van Kemenade FJ, Tan MCCA, Rotteveel FTM, Zandberg P, Butters TD, Miedema F, Ploegh HL, van Boeckel CAA. Synthesis of oxygen-substituted N-alkyl 1-deoxynojirimycin derivatives: azasugar α-glucosidase inhibitors showing antiviral (HIV-1) and immunosuppressive activity. Recl Trav Chim Pays-Bas 1994; 113:507-516.
10. Ermert P, Vasella A. Synthesis of a glucose-derived tetrazole as a new β-glucosidase inhibitor. A new synthesis of 1-deoxynojirimycin. Helv Chim Acta 1991; 74:2043-2053.
11. Altenbach HJ, Himmeldirk K. Stereocontrolled synthesis of 1,5-dideoxy-1,5-imino-allitol (1-deoxy-allonojirimycin) from serine. Tetrahedron: Asymmetry 1995; 6:1077-1080.
12. Van den Broek LAGM, Vermaas DJ, Heskamp BM, van Boeckel CAA, Tan MCAA, Bolscher JGM, Ploegh HL, van Kemenade FJ, de Goede REY, Miedema F. Chemical modification of azasugars, inhibitors of N-glycoprotein-processing glycosidases and of HIV-1 infection. Review and structure-activity relationships. Recl Trav Chim Pays-Bas 1993; 112:82-94.

13. Hughes AB, Rudge AJ. Deoxynojirimycin: synthesis and biological activity. Nat Prod Rep 1994; 135-162.
14. Moremen KW, Trimble RB, Herscovics A. Glycosidases of the asparagine-linked oligosaccharide processing pathway. Glycobiology 1994; 4:113-125.
15. Jacob GS, Bryant ML. Iminosugar glycosylation inhibitors as anti-HIV agents. Persp Drug Discov Design 1993; 1:211-224.
16. Winchester B, Fleet GWJ. Amino-sugar glycosidase inhibitors: versatile tools for glycobiologists. Glycobiology 1992; 2:199-210.
17. Tan A, van den Broek L, Bolscher J, Vermaas DJ, Pastoors L, van Boeckel C, Ploegh H. Introduction of oxygen into the alkyl chain of N-decyl-dNM decreases lipophilicity and results in increased retention of glucose residues on N-linked oligosaccharides. Glycobiology 1994; 4:141-149.
18. Tatsuta K, Miura S, Ohta S, Gunji H. Total synthesis of de-branched Nagstatin and its analogs having glycosidase inhibiting activities. Tetrahedron Lett 1995; 36:1085-1088.
19. Elbein AD, Tropea JE, Mitchell M, Kaushal GP. Kifunensine, a potent inhibitor of glycoprotein processing mannosidase I. J Biol Chem 1990; 265:15599-15605.
20. Kayakiri H, Takase S, Shibata T, Okamoto M, Terano H, Hashimoto M. Structure of kifunensine, a new immunomodulator isolated from an actinomycete. J Org Chem 1989; 54:4015-4016.
21. Kitahara T, Suzuki N, Koseki K, Mori K. Synthesis of (-)-allosamizoline, the pseudoaminosugar moiety of allosamidin, a chitinase inhibitor. Biosci Biotech Biochem 1993; 57:1906-1909.
22. Wong CH, Halcomb RL, Ichikawa Y, Kajimoto T. Enzyme in der organischen Synthese: das Problem der molekularen Erkennung von Kohlenhydraten (Teil 2). Angew Chem 1995; 107:569-593.
23. Kobayashi Y, Miyazaki H, Shiozaki M. Synthesis and absolute configuration of trehazolin aminocyclitol moiety. Tetrahedron Lett 1993; 34:1505-1506.
24. Kesselmans RPW, Wijnberg JBPA, Minnaard AJ, Walinga RE, de Groot A. Synthesis of all stereoisomers of eudesm-11-en-4-ol. 2. Total synthesis of selin-11-en-4α-ol, intermedeol, neointermedeol, and paradisiol. First total synthesis of amiteol. J Org Chem 1991; 56:7237-7244.
25. McCaig AE, Wightman RH. Hydroxylated pyrrolizidines and indolizidines; syntheses using cycloaddition reactions of functionalized cyclic nitrones. Tetrahdron Lett 1993; 34:3939-3942.
26. Herczegh P, Kovacs I, Szilagy L, Varga T, Dinya Z, Sztaricskai F. Cycloaddition reactions of carbohydrate derivatives. Part IV. Synthesis of a tetrahydroxyindolizidine through a cyclic nitrone prepared from D-xylose. Tetrahedron Lett 1993; 34:1211-1214.

27. Tan A, van den Broek L, van Boeckel S, Ploegh H, Bolscher J. Chemical modification of the glucosidase inhibitor 1-deoxynojirimycin. J Biol Chem 1991; 266:14504-14510.
28. Wuensche C, Benninghoven A, Eicke A, Heinen HJ, Ritter HP, Taylor LCE, Veith J. Comparison of 'soft ionization' techniques with electron impact mass spectrometry for desoxinojirimycin and folic acid derivatives. Org Mass Spectrom 1984; 19:176-182.
29. Khanna IK, Mueller RA, Weier RM. New deoxynojirimycin derivs. - active against lentiviruses e.g. 1,5-(2-ethyl-butyl-imino)-1,5-di:deoxy-2-C-methyl-glucitol. GD Searle & Co. EP566556 1993.
30. Johnson CR, Golebiowski A, Braun MP, Sundram H. Chemoenzymatic synthesis of 1,3-dideoxynojirimycin. Tetrahedron Lett 1994; 35:1833-1834.
31. Ichikawa M, Igarashi Y, Ichikawa Y. Facile synthesis of glucose-type 1-N-iminosugars: new inhibitor of glycolipid biosynthesis. Tetrahedron Lett 1995; 36:1767-1770.
32. De Crescenzo Monsanto EP481950.
33. Armone A, Brave P, Donadelli A, Resnati G. Asymmetric synthesis of fluorinated analogues of 1-deoxynojirimycin. J Chem Soc Chem Commun 1993; 984-986.
34. Lee CK, Jiang H. Asymmetric synthesis of 1,3,6-trideoxy-3,6-difluoronojirimycin. J Carbohydr Chem 1995; 14:407-416.
35. Furneaux RH, Mason JM, Tyler PC. The chemistry of castanospermine, part II: synthesis of deoxyfluoro analogues of castanospermine. Tetrahedron Lett 1994; 35:3143-3146.
36. Holt KE, Leeper FJ, Handa S. Synthesis of β-1-homomannojirimycin and β-1-homomannojirimycin using the enzyme aldolase. J Chem Soc Perkin Trans I 1994; 231-234.
37. Kilonda A, Compernolle F, Toppet S, Hoornaert GJ. Synthesis of 6-azido and 6-amino analogues of 1- deoxynojirimycin. J Chem Soc Chem Commun 1994; 2147-2148.
38. Fowler PA, Haines AH, Taylor RJK, Chrystal EJT, Gravestock MB. Synthesis and biological activity of acyclic analogues of nojirimycin. J Chem Soc Perkin Trans I 1994; 2229-2235.
39. Kilonda A, Compernolle F, Toppet S, Hoornaert GJ. The synthesis of 7-carbonyl homologues of 1-deoxynojirimycin. Tetrahedron Lett 1994; 35:9047-9050.
40. Meiji Seika Kaisha. New deoxynojirimycin derivs. which inhibit glycosidase - comprise e.g. 1,5-butyl:imino-1,5,7-trideoxy-D-glycero-D-gluco:heptitol, useful for treating HIV infection. JP05043545 1993.
41. Berger A, Dax K, Gradnig G, Grassberger V, Stutz AE, Ungerank M, Legler G, Bause E. Synthesis and biological activity of C-6 modified

derivatives of the glucosidase inhibitor 1-deoxynojirimycin. Bioorg Med Chem Lett 1992; 2:27-32.
42. Martin OR, Saavedra OM. Concise chemical synthesis of β-homonojirimycin and related compounds. Tetrahedron Lett 1995; 36:799-802.
43. Jespersen TM, Bols M, Sierks MR, Skrydstrup T. Synthesis of isofagomin, a novel glycosidase inhibitor. Tetrahedron 1994; 50:13449-13460.
44. Burgess K, Chaplin DA, Elbein AD, Zeng Y. Preparation of a deoxynojirimycin analog containing an imidazole ring. Heterocycles 1994; 37:673-677.
45. Kang MS, Liu PS, Bernotas RC, Harry BS, Sunkara PS. Castanospermine analogues: their inhibition of glycoprotein processing α-glucosidases from porcine kidney and $B_{16}F_{10}$ cells. Glycobiology 1995; 5:147-152.
46. Wong CH, Provencher L, Porco Jr. JA, Jung SH, Wang YF, Chen L, Wang R, Steensma DH. Synthesis and evaluation of homoazasugars as glycosidase inhibitors. J Org Chem 1995; 60:1492-1501.
47. Khanna IK, Koszyk FJ, Stealey MA, Weier RM, Julien J, Mueller RA, Rao SN, Swenton L. Nucleophilic opening of N-carboalkoxy-2,3-anhydro-1-deoxymannojirimycin. A useful method for the syntheses of 2-, 3- and 2,3-disubstituted 1-deoxynojirimycin analogs. J Carbohydr Chem 1995; 14:843-878.
48. Robinson KM, Rhinehart BL, Ducep JB, Danzin C. Intestinal disaccharidase inhibitors. Drugs Fut 1992; 17:705-720.
49. Kayakiri H, Nakamura K, Takase S, Setoi H, Uchida I, Terano H, Hashimoto M, Tada T, Koda S. Structure and synthesis of nectrisine, a new immunomodulator isolated from a fungus. Chem Pharm Bull 1991; 39:2807-2812.
50. Tropea JE, Kaushal GP, Pastuszak I, Mitchell M, Aoyagi T, Molyneux RJ, Elbein AD. Mannostatin A, a new glycoprotein-processing inhibitor. Biochemistry 1990; 29:10062-10069.
51. Farr RA, Peet NP, Kang MS. Synthesis of 1S,2R,3S,4R,5R-methyl[2,3,4-trihydroxy-5-(hydroxymethyl)cyclopentyl]amine: a potent α-mannosidase inhibitor. Tetrahedron Lett 1990; 31:7109-7112.
52. Ando O, Nakajima M, Hamano K, Itoi K, Takahashi S, Takamatsu Y, Sato A, Enokita R, Okazaki T, Haruyama H, Kinoshita T. Isolation on trehalamine, the aglycon of trehazolin, from microbial broths and characterization of trehazolin related compounds. J Antibiotics 1993; 46:1116-1125.
53. Truscheit E, Frommer W, Junge B, Muller L, Schmidt DD, Wingender W. Chemistry and biochemistry of microbial α-glucosidase inhibitors. Angew Chem 1981; 20:744-761.

54. Vertesy L, Fehlhaber HW, Schulz A. Der Trehalaseinhibitor Salbostatin, ein neuartiger Metabolit aus Streptomyces albus, ATCC 21838. Angew Chem 1994; 106:1936-1937.
55. Horii S, Fukase H, Matsuo T, Kameda Y, Asano N, Matsui K. Synthesis and α-D-glucosidase inhibitory activity of N-substituted valiolamine derivatives as potential oral antidiabetic agents. J Med Chem 1986; 29:1038-1046.
56. Van den Broek LAGM, Kat-van den Nieuwenhof MWP, Butters TD, van Boeckel CAA. Synthesis of α-glucosidase I inhibitors showing antiviral (HIV-1) and immunosuppressive activity. J Pharm Pharmacol In press.
57. Knapp S, Dong Y, Rupitz K, Withers SG. β-Glucosidase inhibitors incorporating pyridinium as a glucosyl mimic. Bioorg Med Chem Lett 1995; 5:763-766.
58. Hiraizumi S, Spohr U, Spiro RG. Characterization of endomannosidase inhibitors and evaluation of their effect on N-linked oligosaccharide processing during glycoprotein biosynthesis. J Biol Chem 1993; 268:9927-9935.
59. Suhara Y, Achiwa K. Disaccharides as endomannosidase inhibitors: syntheses of α-homomannojirimycin and β-homomannojirimycin linked to D-glucose and D-mannose. Chem Pharm Bull 1995; 43:414-420.
60. Liu PS. Total synthesis of 2,6-dideoxy-2,6-imino-7-O-β-D-glucopyranosyl-D-glycero-L-gulo-heptitol hydrochloride: a potent inhibitor of α-glucosidases. J Org Chem 1987; 52:4717-4721.
61. Suzuki K, Hashimoto H. Synthesis of azapyranosyl thioglycoside: novel pseudo-disaccharide having an azasugar residue at the non-reducing end. Tetrahedron Lett 1994; 35:4119-4122.
62. Johnson CR, Miller MW, Golebiowski A, Sundram H, Ksebati MB. Synthesis of aza-C-disaccharides - a new class of sugar mimics. Tetrahedron Lett 1994; 35:8991-8994.
63. Jespersen TM, Dong W, Sierks MR, Skrydstrup T, Lundt I, Bols M. Isofagomin, ein wirksamer neuer Glycosidaseinhibitor. Angew Chem 1994; 106:1858-1860.
64. Knapp S, Purandare A, Rupitz K, Withers SG. A (1-4)-"Trehazoloid" glucosidase inhibitor with aglycon selectivity. J Am Chem Soc 1994; 116:7461-7462.
65. Knapp S, Choe YH, Reilly E. Amidine pseudodisaccharides. Tetrahedron Lett 1993; 34:4443-4446.
66. Cottaz S, Brimacombe JS, Ferguson MAJ. An imino-linked carba-disaccharide α-D-mannosidase inhibitor. Carbohydr Res 1993; 247:341-345.

67. Lehmann J, Rob B. Two cationic analogues of disaccharide heterolysis. Potential epitopes for preparing antibodies with saccharidase activity. Tetrahedron: Asymmetry 1994; 5:2255-2260.
68. Bleriot Y, Dintinger T, Guillo N, Tellier C. Synthesis of an amidine pseudo-1(1-6)-dimannoside and evaluation as a glycosidase inhibitor. Tetrahedron Lett 1995; 36:5175-5178.
69. Uchida C, Kitahashi H, Watanabe S, Ogawa S. Synthesis of trehazolin analogues containing modified sugar moieties. J Chem Soc Perkin Trans I 1995; 1707-1717.

12
Sugar-Modified Pyrimidine Nucleoside Analogues with Potential Antiviral Activity

Stanislas Czernecki and Jean M. Valery
Université Pierre et Marie Curie
Paris, France

I. INTRODUCTION

As far as chemotherapy is concerned, one of the most striking advances in medicine during the twentieth century has been the discovery of antibiotics which are able to inhibit the proliferation of bacteria in an infected organism without or with limited detrimental effect against the host. This was made possible because antibiotics interact with an enzyme present in bacteria but not in human beings.

Comparatively, progress in the treatment of viral infections has been slower and limited, presumably because it is difficult to distinguish viral functions from those of the host. However, in recent years, better knowledge in the field of cellular biology and more specifically, viral replication has made possible the identification of compounds that might interfere selectively with viral functions. Among them, modified nucleosides are of prime importance and almost all antiviral drugs employed at this time are modified nucleosides [1].

Since the discovery of the Human Immunodeficiency Virus (HIV) as the etiologic agent of AIDS [2] intense efforts have been devoted to the synthesis and biological evaluation of compounds with potential anti-HIV activity. Since the virus is a retrovirus, its replicative cycle involves a reverse transcriptase (RT) which is a target of choice for the design of anti-HIV agents. For that purpose, modifications of natural nucleosides were made in the carbohydrate functionalities, in the heterocyclic base or in both. Changes in stereochemistry to obtain non-natural derivatives (α-D, α–L and β-L) were also performed. Several nucleosides with a modified sugar moieties have proved to be selective inhibitors of HIV replication [3]. Among them 3'-azido-3'-deoxythymidine (AZT) was the first to receive wide clinical usage. Other 2',3'-dideoxynucleosides were later reported to inhibit the infectivity and cytopathology of HIV, namely 3'-azido-2',3'-dideoxy-uridine (CS 87) [4] and 2',3'-didehydro-2',3'-dideoxy-thymidine (d4T) [3]. Recently 2',3'-didehydro-2'-3'-dideoxy-cytidine (ddC) and 2',3'-didehydro-2'-3'-dideoxy-inosine (ddI) were approved for clinical use in the chemotherapy of HIV infections, albeit with more restrictions than for AZT. At the same time d4T also entered clinical trials.

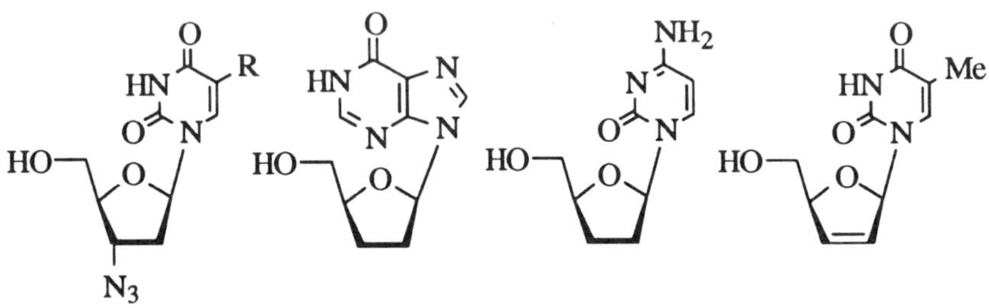

R = Me , AZT **1**
R = H, CS 87 **2** ddI **3** ddC **4** d4T **5**

It is generally accepted that the mechanism of action of these modified nucleosides involves phosphorylation by a kynase of the host cell and incorporation by RT in the growing DNA. The lack of a 3'-hydroxyl makes the chain elongation of DNA impossible.

These two requirements explain why the establishment a structure-activity relationship is very difficult in this field although thousands of modified nucleosides have been evaluated.

Since the more active derivatives contain a modified carbohydrate moiety, this chapter will review efficient syntheses of these kinds of modified nucleosides. In the first part, syntheses of active pyrimidine derivatives will be described. They have in common, the use of a 5'-protected 2,3'-anhydro derivative as an intermediate which allowed stereocontrolled introduction of a hetero-substituent in the 3'-α position of the pentofuranose ring of thymidine and 2'-deoxyuridine.The key step of this sequence of reactions is the easy and very efficient preparation of **8** and **9** resulting from a judicious choice of the protecting group of the primary hydroxyl group.

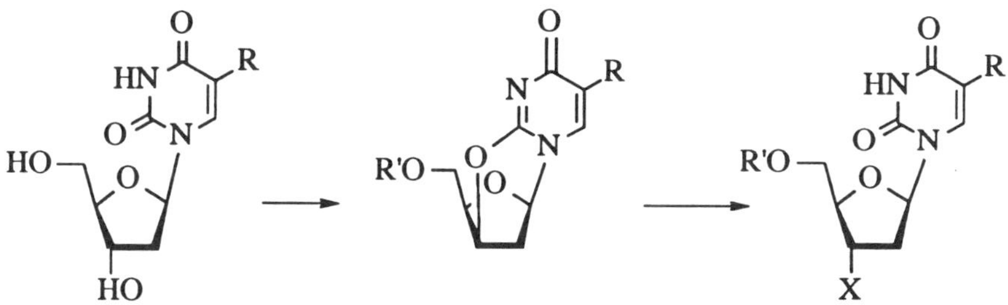

R = CH_3, thymidine **6**
R = H, 2'-deoxyuridine **7**

8 R = CH_3
9 R = H

R' = H, protective group

The second part of the chapter will be devoted to the design and preparation of new modified nucleosides as potential anti-HIV agents focusing on the use of anhydro derivatives as intermediates.

II. PREPARATIONS OF ACTIVE MODIFIED NUCLEOSIDES

1. Preparation of 3'-azido-3'-deoxythymidine (AZT)

Although AZT was recently the first molecule to be employed in AIDS chemotherapy, the first synthesis of this molecule was reported thirty years ago (Scheme 1) [5-6].

The problem of substitution of thymidine 3'-OH by an azido group with retention of configuration was solved by effecting a double inversion at C-3' after protection of the primary hydroxyl group. If the second was clearly an S_N2 reaction, the first was one of the first examples of participation of the heterocyclic base involving a 2,3'-anhydro intermediate which was not isolated, but opened *in situ* by nucleophilic attack of HO^- at C-2 of the base.

In fact, when the reaction was carried out at room temperature, this 2,3'-anhydro derivative was isolated and treatment with sodium hydroxide afforded the desired D-*threo* derivative. Although each step was efficient, a 24% overall yield was obtained in this sequence of reactions.

In order to diminish the number of steps, shorter routes to the key 2,3'-anhydro derivative were studied as well as its opening with azide ion. Kowollik *et al.* reported later a one step procedure in which thymidine was heated at 70°C in the presence of 2-chloro-1-diethylamino-1,1,2-trifluoroethane to afford the 2,3'-anhydro derivative **8** in 40% yield (Scheme 2) [7].

Opening by lithium azide in the presence of a weak acid afforded AZT in two steps (22% overall yield). Difficulties in up-scaling the first step and in purification of AZT limited the use of this approach. Recently the 2,3'-anhydro derivative **8** was obtained by heating thymidine at 150 °C in the presence of diethyl sulphite[8].

Scheme 1

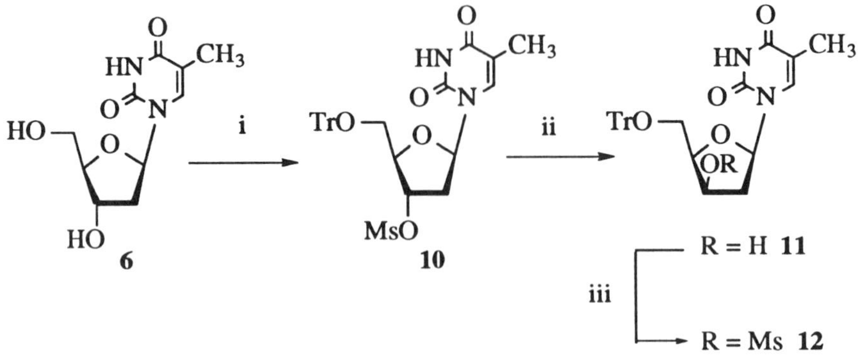

Reagents and conditions: i) TrCl, pyr. then MsCl (65 %); ii) NaOH, 90 % EtOH, reflux (75 %); iii) MsCl, pyr. (80 %); iv) LiN_3, DMF (no purification); v) 80 % AcOH (62 %).

Scheme 2

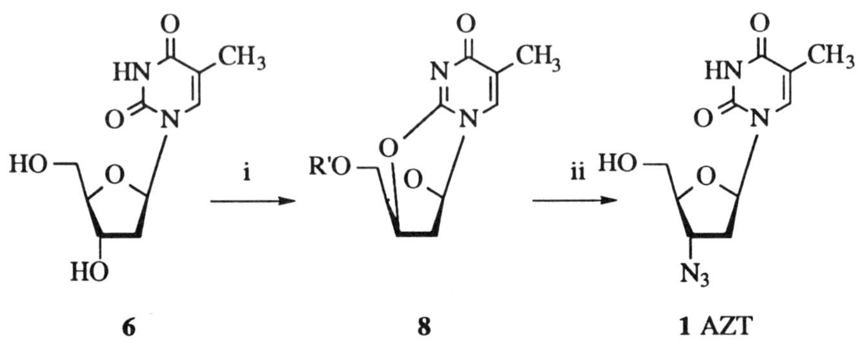

Reagents and conditions: i) $ClCHFCF_2NEt_2$, DMF, 70 °C (40 %); ii) LiN_3, NH_4Cl, 130 °C (55 %).

Owing to the very low solubility of compound **8** in most organic solvents its use as a 5'-protected derivative would be more suitable for further transformations. Along this line, the 5'-trityl derivative **8a** was prepared by reaction of the 5'-trityl-3'-mesylate **10** with a non nucleophilic base such as potassium phthalimide [9]. After opening with sodium azide and detritylation,

AZT was obtained in 33% overall yield (Scheme 3).

Scheme 3

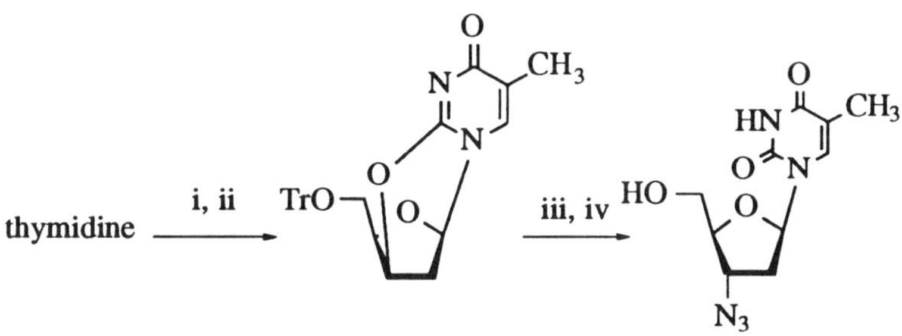

Reagents and conditions: i) TrCl, pyr. then MsCl, pyr. (84 %); ii) excess potassium phthalimide, DMF, 95 °C (71 %); iii) NaN_3, DMF-H_2O, reflux; iv) 80 % AcOH, 90 °C (33 %).

Taking account of these results, we decided to employ an ester group for the protection of the primary hydroxyl group. Rather than trying a selective esterification with an acylating agent we turned our attention to the Mitsunobu reaction [10]. This reaction, which is finding more and more applications in the carbohydrate field, allows the preparation of esters and phenoxy ethers from alcohols. The reaction is much slower with secondary than with primary hydroxyls. So, we expected that a selective protection of C-5' hydroxyl could be achieved, followed by an intramolecular reaction with the phenolic tautomeric form of the heterocyclic base. A 5'-protected anhydro derivative would be obtained in one step.

When a solution of thymidine (R=Me) in dimethylformamide (DMF) was treated with diethylazodicarboxylate (DEAD, 1.5 equiv.) in the presence of triphenylphosphine (Ph_3P, 1.5 equiv.) and benzoic acid (PhCOOH, 1.5 equiv.) consumption of the starting nucleoside was complete within 15 min. Rather than isolating the intermediate 5'-ester **13** it was found more convenient to perform *in situ* a further intramolecular Mitsunobu reaction (Scheme 4) by addition of Ph_3P (1.5 equiv.). After stirring the reaction mixture for 30-60 min. at room temperature the benzoylated 2,3'-anhydro derivative **8b** was precipitated by pouring the mixture into diethyl ether. Filtration and drying gave **8b** (R = Me) in 86% yield and good purity [11].

Scheme 4

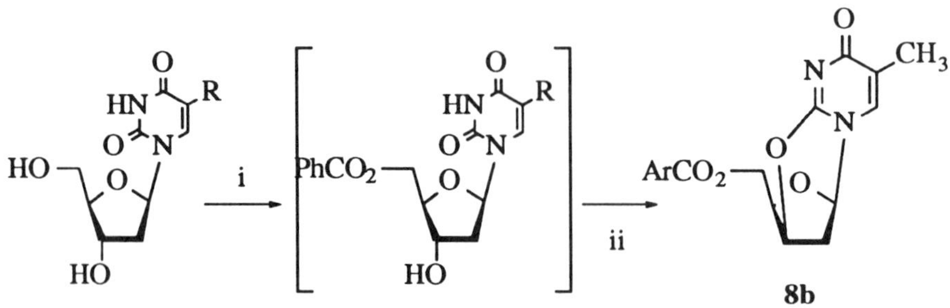

13 not isolated

Reagents and conditions: i) DEAD, PPh_3, $PhCO_2H$, DMF; ii) DEAD, PPh_3, DMF.

This tandem Mitsunobu sequence was shown to be also efficient when various aromatic carboxylic acids were employed (Table 1) [11] or starting from 2'-deoxyuridine (R= H).

Table 1. Preparation of Protected 2,3'-Anhydro Derivatives of Thymidine

Ar	Product	Isolated yield %	M. p. °C
Phenyl	**8b**	86	242
4-Methoxyphenyl	**8c**	86 (76[a])	260
2-Furyl	**8d**	75	260
4-Nitrophenyl	**8e**	66[a]	>300
3,5-Dinitrophenyl	**8f**	90	258
3,4-Dimethoxyphenyl	**8g**	84	240
2-Bromophenyl	**8h**	65	238
2,6-Dichlorophenyl	**8j**	76	222

[a] The product was isolated by cooling the reaction mixture and filtration of the resulting precipitate.

These 5'-*O*-aroyl-2,3'-anhydro-thymidine derivatives were employed as precursors for AZT synthesis. For this purpose the azido group was introduced at the C-3' α position by opening the anhydro ring by an azide ion. At this stage the use of 5'-*O*-benzoyl was shown to afford a mixture containing a moderate (c.a. 10%) amount of free AZT and benzoic acid as by-products. These impurities were considered to arise from nucleophilic cleavage of the ester and furthermore necessitated a more complex final purification resulting in lower overall yield. This unfavorable side reaction was easily suppressed by using *p*-methoxy-benzoyl to protect the 5'-hydroxyl group. In the presence of the electro-donating substituent, the ester did not react with the azide ion, thus allowing the desired opening of the anhydro ring to occur in nearly quantitative yield [12].

The final deprotection was then easy to perform under mild basic conditions (Scheme 5) which are particularly well fitted for the synthesis of dideoxynucleosides which are rather acid sensitive species. This sequence of reactions afforded AZT in 70% yield from thymidine.

Scheme 5

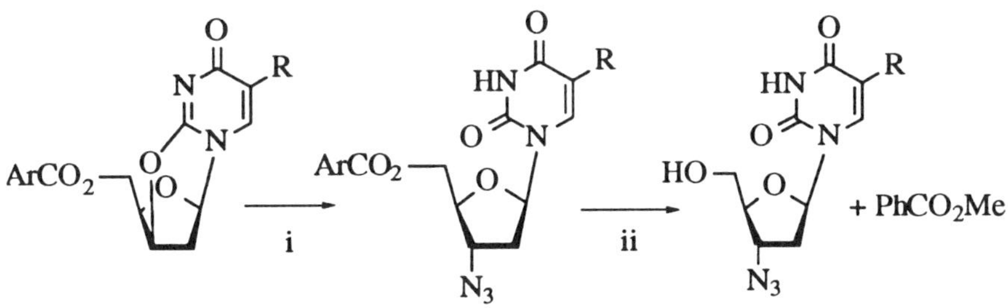

8c Ar = 4-MeOC$_6$H$_4$
R = Me

14

1 R = Me (AZT)
2 R = H (CS 87)

Reagents and conditions: i) NaN$_3$, DMF (90 %); ii) 1.15 eq. NaOMe, MeOH then IRN 77 resin [H$^+$ form] (94 %).

Final deprotection under mild basic conditions due to the use of benzoyl to protect the 5'-hydroxyl group allowed the development of new, high yield synthetic sequences to simplify the preparation of 2',3'-dideoxy-cytidine (ddC, **4**) and 2',3'-didehydro-2',3'-didehydro-2',3'-dideoxy-thymidine (d4T, **5**).

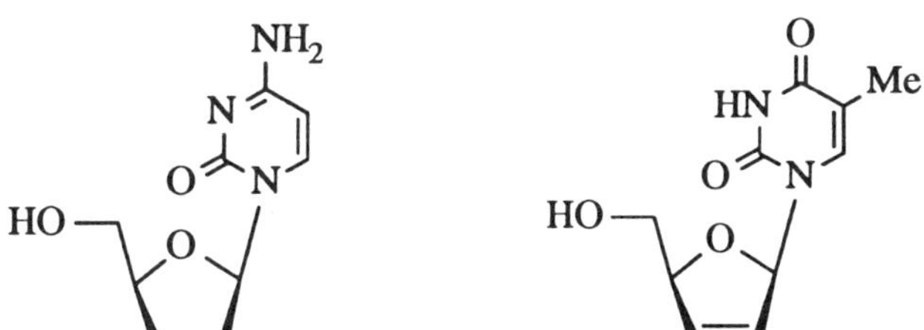

ddC **4** d4T **5**

These compounds, known for a long time, have recently been approved for clinical use in anti-HIV chemotherapy [3].

2. Preparation of 2',3'-didehydro-2',3'-dideoxy-cytidine (ddC)

Horwitz *et al.* first obtained ddC **4** starting from 2'-deoxycytidine [13]. As shown in Scheme 6 the key intermediate of this synthesis was the 3',5' oxetane **18**. Reaction with a strong base in DMSO afforded 2',3'-dideoxy-2',3'-didehydro-cytidine **19** hydrogenation of which over palladium-charcoal catalyst yielded **4** in 19 % overall yield.

Scheme 6

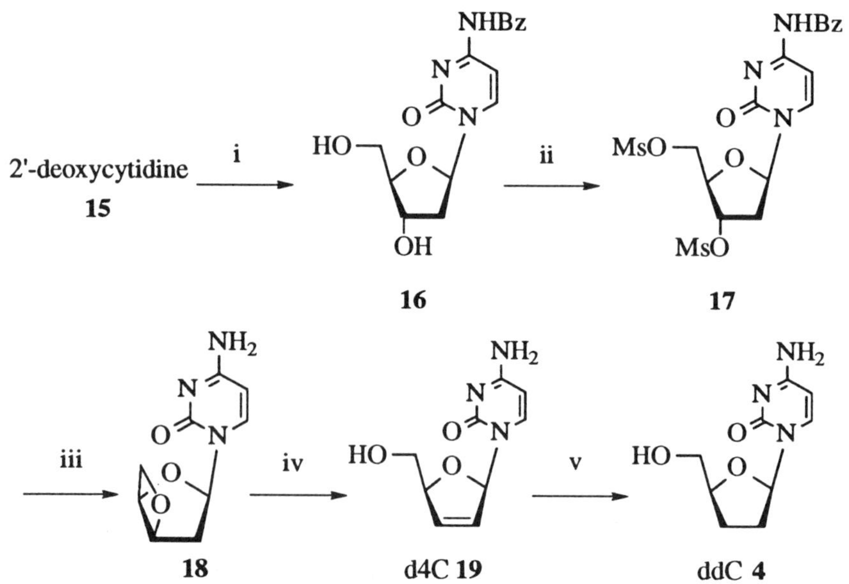

Reagents and conditions: i) Bz_2O, DMF (83 %); ii)MsCl, pyr. (91 %); iii) NaOH, EtOH - H_2O, reflux (58 %); iv) t-BuOK, DMSO, room temp. (61 %); v) H_2, Pd/C, EtOH (72 %).

A similar strategy was used in a number of preparations of ddC which proceeded *via* the didehydro-dideoxy parent nucleoside (d4C) although this intermediate was obtained by different pathways. Representative sequences started from cytidine [14-16] as well as from 2'-deoxy-cytidine [17].

Other synthetic routes have been designed to avoid unstable d4C derivatives as intermediates. Some of them were designed in order to offer the possibility of preparing various dideoxynucleosides besides ddC. These methods involved a coupling reaction between an activated nucleic base and a suitable dideoxypentofuranose synthon which could be obtained either from L-glutamic acid [18], 2-deoxy-D-ribose [19] or even from an achiral precursor [20]. A major drawback of these approaches is the production of anomeric mixtures due to the lack of stereoselectivity during the condensation between 2,3-dideoxysugar derivatives and heterocycles. Although this problem has been recently overcome by the introduction of a phenylselenyl group at the 2-α position of the 2,3-dideoxyfuranose precursor, [21] the preparation of this kind of compound required rather sophisticated reagents and materials which limits the possibilities of scale-up.

For these reasons the transformation of available nucleosides was preferred for the preparation of a precise target. Some straightforward routes have been accomplished: the deoxygenation of the 3'-position of 2'-deoxycytidine was performed either by a Barton reaction [22] or by hydrogenolysis of a 2',3'-dideoxy-3'-iodo-cytidine derivative [23]. ddC **4** was also obtained from 2',3'-dideoxy-uridine (ddU) by amination at C-4 of the uracil ring. This transformation was carried out *via* the 4-triazolyl, [24] the 4-chloro [25] or the 4-thio-pyrimidinone [26] derivatives.

For that purpose we considered as an attractive alternative the use of the readily available 2,3'-anhydro-5'-*O*-benzoyl-2'-deoxy-uridine **9a** which was obtained in 82% yield from 2'-deoxy uridine by the tandem Mitsunobu sequence described previously. Indeed this kind of protected anhydro derivative has proven to be a precursor of 3'-halogeno-2',3'-dideoxy-uridine derivatives. The action of hydrogen bromide was described as affording the 3'-bromo-*erythro* derivative in 60% yield, [27] but the 3'-iodo required for hydrogenolysis could not be prepared efficiently with excessively acidic hydrogen iodide. The use of freshly prepared anhydrous magnesium iodide allowed us the clean opening of the anhydro ring and the *erythro* -3'-iodo derivative **21** was obtained in more than 80% yield (Scheme 7).

Scheme 7

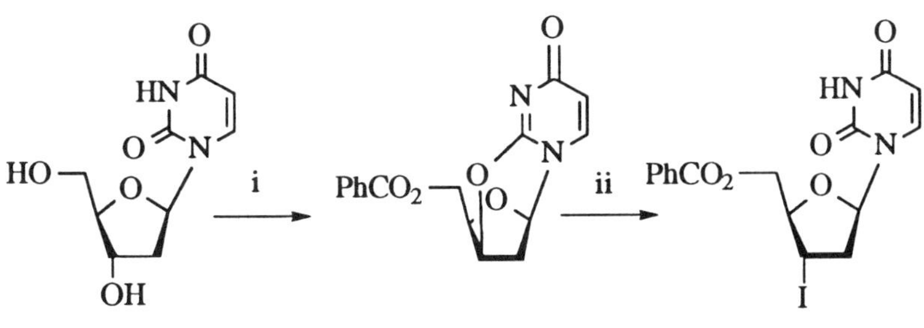

20 **9a** **21**

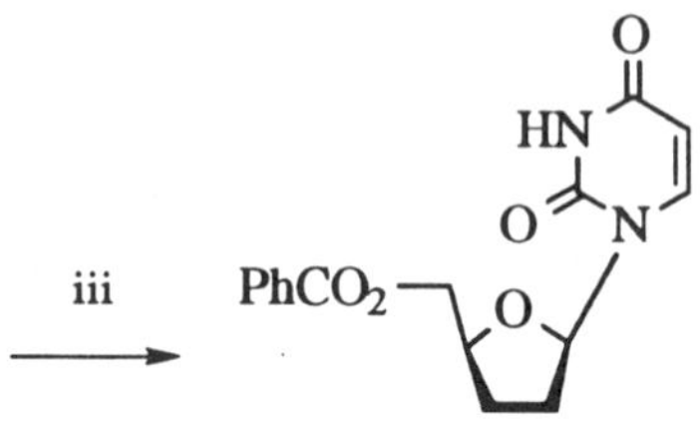

22

Reagents and conditions: i) DEAD, PPh_3, $PhCO_2H$, DMF then DEAD, PPh_3 (78 %); ii) MgI_2, toluene (83 %); iii) H_2, Raney Nickel, AcOEt (84%).

Further deiodination was performed by hydrogenation using the inexpensive widely used Raney nickel as catalyst. This material was found to be superior to the palladium based systems which were previously used in similar reactions because Raney nickel was an efficient scavenger of the co-produced hydrogen iodide. The resulting isolation procedure was exceedingly easy and the benzoylated ddU **22** was obtained in more than 55% overall yield from 2'-deoxyuridine. Deprotection of the 5'-hydroxyl group was not effected at this stage. It was found more convenient to proceed directly to the amination of the nucleic base. This transformation was carried out *via* the corresponding 4-tosylate (Scheme 8). This method was revealed as both easy to perform and efficient for introducing the methoxyamino group at C-4 of the uracil ring [29].

Scheme 8

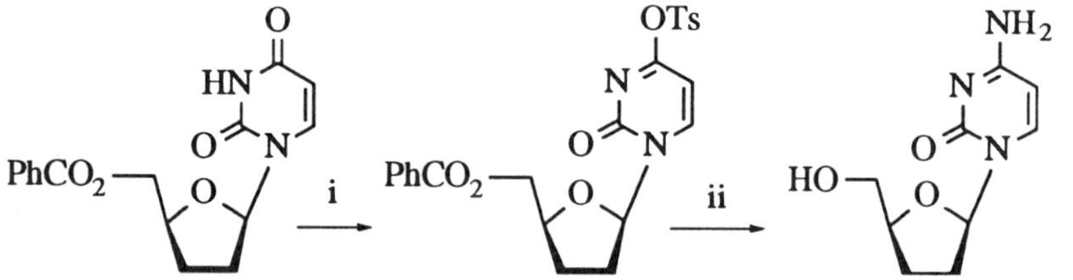

22 23 ddC 4

Reagents and conditions: i) TsCl, K_2CO_3, MeCN; ii) NH_4OH, H_2O / dioxane (80 %).

The unstable tosylate **23** was not isolated but reacted directly with aqueous ammonia at 55 °C. This resulted in simultaneous introduction of the amino group at C-4 and deprotection of the hydroxyl group. The separation of free ddC **4** from the by-products was done by treatment with ion exchange resin. This route required only five steps from 2'-deoxyuridine and afforded the target compound **4** in 43 % overall yield.

3. Preparation of 2',3'-didehydro-2',3'-dideoxy-thymidine (d4T)

Shortly after ddC **4**, the unsaturated nucleoside 1(2,3-dideoxy-β-*glycero*-pent-2 enofuranosyl)thymine (2',3'-didehydro-2',3'-dideoxy-thymidine, d4T, **5**) had been reported to have a potency against HIV comparable to AZT and a lower cytoxicity [30]. This compound has been recently approved for clinical use [3].

However the ready availability of d4T has been a problem because of the difficulties encountered in its synthesis. Recently several groups have addressed this problem starting either from non-nucleosidic [21, 31-34] on nucleosidic [30, 35-37] precursors. As pointed out above about the synthesis of ddC the first routes suffered from difficult multistep preparations of the suitability functionalized pentofuranose synthon. The second strategy, the use of base mediated elimination reactions for introducing the double bond into the sugar moiety of thymidine, [30, 38] gave poor yields due to the instability of d4T under strongly basic conditions [39].

Except for [37] in which the 2',3'-unsaturation was formed by treatment of 5'-*O*-triphenyl methyl-2,3'-anhydrothymidine with sodium in HMPA-THF at 70 °C, in most of recent approaches the double bond was created by elimination of a selenoxide [35-36]. This reaction which is well documented, [40] can be carrried out in particularly mild conditions but the problem lies in the preparation of the required phenylseleno nucleoside derivatives. The introduction of the phenylselenyl group in the carbohydrate residue was generally achieved by nucleophilic substitution of methylsulfonate, [35] or opening of 2,3'-anhydro nucleosides [36-41] as well as 2,2'-anhydro nucleosides [32] with phenyl selenide anion generated by reduction of dipenyldiselenide by sodium in HMPA-THF [36] or lithium aluminium hydride [32,35-36, 41]. One disadvantage of this method is the fact that in opening an anhydro derivative by phenylselenide anion a by-product is formed by nucleophilic attack at C-2 of the heterocyclic base [36,41]. Another drawback is the limited choice of hydroxyl protecting groups that are compatible with the basic and reductive conditions employed.

These difficulties were overcome by promoting an electrophilic assistance process for the opening of a 5'-protected 2,3'-anhydro derivative. Due to its own acidity phenylselenol itself was expected to be suitable. Since no basic or reductive conditions were required, we started from the 5'-*O*-benzoyl-2,3'-anhydro-thymidine derivative **8b** prepared from thymidine as outlined above.

Treatment of **8b** with phenyl selenol (1.5 equiv.) in DMF (1 h at reflux temperature) yielded the phenylseleno nucleoside **24** in high yield (Scheme 9) [42].

Oxidation under mild basic conditions (H_2O_2, pyridine) of such phenylseleno nucleosides was reported to be accompanied by reverse formation of the 2,3'-anhydro derivatives due to intramolecular nucleophilic substitution

[36]. Consequently we carried out the oxidation with H_2O_2 in the presence of acetic acid [40]. The resulting transient selenoxide spontaneously decomposed and protected d4T **25** was formed as the sole product in 88% yield.

The final deprotection of the 5'-hydroxyl group was possible under mild basic conditions which are the most favorable ones due to the instability of d4T **5** towards acidic reagents [30,36]. This is another advantage of the sequence outlined in Scheme 9 which allows the preparation of d4T in 67.5% overall yield from thymidine [42].

Scheme 9

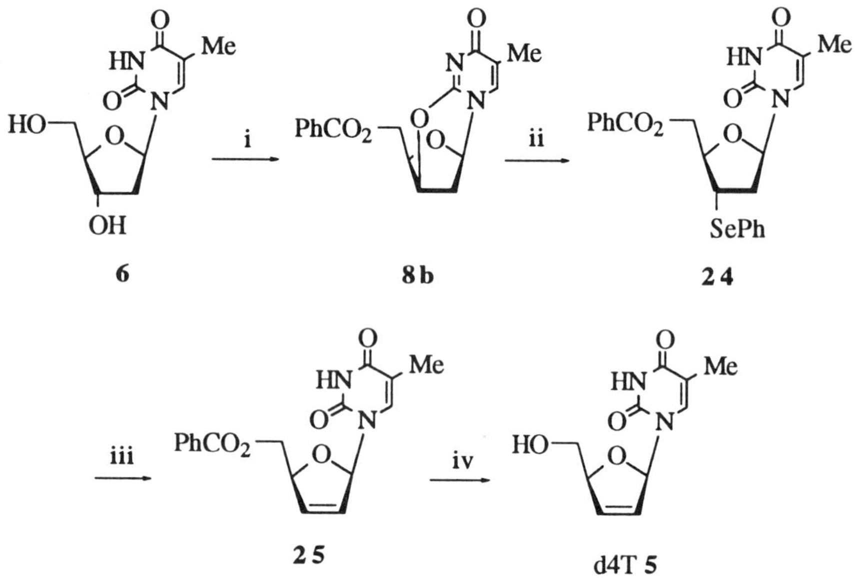

Reagents and conditions: i) DEAD, PPh_3, $PhCO_2H$, DMF then DEAD, PPh_3 (86 %); ii) PhSeH, DMF, reflux (96 %); iii) H_2O_2, AcOH, THF (88 %); iv) NaOMe 1.15 equiv., MeOH then H^+ resin (93 %).

III. SYNTHESES OF POTENTIAL ANTI-HIV AGENTS

In an attempt to understand the structural requirements for anti-HIV activity and in order to find more active and less toxic compounds, many 2',3'-dideoxynucleosides have been targeted towards reverse transcriptase inhibition. [43]. These compounds are of interest because they prove that a simple chemical modification in the sugar moiety can sometimes convert a normal substrate for DNA synthesis into a derivative with a potent activity against the virus.

Among the number of modified nucleosides which were prepared, some had been designed in order to associate the features which were present in known anti-HIV substances and could be responsible for their activity. Since they were the more active ones, pyrimidine nucleosides bearing an azido function in the sugar moiety were chosen as model compounds and a small modification was introduced in the structure. Compounds **26-32** have been designed as analogs of AZT and d4T. Some of them contain an azido group and a double bond, both functionalities which are present in AZT and d4T respectively.

Others ones were more simple analogues of AZT and their syntheses will be described at the end of this chapter.

Scheme 10

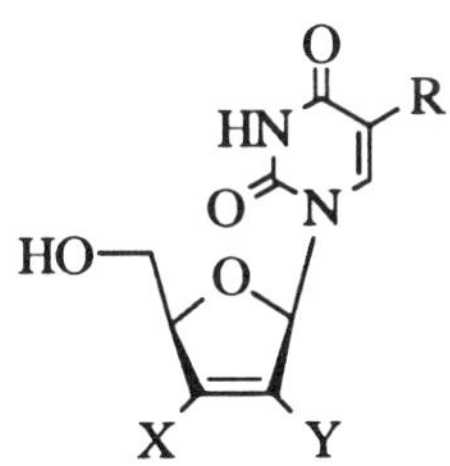

26 R = H, X = H, Y = CH_3
27 R = H, X = H, Y = CH_2N_3
28 R = CH_3, X = CH_3, Y = H
29 R = H, X = CH_3, Y = H
30 R = CH_3, X = CH_2N_3, Y = H
31 R = H, Y = CH_2N_3, Y = H

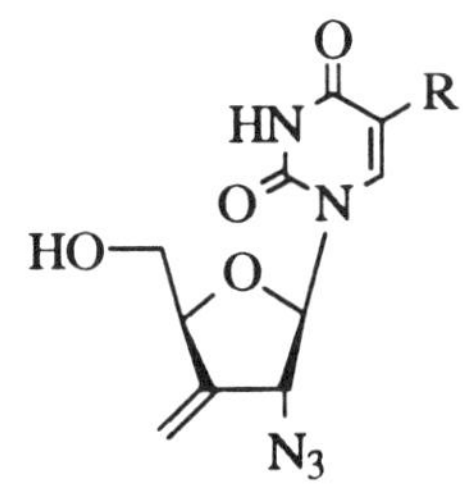

32 R = CH_3

A. Branched-Chain Nucleosides Derivatives

1. Branched-chain nucleosides at C-2'

Compounds **26** and **27** were prepared by modification of uridine *via* a common intermediate (Scheme 12) [46-47]. This compound was obtained from uridine by selective protection of 3'- and 5'- hydroxyl groups followed by oxidation and Wittig reaction on the stable 2'-keto nucleoside. Further deprotection-protection afforded **36** in good yield (Scheme 11).

Scheme 11

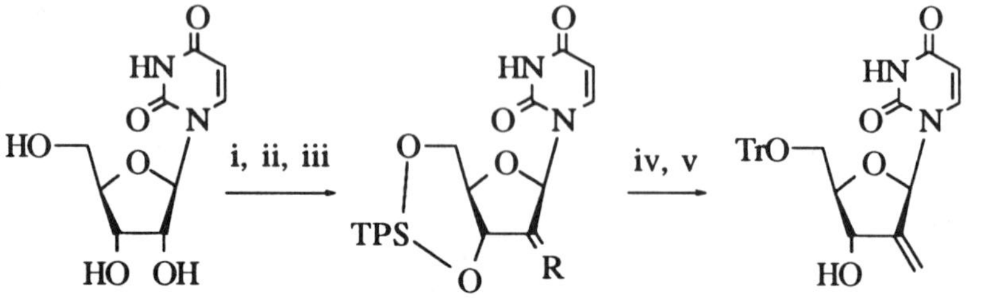

33 34 R = O 35 R = CH_2 36

TPS = 1,1,3,3-tetraisopropyl-1,3-disiloxanyl

Reagents and conditions: i) $TPSCl_2$, pyr.; ii) PDC, AcOH, molecular sieves, CH_2Cl_2 (76 % for two steps); iii) $Ph3P^+Me$, *s*-BuLi, THF, -78 °C (80 %); iv) $Et_3N.HF$, MeCN; v) Ph_3CCl, DMAP, pyr. (92 % for two steps).

Scheme 12

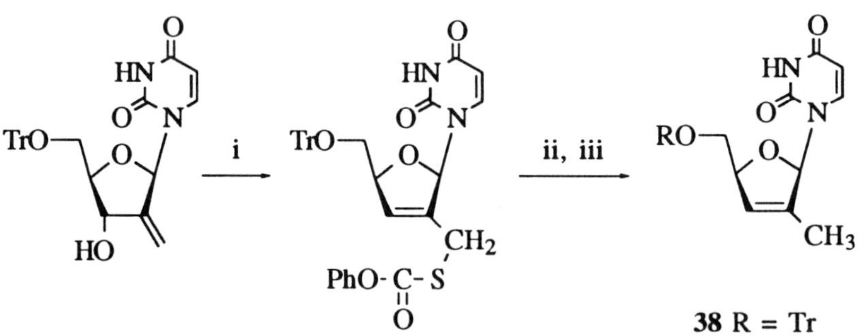

36 37 38 R = Tr 26 R = H

Reagents and conditions: i) PhO-CS-Cl, pyr. (70 %); ii) Bu_3SnH, AIBN, toluene, 100 °C (50 %); iii) $BF_3.OEt_2$, Et_3SiH, MeCN, -40 °C.

During the transformation of the 2'-OH of **36** into its phenoxythiocarbonyl derivative, an allylic rearrangement occurred and the 2',3'-unsaturated nucleoside **37** was obtained. Desulfurization of the latter under Barton's conditions afforded **38** which was deprotected to give the final product **26** (Scheme 12). Detritylation of **38** under classical acidic conditions was not possible because the allylic glycosidic bond was also cleaved. When reducing conditions (Et_3SiH, $BF_3.OEt_2$) were employed at low temperature, the target molecule **26** was obtained.

The intramolecular Mitsunobu reaction carried out with **36** as described above afforded the 2,3'-anhydro derivative **39** (Scheme 13). This compound was very stable and its rather difficult opening with lithium azide resulted in the formation of **40** by S_N2' reaction among other unidentified nucleoside derivatives. Detritylation was achieved as for **26** and **27** was obtained in acceptable yield.

Scheme 13

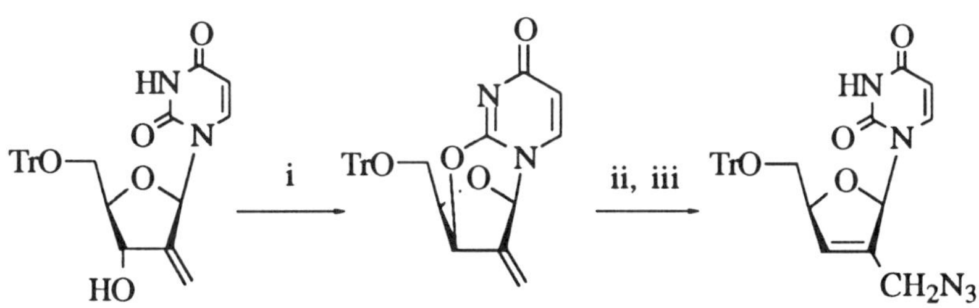

40 R = Tr
27 R = H

36 **39**

Reagents and conditions: i) PPh_3, DEAD, DMF (70 %); ii) LiN_3, DMF, 100 °C (12 %); iii) $BF_3.OEt_2$, Et_3SiH, MeCN, -40 °C (68 %).

2. Branched-chain nucleosides at C-3'

For the synthesis of C-3' branched-chain nucleosides, rather than following a linear route by modifying a nucleoside, another strategy was chosen because it was anticipated that, during the Wittig reaction with a 3'-keto nucleoside a retro-Michaël reaction could happen.

Thus a suitably functionalized and protected pentofuranosyl synthon **41** was prepared to couple with activated thymine or uracil. Further modifications of functionalities would afford the target molecules (Scheme 14).

Scheme 14

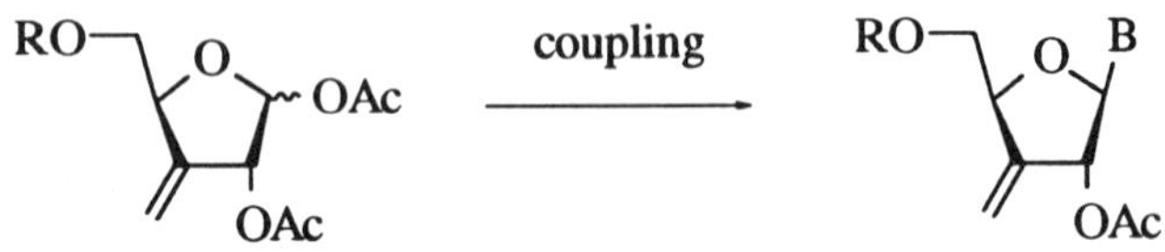

41 R = *p*-$MeOC_6H_4$ B = Thymine or Uracil

Due to the variety of experimental conditions which were employed during the synthesis, a permanent protective group was needed for the protection of the primary hydroxyl group. As the benzyl group could not be used the *p*-anisyl group was chosen. An acetate group was present at C-2 to control the stereochemical outcome of the coupling with the heterocyclic base and afford the β anomer.

This synthon **41** was readily prepared from commercially available 1,2-*O*-isopropylidene-α-D-xylofuranose **42** according to Scheme 15 [44].

Scheme 15

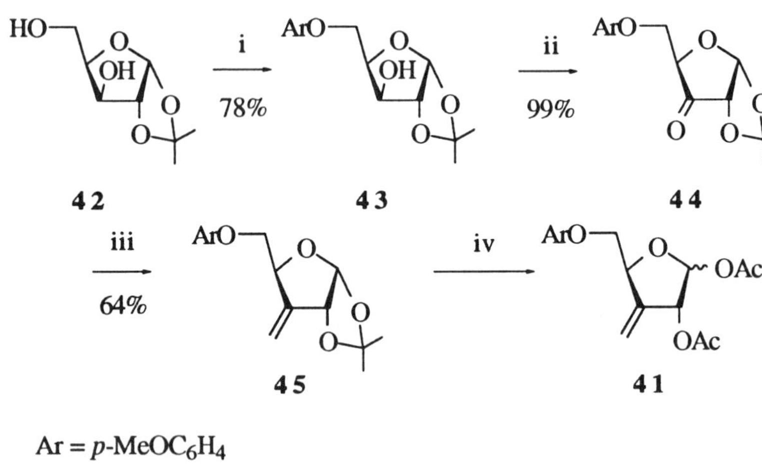

Ar = *p*-$MeOC_6H_4$

Reagents and conditions: i) p.$MeOC_6H_4OH$, PPh_3, DEAD, THF (78 %); ii) PDC, AcOH, molecular sieves, CH_2Cl_2 (99 %); iii) Ph_3P^+Me, I^-, s-BuLi, THF, -30 °C (64 %); iv) Amberlite H^+, AcOH - H_2O then Ac_2O, pyr.

The *p*-anisyl ether was selectively introduced by a Mitsunobu reaction on **42** involving *p*-methoxy-phenol. After oxidation, methylenation and acetolysis of the acetal group, an α/β mixture of **41** was obtained and employed in the coupling with activated thymine and uracil (Scheme 16). Due to the presence of the *p*-anisyl group, tin tetrachloride was not used as catalyst and trimethylsilyl trifluoromethanesulfonate was employed according to Vorbrüggen [45]. The protected 3'-deoxy-3'-methylidene nucleosides **46** were deacetylated with a catalytic amount of potassium cyanide in methanol (Scheme 16). After evaporation the crude products were dissolved in water and nonpolar impurities were extracted with toluene. After neutralization (ion exchange resin) the aqueous phase containing the nucleoside was evaporated to afford **47a** and **47b** in 78 % and 63% yield from **41** respectively [46].

Scheme 16

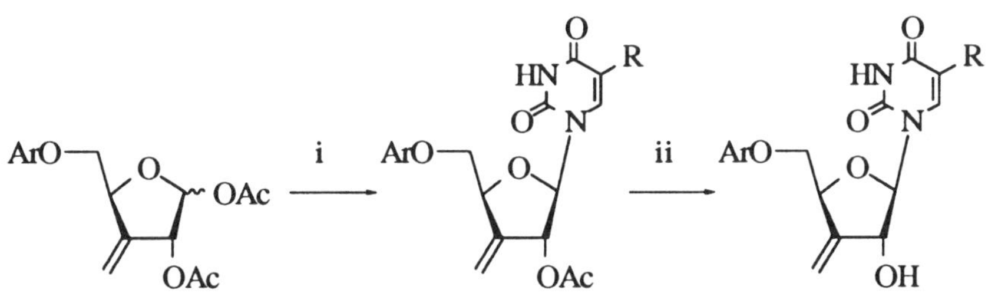

41 **46** **47a** R = Me
47b R = H

Reagents and conditions: i) thymine (or uracil), Me_3SiCl, HMDS, TMSOTf, 1,2-dichloroethane, 60 °C, 2 h; ii) KCN, MeOH.

As observed before in the preparation of C-2' branched nucleosides, an allylic rearrangement occurred during transformation of **47a,b** into their (phenoxycarbonyl)thio derivative and the 2',3'-unsaturated nucleosides **48a** and **48b** were obtained. Barton deoxygenation yielded **49a** and **49b** which were deprotected to give the final products **28** and **29** by oxidative cleavage of the *p*-anisyl group with cerium ammonium nitrate. Partial hydrolysis of the allylic glycosidic bond was observed during this step (Scheme 17).

Scheme 17

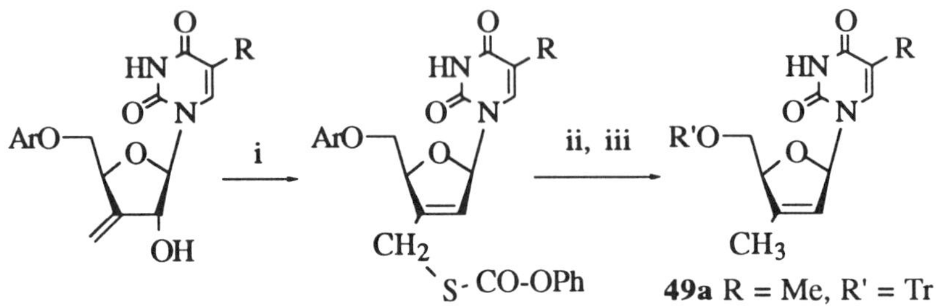

49a R = Me, R' = Tr
49b R = H, R' = Tr
28 R = CH_3, R' = H
29 R = H, R' = H

47a R = Me
47b R = H

48a R = Me
48b R = H

Reagents and conditions: i) PhO-CS-Cl, DMAP, pyr.; ii) *n*Bu_3SnH, AIBN, toluene, 100 °C (for two steps: 75 % for R = Me, 67 % for R = H); iii) CAN, MeCN / H_2O, 0 °C (55 % for R = Me, 20 % for R = H).

Introduction of the azido group in the sugar moiety was achieved by opening of 2,2'-anhydro derivatives which were obtained from **47a,b** by intramolecular Mitsunobu reaction. As outlined in Scheme 18 an allylic rearrangement occurred during the opening of the anhydro ring by lithium azide. Final deprotection afforded the target molecules **30** and **31** [46].

Scheme 18

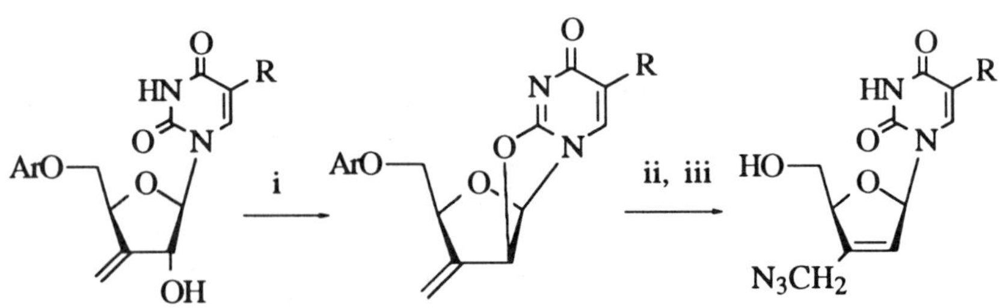

47a R = Me
47b R = H

50 R = Me
50b R = H

30 R = Me
31 R = H

Reagents and conditions: i) DEAD, PPh_3, DMF (70 % for R = Me, 94 % for R = H); ii) LiN_3, DMF; iii) CAN, MeCN / H_2O, 0 °C (for two steps: 46 % for R = Me, 19 % for R = H).

For the preparation of **32** the required introduction of the azido group at C-2' was achieved by palladium-catalyzed substitution of the allylic acetate of **46a** (Scheme 19) [47]. It is well established that this reaction occurs at the acetoxy bearing carbon with retention of configuration. The regio- and stereoselectivity of the reaction were determined by ^{1}H-NMR spectroscopy.

Scheme 19

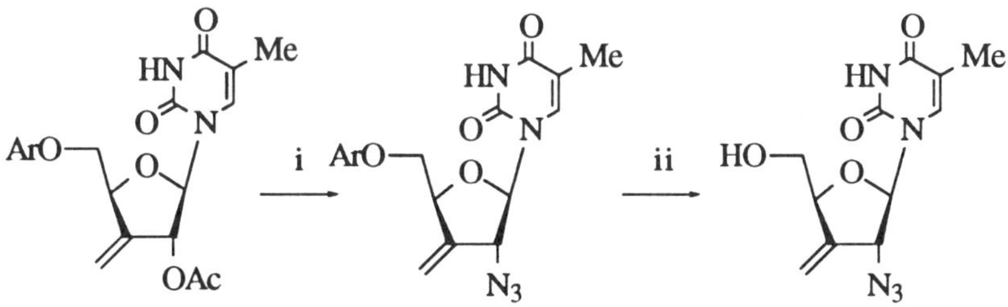

46a **51** **32**

Reagents and conditions: i) $Pd(OAc)_2$, PPh_3, NaN_3, THF / H_2O (65 %); ii) CAN, MeCN / H_2O (28 %).

B. Analogs of AZT

1. Stereoisomer at C-4'

Although AZT was the first modified nucleoside to be employed in the treatment of patients with AIDS, its mechanism of action is not fully understood. Some structure-activity relationship studies with AZT analogs against HIV are already available for base modified nucleosides **52**, 2'-azido-2',3'-dideoxy derivative **53** and stereoisomer **54** in which the azido group is *cis* to the thymine.

The synthesis of the C-4' epimer **55** of AZT was undertaken (Scheme 20). In this derivative the relative positions of the thymine base and the azido group are the same as in AZT and C-4' has the opposite configuration.

Scheme 20

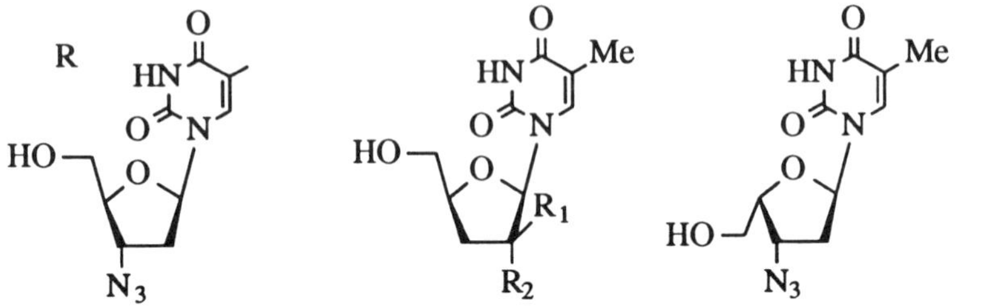

R = Me, AZT
52 R = F, Cl, Br, I

53 $R_1 = H, R_2 = N_3$
54 $R_1 = N_3, R_2 = H$

55

Since L-nucleosides are not naturally occurring, we decided to develop a versatile methodology which could be employed for the synthesis of other L-nucleoside analogs. For that purpose, L-arabinose, the sole L-pentose abundant in nature, was used as starting material. Rather than coupling thymine with a 2-deoxy-sugar derivative prepared from L-arabinose, which would afford an α/β mixture of nucleosides, we decided to employ a synthetic equivalent such as **56** (Scheme 21). In this synthon, a participating group is suitably positioned at C-2 to favor the formation of the α-L-nucleoside which could be transformed into the desired 2'-deoxy-α-L-nucleoside. Furthermore, in L-arabinose, the stereochemistry at C-3 would allow introduction of the azido group in the desired configuration by an S_N2 reaction after protection of the 5'-OH and activation of 3'-OH.

Scheme 21

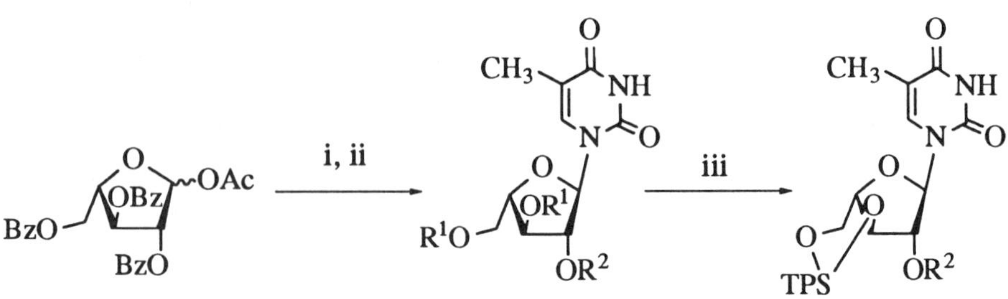

56

57 $R^1 = R^2 = Bz$
58 $R^1 = R^2 = H$

59 $R^2 = H$

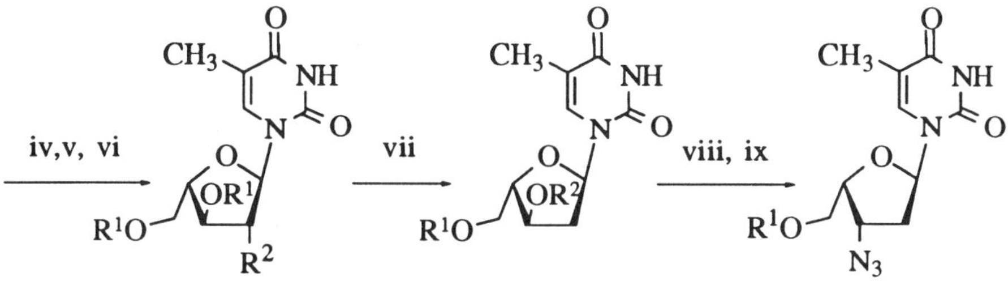

60 R^1 = TPS,
R^2 = OCSOPh
61 R^1 = TPS, R^2 = H
62 R^1 = H, R^2 = H

63 R^1 = Tr, R^2 = Ms

64 R^1 = Tr
55 R^1 = H

Reagents and conditions: i) thymine, TMSCl, HMDS, $SnCl_4$, CH_3CN, reflux (60 %); ii) NaOMe, MeOH (80 %); iii) $TPSCl_2$, pyr. (quant.); iv) PhOCSCl, DMAP, pyr. (68 %); v) $n$$Bu_3SnH$, AIBN, toluene, reflux (83 %); vi) $n$$Bu_4NF$, THF (41 %); vii) Ph_3CCl, pyr., 100 °C then CH_3SO_2Cl, -10 °C (48 %); viii) LiN_3, DMF, reflux (82 %); ix) 80 % AcOH, reflux, 15 min. (64 %).

An α/β mixture of **56** was readily obtained in two steps from L-arabinose [48] and directly condensed with activated thymin [45] to give exclusively the α anomer **57**. Debenzoylation with sodium methoxide in methanol afforded **58**.

Differentiation of the 2'-OH was achieved by selective protection of the two other hydroxyl groups with 1,3-dichloro-1,1,3,3-tetraisopropyldisiloxane [49] to form **59**. Barton deoxygenation [50] via the *O*-phenylthiocarbonate **60** was straightforward and **62** was obtained after desilylation and its structure confirmed by comparison with the known α–D anomer [51].

Selective protection of the 5'-OH and activation of the 3'-OH present in **62** were achieved in one pot. Nucleophilic displacement of the methylsulfonyl group of the product **63** with lithium azide was clean and the protected azide **64** was the sole product formed. It was detritylated under acidic conditions to give **55**. This compound was simultaneously obtained by another group using an analogous methodology [52].

2. Non-polar C-analog of AZT

As part of the structure-activity relationship studies directed towards determination of the precise role of the 3'-azido group in the exceptional anti-HIV activity of AZT, some 3'-carbon substituted 2',3'-dideoxy nucleosides were prepared as non-polar analogs of AZT [53-55].

Among them 3'-cyanomethyl-2',3'-dideoxy-thymidine **65** and 2',3'-dideoxy-3'(2-propynyl)thymidine **66** were designed because of the claimed isosteric relationship of their carbon side chain with an azido group (Scheme 22).[54,55]

We thought that the 1,2-propadienyl (allenyl) moiety would better fit this criterion and we decided to prepare the 3'-α-C-allenyl-2',3'-dideoxy-thymidine (**67**) as a novel non-polar analog of AZT.

Scheme 22

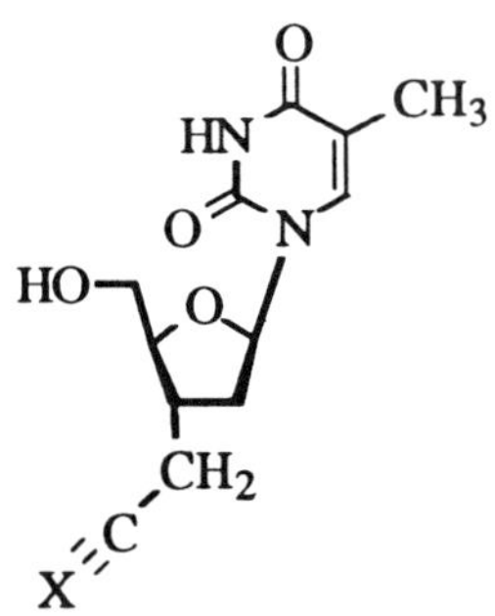

65 X = N
66 X = CH

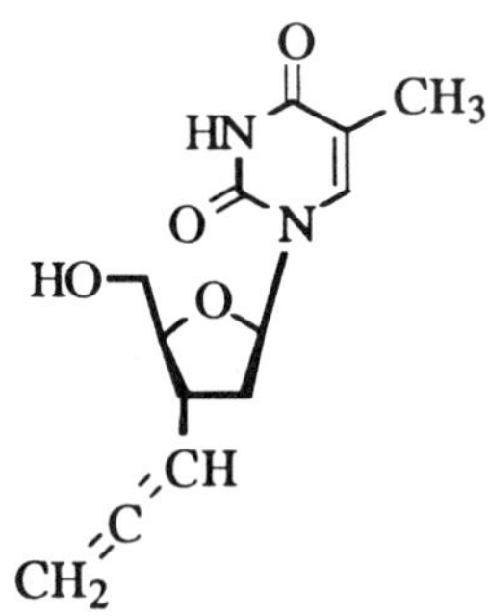

67

Rather than using ionic reactions for our purpose we turned our attention to free radical methodology. It was already known that reaction between triphenylprop-2-ynyl stannane **68** and alkyl bromides, or preferably iodides, in the presence of AIBN as radical initiator resulted in the formation of substituted allenes [56].

As mentioned above in the 2'-deoxyuridine series, the reaction between magnesium iodide and the 5'-*O*-benzoyl anhydro derivative **8b** afforded in good yield the 3-iodo derivative **69** (Scheme 23).

Further radical reaction of **68** with **69** was performed under different conditions and two compounds **70** and **71** were obtained in similar proportions. The use of triethylborane as an initiator was much more efficient than the thermal initiation with AIBN and the reaction could be conducted at room temperature [57].

The major product was identified as 3'-α-C-allenyl 5'-O-benzoyl-2',3'-dideoxythymidine (**71**). The configuration at C-3' was established by X-ray spectroscopy. Usual debenzoylation afforded the target molecule **67**.

Scheme 23

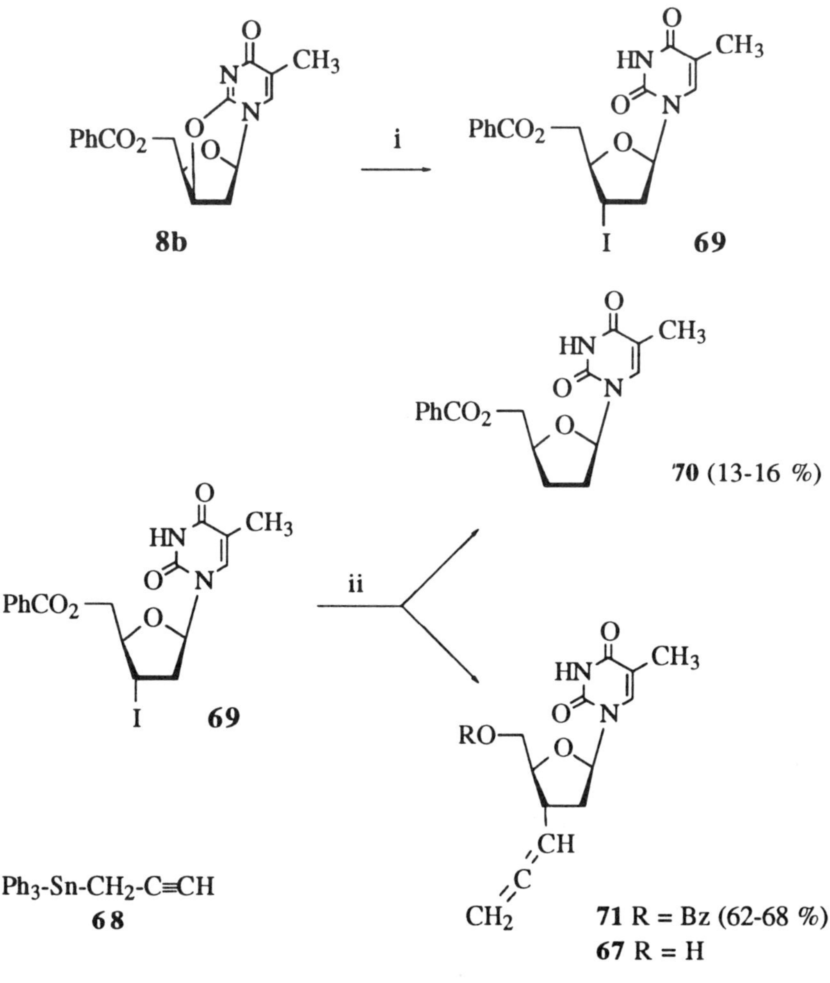

Reagents and conditions: i) MgI_2, 2 equiv., toluene, 110 °C (75 %); ii) **68**, AIBN, 80 °C or Et_3B, 25 °C.

IV. CONCLUSION

The search for improved antiviral compounds with utility in AIDS therapy is confounded by the unusual sensitivity of biological action to chemical structure. Even though a large number of compounds form the intense effort of research directed towards the synthesis of novel anti-AIDS drugs, new synthtic methodologies emerged although most active compounds are nucleosides containing modified carbohydrate moieties. The need therefore for efficient, stereocontrolled synthetic routes to potentially active compounds is critical.

Two general areas are described in this chapter, the synthesis of active pyrimidine compopunds utilizing a 5'-protected 2,3'-anhydro derivative and the design and preparation of new modified nucleosides as potential anti-HIV agents focusing on the use of anhydro derivatives as intermediates. The key to the synthesis of active pyrimidine compounds is deccribed as the easy and efficient synthesis of the 2,3'-anhydro compounds resulting from a judicious choice of the 5'-protecting group. Exploitation of the interaction between the pyrimidine base and the sugar to allow stereocontrolled transformations of the carbohydrate moiety permitted very efficient syntheses including those of AZT and d4T in 73% and 64% overall yield from thymidine respectivelly.

Conditions are described for the synthesis of the inmportant C-2' and C-3' branchged chain nucleoside analogs, the latter utilizing an efficient pantafuranosyl synthon rather than the more traditional linear route. Other important compounds which are also described include derivatives of the C-4' stereoisomer and the non-polar 3'-α-C-allenyl azido isostere of AZT.

REFERENCES

1. Perigaud C, Gosselin G, Imbach J-L. Nucleoside analogues as Chemotherapeutic Agents: A Review. Nucleosides & Nucleotides 1992; 11: 903-945.
2. Barré-Sinoussi F, Chermann JC, Rey F, Nugeyre MT, Chamaret S, Gruest J. Dauget C, Axler-Blin C, Vezinet-Brun F., Rouzioux C, Rozenbaun W, Montagnier L. Isolation of a T-lymphotropic Retrovirus from a Patient at Risk for Acquired Immune Deficiency Syntdrome (AIDS). Science 1983; 220:868-871.
3. De Clercq E. Toward Improved Anti-HIV Chemotherapy:Therapeutic Strategies for Intervention with HIV Infections. J. Med. Chem. 1995; 38:2491-2517.
4. Nasr M, Litterts C, Mac Gowan J. Computer-assisted Structure-activity Correlations of Dideoxynucleoside Analogs as Potential Anti-HIV Drugs. Antiviral Res. 1990; 14:125-148.
5. Michelson AM, Todd AR. Deoxyribonucleosides and Related Compounds. Part. V. Cyclothymidine and other Thymidine Derivatives. J Chem Soc. 1955; 816-823.
6. Horwitz JP, Chua J, Noel M. Nucleosides. V. The Monomesylates of 1-(2-deoxy-β-D-lyxofuranosyl)thymine. J.Org Chem. 1964; 29:2076-2078.
7. Kowollik G, Gaertner K, Langen P. Eine Einfache Methode zur Direkten Darstellung von O^2.3'-Cyclo-2'-desoxy-Nucleosiden. Tetrahedron Lett.

1969; 3863-3865.
8. Rao TS, Reese CB. A Novel One-step Procedure for the Conversion of Thymidine into 2,3'-Anhydrothymidine. J Chem Soc Chem Commun. 1989; 997
9. Glinsky PG, Khan MS, Kalamas RL. Nucleotide Synthesis. IV. Phosphorylated 3'-amino-3'-deoxythymidine and 5'-amino-5'-deoxythymidine and Derivatives. J Org Chem. 1973; 38:4299-4305.
10. Mitsunobu O. The Use of Diethyl Azodicarboxylate and Triphenylphosphine in the Synthesis and Transformation of Natural Products. Synthesis 1981; 1-28.
11. Czernecki S, Valéry JM. One-step Conversion of Thymidine in 2,3'-Anhdydro Derivatives. J Chem Soc Chem Commun. 1990; 801-802.
12. Czernecki S, Valéry JM. An Efficient Synthesis of 3'-Azido-3'deoxythymidine (AZT). Synthesis 1991; 239-240.
13. Horwitz JP, Chua J, Noel M, Donatti JT. Nucleosides. XI. 2',3'-Dideoxycytidine. J Org Chem. 1967; 32:817-818.
14. Kawana M, Yamasaki N, Nishikawa M, Kuzuhara H. The synthesis of 2',3''-dideoxycytidine and its 2'-azido Analogue. Chemistry Lett. 1987; 2419-2422.
15. Chu CK, Bhadti VS, Doboszewski B, Gu ZP, Kosugi Y, Pullaiah KC, Van Roey P. General Syntheses of 2'3'-dideoxynucleosides and 2',3'-didehydro-2'3'-dideoxynucleosides J Org Chem. 1989; 54:2217-2225.
16. Manchand PS, Belica PS, Holman M J, Huang TN, Maehr H, Tam SYK, Yang RT. Syntheses of the Anti-AIDS Drug 2',3'-Dideoxycytidine from Cytidine. J Org Chem. 1992; 57:3473-3478.
17. Bhat V, Ugarkar BG, Grimm K, Stocker E, Domenico PA, Sayeed VA, Kosora N. An Improved Synthesis of 2'3'-dideoxycytidine. Nucleosides & Nucleotides 1990; 9:1061-1065.
18. Farina C, Benigni DA. A New Synthesis of 2',3'-Dideoxynucleosides for AIDS Chemotherapy. Tetrahedron Lett. 1988; 29:1239-1242. See also: Okabe M, Sun RC, Tam SYK, Todaro LJ, Coffen DK Synthesis of Dideoxynucleosides ddC and CNT from Glutamic Acid, Ribonolactone and Pyrimidine Bases. J Org Chem. 1988; 53:4780-4786.
19. Motawia MS, Pedersen EB. A New Route to 2',3'-dideoxycytidine. Liebigs Ann Chem. 1990; 599-602.
20. Jung ME, Castro C, Gardiner JM. Rapid Synthesis of 2',3'-dideoxycytidine (ddC) from a simple achiral precursor. Tetrahedron Lett. 1991; 32:5717-5720.
21. Beach JW, Kim HO, Jeong LS, Nampalk S, Islam Q, Ahn SK, Babu JR, Chu CK. A Highly Stereoselective Synthesis of Anti-HIV 2',3'-dideoxy-and 2',3'-didehydro-2',3'-dideoxynucleosides. J Org Chem. 1992; 57:3887-3894.

22. Prisbe EJ, Martin JC. A Novel and Efficient Preparation of 2'3'-dideoxynucleosides. Synthetic Commun. 1985; 15:401-409.
23. Kaskar B, Markovac A. A New Synthesis of 2',3'-Dideoxycycidine. J Heterocyclic Chem. 1989; 26:1531-1533.
24. Lin TS, Chen MS, McLaren C, Gao YS, Ghazzouli I, Prusoff WH. Synthesis and Antiviral Activity of Various 3'-Azido, 3'-Amino, 2',3'-unsaturated, and 2',3'-dideoxy Analogues of Pyrimidine Deoxyribonucleosides against Retroviruses. J MedChem. 1987; 30:440-444.
25. Kaneko M, Kimura I, Kamogari M. Preparation of 2',3'-dideoxyctidine Derivatives as Drugs for treatment of AIDS. Jpn. Kokai Tokkyo Koho JP 63, 284, 193 [88, 284, 193] Chem. Abstract 1989; 111:23917d.
26. Yoshioka H, Ishida S, Kojima E, Murakami K. Preparation of 2',3'-dideoxycytidine Derivatives as Antiviral Agents. Jpn. Kokai Tokkyo Koho JP 01, 100, 191 [89, 100, 191]. Chem. Abstract 1989; 111:154310.
27. Joecks A, Köppel H, Schleinitz KD, Cech D. N. M. R. Spectroscopical Investigations on the Conformational Behaviour of some 2'- and 3'-Halogen-substituted Pyrimidine Nucleosides. J Prakt Chem. 1983; 325:881-1048.
28. Chu CK, Schinazi RF, Ahn MK, Ullas GV, Gu ZP. Structure-Activity Relation ships of Pyrimidine Nucleosides as Antiviral Agents for Human Immunodeficiency Virus Type 1 in Peripheral Blood Mononuclear Cells. J Med. 1989; 32:612-617.
29. Czernecki S, Le Diguarher T, Valéry JM. Synthesis and Anti-HIV-1 Activity of Base Modified Analogues of 3'-Azido-2',3'-Dideoxy thymidine (AZT). Nucleosides & Nucleotides 1993, 12:369-380.
30. Mansuri MM, Starret JE, Ghazzouli I, Hitckock M JM , Sterzycki RZ, Brankovan V, Lin TS, August EM, Prusoff WH. Sommadossi JP, Martin C. 1-(2,3-dideoxy-b-D-glycero-pent-2-enofuranosyl)thymine. A Highly Potent and Selective Anti-HIV Agent. J.Med Chem. 1989; 32:461.
31. Wilson LJ, Liotta DA. A General Method for controlling Glycosylation Stereochemistry in the synthesis of 2'-deoxyribose Nucleosides. Tetrahedron Lett. 1990; 31:1815-1818.
32. Vargeese C, Abushanab E. A Pratical and Stereospecific Approach to the Synthesis of 3'-deoxy-2',3'-didehydro-thymidine (d4T). Nucleosides & Nucleotides 1992; 11:1549-1559.
33. Kim CV, Misco PF. Facile, highly Stereoselective Synthesis of 2',3'-di-deoxy- and 2',3'-didehydro-2',3'- dideoxy Nucleosides via a Furanoid Glycal Intermediate. Tetrahedron Lett. 1992; 33:5733-5736.
34. Lipshutz BH, Stevens KL, Lowe RF. A Novel Route to the anti-HIV Nucleoside d4T. Tetrahedron Lett. 1995; 36:2711-2712.
35. Vial JM, Agback P, Chattopadhyaya J. A New Synthesis of 1-(2,3-di-

deoxy-β-D-glycero-pent-2-enofuranosyl)-Thymine. A highly Potent and Selective Anti-HIV Agent. Nucleosides & Nucleotides 1990; 9:245-258.
36. Cosford NDP, Schinazi R. Selenium Nucleophiles for the Preparation of Antiviral Nucleosides. J Org Chem. 1991; 56:2161-2165.
37. Cosford NDP, Schinazi RF. A Short Synthesis of 2',3'-didehydro-3'-deoxythymidine. Nucleosides & Nucleotides 1993; 12:149-155.
38. Martin JC, Mansuri MM, Starett JE, Sommadossi JP, Brankovan V, Ghazzouli I, Ho MT, Hitchcock MJM. Nucleosides & Nucleotides 1989; 8:841.
39. Horwitz JP, Chua J, Da Rooge MA, Noel M, Klundt I. Nucleosides IX. The formation of 2',3'-Unsaturated Pyrimidine Nucleosides *via* a Novel β-Elimination Reaction. J Org Chem. 1966; 31:205-211.
40. Grieco PA, Migashita M. Organoselenium Chemistry. α-Phenylseleno Lactones. A new General Route to the Synthesis of Fused α-Methylene Lactones. J Org Chem. 1974; 39:120-122.
41. Haraguchi K, Tanaka H, Maeda H, Itoh Y, Saito S, Miyasaka T. Selenoxide Elimination for the Synthesis of Unsaturated-Sugar Uracil Nucleosides. J Org Chem. 1991; 56:5401-5408.
42. Czernecki S, Valéry JM. Efficient Transformation of Thymidine into 2',3'-didehydro-2',3'-dideoxy-thymidine (D4T) Involving Opening of a 2,3'-Anhydro Derivative by Phenylselenol. Nucleosides & Nucleotides 1995; 14:1227-1232.
43. Herdewijn P, De Clercq E. Dideoxynucleoside Analogues as Inhibitors of HIV Replication. In:De Clercq E. ed. Design of Anti-AIDS Drugs. Amsterdam: Elsevier, 1990:141-174.
44. Czernecki S, Ezzitouni A, Krausz P. Synthesis of 3'-deoxy-2',3'-didehydro-3'-methylthymidine. Synthesis 1990; 651-653.
45. Vorbrüggen H. Bennua B. Nucleoside Syntheses, XXV. A New Simplified Nucleoside Synthesis. Chem Ber. 1981; 114:1279-1286.
46. Czernecki S, Ezzitouni A. Synthesis of Branched Nucleosides Closely Related to AZT Involving SN_2' Opening of Anhydronucleosides. Tetrahedron Lett. 1993; 34:315-318.
47. Czernecki S, Ezzitouni. A. Synthesis of Various 3'-Branched 2',3'-Unsaturated Pyrimidine Nucleosides as Potential Anti-HIV Agents. J Org Chem. 1992; 57:7325-7328.
48. Czernecki S, Le Diguarher T. Synthesis of 1-(3'-azido-2',3'-dideoxy-α-L-threo-pentofuranosyl)thymine as a Potential Anti-HIV Agent. Synthesis 1991; 683-686.
49. Markiewicz WT. Tetraisopropyldisiloxane-1,3-diyl, a Group for Simultaneous Protection of 3'- and 5'-Hydroxy Functions of Nucleosides. J Chem Research (S) 1979;24-25.
50. Barton DHR, McCombie SW. A New Method for the Deoxygenation of

Secondary Alcohols. J Chem Soc Perkin Trans. I 1975; 1574-1585.

51. Palomino E, Meltsner BR, Kessel D, Horwitz JP. Synthesis and in Vitro Evaluation of Some Modified 4-Thiopyrimidine Nucleosides for Prevention or Reversal of AIDS-Associated Neurological Disorders. J Med Chem. 1990; 33:258-263.
52. Génu-Dellac C, Gosselin G, Imbach J-L. Synthesis of New 2'-Deoxy-3'-substituted-a-L-threo-pentofuranonucleosides of Thymine as Potential Antiviral Agents. Tetrahedron Lett. 1991; 32:79-82.
53. Chu CK, Doboszewski B, Schmidt W, Ullas GV. Synthesis of Pyrimidine 3'-Allyl-2',3'-dodeoxyribonucleosides by Free Radical Coupling. J Org Chem. 1989; 54:2767-2769.
54. Fiandor J, Huryn DM, Suboski B, Todaro LJ, Tam S. Synthesis of 3'-C-Substituted-2',3'-dideoxynucleosides as Potential Anti-AIDS Agents. Nucleosides & Nucleotides 1989;8:1107-1108.
55 Fiandor S, Tam SY. Synthesis of 3'-Deoxy-3'-(2-propynyl)thymidine and 3'-Cyanomethyl-3'-deoxythymidine, Analogs of AZT. Tetrahedron Lett. 1990;31:597-600.
56. Baldwin JE, Adlington RM, Basak A. Allene Transfer Reactions. A New Synthesis of Terminal Allenes. J Chem Soc Chem Commun. 1984; 1284-1285.
57. Becouarn S, Czernecki S, Valéry JM. Radical Mediated Synthesis of 3'-α-C-allenyl-2',3'-Dideoxythymidine as a Non-polar Analogue of AZT. Tetrahedron Lett. 1995;36:873.876.

13
Carbohydrate Templates for the Synthesis of 3′-Heteronucleosides

J. Warren Beach
University of Georgia
Athens, Georgia

I. INTRODUCTION

The search for compounds useful for the treatment of viral diseases such as AIDS and Hepatitis B has lead to the design and synthesis of a number of novel nucleoside analogues. In general, these compounds are modified natural 2'deoxynucleoside analogues where the 3'hydroxyl group has been replaced or eliminated entirely. Those that have successfully negotiated the clinical trial maze include AZT (3'-azidothymidine), DDC (2',3'-dideoxycytidine), DDI (2',3'-dideoxyinosine) and D_4T (2'3'-didehydro-3'deoxythymidine) (Figure 1). In 1989 a new class of nucleoside analogues appeared in which the 3'-methylene of the dideoxy nucleosides (Figure 1) was replaced with a sulfur or an oxygen atom. These compounds were first synthesized by Belleau et al.[1] (O and S) and Norbeck et al. [2] (O) as racemic mixtures. In most biologically active compounds that are racemates, either all the activity lies in one enantiomer or both enantiomers have approximately equal activity but one enantiomer is more toxic.

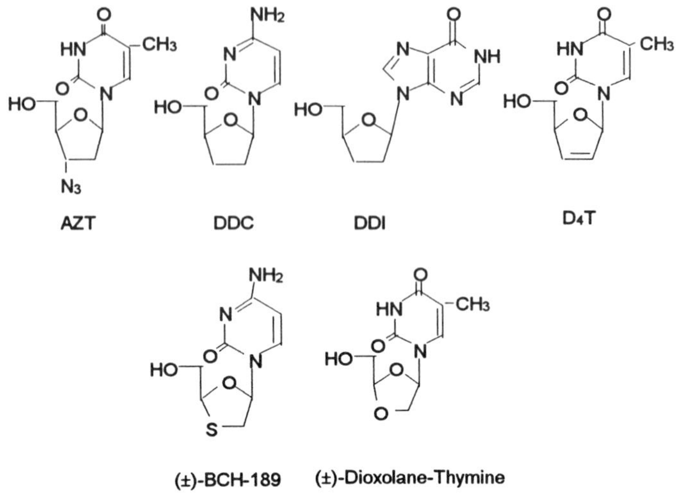

Figure 1. Structures of anti-HIV nucleosides

The initial reports of this new class of compounds set off a flurry of activity to synthesize or separate the individual enantiomers. Old paradigms die hard most of this activity was concentrated of the synthesis of the D-like enantiomer. The result from biological evaluation of the D enantiomer revealed a new area to the nucleoside chemist, that of L-nucleosides analogues as antiviral agents.

The purpose of this chapter is to review the strategies for the stereoselective synthesis and chiral resolution of the D and L enantiomers of the oxathiolane and dioxolane nucleosides and to review their biological activity, their structure activity relationships, their biochemical pharmacology and their cellular metabolism.

II. SYNTHESIS OF THE DIOXOLANE NUCLEOSIDES

It was determined that the dioxolane ring system of a 1,6-anhydrohexose would serve well as a chiral template for the dioxolane nucleosides. Thus 1,6-anhydromannose [3,4] was chosen for its ease of large-scale preparation and the configuration of its hydroxyl groups at 2,3 and 4 (2,3 *cis* and 3,4 *trans*) which allows for selective protection prior to oxidative cleavage (Figure 2).

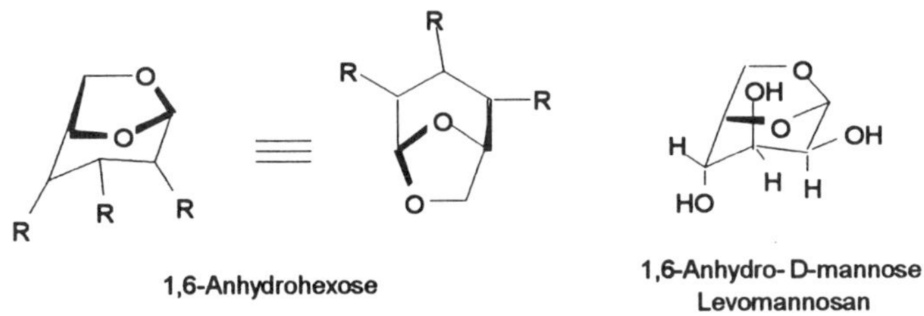

Figure 2. 1,6-Anhydro-D-mannose

1,6-Anhydro-D-mannose or levomannosan was originally synthesized by Hudson and co-workers [3] by the destructive distillation of ivory nut turnings and more recently on a preparative scale by Fraiser-Reid and co-workers [4] from D-mannose. In the later process the levomannosan is isolated as its 2,3-isopropylidene derivative (**1**). Benzoylation of the 4 position to give **2** (Scheme 1) [5] followed by removal of the isopropylidene group using dilute H_2SO_4 in aqueous dioxane at 75°C, gave the diol **3**.

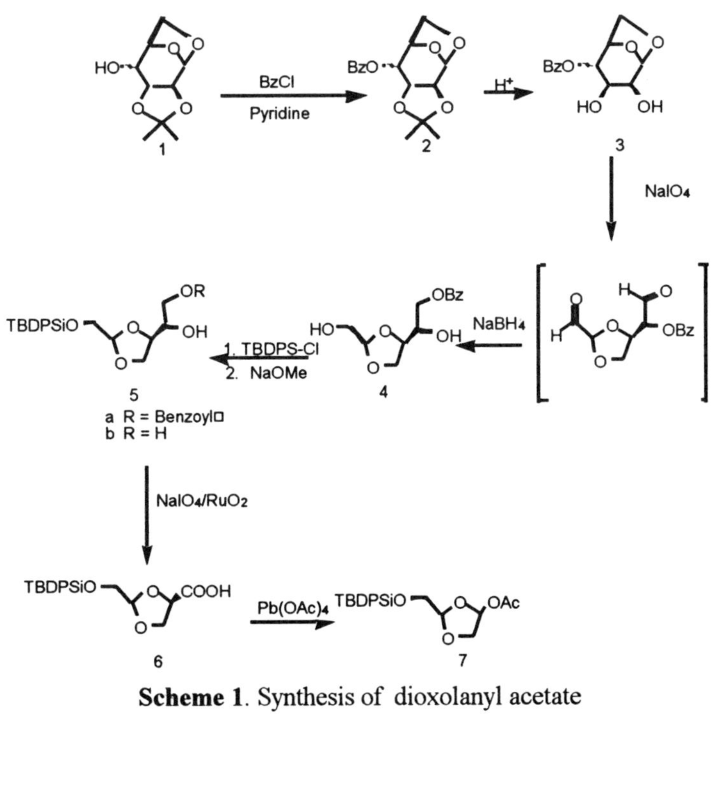

Scheme 1. Synthesis of dioxolanyl acetate

Treatment of the diol with $NaIO_4$ in aqueous methanol, followed by reduction with $NaBH_4$, gave the dioxolane derivative **4**. Note that the benzoate migrated from the secondary to the primary alcohol under the conditions of the reaction. This fortunate occurrence allowed the selective protection of the remaining primary alcohol with t-butyldiphenylsilyl chloride to give **5a**, which was deprotected using NaOMe in MeOH to give **5b**. The diol **5b** was converted to the acid by a process similar to that described by Nunez and Martin[6] using $NaIO_4$ and RuO_2 in a two-phase reaction system. The acid **6** was subjected to oxidative decarboxylation using $Pb(OAc)_4$/pyridine in ethyl acetate as described by Dhavale *et al.* [7] giving the acetate **7**. It was later found that dry THF was a better solvent, giving faster reactions and better yields.

Condensation of the acetate **7** with silylated thymine (Scheme 2) using TMSOTf as Lewis acid gave a 1:1.5 α:β mixture of the protected nucleosides **8** and **9**, which were separated by column chromatography over silica gel.

Deprotection with TBAF gave the desired free α (**10**) and β (**11**) dioxolane-T[5].

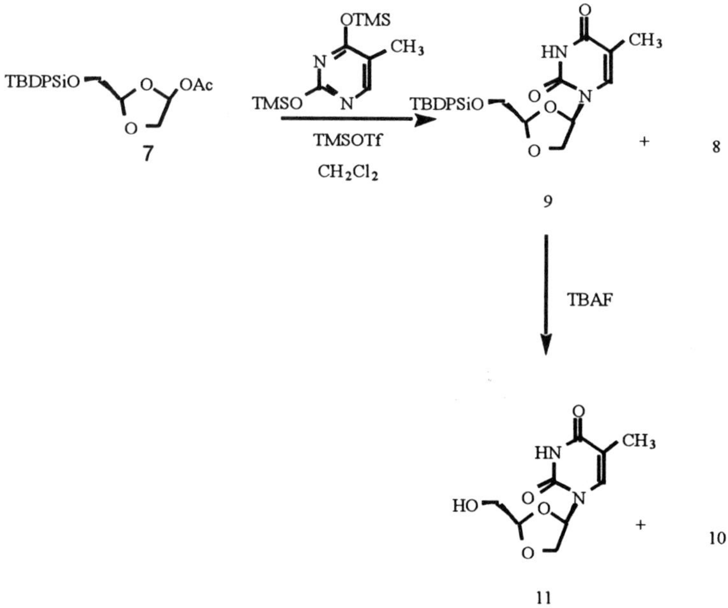

Scheme 2. Synthesis of (-)-dioxolane-T

The 1-acetate was also condensed with other pyrimidine and purine bases in order to explore the structure-activity relationships of this class of nucleosides [8,9]. The α and the β anomers were subjected to NOE studies in order to confirm the anomeric assignments due to the unusual activity observed for the α anomers of this series .

In the search for an L-sugar that would serve as a chiral template for the synthesis of both the dioxolane as well as oxathiolane nucleosides, L-mannose was considered. L-mannose, however, suffers from several disadvantages in that it is not readily available or easily synthesized from readily available starting materials.

The solution was found in the 5 "epimer" of D-mannose, namely L-gulose **17** (Scheme 3), which can be readily obtained from commercially available L-gulono-γ-lactone **14** [10].

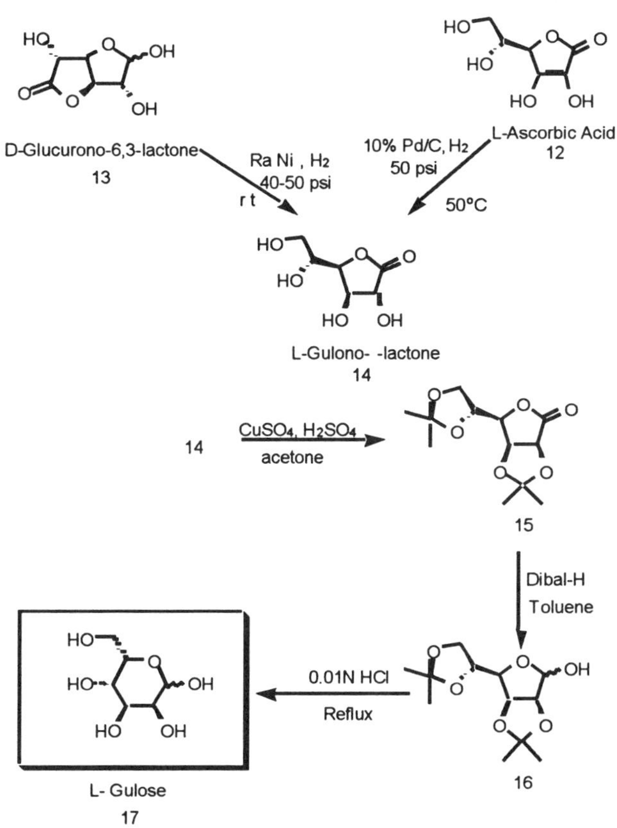

Scheme 3. Synthesis of L-gulose

The lactone may also be obtained by catalytic hydrogenation of L-ascorbic acid [11] **12** over 10% Pd/C at 50°C or D-glucurono-6,3-lactone [12] **13** over Raney nickel at room temperature. The latter method was found to be more efficient in that the nickel catalyst could be reused for up to one kilogram of starting material without noticeable changes in yield or reaction length. The complete reaction mixture was decanted from the catalyst, passed through a strongly acidic ion-exchange column to remove nickel ion and the eluant concentrated to give crystalline L-gulono-γ-lactone **14** in 85-90% yield. The lactone was isopropylidenated using acetone/$CuSO_4$/H_2SO_4 to give the crystalline di-*O*-isopropylidene derivative [10] **15** that was reduced with Dibal-H in toluene at -78°C to give the di-*O*-isopropylidene-L-gulofuranose **16** as a crystalline solid. This process could be run on a 250 gm scale with little difficulty.

To obtain L-gulose itself, the di-*O*-isopropylidene derivative was hydrolyzed with 0.01 N HCl at 100°C for approx. 2 h followed by cooling, neutralization with basic ion-exchange resin in the bicarbonate form, filtration and concentration to give L-gulose **17** as a syrup, which was crystallized from 95% ethanol [13].

L-Gulose **17** was converted the anhydrosugar directly by heating 2,3,5,6-di-*O*-isopropylidene-L-gulose **16** in 0.5 N HCl [14] (Scheme 4). This process causes deprotection, followed by anhydro formation. 1,6-Anhydro-L-gulose **18** may be obtained after chromatography in a crystalline form in approximately 45% yield.

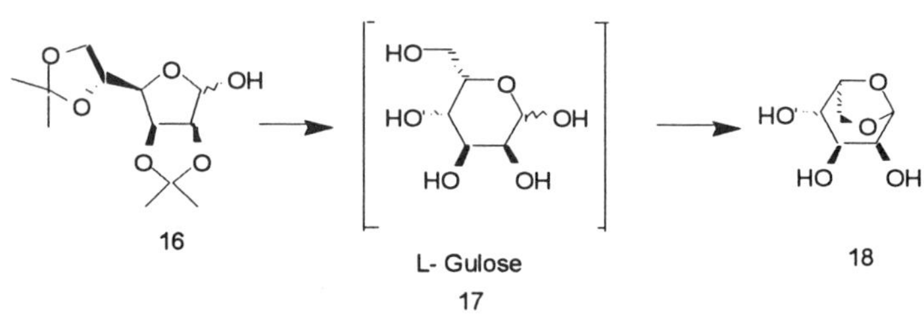

Scheme 4. Synthesis of 1,6 anhydro-L-gulose

Direct cleavage [15] of the cis diols of **18** with $NaIO_4$, followed by $NaBH_4$ reduction and reaction with p-toluenesulfonic acid in acetone, gave the isopropylidene derivative **19** (Scheme 5). Protection of the remaining hydroxyl as its benzoate, followed by removal of the isopropylidene group, gave the protected dioxolane diol **20**.

Treatment of this material with $NaIO_4$/RuO_2 gave the acid **21** which was converted to the acetate **22** as before with lead tetraacetate/pyridine in THF. This chiral acetate was used as before for condensation with various heterocyclic bases [16].

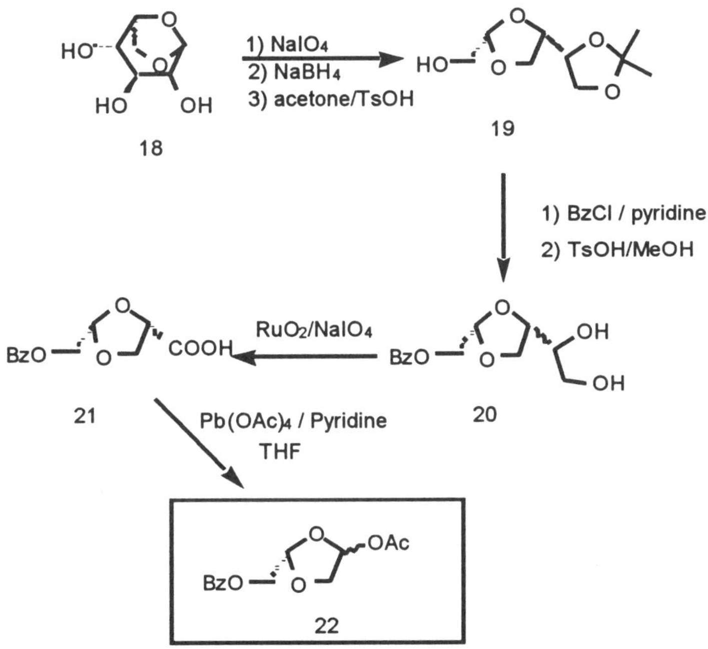

Scheme 5. Synthesis of L-dioxolanyl acetate

III. SYNTHESIS OF OXATHIOLANE NUCLEOSIDES

Another important member of this class of nucleoside analogues are those that contain an oxathiolane ring system, in particular, the cytosine derivative, BCH-189 [1]. This compound shows potent anti-HIV activity [1] as well as anti-HBV activity [17]. Again it was observed that the five-membered ring system of a thioanhydrohexose could be used to fix the stereochemistry of what would become the 4' position in the final nucleoside. Two such sugars were chosen as candidates for use as a chiral template due to the cis relationship of two adjacent hydroxyls and their low cost. These were unknown 1,6-thioanhydro-D-mannose [18] and the known 1,6-thioanhydro-D-galactose [19] (Figure 3). The synthesis of 1,6-thioanhydro-D-mannose was accomplished in four steps [18] (Scheme 6). Selective tosylation of D-mannose (**23**) at the 6 position, using a modification of the procedure for 1,6-anhydro-D-mannose, followed by acetylation, gave 1,2,3,4,-tetra-*O*-acetyl-6-*O*-tosyl-D-mannose (**24**) as a foam, which, without further purification, was treated with HBr/acetic acid using acetic acid as solvent, to give the bromo sugar **25**.

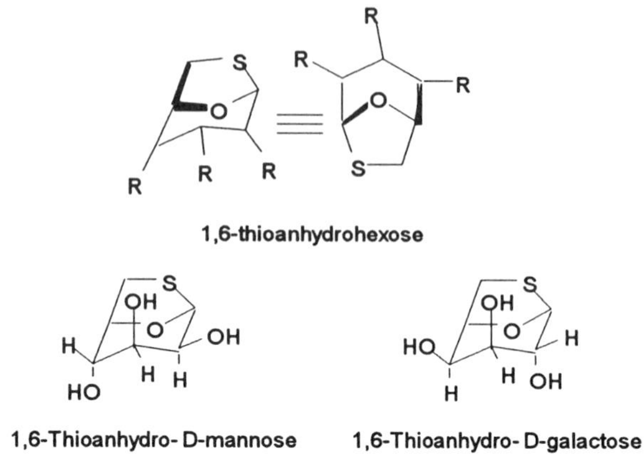

Figure 3. 1,6-Thioanhydrohexoses

Reaction of **25** with 3 equivalent of potassium ethyl xanthate in DMF at 0°C, followed by stirring overnight at room temperature, gave 2,3,4, tri-*O*-acetyl-1,6-thioanhydro-**D**-mannose (**26**), which was not isolated but subjected to deprotection using NH_4OH in MeOH. Concentration of the reaction mixture followed by chromatography over silica gel gave the desired product **27** in approximately 40% yield. The classical method for thioanhydro formation [20] involves using 1.5 to 2 equivalent of potassium ethyl xanthate in refluxing acetone, which produces the xanthate intermediate. This material is isolated, usually by crystallization, followed by treatment with NaOMe, which hydrolyzes the xanthate as well as the acetates, producing a sulfide anion that attacks the 6 position, displacing the tosyl group and forming the thioanhydro ring. The majority, if not all, previously known thioanhydro sugars have an α hydroxyl group and thus an α acetate at the 2 position. This assists in the formation of the β xanthate and the thioanhydro sugar. However, in the case of mannose, the 2 position is β, and thus little if any of the desired β xanthate could form when using the above conditions. This was observed in our initial experiment on the thioanhydro formation where we could only obtain approximately 5% of the desired product. Subsequently, we reasoned that by using a solvent that would, 1) dissolve potassium ethyl xanthate at less than room temperature and 2) be conducive to an S_N2 type reaction, the yield would improve. We assumed that the major anomer of the bromo sugar was α due to electronic effects and the presence of the β-acetate. DMF was chosen based on the yield it gave in intial experiments as well as ease of workup.

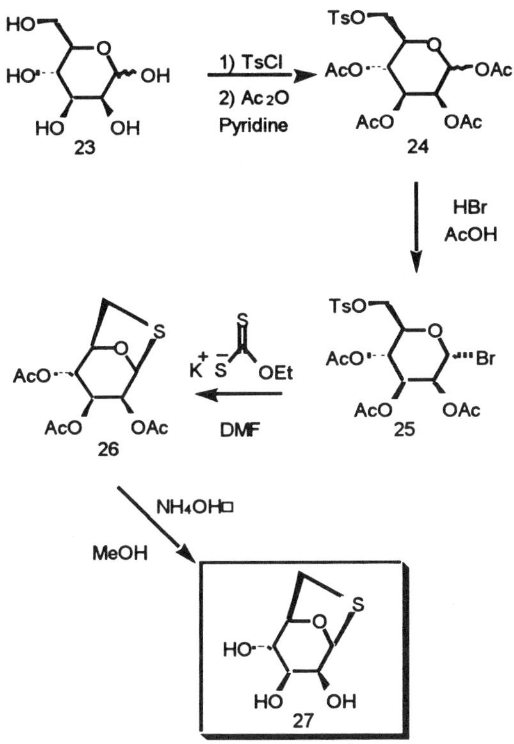

Scheme 6. Synthesis of 1,6-thioanhydro-D-mannose

The second major modification of the classical procedure was the use of 3 equiv of potassium ethyl xanthate. It was found by accident that when more than two equivalent of the xanthate were used, some thioanhydrosugar was formed; as more xanthate was added a proportional increase in thioanhydro compound was observed. At slightly more than 3 equivalent, the intermediate xanthate had completely disappeared giving the thioanhydro as the major product, with a minor amount of dixanthate. A probable mechanism is shown in Scheme 7.

Isopropylidenation of the 2,3-cis diol of **27** (Scheme 8), followed by protection of 4 position with benzoyl chloride, gave **28,** which was deprotected using dilute H_2SO_4 in aqueous dioxane at 75-80°C, furnishing the desired intermediate **29**. Treatment of **29** with $Pb(OAc)_4$ in ethyl acetate gave the dialdehyde, which, after partial purification to remove lead salts, was reduced with $NaBH_4$ in MeOH. As was seen in the dioxolane series earlier, the benzoyl group underwent a secondary-to-primary migration to give **30**. Selective protection of the remaining primary alcohol with t-butyldiphenylsilyl chloride and removal of the benzoate protection gave the diol **31**. Treatment of **31** with $Pb(OAc)_4$ in ethyl acetate gave the aldehyde, which was not isolated, but was further oxidized to the acid **32** using $NaClO_2$ in aqueous t-butanol [21].

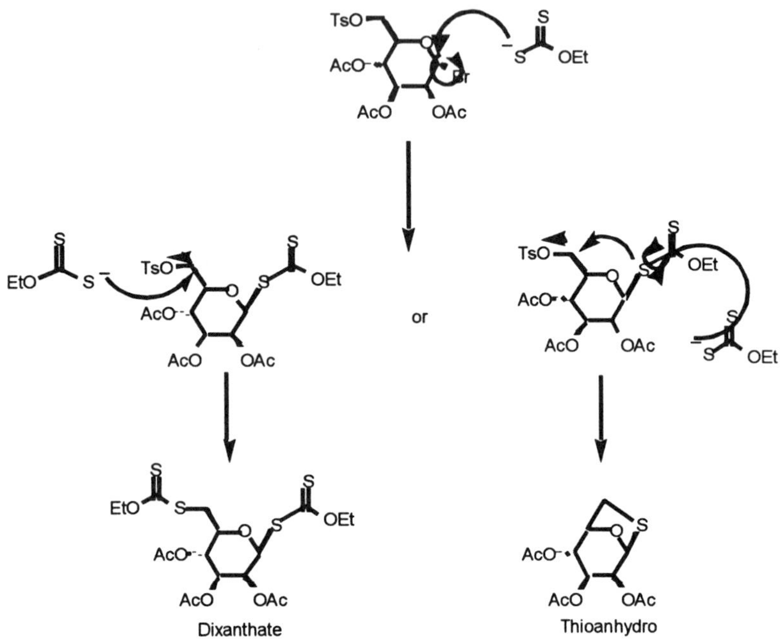

Scheme 7. Possible mechanism of direct thioanhydro formation

Unfortunately, this also oxidized the sulfur atom of the ring to a mixture of endo- and exo-sulfoxides. It was found that the sulfoxides would not undergo oxidative decarboxylation; therefore, the mixture of sulfoxides was esterified using dimethyl sulfate/K_2CO_3/acetone at room temperature to give **33**. Treatment of **33** with the dichloroborane-dimethylsufide complex [22] in dry THF at 0°C cleanly converted the

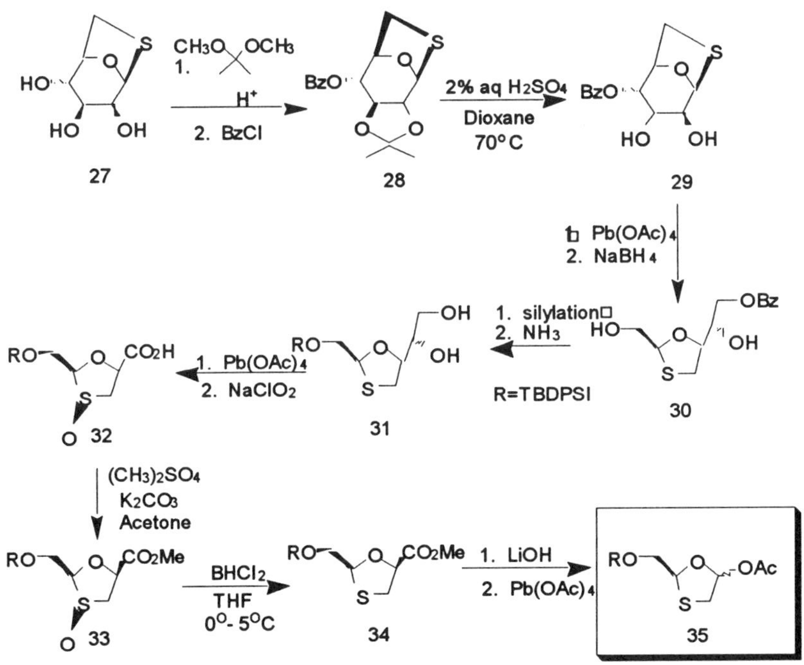

Scheme 8. Synthesis of oxathiolanyl acetate

mixture to the sulfide ester **34**, which was hydrolyzed to the acid with LiOH in 4:1 THF/H_2O [2] and converted to the key intermediate acetate **35** by treatment with lead tetraacetate/pyridine [2,7] in ethyl acetate. Later THF was found to give faster reactions and homogeneous solutions. This may be due to the fact that oxygen tends to inhibit the reaction, as the THF was freshly distilled under nitrogen from Na/benzophenone. Condensation of the acetate with silylated N^4-acetyl cytosine gave a 2:1, β:α mixture of **36** and **37** (Scheme 9), which were separated by chromatography and subjected to deprotection with methanolic ammonia and TBAF to give the final free (+)-β **38** and (-)-α **39** BCH-189. These compounds were subjected to chiral HPLC analysis to confirm their enantiomeric purity, and both were found to be homogeneous.

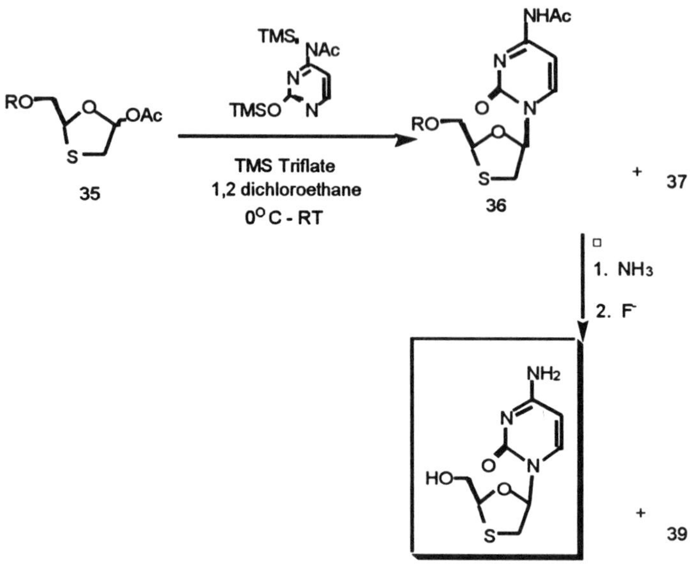

Scheme 9. Synthesis of (+)-BCH-189

An alternative and more efficient synthesis [19] of the (+)-β isomer of BCH-189 **38** uses 1,6-thioanhydro-D-galactose as the starting material. 1,6-Thioanhydro-D-galactose was originally prepared by Whistler and Sieb [20b] (Figure 4) and is easily scaled to large quantities.

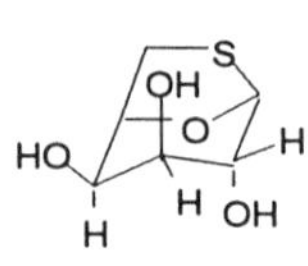

Figure 4. 1,6-Thioanhydro-D-galactose

The key to the efficiency of this process (Scheme 10) is the selective cleavage of the cis diol of **40** without protection followed by reduction with $NaBH_4$. The triol produced without isolation is then protected as its isopropylidene **41** and isolated.

The diol, which is a side product of the $NaIO_4$ reaction is washed out of the reaction mixture during work up. The remaining primary hydroxyl group of **41** is benzoylated, and the isopropylidene function is removed. The diol produced is cleaved and reduced to the alcohol **42**, which is protected with t-buyldiphenylsilyl chloride. This fully protected oxathiolane **44a** was then hydrolyzed with NaOMe to give the alcohol **44b**. Oxidation of **44b** with pyridium dichromate in DMF [23] gave the acid **45** without oxidation of the ring sulfur atom.

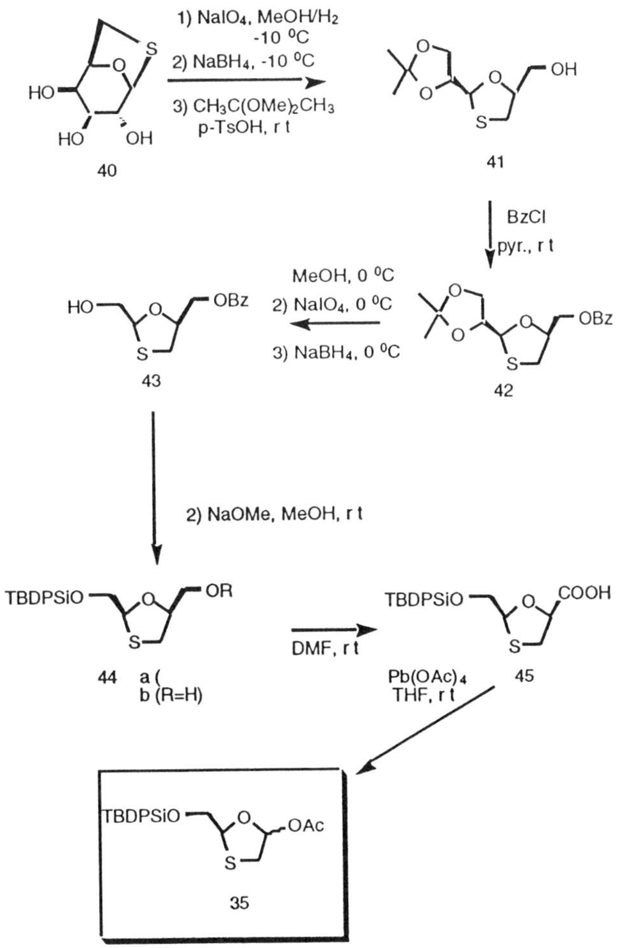

Scheme 10. Alternative synthesis of **35** from 1,6-thioanhydro-D-galactose.

Oxidative decarboxylation as before with lead tetraacetate/pyridine in THF gave the acetate **35**, which could be used for condensation as before.

The (+)-β-BCH-189 (**39**) was tested against HIV in peripherial blood mononuclear (PBM) cells and found to be a potent inhibitor; however, it was found to be less potent than the racemic mixture [18]. Racemic BCH-189 had also been shown to be a potent inhibitor of hepatitis B virus (HBV) in 2.2.15 cells [17]. When the (+)-β-isomer **39** was tested in this system, it was also found to be less active than the racemate. Thus the L-like enantiomer must be the more potent species in the oxathiolane.

As the with dioxolane nucleosides L-gulose was chosen for the chiral template. The advantage of this compound in the thioanhydrosugar synthesis is due to the arrangement of the alcohols the acetate would tend to the direct the nucleophile to the desired side, thus improving the yield.

As with D-mannose, the L-gulose [24] **17** was selectively to tosylated, followed by acetylation, to obtain 1,2,3,4-tetra-*O*-acetyl-6-*O*-tosyl-L-gulopyranose **46** (Scheme 11). Treatment of this material with HBr/AcOH in acetic acid at room temperature gave the glycosyl bromide **47** as a foam, which was reacted with 3 equiv of potassium ethylxanthate in refluxing acetone to give the 2,3,4-tri-O-acetyl-1,6-thioanhydro-L-gulose, which was not isolated but subjected to hydrolysis using NH_4OH in MeOH to give, after column chromatography, 1,6- thioanhydro-L-gulose **48** in approximately 75% yield from L-gulose.

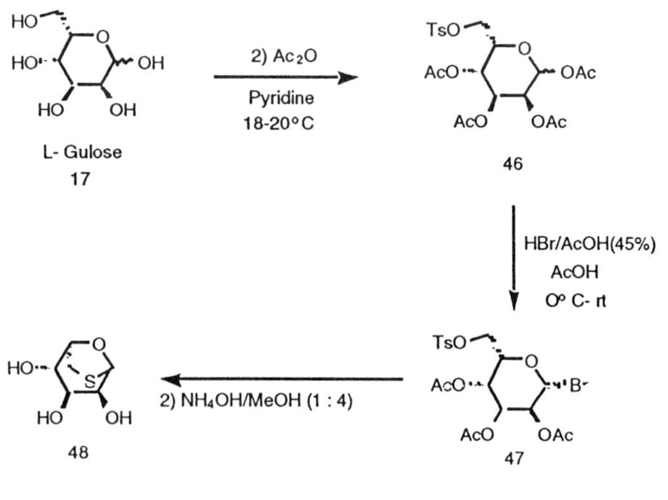

Scheme 11. Synthesis of 1,6-thioanhydro-L-gulose

The direct-cleavage methodology, developed for 1,6-thioanhydro-D-galactose was applied to 1,6-thioanhydro-L-gulose (Scheme 12). Treatment of the thioanhydro-L-gulose **48** with $NaIO_4$ followed by reduction of the resulting aldehyde with $NaBH_4$ gave the triol which was not isolated as such but converted to its isopropylidene **49** that was purified. Protection of the remaining alcohol with t-butyldiphenylsilyl chloride, followed by removal of the isopropylidene group, gave the protected oxathiolane diol **50**. Treatment of this material with lead tetraacetate to give the aldehyde, followed by further oxidation with pyridium dichromate in DMF, gave the acid **51**. Oxidative decarboxylation using lead tetraacetate/pyridine in THF gave the chiral acetate **52**. Condensation of **52** with silylated N^4-acetyl cytosine in the presence of TMSOTf/1,2-DCE gave, as before, a 1:2 α:β mixture of BCH-189. The α,β mixture was separated and treated with methanolic ammonia and TBAF to give the individual (+)-α and (-)-β BCH-189.

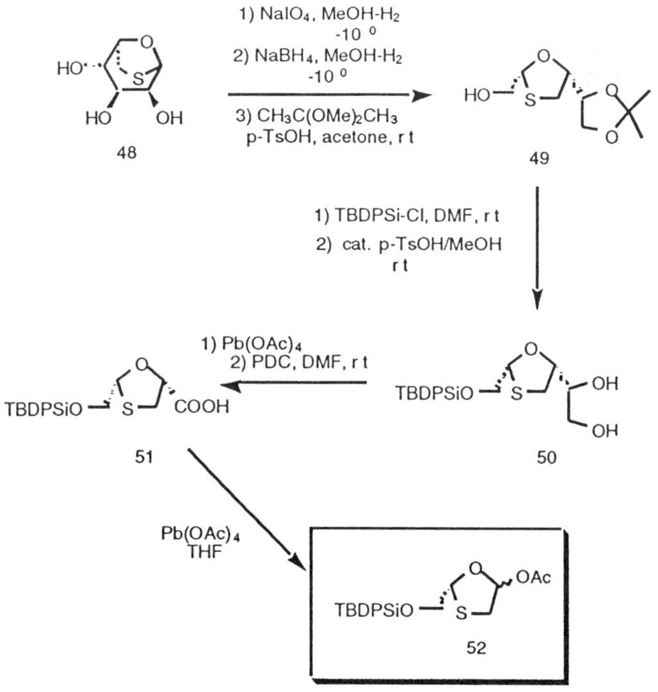

Scheme 12. Synthesis of L-oxathiolanyl-acetate

Chiral HPLC showed that again these were homogeneous and enantiomerically pure. The (+)-α and (-)-β anomers were evaluated against HIV-1 in PBM cell [25] in comparison to racemic and the (+)-β BCH-189, which indicated the this L-like enantiomer was the most potent. The (-)-β BCH-189 was also found to be the most potent compound against HBV [26].

IV. ALTERNATE SYNTHETIC ROUTES TO THE CHIRAL 3'HETERONUCLEOSIDES

A. Dioxolane Nucleosides

Another approach reported by Mansour and co-workers [27] employs L-ascorbic acid (Scheme 13) not so much as a chiral template but as a chiral auxiliary. L-ascorbic acid was used to produce a set of diastereomer which could be separated. The chiral center of L-ascorbic acid was, however, later destroyed on oxidative decarboxylation. L-ascorbic acid was condensed under acidic conditions with benzyloxyacetaldehyde dimethylacetal to give a diastereomeric mixture (**53** and **54**).

This mixture could be separated by fraction crystallization to yield the individual diastereomer in high **de**. Oxidation degradation using basic H_2O_2 to give the α-hydroxy acid **55** followed by Wolfe oxidation($RuCl_3$/NaOCl) under controlled pH conditions produced the dioxolanyl carboxylic acid **56**, diastereomeric with that produced from anhydromannose. Oxidative decarboxylation using $Pb(OAc)_4$ in the presence of pyridine gave an anomeric mixture of acetates **57**. The acetates were condensed with silyated N-acetylcytosine using TMSOTf as the Lewis acid to give an anomeric mixture of the α and β dioxolanyl cytidines (BCH 203 and BCH204 respectively). In this way both the D and L nucleosides could be produced.

B. Oxathiolane Nucleosides

Another application of chiral templates involves the use of (+) thiolactic acid [28] as the chiral starting material. This method is similar to that used for the synthesis of the dioxolanyl nucleosides which employed L-ascorbic acid. (+)-Thiolactic acid (Scheme 14) was condensed with benzoylacetaldehyde in the presence of $BF_3{\cdot}Et_2O$ to give a diastereomeric mixture of oxathiolanyl carboxylic acids(**58** and **59**) in a 1:2 ratio produced from one common starting material.

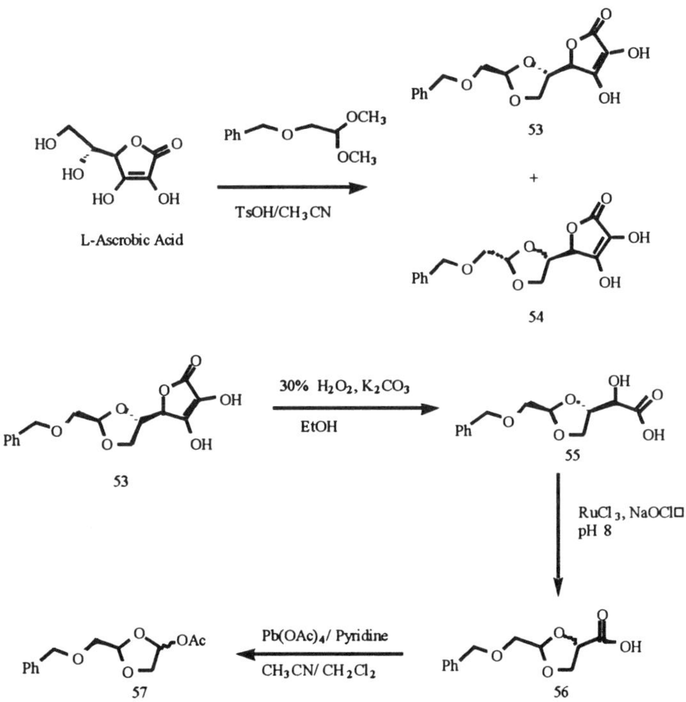

Scheme 13. Synthesis using L-Ascrobic Acid as starting material

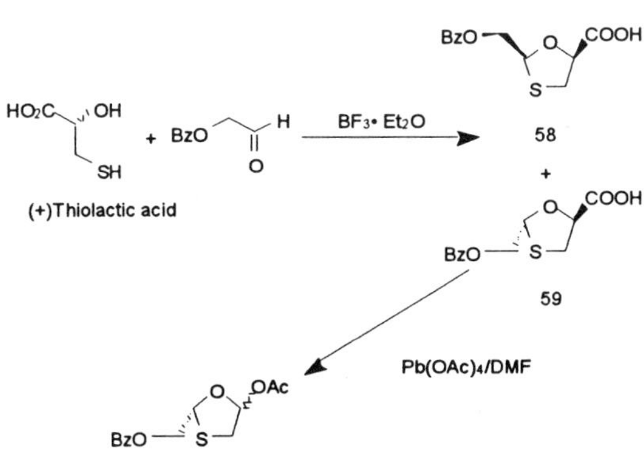

Scheme 14. Synthesis using (+)-Thiolactic acid as a chiral template

This mixture was separated by chromatography. The configurationally correct acid was then subjected to oxidative decarboxylation using $Pb(OAc)_4$ in DMF to produce an anomeric mixture of 5-acetates (**60**). This material is similar to **52** derived from L-Gulose, differing only in the protecting group at the 2-hydroxymethyl. The acetates were condensed with persilyated cytosine using TMS iodide as the Lewis acid to obtain an α/β (1:1.3) mixture. This may, as with L-ascorbic acid, be thought of a chiral auxillary rather than a chiral template method because the orginal chiral center is destroyed upon oxidation decarboxylation.

An alternative to the application of a chiral template as presented above to produce the oxathiolane "sugar" intermediate involves use of either chiral esters of the "sugar" or the nucleoside derivatives. The presence of a chiral center or centers of known configuration produces sets of diastereomers which allows for the resolution of individual enantomers. Examples of compounds which have been used in this way include (Figure 5) menthol esters **61** [29] and methyl lactate esters **62** [28].

A similar process involves the use of R and S α-methylbenzylamine [30] to form diastereomeric salts of oxathiolane 2-carboxylic acid. The separated salts were then reduced, protected and as before condensed to give the nucleoside deviratives.

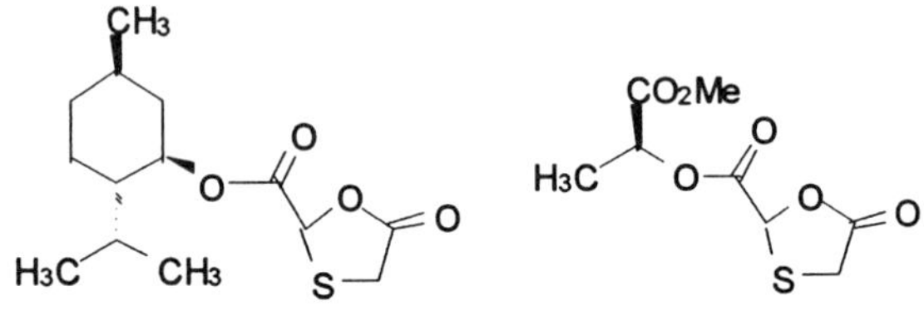

61 62

Figure 5. Chiral esters of oxathiolane intermediates

Another method which has found application in this area is the use of enzymes for the resolution of enantiomers. Those which have been sucessfully applied (Figure 6) include esterases and lipases derived from microorganisms and mammals [31] and 5'-nucleotidase derived from *Crotalus atrox* venom [32].

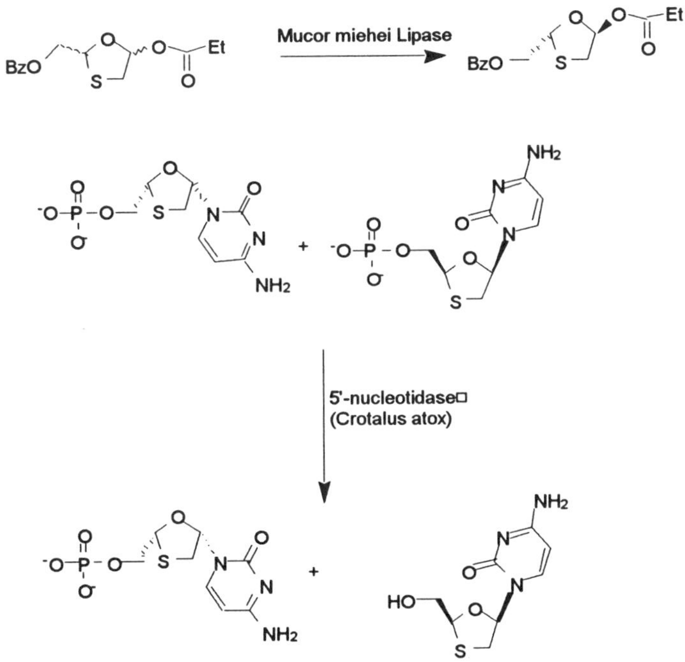

Figure 6. Examples of enzymatic resolution of final products and intermediates

This methodology has been used to resolve the enantomers of the "sugar" prior to condensation as well as to resolve the (+) and (-) enantomers of the cis oxathiolane nucleosides. The latter seemd to have greater utility. It has been found that the condensations carried out using strong Lewis acids such as tin tetrachloride and titanium tetrachloride give high yields of the desired cis isomer. However, this methodology also causes racemization if enantomerically pure oxathiolane "sugar" is used. By first condensing racemic "sugar" with base to give the (±) cis isomer followed by resolution , an effcient synthesis of both enantomers can be achieved.

V. BIOCHEMICAL PHARMACOLOGY AND METABOLISM

The greater potency and lower toxicity of (-) 3TC ((+)-BCH-189) and (-) FTC (5-Fluro derivative of 3TC), which possess the unnatural L configuration, over the (+) isomers is not easily explained. It has been shown that (+)3TCTP is actually a five times more potent inhibitor of purified reverse transcriptase (RT) than the (-) isomer

[33]. The best explanation for the higher potency of the (-) isomers over the (+) isomers of both compounds lies in their catabolism and anabolism. Both isomers of 3 TC [33] are converted to their triphosphates,apparently by deoxycytidine kinase.For both isomers the triphosphate comprises approximately 40% of the total phosphorylated products (mono, di and triphosphates) detected in mock infected and HIV infected PBL cells. The $T_{½}$ of (-) 3TCTP is ~13.5hr is this cell system whereas the $T_{1/2}$ for the (+) isomer is ~3.5 (mock infected) and 6 h (HIV infected). Furthermore the (-) isomer is phosphorylated faster than the (+) isomer [33]. A similar result has been found for L-DDC verses D-DDC. Unlike AZT, which increases dCTP and decreases dGTP levels, 3TC has no effect on other deoxynucleotide levels [33]. The (-) isomer of 3TC is not a substrate for cytidine deaminase whereas the (+) is a substrate [25,26,33]. Neither the (+) or the (-) isomer is a substrate for phosphorylase [33]. In addition (+)3TCTP is a more potent competitive inhibitor of dCTP than a (-) isomer against all DNA polymerase especially γ which is required for mitochondrial DNA synthesis and is associated with the development of peripheral neuropathy[34]. It has also been shown that cytoplasmic 3' to 5' exonuclease will remove (+) 3TC but not the (-) isomer from DNA These pieces of evidence explain the increased potency and lower toxicity of the (-) isomer over the (+) isomer of 3TC [34]. The (-) isomer is phosphorylated faster and is not converted into the inactive uridine derivative by deaminase as the (+) isomer is . The finding that the (-) isomer does not interfere with DNA polymerase γ and thus may not cause peripheral neuropathy will give this compound a clinical advantage over compounds such as DDC.

For the 5-fluoro analogue of 3TC, FTC, similar results were found in [35,36]. The (-) isomer is phosphorylated faster than the (+) isomer. The (-) isomer is also not a substrate for deaminase whereas the (+) isomer is. This would lead to the production of a potentially toxic metabolite.

Although the complete series of both the D and L oxathiolane nucleosides have been synthesized and evaluated for activity, the metabolism of these other analogues has not been extensively studied.

VI. STRUCTURE-ACTIVITY RELATIONSHIPS

Table 1 [8,9,16] shows the comparison of the EC_{50} for anti-HIV activity in PBM cells for the dioxolane nucleosides. Only α-anomers which showed significant activity have been included in the table. The most potent compounds of the dioxolane series were the L enantiomers of the cytosine derivatives. Hydrogen or fluorine at the 5 position gave the most potent compounds. The D-enantiomer of thymine was found to be more potent than the L enantiomer by a factor of 10.
However neither was as potent as the reported value of the racemic (0.09 μM). This observation remains to be explained.

The D enantiomers of the purine derivatives are in general more potent than the L enantiomer with the exception of the 2,6-diaminopurine. The L enantiomer was not only more potent than the D-enantiomer but was also the most potent of the purine series (D or L).

Table 2 [37,38] shows the comparison of the EC_{50} for anti-HIV activity in PBM cells for the oxathiolane nucelosides. Again only α-anomers which showed significant activity have been included in the table. In general, the 4-amino pyrimidine (cytosine) derivatives are more potent than the corresponding 4-oxo (uracil, thymine) derivatives. Of these derivatives, when the 5 substituent is small (H or F) the L enantiomer is more active. The D enantiomers are more active for large substituents at the 5 position with the 5-bromo being the most active followed by the 5 methyl and 5 chloro derivatives. Interestingly, this trend reverses for the 5 iodo where the L enantiomer is more active than the D.

For the purine derivatives, the D enantiomers are more active, in general, than the L enantiomers with hypoxanthine derivative followed by the adenine and 6-chloropurine derivatives showing the most activity. The D-enantiomers of guanine and 2-amino-6-chloropurine also showed fairly good activity.

It must be kept in mind that these compounds as well as all anti-HIV nucleosides are prodrugs. That is they must be activated to the triphosphate level before they can inhibit reverse transcriptase or targets in the infected cell. An idea of the selectivity of the various kinases and other catabolic and anabolic enzymes may be obtained by comparing the activity of the L-nucleoside versus the D-nucleoside.

It appears that in general (there are some exceptions) the pyrimidine metabolic enzymes are less selective (except for L-nucleosides) than the purine metabolic enzymes.

Table 1 Anti-HIV Activity of the Dioxolane Nucleosides in PBM cells [8,9,16]

Base	**configuration**	**D series** EC_{50} (μM)	**L series** EC_{50} (μM)
cytosine	β	0.016	0.002
cytosine	α	2.4	1.3
uracil	β	>100	nd
thymine	β	0.39	4.8
5-fluorouracil	β	69.6	nd
5-chlorouracil	β	6.8	nd
5-bromouracil	β	9.3	nd
5-iodouracil	β	>100	nd
5-methylcytosine	β	>100	45.9
5-methylcytosine	α	nd	18.9
5-fluorocytosine	β	nd	0.0012
5-fluorocytosine	α	nd	0.063
5-chlorocytosine	β	nd	34.3
5-bromocytosine	β	nd	1.8
5-iodocytosine	β	nd	0.41
5-iodocytosine	α	nd	28.8
6-chloropurine	β	22.8	34.7
6-chloropurine	α	11.1	1.27
adenine	β	0.5	3.8
adenine	α	6.2	29.0
N-methyladenine	β	14.3	62.6
N-methyladenine	α	30.3	nd
hypoxanthine	β	5.0	>100
2-fluoroadenine	β	0.3	1.6
2-fluoroadenine	α	2.5	23.7
2-NH_2-6-Cl-purine	β	0.09	13.4
2-NH_2-6-Cl-purine	α	23.1	8.1
guanine	β	0.03	17.5
2,6-diaminopurine	β	0.7	0.014

Table 2 Anti-HIV Activity of the Oxathiolane Nucleosides in PBM cells [37,38]

Base	**configuration**	**D series** $\mathbf{EC_{50}}$ (μM)	**L series** $\mathbf{EC_{50}}$ (μM)
cytosine	β	0.21	0.0018
cytosine	α	>100	10.1
uracil	β	94.7	11.7
uracil	α	>100	32.8
thymine	β	11.6	34.3
thymine	α	>100	4.4
5-fluorouracil	β	6.3	>100
5-fluorouracil	α	2.3	>100
5-chlorouracil	β	30.4	>100
5-bromouracil	β	>100	>100
5-iodouracil	β	29.3	92.9
5-methylcytosine	β	0.172	1.90
5-methylcytosine	α	>100	0.45
5-fluorocytosine	β	0.38	0.0013
5-fluorocytosine	α	77.5	0.43
5-chlorocytosine	β	0.28	31.8
5-bromocytosine	β	0.012	2.51
5-iodocytosine	β	1.9	0.14
5-iodocytosine	α	0.28	>100
6-chloropuirne	β	0.48	1.44
6-chloropurine	α	2.7	2.75
adenine	β	0.28	1.01
adenine	α	11.9	78.9
N-methyladenine	β	1.1	15.0
N-methyladenine	α	11.7	13.0
hypoxanthine	β	0.05	>100
2-fluoroadenine	β	2.8	16.3
2-fluoroadenine	α	61.9	15.8
2-NH_2-6-Cl-purine	β	0.52	9.8
2-NH_2-6-Cl-purine	α	5.6	42.1
guanine	β	0.83	10.2
guanine	α	7.62	>100

REFERENCES

1. Belleau B, Dixit D, Nguyen-Ba N, Kraus JL. Fifth International Conference on AIDS, Montreal, Canada, June 4-9, 1989, paper no. TCO1.
2. Norbeck DW Spanton S Broder S Mitsuya H ((-1-[2β,4β)-2 hydroxymethyl-4-dioxolane thymine). A new 2',3'-dideoxy nucleoside prototype with in vitro activity against HIV. Tetrahedron Lett. 1989 30: 6263.
3. Knauf AE, Hann RM , Hudson CS. D-Mannosan <1,5>β<1,6> or Levomannosan. J. Am. Chem. Soc. 1941 63:1447.
4. Zottola MA Alonso R Vile GD Fraser-Reid B. A practical, efficient,large-scale synthesis of 1,6-anhydrohexopyranoses. J. Org. Chem. 1989 54: 6123 5.
5. Chu CK, Ahn SK, Kim HO, Beach JW, Alves AJ, Jeong LS, Islam Q, VanRoey P, Schinazi RF Asymmetric synthesis of enantiomerically pure (-) - (1'R, 4'R) - dioxolane thymine and its anti-HIV activity. Tetrahedron Lett. 1991 32:3791.
6. Nunez MT, Martin VS. Efficient oxidation of phenyl group to carboxylic acids with rutherniun tetroxide. A simple synthesis of (R)-γ-caprolactam, the pheromone of Trogoderma granarium. J. Org. Chem. 1990 55:1928.
7. Dhavale D, Taghavini E, Trombini C, Umani-Romchi A. A novel synthetic equivalent of differentially protected tartaric aldehyde. A simple route to useful C-4 chiral synthesis. Tetrahedron Lett. 1988 29:6123.
8. Kim HO, Ahn SK, Alves AJ , Beach JW, Jeong LS,Choi BG, Van Roey P, Schinazi RF, Chu CK Asymmetric synthesis of 1,3-dioxolane pyrimidine nucleosides and their anti-HIV Activity. J. Med. Chem. 1992 35:1987.
9. Kim HO, Schinazi RF, Nampalli S, Shanmuganathan K, Cannon DL, Alves AJ, Jeong LS, Beach JW, Chu CK. 1,3-Dioxolanylpurien Nucleosides (2R,4R) and (2R,4S) with selective Anti-HIV-1 activity in Human Lymphocytes. J. Med. Chem. 1993 36:30.
10. Lerner LM, Kohn BD, Kohn P. Preparation of nucleosides via isopropylidene sugar derivatives III. J. Org. Chem. 1968 33:1780.
11. Andrews G, Crawford TC, Bacon BE. Stereoselective, catalytic reduction of L-ascorbic Acid. A convent synthesis of L-gulono-1,4-lactone. J. Org. Chem. 1981 46:7976.
12. Ishedate M, Mai YI, Hirasaka Y, Umemoto K, A new action of ion-exchange resins on the lactone ring of some carbohydrates. Chem.

Pharm. Bull. 1965 13:173.

13. Evans ME, Parrish FW. A simple synthesis of L-gulose. Carbohydr. Res. 1973 28: 359.
14. Stewart LC, Richtmyer NK. Transformation of D-gulose to 1,6 anhydro-β-D-gulopyranose in acid solution. J. Am. Chem. Soc. 1955 77:1021.
15. Kim HO, Shanmuganathan K ,Alves AJ, Jeong LS, Beach JW, Schinazi RF, Chang C-N, Cheng Y-C, Chu CK Potent anti-HIV and anti-HBV activities of (-)-L-β-Dioxolane-C and (+)-L-β dioxolane-T and their asymmetric synthesis. Tetrahedron Lett. 1992 33:6899.
16. Kim HO, Schinazi RF, Shanmuganathan K , Jeong LS, Beach JW, Nampalli S, Cannon DL, Chu CK. L-β-(2S, 4S) and L-α-(2S,4R)-dioxolanyl nucleosides as potential anti-HIV agents: asymmetric synthesis and strucutre-activity relationships. J. Med.Chem. 1993 36:519 17.
17. Doong S-L, Tsai C-H, Schinazi RF, Liotta DC, Cheng Y-C. Inhibition of the replication of hepatitis B virus in vitro by 2',3' dideoxy-3'thiacytidine and related analogues. Proc. Natl. Acad. Sci, USA 1991 88:8495.
18. Chu,CK Beach JW, Jeong LS, Choi BG, Comer FI, Alves AJ , Schinazi RF. Enantiomeric Synthesis of (+) BCH-189. [(+)-2S,5R)-1-[2-(Hydroxymethyl)-1,3-oxathiolane-5-yl) cytosine] from D-mannose and its anti-HIV activity. J. Org. Chem. 1991 56:6503.
19. Jeong LS, Alves AJ, Carrigan SW, Kim HO, Beach JW, Chu CK. An Efficient synthesis of enantiomerically pure (+) (2S,5R)-1-[2-(hydroxymethyl)-1,3 oxathiolane-5-yl] cytosine [(+) BCH-189] from D-galactose. Tetrahedron Lett. 1992 33: 595.
20. a) Akagi M, Tejima S, Haga M. Biochemical studies in thio sugars IV.Synthesis of 1,6-anhydro-1,6-sulfide-β-D-glucopyranose (thiolevo glucosan) and 6-deoxy-6-mercapto-1-thio-D-glucose.Chem. Pharm. Bull. 1963 11: 58.
    b) Whistler RL, Seib PA. Alkaline degradation of 6-thio derivatives of D-glucose and D-galactose. Carbohydr. Res. 1966 2:93.
21. Kraus GA, Taschner MJ. Model studies for the synthesis of quassinoids. Construction of the BCE ring system. J. Org. Chem. 1980 45:1175.
22. Brown HC, Ravindran N. An exceptionally rapid and selective deoxygenation of aliphatic sulfoxides to sulfides under mild conditions with a new reducing agent,dichloroborane. Synthesis 1973 42 23. Corey EJ, Schmidt G. Procedures for the oxidation of alcohols involving pyridinium dichromate in aprotic media. Tetrahedron Lett. 1979 379.
24. Beach JW, Jeong LS, Alves AJ, Pohl D, Kim HO, Chang C-N, Doong SL,

Schinazi RF, Cheng Y-C, Chu CK. Synthesis of enantiomerically pure 2'R,5'S-(-)-1-[2-(Hydroxymethyl)-oxathiolane-5-yl] cytosine as a potent antiviral agent against hepatitis B virus (HBV) and human immunodeficiency virus (HIV). J. Org.Chem. 1992 57:2217.

25. Schinazi RF, Chu CK, Peck A, McMillan R, Mathis R, Cannon D, Jeong LS, Beach JW, Choi WB, Yeola S, Liotta DC. Activities of the four optical isomers of 2',3'-dideoxy 3' thiacytidine (BCH-189) against human immunodeficiency virus type 1 in human lymphocytes. Antimicrob. Agents Chemother. 1992 36:672.
26. Chang CN,Doong S-L , Zhou JH, Beach JW, Jeong LS, Chu CK, Tsa CH, Cheng Y-C. Deoxycytidine deaminase resistent stereoisomer is the active form of (±) 2',3'-dideoxy 3'-thiacytidine in the inhibition of hepatitis B virus replication. J. Biol.Chem. 1992 267:13938.
27. Belleau BR, Evans CA, Tse HLA, Jin H, Dixit DM, Mansour TS Oxidative Degradation Of L-Ascrobic Acid Acteals to 2',3'-Dideoxy-3'-Oxaribofuranosides. Synthesis Of Enatiomerically Pure 2',3'-Dideoxy-3'-Oxacytidine Stereomers As Potential Antiviral Agents. Tetrahedron Lett. 1992 33:6949.
28. Humber DC, Jones MF, Payne JJ, Ramsay MVJ Expeditious Preparation of (-)-2'-Deoxy-3'-Thiacytidine (3TC). Tetrahedron Lett. 1992 33:4625.
29. Siddiqui MA, Jin H, Evans CA, Dimarco MP, Tse HLA, Mansour TS. (1'R,2'S,5'R)-Menthyl-(5R)-acetoxy-1,3-oxathiolan-(2R)-Carboxylate: Synthesis and Isomeric Purity Determination by HPLC. Chirality 1994 6:156.
30. Kraus J-L, Attardo G New Optically Active 1,3-Oxathiolanes. Chirality 1993 5:97.
31. Cousins RPC, Mahmoudian M, Youds PM Enzymic Resolution of Oxathiolane Intermediates- An alternative Approach to the Anti-viral Agents Lamivudine (3TC). Tetrahedron Asymmetry 1995 6:393.
32. Storer R, Clemens IR, Lamont B, Noble SA, Williamson C, Belleau B The Resolution and Absolute Stereochemistry of cis-1-[2-(hydroxylmethyl)-1,3-oxathiolan-5-yl]cytosine(BCH189):equipotent anti-HIV Agents. Nucleosides and Nucleotides 1993 12:225.
33. Cammack N, Rouse P, Marr CLP, Reid PJ, Boehme RE, Coates JAV, Penn CR, Cameron JM Cellular metabolism of (-) enantiomeric 2'-deoxy- 3'-thiacytidine. Biochem. Pharmacol. 1992 43:2059.
34. Shalski V, Chang C-N, Dutschman G, Cheng Y-C The biochemical basis of the differential anti-human immunodeficiency virus activity of two cis enantiomers of 2',3'-dideoxy-3'-thia cytidine. J. Biol. Chem. 1993 268:23234.

35. Wilson JE, Martin J L, Borroto-Esoda K, Hopkins S, Painter G, Liotta DC, Furman PA. The 5'-triphospahtes of the (-) and (+) enantiomers of cis-5-fluoro-1-[2-(hydroxymethyl)-1,3-oxathiolan-5-yl]cytosine equally inhibit human immunodeficiency virus type 1 reverse transcriptase. Antimicrob. Agents and Chemo. 1993 37:1720.
36. Furman PA, Davis M, Liotta DC, Paff M, Frick LW, Nelson DJ, Dornsife RE, Wurster JA, Wilson LJ, Fyfe JA, Tuttle JV, Miller WA, Condreay L, Averett DR, Schinazi RF, Painter GR The Anti-hepatitis B virus activities, cytotoxicities and anabolic profiles of the (-) and (+) enantiomers of cis-5-fluoro-1-[2-(hydroxymethyl)-1,3-oxathiolan-5-yl]cytosine Antimicrob. Agents and Chemo.1992 36:2686.
37. Jeong LS, Schinazi RF, Beach JW,Kim HO, Nampalli S, Shanmuganathan K, Alves AJ, McMillan A, Chu CK, Mathis R Asymmetric sytnhesis and biological evaluation of L-β-(2R, 5S) and L-α-(2R,5R)-1,3-oxothiolanyl-pyrimidine and -purine nucleosides as potential anti-HIV agents J. Med. Chem. 1993 36:182.
38. Jeong LS, Schinazi RF, Beach JW, Kim HO, Shanmuganathan K, Nampalli S, Alves AJ, Chun MW, Chung WK, Choi BG, McMillan A Chu CK Structure-activity relationships of D-β-(2S, 5R) and D-α-(2S,5S)-1,3-oxothiolanyl nucleosides as potential anti-HIV agents J. Med. Chem. 1993 36:2627.

14
Sugar-Modified Anthracyclines: A Search for Drugs Able to Overcome Multidrug Resistance

Waldemar A. Priebe and Roman Perez-Soler
University of Texas
Houston, Texas

I. INTRODUCTION

Doxorubicin (DOX) is an effective anticancer agent with a broad spectrum of activity [1-3]. It is primarily used to treat leukemias, lymphomas, breast carcinomas, and sarcomas [4]. DOX's clinical application, however, is limited by its significant cardiotoxicity and by the fact that often, despite initially effective systemic chemotherapy, tumors develop resistance to it (acquired resistance). These limitations, combined with the existence of tumors resistant de novo to chemotherapy, were a stimulus to the design and synthesis of novel anthracyclines.

Over the last 20 years, numerous modifications of DOX's aglycon portion and carbohydrate moiety have been made; however, they have not resulted in an agent clearly superior to DOX. The other anthracyclines that have been approved for clinical use are closely related to daunorubicin (DNR) or DOX and are generally equally ineffective against resistant tumors. One such drug is epirubicin; although not approved for use in the United States, it is being used in Europe, where its clinical use in recent years have increased substantially. Epirubicin differs from DOX only in one position at the sugar portion: it is a 4'-epimer of DOX. The other compound, idarubicin, an analog closely related to DNR, is modified at the aglycon portion: its methoxy group at C-4 is replaced with hydrogen.

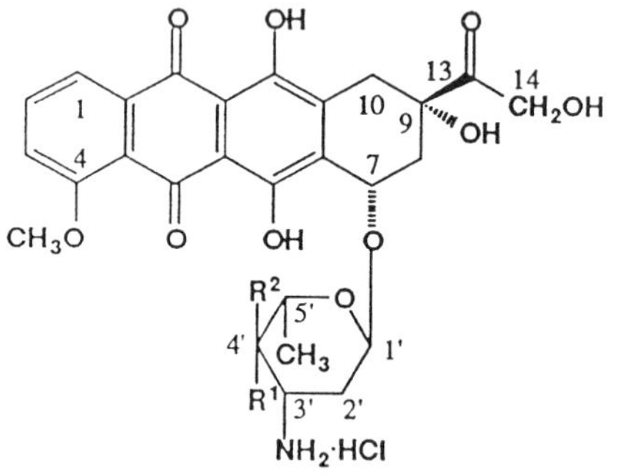

Doxorubicin (DOX) - R^1 = OH, R^2 = H
Epirubicin - R^1 = H, R^2 = OH

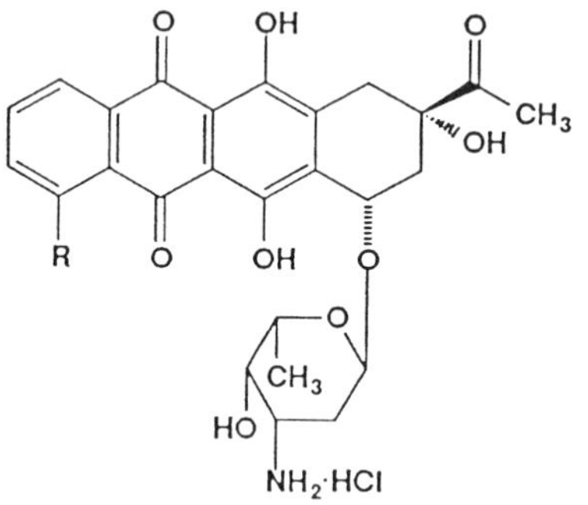

Daunorubicin (DNR) - R = OCH_3
Idarubicin - R = H

Although the sugar portion itself does not exhibit antitumor properties, it is an important part of anthracycline antibiotics. This has been revealed, in part, by evaluations of the aglycon portion. Such studies showed that, although some antitumor activity can be detected, the activity is too low for aglycons to be considered potential candidates for drug development. From our recent studies, it is evident that sugars play an important role in the binding of anthracyclines to DNA. [5, 6] For example, removal of the daunosamine reduces binding free energy by 2 kcal mol-1 and consequently decreases by about 100-fold the DNA binding constant. It is also very clear that the sugar's orientation is critical and that alteration of the configuration at the anomeric center can drastically reduce drug-DNA binding. In fact, the DNA binding of the ?-anomer of DOX is less favorable than the binding of the aglycon itself. These observations correlate with the biological activity of these compounds and highlights the importance of the sugar part. The ability of anthracyclines to bind to DNA is associated with their ability to inhibit topoisomerase II (topo II), which is currently recognized as the most important mechanism of antitumor action of this group of drugs. Thus, in effect, the sugar part seems to be a key substructure whose modification can alter anthracyclines' ability to inhibit topo II and probably other biochemical interactions important for biological activity.

Multidrug resistance (MDR) is probably the most studied mechanism of resistance and has been shown to be clinically relevant. Therefore, in our approach we have focused on designing anthracyclines active against multidrug resistant tumors. The mechanism of resistance called typical or classic MDR

[7-9] is associated with overexpression of the *MDR*1 gene and production of the transmembrane glycoprotein P-glycoprotein (P-gp). P-gp, by acting as an energy-dependent efflux pump, decreases intracellular drug accumulation of structurally and mechanistically unrelated drugs like DOX, DNR, mitoxantrone, vincristine, vinblastine, vindesine, etoposide (VP-16), teniposide (VM-26), dactinomycin (actinomycin D), gramicidin D, plicamycin (mithramycin), mitomycin C, trimetrexate, and paclitaxel.

In our approach, we hoped to identify structural elements responsible for the affinity of anthracyclines to P-gp, dissect those elements from the structural factors responsible for inhibition of topo II, and then design drugs that would be retained by MDR cells and would be able to inhibit topo II, expecting in effect to obtain drugs highly cytotoxic to MDR cells.

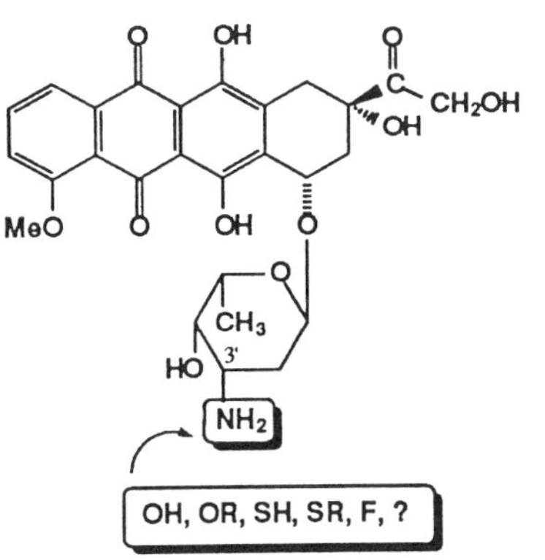

Figure 1. Replacement of basic nitrogen with other heteroatom

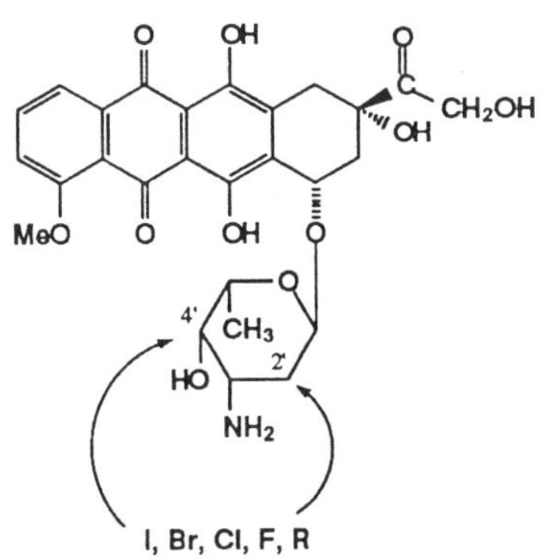

Figure 2. Introduction of electron-withdrawing substituents at C-2' or at C-4'

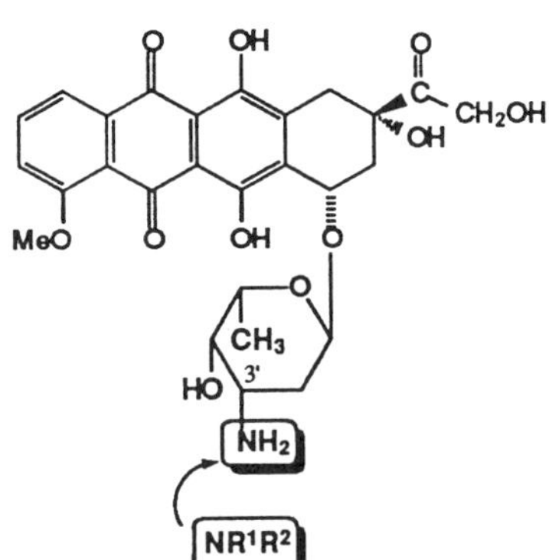

Figure 3. Direct derivatization of basic amine

In our studies, we hoped to identify and learn to manipulate the "molecular switch" in the sugar moiety that "turns off" P-gp-mediated efflux while leaving at least undisturbed the "switch" that "turns on" the inhibition of topo II. This concept of anthracycline design is discussed in more detail in a recent review [10].

In our search for those "molecular switches," our departure point was the hypothesis that the basic amine in the sugar moiety of anthracyclines is an important factor facilitating P-gp-mediated transport [11-16]. Thus, decreasing basicity would be the "off" switch for P-gp-mediated efflux. Even though it was difficult to imagine that one modification would completely turn off drug transport by such a relatively nonspecific transport system, we believed that even partial circumvention of P-gp might sufficiently increase cytotoxicity against MDR cells, and so our hypothesis became a cornerstone of the design of first-generation MDR-active anthracyclines.

To effectively use decreased basicity as an element of the drug design, it was important for us to assess the biological role of the amine group and to examine to what extent the amine group is essential for the antitumor activity of DOX. To achieve these goals, we decided to select a series of nonbasic or decreased-basicity analogs for comparative studies with parent aminated anthracyclines.

From the chemical point of view, the basicity of doxorubicin can be modulated by modifying the daunosamine moiety in several different ways. First, the basic nitrogen atom in the 3'-amine can be replaced with other nonprotonable heteroatoms like oxygen or sulfur (Figure 1). Second, the amino group can be preserved while electron-withdrawing substituents at C-2' or C-4' are introduced to decrease the basicity of the amino group (Figure 2). In the third option, the basic amine can be directly derivatized to the nonbasic compounds (Figure 3). We have decided to focus on the first two options, rather than on the third one, because it was shown previously that acylation of the amino group leads to significantly less cytotoxic analogs. Therefore, such modifications, besides decreasing basicity, probably lead to decreased topo II inhibition as well.

II. SYNTHESIS OF 3'-DEAMINO-3'-HYDROXYANTHRACYCLINES

The simplest and most radical approach seemed to be replacement of amino group with hydroxyl. This modification was favored by us because it has some roots in our earlier efforts, when a more random approach to drug synthesis was used. The first synthesized 3'-hydroxyanthracyclines were not good candidates for drug development [17-21], most probably because the aglycons used in those early syntheses (ε-rhodomycinone, ε-pyrromycinone, and daunomycinone) were different from adriamycinone, the aglycon present in

DOX. On the contrary, we were the first group to synthesize a 3'-hydroxy analog containing adriamycinone as an aglycon, and the initial biological evaluations clearly indicated that the 3'-deamino-3'-hydroxydoxorubicin (**6**, hydroxyrubicin) has substantial antitumor activity [22].

The synthesis of hydroxyrubicin was complicated by the fact that typical transformation from the daunorubicin series to the DOX series, which includes bromination at C-14 and hydrolysis, could not be used to synthesize deaminated analogs. Instead of formation of a 14-bromo intermediate, we have observed decomposition of substrate and glycosidic cleavage.

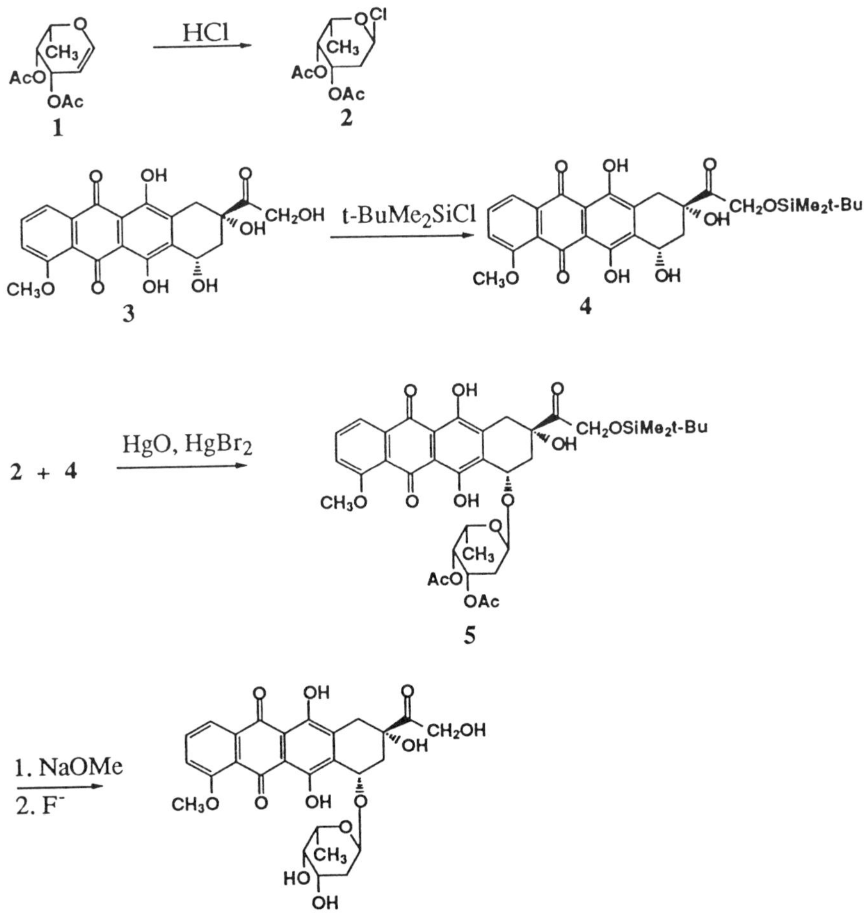

6 - Hydroxyrubicin

Thus, we have initiated the use of adriamycinone selectively silylated at the primary hydroxyl at C-14 (**4**) as a substrate. Such an aglycon was obtained from daunomycinone by bromination, hydrolysis, and selective silylation of 14-OH.

In brief, synthesis proceeded as follows: **4** was coupled via Koenigs-Knorr reaction with glycosyl chloride **2** obtained by electrophilic addition of dry hydrogen chloride to 3,4-di-*O*-acetyl-L-fucal (**1**) to give α anomer **5** with good yield. Compound **5** was easily purified since no β anomer was detected in the reaction mixture. Hydroxyrubicin (**6**) was then obtained from **5** in a two-step deblocking procedure: **5** was first deacetylated with sodium methoxide and then desilylated with tetrabutylammonium fluoride in tetrahydrofuran solution to give the desired hydroxyrubicin **6** [22].

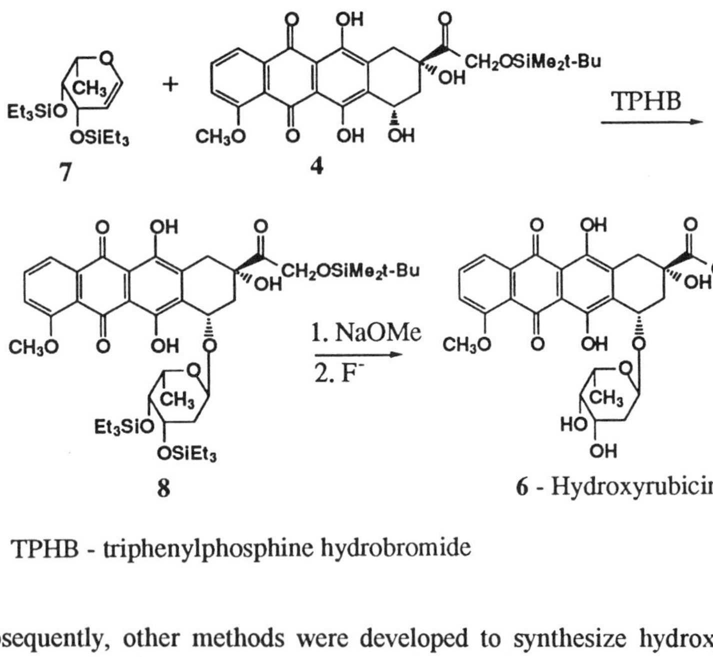

7 **4** **8** **6** - Hydroxyrubicin

TPHB - triphenylphosphine hydrobromide

Subsequently, other methods were developed to synthesize hydroxyrubicin. The most straightforward starts from 3,4-di-*O*-triethylsilyl-L-fucal (**7**) and 14-*O*-silylated adriamycinone (**4**) [23]. In brief, one-step electrophilic addition of aglycon **4** to L-fucal **7** in the presence of triphenylphosphine hydrobromide (TPHB), which gives **8** with good yield, was followed by one-step desilylation to free hydroxyrubicin **6**.

The 3'-deamino-3'-hydroxy analogue of another important anthracycline antibiotic, epirubicin, was also prepared by us [24]. The synthetic approach here was similar to that used for hydroxyrubicin, but instead of 3,4-di-*O*-acetyl-L-fucal (**1**), we used 3,4-di-*O*-acetyl-L-rhamnal (**9**) as a substrate. L-Rhamnal **9** reacted with hydrogen chloride gave glycosyl chloride **10,** which then was coupled with 14-*O*-*tert*-butyldimethylsilyl-adriamycinone (**4**) under Koenigs-Knorr conditions. In contrast to the reaction from glycosyl chloride **2** (α-L-lyxo configuration), reaction from glycosyl donor **10** (α-L-arabino configuration) led to an inseparable mixture of α and β anomers. In fact, the anomers could only be partially separated after deacetylation. To increase stereoselectivity, we took an alternative approach using silver triflate as the promoter of glycosidation to isolate α anomer **11**, although with low yield. Using the same deblocking procedures as previously for hydroxyrubicin, we have successfully prepared hydroxyepirubicin **12** from compound **11**. When compared to antitumor activity, hydroxyrubicin in vivo was consistently more active and also less toxic than hydroxyepirubicin [22, 24].

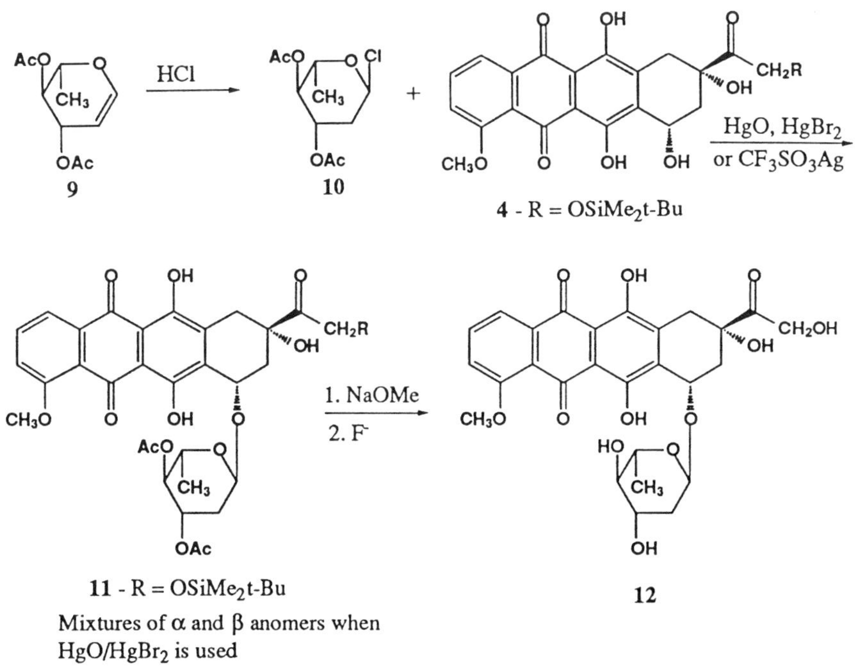

Whereas L-rhamnal **13** and L-fucal were easily accessible from the natural sugars L-rhamnose and L-fucose, respectively, the starting glycal **17** in the synthesis of another 3'-deaminated anthracycline, the 3'-hydroxyesorubicin (**22**)

had not been described before, and our attempts to prepare it using known procedures failed.

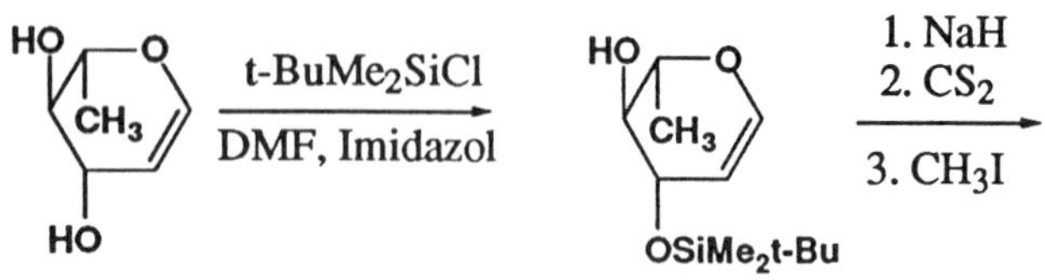

13 **14**

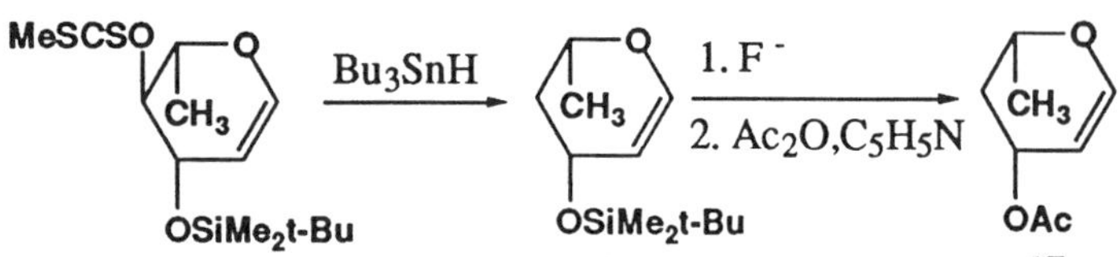

15 **16** **17**

Consequently, we have developed a novel method for preparing the required 3-*O*-acetyl-4-deoxy-glycal **17**. The five step procedure involves the free radical deoxygenation of properly blocked L-rhamnal **15** or L-fucal [25]. Critical for the success of this synthesis is the elimination of acetate as a blocking group and the introduction of *t*-butyldimethylsilyl to block hydroxyl at C-3 (compounds **14** and **15**). The glycal **17** we obtained was reacted with hydrogen chloride to glycosyl chloride **18**.

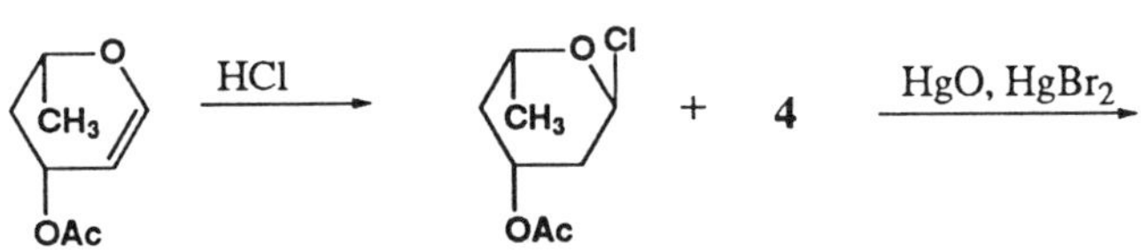

17 **18**

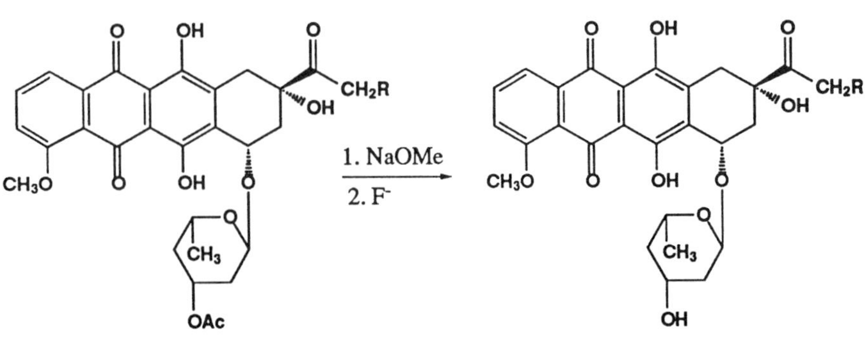

19 - R = H
20 - R = OSiMe$_2$t-Bu

21 - R = H
22 - R = OH

Subsequent coupling of **18** with daunomycinone or 14-*O-tert*-butyldimethylsilyl-adriamycinone (**4**) in methylene dichloride solution in the presence of mercuric bromide, yellow mercuric oxide, and 3 Å molecular sieves gave daunorubicin and doxorubicin analogs **19** and **20**, respectively. Removal of acetate from **19** gave 4'-deoxy-3'-hydroxydaunorubicin (**21**), and the two-step deprotection of **20** gave 3'-hydroxy-esorubicin (**22**).

III. HOW THE REPLACEMENT OF 3'-AMINE WITH HYDROXYL AFFECTS BIOLOGICAL ACTIVITY

To evaluate the role of the amino group in and to study the effects of the basicity of the sugar moiety on the toxicity and antitumor activity of anthracycline antibiotics, as well as to assess the potentials of deaminated anthracyline analogues as new anticancer drugs, we have selected for comparative studies hydroxyrubicin (**6**). This analogue has the 3'-amine replaced by hydroxyl group, but in all other structural features and configurations it is identical to DOX. In our series of studies of hydroxy rubicin and DOX, [11, 12, 14, 15] we have shown that hydroxyrubicin is similar or superior to DOX in its *in vitro* cytotoxicity against P-388, L-1210, and M-5076 cells, as determined by MTT assay, and against 8226 and CEM cells, as determined by growth inhibition assay. For example, hydroxyrubicin was 5- and 13-fold more effective than DOX in inhibiting the growth of multidrug-resistant CEM (CEMvbl) and 8226 (8226R) cells, respectively. Hydroxyrubicin was equally cytotoxic against KB-3-1 and multidrug-resistant KB-V-1 cells (resistant index [RI] of 1.1 for hydroxyrubicin vs. RI of >50 for DOX). Flow cytometry studies of cellular uptake and retention of hydroxyrubicin and DOX in parent and multidrug-resistant 8226 cells showed that, in sensitive 8226 cells, the 2-h uptake and retention of DOX were similar to or higher than those of hydroxyrubicin. In 8226R cells, the uptake and retention of hydroxyrubicin were about 3-fold higher than those of DOX. Together, these data suggest that the lack of a basic amino group in hydroxyrubicin is responsible for its increased cellular uptake and retention in multidrug-resistant cells, possibly because of the reduced interaction of hydroxyrubicin with the multidrug transporter P-gp. Moreover, the increased cellular uptake and retention of hydroxyrubicin correlated with its partial or total lack of cross-resistance of hydroxyrubicin with its parent compound, DOX. Hydroxyrubicin's activity against MDR tumors was confirmed *in vivo* against P388 leukemia cells resistant to DOX. Hydroxyrubicin at the optimal dose (37.5 mg/kg i.p. on day 1) had significant activity, whereas DOX (10 mg/kg i.p. on day 1) was inactive (T/C 163-200 vs. 118-120; data from two experiments) (T/C is a ratio of the test median survival time to median survival time of the control x 100%) [15].

Hydroxyrubicin was also less toxic in mice. Its acute LD50 was about 3-fold higher than that of DOX (79.1 mg/kg versus 25.7 mg/kg), and at equitoxic doses it was less cardiotoxic, as assessed by the Bertazzoli test. The lack of basic amine function was also apparently responsible for the decreased affinity of hydroxyrubicin to negatively charged cardiolipin, which might explain in part the lower cardiotoxicity of hydroxyrubicin versus DOX [15].

Like hydroxyrubicin, both epihydroxyrubicin **12** and 3'-hydroxyesorubicin **22** show substantial antitumor activity. In in vitro studies, the cytotoxic potential of 3'-hydroxyesorubicin **22** was similar to that of DOX, and **22** showed higher activity against P388 leukemia in vivo than did DOX. In fact, in two experiments, the T/C for **22** was >600 with 100% and 90% long-term survivals noted, respectively. In comparison, the T/C for DOX at its maximum tolerated dose of 10 mg/kg was 172 and 300, respectively, with 0% and 20% long-term survivals, respectively [27].

An interesting lipophilic analogue of hydroxyrubicin **WP474** was obtained by acylation of 14-hydroxyl of **6** with palmitoyl chloride [26]. 7-*O*-(2,6-Dideoxy-α-L-lyxo-hexopyranosyl)14-*O*-palmitoyladriamycinone (**WP474**) was obtained in 32% yield.

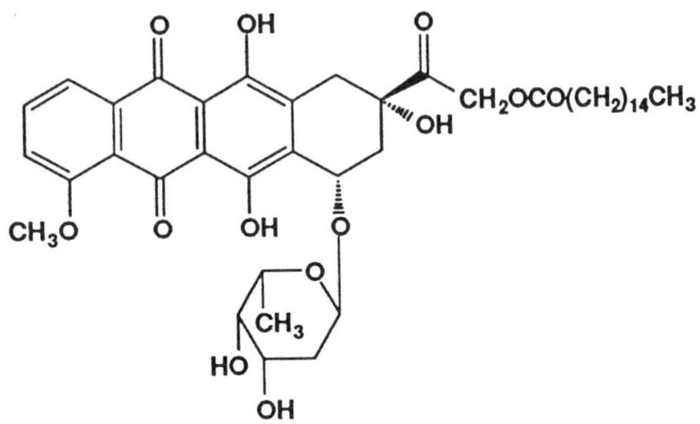

WP474

The antitumor activity of this analogue was assessed in vivo as a liposome-formulated drug. Particularly interesting was its good activity in vivo against M-5076 reticulosarcoma cells when drug was administered i.v.: DOX was basically inactive at its maximum tolerated dose, whereas **WP474** showed T/C 175 at 20 mg/kg. The ester **WP474** was also highly active against L1210 lymphoid leukemia (T/C > 600 at 60 mg/kg vs. T/C 337 at 10 mg/kg for DOX) [26].

In summary, the presence of the amino group at position 3' seems unessential to DOX's biological activity. In addition, removal of the basic amine seems to reduce general toxicity and cardiotoxicity. Thus, these studies assured us that

3'-hydroxyanthracyclines should be evaluated further and that, based on our hypothesis, we should focus on evaluating their potential activity against MDR tumors.

IV. ANTHRACYCLINES WITH 3'-AMINE REPLACED WITH MERCAPTO GROUP

Our finding that the 3'-amine is not essential for the high antitumor activity of DOX moved us to explore other hetero atoms and to synthesize anthracyclines whose basic amine is replaced with a mercapto group. Our progress towards that goal was slowed, however, by the lack of adequate synthetic methods for the preparation of 3-mercapto sugars.

During our previous studies, we tried to prepare 3-mercapto sugars using Ferrier rearrangement reaction of glycals. This reaction worked well with mercaptans and led to respective 3-thioalkyl sugars [28]. However, when 3,4-di-*O*-acetyl-L-fucal was reacted with thioacetic acid or potassium thioacetate in the presence of Lewis acid ($BF_3 \cdot Et_2O$), primarily 2,3-unsaturated sugars were obtained with no observed formation of 3-*S*-acetyl-glycals [29]. Even though these results nicely confirmed our explanation of regioselectivity of Ferrier rearrangements in glycals, [28, 29] this was a small comfort to us.

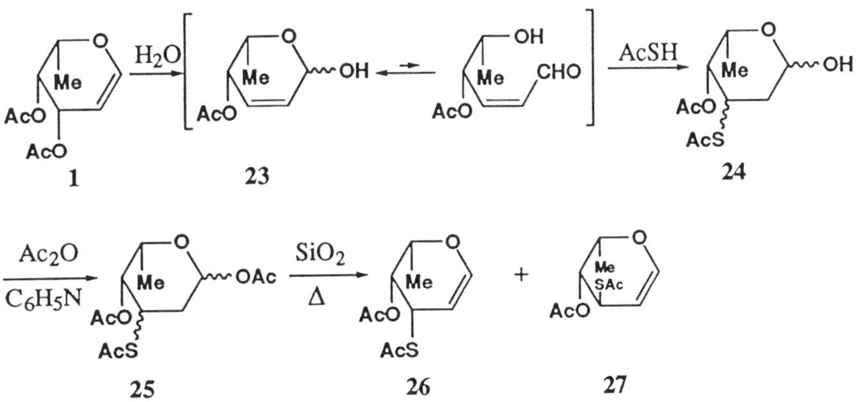

Therefore, to bypass glycals'-based regioselectivity, we have based our approach on conjugate addition reaction to hex-2-enopyranoses.29 Hex-2-enopyranoses can be easily prepared from their respective glycals.

Thus, heating 3,4-di-*O*-acetyl-L-fucal (**1**) in water gave 4-*O*-acetyl-2,3,6-trideoxy-L-threo-hex-2-enopyranose (**23**). Reaction of **23** with thioacetic acid in turn gave a mixture of products **24**, which was acetylated to 1,4-di-*O*-acetyl-3-*S*-acetyl-3-thio-hexopyranoses **25**. The 3-epimers were separated only after 3-

S-acetyl-2,3,6-trideoxy-L-lyxo- and -L-xylo-hex-1-enitols (**26**, **27**) were obtained by refluxing **25** in xylene-containing silica gel for 1 h. The desired 4-*O*-acetyl-3-*S*-acetyl-1,5-anhydro-3-thio-2,3,6-trideoxy-L-lyxo-hex-1-enitol (**26**), a crystalline compound with a melting point of 95°C, was treated with dry HCl in benzene to give respective glycosyl chloride.

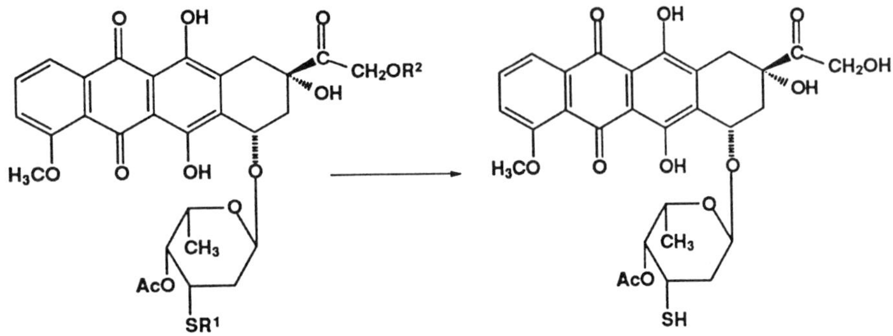

28 -R^1 = Ac; R^2 = OSit-BuMe$_2$
29 -R^1 = H; R^2 = OSit-BuMe$_2$

30

Coupling of excess glycosyl chloride with 14-*O*-*t*-butyldimethylsilyl-adriamycinone in dichloromethane in the presence of $HgBr_2$, HgO, and 3 Å (600 mesh) molecular sieves gave 3'-*S*-acetyl DOX derivative **28** in 78% yield. Deblocking the thiol group gave the 3'-mercaptodoxorubicin analog **29**. However, the 4-*O*-acetyl group was stable in basic conditions and resisted conventional deblocking procedures that can be used without causing decomposition of the antibiotic molecule. Despite that problem, **29** was desilylated to 3'-mercaptodoxorubicin **30**.

The 4'-Acetylated analog **30** was tested in vitro and showed cytotoxic potential similar to that of DOX [29]. Identical conditions were used to prepare 3'-*S*-acetyldaunorubicin [29]. Recently, by using different blocking groups at C-4' we were able to prepare a fully unblocked 3'-mercaptodoxorubicin. Even though the results of their biological evaluation are not yet available, these analogs deserve full evaluation to determine whether the unique chemical properties of the mercapto group can significantly differentiate biological activity of mercapto analogs from that of DOX.

V. ANTHRACYCLINES SUBSTITUTED WITH HALOGEN AT C-2'

Our initial hypothesis predicted that anthracyclines having a nonbasic or significantly less basic amine than either DOX or DNR should also be less likely substrates for P-gp-mediated efflux. To explore this part of our hypothesis, we proposed to alter the basicity of an amino group by introducing halogen at C-2'. Alteration of basicity of the amino group strongly affects equilibrium between charged and uncharged species in physiological solution. In fact, introduction of halogen at C-2' was expected to alter pKa from ~8 for natural antibiotics to ~6 for 2'-halo-substituted compounds. Thus, in contrast to DOX where more than 90% of drug, is in physiological pH in charged form, less than 10% of drug was in protonated form for 2'-halo-3'-amino analogs. These changes of pKa can affect the rate of cellular uptake of the drug through a passive diffusion process and its intracellular distribution, as well as interaction with macromolecules.

Interestingly, the electron-withdrawing properties of halogen at C-2', besides decreasing the basicity of the amino group, also affected properties of the glycosidic bond, a crucial element of the anthracycline structure. Thus, in an attempt to dissect and test both effects, we synthesized both 3'-amino and 3'-deamino-3'-oxy analogs with halogen at C-2'. It was also important to prepare both groups of analogs because such compounds were not known and the biological effects of the substitution at C-2' were never tested. Thus, in a series of 2'-halogenated anthracyclines prepared by us, [30-39] we demonstrated that halogen at C-2' is important for biological activity. Several years later, other groups prepared 2'-halo analogs of anthracyclines and assessed their activity [40-45]. Together, these experiments confirmed that a properly oriented halogen at C-2' can be used to decrease the basicity of an amino group without adversely affecting biological activity.

VI. 3'-AMINO-2'-HALO ANTHRACYCLINES WITH DECREASED BASICITY

In our synthetic strategy, we have used electrophilic addition to glycals as the most direct way to prepare 2'-halo-glycosides. Thus, we needed to developed efficient methods for the preparation of 3-amino glycals from respective amino sugars. Initially, we prepared them from acetylated 2-deoxy-hexopyranose by elimination of acetic acid on silicagel [46] and more recently by a more efficient method that allowed us to obtain glycal in a two-step one-pot reaction by, first, generating of glycosyl bromide in a mild condition and then subsequently eliminating HBr with *N,N*-diisopropylethylamine [47,48].
Reaction of the obtained daunosaminal **31** with daunomycinone (**33**) in the presence of NIS led to one main product **34** in an α-L-*talo* configuration. Deblocking the hydroxyl group at the 4' position with sodium methoxide in **34** was followed by removal of the trifluoroacetyl group with 1M sodium hydroxide solution.

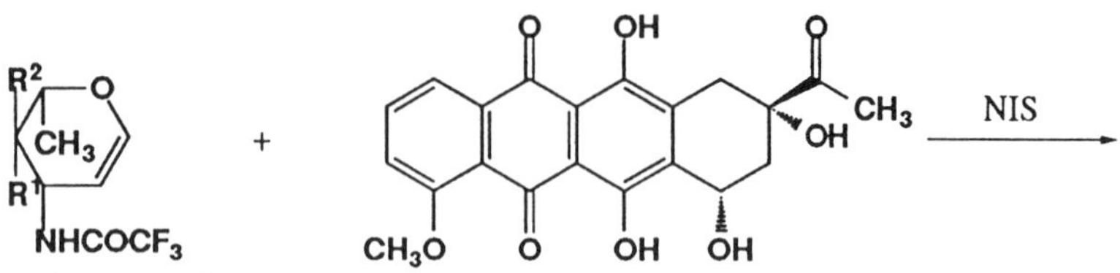

31 - R[1]=OAc; R[2]=H
32 - R[1]=H; R[2]=OAc

33

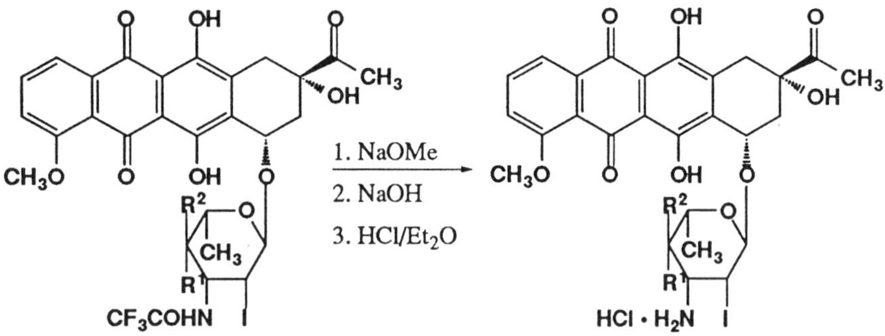

34 - R[1]=OAc; R[2]=H
35 - R[1]=H; R[2]=OAc

36 - R[1]=OH; R[2]=H
37 - R[1]=H; R[2]=OH

The final product, the 2'-iodo-daunorubicin **36**, was obtained and characterized as hydrochloride by precipitation from methanol/ethyl ether solution.36 We have followed then with synthesis of the 4'-epimer **37**. Starting from glycal **32** in an arabino configuration and using similar conditions, the 2'-iodo-4'-epidaunorubicin **37** was prepared with a good yield.

Biological evaluation of **36** and **37** indicated that trans orientation of 4'-OH to the axially spaced 2'-iodine is associated with higher antitumor activity in vivo. These observations focused our further efforts on 2'-bromo-4'-epi analogs having the α-L-*manno* configuration rather than on analogs having the α-L-*talo* configuration.

Our approach to the preparation of 2'-bromo anthracyclines differed from that of 2'-iodo analogues because we were unable to achieve direct glycosidation with NBS, like we did with NIS. Instead, we used the two-step approach. First, simple bromination gave us a mixture of glycosyl bromides, which were separated by column chromatography; then, in a second step, glycosyl bromides with a defined configuration at C-2' were coupled with daunomycinone.

This approach was complicated by the fact that bromination of acosaminal **38** in dichloromethane gave a mixture of glycosyl bromides **39** and **40** epimeric at C-2 with *manno* isomer as a minor product (ratio of *manno* to *gluco,* 3:7), whereas from our biological evaluation, it was clear that the equatorial orientation of halogen (*gluco* isomer) strongly reduces cytotoxic potential in vitro and antitumor activity in vivo. Thus, to solve the problem of low yield of the desired *manno* glycosyl bromide **39**, we undertook separate investigations into factors influencing the stereochemistry of electrophilic addition of bromine to acosaminal.

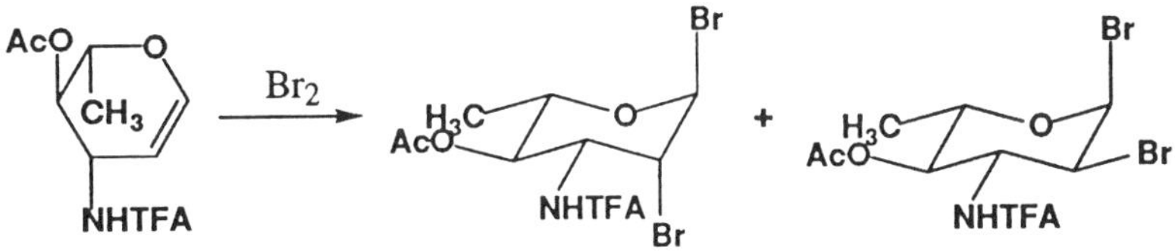

38 **39** - α-*manno* **40** - α-*gluco*

Interestingly, we have found that the presence of iodine during the bromination reaction alters stereochemical outcome. Using this method we were able to reverse the ratio of *manno* to *gluco* from 3:7 to 6:4 and to isolate α-*manno* glycosyl bromide as the main product [49, 50]. In addition, a 6:4 mixture of glycosyl bromides could be further enriched in the α-*manno* configuration by crystallization. Therefore, the combination of a novel bromination procedure and subsequent crystallization allows for coupling reactions to be performed on pure or highly enriched glycosyl bromides.

The required 2'-bromo-α-L-*manno* daunorubicin was obtained as the only product in very good yield by using silver trifluoromethanesulfonate as a promoter [49,50]. In contrast, the coupling of glycosyl bromide **40** in the *gluco* configuration led to a mixture of α and β anomers. All 2'-bromo-analogs isolated from coupling reactions were deblocked with sodium methoxide and then with 1M sodium hydroxide. Amino groups of these unblocked 2'-halo analogs were significantly less basic than those in DOX and DNR; however, ammonium salts were formed easily and compounds were water soluble. Thus, final products having the α-L-*manno* (**WP401**), α-L-*gluco* (**WP400**), or β-L-*gluco* (**WP402**) configuration were isolated and characterized as hydrochlorides.

Comparison of cytotoxic potentials in L1210 and M5076 cell lines revealed that **WP401** was as cytotoxic or more cytotoxic than DOX. However, **WP400** and **WP402** were clearly less cytotoxic than DOX and **WP401**. Thus, it is apparent from these experiments that the orientation of the halogen in **WP400** and **WP402** is responsible for the diminished cytotoxicity. Currently, we are investigating how the equatorial (2'*S*) orientation of halogen can decrease cytotoxicity of the molecule when compared to 2'-unsubstituted compound and how axially oriented halogen and potentially other substituents can positively affect interaction with such molecules as DNA or topoisomerases.

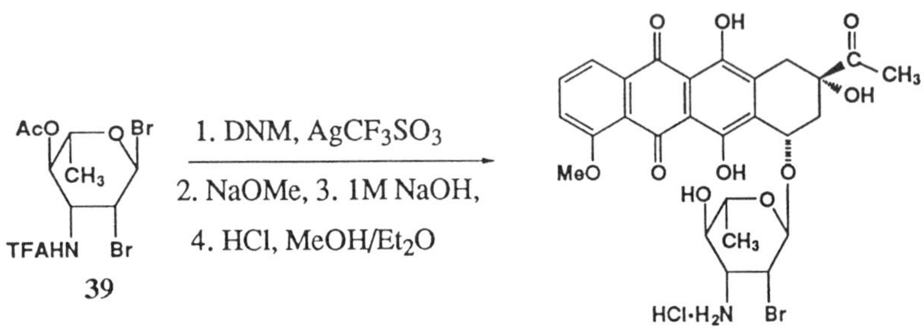

41 (WP401)

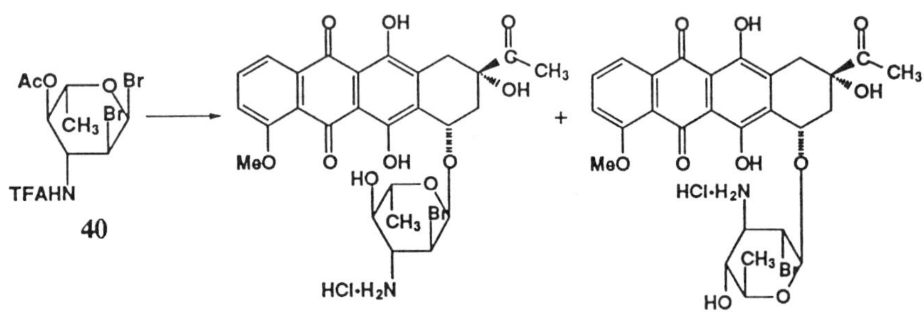

42 (WP400) 43 (WP402)

The same effects of halogen orientation at C-2' on biological activity were noticed by us in the 3'-deamino anthracyclines (discussed in detail in the next chapter; see Figure 4) and by others in the 2'-fluoro substituted anthracyclines [51-53].

Comparison in vivo of 2'-iodo analogues **36** and **37** revealed that compound **37** having an α-L-manno configuration (epi orientation of hydroxyl at C-4') displayed higher activity than analogue **36** having an α-L-talo configuration (orientation of hydroxyl at C-4' identical to that of DOX and DNR). These data indicated that epimerization at C-4' has a positive effect on antitumor activity and that a compound having α-L-manno is more active than its parent drug

DNR, whereas the analogue having an α-L-talo configuration (**36**) showed activity similar to DNR. Confirmation of that observation and another indication that the nature of a halogen at C-2' might not be important to biological activity came from the evaluation of α-L-talo analogues fluorinated at C-2'. Both 2'-fluorodaunorubicin [54] and 2'-fluorocarminomycin [55] displayed activity against sensitive tumors similar to their parent 3'-amino drugs.

Having established the structure-activity relationship for the 2'-halo-3-amino anthracyclines, we returned to our original task of evaluating our initial hypothesis and checking if in fact the reduced basicity can increase activity against MDR cells. Both **WP401** and less toxic **WP402** were compared with DOX against sensitive human carcinoma KB-3-1 and childhood leukemia CEM and against their multidrug-resistant counterparts KB-V-1 and CEM-VBL cell lines. Both compounds with reduced basicity displayed significantly reduced RIs. **WP401** displayed the same potency as DOX against CEM cells and was significantly more cytotoxic against the resistant CEM-VBL cells (RI of 10). **WP401** showed better activity against sensitive KB-3-1 cells than DOX and was more than 20-fold more cytotoxic than DOX against KB-V-1 MDR cells. Within its own range of cytotoxic potentials, a less potent **WP402** analogue also displayed a significantly lower RI of 25 than DOX, which showed an RI of 356.

In summary, these results clearly indicate that by reducing the basicity of the amino group, significantly better activity against MDR cells was achieved. Thus, these experiments provided additional support for our hypothesis that reduction of basicity will lead to circumvention of P-gp-mediated efflux.

VII. 3'-HYDROXY ANTHRACYCLINES HALOGENATED AT THE C-2' POSITION

The positive biological effects resulting not only from decreased basicity but also from the presence of a 2'*S*-oriented halogen indicated that compounds with the 3'-amino group removed and replaced with a nonbasic substituent and having a halogen at C-2' might have a unique activity. To analyze their potential and their structure-activity relationship, we have synthesized and examined a series of 2'-iodo, 2'-chloro-, and 2'-bromo-3-deaminated anthracyclines. The 2'-iodo analogs were prepared in a one-step reaction by reacting daunomycinone in the presence of *N*-iodosuccinimide (NIS) with 3,4-di-*O*-acetyl-L-fucal (**1**) to give 2'-iodo-3',4'-diacetoxy-α-L-*talo*-daunorubicin **44** and subsequently with 3,4-di-*O*-acetyl-L-rhamnal (**9**) to give two 2'-iodo trans addition products with (**45**) and β-L-*gluco* (**46**) configurations [32, 33, 35, 38].

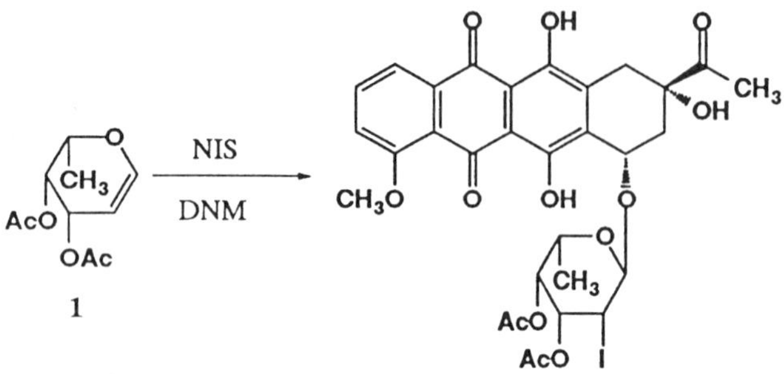

44

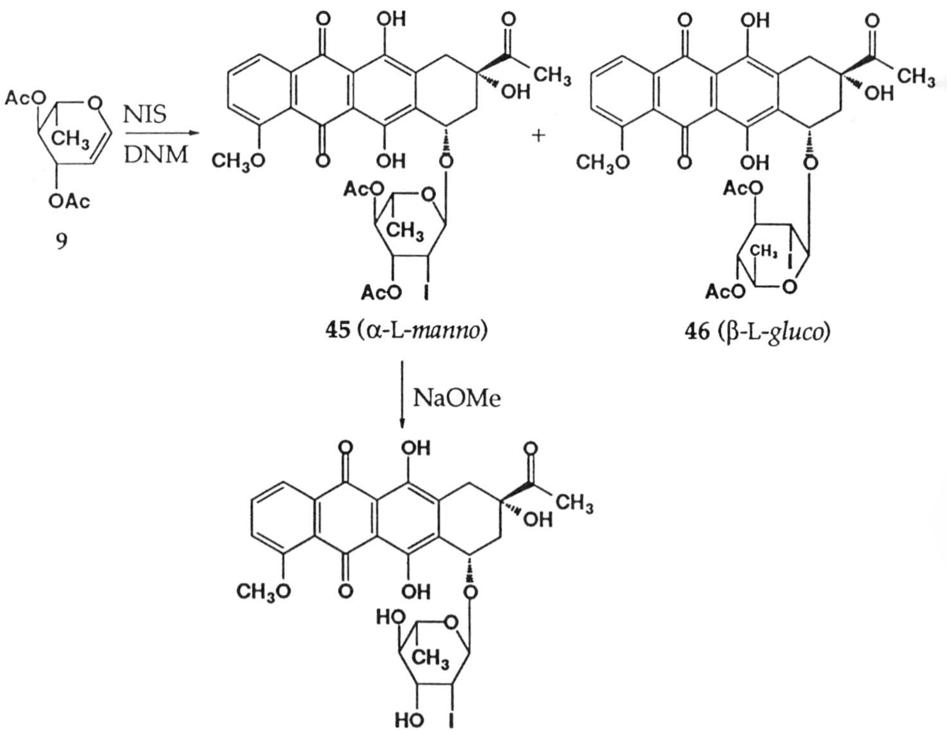

45 (α-L-*manno*) **46** (β-L-*gluco*)

47 (WP8)

Our biological evaluation indicated that, like the 2'-halo-3'-amino series, the compound having the α-L-*manno* configuration (**45**) was significantly more active against P-388 leukemia (T/C 247 at 50 mg/kg and 208 at 25 mg/kg) than DNR. **45** also showed good activity against L-1210 lymphoid leukemia (T/C 196 at 25 mg/kg) and Lewis lung carcinoma (T/C 187 at 25 mg/kg) and significant activity against B-16 melanoma (T/C 218 at 25 mg/kg). Again, similar to the 2'-halo-3'-amino series, compound **44** with the talo configuration was less efficacious than **45** and had activity comparable to that of daunorubicin (T/C 162), whereas the compound with the β-L-*gluco* (**46**) configuration was inactive up to the highest tested dose of 50 mg/kg.

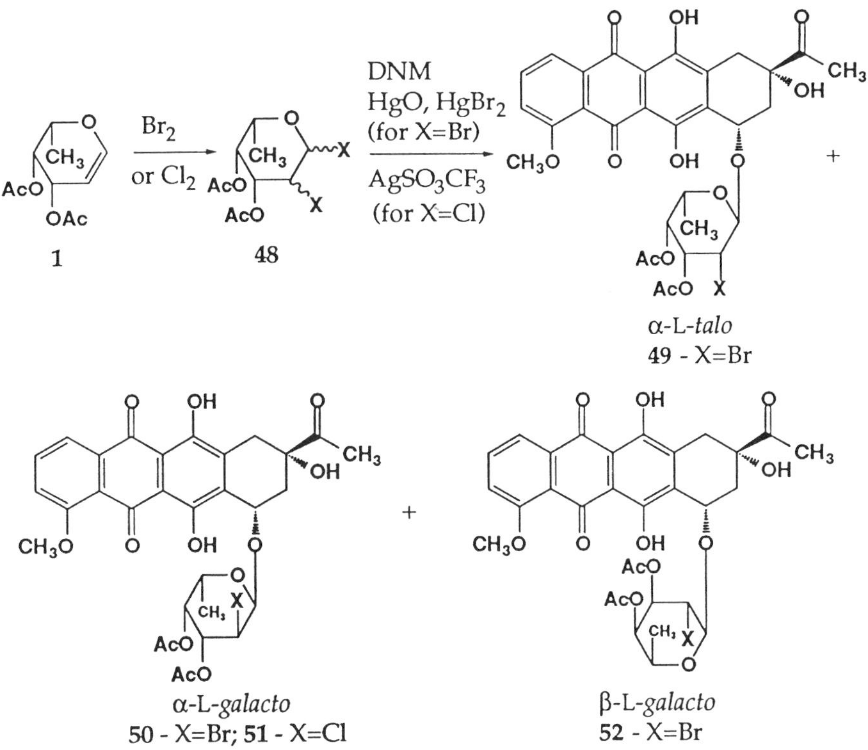

1 48 α-L-*talo* 49 - X=Br

α-L-*galacto* 50 - X=Br; 51 - X=Cl

β-L-*galacto* 52 - X=Br

As expected, the deblocking of hydroxyl groups at C-3' and C-4' led to compounds with increased potency. For example, the unblocked 2'-iodo-2,6-dideoxy-?-L-*manno*-daunorubicin **WP8** (**47**) was more potent in vivo against P-338 than its 3,4-di-*O*-acetyl derivative **45** (at a dose of 3.12 mg/kg, the T/C range was 144-178); however, efficacy remain unchanged (T/C of 222 at 12.5

mg/kg and 200 at 6.25 mg/kg for **WP8** vs T/C 247 at 50 mg/kg and 208 at 25 mg/kg for **45**).

These positive results with 2'-iodo-3'-oxy-anthracyclines prompted synthesis of 2'-bromo and 2'-chloro analogues. Our synthetic approach was straightforward: both 3,4-di-*O*-acetyl-L-fucal (**1**) and 3,4-di-*O*-acetyl-L-rhamnal (**9**) were brominated or chlorinated, and the mixture of glycosyl halides was then coupled with daunomycinone, followed by careful chromatography. Thus, in a reaction starting with bromination of L-fucal **1**, a mixture of glycosyl bromides **48** was coupled with daunomycinone under Koenigs-Knorr conditions (HgO, $HgBr_2$) to 2'-bromo-daunorubicin analogues having α-L-*talo* (**49**), α-L-*galacto* (**50**), and β-L-*galacto* (**52**) configurations in the ratio 3: 6: 2 [39]. Reaction starting from bromination of 3,4-di-*O*-acetyl-L-rhamnal **9** also resulted in a mixture of products whose separation gave pure α-*manno* analogue **54** (13%), α-*gluco* isomer **56** (28%), and β-*gluco* isomer **58** (27%).

In contrast to bromination, the chlorination of 3,4-di-*O*-acetyl-L-fucal (**1**) gave an L-*galacto* isomer as the main component. Thus, coupling with daunomycinone in the presence of silver triflate gave α-L-galacto-daunorubicin analog **51** in 44% yield. Chlorination of 3,4-di-O-acetyl-L-rhamnal (**9**) gave a mixture of L-*gluco* and L-*manno* isomers in a 4:1 ratio. Silver triflate-promoted coupling of the reaction mixture with daunomycinone gave α anomers with L-*manno* (**55**) and L-*gluco* (**57**) configurations.

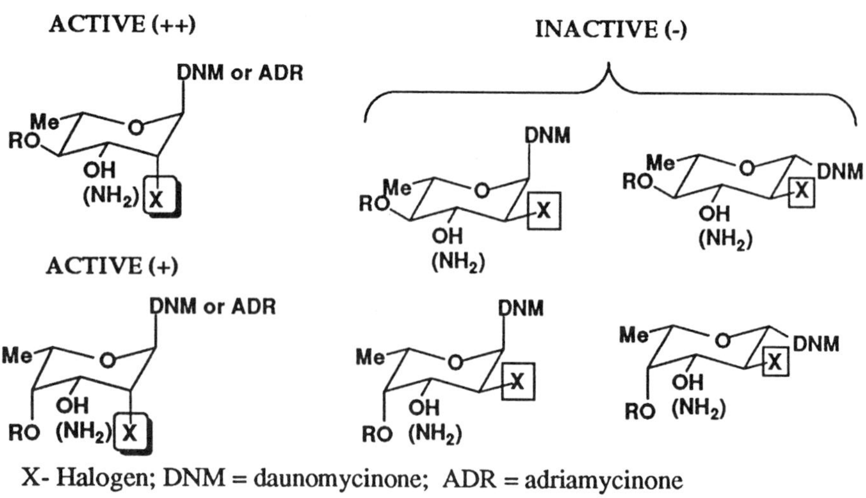

X- Halogen; DNM = daunomycinone; ADR = adriamycinone

Figure 4. Structure-activity relationship of 2'-halogenated anthracyclines.

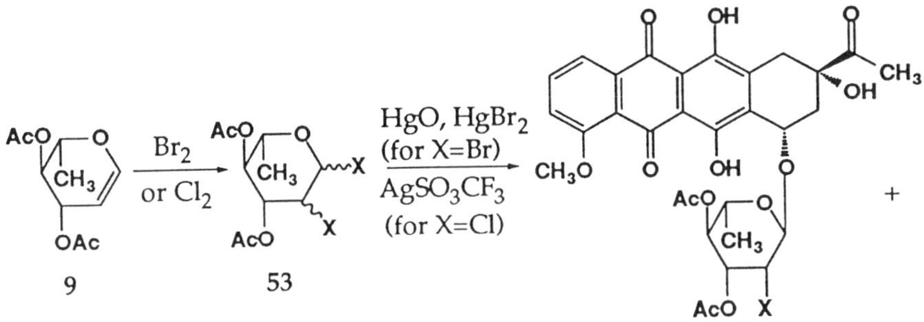

9 53

α-L-*manno*
54 - X=Br; 55 - X=Cl

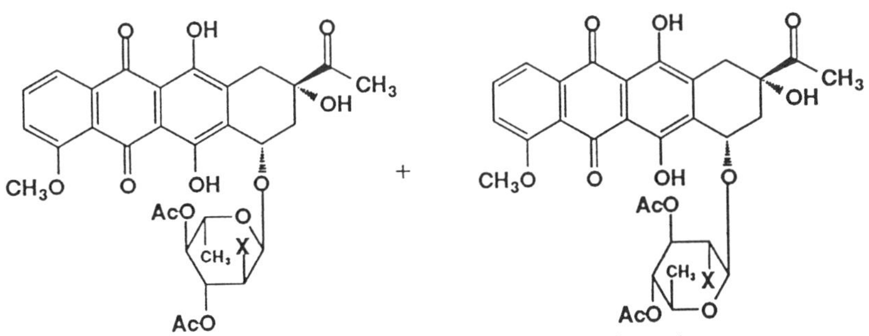

α-L-*gluco*
56 - X=Br; 57 - X=Cl

β-L-*gluco*
58 - X=Br

The structure-activity relationships observed for the bromo and chloro series were identical to the one established for iodo anthracyclines (Figure 4). Thus, the 2'-bromo-α-L-*manno*-daunorubicin **54** was significantly more active than the 2'-bromo-α-L-*talo*-daunorubicin **49**. Analogue **54** with an α-L-*manno* configuration showed in vivo activity against murine P-388 leukemia (T/C 245 at 25 mg/kg) when first injected on day 5, and in separate experiment (when injected on days 1, 5, and 9) showed T/C 278 at 25 mg/kg and T/C 229 at 12.5 mg/kg. Conversely, for α-L-*talo*-daunorubicin **49**, the best noted T/C was 181 at 83 mg/kg. Like bromo analogue **54**, the 2'-chloro-daunorubicin (**55**) with an L-*manno* configuration showed good antitumor activity in vivo against P-388 leukemia (T/C 248 at 25 mg/kg). Finally, as expected, all 2'-bromo and 2'-chloro analogs with L-*gluco* (**56**, **57**, **58**) and L-*galacto* (**50**, **51**, **52**) configurations displayed no activity in vivo at doses up to 50 mg/kg against P-388 leukemia cells, further confirming the strong deactivating effect of equatorially oriented halogen at C-2'.

VIII. SELECTION OF ANNAMYCIN AND ITS PRECLINICAL EVALUATION

The studies presented above have established a well-supported structure-activity relationship for 2'-halo- and most likely for the other 2'-substituted anthracyclines. Consequently, it became obvious to us that other aglycons should be tried to refine the 3'-deaminoanthracyclines' activity against MDR tumors and to increase its compatibility with drug delivery-enhancing tumor targeting. In the selection process, which is discussed in more detail elsewhere,16 we have focused on 4-demethoxyadriamycinone, an aglycon similar to that in the clinically used drug idarubicin and hydroxylated at C-14. This aglycon was coupled with 3,4-di-*O*-acetyl-L-rhamnal to produce, after deblocking, a compound called Annamycin [10,16,35,56-63]

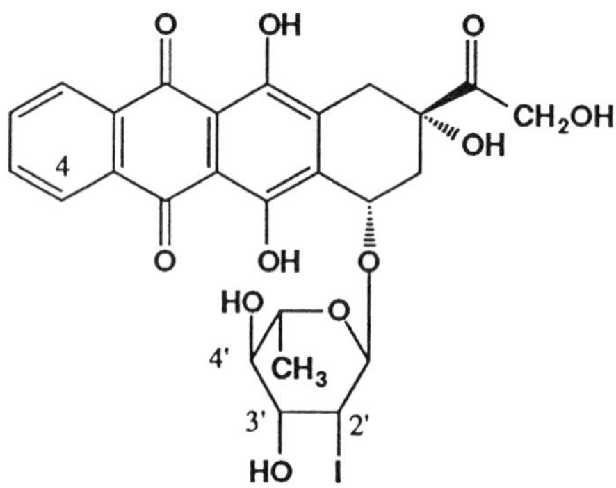

Annamycin

Interestingly, this analogue incorporates not only changes imposed by our hypothesis, such as decreased basicity (3'-hydroxy group) and introduction of an axially oriented halogen at C-2', but also components of the new generation of anthracyclines in clinical use. Thus, it is demethoxylated at C-4 like idarubicin and has an inverted configuration at C-4' like epirubicin. Another important reason for the selection of Annamycin for further development was its ability to form a stable liposomal preparation [57, 58, 64]. Below, we shortly summarize the results of the preclinical studies of Annamycin in its free and liposomal forms.

Comparison in vitro of Annamycin (F-Ann) with its liposomal form (L-Ann) in sensitive and MDR cell lines (KB, KB-V-1, P388, P388/DOX, CEM, CEM/VBL, 8226, 8226/DOX) indicated no difference between the two [57]. Both F-Ann and L-Ann were more cytotoxic than DOX in all MDR cell lines, which clearly indicates the potential of Annamycin against MDR tumors. Owing to the higher uptake by resistant P388/DOX cells, Annamycin activity against MDR cell lines was higher than DOX. Furthermore, the efflux pattern of Annamycin suggested that it is not P-gp mediated, as supported by the fact that efflux and uptake were not affected by resistance-reversing agents like verapamil [59]. To illustrate the observed differences in vitro between Annamycin and DOX, let us point out that Annamycin was as cytotoxic as DOX against sensitive P388 cells but 50 to more than 100 times more cytotoxic against resistant P388/DOX cells [59].

Pharmacological and biochemical evaluations revealed that Annamycin induced single-strand DNA breaks, double-strand DNA breaks, and DNA-protein cross-links in both sensitive and resistant cells at much lower concentrations that DOX [59]. Interestingly, Annamycin also showed a significantly higher ability to induce DNA degradation and programmed cell death (apoptosis) than did DOX [61].

The important part of our studies was confirmation in vivo of the observed in vitro activity. Also, as expected, we demonstrated the advantage of liposomal formulation. In one of the experiments, we showed that liposomes increased tumor targeting. In mice bearing advanced subcutaneous B16 melanoma tumors, tumor uptake studies of F-Ann, L-Ann, and DOX showed that levels of L-Annamycin in tumor were 10 times higher than those of DOX, whereas levels of F-Ann were 5 times higher than those of DOX [56].

Further studies of F-Ann, L-Ann, and DOX in vivo in subcutaneous KB-3-1 and KB-V-1 (MDR tumor) human xenografts revealed that L-Ann was clearly superior to DOX, though F-Ann was also significantly more active than DOX [57]. These results confirmed the activity of Annamycin in vivo against KB-V-1 MDR tumors.

The survival rates in other in vivo models like M5076 reticulosarcoma and Lewis lung carcinoma ranged from 39% to 45% for F-Ann and 63% to 85% for L-Ann, whereas DOX was inactive or marginally active in these models [57]. These experiments illustrate again our decision to use a drug delivery system and demonstrate the potential of the "double-advantage approach," a novel strategy reviewed by us in detail [16].

In summary, Annamycin appears to be a drug significantly different from DOX, with its own unique activity profile, a definitely broader spectrum of activity, and lower cardiotoxicity. Especially important seems to be its activity against resistant tumors. Even though the decreased basicity of the sugar portion is not the only element of Annamycin's structure that contributes to the observed increase in activity against resistant tumors, we believe that it is the main factor. Thus, future clinical studies of Annamycin in humans should determine if our hypothesis has clinical relevance and if decreasing drug's basicity can be used to design other clinically useful drugs. Annamycin is currently undergoing Phase I clinical studies at The University of Texas M. D. Anderson Cancer Center, which are expected to be finished in early 1996.

REFERENCES

1. Arcamone F. Doxorubicin Anticancer Antibiotics. In: DeStevens G, ed. Medicinal Chemistry. Vol. 17. New York: Academic Press, 1981.
2. Lown JW. Anthracycline and anthracenedione-based anticancer agents. Bioactive Molecules. Vol. 6. Amsterdam; Elsevier Science Publishers B. V., 1988.
3. Priebe W. Anthracycline Antibiotics: Novel Analogues, Methods of Delivery, and Mechanisms of Action. Washington, DC: American Chemical Society, 1995.
4. DeVita JVT, Hellman S, Rosenberg SA. Cancer: Principles and Practice of Oncology. Philadelphia: J. B. Lippincott, 1993.
5. Chaires JB, Priebe W, Graves DE, Burke TG. Dissection of the free energy of anthracycline antibiotic binding to DNA: electrostatic contributions. J Am Chem Soc 1993; 115:5360-5364.
6. Chaires JB, Satyanarayana S, Suh D, Fokt I, Przewloka T, Priebe W. Parsing the Free Energy of Anthracycline Antibiotic Binding to DNA. Biochemistry 1996. In press.
7. Bradley G, Juranka PF, Ling V. Mechanism of multidrug resistance. Biochim. Biophys. Acta 1988; 948:87-128.
8. Endicott JA, Ling V. The biochemistry of P-glycoprotein-mediated multidrug resistance. Annu Rev Biochem 1989; 58:137-171.
9. Pastan I, Gottesman M. Multiple-drug resistance in human cancer. N Engl J Med 1987; 316:1388-1393.
10. Priebe W. Mechanism of Action-Governed Design of Anthracycline Antibiotics: A "Turn-Off/Turn-On" Approach. Current Pharmaceutical Design 1995; 1:51-68.
11. Priebe W, Van NT, Perez-Soler R. Biological effects of deamination at C-3': Evaluation of 3'-hydroxy-doxorubicin (Hydroxyrubicin). J Cancer Res Clin Oncol 1990; 116 (Suppl., Part I):439.
12. Priebe W, Van NT, Neamati N, Grynkiewicz G, Perez-Soler R. Non-cross-resistant anthracyclines with reduced basicity in sugar moiety. Proceedings of the Third Internatational Symposium on Molecular Aspects of Chemotherapy. Gdansk, Poland, 1991.
13. Priebe W, Van NT, Perez-Soler R. Activity of 3'-deaminated anthracyclines against MDR cells. Proc Am Assoc Cancer Res, 1991; 82:2245.
14. Priebe W, Van NT, Perez-Soler R. Reduction of basicity of anthracyclines as a method to overcome multidrug resistance. Proc Am Chem Soc Meet, San Francisco, 1992; 203: MEDI-40.
15. Priebe W, Van NT, Burke TG, Perez-Soler R. Removal of the basic center from doxorubicin partially overcomes multidrug resistance and decreases cardiotoxicity. Anti-Cancer Drugs 1993; 4:37-48.

16. Priebe W, Perez-Soler R. Design and tumor targeting of anthracyclines able to overcome multidrug resistance: A double-advantage approach. Pharmacol Ther 1993; 60:215-234.
17. El Khadem HS, Swartz DL, Cermak RC. Synthesis of ε-rhodomycinone glycosides. J Med Chem 1977; 20:957-60.
18. Fuchs E-F, Horton D, Weckerle W. Synthesis of 7-*O*-(2,6-dideoxy-α-L-*lyxo*-hexopyranosyl)daunomycinone, a functional analog of daunorubicin. Carbohydr Res 1977; 57:C36-C39.
19. El Khadem HS, Swartz DL. Synthesis of 2-deoxy-L-fucopyranosyl-ε-pyrromycinone and 2-deoxy-D-erythro-pentopyranosyl dauno-mycinone, -carminomycinone, and -ε-pyrromycinone. Carbohydr Res 1978; 65:C1-C2.
20. El Khadem HS, Swartz DL. Synthesis of 2'-deoxy-L-fucopyranosylcarminomycinone and-ε-pyrromycinone as well as 2'-deoxy-D-erythro-pentopyranosyldaunomycinone, -carminomycinone, and -ε-pyrromycinone. J Med Chem 1981; 24:112-115.
21. Fuchs E-F, Horton D, Weckerle W, Winter-Mihaly E. Synthesis and antitumor activity of sugar-ring hydroxyl analogues of daunorubicin. J Med Chem 1979; 22:406-411.
22. Horton D, Priebe W, Varela O. Synthesis and antitumor activity of 3'-deamino-3'-hydroxydoxorubicin. A facile procedure for the preparation of doxorubicin analogs. J Antibiot 1984; 37:853-858.
23. Priebe W, Skibicki P, Krawczyk M, Grynkiewicz G. Stereoselective synthesis of anthracyline antibiotics. Proc Am Chem Soc Meet, San Diego, CA, 1994; 207:CARB-79.
24. Horton D, Priebe W, Varela O. 3'-Deamino-4'-epi-3'-hydroxydaunorubicin and -doxorubicin. Synthesis and antitumor activity. J Antibiot 1984; 37:1635-41.
25. Priebe W, Neamati N. Synthesis of optically active 3,4-dihydro-4-oxy- 2H-pyrans, Proceedings of the XIVth IUPAC International Carbohydrate Symposium, Stockholm, Sweden, 1988.
26. Perez-Soler R, Priebe W. Liposomal formulation and antitumor activity of 14-O-palmitoyl-hydroxyrubicin. Cancer Chemother Pharmacol 1992; 30:267-271.
27. Priebe W, Neamati N, Perez-Soler R. 3'-Hydroxyesorubicin. Synthesis and antitumor activity. J Antibiot 1990; 43:838-846.
28. Priebe W, Zamojski A. The acid-catalyzed reaction of thiols with alkyl 2,3-dideoxy-glyc-2-enopyranosides or glycals. Tetrahedron 1980; 36:287-297.
29. Priebe W, Grynkiewicz G, Neamati N, Perez-Soler R. A facile method for preparation of 3-thio-sugars and 3-thio-glycals: Synthesis of 3'-mercapto-3'-deamino-doxorubicin. Tetrahedron Lett 1991; 32:3313-3316.
30. Horton D, Priebe W. Novel 2'- and 14-substituted adriamycin analogs. Proc

Am Chem Soc Meet, San Francisco, CA, 1980; 180:CARB-22.
31. Horton D, Priebe W. Glycon and C-14 modified, antitumor doxorubicin analogs. Proc Am Chem Soc Meet, New York, 1981; CARB-11.
32. Horton D, Priebe W. Glycon- and C-14 modified, antitumor doxorubicin analogs. In: El Khadem HS, ed. Anthracycline Antibiotics. New York: Academic Press, Inc., 1982:197-224.
33. Horton D, Priebe W, Varela O. Novel daunorubicin analogs having substituents modifying the properties of the glycosidic bond. Proceedings of the XIth International Carbohydrate Symposium, Vancouver, British Columbia, Canada, 1982.
34. Priebe W, Horton D, Wolgemuth RL. 14-Acyloxy-2'-haloanthracycline anticancer antibiotics. European Patent EP116,222. August 22, 1984.
35. Horton D, Priebe W. 2'-Halo derivatives of daunomycin, desmethoxy daunomycin, adriamycin, carminomycin. U.S. Patent 4,427,664. January 24, 1984.
36. Horton D, Priebe W. 3'-Amino-2'-halo-anthracycline antibiotics. U.S. Patent 4,562,177. December 31, 1985.
37. Horton D, Priebe W. 4-Demethoxy-3'-desamino-2'-halo-anthracycline and pharmaceutical composition containing same. U.S. Patent 4,537,882. August 27, 1985.
38. Horton D, Priebe W. Oxyhalogenation of glycals for the synthesis of antitumor active 2'-halo daunorubicin analogs. Carbohyd Res 1985; 136:391-396.
39. Horton D, Priebe W, Varela O. Synthesis and antitumor activity of 2'-bromo- and 2'-chloro-3'-acetoxy daunorubicin analogs. Carbohyd Res 1985; 144:305-315.
40. Tsuchiya T, Takagi Y, Ok KD, et al. Syntheses and antitumor activities of 7-O-(2,6-dideoxy-2-fluoro-α-L-talopyranosyl)-daunomycinone and -adriamycinone. J Antibiot 1986; 39:731-733.
41. Tsuchiya T, Takagi Y, Umezawa S, et al. Synthesis and antitumor activities of 14-O-acyl derivatives of 7-O-(2,6-dideoxy-2-fluoro-α-L-talopyranosyl)adriamycinone. J Antibiot 1988; 41:988-991.
42. Umezawa H, Umezawa S, Tsucha O, Takeuchi T, Takagi Y. Preparation of 7-O-(2,6-dideoxy-2-fluoro-α-L-talopyranosyl)daunomycinone and -adriamycinone as antitumor agents. Jpn Kokai Tokkyo Koho, 1987.
43. Umezawa H, Umezawa S, Tsuchiya O, et al. Preparation and testing of 7-O-(2,6-dideoxy-2-fluoro-?-L-talopyranosyl)adriamycinone 14-O-carboxylic acid esters as antitumor agents. Jpn Kokai Tokkyo Koho, 1988.
44. Gerken M, Blank S, Kolar C, Hermentin P. Synthesis of 2'-iodo analogs of β rhodomycins. J Carbohydr Chem 1989; 8:247-254.
45. Florent JC, Genit A, Monneret C. Synthesis and antitumor activity of new

anthracyclines. J Antibiot 1989; 12:1823-1830.
46. Horton D, Priebe W, Sznaidman M. Preparation of 4-O-acetyl-1,5-anhydro-2,3,6-trideoxy-3-trifluoroacetamido-L-lyxo-hex-1-enitol, a key intermediate in synthesis of daunosamine glycosides. Carbohyd Res 1989; 187:145-147.
47. Priebe W, Grynkiewicz G, Krawczyk M, Fokt I. A facile one-pot synthesis of glycals and 2-acetoxy-glycals. Proceedings of XVIIth International Carbohydrate Symposium, Ottawa, Canada, 1994.
48. Priebe W, Grynkiewicz G, Krawczyk M, Fokt I. Improved synthesis of glycals and 2-acetoxy-glycals. Proc Am Chem Soc Meet, 1994; 207:CARB-78.
49. Priebe W, Neamati N, Grynkiewicz G, Perez-Soler R. 2'-Bromo-anthracyclines. Proc Am Chem Soc Meet, Atlanta, 1991; 201:CARB-11.
50. Priebe W, Neamati N, Grynkiewicz G, Van NT, Burke TG, Perez-Soler R. 2'-Bromo anthracyclines. Synthesis and non cross resistant properties. Proc Am Assoc Cancer Res, 1992; 83:3332.
51. Tsuchiya T, Takagi Y. Synthesis and biological activities of fluorinated daunorubicin and doxorubicin analogues. In: Priebe W, ed. Anthracycline Antibiotics. Novel Analogues, Methods of Delivery, and Mechanisms of Action. Washington, DC: American Chemical Society, 1995:100-114.
52. Castillon S, Dessinges A, Faghih R, Lukacs G, Olesker A, Ton TT. Synthesis of 2'-C-β-fluorodaunomycin: an example of configurational retention in fluorodehydroxylation with diethylaminosulfur trifluoride. J Org Chem 1985; 50:4913-4917.
53. Baer HH, Siemsen L. Synthesis and biological activity of (*S*)-2'-fluorodaunorubicin. Can J Chem 1988; 66:187-90.
54. Takagi Y, Park H, Tsuchiya T, et al. Syntheses and antitumor activities of 7-O-(3-amino-2,3,6-trideoxy-2-fluoro-α-L-talopyranosyl) daunomycinone and -adriamycinone. J Antibiot 1989; 42:1315-17.
55. Baer HH, Mateo FH. Synthesis and biological activity of (*R*)-2'-fluorocarminomycin. Can. J. Chem. 1990; 68:2055.
56. Zou Y, Priebe W, Ling YH, Perez-Soler R. Organ distribution and tumor uptake of Annamycin, a new anthracycline derivative with high affinity for lipid membranes, entrapped in multilamellar vesicles. Cancer Chemother Pharmacol 1993; 32:190-196.
57. Zou Y, Ling YH, Van NT, Priebe W, Perez-Soler R. Antitumor activity of free and liposome-entrapped Annamycin, a lipophilic anthracycline antibiotic with non-cross-resistance properties. Cancer Res 1994; 54:1479-1484.
58. Zou Y, Ling YH, Reddy S, Priebe W, Perez-Soler R. Effect of vesicle size and lipid composition on the *in vivo* tumor selectivity and toxicity of the

non-cross resistant anthracycline Annamycin incorporated in liposomes. Int J Cancer 1995; 61:666-671.

59. Ling YH, Priebe W, Yang LY, Burke TG, Pommier Y, Perez-Soler R. *In vitro* cytotoxicity, cellular pharmacology, and DNA lesions induced by Annamycin, an anthracycline derivative with high affinity for lipid membranes. Cancer Res 1993; 53:1583-1589.
60. Ling YH, Priebe W, Perez-Soler R. Intrinsic cytotoxicity and reversal of multidrug-resistance by monensin in KB parent and MDR cells. Int J Oncolog 1993; 3:971-977.
61. Ling YH, Priebe W, Perez-Soler R. Apoptosis induced by anthracycline antibiotics in P388 parent and MDR cells. Cancer Res 1993; 53:1845-1852.
62. Ling YH, Zou Y, Van NT, Priebe W, Perez-Soler R. Partial circumvention of multi-drug resistance by Annamycin is associated with comparable inhibition of DNA synthesis in the nuclear matrix of sensitive and resistant cells. Int J Cancer 1995; 61:402-408.
63. Priebe W, Skibicki P, Varela O, et al. Non-cross-resistant anthracyclines with reduced basicity and increased stability of the glycosidic bond. In: Priebe W, ed. Anthracycline Antibiotics. Novel Analogues, Methods of Delivery, and Mechanisms of Action. Washington, DC: American Chemical Society, 1995:14-46.
64. Perez-Soler R, Priebe W. Anthracycline antibiotics with high liposome entrapment: structural features and biological activity. Cancer Res 1990; 50:4260-4266.

15
Lincomycin: An Organic Chemistry Perspective

Osman Achmatowicz and Barbara Szechner
Pharmaceutical Research Institute
Warsaw, Poland

1. Introduction

Lincomycin (**1**) is a member of a unique family of antibiotics characterized by the molecular structure comprised of alkyl 6-amino-6,8-dideoxy-1-thio-D-*erythro*-D-*galacto*-octopyranoside connected by an amide bond with a substituted L-proline moiety.

Structural and configurational details of the sugar portion as well as the substitution pattern of the amino acid proved to be essential for the antibacterial activity of the antibiotic.

Lincomycin exhibited antibacterial activity similar but not identical to the macrolide antibiotics and possessed good *in vitro* and *in vivo* potency against a variety of aerobic and anaerobic gram-positive micro-organisms. It was not cross-resistant with known antibiotics and possessed a low level of toxicity. The antibacterial activity of lincomycin is based on its ability to inhibit ribosomal protein synthesis, while allowing RNA and DNA synthesis to continue.

Lincomycin acts by specifically blocking a peptidyl transferase site on the 50S ribosomal subunit and affecting the process of peptide chain initiation [1-3]. A theoretical model which rationalized structure-activity relationships of lincomycin and related antibiotics was presented [1].

Lincomycin may be administered intravenously or by the oral route and is well tolerated by animals and humans. There was no evidence indicating life-endangering toxicity in clinical use of lincomycin. Preparations containing lincomycin are widely used in human [5] and veterinary [6] diseases.

Early reports of experimental studies and clinical experience with lincomycin in treatment of humans and animals were comprehensively presented in the monograph by Herrel [7]. Lincomycin was thereafter modified chemically on one hand to better understand structure-activity relations and on the other to prepare analogues of enhanced potency and broadened antibacterial spectrum.

As with any natural product endowed with significant biological activity and novel structure, lincomycin inspired numerous synthetic investigations. The molecular structure of the lincomycin sugar portion, commonly named lincosamine, with a "density" of chiral centers located in the cyclic and acyclic portion presented a formidable synthetic problem. Efforts to solve this problem included total and partial, starting from sugar precursor, synthetic approaches which became part of a broader effort to identify new reactions and strategies of general utility [8].

This ultimately contributed to the development of monosaccharide chemistry and new methodologies in stereoselective organic synthesis.

The present review briefly describes derivatives and analogues of lincomycin (**1**). However, it is mainly devoted to synthetic endeavors which evolved from lincomycin chemistry, in particular syntheses of the sugar portion lincosamine.

2. Isolation and Structure

2.1. Isolation

The discovery of lincomycin (**1**) in the fermentation broth of *Streptomyces lincolnensis var. lincolnensis*, a new species of streptomycete found in a soil sample from the vicinity of Lincoln, Nebraska was published in 1962 [9]. In the same year, lincomycin was isolated, its chemical and physical characteristics [10] as well as *in vitro* and *in vivo* evaluation [11] were reported.

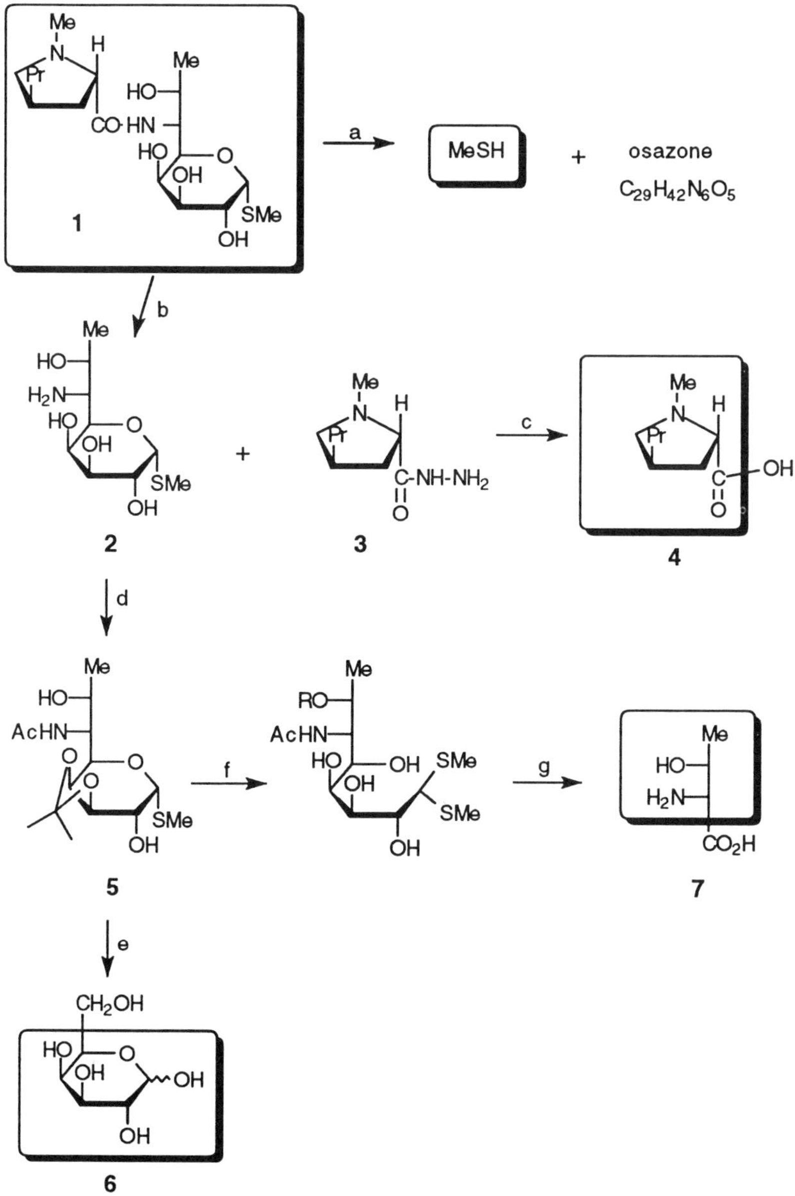

a. HCl/H_2O; b. H_2N-NH_2·H_2O; c. HCl/H_2O, Δ; d. Ac_2O, Me_2C=O - H^+; e. i. b, ii. HIO_4, iii. $NaBH_4$, iv. H_3O^+; f. i. MeSH/H^+, ii. $NaIO_4$ - $KMnO_4$.

Scheme 1

2.2. Structure and configuration

The structure and absolute configuration of lincomycin (**1**) were established with the aid of classical chemical methods [12-15] supported by the spectrometric data of the parent antibiotic and its degradation products [16]. Lincomycin had molecular formula $C_{18}H_{34}N_2O_6S$ and contained one basic function with $pK_a = 7.5$. The ^1H NMR spectrum revealed the presence of one N-methyl and two C-methyl groups. The IR spectrum indicated a monosubstituted amide (1640, 1530 cm^{-1}). Key transformations which led to the lincomycin structure elucidation were as follows. Acidic hydrolysis liberated methanethiol and a second fragment, a reducing sugar isolated as the phenylosazone (Scheme 1). Hydrazinolysis cleaved lincomycin into a sugar portion, methyl α-thiolincosaminide (**2**) and amino acid isolated as a hydrazide **3**, respectively. Acid hydrolysis of hydrazide **3** freed an amino acid $C_9H_{17}NO_3$. Its structure and absolute configuration as *trans*-L-4-propylhygric acid (**4**) was established by the synthesis from 4-keto-L-proline **8** (Scheme 2).

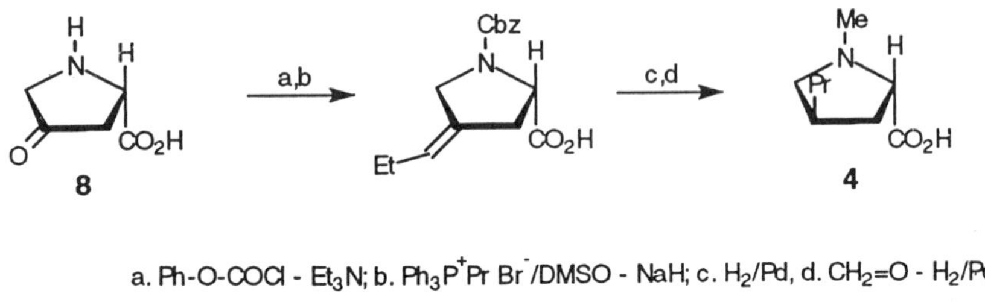

Scheme 2

The *galacto* configuration of the pyranoside moiety and location of the amino function in the sugar portion were suggested by the ^1H NMR spectra of its *N*-acetyl and pentaacetyl derivatives. *Galacto* configuration was demonstrated by the oxidative cleavage of the side chain two carbons and isolation of D-galactose (**6**) (as α -methylphenylhydrazone). On the other hand, formation of D-*allo*-threonine (**7**) (isolated as *N*-2,4-dinitrophenyl derivative) as a result of selective oxidative degradation of the pyranoside moiety in protected α-thiolincosaminide (**5**) demonstrated the structure and D-*erythro* absolute configuration of the side chain (Scheme 1).

2.3. Three-dimensional structure

The crystal structure and molecular conformation in the solid state was determined for lincomycin hydrochloride monohydrate by single crystal X-ray analysis [17]. This result furnished unequivocal confirmation of the parent antibiotic structure **1** deduced from the chemical and spectrometric evidence. The crystal structure of lincomycin hydrochloride is characterized by a network of hydrogen bonds, including the chloride ion hydrogen bonded to the amide nitrogen and not to the proline ring nitrogen atom. In the solid state there occurs a hydrogen bond between hydrogen atom at N-1' and the amide oxygen atoms.

The three-dimensional structure in solution of lincomycin (**1**) and its protonated form has been examined in detail with the aid of spectrometric (IR, ^{1}H and ^{13}C NMR) and theoretical methods.

The IR spectra of lincomycin in various inert solvents (CCl_4, $CCl_2{=}CCl_2$, dioxane, $CHCl_3$) indicated formation of associates due to the intermolecular hydrogen bonds. In dilute solutions intermolecular hydrogen bonds stabilized the rigid molecular conformation [18] The rigidity of the lincomycin molecule was also deduced from the ^{13}C spin-lattice relaxation times (T_1) of lincomycin hydrochloride [19]. The similarity of T_1 values for the pyranoside ring and the side chain carbon atoms indicated isotropical tumbling of the sugar portion without free rotation around C-6 — C-7 bond. The pyrrolidine ring ^{13}C atoms were characterized by larger nT_1 (n-number of directly bonded hydrogen atoms) values thus suggesting its increased mobility with respect to the sugar moiety. Analogous data obtained for clindamycin (**50**) hydrochloride demonstrated conformational similarity in solution of both antibiotics [19]. Theoretical studies on the lincomycin conformation in conjunction with advanced NMR data analysis resulted in defining the preferred three-dimensional structure of the antibiotic in solution.

Ab initio Hartree-Fock SCI calculations of the several molecular species used as molecular fragments to model the pyrrolidine, amide and sugar fragments of lincomycin led to identification of the energetically preferred conformation in solution of the free antibiotic and its protonated form [20]. The absolute minimum for the free antibiotic corresponded to the conformation in which a hydrogen bond occurred between the amide hydrogen and the pyrrolidine nitrogen atom forming a planar five-membered ring structure. For the protonated form the absolute minimum appeared in the conformation similar to the one found in the solid state [17] with hydrogen bond between the quaternary amine hydrogen and amide oxygen atom also forming a planar five-membered ring structure. In both conformations the position of hydrogens: HC-5 — C-6-H and H—N — C-4—H remain essentially *anti*-periplanar whereas

H—C-5 — C-6—H *syn*-clinal.

In another theoretical study, based on the classical potential energy calculations [21] the potential energy contour plot for the rotations around C-5 — C-6, C-6 — C-7 and C-6 — N bonds was obtained. The result indicated the energetically preferred conformation in agreement with that previously [20] deduced. The same authors [21] carried out an extensive analysis of one and two-dimensional ^{1}H NMR spectra of lincomycin. Coupling constants and nuclear Overhauser enhancement (nOe) supported the earlier theoretically identified conformation. Three-dimensional structures of lincomycin-related antibiotics: *N*-demethylclindamycin (**80**) and two its cyclic derivatives **91** and **92** were also studied recently by ^{1}H and ^{13}C NMR spectrometry [22] and the homology of their conformations was established.

The least-energy conformations of lincomycin (**1**) and its protonated form obtained by us by molecular mechanics calculations* are shown in Fig.1. They are in accord with energetically preferred conformations found in the theoretical and spectrometric studies [20-21].

Because energetically favoured three-dimensional structures of free and protonated lincomycin corresponded to two distinctly different conformations around the amide bond, it followed that depending on pH of solution a varying conformational equilibrium should occur. It should be added that kinetic

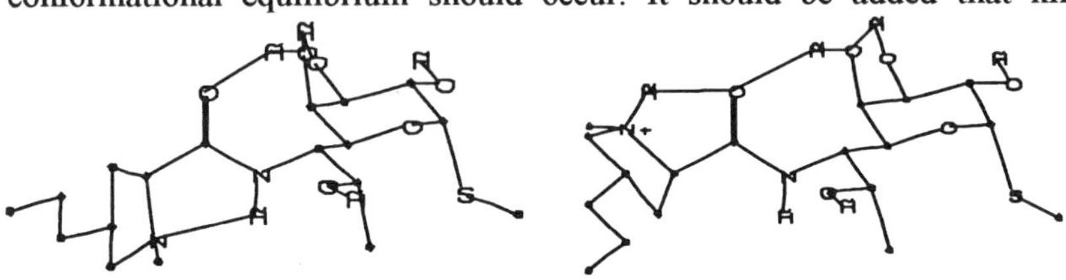

A B

Figure 1. The stick model of **1** (A) and **1**·HCl (B) in the least-energy conformation.

measurements of the lincomycin activity dependance on solution pH indicated that it is active as the free base [23].

Molecular orbital calculations of model structures established that HOMO and LUMO of lincomycin molecule were localized in the amide group [20]. Thus the nucleophilic as well as electrophilic reactivity of the antibiotic are connected primarily with the amide fragment of the molecule. From the above data it follows that it is also the portion of the molecule most sensitive to the

---

*MMX force field, Serena Software, P.O.Box 307, Bloomington, IN 47402-3076.

conformational and consequently steric and stereoelectronic effects following variation of the pH. Although the foregoing structural data permitted the explanation of much of the experimental evidence of lincomycin activity, further studies are required to gain a deeper insight into the relationship between its structure and mode of action at the molecular level.

3. Lincomycin-Related Antibiotics

3.1. Lincomycins

Several lincomycin-related antibiotics have been isolated with varying degrees of biological activity. In comparison to lincomycin (**1**) none of them was more active nor was the spectrum of activity significantly different. 4'-Depropyl-4'-ethyllincomycin (lincomycin B) (**9**) [24] was obtained from the fermentation of *Streptomyces lincolnensis var. lincolnensis* in addition to lincomycin.

New antibiotics related to lincomycin (Fig. 2) were produced by *Streptomyces lincolnensis var. lincolnensis* when various chemicals were added in the culture media. Thus, addition of DL-ethionine led to isolation of 1-demethylthio-1-ethyl-thiolincomycin (lincomycin C)(**10**) [25] and 1'-demethyl-1-demethylthio-1'-ethyl-1-ethylthiolincomycin (lincomycin S) (**11**) [26]. Upon addition of methyl thiolincosaminide (**2**) or methylation inhibitors the same strain produced in addition to lincomycin (**1**) *N*-demethyllincomycin (lincomycin D) (**12**) [27][28]. Introduction of ethyl thiolincosaminide into the fermentation broth yielded 1'-demethyl-1-demethylthio-1-ethylthiolincomycin **13** [29].

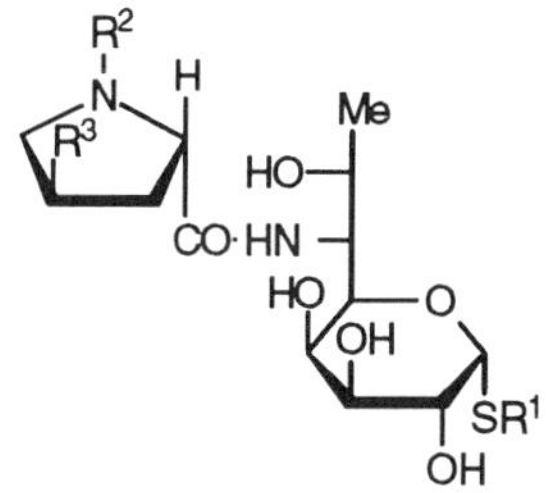

	R^1	R^2	R^3	
9	Me	Me	Et	lincomycin B
10	Et	Me	Pr	lincomycin C
11	Et	Et	Pr	lincomycin S
12	Me	H	Pr	lincomycin D
13	Et	H	Pr	

Figure 2

3.2. Celesticetins

Antibiotics structurally related to lincomycin were also produced by other species of *Streptomycete*. Celesticetin (**14**) is an antibiotic produced by

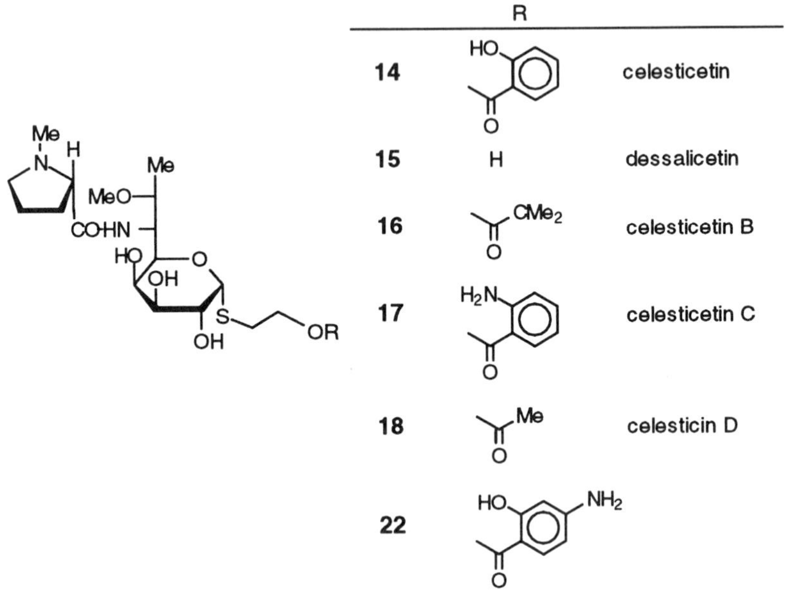

Figure 3

Streptomyces caelestis. Its isolation by Hoeksema et at. [30] preceded the discovery of lincomycin (**1**) however the structure remained unknown. Taking advantage of the experience gained in the structural studies on lincomycin Hoeksema [31][32] assigned structure **14** to celesticetin. Re-examination of the culture of *S. caelestis* led to the isolation [33] of four celesticetin-related antibiotics (Fig. 3): dessalicetin (**15**), celesticetin B (**16**), celesticetin C (**17**) and celesticetin D (**18**). It was found that mutants of *S. caelstin* are capable of producing antibiotics different from those produced by the parent culture. Argoudelis et al. isolated 7-O-demethylcelesticetin (**19**) [34], *N*-demethylcelesticetin (**20**) and N-demethyl-7-O-demethylcelesticetin (**21**) [35] from mutant strain (Fig. 4). Celesticetins proved to be less effective than lincomycin when assayed *in vitro* and *in vivo* against a spectrum of microorganisms.

It should be mentioned that addition of aromatic acids to the culture medium of *Streptomyces caelestis* induces production of antibiotics different from those normally obtained from this organism. Argoudelis et al. [36] examined the effect of aromatic acids and *e.g.* isolated 2'-(4-aminosalicylate) **22** resulting from the addition of 4-aminosalicylic acid to the culture medium.

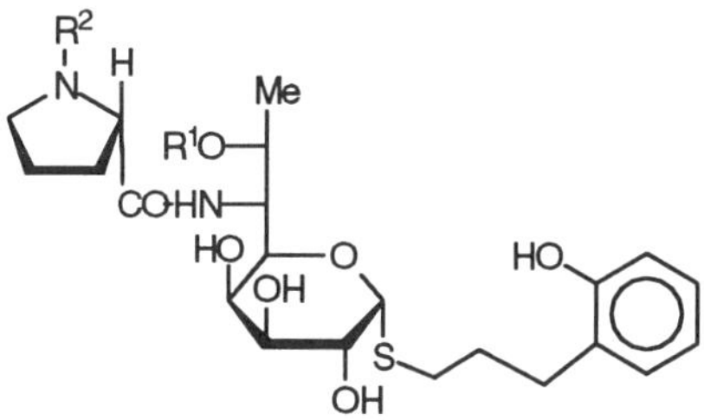

19 $R^1 = H$, $R^2 = Me$

20 $R^1 = Me$, $R^2 = H$

21 $R^1 = H$, $R^2 = H$

Figure 4

4. Chemical Modification of Lincomycin

Studies involving modification of lincomycin (**1**) were developed with two related goals in mind and consequently followed separate paths. The first, preparation of derivatives of lincomycin (**1**) was aimed at obtaining compounds with enhanced pharmaceutical acceptance, improved taste and absorption following oral administration, lower tissue irritability etc. Antibacterial potency of such derivatives depended on the activity of the parent antibiotic. The other goal, discovery of compounds with advantageously altered biological properties such as antibacterial spectrum, potency, rate of absorption, tissue distribution, excretion and metabolism, called for modification of the lincomycin molecular structure.

This distinction is convenient also from the chemical point of view because the first objective required preparation of derivatives easily hydrolysed *in vivo* like esters or carbonates, whereas the second entailed synthetic operations bringing about changes in one or more substituents in the sugar or amino acid portion of lincomycin.

4.1. Derivatives

Synthesized derivatives of lincomycin included esters of phosphoric acid and C_2 to C_{16} aliphatic acids, carbonates and esters of retinoic and polycyclic aromatic carboxylic acids. Methodology for ester (carbonate) group introduction at positions C-2, C-3, C-4 and C-7 are described below.

Monoesters

Lincomycin 2-phosphate, 2-acylate and 2-carbonate

Preparation of monoesters in a regioselective manner with four secondary hydroxyls in the lincomycin (**1**) molecule required their effective differentiation. Methodology based on protecting groups was developed by Morozowich et al.

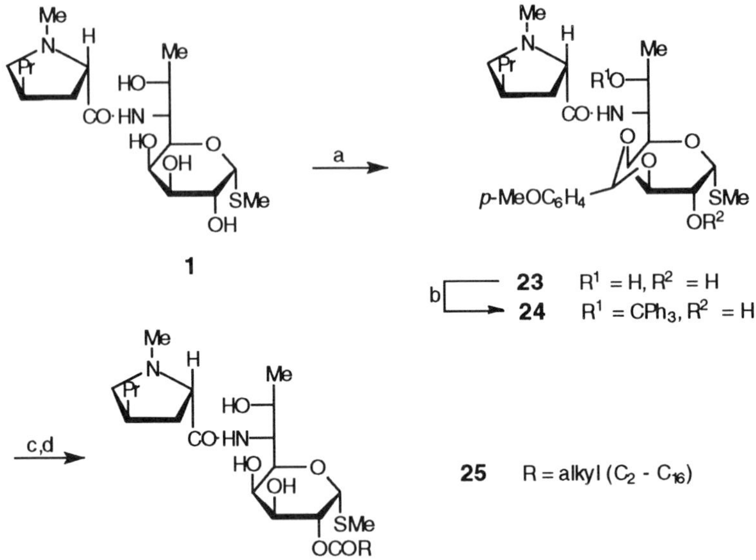

a. *p*-MeOC$_6$H$_4$CHO/H$^+$; b. Ph$_3$CCl - NEt$_3$/acetone; c. R-COCl/Pyr; d. 80% AcOH/110° C

Scheme 3

for the synthesis of 2-phosphate [37], 2-acylates and 2-carbonates [38]. Reaction of lincomycin (**1**) with anisaldehyde gave the 3,4-*O*-anisylidene derivative **23** (Scheme 3) [39] which when treated with trityl chloride and triethylamine in acetone provided selectively 7-*O*-trityl-3,4-*O*-anisylidenelincomycin (**24**). Esterification of position C-2 with a given acyl chloride or chlorocarbonate followed by the removal of protecting groups afforded the desired 2-ester **25**. In this way a number of 2-*O*- substituted lincomycins have been obtained.

Lincomycin 3- and 4- acylates

Employing conditions favoring acylation of less hindered hydroxyls, i.e. catalysis with sterically bulky, strongly basic tertiary amine, 3-propionate **26** and 3-hexanoate **27** of lincomycin were prepared (Scheme 4).

Acyl groups in lincomycin esters isomerized readily between C-2, C-3 and C-4 positions. The process was most rapid between *cis* hydroxyls at C-3 and C-4, favoring migration to C-4 [40]. Taking advantage of this property 4-hexanoate **28** was obtained from the available C-3 isomer **37** by basic or acidic rearrangement (Scheme 4).

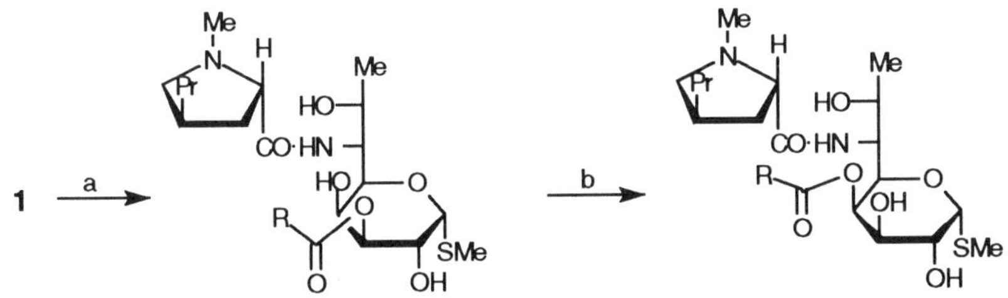

26 R = Et
27 R = C_5H_{11}
28 R = C_5H_{11}

a. $(EtCO)_2O$ or $(C_5H_{11}CO)_2O$/base; b. Et_3N

Scheme 4

Lincomycin 7-acylates and 7-carbonates

To obtain monoesters of the C-7 hydroxyls protection of all other hydroxyls was necessary. This was accomplished by preparing tetra(trimethylsilyl) lincomycin (**29**) and then performing selective acid hydrolysis of the silyl group at position C-7. Acylation of trisilyl derivative **30** gave ester **31** which after removal of the silyl protecting groups afforded lincomycin 7-acylates or 7-carbonates **32** [41] (Scheme 5).

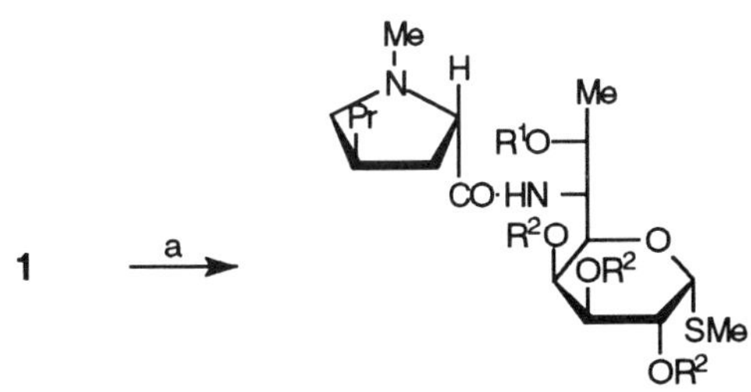

29 $R^1 = R^2 = SiMe_3$
b → **30** $R^1 = H$, $R^2 = SiMe_3$
c → **31** $R^1 = COR$ [or C(=O)OR], $R^2 = SiMe_3$
d → **32** $R^1 = COR$ [or C(=O)OR], $R^2 = H$ R = alkyl

a. Me_3SiCl - NEt_3; b. dilut. HCl; c. RCOCl [or RO(C=O)Cl]/Pyr; d. 80% AcOH.

Scheme 5

Using 3,4-*O*-anisylidenelincomycin (**23**) as a starting material, selective acylation at position C-7 via compound **24** was also accomplished [41].

Diesters

Sinkula and Lewis [42] prepared a series of 2,7-dicarbonates with a goal of obtaining products sufficiently insoluble in aqueous media for a pleasant tasting suspension of lincomycin (**1**) for pediatric use. In their approach 3,4-*O*--anisylidene lincomycin **23** was used as a starting material. Upon acylation with a slight excess of alkyl chloroformate and subsequent acidic hydrolysis of the acetal group the desired lincomycin 2,7-dialkylcarbonate **33** was obtained.

23

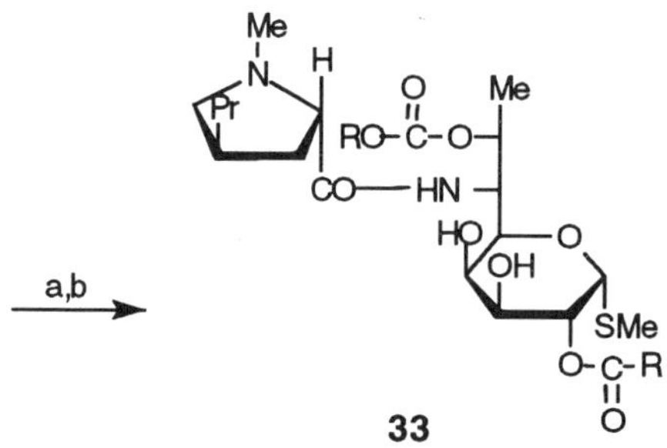

33

a. RO(C=O)Cl - Pyr; b. 80% AcOH

Scheme 6

Antibacterial assays of the four, 2-, 3-, 4- and 7-monohexanoates of lincomycin showed that antibacterial potency *in vivo* and *in vitro* diminishes in the following order: 2 > 3 ≈ 7 > 4. Multiple esters (tri- and tetraacylates) of lincomycin (**1**) possessed low *in vivo* and *in vitro* antibacterial activity [43] and were of minor interest.

In the patent literature, esters of lincomycin (**1**) are reported as components in the formulations for treatment of dermatosis, especially acne [44], cosmetic products [45] and products for reducing hair loss [46]. They are comprised of esters of retinoic acid at position C-2, C-3 and C-7 and esters at C-3 of various aromatic, polycyclic and heterocyclic, carboxylic acids. Representative examples of acyl groups are shown in Scheme 7. In these and other patents [47] [48] analogous esters of clindamycin (**50**) with similar pharmaceutical applications were described.

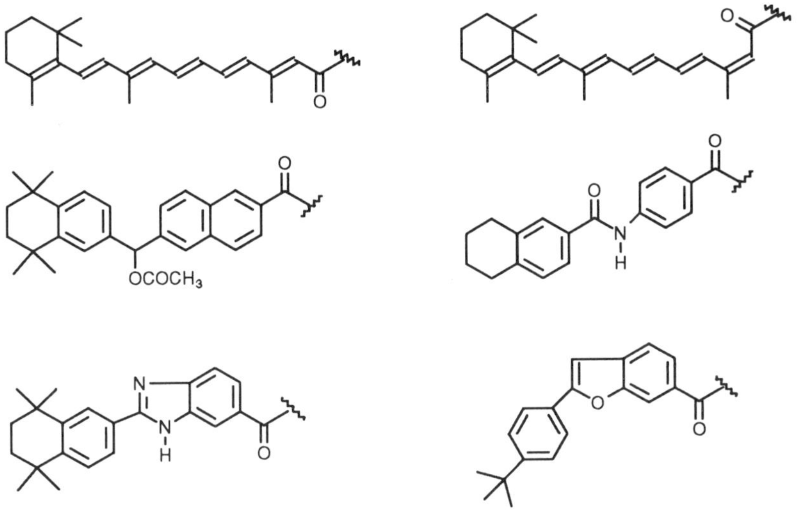

Scheme 7

4.2. Analogues

Chemical modification of the lincomycin (**1**) molecule which introduced variation in one or more substituents in the sugar, amino acid or aglycone fragment, leaving the skeleton of the parent antibiotic unchanged, led to compounds referred to as lincomycin analogues. Their syntheses were developed according to two different methodologies. In one general approach chemical modification was carried out in the sugar or/and amino acid moiety, on the otherwise intact molecule of lincomycin (**1**). The other methodology relied on the easy cleavage of the lincomycin amide bond, yielding methyl thiolincosaminide (**2**) and *trans*-1-methyl-4-propyl-L-proline (**4**) (Scheme 1). Modification of the sugar and/or amino acid portion and subsequent recombination, by the use of one of the methods for amide bond formation, was the basis of the second approach to the lincomycin analogues.

4.2.1 Modification in the sugar moiety

Synthesis of analogues with D-*gluco* and D-*ido* configuration [49] as well as 2-*O*-methyl- and 2-deoxy- analogues [50] are presented to illustrate the two described routes to synthetic lincomycins.

Treatment of lincomycin (**1**) hydrochloride with butanoic anhydride in pyridine gave preferentially 2,3,7-triester **34**. The hydroxyl at C-4 in **35** which was activated with the methanesulfonyl group was displaced with inversion of configuration by benzoate anion. Removal of the acyl groups gave the D-*gluco*-analogue **36** (Scheme 8) [49].

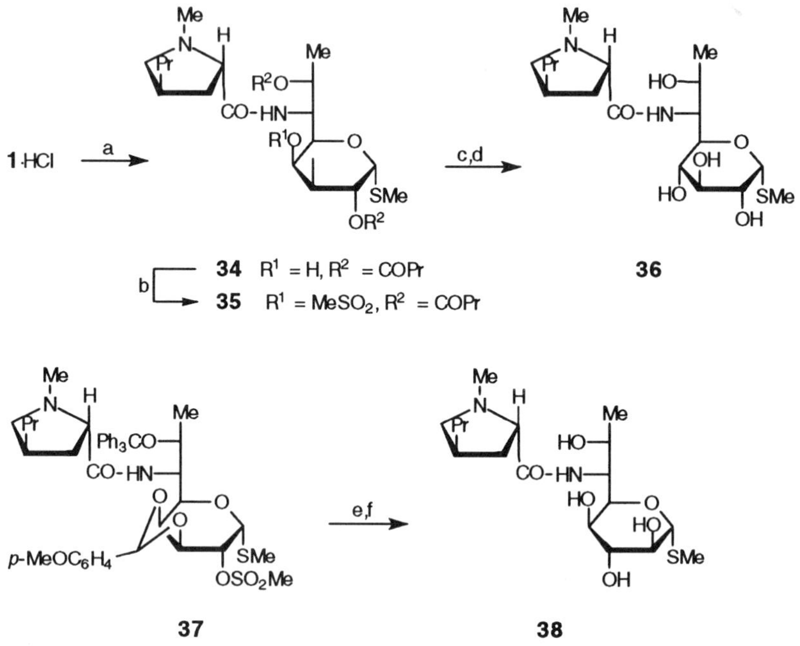

a. $(PrCO)_2O$ - Pyr; b. $MeSO_2Cl$ - Pyr; c. $PhCO_2Na$/DMF, 150° C; d. MeO^-/MeOH; e. 80% AcOH, 110° C; f. NaOH/H_2O.

Scheme 8

The D-*ido* stereoisomer **38** was obtained via the 2-*O*-methanesulfonyl ester **37** which was prepared according to the procedure described (vide supra) for other 2-acylates of lincomycin (**1**). Acid hydrolysis of the acetal grouping in ester **37** followed by treatment with aqueous sodium hydroxide, via D-*talo*-epoxide, afforded D-*ido*- analogue **38** [49].

2-*O*-Methylation of lincomycin (**1**) would result in concomitant quaternization of the proline nitrogen atom. To avoid this unwanted course of the reaction methodology based on splitting the antibiotic into the amino acid and sugar fragments, followed by modification and recombination, was employed. Methyl thiolincosaminide (**2**) available by hydrazinolysis of lincomycin (**1**) was treated with acetic anhydride to give the *N*-acetyl derivative **39**. Protection of hydroxyls at C-3 and C-4 by isopropylidene group and *O*-methylation of acetonide **40** with 1 equivalent of potassium *tert*-butoxide and methyl iodide afforded, after removal of protecting groups, the aminosugar **41**. Coupling with *trans*-1-methyl-4-propyl--L-proline (**4**) via its mixed carbonate gave 2-*O*-methyllincomycin (**42**) (Scheme 9) [50].

Starting from the methyl *N*-acetylthiolincosaminide **39**, according to the typical sugar chemistry procedures shown in Scheme 10, peracetylated lincosaminal **45** was obtained via the 1-bromo derivative **44**. Addition of hydrogen bromide to the unsaturated sugar **45**, exchange of bromine atom in **46** for thiol and

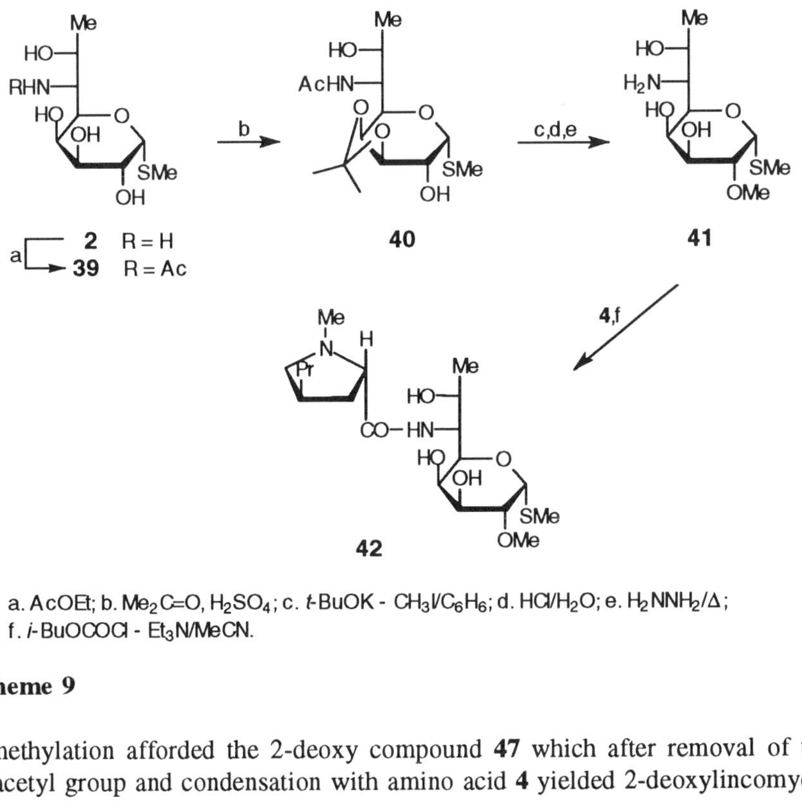

a. AcOEt; b. $Me_2C{=}O$, H_2SO_4; c. *t*-BuOK - CH_3I/C_6H_6; d. HCl/H_2O; e. H_2NNH_2/Δ; f. *i*-BuOCOCl - Et_3N/MeCN.

Scheme 9

S-methylation afforded the 2-deoxy compound **47** which after removal of the *N*-acetyl group and condensation with amino acid **4** yielded 2-deoxylincomycin **48** [50] (Scheme 10).

At this point it has to be noted that altering the configuration at any carbon atom of the pyranoside moiety, 2-*O* substitution as well as removal of C-2 hydroxyl group, led to lincomycin analogues virtually devoid of antibacterial activity. It was also found that replacement of the methylthio substituent by a hydroxyl group or hydrogen atom, inversion of configuration at anomeric center or oxidation to sulfoxide in all cases gave products showing very low activity [43]. Biologically active and therefore most studied synthetic lincomycin analogues are those with modifications at C-7 in the sugar portion.

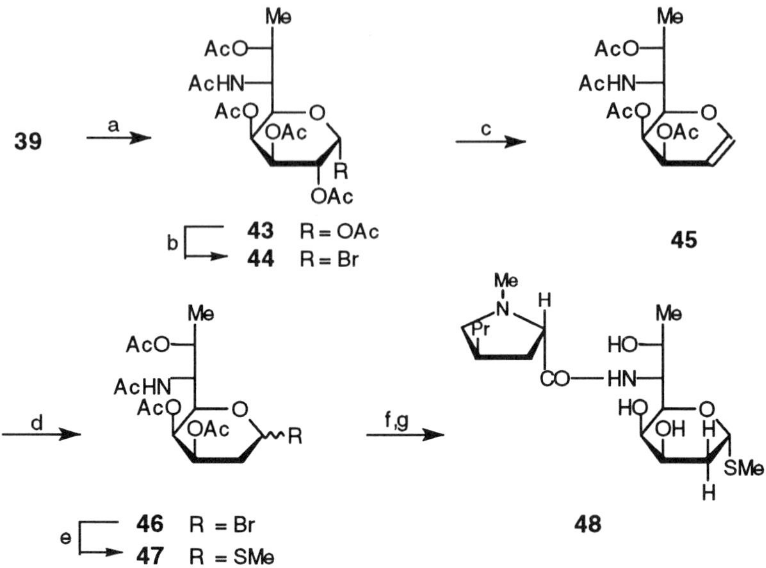

a. Ac_2O; b. HBr - AcOH; c. Zn - AcOH; d. HBr/C_6H_6; e. $(H_2N)_2C{=}S/\Delta$ then K_2CO_3, $NaHSO_3$, MeI; f. H_2NNH_2/Δ; g. **4**, *i*-BuOCOCl - Et_3N/MeCN.

Scheme 10

C-7 substituted lincomycin analogues

When lincomycin (**1**) was treated with thionyl chloride in carbon tetrachloride solution, substitution of the hydroxyl at C-7 occurred with inversion of

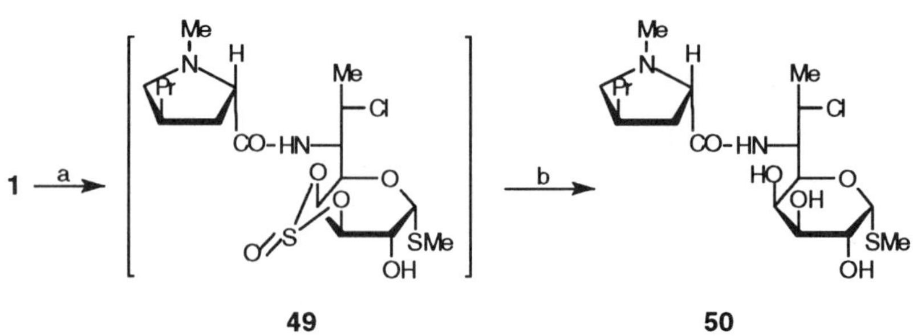

a. SO_2Cl/CCl_4; b. H_2O, Na_2CO_3.

Scheme 11

configuration yielding (7*S*)-7-chloro-7-deoxylincomycin **50** which was given the common name clindamycin (Scheme 11) [51]. The selectivity of the reaction was explained by the formation of the 3,4-cyclic sulfite **49** which was hydrolysed during the workup of the reaction mixture. An alternate method for the preparation of clindamycin (**50**) involved the use of triphenylphosphine dichloride or triphenylphosphine and CCl_4. The latter procedure using CBr_4 or CI_4 in acetonitrile afforded (7*S*)-7-bromo- and (7*S*)-7-iodo-7-deoxylincomycin [51].

Reduction of clindamycin with zinc in water solution gave 7-deoxylincomycin **51** (Scheme 12).

50

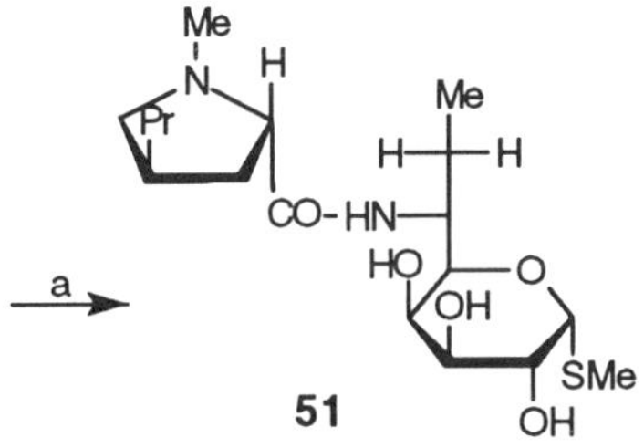

51

a. Zn - H_2O

Scheme 12

Selective oxidation of the hydroxyl at C-7 in 3,4-*O*-isopropylidenelincomycin (**52**) followed by reduction of the carbonyl group in ketone **53** and hydrolysis led to 7-epi-lincomycin **54**. From the latter, (7*R*)-7-chloro-7-deoxylincomycin (7-epi-clindamycin) **55** was obtained [51] (Scheme 13). Clindamycin (**50**) was also obtained by recombination methodology. Treatment of methyl thiolincosaminide (**2**) with thionyl chloride gave the (7*S*)-7-chloro- derivative **56** with a substituted C-7 hydroxyl. Condensation with *trans*-4-L-hygric acid (**4**)

2

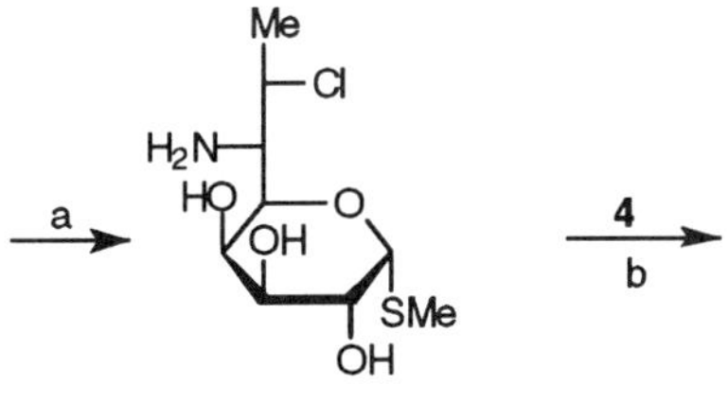

50

56

a. SO_2Cl_2/CCl_4, Na_2CO_3; b. *i*-BuOCOCl, Et_3N.

Scheme 14

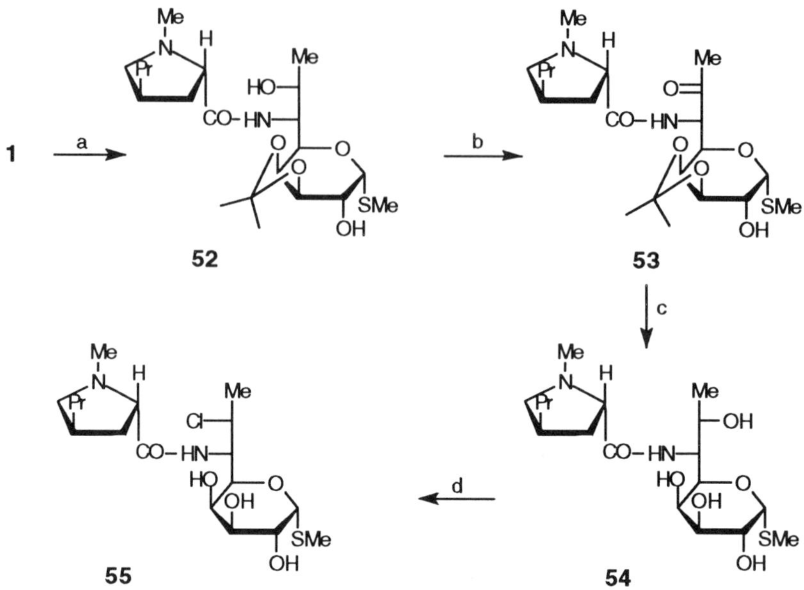

a. Me_2CO/H^+ ; b. [O]; c. 1. $NaBH_4/EtOH$, 2. HCl/H_2O, d. $SOCl_2/CCl_4$ - Na_2CO_3.

Scheme 13

afforded clindamycin **50** (Scheme 14) [51].

Recently, in the improved procedures for preparation of clindamycin (**50**), chlorination of lincomycin (**1**) was carried out with Vilsmayer-Haack reagent ($POCl_3$ - DMF) [52] or amide chlorides formed *in situ* from DMF and thionyl chloride or phosphorous pentachloride as well as *N*-formylpiperidine with phosgene [53].

The comparison of the *in vitro* activities (standard curve assay against *Sarcoma lutea*) of lincomycin (**1**) and its C-7 analogues is shown below:

lincomycin (**1**)	1.0
7-epi-lincomycin (**54**)	0.5
7-deoxylincomycin (**51**)	0.10
7-ketolincomycin (**53**)	0.02
clindamycin (**50**)	4.0
(7*R*)-epi-clindamycin (**55**)	1.2
(7*S*)-bromo-7-deoxylincomycin (**55**)	8.0

It was apparent that clindamycin (**50**) possessed significantly higher activity than lincomycin (**1**). (7*S*)-7-Bromo-analogue proved to be even more active. This result demonstrated that modifications of lincomycin at C-7 of its sugar portion may give biologically interesting analogues and consequently stimulated synthetic efforts towards other C-7 analogues. On the other hand because clindamycin (**50**) itself became a clinically important antibiotic, synthesis of various esters, analogous to those obtained from lincomycin were reported [44] [45] [46] [47] [48] [54] in the patent literature.

Other modifications at C-7 in the sugar portion of lincomycin were pursued according to both mentioned methodologies ("intact" and "recombination"). In the first approach clindamycin (**50**) was used as a starting material. Reaction of **50** with nucleophilic reagents: sodium azide or cyanide proceeded with inversion of configuration yielding 7-substituted-7-deoxylincomycin **57** and **58**, respectively (Scheme 15) [43]. Hydrogenation of 7-azido-7-deoxylincomycin (**58**) afforded the (7*R*)-7-amino analogue **59**.

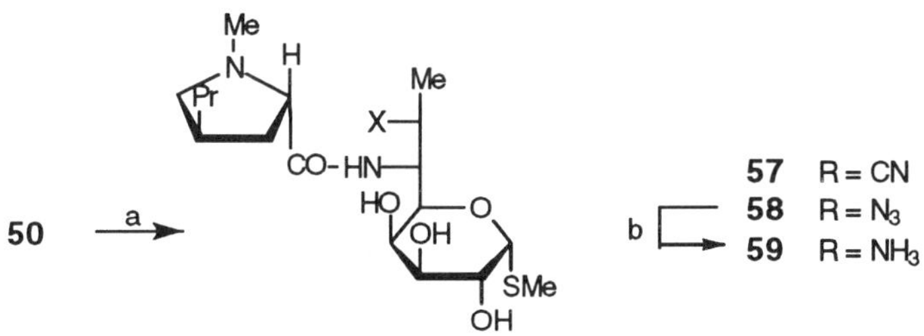

a. NaCN/DMF or NaN_3/DMF; b. H_2/Pd.

Scheme 15

In the "recombination" approach methyl (7*S*)-7-chloro-7-deoxythiolincosaminide **56** was used as a starting material. Treatment of the chloro derivative **56** with

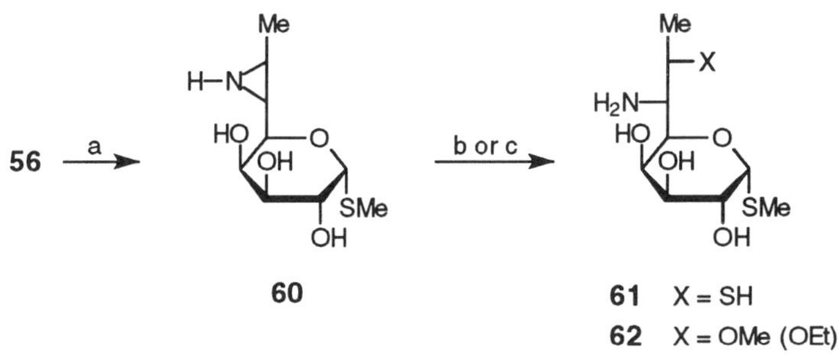

a. DBU; b. H_2S/*i*-PrOH; c. MeOH (EtOH) - AcOH.

Scheme 16

base gave epimine **60**. Opening of the aziridine ring in **60** with hydrogen sulphite or alcohols in acidic medium afforded (7*R*)-thiol **61** and (7*R*)-alkoxy **62** compounds [55], respectively (Scheme 16).

Bannister [56] extensively studied *S*-alkylation of sulfides, mono and dithioacetals, methoxymethyl sulfides and disulfides with *N*-acetylepimine **63** (Scheme 17). The amide bond was cleaved in the (7*S*)-thiosubstituted sugars **64** obtained in these reactions, and the free aminosugars were coupled with an amino acid **4** to produce lincomycin thio-analogues **66** (Scheme 17).

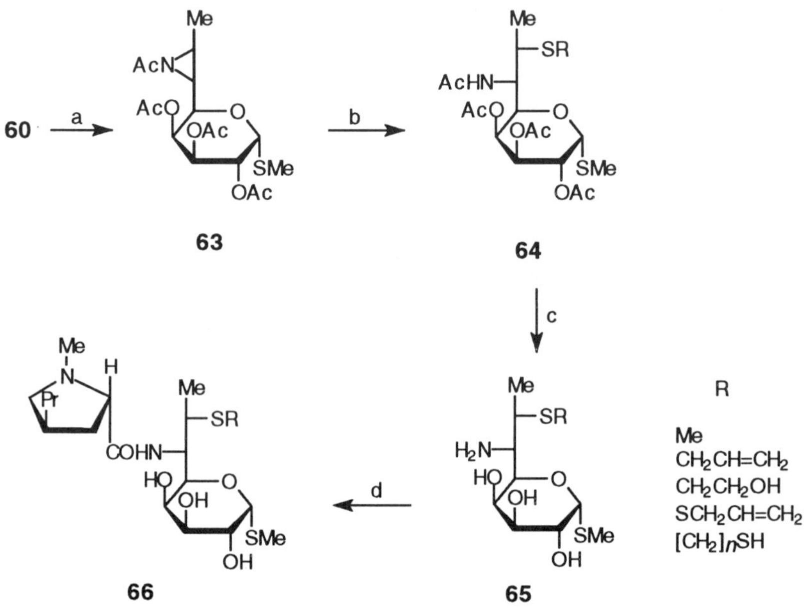

a. Ac_2O - Pyr; b. sulfur nucleophiles: RSH or RSR or $HSCH_2CH_2OH$ or (1,3-dithiolane) /AcOH; c. $H_2NNH_2{\cdot}H_2O$, Δ; d. **4**, *i*-BuOCOCl - Et_3N.

Scheme 17

4.2.2. Modification of the aglycon

Celesticetin (**14**) and other antibiotics produced by *Streptomyces caelestis* and structurally related to lincomycin (**1**), contained substituents at the anomeric carbon atom other than methylthio. Since they exhibited significant activity a series of 1-demethylthio-1-alkylthiolincomycins were prepared from lincomycin (**1**) for structure - activity relationship studies. Modification of the C-1 substituent was carried out by two routes. In one method the intermediate dialkyldithioacetal **67**, obtained by reacting lincomycin (**1**) with a given thiol, was cyclized by treatment with acid to yield the desired analogue **70**. The process is complicated by the formation of three possible dialkyldithioacetals: **67, 68** and **69** (Scheme 18).

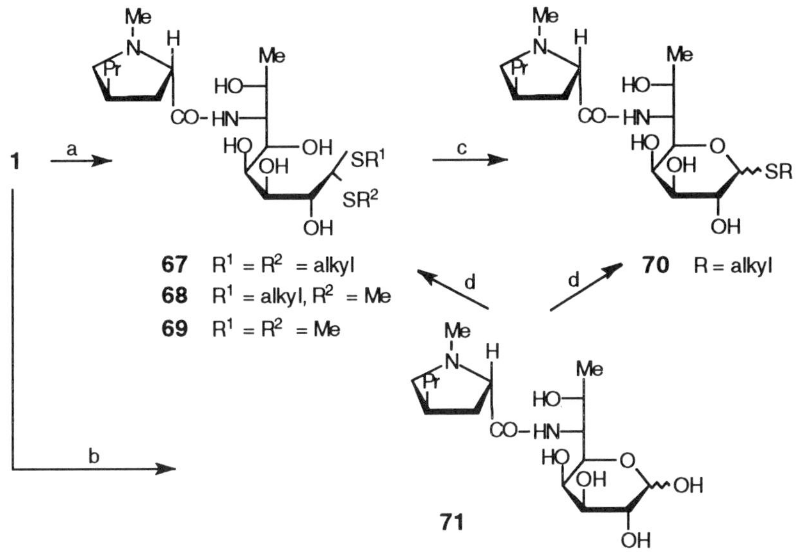

a. RSH - H_3O^+; b. Br_2 - H_2O; c. HCl, Δ; d. RSH - HCl.

Scheme 18

In the other method which avoided this difficulty, lincomycin (**1**) is brominated in aqueous solution to give lincomycose **71**. Reaction of the free sugar with alkanethiol and acid gave a mixture of anomeric thioglycosides **70** (Scheme 18). In both approaches separation of resulting stereoisomers of **70** was necessary [43].

4.2.3 Modification of the amino acid fragment

The synthesis and antibacterial activity of lincomycin (**1**) and clindamycin (**50**) analogues with modified amino acid portions were extensively studied. Analogues having variously altered groups at C-4' and/or N-1' of the proline moiety were obtained as well as analogues in which different amino acids were introduced in place of proline derivatives.

4'-Alkyl-4'-depropyllincomycins

A versatile method of obtaining lincomycin (**1**) or clindamycin (**50**) analogues modified at N-1' and C-4' positions was based on the cleavage - recombination methodology. Thus methyl thiolincosaminide **2** was condensed (Scheme 20) with a series of *N*-protected amino acids **74** prepared according to the synthetic route outlined in Scheme 19. 1-Carbobenzoxy-4-keto-L-proline **72** reacted with tripehenylalkylidenephosphorane to give the unsaturated acid **73**.

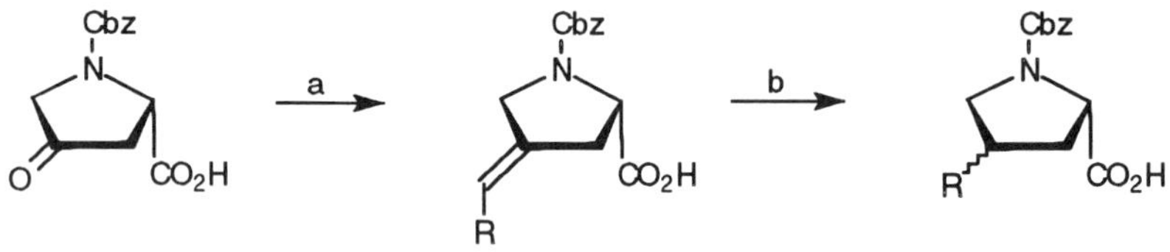

72 **73** R = alkyl **74** R = alkyl

a. DMSO - NaH, $R\cdot Ph_3P^+ Br^-$; b. H_2 - Pt - Dowex 1/MeOH.

Scheme 19

Careful catalytic reduction of **73** yielded a mixture of *cis* and *trans* saturated acids **74**. Epimers **74**, as a mixture, were condensed via isobutyl carbonates with methyl thiolincosaminide **2**. Catalytic removal of the carbobenzoxy group followed by reductive *N*-alkylation with formaldehyde (or acetaldehyde) afforded an epimeric mixture of lincomycin analogues **77**. *Cis* and *trans* isomers were separated by chromatography [57] (Scheme 20).

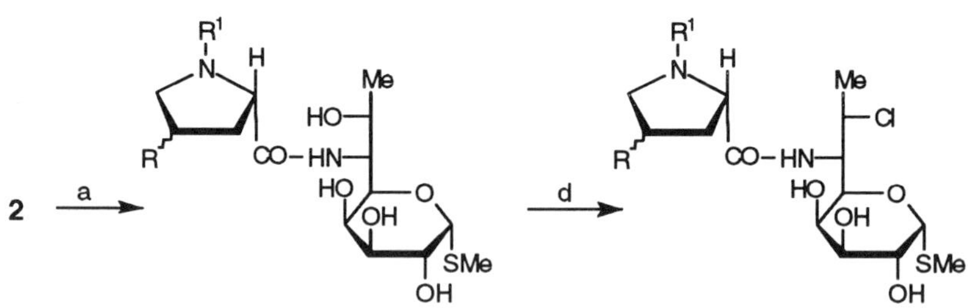

75 R^1 = Cbz, R = alkyl **78** R^1 = Me (or Et), R = alkyl
b → **76** R^1 = H, R = alkyl
c → **77** R^1 = Me (or Et), R = alkyl

a. **74**, *i*-BuOCOCl - Et_3N/MeCN; b. H_2 - 10% Pd-C/MeOH; c. CH_2O or MeCHO - H_2 - 10% Pd-C/MeOH-H_2O; d. $SOCl_2$.

Scheme 20

Clindamycin analogues **78** were obtained by thionyl chloride treatment of corresponding lincomycin derivatives [57] (Scheme 20). In a similar manner using as an amino acid component 1-methyl-4-ethoxy-L-proline lincomycin, analogues with *cis* and *trans* 4'-alkoxy substituents instead of alkyl groups were prepared [58].

Modification of the N-1' alkyl substituent could also be performed on the intact molecule of lincomycin (**1**) or clindamycin (**50**). Birkenmeyer and Dolak [59] accomplished *N*-demethylation of both antibiotics **1** and **50** by treatment in aqueous solutions with a stream of oxygen or air in the presence of a platinum catalyst (Scheme 21). The resulting *N*-demethyl compounds **79** and **80** could serve as substrates for the preparation of variously *N*-alkylated analogues.

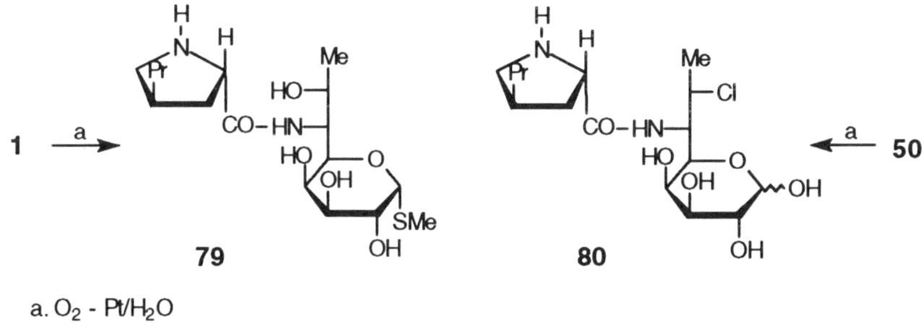

Scheme 21

4.2.4. Modification in the amino acid and sugar moieties

Alterations of the amino acid portion of clindamycin (**50**) which involved N-1' and C-4' alkyl substituents have been described earlier. A group of compounds was also prepared in which the amino acid fragment of clindamycin was replaced with cyclic amino acids with varying ring size, increased or decreased by one, two or three methylene groups [60].

In the six-membered series the synthesis of clindamycin analogues was carried out as shown in Scheme 22. The amino acids **81** after conversion to mixed carbonates **82** were coupled with methyl (7*S*)-7-chloro-7-deoxylincosaminide (**56**) to give aromatic amides **83**. Hydrogenation in the presence of platinum oxide yielded a mixture of diastereoisomeric products **84** and **85** which were separated by crystallization and chromatography. Absolute configurations of the amino acid portions of the isomeric pair **84** (R = Et) and **85** (R = Et) were established by X-ray analysis. For the other compounds of the series it was deduced on the basis of homology of their mobility in TLC. Reductive alkylation with formaldehyde or oxirane afforded *N*-substituted analogues **86** and **87**. Among the compounds in this series, pirlimycin **85** (R = Et) exhibited the most promising biological activity.

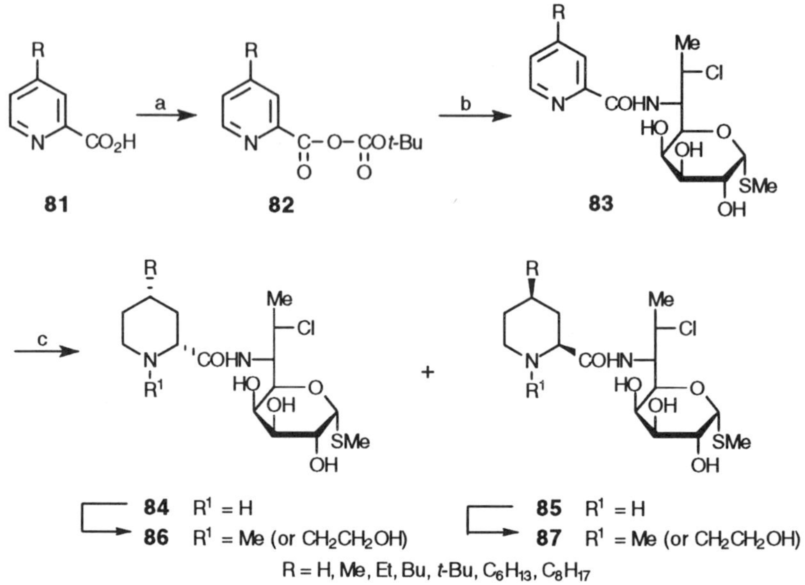

a. *i*-BuOCOCl - Et_3N/MeCN; b. **56**/Me_2CO - H_2O - MeCN; c. H_2 - PtO_2 /HCl/H_2O - MeOH.

Scheme 22

Condensation of methyl (7*S*)-7-chloro-7-deoxylincosaminide (**56**) with appropriate *N*-protected amino acids via their mixed carbonates, yielded clindamycin analogues **88** with various ring sizes of the amino acid portion after removal of the protecting groups (Scheme 23).

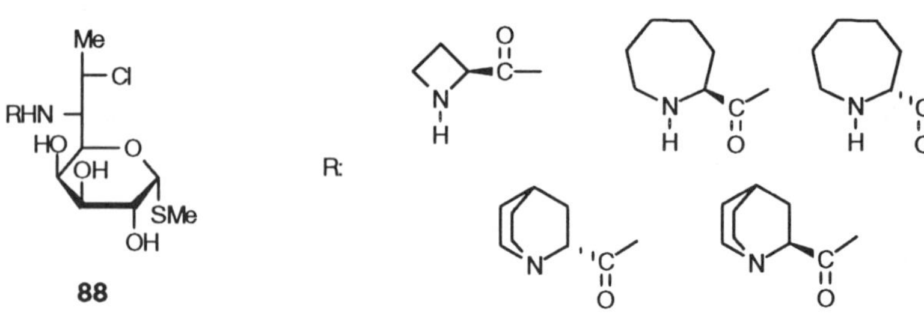

Scheme 23

Synthesis of 1'-demethyl-4'-depropyl-(4'*R*)- and (4'*S*)-pentylclindamycin (**78**, R' = H, R = C_5H_{11}), the antimalarial agents, based on the racemic 4'-pentylproline and the condensation - resolution sequence, was reported [61] [62].

4.2.5. Miscellaneous modifications

Reacting lincomycin (**1**) or clindamycin (**50**) peracetate with phosphorous pentasulfide, followed by hydrolysis of acetyl groups yielded thioamide analogues **89** and **90**, respectively (Scheme 24) [55].

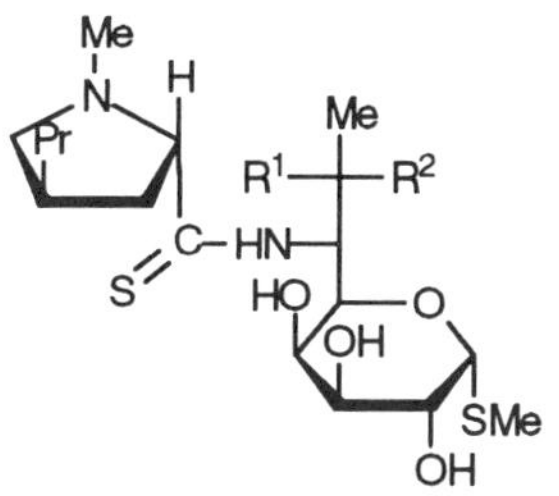

89 R^1 = OH, R^2 = H

90 R^1 = H, R^2 = Cl

Scheme 24

Base catalyzed condensation of 1'-demethyllincomycin (**79**) or 1'-demethylclindamycin (**80**) with formaldehyde or benzaldehyde led to the imidazolinone derivatives **91** and **92** (Scheme 25) [63]. In view of the biological activity of compounds **91** and **92** their three-dimensional structures in solution was studied using ^{1}H and ^{13}C NMR spectrometry [22].

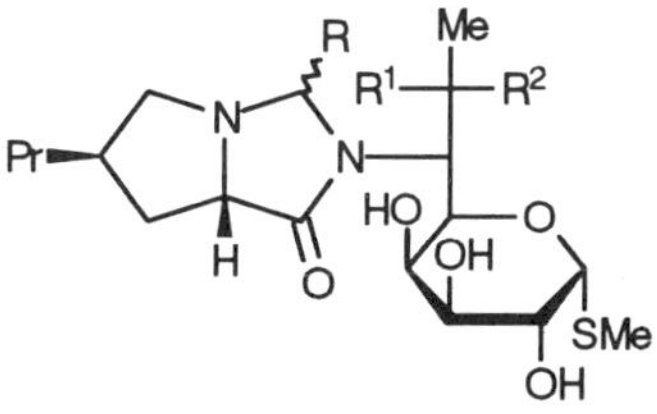

91 R^1 = OH, R^2 = H
92 R^1 = H, R^2 = Cl
} R = H, Ph

Scheme 25

5. SYNTHESIS FROM SUGAR PRECURSORS

Numerous efforts directed towards synthesis of lincomycin (**1**) or its sugar portion lincosamine (**93**) (Fig. 5) concentrated on partial synthesis. This approach depended on a chain elongation of the common aldohexose and introduction of functional groups into the resulting side chain. For a starting monosaccharide the obvious choices were variously protected derivatives of D-galactose (**6**). The advantages of employing them as substrates were enantiopurity and the stereochemistry which appeared in pyranoside moiety of lincomycin (**1**). Several methods of chain extension and functional group introduction have been developed. In some methodologies a high level of stereochemical control has been achieved.

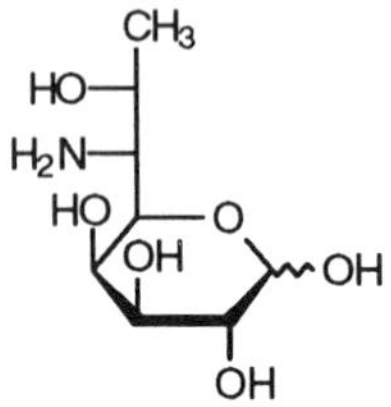

93

Figure 5

In this section all attempts to construct the carbon frame-work of the aminooctose portion of lincomycin (**1**) are discussed, i.e. synthesis of lincosamine (**93**) or its derivatives and diastereoisomers. In the syntheses aimed at methyl thiolincosaminide (**2**) the problem of the efficient introduction of the anomeric methylthio group was also addressed. Completion of lincomycin (**1**) synthesis required *N*-acylation of methyl thiolincosaminide (**2**) with *trans*-1-methyl-4--propyl-L-proline (*trans*-L-4-propylhygric acid) (**4**). However synthesis of methyl thiolincosaminide (**2**) may be regarded as tantamount to the total synthesis of lincomycin (**1**) since condensation of **2** with amino acids yielding an amide bond has been reported in connection with structure elucidation of lincomycin and preparation of its analogues [57].

All immediate substrates for chain elongation, aldehydes **94** and **95** as well as imine **96** and nitro compound **97** (Scheme 26), were obtained from D-galactose (**6**). Introduction of the methylthio group in one case was performed at the very beginning of the synthesis prior to the chain elongation step. In most syntheses it was left to the last steps of the synthetic route. Chain extension of the aldohexose moiety was carried out either in one step by addition of the two-carbon unit, or in a step-wise manner by two consecutive additions of one carbon units.

Syntheses are grouped into two-carbon and double one-carbon extension methodologies and are discussed separately. Two-carbon and one-carbon elongation of benzylimine **96** are reported in separate sections.

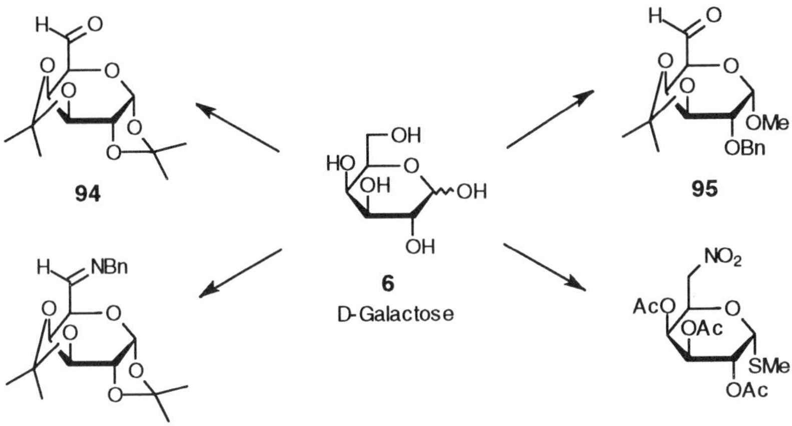

Scheme 26

5.1. Two-Carbon Chain Extension

The first synthesis of lincomycin (**1**) from D-galactose (**6**) was accomplished by Magerlain [64]. In his strategy introduction of the *S*-methyl group preceded construction and functionalization of the side chain. D-Galactose (**6**) was reacted with methanethiol to give methyl 1-thio-α-D-galactoside (**98**) (Scheme 27). Selective tosylation of the C-6 hydroxyl in the latter followed by acetylation, treatment with sodium iodide and then sodium nitrite afforded the nitro compound **97**. Two-carbon chain elongation was effected by condensation of **97** with acetaldehyde which led to a mixture of diastereoisomeric nitrooctoses **99**. Lithium aluminum hydride reduction of **99** afforded a mixture of aminooctoses **100** from which, after acetylation, methyl pentaacetyl-α-thiolincosaminide **101** was isolated. Aminosugars **100** were also condensed with *trans*-4-propyl-L-hygric acid (**4**) to afford, after careful chromatography, lincomycin (**1**). In the corresponding reaction, products methyl 7-epi-thiolincosaminide and 7-epi-lincomycin (**54**) were identified.

5.1.1. Syntheses from α-D-*galacto*-hexodialdo-1,5-pyranose 94

Szarek et al. [65] [66] were the first to employ dialdopyranose (**94**), the most commonly used starting material for lincomycin (**1**) synthesis. In their approach, the introduction of methylthio group was left to the later stages of synthesis. Chain elongation of aldehyde **94** was carried out by a Wittig reaction with ethylidenetriphenylphosphorane, addition of vinylmagnesium bromide and

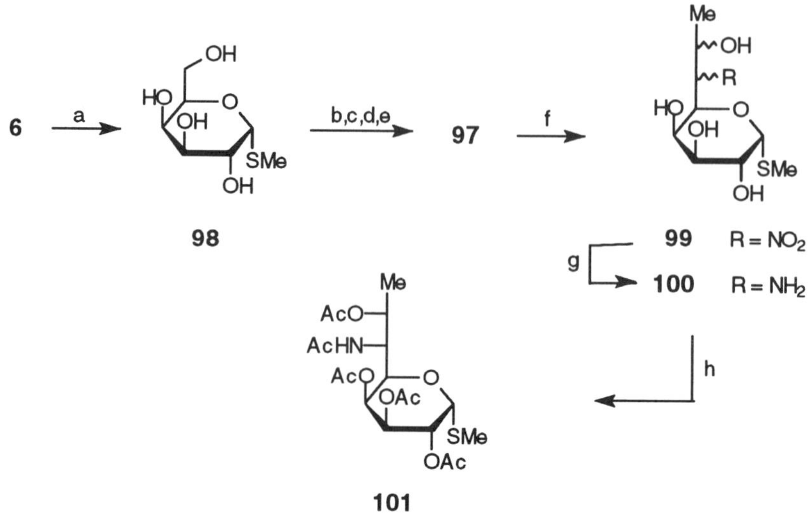

a. MeSH/HCl; b. TsCl-Pyr; c. Ac_2O-Pyr; d. NaI/acetone, 125°C; e. $NaNO_2$/DMF;
f. MeCHO - MeO^-/MeOH; g. LAH/THF; h. Ac_2O-Pyr.

Scheme 27

condensation with nitromethane yielding octenose **102**, a mixture of epimeric hydroxyoctenoses **103** and a mixture of diasteroisomeric nitrooctoses **104**, respectively [65] [66] (Scheme 28). Szarek developed several methods for introduction of functional groups into a side-chain of intermediates **102, 103** and **104**. These methodologies will now be discussed.

cis-Hydroxylation of octenose **102** with aqueous potassium permanganate proceeded with remarkable selectivity yielding D-*erythro*-diol **105** as a single product [67]. On the other hand Markovnikov hydration of the double bond in the major isomer of hydroxyoctenoses **103** with mercuric acetate - sodium borohydride reagent yielded a mixture of *vic*-diols **106** (Scheme 28). Selective benzoylation of **105** C-7 hydroxyl followed by oxidation with DMSO - DCC or ruthenium tetroxide afforded ketone **107** (Scheme 29). Treatment of the latter with hydroxylamine, debenzoylation and catalytic reduction of the resulting oxime **108** gave, after *N*-acetylation, a mixture of epimeric at C-6 acetamidoalcohols from which a small yield of the desired D-*erythro* stereoisomer **109** was isolated. Acetamidoalcohol **109** was reacted with methanethiol and hydrochloric acid to give dimethyl dithioacetal **110** which under acidic conditions eliminated methanethiol with concomitant cyclization. Hydrazinolysis of **111** gave a small yield of methyl α-thiolincosaminide **2** [66].

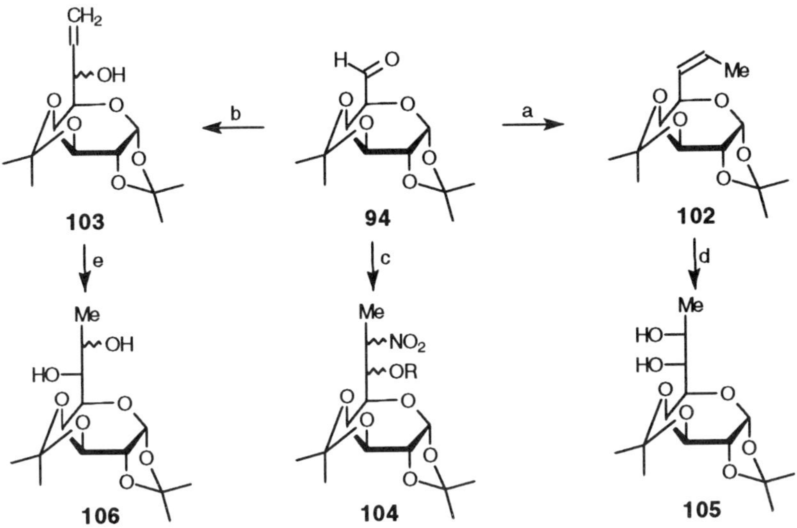

a. $Ph_3P{=}CHMe$; b. $CH_2{=}CHMgBr/THF$; c. $EtNO_2$ - MeO^- /MeOH; d. $KMnO_4/H_2O$;
e. $Hg(OAc)_2/H_2O$ - THF, then $NaBH_4$.

Scheme 28

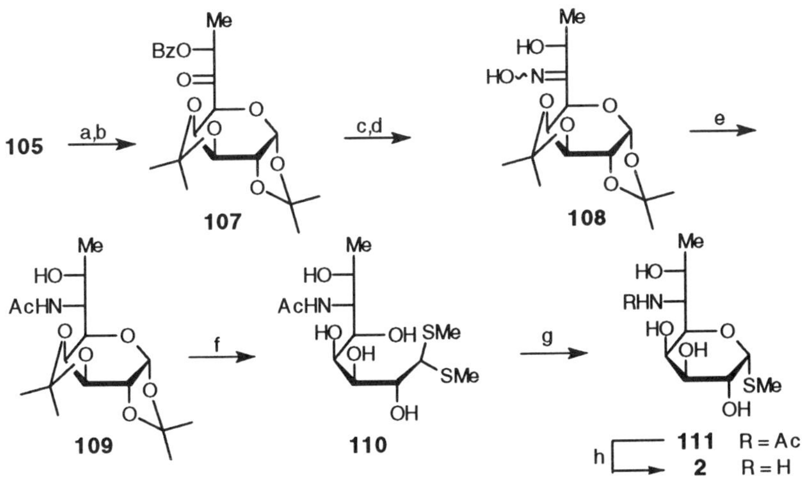

a. BzCl - Pyr; b. DMSO - Pyr·HCl - DCC; c. $H_2NOH{\cdot}HCl$ - Pyr/EtOH; d. MeO^- /MeOH,
e. 1. Ni - H_2, 2. Ac_2O - Pyr; f. MeSH - HCl/H_2O; g. conc. HCl/H_2O; h. $H_2N\text{-}NH_2{\cdot}H_2O$ - Δ.

Scheme 29

The acetyl derivative of the nitroalcohol **104** on elimination of acetic acid gave a mixture of *cis-* and *trans-*nitroolefins **112**. Addition of ammonia to the *trans-*nitrooctenose **112** led to the 1:1 mixture of *vic-*nitroamines **113** (Scheme 30). Their acylation and oxidative denitration with potassium permanganate afforded a mixture of aminoketones **114** and **115** which were separated and reduced to yield *vic-*acetamidoalcohol **116** and **109**, respectively. The D-*erythro* isomer **109** on treatment with hydrazine afforded 1,2;3,4-*O*-diisopropylidene-α-D-lincosamine **117** [66] [68].

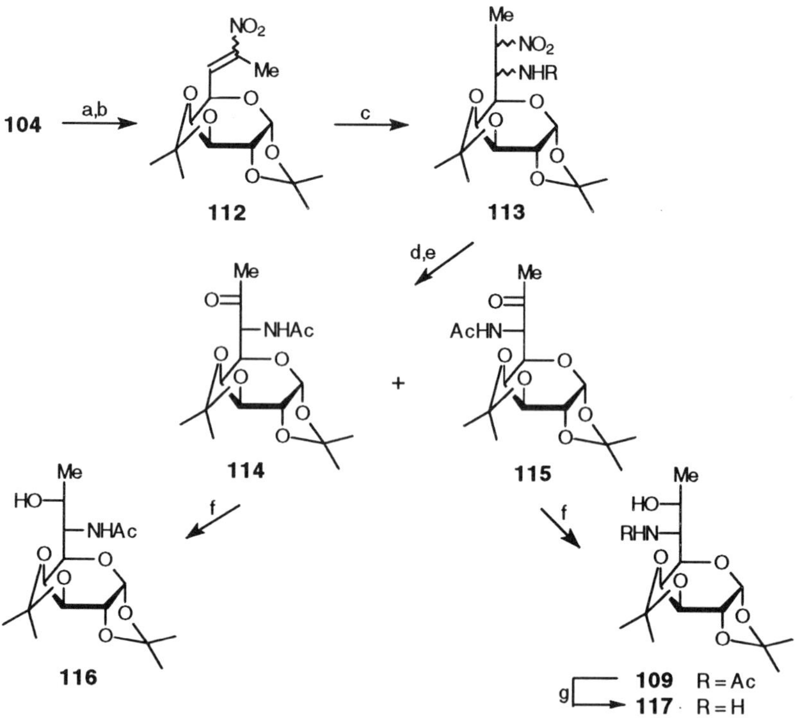

a. Ac_2O - Pyr; b. Et_3N - C_6H_6, Δ; c. NH_3 - MeOH; d. Ac_2O - Pyr;
e. $KMnO_4$ - H_2O; f. $NaBH_4$ - EtOH; g. H_2NNH_2

Scheme 30

Alternatively functionalization of the side-chain in nitrooctenose **112** was carried out via epoxide **118** (Scheme 31) [69]. Treatment of **112** with alkaline hydrogen peroxide gave a mixture of epimeric nitroepoxides **118** and **119**. Separation and splitting off the oxirane ring with benzylamine in the stereoisomer **118** led to aminoketone **120** from which after reduction of the carbonyl group, catalytic removal of benzyl group and acetylation yielded the lincosamine derivative **109**. 7-Epi-lincosamine **121** was also isolated as a side-product.

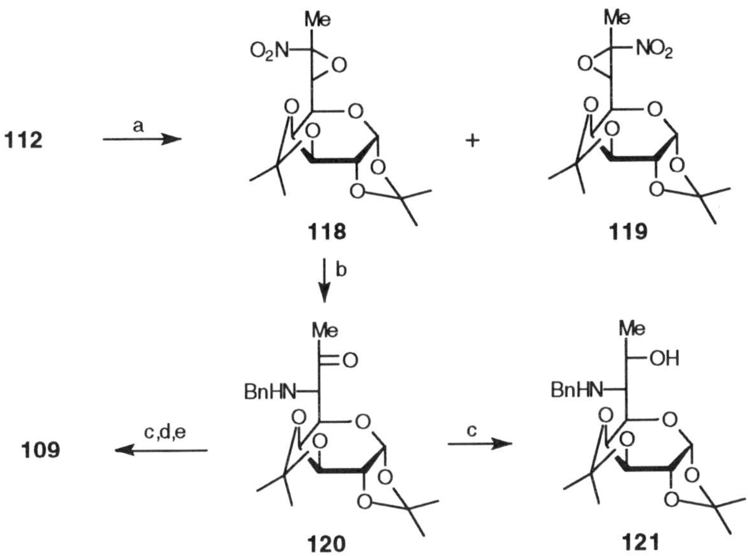

a. H_2O_2 - OH^-; b. $BnNH_2$/DMF; c. $NaBH_4$ - EtOH; d. 5% Pd/C - H_2; e. Ac_2O - Pyr.

Scheme 31

The other method for introduction of an amine function into the nitroepoxide **118** consisted of opening the oxirane ring with azide instead of benzylamine (Scheme 32) [69]. Thus treatment of **118** with sodium azide in DMF gave a mixture of azidoketones **122** and **123**. Step-wise, one pot sodium borohydride reduction of the carbonyl and azido group in the major azidoketone **122** afforded a mixture of epimeric aminoalkohols **117** and **124** from which after acetylation, the *N*-acetyl derivatives of lincosamine **109** and its C-7 epimer **125** were obtained. Diol **105**, an intermediate in the synthesis of methyl α-thiolincosaminide (**2**), developed by Szarek et al. [66] was obtained from aldehyde **94** also in a different manner. Géro et al [70] *C*-acylated the carbanion derived from 2-methyl-1,3-dithiane (**126**) with aldehyde **94** to obtain epimeric dithioacetals **127** and **128** (Scheme 33). Hydrolysis of the dithioacetal and reduction of the carbonyl group in the D-*glycero* ketoalcohol **129** led to the D-*erythro* diol **105**, precursor of lincosamine (**93**) (cf. Scheme 29).

Carbanion chemistry was also used by Hoppe and Schölkopf [71] for the two-carbon elongation of aldehyde **94**. Treatment of **94** with 1-ethoxyvinyllithium yielded a mixture of epimeric at C-6 alcohols **130** (Scheme 34). Activation of the C-6 hydroxyl as mesylates **131** and subsequent nucleophilic substitution with azide followed by hydrolysis of the vinyl ether afforded a 1:1 mixture of L- and

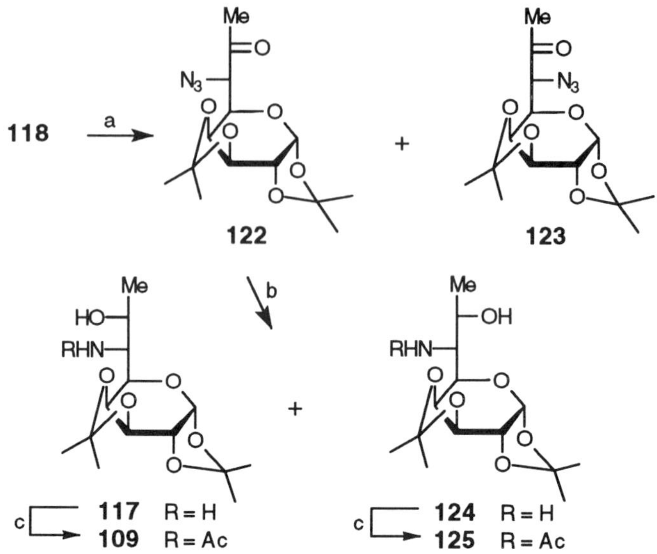

a. NaN_3 - DMF; b. $NaBH_4$ - *i*-PrOH; c. Ac_2O - Pyr.

Scheme 32

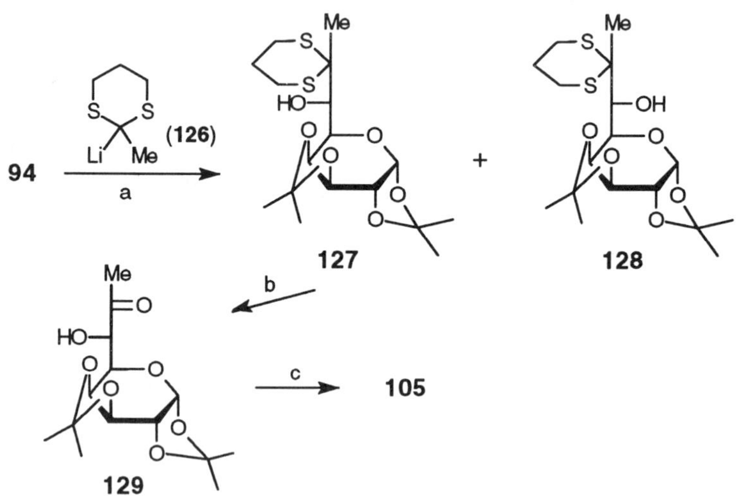

a. THF or HPT; b. $CaCO_3$ - $HgCl_2$/MeOH; c. $NaBH_4$

Scheme 33

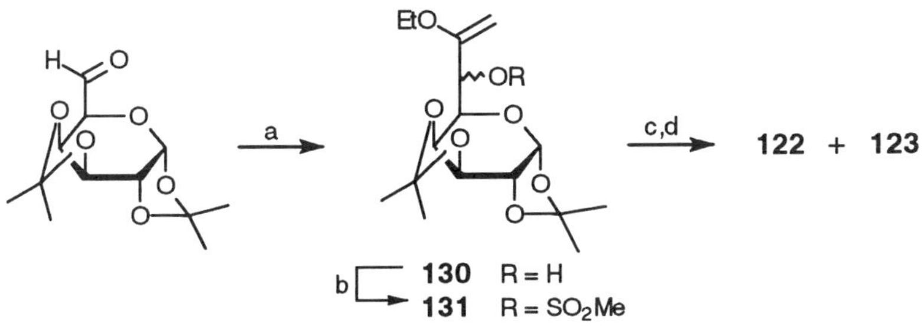

a. EtO-CH=CH_2, BuLi/THF; b. $MeSO_2Cl$ - NEt_3/CH_2Cl_2; c. LiN_3 - HMPT, 100° C; d. HCl - MeOH.

Scheme 34

D-*glycero* ketoazides **122** and **123**. Fortunately base-catalysed equilibration led to 6:1 mixture enriched in the D-*glycero* isomer **122**, an advanced intermediate in the lincosamine (**93**) synthesis (cf. Scheme 32).

David et al. [72] accomplished two-carbon extension of the acid chloride **132** prepared from D-galactose (**6**). Reaction of **132** with diazoethane afforded epimeric diazoketones **133** which after acetolysis yielded epimeric acetates **134** and **135** in a moderate yield. Oximation, reduction and acetylation of the epimer **134** corresponding to Szarek's ketobenzoate **107**, formed the lincosamine derivative **137** (Scheme 35).

Diazoketone **133** also served as a substrate for the synthesis [73] of 1,2;3,4-di-*O*-isopropylidene-*N*-acetylcelestosamine (**141**), the sugar portion of the celesticetin (**14**), an antibiotic related to lincomycin. Thus treatment of diazoketone **133** with methanol in the presence of boron trifluoride gave ketoethers **138** and **139**. The one with the D-*glycero*-D-*galacto* configuration (**138**) was transformed into the oxime **140** which after reduction and acetylation afforded celestosamine derivative **141** (Scheme 35).

Acetylene sugars

Ethynylation of aldehyde **94** extended a sugar chain by two carbons and at the same time introduced a terminal acetylene function with versatile synthetic potential. Horton et al. [74] amongst other substrates reacted aldehyde **94** with ethynylmagnesium bromide to obtain a 1:1 mixture of D- and L-*glycero* acetylene alcohols **142** and **143** (Scheme 36).

Separation of the epimers (**142, 143**) and partial reduction of the triple bond yielded D-*glycero* or L-*glycero* allylic alcohols **103** which were previously utilized by Szarek et al. [66] in the synthesis of lincosamine (**93**).

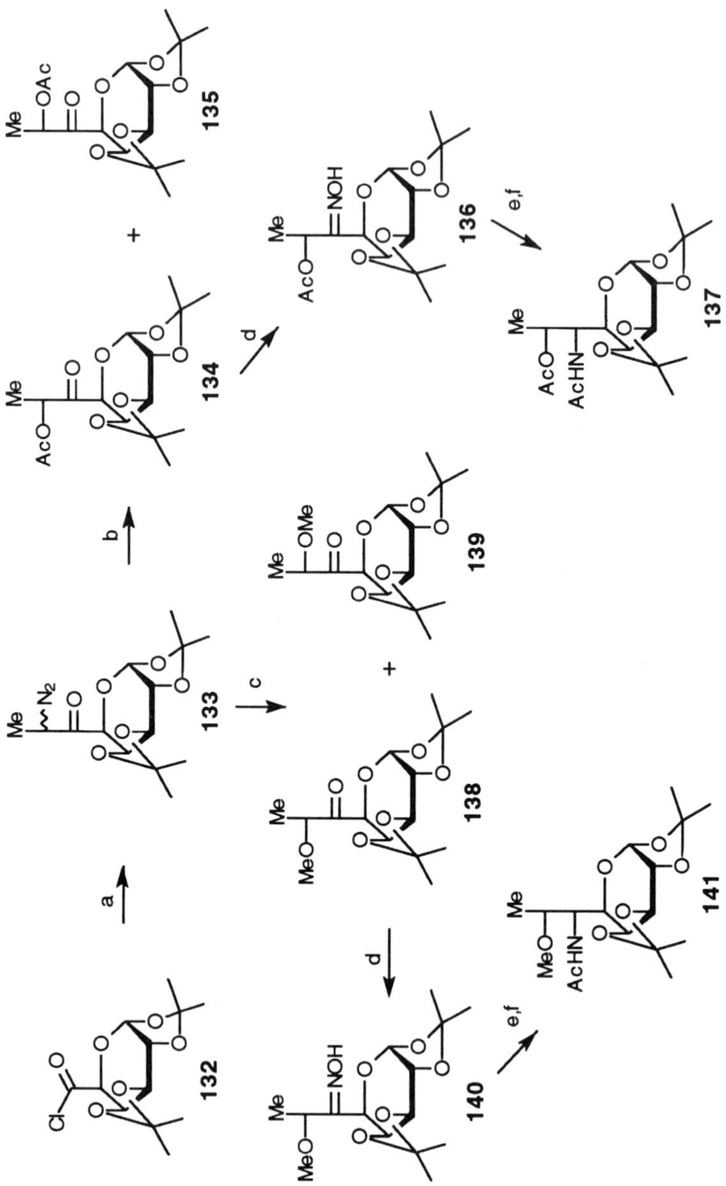

a. EtN_2; b. AcOH; c. MeOH - BF_3; d. $H_2NOH{\cdot}HCl$, Pyr; e. $LiAlH_4$ - THF; f. Ac_2O - Pyr.

Scheme 35

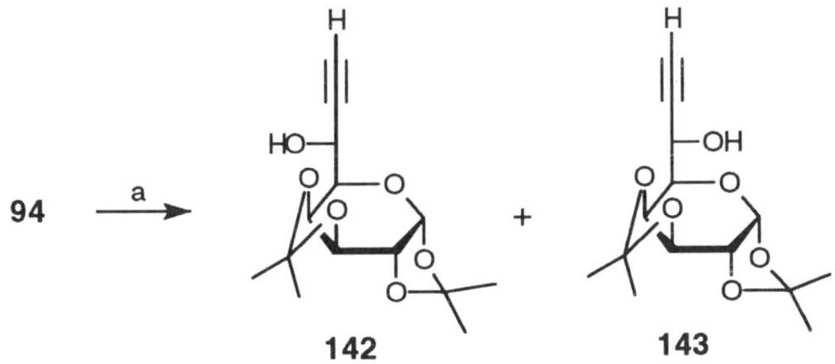

a. EtMgBr, HC ≡CH, THF.

Scheme 36

L-*Glycero* allylic alcohol **103** was obtained also by Brimacombe et al. [75]. In connection with exploration of synthetic routes to various octoses

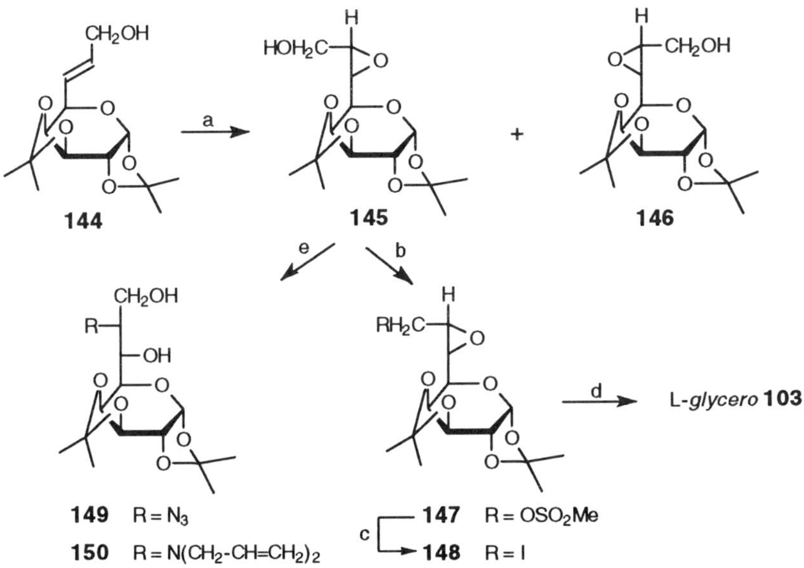

a. *m*-$ClC_6H_4CO_3H/CH_2Cl_2$; b. $MeSO_2Cl$ - Pyr; c. NaI/acetone; d. Zn - Cu/EtOH, Δ; e. $Ti(O\text{-}i\text{-}Pr)_4$ - Me_3SiN_3/C_6H_6 or $(CH_2{=}CH\text{-}CH_2)_2NH$ - $Ti(O\text{-}i\text{-}Pr)_4$.

Scheme 37

[75,76] they studied stereoselectivity of epoxidation and *cis*-hydroxylation of octenopyranose **144**. Oxidation of **144** with *m*-chloroperbenzoic acid yielded

a mixture of D-*threo* **145** and L-*threo* **146** epoxides (Scheme 37). From the major product **145** via mesylate **147**, epoxy-iodide **148** was obtained. Reductive elimination of the latter gave the L-*glycero* alcohol **103**.

Stick et al. [77] attempted transformation of epoxide **145** into lincosamine (**93**). Opening of the oxirane ring in **145** with nitrogen nucleophiles, trimethylsilyl azide - titanium tetraisopropoxide, or di(allyl)amine led exclusively to the products **149** or **150** with nitrogen substituents at C-7 (Scheme 37). Also, the oxyamination reaction of alkene **102** [78] carried out by Australian authors [77] as a mean of direct introduction of amino and hydroxy groups gave a product **151** with unwanted regiochemistry (Scheme 38).

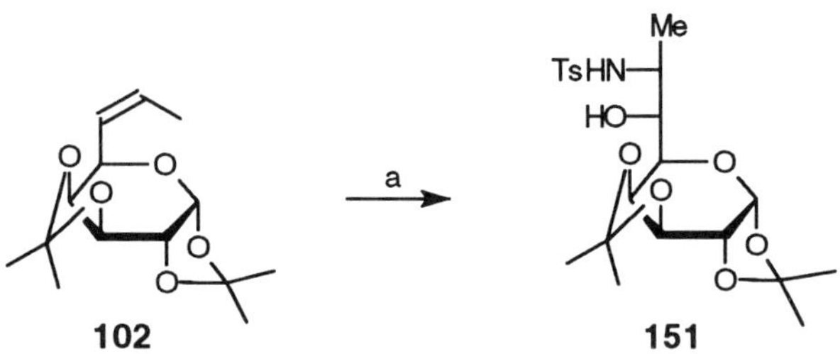

a. Chloramine T - OsO_4.

Scheme 38

Significant progress in the methodology based on acetylene sugars was achieved by Czernecki and Valery [79], who reported on the stereoselective formation of the C-6 chiral center in the ethynylation reaction of aldehyde **94**. Their success depended on the choice of the ethynylation reagent. Thus, reaction of **94** with trimethylsilylacetylene in the presence of magnesium bromide afforded, after desilylation, L-*glycero* alcohol **143** as a single product. Smooth inversion of configuration at C-6 accomplished by the Mitsunobu reaction (Ph_3P - DEAD - *p*-$O_2NC_6H_4CO_2H$) yielded epimer D-*glycero* **142** [80] (Scheme 39). Nucleophilic substitution of the triflate **152** with benzylamine proceeded with clean inversion of configuration at C-6 to give amine **153**. After protection of amino group, hydration of the triple bond under mild conditions (mercuric acetate) followed by hydrogenolysis of the carbobenzyloxy group gave the aminoketone **120** which was previously [69] transformed into the substituted lincosamine (Scheme 39). Alternatively, amine **153** was hydrated in the above-mentioned conditions to give acetamidoketone **155** which was also used before as an intermediate in the synthesis of lincosamine (**93**). Hence stereoselective synthesis of the acetylene sugar introduced stereocontrol to formerly developed [66][69] route to **93**.

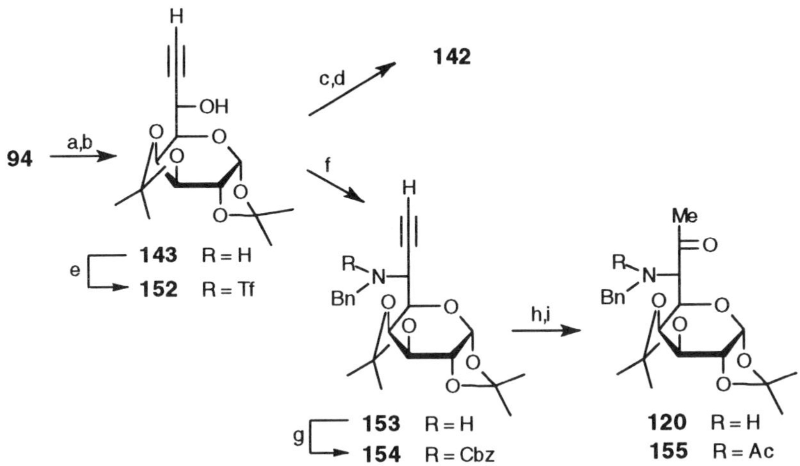

a. H-C≡ C-$SiMe_3$ - $MgBr_2$/ether; b. Et_3N, 3HF - MeCN; c. PPh_3 - DEAD - *p*-$O_2NC_6H_4CO_2H$; d. K_2CO_3/MeOH; e. CF_3SO_2Cl - Pyr; f. $BnNH_2$; g. CbzCl; h. 1.$Hg(OAc)_2$/AcOEt, 2. H_2S; i. H_2/Pd - C

Scheme 39

Employing 2-thiazolylmethylenetriphenylphosporane (**157**) Dondoni developed an efficient general technology for a two-carbon homologation: amination of aldehydes. Its application to aldehyde **94** led to derivatives of aminooctoses **161** and **162** remotely related to lincosamine (**93**). Key steps of the synthesis were as follows. Olefination of aldehyde **94** with ylide **157** generated in situ from the phosphonium salt **156** followed by iodine catalyzed equilibration of the *E*:*Z* (1:3) mixture of alkenes afforded *E*-alkenylthiazole **158** (Scheme 40). Michael addition of benzylamine to the iodide **159** and one pot acetylation - deblocking sequence gave the acetylamido dialdose **160**. Unfortunately, the mode of addition, nucleophile (benzylamine) attack on the double bond of iodide **159** from the side

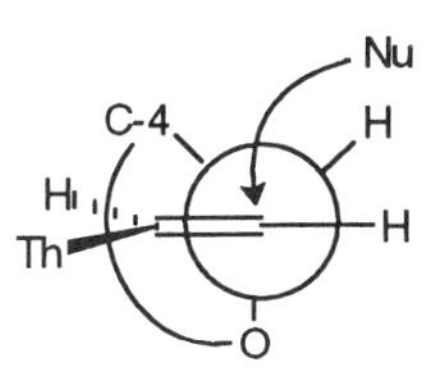

Figure 6

ring as depicted by the Newman projection (Fig. 6) led to the configuration (6*S*) opposite to the pyranose ring as depicted by the Newman projection (Fig. 6) led to the configuration (6*S*) opposite to that of lincosamine (**93**).

Reduction of the aldehyde **160** and reductive removal of the C-8 hydroxyl in the resulting alcohol **161** afforded aminooctose **162**.

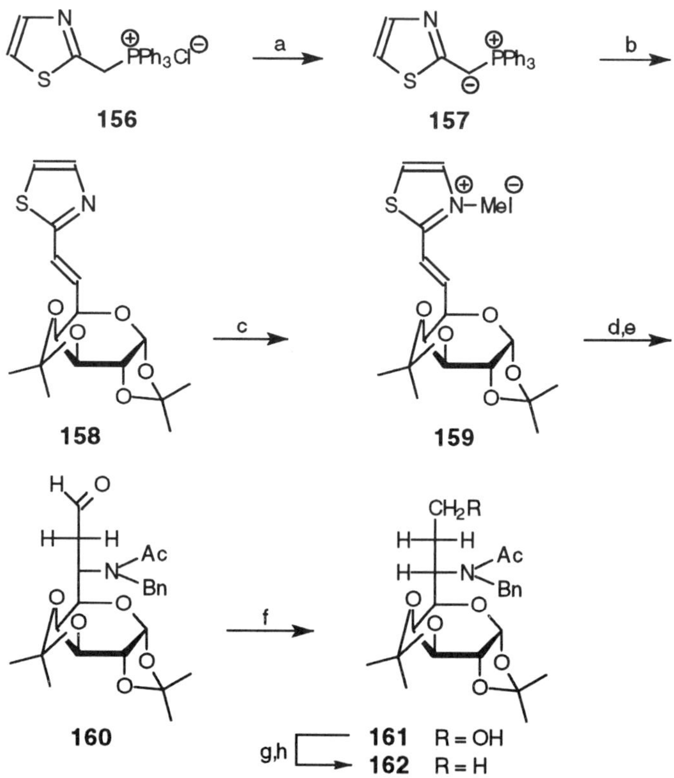

a. *t*-BuOK; b. i. **94**, ii. I_2; c. MeI - MeCN; d. $BnNH_2$ - MeOH; e. i. Ac_2O - Pyr; ii. $HgCl_2/H_2O$ - MeCN; f. $NaBH_4$ - MeOH; g. i. TsCl - Pyr, ii. $LiAlH_4$

Scheme 40

Analogues of lincosamine and methyl thiolincosaminide

Tronchet and Massoud [82] described the chain elongation of aldehyde **94** in the Wittig reaction with various stabilized phosphoranes, e.g **163** yielding octenopyranose **164** (Scheme 41). The conjugate addition of ammonia to **164** afforded epimine **165** in good yield as a mixture of *cis* and *trans* isomers. Opening of the aziridine rings in *cis*-**165** and *trans*-**165** gave epimeric 6-epi-lincosamine analogues **167** and **168**, respectively, with carboxymethyl groups instead of C-8 methyls. Oxidation of acetamidoalcohol **167** or **168** and epimerization of the resulting acetamidoketone **169** led to ketoester **170** with (*R*) configuration at C-6 (Scheme 41).

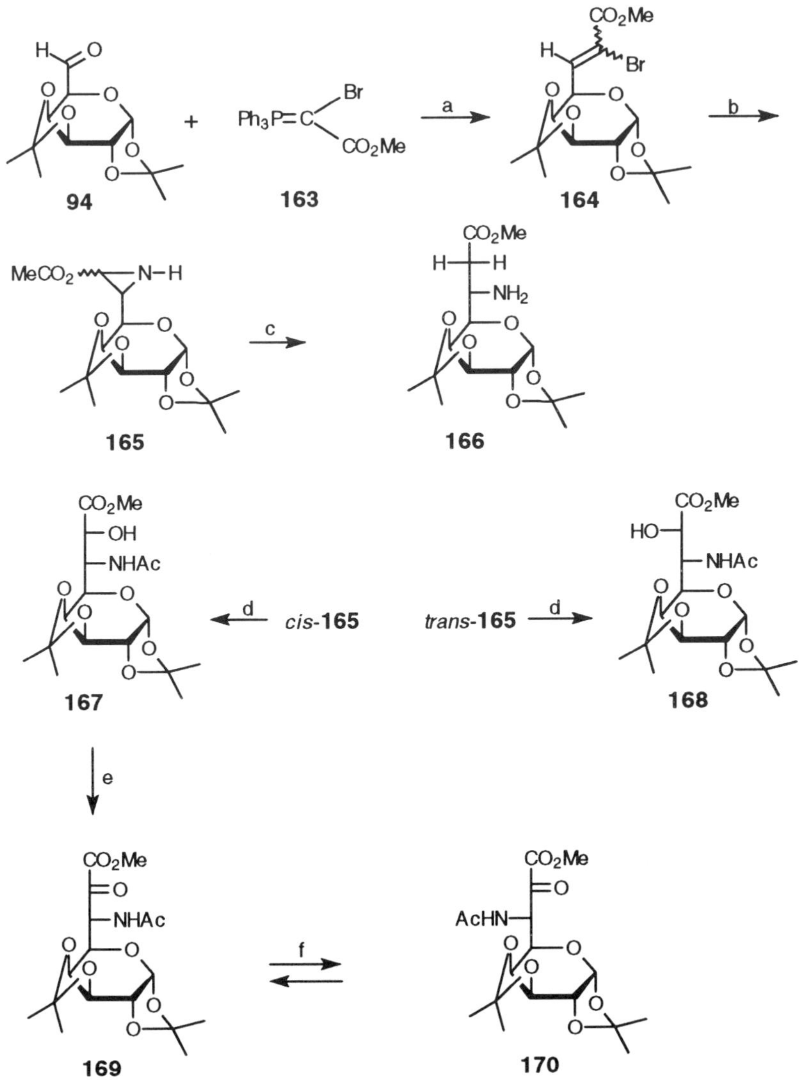

a. C_6H_6; b. NH_3 - MeOH; c. H_2/RaNi; d. MeCOBr - Et_3N (trace); e. CrO_3 - Pyr; f. base.

Scheme 41

Methylthio aldehyde **174** was prepared by the same authors [83] starting from D-galactopyranosyl bromide **171** as shown in Scheme 42. Reaction of **171** with thiourea followed by hydrolysis and methylation afforded a mixture of two methyl thiogalactosides **172** and **173**. Chromatographic separation of the major β-anomer, removal of acetyl groups and protection of C-3 and C-4 hydroxyls was followed by oxidation of the primary hydroxyl group yielding aldehyde **174**. It's Wittig reaction with phosphorane **163** gave the eight-carbon methyl thiopyranoside **175**, opening a route to the C-8 modified methyl β-D-thiolincosaminide.

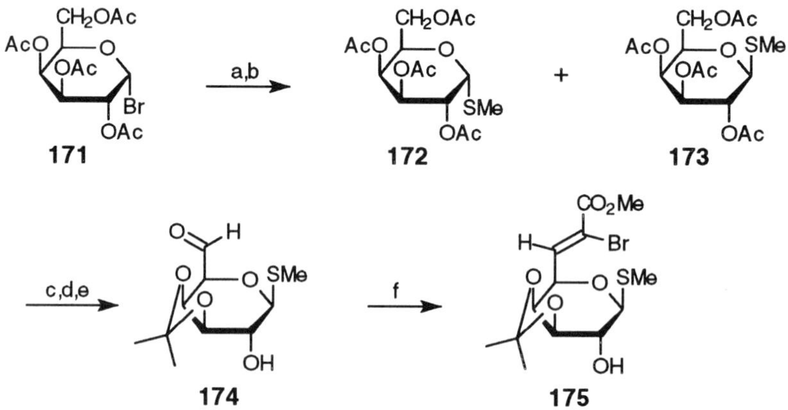

a. $S{=}C(NH_2)_2$, HMPT; b. i. H_2O - K_2CO_3, ii. $NaHSO_3$, iii. MeI; c. MeO^- -MeOH;
d. $(CH_3)_2C{=}O$ - $CuSO_4$; e. CrO_3 - Pyr/CH_2Cl_2; f. $Ph_3P{=}C(Br)CO_2Me$ (**163**)/C_6H_6

Scheme 42

5.1.2. Syntheses from 6-benzylimino-α-D-galactopyranose 96

A novel approach to lincosamine (**93**) synthesis was proposed by Fukumaru and Atsumi [84] who introduced as a starting material aldimine **96**, readily available by condensation of aldehyde **94** and benzylamine. The obvious benefit of employing aldimine **96** in the synthesis was the presence of the amino function at C-6 already in the substrate. Alkylation of **96** with vinylmagnesium bromide led to a mixture of epimeric unsaturated amines **176** (Scheme 43). After acetylation and debenzylation the L-*glycero* **178** and D-*glycero* **179** epimers were separated. Hydration of the major L-*glycero* isomer **178** afforded the derivative of the D-*threo*-D-*galacto*-octose **116** (cf. Scheme 30).

Alternatively, a mixture of epimers **177** was separated and ozonolysis of each isomer yielded aminoaldehydes **180** and **181**, respectively. Addition of methylmagnesium bromide to the L-*glycero* aminoaldehyde **181** afforded

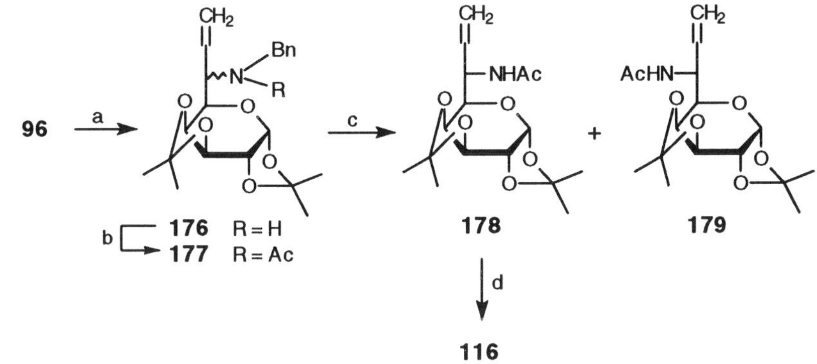

a. CH_2=CH-MgBr/THF; b. Ac_2O; c. Na-NH_3/THF; d. $HgAc_2$/H_2O-THF, then $NaBH_4$.

Scheme 43

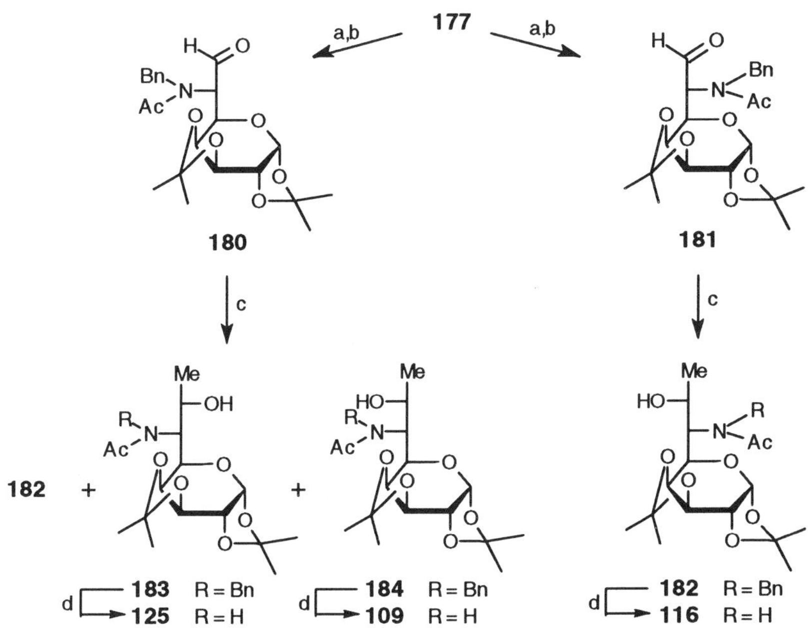

a. chromatography; b. O_3/CH_2Cl_2, then Zn/AcOH; c. CH_3MgBr/THF; d. Na - NH_3/THF.

Scheme 44

stereoselectively D-*threo*-acetamidoalcohol **182** (Scheme 44). On the other hand reaction of the same Grignard reagent with D-*glycero* isomer **180**, a potential precursor of lincosamine (**93**) was not stereoselective. The addition proceeded with epimerization at C-6 yielding three products with D-*threo* **182**, L-*threo* **183** and D-*erythro* **184** configurations of the side chain. Base-catalyzed epimerization of the C-6 stereogenic center in acetamidoaldehydes **180** and **181** was confirmed in the independent experiments [85]. Removal of the benzyl groups in compounds **184**, **183** and **182** gave derivatives of lincosamine **109** and of its diastereoisomers **125** and **116**, respectively.

This methodology was extended by employment of another Grignard reagent [86]. Addition of 1-methylvinylmagnesium bromide to aldimine **96** yielded alkenylamine as a 4:1 mixture of L-*glycero* **185** and D-*glycero* **186** epimers (Scheme 45). Separate ozonolysis of **185** and **186** followed by sodium borohydride reduction of the resulting ketones **187** and **188** afforded in each case a mixture of epimeric aminooctoses: 3:1 D-*threo* **189** and L-*erythro* **190** as well as 4:1 D-*erythro* **191** and L-*threo* **192**. Debenzylation of compounds **191**, **189**, **190** and **192** gave derivatives of lincosamine **109** and of its diastereoisomers **116**, **193** and **125**, respectively.

The expediency of the Fukumaru and Atsumi methodology based on utilization of a substrate which already possessed an amino function at C-6 was marred by the lack of stereocontrol in creating new chiral centers. However, in other approaches highly stereoselective two-carbon and one-carbon additions to aldimine **96** were achieved.

Bose et al. [87] accomplished a two-carbon elongation of aldimine **96** skeleton with retention of the amino function at C-6 by utilizing β-lactam chemistry. Thus [2 + 2] cycloaddition of methoxyketene, generated in situ from methoxyacetyl chloride and triethylamine, to the C=N bond of **96** proceeded stereoselectively yielding β-lactam **194** as a single product (Scheme 46). The absolute configuration of the four-membered heterocycle was established by the degradation of **194** to the known *cis*-β-lactam derivative **199**. Methanolysis followed by benzylation gave ester **195**. Lithium aluminum hydride reduction of **195** furnished the aminoalcohol **196** which via chloride **197** was reduced to the derivative of the 6-epi-celestosamine **198**, a sugar portion of celesticetin (**14**) the antibiotic related to lincomycin (**1**), which was discussed earlier.

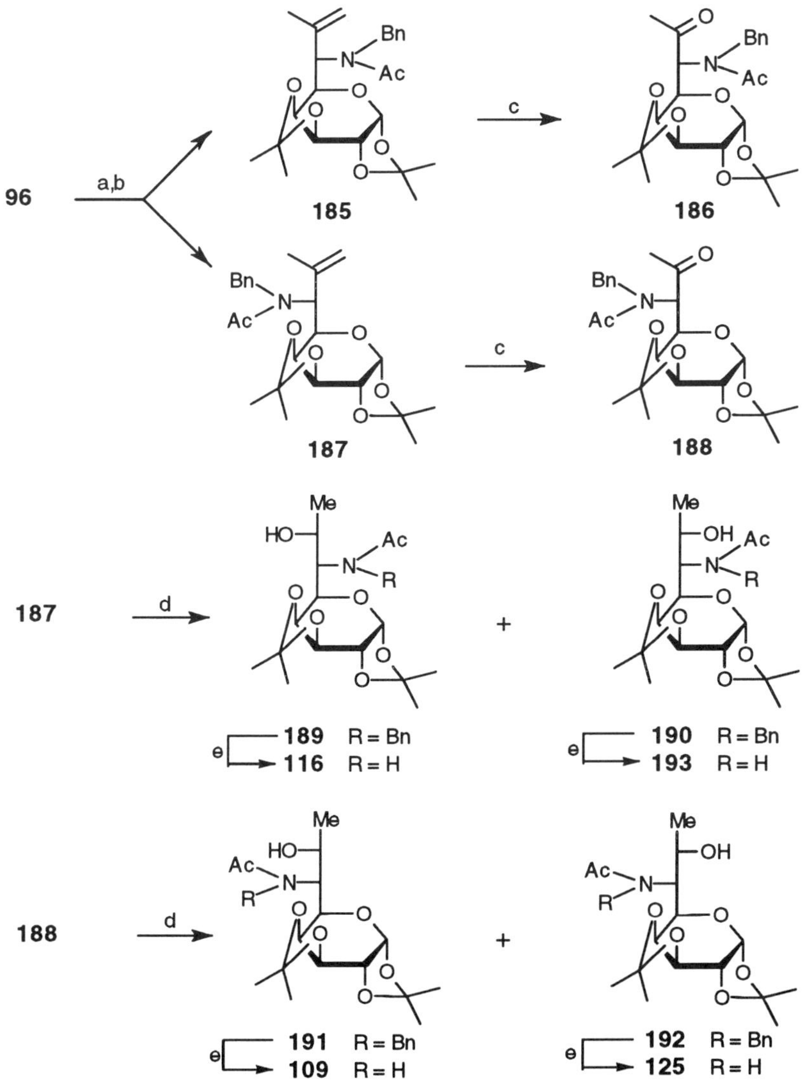

a. $CH_2=C(CH_3)MgBr/THF$; b. Ac_2O; c. O_3/CH_2Cl_2, then Zn/AcOH; d. $NaBH_4$ - EtOH; e. H_2 - Pd/C

Scheme 45

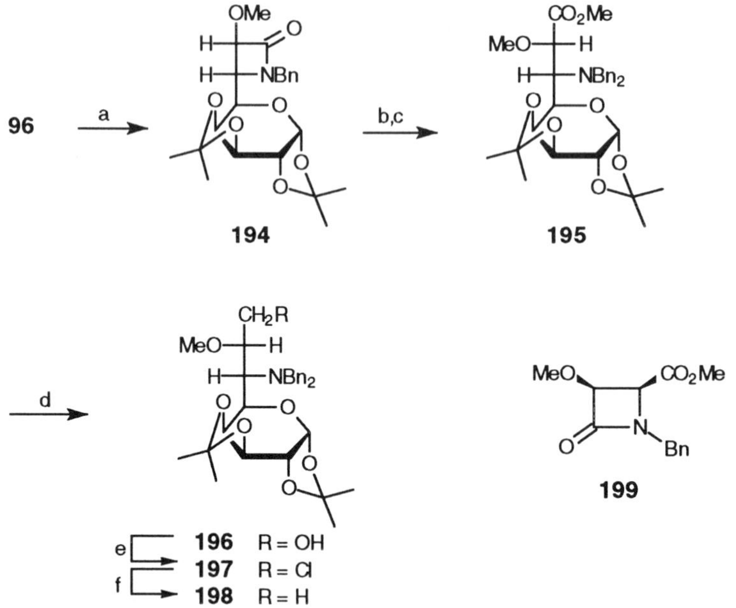

a. MeO-CH_2-COCl - Et_3N/CH_2Cl_2; b. MeOH - MeO^- ; c. BnBr; d. $LiAlH_4$, e. $MeSO_2Cl$ - Pyr, 0°C; f. $LiAlH_4$/THF.

Scheme 46

5.2. One-carbon chain extension

Whereas two-carbon extensions of sugar chain aimed at constructing the carbon skeleton of lincosamine (**93**) were studied extensively, only a few approaches based on one-carbon elongation were reported. They are however of interest because, using aldimine **96** as a substrate, stereocontrolled syntheses of lincosamine (**93**) precursors were recently reported.

The first method of the step-wise construction of lincosamine (**93**) skeleton, with concomitant introduction of appropriate functional groups in the side chain, started from aldehyde **94**. Saeki and Ohki [88] converted **94** into a mixture of D-*glycero* **200** and L-*glycero* **201** cyanohydrins which after tosylation were separated into pure epimers **202** and **203** (Scheme 47). Treatment with lithium aluminum hydride of D-*glycero* tosylate **203** followed by acetylation afforded the *N*-acetylated epimine **204**. Opening of the aziridine ring in **204** with acetic acid gave acetamidoacetate **205**, which after selective deacetylation was oxidized to yield aldehyde **206**. Addition of methylmagnesium bromide to the latter gave

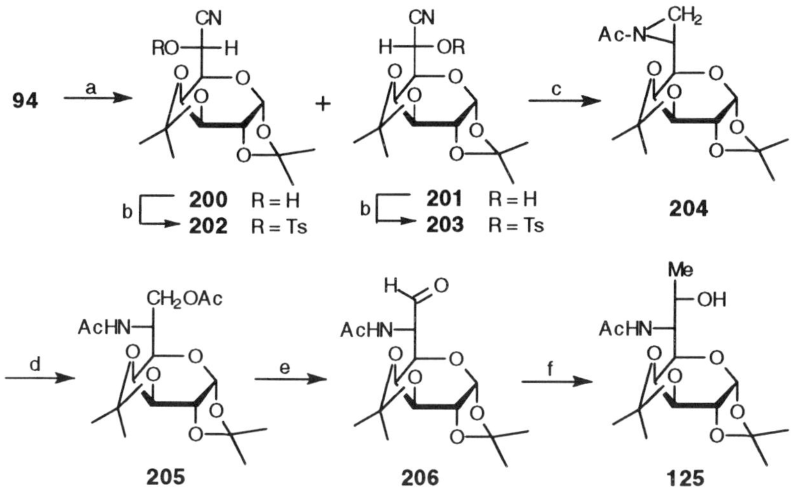

a. NaCN/MeOH - H_2O; b. TsCl - Pyr; c. 1. $LiAlH_4$/ether, 2. Ac_2O; d. AcOH - Δ;
e. 1. MeO^-/MeOH, 2. DCC - Me_2SO - 1M H_3PO_4; f. MeMgBr/THF.

Scheme 47

L-*threo*-aminooctose **125**. The D-*erythro* configuration in the side chain was achieved in the oxidation - reduction cycle yielding a derivative of *N*-acetyllincosamine **109** (cf. Scheme 30 and 32). The same sequence beginning with L-*glycero* cyanohydrin **200** led to the mixture of aminooctoses **116** and **193**, diastereoisomeric with lincosamine (**93**). The synthesis of lincosamine was accomplished, however stereocontrol of individual steps was low or just opposite to that wanted; in effect inversion of configuration was required (**125** → **105** → **109**).

High stereoselectivity albeit opposite to the one required for lincosamine (**93**) synthesis was achieved by Czernecki et al. [89] in the cyanoamination reaction of aldehyde **94**. One pot reaction of **94** with sodium hydrogen sulfite, benzylamine and sodium cyanide yielded epimeric at C-6 cyano-amines **207** and **208** in a 10:1 ratio (Scheme 48). Diastereoselectivity of the addition was rationalized in terms of the preferred attack of cyanide ion on the less hindered face of the >C=N double bond in the depicted conformation of the iminium intermediate **209** (Figure 7).

To raise the yields of D-*glycero* aminonitrile **208** an indirect route was explored. A mixture of cyanohydrins **209** and **210** was treated with triflic anhydride to give trifluoromethanesulfonates **211** and **212** in a 3:1 ratio. Nucleophilic substitution at C-6 with benzylamine proceeded with some

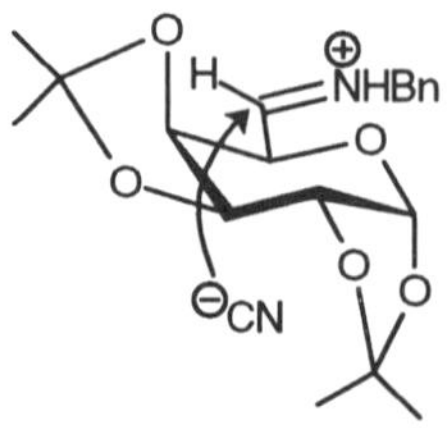

Figure 7

equilibration because the resulting aminonitriles **207** and **208** were obtained in the ratio 3:7. Again the D-*glycero* isomer required for lincosamine (**93**) synthesis was a minor product (Scheme 48).

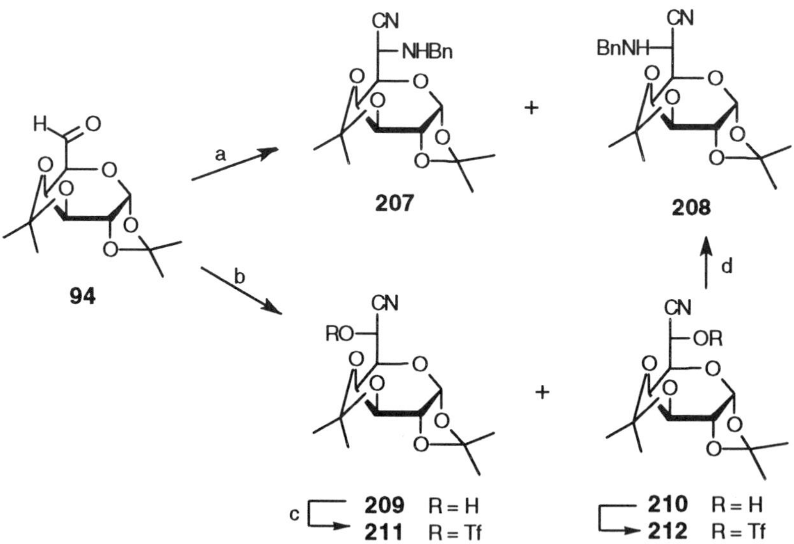

a. $NaHSO_3$, $BnNH_2$, $NaCN/H_2O$ - EtOH; b. NaCN/AcOH - H_2O;
c. $(CF_3SO_2)_2O$ - Pyr/CH_2Cl_2; d. $BnNH_2$.

Scheme 48

An addition of 2-metalated thiazoles to formyl groups was developed by Dondoni [90] into an elegant method of one-carbon homologation of aldehydes affording α-hydroxyaldehydes in a stereocontrolled fashion. Addition of 2-metalated thiazoles to nitrones [91] extended this methodology to the synthesis of α-aminoaldehydes. Using nitrone **213** obtained by condensation of hydroxylamine with aldehyde **94** as starting material, two epimeric amines, either **214** or **215**, depending on reaction conditions, were obtained. Amine **215** was subsequently converted into an advanced intermediate in the lincosamine (**93**) synthesis (Scheme 49). Control of the diasteroselectivity of the

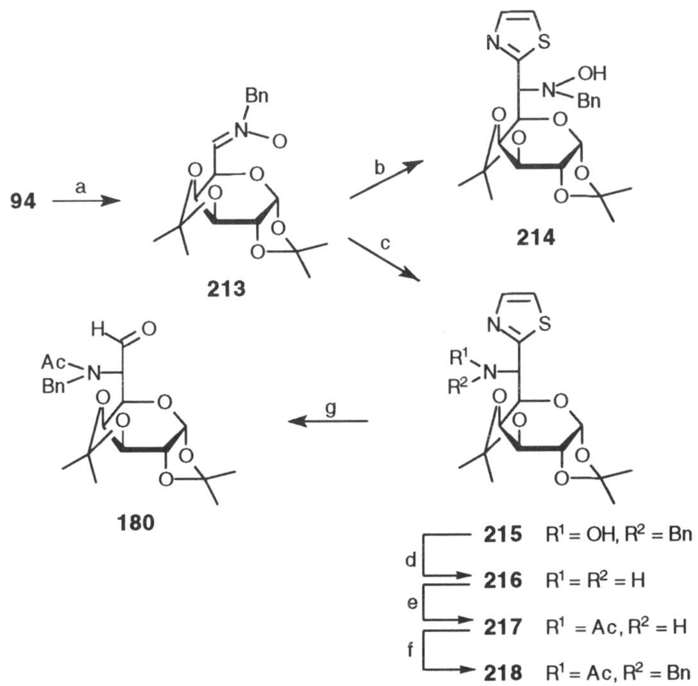

a. $BnNHOH/CH_2Cl_2$; b. 2-lithiothiazole, $ZnBr_2/Et_2O$, -80°C; c. 2-lithiothiazole - Et_2ClAl/Et_2O, -10°C; d. $TiCl_3$, $MeOH$-H_2O, RT; e. Ac_2O-Pyr-DMAP; f. NaH, DMF, BnBr, 0°C; g. CF_3SO_3Me - MeCN, then $NaBH_4$ - MeOH, then CuO, $CuCl_2$ - MeCN - H_2O

Scheme 49

2-lithiothiazole addition was achieved by complexing nitrone **213** with an appropriate Lewis acid. Thus reacting zinc bromide or magnesium bromide with **213** followed by addition of 2-lithiothiazole gave *syn*-adduct **214** with (*R*) configuration at the new stereocenter as the main reaction product (*syn:anti* = 9:1). Complexation of **213** with diethyl aluminum chloride or titanium tetrachloride led to the *anti*-addition affording as a major product (6*S*)-amine **215** (*anti:syn* = 9:1) (Scheme 49).

To rationalize this remarkable change of diastereoselectivity, two different conformations depending on the Lewis acid ($MgBr_2$ or Et_2AlCl) were postulated. The attack of nucleophile from the less hindered face of the C=N double bond in conformation A gave the *syn*-adduct **214** whereas in the case of conformation B, led to the *anti*-adduct **215** (Figure 8).

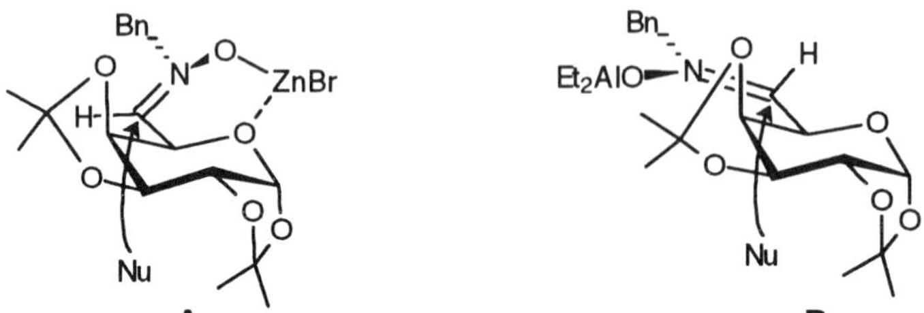

A

B

Figure 8

Anti-adduct **215** after deoxygenation with concomitant debenzylation by titanium trichloride produced amine **216** which was acetylated and benzylated to yield derivative **218**. The thiazole ring was removed by treatment with methyl trifluoromethanesulfonate followed by sodium borohydride and then CuO and $CuCl_2$ liberating the formyl group. The obtained aminoaldehyde **180** was used as a precursor of lincosamine (**93**) [84].

Recently van Boom et al. [92] reported on yet another elegant method for stereoselective one-carbon extension of the benzylimine **96** side-chain. A key step which was susceptible to stereocontrol in their approach consisted in the addition of precomplexed [{dimethylphenylsilyl}methyl]magnesium chloride (**219**) (Scheme 50) to **96**. Precomplexation of the Grignard reagent **219** with cerium(III) chloride followed by addition of **96** resulted in a *syn*-adduct **220** as a sole product. The (*R*) configuration of the newly-created stereogenic center at C-6 was established by conversion of **220** into alcohol **222**, reported precursor of destomic acid.

Precomplexation of Grignard reagent **219** with cuprous iodide and boron trifluoride dramatically changed the steric course of the reaction with benzylimine **96**. The *anti*-adduct **221** with (6*S*) configuration at the new chiral center was obtained as a sole product. Product **221** after acetylation and oxidative removal of the silyl substituent yielded the aminoalcohol **223** which after Swern oxidation afforded the aminoaldehyde **180**, a precursor of lincosamine (**93**) [84].

5.3 Intramolecular nitrogen atom delivery

Based on the syntheses of the lincosamine (**93**) discussed so far, which rely on the eight-carbon framework obtained from D-galactose (**6**) it is apparent that difficulties encountered in the stereocontrolled functionalization of the side-chain were not completely resolved. In particular the regio- and stereoselective

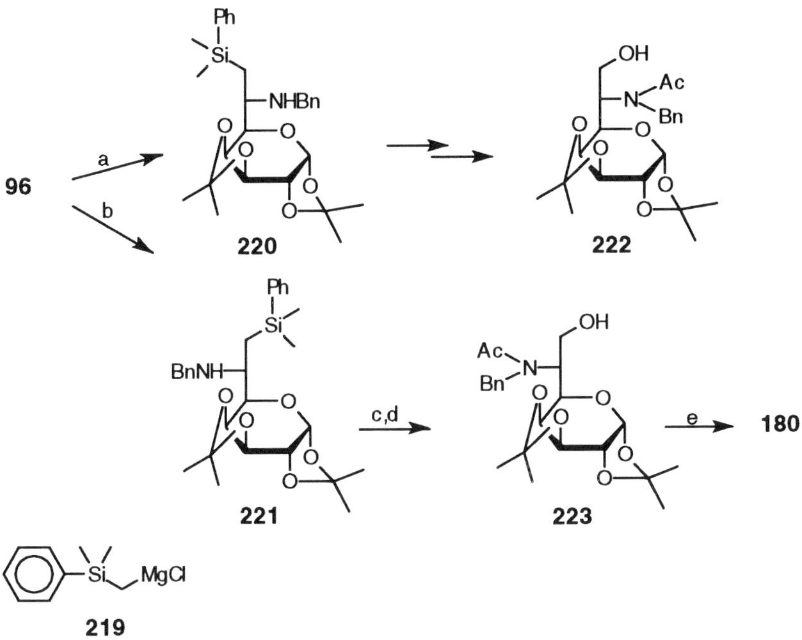

a. **219**, $CeCl_3/Et_2O$ - THF, -78° C; b. CuI, $BF_3{\cdot}Et_2O/Et_2O$, -40° C; c. Ac_2O - Pyr; d. KBr, $MeCO_3H$; e. DMSO - $(COCl)_2$ - Et_3N.

Scheme 50

introduction of the amino function at C-6 remained a vexing problem. A conceptually novel approach to its solution has been suggested independently by two groups; Knapp and Kukkola [93] as well as Stick et al [77]. Their idea was based on the methodology of intramolecular nitrogen delivery [94] and observations made by Danishefsky et al. [95] on the facile interaction of the substituent at C-4 with the C-6 functional group. Consequently substituents

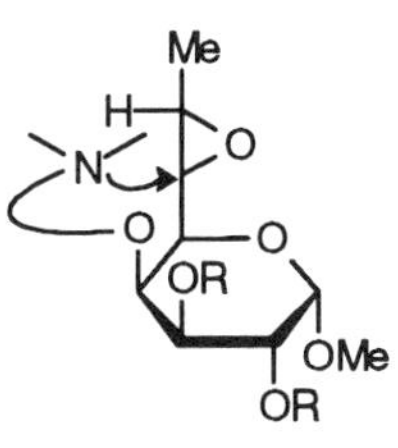

Figure 9

carrying a nitrogen atom tethered to the C-4 hydroxyl were examined as nucleophiles for the intramolecular opening of the oxirane ring in D-*threo* epoxide (Fig. 9) and introduction of the amino function, or its precursor, at C-6 with full control of regio- and stereochemistry.

Octenose **227** was obtained from methyl α-D-galactopyranoside (**224**) by Stick et al. [77] as well as Knapp and Kukkola [93] in a similar manner, differing by the use of protecting groups and reagents. Both groups employed triethyl phosphonoacetate in Emmons - Wadsworth reaction to elongate the carbon chain of aldehyde **226**. One of the syntheses [77] is shown in Scheme 51.

Direct epoxidation of alkene **227** with *m*-chloroperbenzoic acid gave a 1:4 mixture of epoxides with preponderance of L-*threo* isomer, which was unsuitable for the purpose in hand. Ultimately both groups carried out the epoxidation of octenose **227** using Danishefsky protocol [95] (vide infra) which gave selectively the desired D-*threo* epoxide **228** (Scheme 51).

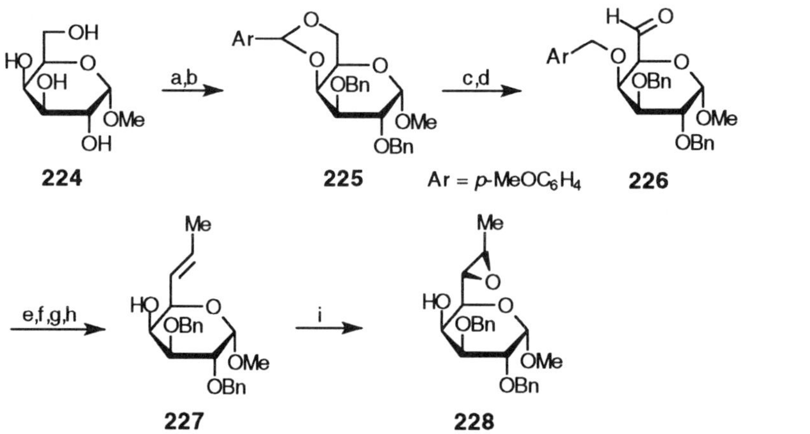

a. p-$MeOC_6H_4CH(OEt)_2$ - PPTS; b. BnBr - NaH/THF; c. $LiAlH_4$ - $AlCl_3/Et_2O$ - CH_2Cl_2; d. $(COCl)_2$ - DMSO - $(i\text{-}Pr)_2NEt/CH_2Cl_2$; e. $(EtO)_2POCH_2CO_2Et$ - LiBr - $(i\text{-}Pr)_2NEt/CH_2Cl_2$; f. DIBAH/THF; g. $[Me_2N^+{=}CCl_2]Cl^-$ - NEt_3/CH_2Cl_2, then $LiAlH_4$/THF; h. CAN/MeCN - H_2O; i. 1.NBS/CH_2Cl_2, 2. DBN/C_6H_6

Scheme 51

Stick et al. [77] examined the preparation and intramolecular nucleophilic reactivity of several derivatives of **228** with C-4 substituents carrying nitrogen atoms. They included trichloroacetamidate **229**, *N*-trimethylsilyltrichloroacetamidate **230**, trichloroacetylcarbamate **231**, carbamate **232** and 2-amino-2-phenylacetate **233**. However, either the introduction of the desired substituent at C-4 failed (**229**, **230**) or derivatives (**231**, **232**, **233**) which proved to be unreactive were obtained. Also preparation of *N*-benzylcarbamate **234** met with failure [93] and *N*-benzoylcarbamate **235** did not react in the expected fashion [93] (Scheme 52).

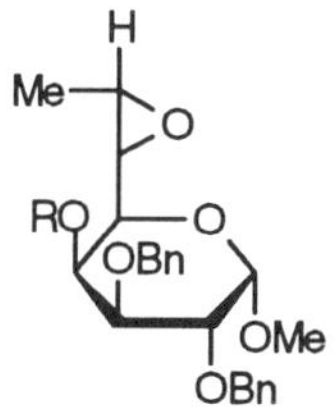

	R
229	$Cl_3C\text{-}C{=}NH$
230	$Cl_3C\text{-}C{=}NSiMe_3$
231	$Cl_3C\text{-}CONHCO$
232	H_2NCO
233	$Ph\text{-}CH(NH_2)CO$
234	BnNH-CO
235	BzNH-CO

Scheme 52

The concept of the intramolecular nitrogen delivery was successfully put to

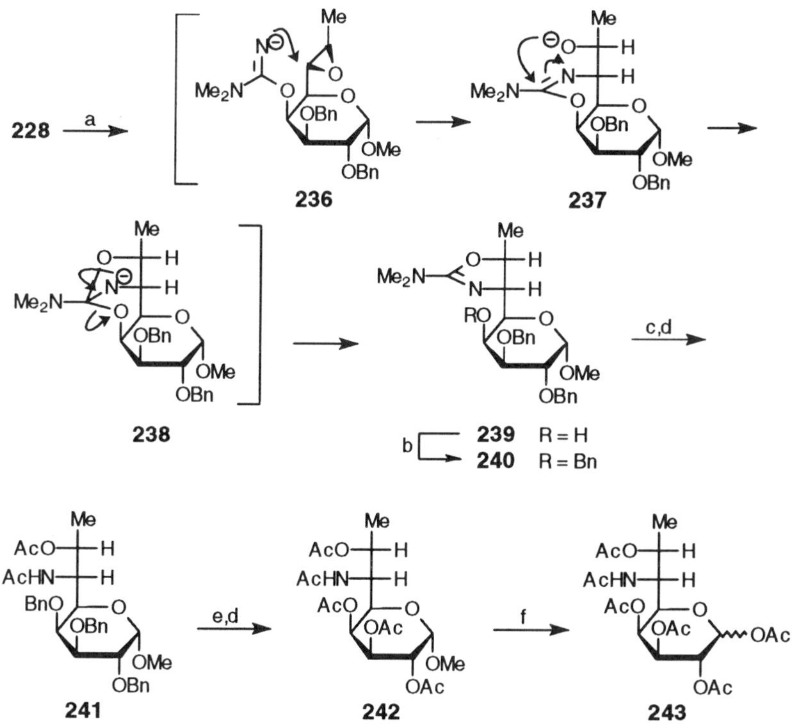

a. NaH, Me_2N-CN, imidazole/DMF; b. BnBr, NaH/DMF; c. KOH, $HOCH_2CH_2OH$ - Δ; d. Ac_2O, Et_3N, DMAP, Pyr; e. H_2/Pd$(OH)_2$/MeOH - EtOAc; f. Ac_2O, H_2SO_4.

Scheme 53

practice by Knapp and Kukkola [93] when they employed *N,N*-dimethylcyanamide as the nitrogen carrying reagent. Reaction of *N,N*-dimethylcyanamide with the anion generated from epoxyalcohol **228** presumably led to an isourea anion **236**, which after intramolecular oxirane splitting and rearrangement of anions **237** to **238** afforded oxazoline **239** (Scheme 53). In this way the crucial problem of lincosamine synthesis, namely setting the stereochemistry of the side-chain, was solved. Protection of the C-4 hydroxyl as benzyl ether **240** followed by hydrolysis and acetylation gave the *N,O*-diacetate **241**. Catalytic removal of benzyl groups and acetylation afforded the pentaacetate of methyl α-lincosaminide **242** in excellent (32% from methyl α-D-galactopyranoside) overall yield. The pentaacetate was also converted into an anomeric mixture of lincosamine peracetates **243**.

Knapp and Kukkola [93] also developed an efficient method for stereocontrolled introduction of the methylthio group at the anomeric position. Reaction of acetate **244** with (methylthio)trimethylsilane catalyzed by trimethylsilyl triflate replaced an anomeric acetate for the methylthio group yielding a 3:1 mixture of α- and β-anomers of thioglycopyranoside **245**. The efficacy of this transformation was improved by recycling of the unwanted β-anomer via acetate **244**. After chromatographic separation α-**245** was hydrolysed to give alcohol **246** which in

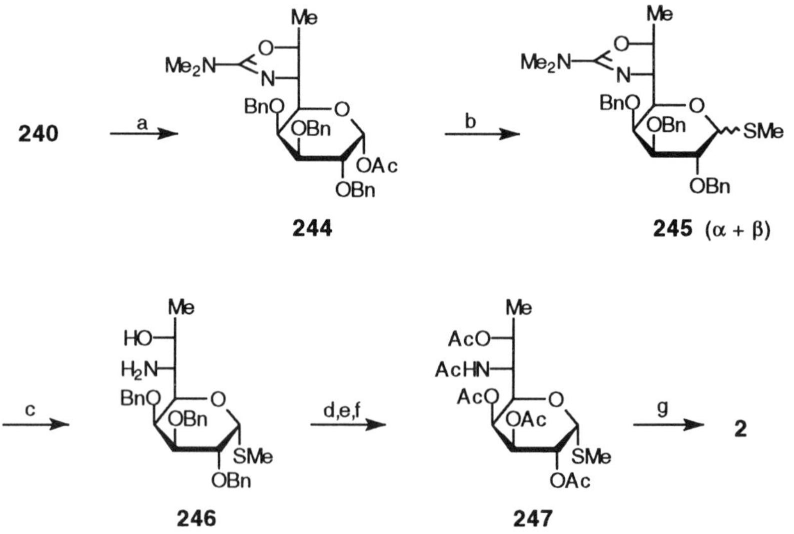

a. Ac_2O, H_2SO_4, -20° - 0°C; b. Me_3Si-SMe, Me_3SiOTf/CCl_4, Δ.; c. KOH, EtOH; d. Ac_2O/MeOH; e. Na - NH_3/THF, -78°C; f. Ac_2O, Et_3N, DMAP, Pyr, g. H_2NNH_2, Δ.

Scheme 54

the sequence of reactions: acetylation, removal of benzyl groups and further acetylation (Scheme 54) was transformed into methyl thiolincosaminide pentaacetate **247**. Hydrazinolysis of **247** yielded methyl α-D-thiolincosaminide (**2**) thus completing its first fully stereocontrolled synthesis starting from D-galactose (**6**).

6. TOTAL SYNTHESIS

Two general approaches towards total synthesis of lincosamine (**93**) were developed. One of them relied on the hetero-Diels-Alder reaction of 1-alkoxy-1,3-butadienes with carbonyl compounds for the construction of pyranose moiety. The other employed an oxidative rearrangement of furyl alcohols into pyran system. Both methodologies were extensively studied in terms of chemistry, regio- and stereochemistry of involved transformations, ultimately furnishing stereocontrolled technologies for construction of the lincosamine (**93**) molecule as well as its diastereoisomers or analogues.

6.1 Hetero-Diels-Alder Approach

In the first reported approach Zamojski et al. [96] employed dihydropyran **250** as a starting material, which was available by the [4+2] cycloaddition of 1-methoxybuta-1,3-diene (**248**) with butyl glyoxylate (**249**) (Scheme 55). Ester **250** was utilized previously as a substrate in the general method of

248 **249** **250**

Scheme 55

monosaccharide synthesis developed by Zamojski et al. [97]. For the purpose of lincosamine (**93**) synthesis the eight-carbon frame-work was obtained from **250** by a Claisen condensation with ethyl propionate followed by the ketone cleavage of ketoester **251** yielding ketone **252** (Scheme 56). The amino function at C-6 was introduced according to the procedure used by Szarek et al [65] [66]. Treatment of **252** with hydroxylamine yielded oxime **253** (mixture of *Z* and *E* isomers), which after lithium aluminum hydride reduction and subsequent acetylation afforded epimeric *N*-acetates **254** and **255**. Functionalization of the dihydropyran ring of the major epimer **254**, whose α-DL-*erythro* configuration was proven at a later stage of synthesis, was carried out according to earlier developed methodology [97]. Epoxidation of **254** with *m*-chloroperbenzoic acid

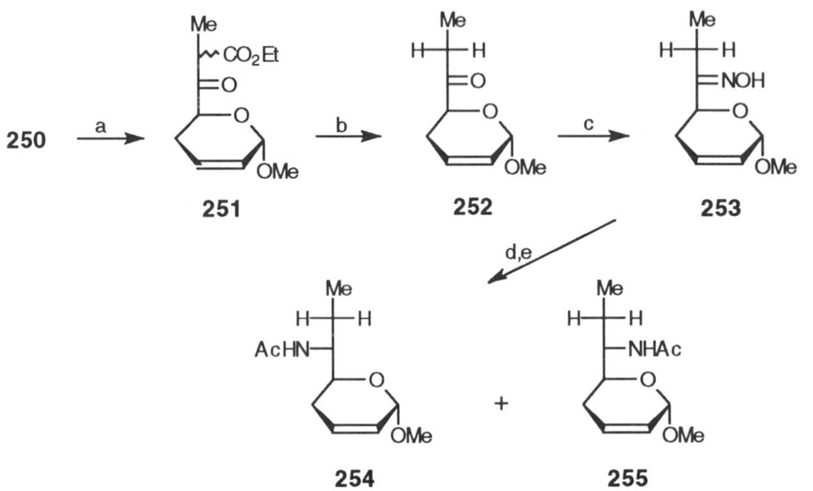

a. $EtCO_2Et$ - EtO^-; b. 5% NaOH - H_2O, Δ; c. H_2NOH - AcOH; d. $LiAlH_4/Et_2O$; e. Ac_2O - Pyr.

Scheme 56*

afforded a 1:1 mixture of epoxides **256** and **257**. After chromatographic separation of the mixture, opening of the oxiran ring in **256** with dimethylamine, an oxidation of amine **258** followed by Cope degradation, octenoside **259** was obtained (Scheme 57).

cis-Hydroxylation of **259** with H_2O_2 - OsO_4 reagent and subsequent acetylation yielded a mixture of two diastereoisomeric acetates **260** and **261**. Their chromatographic separation furnished the target tetraacetate of methyl 7-deoxy-α-DL-lincosaminide (**260**). The configuration of **260** was unequivocally confirmed by comparison, via peracetate, with an authentic sample obtained from lincosamine (**93**). In an analogous manner epoxide **257** was transformed into alcohol **262**. Inversion of configuration at C-2 in **262** in the oxidation - reduction cycle (Scheme 57) raised the yield of the target acetate **259**.

To permit introduction of an oxygen function at C-7 functionalization of the dihydropyran ring in ketone **252** was carried out with a protected carbonyl group in the side-chain. The α-DL-*erythro* configuration of the dihydropyran ring with carbonyl group protected as dimethyl ketal was obtained according to the procedures outlined already in Scheme 57 yielding ketal **263** [98]. Mild

*In Scheme 56 and all other showing racemic compounds for the sake of simplicity only enantiomers corresponding to D-series are depicted.

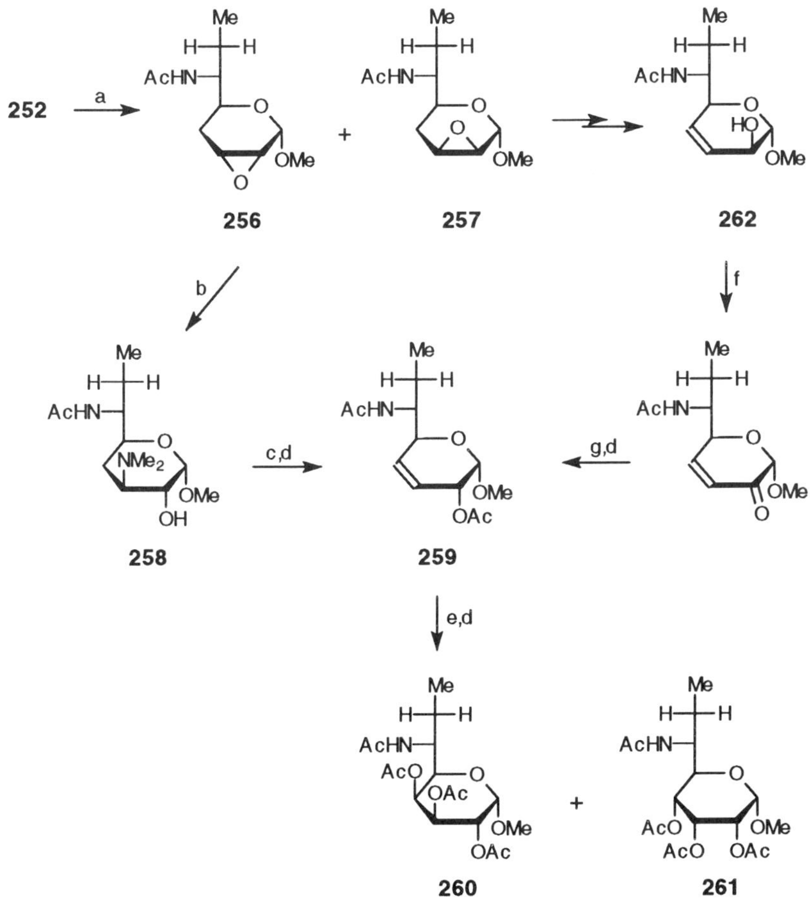

a. m-$ClC_6H_4CO_3H/CH_2Cl_2$; b. $HNMe_2/H_2O$; c. 30% H_2O_2, Δ; d. Ac_2O - Pyr;
e. 6% H_2O_2/t-BuOH - OsO4; f. MnO_2/CH_2Cl_2; g. $NaBH_4$/MeOH.

Scheme 57

hydrolysis of **263** gave octenulose **264** which was *cis*-hydroxylated to afford a mixture of diastereoisomeric products from which after acetylation, α-DL-*galacto* -ketone **265** was isolated [99] (Scheme 58).

Treatment of **265** with lead tetraacetate in the presence of boron trifluoride furnished mixture of tetraacetates **266** epimeric at C-7. The relative configuration at C-7 in **266** was not assigned, however it should be pointed out that ketone **266** closely resembles the key intermediate used previously [66] in the synthesis of methyl lincosaminide (**93**). Because of modest yields and/or lack of stereocontrol of the synthetic steps (Scheme 58 shows only desired stereoisomers) this approach was not pursued further.

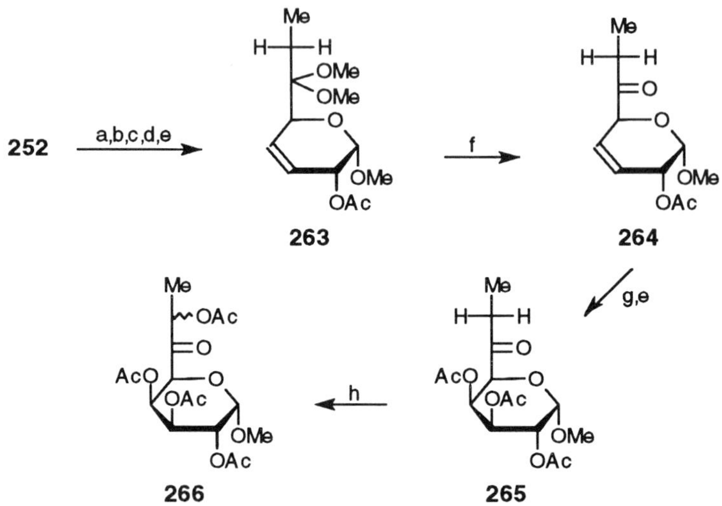

a. m-$ClC_6H_4CO_3H/CH_2Cl_2$; b. $HNMe_2/H_2O$; c. $HC(OME)_3$ - HCl/MeOH; d. 30% H_2O_2, Δ; e. Ac_2O - Pyr; f. 1% CF_3CO_2H/THF - H_2O; g. 6% H_2O_2 - OsO4/*t*-BuOH; h. $Pb(OAc)_4$ - $BF_3 \cdot Et_2O/C_6H_6$ AcOH, Δ.

Scheme 58

An impressive extension of the scope of hetero-Diels-Alder reactions was achieved by Danishefsky and DeNinno [100]. This was accomplished by introduction of 1,3-dienes bearing 2 or 3 oxygen functions which withstood mild Lewis acid catalysis and the discovery that α,β-unsaturated aldehydes undergo Lewis acid catalyzed [4+2] cycloaddition at the carbonyl group (unlike thermal reaction when addition to the C=C bond prevailed). These findings allowed them to assemble the eight-carbon framework required for lincosamine (**93**) synthesis with conveniently located functional groups.

Thus reaction of diene **267** and crotonaldehyde (**268**) in the presence of $BF_3 \cdot Et_2O$ at -78° C afforded, after treatment with trifluoroacetic acid , pyranone **269** (Scheme 59) [95] [101]. Stereoselective reduction of the carbonyl group in **269** and benzoylation led to dibenzoate **270**. Treatment of **270** with *m*-chloroperbenzoic acid in methanol and subsequent benzoylation gave tribenzoates **271** as a mixture of α- and β- anomers. Stereoselective epoxidation of β-**271**, in the proper configurational sense, was attained via bromohydrine **272**. Treatment of β-**271** with *N*-bromosuccinimide in wet acetic acid gave **272** as a sole product, which when reacted with 1,5-diazabicyclo[4.3.0]non-5-ene (DBN)

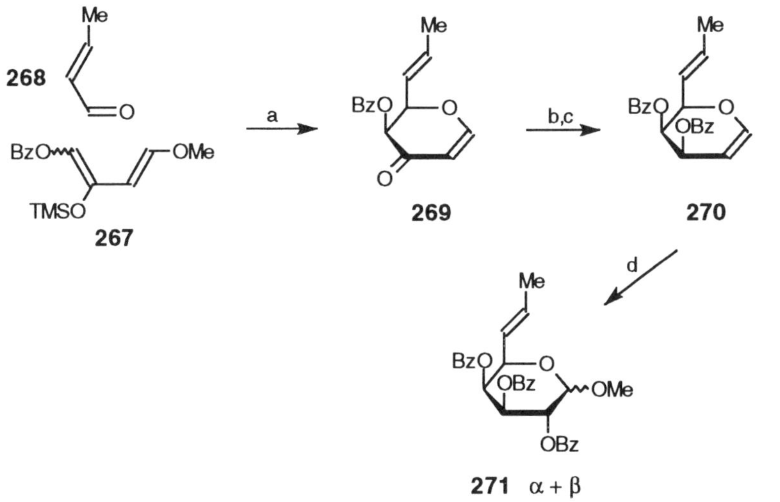

a. $BF_3 \cdot Et_2O/CH_2Cl_2$, -78° C, then CF_3CO_2H; b. $CeCl_3 \cdot 7H_2O$ - $NaBH_4$/MeOH, EtOH, -78°C; c. BzCl - NEt_3 - DMAP/CH_2Cl_2; d. *m*-$ClC_6H_4CO_3H$/MeOH.

Scheme 59

afforded epoxide **273** (Scheme 60). Reaction of **273** with tetrabutylammonium azide in the presence of trimethylsilyl azide followed by trifluoroacetic acid gave the azidohydrine which was mesylated to give **274**. Treatment of **274** with trimethyl phosphite and then sodium hydride gave phosphorylaziridine **275**. A change in protecting groups in **275** to avoid an unwanted reactions between the C-4 ester and C-6 center resulted in carbonate **276**. Acetolysis of **276** proceeded regioselectively yielding phosphoramidoacetate **277** as a single product. Methanolysis of the acyl groups in **277** followed by acetylation afforded the target methyl *N,O*-pentaacetyl-β-DL-lincosaminide (**278**).

Introduction of the side-chain functional groups with full stereocontrol relative to the stereocenters in the pyran ring was one of the key steps of Danishefsky's synthesis of lincosamine and other higher sugars [100]. Nevertheless implementation of the C-6 amino and C-7 hydroxy substituents required a rather circuitous route and moreover the final product was obtained in racemic form. To obviate these shortcomings a chiral pool methodology in the hetero-Diels-Alder reaction was studied by Jurczak et al. [102]. In this approach α-amino-β-hydroxyaldehydes, readily available from natural α-amino-β-hydroxy acids, were utilized as dienophiles [103].

Jurczak et al. demonstrated [104] [105] that *O*-protected *N*-carbobenzoxy-D-*allo*-threoninal **280** and *N*-carbobenzoxy-D-threoninal **281** could be utilized as dienophiles in the Danishefsky type cycloaddition. Reaction

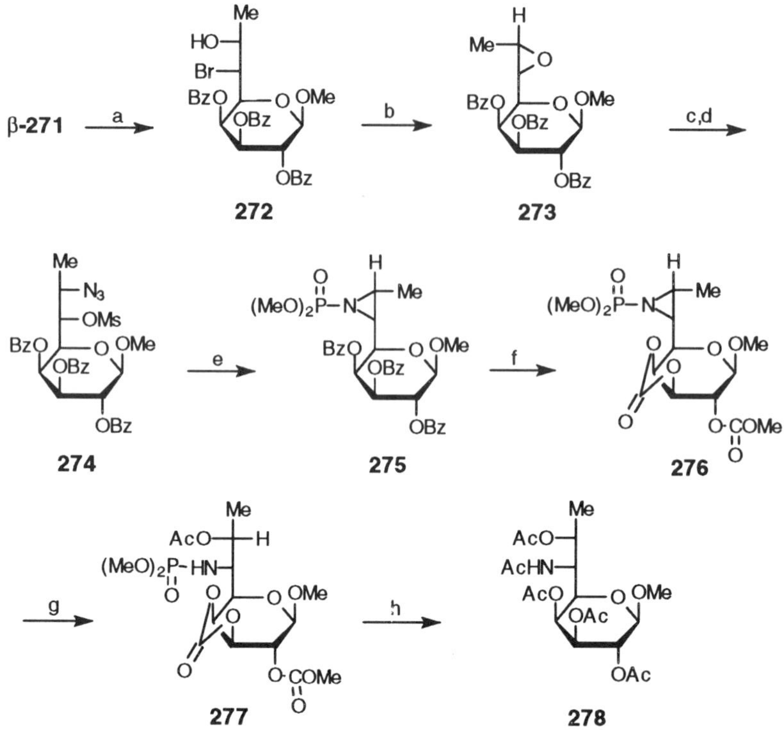

a. NBS - AcOH-H_2O; b. DBN/C_6H_6; c. Bu_4N-N_3 - Me_3SiN_3/PhMe; d. $MeSO_2Cl$ - Et_3N/CH_2Cl_2; e. 1. $P(OMe)_3$, 2. NaH/THF; f. 1. K_2CO_3 - MeOH, 2. Im_2CO/THF; g. AcOH, h. 1. K_2CO_3/MeOH, 2. Ac_2O - Pyr.

Scheme 60

of **280a** or **280b** and **281a** or **281b** with 1-benzoyloxy-2-*tert*-butyldiphenyl-silyloxy-4-methoxybuta-1,3-diene (**279**) in the presence of 1 equiv. of $ZnBr_2$ followed by treatment with trifluoroacetic acid gave in each case mixtures of cycloadducts in the ratios shown in Scheme 61.

The steric course of the reaction was interpreted in terms of the following model; the prevalence of α-chelating interactions in the zinc bromide catalysed cycloaddition favored conformation A (Fig.10) of the dienophile for which *threo*-selectivity ensued and consequently adducts **284**, **285**, **286** and **287** were obtained. Only in the case of *allo*-threoninal **280b** with the strongly chelating benzyloxymethyl group β-chelation increased the population of conformation B (Fig. 10) which yielded *erythro*-adducts **282b** and **283b** albeit as minor products (Scheme 61). Unlike the analogous cycloaddition of crotonaldehyde (**268**) and diene **267** [95] (Scheme 59) stereocontrol of the C-4 center was not achieved.

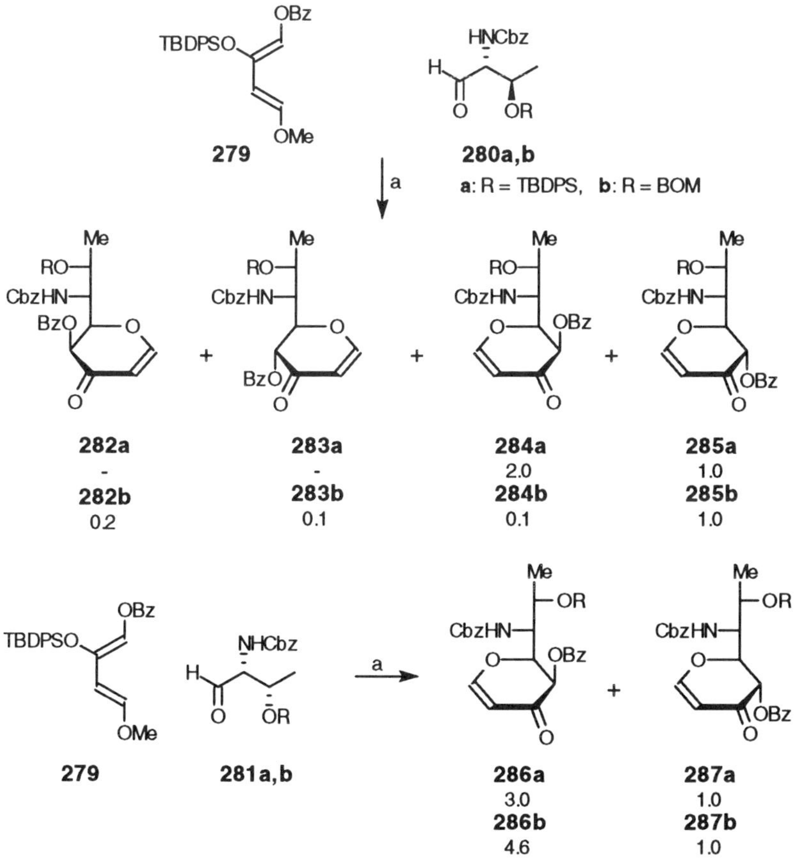

a. 1. $ZnBr_2$/THF, 2. CF_3CO_2H.

Scheme 61

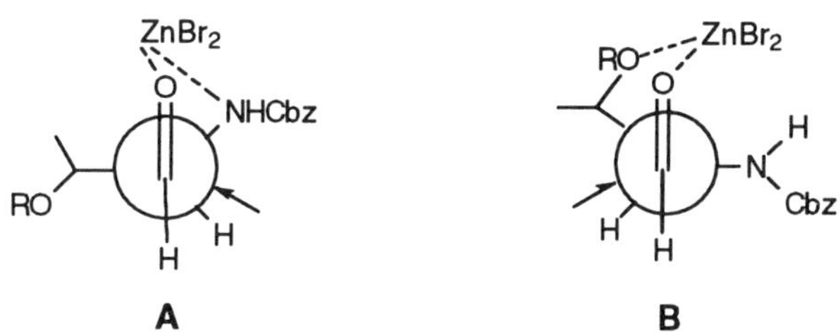

Figure 10

Functionalization of the dihydropyran ring of the major cycloadducts 284a and 285b leading to the L-*galacto* configuration was carried out according to the Danishefsky procedure [95]. Reduction of the carbonyl groups in **284a** and **285b** followed by debenzylation and acetylation gave diacetates **288** and **289** which were separated. Treatment of **288** with *m*-chloroperbenzoic acid in methanol and acetylation afforded glycoside **290**, which was transformed into the pentaacetyl derivative of L-aminooctose **291**. In an analogous manner the pentaacetyl drivative of L-aminooctose **292** was obtained from the dihydropyranone **286a**.

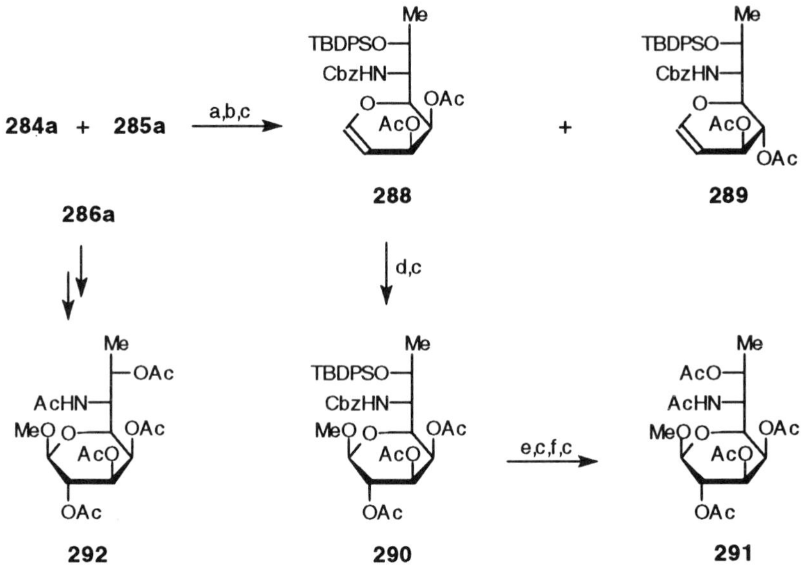

a. $NaBH_4$, $CeCl_3{\cdot}H_2O/MeOH$; b. NH_3 - MeOH; c. Ac_2O, Et_3N, $DMAP/CH_2Cl_2$; d. m-$ClC_6H_4CO_3H$ - MeOH; e. H_2 - Pd - C/EtOAc; f. Bu_4NF/THF.

Scheme 62

Threo-selectivity prevailed in the zinc bromide catalyzed cycloadditon which formed adducts with (5*S*) configuration, i.e. precursors of the sugars belonging to the L-series. In search for a stereocontrol favoring *erythro*-selectivity Jurczak and Gołębiowski examined high-pressure cycloaddition of D-*allo*-threoninals **280a** and **280b** as well as D-threoninals **281a** and **281b** with 1-methoxybuta-1,3-diene **(293)** [106]. The reaction was carried out at 50°C and 13 kbar in ether solution in the presence of $Eu(fod)_3$ as a mild Lewis catalyst. Equilibration of the crude reaction products yielded the thermodynamically more stable *trans*-adducts **294**, **295** and **296**, **297** in the ratios shown in Scheme 63.

Under the high-pressure conditions *threo*-selectivity predominated, as in zinc

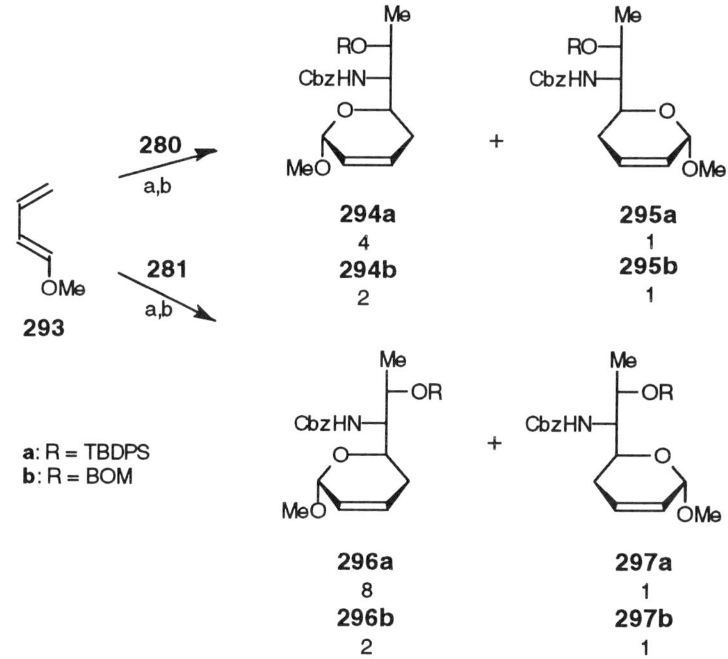

a. Eu(fod)$_3$/ether, 15 kbar, 50° C; b. PPTS/MeOH.

Scheme 63

bromide catalysed cycloadditions; however a modest control favoring *syn* addition was achieved. Replacement of the bulky silyl by the strongly-chelating benzyloxymethyl group (**280b** and **281b**), presumably due to the β-chelating effect which shifted equilibrium towards the conformation B (Fig. 10), gave rise to relatively higher yields of *erythro*-adducts **295b** and **297b** with (5*R*)-configuration.

While yielding much in the way of interesting chemistry and opening routes to various aminosugars, none of these efforts succeeded in forming the C-5 chiral center selectively with (*R*) configuration corresponding to that occuring in lincomycin (**1**).

Syntheses from furan compounds

A general method for monosaccharide synthesis from furan compounds [107] is particularly well suited for the synthesis of higher-carbon sugars [108]. Because stereocenters at C-1 and C-2 of the furan compound side-chain correspond to C-5 and C-6, respectively, of the monosaccharide **299** (Scheme 64) the relative configuration of its side-chain and pyranose ring is controlled by the stereochemistry of the starting material.

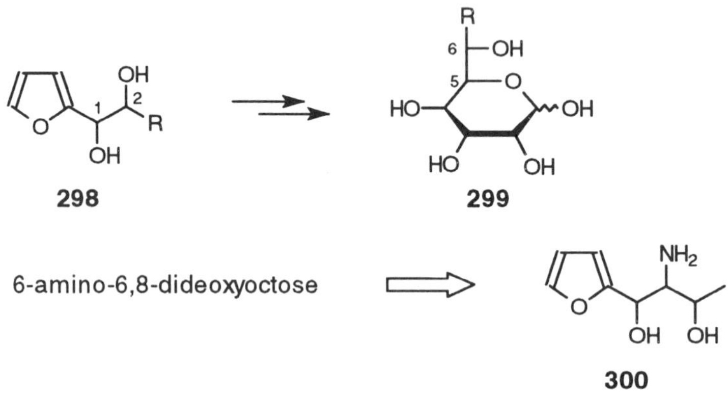

Scheme 64

A prerequisite for the success of this methodology is the availability of appropriate homochiral furan compounds. 2-Amino-1(2-furyl)-butane-1,3-diols (**300**) were required for the synthesis of lincosamine (**93**) or other 6-amino-6,8-dideoxyoctoses (Scheme 64). Enantiopure compounds **300** could be obtained by the addition of furan to the suitable four-carbon chiral synthon.

In the synthesis directed towards lincosamine (**93**) we utilized, as a chiral substrate, the lithium salt of oxazoline **302** obtained from D-threonine (**301**) [109]. The sequence of reactions involved two inversions, one at each stereocenter, hence the L-*threo* configuration of **302** (Scheme 65). Reaction of 2-furyllithium (**303**) with salt **302** afforded the expected ketone **304** in 30% yield. Reduction of **304** with lithium aluminum hydride gave alcohols **305** and **306** in a 3:1 ratio. After chromatographic separation of the mixture and acid hydrolysis of the oxazoline rings, both alcohols yielded *N*-benzoyl derivatives **307** and **308**, respectively. The furan compound **307** could be used in the synthesis of L-lincosamine but in view of low yields in the condensation step and unwanted diastereoselectivity of the carbonyl reduction this approach was not pursued.

Gołębiowski and Jurczak [103] reported on the highly diastereoselective addition of 2-furyllithium (**303**) to *N,N*-diprotected D-*allo*-threoninal **309** yielding almost pure adduct **310**. Alcohol **310** was rearranged into ulose **311**, an advanced intermediate in the synthesis of lincosamine (**93**) (Scheme 66).

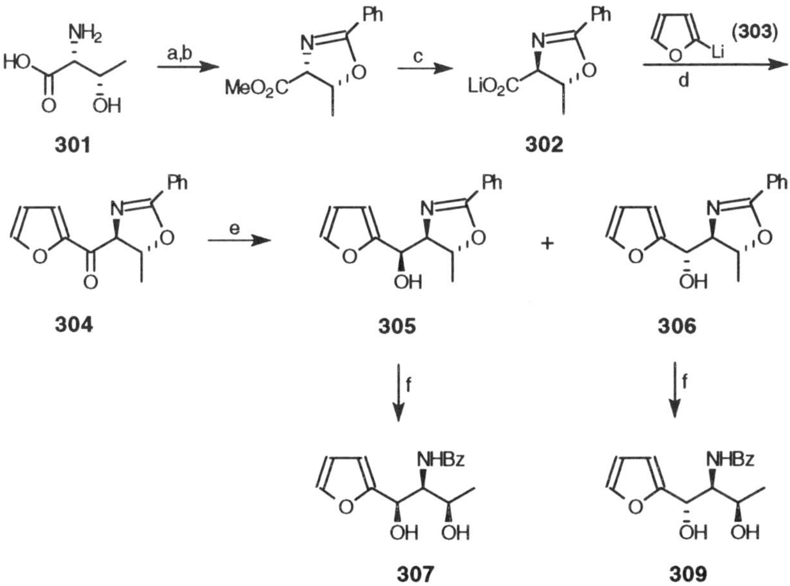

a. MeOH, $SOCl_2$, Δ; b. BzCl, NEt_3 then $SOCl_2$; c. EtO^-Li^+/EtOH; d. furan - BuLi/ether - THF; e. $LiAlH_4$/ether; f. 4% HCl/THF - H_2O

Scheme 65

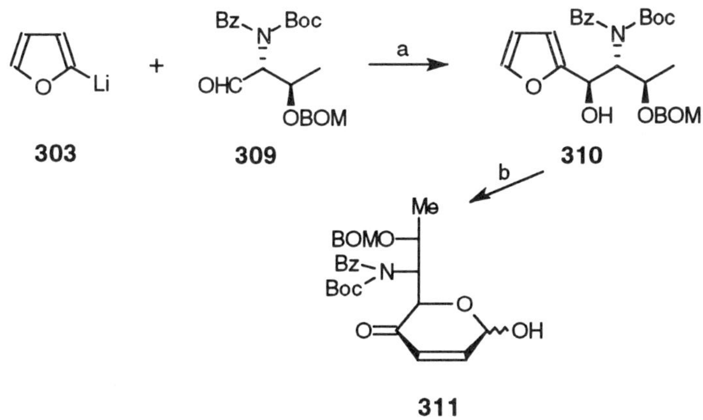

a. THF; b. Br_2/MeCN - H_2O.

Scheme 66

Thus the judicious choice of *N*- and *O*-substituents led to high stereoselectivity of the addition [103]. Notwithstanding this result we looked for protecting groups which could control stereoselectivity of dihydropyran ring functionalization in the desired sense i.e. leading to the *galacto* configuration. Based on the molecular modeling, we have employed *N*-benzenesulfonyl-*N,O*-isopropylidene-D-threoninal (**312**) and the corresponding D-*allo*-threoninal (**313**). Aldehydes **312** and **313** were obtained from D-threonine (**301**) following known procedures as shown in Scheme 67.

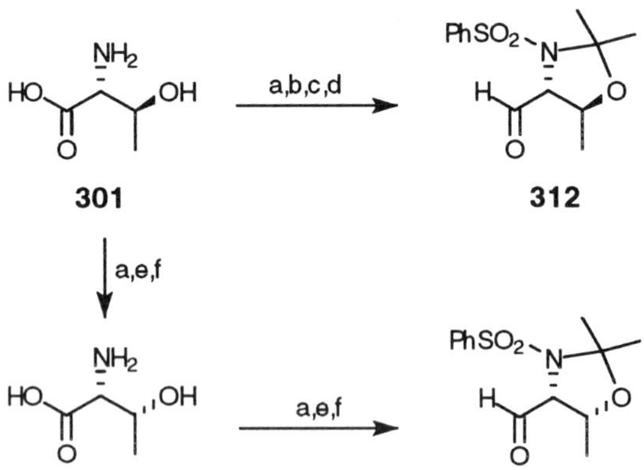

301 312 7 313

a. MeOH, $SOCl_2$, Δ; b. $PfSO_2Cl$, NEt_3; c. $Me_2C(OMe)_2$, Me_2CO, *p*-TsOH; d. DIBAL-H, PhMe, -70° C; e. BzCl, NEt_3; f. $SOCl_2$ then 10% HCl, Δ.

Scheme 67

Reaction of 2-furyllithium (**303**) with D-threoninal **312** and D-*allo*-threoninal **313** gave, with good to excellent yields (78 - 90%) in each case a pair of alcohols **314**, **315** and **316**, **317**, respectively [110] (Scheme 68). The product ratio varied considerably depending on the reaction conditions. For D-threoninal **312** either *syn* (conditions a) or *anti* (conditions b) addition was attained leading with satisfactory selectivity to alcohols **314** and **315**, respectively. In the case of D-*allo*-threoninal **313** only *syn* addition (conditions a) was achieved with high selectivity yielding alcohol **317**.

Gross structure and configuration of the new stereocenter at C-1 in compounds **314**, **315**, **316** and **317** was established by chemical and spectrometric methods. Since the aminodiol was to be used as a substrate in lincosamine (**93**) synthesis its structure and that of its C-1 epimer were confirmed by X-ray single crystal analysis [111].

Results of furyllithium addition can be rationalized in terms of the α-chelation-controlled conformation A and Felkin-Anh conformation B (Fig. 11). In the presence of the chelating agents threoninal **312** or *allo*-threoninal **313** have reacted predominantly in conformation A via addition to the less hindered *re* face leading to aminodiols **315** and **317**, respectively. Under non-chelating conditions for threoninal **312**, conformation B predominated and under *si*-attack yielded butanediol **314**. In the case of *allo*-threoninal **313** rotamer B, due to the presence of the *N,O*-isopropylidene moiety had a steric interaction between the C-4 methyl and carbonyl group. Consequently conformation B was destabilized and even in

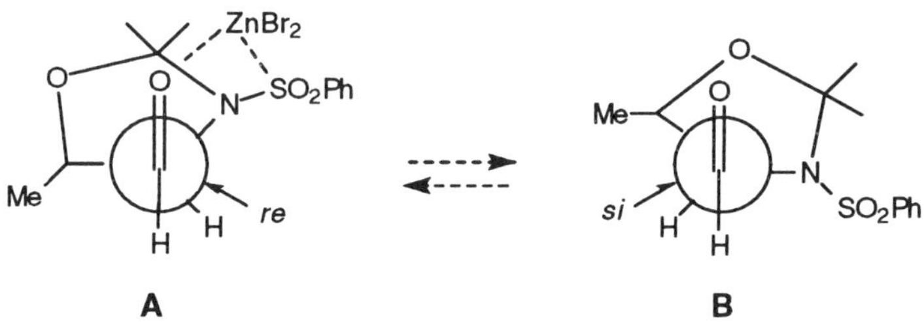

Figure 11

the absence of a chelating cation was in equilibrium with conformation A (Figure 11). Therefore an attack on the *re* and *si* face of the carbonyl group was equally probable, leading to approximately equal amounts of **316** and **317** (conditions c, Scheme 68).

Derivatives of 1-(2-furyl)-2-aminobutane-1,3-diols **314, 315, 316** and **317** were oxidized with *m*-chloroperbenzoic acid and then methylated to give methyl α-ulosides **318, 319, 320** and **321**, respectively (Scheme 68) [112], advanced enantiopure intermediates for the synthesis of D and L-6-amino-6,8-dideoxy- octoses. The absolute configuration of uloside **320**, the appropriate chiron for lincosamine (**93**) synthesis was confirmed by single crystal X-ray analysis. X-ray data also revealed a solid state molecular conformation of **320** consistent with conformation in solution indicated by molecular modeling [112] and ^{1}H NMR spectrometry. Its important feature was the shielding of the β-side of the dihydropyran ring by the oxazolidine moiety on which depended stereoselectivity of the following synthetic steps. Reduction of the carbonyl in **320** using sodium borohydride - cerium chloride reagent gave alcohol **322** resulting from the attack of the hydride on the α-face of carbonyl group. The foregoing stereoselectivity is opposite to that observed so far in the reduction of α-4-pyranosides [113]. Also epoxidation of the double bond in alcohol **322** afforded the anhydrosugar **323** with high stereoselectively as the sole product. Opening of the epoxide ring with 1 M perchloric acid proceeded with full regioselectivity and was accompanied by splitting of the oxazolidine ring yielding, after acetylation, tetraacetate **324**. Finally, the benzenesulfonyl group

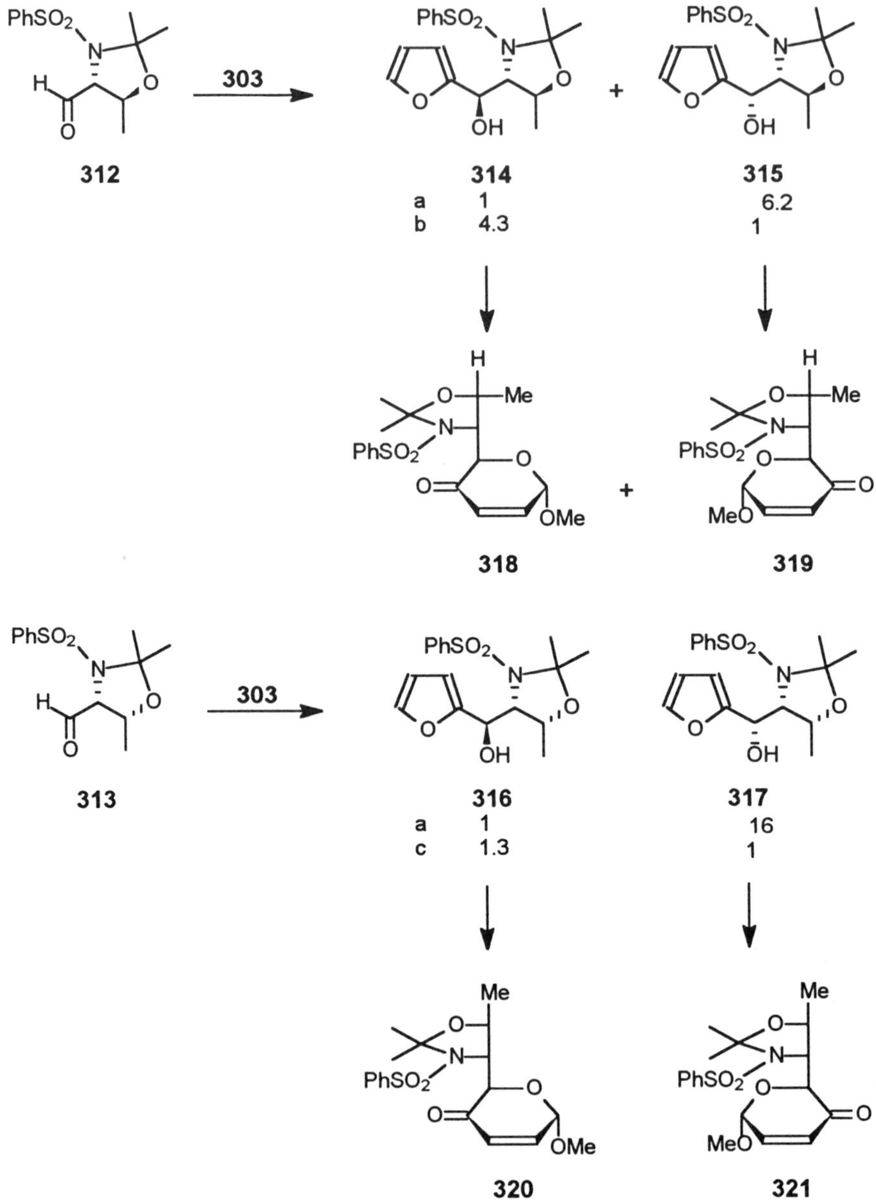

a. $ZnBr_2$/ether; b. ether, -70° C; c. THF/ether/hexane, - 70° C;
d. *m*-$ClC_6H_4CO_3H$ - $NaHCO_3$/CH_2Cl_2; e. MeI - Ag_2O.

Scheme 68

was removed with sodium in liquid ammonia yielding, after acetylation, pentaacetyl-α-D-lincosaminide (**242**).

Taking into account that D-*allo*-threonine (**7**), the primary source of chirality, could be obtained from crotyl alcohol by use of asymmetric Sharpless epoxidation [114], this result completed the first total synthesis of D-lincosamine (**93**).

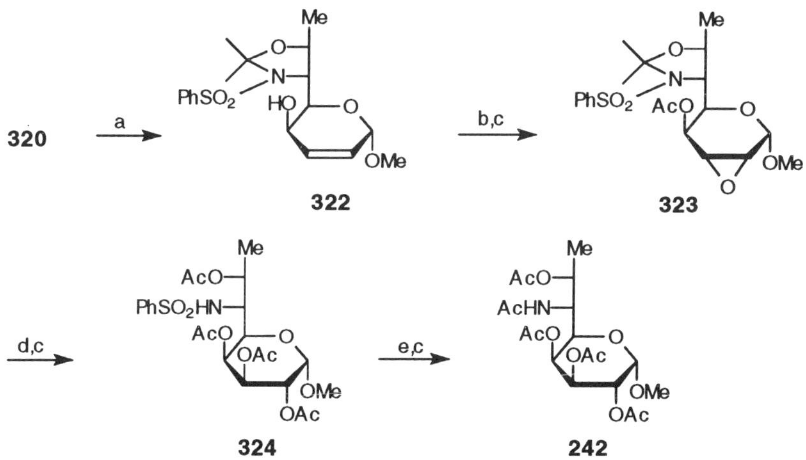

a. $NaBH_4$, $CeCl_3{\cdot}7H_2O$; b. 60% H_2O_2, MeCN; c. Ac_2O - Pyr - DMAP;
d. $HClO_4$/THF-H_2O, 40° C; e. Na/NH_3, -70° C - RT.

Scheme 69

REFERENCES

1. Cundlife E. Antibiotics and Polyribosomes II. Some Effects of Lincomycin, Spiramycin and Streptogramin A in vivo. Biochem. 1969;8: 2063-2066.
2. Reusser F. Effect of lincomycin and clindamycin on peptide chain initiation. Antimicrob Agents Chemother. 1975; 7: 32-37.
3. Kucers A, Bennett NMCK. The use of antibiotics. 3rd ed, W Heinemann Medical Books Ltd, London, 1979.
4. Chaney BV. Ab initio calculations on large molecular fragments. Structural correlations between natural substrates moieties and some antibiotic inhibitors of peptidyl transferase. J Med Chem. 1974; 17: 590-599.
5. US Pharmacopeia 1995, The United States Pharmacopeial Convention, Inc. 1994: 890-891.
6. British Pharmacopoeia (Veterinary) 1993. EMSO, London 1993: 74-75.
7. Herrel WE. Lincomycin, Modem Scientific Publications, Inc., Chicago, 1969.
8. Golebiowski A, Jurczak J. Total synthesis of lincomycin and related antibiotics. n: Recent Progress in the Chemical Synthesis. Lukacs G, Ohno M. eds. Springer Verlag Berlin Heidelberg, 1990: 365 - 385.
9. Mason DJ, Dietz A, Deboer C. Lincomycin, a new antibiotic. I. Discovery and biological properties. Antimicrob Agents Chemother. 1962: 554-559.
10. Herr RR, Bergy ME. Lincomycin, a new antibiotic.II. Isolation and characterization. Antimicrob Agents Chemother. 1962: 560-564.
11. Lewis C, Clapp HW, Grady JE. *In vitro* and *in vivo* evaluation of lincomycin, a new antibiotic. Antimicrob Agents Chemother. 1962: 570-582.
12. Hoeksma H, Bannister B, Birkenmeyer RD, Kagan F, Magerlein BJ, MacKellar FA, Schroeder W, Slomp G, Herr RR. Chemical studies on lincomycin. I. The structure of lincomycin. J Am Chem Soc. 1964; 86: 4223 -4224.
13. Heff RR, Slomp G. Lincomycin. II. Characterization and gross structure. J Am Chem Soc. 1967; 89: 2444-2447.
14. Schroeder W, Bannister B, Hoeksema H. Lincomycin. III. The structure and stereochemistry of the carbohydrate moiety. J Am Chem Soc. 1967; 89: 2448 - 2453.
15. Megerlein BJ, Birkenmeyer RD, Heff RR, Kagan F. Lincomycin. V. Amino acid fragment. J Am Chem Soc. 1967; 89: 2459-2464.

16. Slomp G, MacLellar FK. Lincomycin. IV. Nuclear magnetic resonance studies on the structure of lincomycin, its degradation products, and some analogs. J Am Chem Soc. 1967: 89: 2454 - 2459.
17. Davis RE, Parthasarathy R. Crystal structure of lincomycin hydrochloride, Acta Crystallogr. 1966; 21: A109-AIIO.
18. Gusakova GV, Denisov GS, Kagan EZ, Smolyanskii Al. Intramolecular hydrogen bonding and conformation of the lincomycin molecule in organic solvents. Antibiot. Khimioter. 1988; 33: 657-660.
19. Mizsak S, Slomp G, Neszmelyi , Gero SD, Lucacs G. Carbon-13 N.M.R. spectral analysis and spin lattice relaxation times of the anfibiotic lincomycin and related compounds. Tetrahedron Lett. 1977; 721-724.
20. Shipman LL, Christoffersen RE, Chaney BV. Ab initio calculations on large molecules using molecular fragments. Lincomycin model studies. J Med Chem. 1974; 17:583-590.
21. Moloney GP, Craik DJ, Iskander MN, Munro SLA. 'H NMR and theoretical studies of the conformation of the antibiotic lincomycin. Aust J Chem. 1990; 43: 535-548.
22. Maloney GP, Craik DJ, Iskander MN. 'H and ^{13}C NMR studies of aminoglycoside antibiotics. Magn Res Chem. 1993; 31: 1077-1084.
23. Heman-Ackah SM, Garret ER. Kinetics and mechanisms of drug action on microorganisms. 13. Comparative studies on action of lincomycin, clindamycin and U 2427A against Escherichia coli. J Med Chem. 1972;15: 152-163.
24. Argoudelis AD, Fox JA, Able TE. A new lincomycin-related antibiotic. Biochemistry. 1965; 4: 698-708.
25. Argoudelis AD, Mason DJ. Studies on the biosynthesis of lincomycin. I. Antibiotic U-11,921, an S-ethyl homolog of lincomycin. Biochemistry. 1965; 4: 704-709.
26. Argoudelis AD, Eble TE, Mason DJ. Studies on the biosynthesis of lincomycin. V. Effect of ethionine on fermentation of S. Lincolnennsis. J Antiobiot. 1979; 23: 1-8.
27. Argoudelis AD, Fox JA, Mason DJ. Studies on the biosynthesis of lincomycin. II. Antibiotic U-11,973, N-demethyl lincomycin. Biochemistry 1965; 4: 710 -713.
28. Argoudelis AD, Johnson LE, Pyke TR. Effect of methylation inhibitors on fermentation of S. Lincolnensis. Production of N-demethyllincomycin. J Antibiot. 1973; 26: 429-463.
29. Argoudelis AD, Fox JA, Mason DJ, Eble TE. New lincomycin-related antibiotics. J Am Chem Soc. 1964; 86: 5044-5045.

30. Hoeksema H, Crum GF, Devries WH. Isolation and purification of celesticetin. Antibiotic Ann. 1954/1955: 837-841.
31. Hoeksema H, Hinman JW. Celesticetin. III. The partial structure. J Am Chem Soc. 1964; 86: 4979-4982.
32. Hoeksema H. Celesticetin. V. The structure of celesticetin. J Am Chem Soc. 1968; 90: 755-757.
33. Argoudelis AD, Brodasky TF. Studies with Streptomyces caelestis. I. New celesticetins. J Antibiot. 1972; 25: 194-196.
34. Argoudelis AD, Coates JH, Lemaux PG, Sebek OK. Antibiotics produced by mutants of Streptomyces caelestis. I. 7-O-demethyl celesticetin and its degradation products. J Antibiot. 1972; 25: 445-455.
35. Argoudelis AD, Coates JH, Lemaux PG, Sebek OK. Antibiotics produced by mutants of Streptomyces caelestis. II. N-Demethylcelesticetin and N-demethyl-7-O-demethylcelesticetin. J Antibiot. 1972; 26: 7-14.
36. Argoudelis AD, Coats JH, Johnson LE. Directed biosynthesis of new celestosaminide antibiotics by Streptomyces caelestis. J Antibiot. 1974; 27: 738-739.
37. Morozowich W, Lamb DJ, Karnes HA, MacKellar FA, Lewis C, Stern KF, Rowe EL. Synthesis and biological activity of lincomycin 2-phosphate. J Pharm Sci. 1969; 58: 1485-1489.
38. Morozowich W, Sinkula AA, MacKellar FA, Lewis C. Synthesis and bioactivity of lincomycin. 2-Monoesters. J Pharm Sci. 1973; 62: 1102-1105.
39. Taraszka MJ, Morozowich W. Synthesis and hydrolysis kinetics of lincomycin acetals. J Org Chem. 1968; 33: 2349-2354.
40. Oesterling TO, Metzler CM. Kinetics of alkaline isomerization and hydrolysis of lincomycin monoesters. J Pharm Sci. 1969; 8: 1389-1392.
41. Sinkula AA, Morozowich W, Lewis C, MacKellar FA. Synthesis and bioactivity of lincomycin 7-monoesters. J Pharm Sci. 1973; 62: 1757-1760.
42. Sinkula AA, Lewis C. Chemical modification of lincomycin: synthesis and bioactivity of selected 2,7-dialkylcarbonate esters. J Pharm Sci.1973; 62: 1757-1760.
43. Magerlein BJ. Modification of lincomycin. Advan Appl Microb. 1971; 14: 185 - 229.
44. Philippe M, Sebag H. Preparation of esters of macrolide and lincosainide antibiotics with polyaromatic acids and pharmaceutical compositions containing them. Euro Patent Appl. EP 315.536, 1989.

45. Philippe M, Sebag H. Preparation of bicyclic aromatic esters of macrolide and lincosainide antibiotics and pharmaceutical and cosmetic compositions containing them. Euro Patent Appi. EP 315.538, 1989.
46. Saint Leger D. Composition based on hydroxypyridone derivatives for reducing hair loss. LTK Patent Appl. GB 2207051, 1989.
47. Philippe M, Sebag H. Preparation of unsaturated cycloaliphatic esters of macrolide antibiotics and lincosamide antibiotics and pharmaceutical and cosmetic compositions containing them. Euro Patent Appl. EP 315535, 1989.
48. Philippe M, Sebag H, Preparation of etretin-related esters of macrolide or lincosamide antibiotics for treatment of acne and pharmaceutical and cosmetic compositions containing them. Euro Patent Appl. EP 315534, 1989.
49. Bannister B. Modification of lincomycin involving the carbohydrate portion. Part H. Analogues with gluco *and* D-ido- stereochemistry. J Chem Soc Perkin 1 1972; 3031-3036.
50. Bannister B. Modification of lincomycin involving the carbohydrate portion. Part I. The 2-O-methyl and 2-deoxy- analogues. J Chem Soc Perkin 1 1972; 3025-3030.
51. Birkenmeyer RD, Kagan F. Lincomycin. XI. Synthesis and structure of clindamycin, a potent antibacterial agent. J Med Chem. 1970; 13: 616-619.
52. Hu S, Zhou H. An improved method for the chlorination of lincomycin. Zhoingguo Yaoke Daxu Xuebao 1992; 23: 248-249; CA 1993; 118: 212723t
53. Ying H, He L, Hong J, Fan Z, Xu Y. Improved process for preparation of clindamycin hydrochloride. Zhongguo Yiyao Gongye Zazhi. 1991; 22: 197-199, CA 1992; 116: 27918f.
54. Tobkes M, Diaz S, Krishnan L. Preparation of clindamycin phosphate. Patent US 5,182,374, 1993.
55. Magerlein BJ, Kagan F. Lincomycin. IX. 7-Thio and thioamido analogs of lincomycin. J Med Chem. 1969; 12: 974 - 977.
56. Bannister B. (7S)-7-Deoxy-7-substituted-alkylthio-lincomycins. S-Alkylation of sulphides by an activated epimine under acidic catalysis: formation of α-acetamido-sulphides. Tetrahedron 1984; 40: 1633 - 1684.
57. Magerlein RJ. Lincomycin. VI. 4'-Alkyl analogs of lincomycin. Relationship between structure and antibacterial activity. J Med Chem. 1967; 10: 355 - 359.
58. Magerlein RJ. Lincomycin. VII. 4'-Depropyl-4'-ethoxylincomycin. J Med Chem. 1967; 10: 1161 - 1163.

59. Birkenmeyer RD, Dolak LA. Lincomycin. XIII. N-Demethylation of lincomycin and its analogs. Tetrahedron Lett. 1970; 5049 - 5051.
60. Birkenmeyer RD, Knoll SJ, Lewis C, Stern KF, Zurenko GF. Synthesis and antimicrobial activity of clindamycin analogues: pirlimycin, a potent antibacterial agent. J Med Chem. 1984; 27: 216 - 223.
61. Magerlein BJ, Kagan F. Lincomycin. VIII. 4'-Alkyl-l'-demethyl-4'-depropylclindamycin, potent antibacterial and antimalarial agents. J Med Chem. 1969; 12: 780 - 784.
62. Magerlein BJ. Lincomycin. 14. An improved synthesis and resolution of the antimalarial agent, l'-demethyl-4'-depropyl-(4'R)- and (4'S) pentylclindamycin hydrochloride. J Med Chem. 1972; 15: 1255-1259.
63. Argoudelis AD, Magerlein BJ. Lincomycin-type compounds. US 3 758 454, 1973.
64. Magerlein BJ. Lincomycin. X. The chemical synthesis of lincomycin. Tetrahedron Lett. 1970; 33 -36.
65. Howarts GB, Szarek WA, Jones JKN. The synthesis of N-acetyllincosamine (6-acetamido-6,8-dideoxy-D-*erythro*-D-galacto-octose, a derivative of the free carbohydrate moiety of lincomycin. Chem Comm. 1969; 1339 -1340.
66. Howarth GB, Szarek WA, Jones JKN. The synthesis of lincomycin. J Chem Soc (C) 1970; 2218 - 2224.
67. Lance DG, Szarek WA, Jones JKN, Howarth GB. Synthesis of 8-deoxy-D-erythro-D-galacto-octose. Determination of the configuration of two octenoses. Can J Chem. 1969; 47: 2871 -2874.
68. Howarth GB, Lance DO, Szarek WA, Jones JKN. Syntheses related to the carbohydrate moiety in lincomycin. Can J Chem. 1969; 47: 75 - 79.
69. Woolard GR, Rathbone EB, Szarek WA, Jones JKN. Syntheses towards the carbohydrate moiety of lincomycin. J Chem Soc Perkin Trans I 1976; 950 - 954.
70. Gateau-Olesker A, Sepulcbre AM, Vass G, Gdro SD. Synthese de sucres a chaine allongde heptose et octose par C-acylation nucleophile. Tetrahedron 1977; 33: 393 - 397.
71. Hoppe 1, Scholkopf U. Lincomycin. Synthese des kohlenhydrat-teils durch nucleophile acylierung mit I-ethoxyvinyllithium. Liebigs Ann Chem. 1980; 1474 - 1476.
72. David SM, Fischer JC. Acetolyse d'une diazocetone derivde d'octose: synthdse d'une N-acetyllincosamine proteg6e, de ses epimeres en C-6 et C-7, et d'une 3-oxetanone a chaine laterale glucidique. Carbohydr Res. 1976; 50: 239 - 246.
73. David SM, Fisher J-M. Synthese, a partir de D-galactose, d'un derivde protege de la celestodamine (6-amino-6,8-didesoxy-7-O-methyl-D-*erythro-D-galacto-octose),* sucre de certain antibiotiques du groupe de la lincomycine. Carbohydr Res. 1974; 38: 147 - 159.

74. Hems R, Horton D, Nakadate M. Synthesis of *D-glycero-D-galacto*-heptitol (perseitol), *L-glycero-D-galacto-heptitol,* and an allenic sugar derivative, from acetylenic precursors. Carbohydr Res. 1972; 25: 205-216.
75. Brimacombe JS, Kabir Abul KMS. Higher-carbon sugars. Part 10. The synthesis and osmylation of 7,8-dideoxy-1,2;3,4-di-O-iso-propylidene-β-L-glycero-D-galacto-oct-7-enopyranose and related studies. Carbohydrate Res. 1988; 174: 37- 45.
76. Brimacombe JS, Kabir AKMS, Bennet F. Higher-carbon sugars. Part 6. The synthesis of some octose sugars *via* the epoxidation of unsaturated precursors. J Chem Soc Perkin Trans I 1986; 1677 - 1680.
77. Engelhardt LM, Skelton BW, Stick RV, Tilbrook DMC, White AH. Approaches to the synthesis of lincosamine, a derived portion of the antibiotic lincomycin. Aust J Chem. 1990; 43: 1657-1680.
78. Stick RV, Tilbrook DMG. The synthesis of some octenoses as potential precursors to lincosamine, a derived portion of the antibiotic lincomycin. Aust J Chem. 1990; 43: 1643 -1656.
79. Czernecki S, Valery JM. A Stereocontrolled synthesis of lincosamine precursor. J Carbohydr Chem. 1990; 9: 767- 770.
80. Jarosz S, Głodek J, Zamojski A. Inversion of configuration at the proparglylic alcohol center in sugar acetylenes. Carbohydr Res. 1987; 163: 289 - 296.
81. Dondoni A. Carbohydrate synthesis via thiazoles. In Modem Synthetic Methods. Scheffold R ed. Verlag Helvetica Chimica, Helvetica Acta Basel 1992; 377 - 398.
82. Tronchet JMJ, Massoud MAM. Glycosylaziridine derivatives. Heterocycles 1989; 29: 419-426.
83. Tronchet JMJ, Massoud MAM. Lincomycin Analogues.II. Chain-extension of methyl 6-aldehydo-3,4-O-isopropylidene-l-thio-β-D-galacto-1,5-pyranoside. Heterocycles 1986; 24: 1265 - 1269.
84. Fukumaru T, Atsunii T, Ogawa T, Matsui M. A new approach to the synthesis of N-acetyllincosamine isomer. Agr Biol Chem. 1973; 37: 2617-2619.
85. Atsumi T, Fukumaru T, Ogawa T, Matsui M. A stereoselective synthesis of N-acetyllincosamine derivatives. Agr Biol Chem. 1973; 37: 2621 2626.
86. Atsumi T, Fukumaru T, Matsui M. A stereoselective synthesis of N-acetyllincosamine isomers. Agr Biol Chem. 1973; 37: 2627 - 2630.
87. Bose AK, Mathur C, Wagle DR, Nagvi R, Manhas MS, Urbanczyk Lipkowska Z. Chiral P-lactams as synthons. Stereospecific synthesis of 6-epi-lincosamine derivative. Heterocycles 1994; 39: 491 - 496.

88. Saeki H, Ohki E. Synthesis of N-acetyllincosamine. Chem Pharm Bull. 1970; 18: 412 - 413.
89. Czernecki S, Valery JM. Stereochemical studies of the synthesis of α-amino-nitriles related to lincosamine. Carbohydr Res. 1988; 184: 121-130.
90. Dondoni A. New organometallic reagents. Use of 2-(trimethylsilyl)-thiazole in acyclic stereoselective strategies. Pure Appl Chem. 1990; 62: 643 - 652.
91. Dondoni A, Junguera F, Merchan FL, Merino P, Tejero T, Stereoselective aminohomologation of chiral α-alkoxy aldehydes *via* thiazole addition to nitrones. Applicafion to the synthesis of N-acetyl-D-mannosamine. Tetrahedron Lett. 1992; 33: 4221 -4224.
92. van Delft FL, de Kort M, van der Marel GA, van Boom JH. Stereocontrolled hydroxymethylafion of carbohydrate imines: formal synthesis of destoniic acid and lincosamine. Tetrahedron Asymm. 1994; 5: 2261 - 2264.
93. Knapp S, Kukkola PJ. Stereocontrolled lincomycin synthesis J Org Chem. 1990; 55: 1632 - 1636.
94. Knapp S, Kukkola SS, Pietranico S. N-Benzoylcarbamate cyclization. Tetrahedron Lett. 1987; 28: 5399 - 5402 and references cited therein.
95. Danishefsky SJ, Larson E, Springer JP. A totally synthetic route to lincosamine: some observations on the diastereofacial selectivity of electrophilic reactions on the double bond of various 5-(1-alkenyl)-arabinopyranosides. J Am Chem Soc. 1985; 107: 1274 -1280.
96. Konował A, Achmatowicz O Jr, Zamojski A. Total synthesis of lincomycin system. Part 1. Rocz Chem. 1976; 50: 879 - 893.
97. Zamojski A, Banaszek A, Grynkiewicz G. The synthesis of sugars from non-carbohydrate substrates. Advan Carbohydr Chem Biochem. 1982; 40: 1-129.
98. Chmielewski M, Achmatowicz O Jr, Zamojski A. Stereoselective synthesis of methyl 3,4,7,8-tetradeoxy-(α- and β-DL-erythro-oct-3-enopyranosid-6-ulose dimethyl ketals. Bull Pol Acad Chem. 1984; 32: 19 - 28.
99. Chmielewski M, Doboszewski B, Achmatowicz O Jr, Zamojski A. The synthesis of lincomycin system. Part III. Bull Pol Acad Chem. 1984; 32: 423 - 431.
100. Danishefsky SJ, DeNinno MP. Totally synthetic routes to the higher monosaccharides. Angew Chem Int Ed Engl. 1987; 26: 15 - 23.
101. Larson ER, Danishefsky SJ. A totally synthetic route to lincosamine. J Am Chem Soc. 1983; 105: 6715 - 6716.

102. Kiciak K, Jacobsson U, Gołebiowski A, Jurczak J. α-Amino acids in the total synthesis of amino sugars. Pol J Chem. 1994; 68: 199 - 238.
103. Gołebiowski A, Jurczak J. Optically active N-protected α-amino aldehydes in organic synthesis. Chem Rev. 1989; 89: 149-164.
104. Gołebiowski A, Jurczak J, Krajewski JW, Galitskii NM, Verenich AJ. Bull Pol Acad Chem. 1989; 37: 307-312.
105. Gołebiowski A, Jurczak J. The cycloaddition reaction of 1-benzyloxy-2-tert-butyldimethylsilyloxy-4-methoxy-1,3-butadiene with N,O-protected D-threoninals and D-allo-threoninals. Tetrahedron 1971; 47: 1045 - 1052.
106. Gołebiowski A, Jurczak J. High-pressure [4+2] cycloaddition of 1-methoxy-1,3-butadiene to N, O-protected D-threoninals and D-allo-threoninals. Tetrahedron 1991; 47: 1037 - 1044.
107. Achmatowicz O Jr, Bukowski P, Szechner B, Zwierzchowska Z, Zamojski A. Synthesis of methyl 2,3-dideoxy-DL-alk-2-enopyranosides from furan compounds. A general approach to the total synthesis of monosaccharides. Tetrahedron 1971; 27: 1973 -1996.
108. Achmatowicz O. An approach to the synthesis of higher-carbon sugars.In: Trost BM, Hutchinson CR, eds. Organic Synthesis Today and Tomorrow. Pergamon Press Oxford and New York 1981.
109. Szechner B. Synthesis and absolute configuration of two diasteroisomeric *(lR,2S,3R)-* and (lS,2S,3R)-2-amino-l-(2-furyl)-1,2,3-butanediols. Tetrahedron 1981; 37: 949 - 952.
110. Szechner B, Achmatowicz O Jr. A totally synthetic route to enantiomerically pure D and L-aminooctoses: stereocontrolled synthesis of methyl α-D-lincosaminide. J Carbohydr Chem. 1992; 11: 401-406.
111. Szechner B, Achmatowicz O Jr, Gałdecki Z, Fruzinski A. Synthesis and absolute configuration of four diastereoisomeric 1-(2-furyl)-2-aminobutane-1,3-diols. Tetrahedron 1994; 50: 7611-7624.
112. Szechner B, Achmatowicz O. A totally synthetic route to enantiomerically pure D and L-6-amino-6,8-dideoxyoctoses: stereocontrolled synthesis of methyl α-D-lincosaminide. Pol J Chem. 1994-. 68: 1149-1160.
113. Achmatowicz O Jr, Burzynska MH. Stereochemistry of lithium aluminium hydride reduction of cyclic enones. Pol J Chem. 1983; 57: 1275 -1282.
114. Pons D, Savignac M, Genet J-P. Efficient synthesis of enantiomerically pure L and D-allothreonine and (S) and (R) isoserines. Tetrahedron Lett. 1990; 31: 5023 - 5026.

16
Fungal (1-3)-β-D-Glucans: Chemistry and Antitumor Activity

Akira Misaki
Faculty of Human Life Science, Osaka City University
Osaka, Japan

Mariko Kakuta
Koman Women's University
Osaka, Japan

I. INTRODUCTION

Investigations of the tumor-necrotizing actions of polysaccharides published in the 1940s were concerned with Shear's polysaccharide prepared from the cells of *Serratia marcescens.* Although Shear's polysaccharide was known to cause intensive necroses of Sacroma 180 and other tumor cells in mice, it was not chemically pure and its severe toxicity prevented further clinical study [1-3]. Since then, however, chemical and pharmacological studies have been focused on the non-cytotoxic, immunomodulating natural carbohydrate polymers occurring mainly in plants, yeasts and fungi. For instance, zymosan, a crude preparation of yeast cell-wall polysaccharide, was shown in the 1950s to dramatically stimulate the immune system of mammals through a mechanism involving activation of the reticulo-endothelial system (RES) which stimulated the host's defense mechanisms against bacterial and fungal infections [4].

Apart from these studies, in China and other Asian regions there have been several studies of traditional folk remedies such as, medicinal herbs, ginseng and a variety of mushrooms. In Japan, the cell-mediated antitumor actions of fungal (1-3)-ß-D-glucans obtained from some edible mushrooms, fruiting bodies of Basidiomycetes, have attracted much chemical and pharmacological attention [5] during the last 25 or 30 years. Consequently, in the 1980s some (1-3)-ß-D-glucans, *e.g.*, lentinan from Shiitake, from the fruiting body of *Lentinuus edodes* and schizophyllan, an extracellular product of *Scizophyllum commune*, have been used as immunopotentiating agents in cancer chemotherapy or in radiotherapy. These antitumor glucans contain a (1-3)-ß-D-glucan backbone with O-6-linked ß-glucosyl branches with a degree of branching (db), the ratio of branching residues to all the (1-3)-linked D-glucose residues of the back bone equal to 1/3. This is essentially the same as the structure of the exoglucan of *Fungi infecti,* later named *Scleotum glucanicum,* [6]. Recent extensive studies on many fungal glucans have demonstrated that there may be a wide range of structural diversity with regard to the arrangements of ß-glucosyl branches, e.g., a highly branched, insoluble glucan (db, 4/5) of *Auriculalaria auricula* and a less branched glucan (db 1/16) of *Ganoderma lucidum,* etc. Recent conformational studies have indicated that the antitumor effects of these ß-glucans are closely related to the triple helical structure of the (1-3)-ß-glucan backbone. In addition, our studies of chemical modification of the glucosyl side chains [7] have suggested that the arrangements of the glycosyl side chains must also play an important role in their host-mediated, tumor-inhibitory actions. In Table 1 tumor growth-inhibitory activities of various types of (1-3)-ß-D-glucans isolated from cell walls and cultural fluids of yeast, fungi, and fruiting bodies of Basidiomycetes (mushrooms), against mouse-implanted Sarcoma 180 tumors are listed.

II. ZYMOSAN AND YEAST CELL WALL GLUCAN

Among various species of yeasts and fungi, *Saccharomyces cerevisiae* has been most extensively investigated chemically and physiologically. This yeast's cell wall contains 1,3- and/or 1,6- glucan (29%) as the skeletal architecture, cell surface mannan (31%), protein (13%), lipid (8.5%) and a small amount of chitin.

Zymosan, first described in 1941 as a crude yeast cell wall preparation, is recognized as a unique immunomodulating agent which stimulates the reticulo-endothelial system (RES) through activation of the complement system [4,5] RES refers to the lympho-reticulo cells of the mammalian defense system, *e.g.*, macrophages, endothelial and reticulum cells, etc. Zymosan is prepared by boiling yeast cells in 0.5M sodium phosphate and exclusively digesting the insoluble residue with trypsin. The resulting insoluble residue, designated zymosan, comprises glucan (58%), mannan (17-22%), protein (13-17%), chitin (~1%) and ash (~3%) [8]. It seems, therefore, that zymosan is a major part of the cell wall or "cell ghost." A series of immunopotentiating effects of zymosan including stimulation of RES host defense mechanism through complement system activation, especially the alternative pathway, was extensively investigated during the 1950s. When zymozan is administered intravenously into rats or mice (approx. 4mg/ 100g body weight) a remarkable effect on RES, as measured by the"carbon clearance test" [9] is observed. This results in stimulation of the host defense mechanism against various pathogenic infections, regression of tumor cells, etc. These immunological effects of zymosan are apparently attributed to activation of macrophages through endocytosis of zymosan, possibly initiated by an interaction with the cell surfaces of macrophages. In this connection, it is interesting that the alkali-insoluble glucan fraction of the yeast cell wall reproducibly stimulated RES, while the soluble mannan was inactive [10].

Since insoluble glucan of the yeast cell wall seemed to be a major component of zymosan, further knowledge of its molecular structure was of great interest. The structure of the glucan of *Saccharomycese cerevisiae*, first named "yeast cellulose" by Salkowski (1894), has attracted the interest of many chemists to gain an understanding of this hot alkali-insoluble, rigid molecule though its molecular structure is still not completely clarified.

Baker's yeast cells were extensively treated with hot alkali (0.75M NaOH, 75C°, 6hr) to remove mannan, glycogen and protein, leaving an alkali-insoluble polysaccharide comprising only D-glucose. Repeated methylations using 30% NaOH and $(CH_3)_2SO_4$ (Haworth method) by Hassid et al. [11] showed that the insoluble glucan contains exclusively (1-3)-linked D-glucose units. In addition Peat et al. [12] demonstrated the presence of β-(1-6) glucosidic linkages by isolation of a series of ß-(1-6)-linked oligosaccharides in the partial acid hydrolysis products (12). It was believed that the β-(1-6)-linked chains must also be implicated in the building unit of the glucan molecule.

Later detailed chemical investigation of the insoluble yeast glucan was carried out by Misaki et al. [13] using a combination of methylation analysis and Smith degradation, involving periodate oxidation, reduction and mild acid hydrolysis. The cell wall glucan of Baker's yeast was prepared by exclusive alkali treatments under N_2, followed by the removal of residual glycogen-like material by repeated treatments with 0.5M acetic acid at 90° until the absence of I_2 coloration, according to the procedure of Bell (1950).

The alkali-insoluble glucan was found to be soluble in DMSO (dimethylsulfoxide); [a] -10_ (DMSO) and its *ir* spectrum (890 cm^{-1}) confirmed the ß-glucosidic linkage. In the methylation analysis, the methylated glucan (DP 410) yielded on acid hydrolysis 2,3,4,6-tetra-, 2,4,6-tri-, and 2,4-di-O-methyl glucose (molar ratio, 1:7:1).

Table 1. ANTITUMOR ACTIVE FUNGAL (1→3)-ß-GLUCANS AGAINST MOUSE-IMPLANTED SARCOMA 180 SOLID TUMOR

Source	Polysaccharide	Structure	Tumor-growth inhibition[a)] Dose mg/kg ×days	Inhibition %	Reference
Cell wall					
Saccharomyces cerevisiae	Yeast glucan DMSO s.	Less branched (1→3)	5×10	68.1	(16)
Elsinoe leucospila	Water s.	Branched, db 1:8	5×10	98.5	(49)
	Alkali s.	Branched, db 2:3	5×10	67.8	
Aureobasidium pululans	Alkali s.	Branched, db 1:5-7	10×10	100.0	(50)
Exopolysaccharide					
[Bacteria]					
Alcaligenes facalis var. myxogenes	Curdlan water ins.	Linear (1→3)	10×10	99.8	(25), (51)
			3×10	40.3	
[Fungi]					
Sclerotium glucanicum	Scleroglucan water s.	Branched, db 1:3	5×10	88.2	(6), (24)
Schizophyllum commune	Schizophyllan	Branched, db 1:3	1×10	95.1	(30-31), (36)
	sonicated	Mr=5×10^5 db 1:3	1×10	95.1	
Pestalotia 815	Pestalotian water-less s.	Highly branched, db 4:6	50×10	75.6	(42)
			5×10	57.3	
			1×10	16.0	
	Glucan polyol	Ox-Red.	5×10	92.5	
			1×10	88.1	

	Debranchewd	Linear (1→3)	5×10	8.0	
Mushroom (Fruiting body, *Basidiomycetes*)					
Poria cocus	Pachyman	Lessbranched, db 1:40[a]	5×10	0	(25), (52)
	polyol of,	Ox-Red.	5×10	96.0	
Coriolus	PSK	(1→3), (1→4)	10×10	90.	(19)
versicolor		Protein (20-38%)	5×10	94.0	
CM101					
Shiitake					
Lentinus edodes	Lentinan	Baranched, db 1:3	1×10	99.2	(21), (22)
Kikurage					
Auricularia	Glucan I water s.	Branched, db 2:3	8×10	96.6	(40)
auriularia	Glucan II	Highly bramched db 3:4	10×10	18.9	
	G II alkali ins.	Ox-Red.	1×10	79.0	
	G II polyol		5×10	97.4	
Ganoderma		(1→3), (1→6)	1×10	100.0	(53)
applanatum					
Ganoderma	HW water s.	Branched, db 1:3	10×10	97.7	(41)
lucidum	CA I alkali s.	Less branched, db 1:17	10×10	59.6	
	CA II	db 1:23	10×10	10.9	
Maetake					
Griforafrondosa	Glucan	6-*O*-branched	4×10	73.4	(54)
Tsao ru or Fukurotake					
Volvariella	Alkali-extracted				(43)
volvacea	Glucan VVG	Branched, db 1:4	1×10	73.2	
			5×10	95.0	
	Hot alkali-extr. VVG	db 1:2	10×10	33.9	
	VVG polyol	Ox-Red.	1×10	100.0	

a) Assayed after 5 weeks, JCL-ICR mice.

db: degree of branching = branch/(1→3)-linked glucose residues of the back bone.

Methylation analysis of the degraded glucan which survived the Smith degradation showed the same O-methylated sugars as from the parent glucan, in a ratio, 1:6:1. These experimental results indicated that the glucan has a branched structure of a ß-(1-3)-linked backbone chain in which every 8 sugar unit is attached with (1-3) linked side chains at the O-6, shown as (1). Fig.1. In the Smith degradation, glycerol and 1-0-ß-glucosylglycerol were produced in addition to the degraded glucan. This suggested that the insertion of the (1-6)-linkage between the (1-3)- linkages is feasible, such as in a -6)-Gp-(1-3)-G-(1-6)Gp-(1- sequence. If consideration is given to the Peat's work which showed that gentiobiose and other ß-(1-6)-linked oligosaccharides are released by partial hydrolysis (12), in addition to structure (1) an alternative structure (2) is possible, where the ß-(1-6)-linked main chain is attached with (1-3)-linked side chains to form a framework. If further consideration is given to the solubility characteristics of the branched glucan, the lack of solubility in water and in hot alkali and the solubility only in a solvent such as DMSO which can disrupt hydrogen bonds, could very well suggest an alternative structural model, either (3) or (4) that would result from combination of (1) and (2).

Concerning the structure of yeast glucan, the investigations of Manners et al. (1973) [14] revealed the occurrence of two types of ß-glucans , *i.e.*, a branched ß-(1-6)-glucan (14) and ß-(1-3)-glucan chains joined by (1-6)-interchain-linkages [15]. The soluble ß-(1-6)-glucan with (1-3)-linked branches was obtained from repeated warm acetic acid extracts of the alkali-insoluble cell glucan [14]. The insoluble glucan (DP 1450) that remained is comprised of long (1-3)-ß-glucan chains, interlinked by (1-6)-linkages (db, 60), shown as (5). There may be still a question, whether the cell wall glucan is made up of two independent ß-(1-3)- and ß-(1-6)-glucans or if the two polysaccharides are covalently linked together to form a homogeneous glucan, from which the ß-(1-6)-glucan segments were liberated during the hot acetic acid extractions.

After Manners' model was proposed, Misaki and coworkers [16] re-examined the fine structure of yeast glucan [16] which was purified from mechanically disintegrated cell walls of Oriental Baker's yeast (*S. cerevisiae.*) by fractionation according to Scheme 1. The cell wall was first treated with protease (pronase) to liberate the surface mannan and the digestion residue was successively extracted with hot water (120°), 1 M NaOH (25°) and then hot alkali (70°). The resulting alkali-insoluble glucan was dissolved in DMSO with aid of ultra-sonication. Addition of water to the DMSO solution afforded the precipitated fraction (HP2) and non-precipitated fraction (HS), both fractions showing a single peak on ultracentrifugation (solvent, DMSO). Methylation analysis by an improved Hakomori method (DMSO and CH_3I) indicated that these glucan fractions possessed essentially the same chemical structure consisting predominantly of (1-3)-linked sugar residues, possibly forming a backbone with slight branches at O-6 [branch point/linear (1-3)-glucosyl residues, 1:13-17.2] (see Table 2). In addition, internal (1-6)-linkages and the presence of sequence, -6)-Gp-(1-3)-Gp-(1-6)-Gp-(1- were confirmed as before [13]. Glycerol and 1-0-ß-glucosylglycerol (2:1) were produced by Smith degradation. The remaining degraded glucan was a slightly branched ß-(1-3) glucan chain with db, 31-32 suggesting that the (1-6)-linkages may be located in the side chains. Interestingly, the structure of the Smith-degraded glucan appears to resemble that of the less branched (1-3)-ß-glucan that remained after removal of the ß-(1-6)-glucan according to Manners et al. [15].

Based on this knowledge, two structural models may be possible. One is an essentially linear ß-(1-3) glucan having ß-glucosyl groups or side chains of ß-(1-3) G-ß-(1-6)G-(1- , shown as (6). An alternative structure (7) could be depicted where infrequently branched (1-3)-linked chains are joined to each other through a (1-6)-linked chain, forming a stable network architecture. Although elucidation of the true structural features of the cell wall glucan of *S. cerevisiae* must wait until more precise data are available, the structure (7) does seem close to the structure of the cell wall R-glucan of *Schizophyllum commune* [16].

Whatever the precise structure of the glucan may be, the distinct immunomodulating actions exerted by zymosan must be closely related to the molecular shape of the cell wall glucan. This was demonstrated by comparing the tumor-inhibitory activities of mannnan and glucan fractions on Sacroma 180 solid tumors, as shown in Table 2. It is obvious that the ß-(1-3)-glucan fractions exhibited remarkable growth-inhibiting activity whereas the mannan fractions are less active. These data support the supposition that the most immunologically active component in zymosan is the cell wall (1-3)-glucan, although yeast mannan may also exert adjuvant action and augmentation functions both in humoral and cellular immunities [17]

III. LENTINAN, SCHIZOPHYLLAN AND OTHER GLUCANS USED AS CLINICAL AGENTS

In the last two or three decades extensive pharmacological investigations have been concerned with the host-intermediated, antitumor active, natural polysaccharides. Consequently, specific fungal ß-(1-3)-linked glucans have been selected as immunotherapeutic agents in cancer chemotherapy or radiotherapy in Japan because of their observed tumor inhibitory activity in experimental animals. Thus, **"krestin"** or **PSK,** a protein-bound polysaccharide of *Coriolus versicolor*, of the polypolraceae family, was first developed as an orally active anti-cancer drug in 1975. Krestin was prepared from the hot water extract of mycelia of *Coriolus versicolor*, "hiratake." It is a ß-glucan protein complex with a molecular weight 100,000.

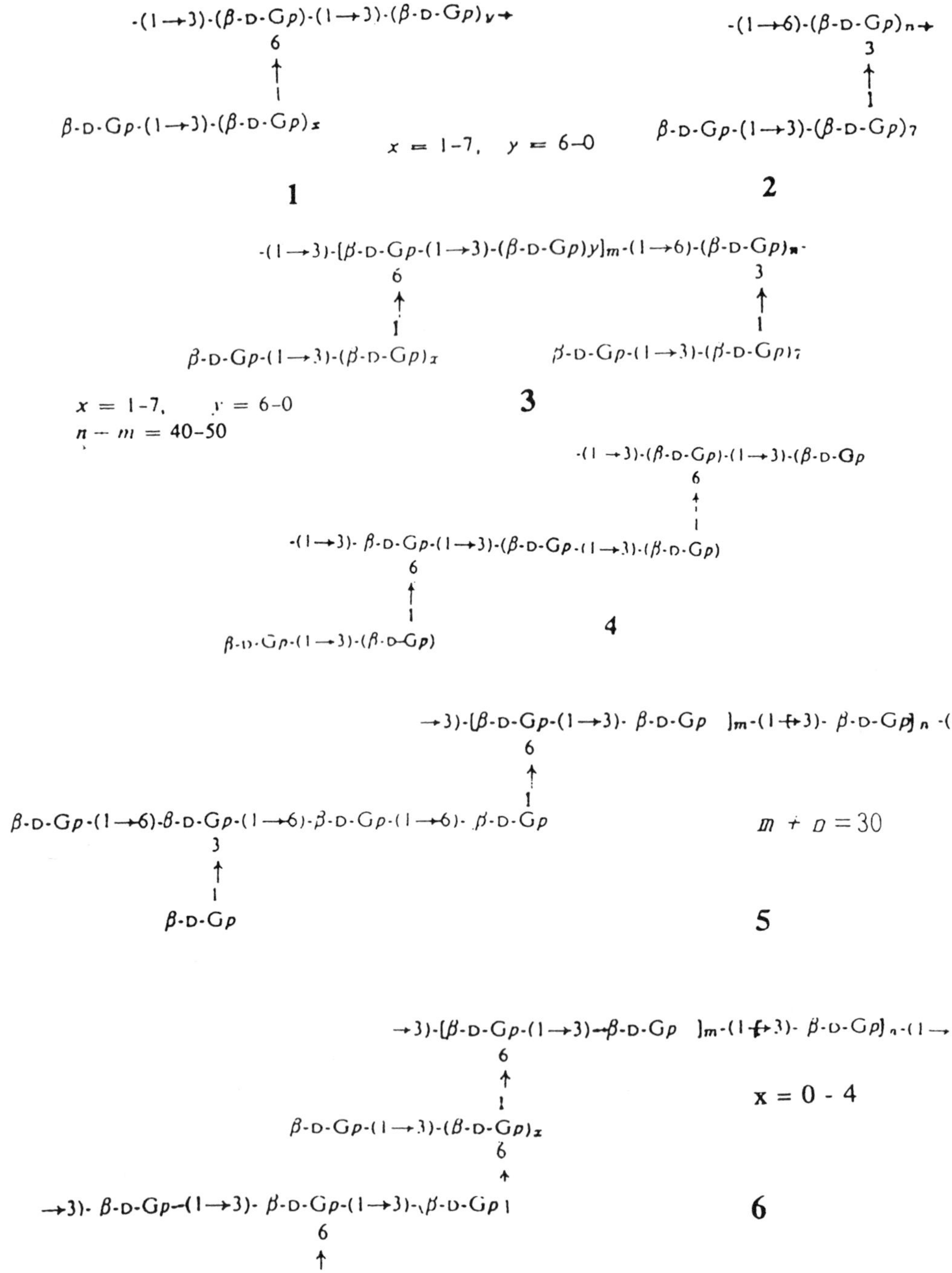

Fig. 1 Possible structural features of yeast cell wall glucan

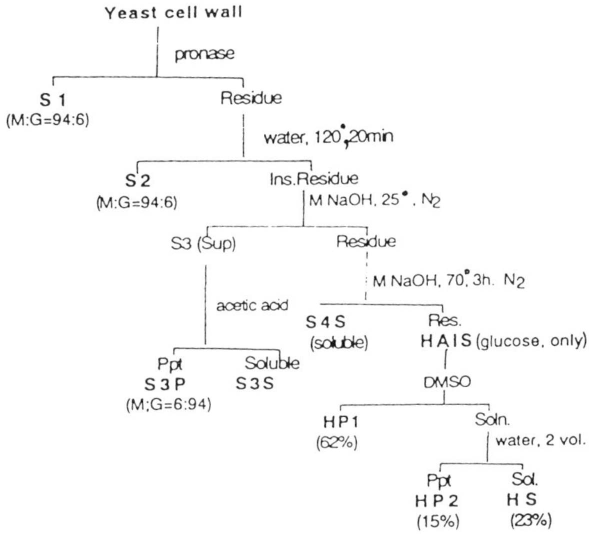

Scheme 1. Fractionation procedure of yeast cell wall

Table 2. Comparison of antitumor activities of the yeast cell wall polysaccharides against sacroma 180 solid tumor[a)]

Fraction	Mode of linkage				Dose mg/kg 10 days	Inhibition %	Tumor regression numbers
Mannan	→M₂→M₂→M₂→M₂→M→ (side chains ↑ M₂, M₃, M₂, M; ↑ M₃, M, M; ↑ M)						
S1					10	35.1	0/5
S2					10	32.3	0/5
S3S					10	58.1	0/5
S4S[b)]					10	85.2	2/5
Glucan	G1-	-3G1-	-6G1-	-3G1-6 ↑			
HAIS	1	15	2	1			
HP1	1	15-16	2	1	10	95.2	2/5
HP2	1	16	2	1	10	88.5	1/5
HS	1	15	2	1	10	91.6	2/5

a) Implanted in ICR-JCL mice. The sample suspended in water was i.p. injected, starting at 0 time. After 5 weeks the animals were killed.
b) Containing 24% glucan.

Unlike other antitumor polysaccharides, the glucan moiety contains a backbone of (1-3) and (1-4)-linked glucose residues with O-6 branched side chains [18].

The peptide moiety contains aspartic acid (13%), glutamic acid (14.5%) and other amino acids [19]. This protein moiety appeared to covalently conjugate with the glucan. Some pharmacological studies indicated that krestin remarkably inhibited the growth of Sarcoma 180 transplanted in ICR-JCL mice. Intraperitoneal injections of 10-100mg/kg resulted in 86-97.6% inhibition of tumor growth. It was also reported that krestin exhibits antitumor activity with oral administration [20]. Therefore krestin has often been used in cancer chemotherapy. Recently, serious concerns have been raised concerning the anti-cancer effect by oral administration of such a high molecular agent.

A. Lentinan

Lentinan and schizophyllan were both licensed in Japan in 1985 and 1986 respectively, as immunostimulants in cancer therapy. Lentinan was isolated from a popular edible mushroom, "shiitake," the fruiting body of *Lentinus edodes*, while schizopyllan was purified from the cultural fluid of *Schizopyllium commune*. Incidentally both lentinan and schizophyllan have essentially the same building units as scleroglucan, an extracellular branched (1-3)-ß-glucan elaborated by *Fungi inperfective* (later named, *Sclerotum glucanicum*), as shown in Fig. 2. Both glucans are capable of stimulating RES, like zymosan, possibly due to their triple helix conformation, unlike laminaran which is believed to assume a single helix conformation.

According to the work of Chihara et al. [21], a large quantity (200kg) of shiitake mushroom was homogenized and extracted with hot water for 8-16 hours at 80-100°C. The crude polysaccharide precipitated with 50% ethanol was fractionated using 20% and 50% acetic acid and 6% NaOH to afford a major polysaccharide, designated lentinan (31g), and another 5 fractions LC-11, LC-12, LC-13, EC-11 and EC-14. Purified lentinan is a water insoluble but alkali soluble ß-D-glucan ([«] +19.5 to +21.5_), and shows a single peak on ultracentrifugation (1% NaOH) corresponding to a molecular weight 10 x 10^5 [22].

These polysaccharide fractions were assayed for antitumor activity using Sarcoma 180 solid tumor implanted into a group of Swiss albino or SWM/Ms mice. The results indicated that among the six polysaccharide fractions lentinan and one other fraction (LC-11) exhibited striking tumor inhibitory activity. Doses of 1 to 5 mg/kg for 10 days resulted in 95-97.5 % tumor inhibition in Swiss albino mice and in four groups of SWM/Ms mice lentinan in 1mg/kg doses for 10 days gave 100% inhibition with complete tumor regression [27].

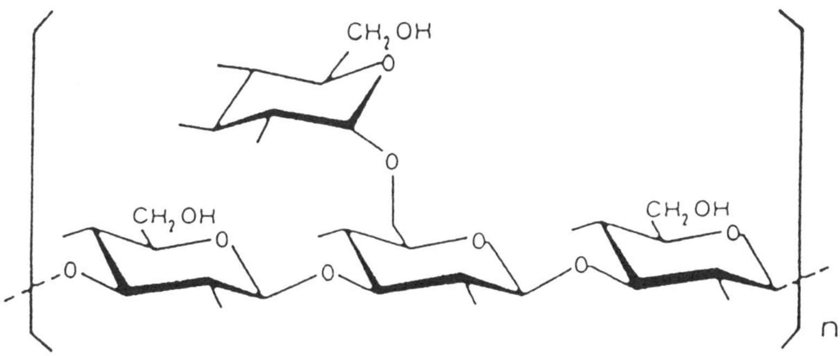

Fig. 2 Repeating unit structure of scleroglucan, lentinan and schizophyllan

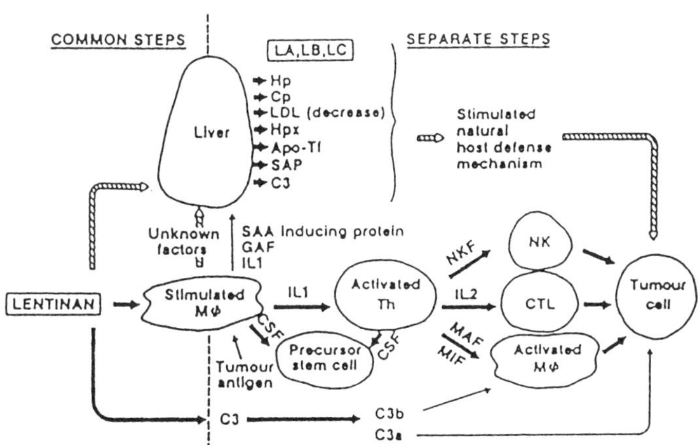

Hp: haptoglobin, Cp: ceruloplasmin, LDL: low density lipoprotein, Hpx: hemopexin, Apo-Tf: apo-transferrin, SAP: serum amyloid P component, SAA: serum amyloid associated, GAF: glucocorticoid antagonizing factor, IL1: interleukin-1, IL2: interleukin-2, NKF: natural killer cell factor, MAF: macrophage activating factor, MIF: migration inhibitory factor, CSF: colony stimulating factor, Th: helper T-cell, CTL: cytotoxic T-lymphocytes.
→ a route stimulated by lentinan, ⇨ a hypothetical route for activation by lentinan.

Fig. 3 Proposed possible mode of action of lentinan (27)

The assay method that was developed to evaluate the tumor inhibitory effect of lentinan has been subsequently employed in many other studies.

1. Assay method for antitumor activity

Swiss albino and SWM/Ms mice (later, ICR-JCL mice were often used) weighing 20 g were inoculated subcutaneously in the right groin with seven-day-old Sarcoma 180 ascites (0.05 ml, approx 0.6 $\times 10^6$ cells). The polysaccharide samples dissolved or suspended in water were intraperitoneally injected daily for 5 or 10 days starting after tumor implantation. After 5 weeks the mice were sacrificed and the tumors weighed. The inhibition (%) was expressed: [(A-B)/A]x100, where A is an average tumor weight of the control group and B is that of the treated group.

2. Structure

The purified lentinan, based on partial acid hydrolysis and enzymatic hydrolysis, was first believed to be a curdlan-like ß-glucan [21] consisting of ß-(1-3)-linked glucose residues. Detailed structural information of lentinan and a depolymerized lentinan (small lentinan: [α] +29° in water) were obtained by methylation, periodate oxidation and Smith degradation. The results indicated that lentinan contains a main chain of ß-(1-3)-linked glucose residues. Two out of every five glucose residues are branched at O-6-positions to have D-glucosyl side chains and a few internal ß-(1-6)-linked side chains may also be present. Nevertheless, such branched structural features appear similar to that of *Sclerotum glucanicum*, as shown in Fig. 1, but not curdlan. Formic acid-deploymerized lentinan (mol.wt., 16200) showed high antitumor activity (2mg/kg, 10 days: 97.6%).

Sasaki et al. [26] compared the antitumor activities of several acid depolymerized lentinan (mol. wt. 3x 10^3 to 10×10^4) and native lentinan to correlate antitumor activity and molecular size of lentinan. The results indicated that glucan fractions with a molecular weight > 150,000 exhibited high activities similar to native lentinan and that decreases in molecular size below a molecular weight of 100,000 might diminish their original antitumor activity.

3. Immunopharmacology

Since lentinan possessed potent antitumor effects on not only allogeneic tumors but also on various syngeneic and even autologous tumors in experimental animals, as shown in Table 3 [27], there have been extensive pharmacological studies in connection with immunomodulation in cancer immunotherapy with particular attention on the distinct antitumor actions and augmentation of immunological systems exerted by lentinan. In clinical use, lentinan was proved to be effective as an immunopotentiator in the prolongation of life and improvement of host immune responses of patients with stomach and colorectal cancer when given in combination with other chemotherapeutic agents (cf. Chihara et al. [27]).

Lentinan is also known to exert immunostimulasting effects on host defense mechanisms against various infections by bacteria, fungi, virus and parasites. This involves stimulation of RES activity, like zymosan, through the alternative complement system pathway leading to a split in C3 into C3a and C3b. The latter acts with other serum components, which in turn interact with other targets resulting in an activation of the phagocytic capacity of macrophages.

The fact that the *in vivo* antitumor activity of lentinan is abolished in thymectomized mice or in nude mice supports the above thymus dependent mechanism. Thus, the antitumor action of lentinan (and schizophyllan) may involve T-cell dependent activation of the immune system, that is, augmentation of the production of IL1, lymphocyte activating factor, which leads to differentiation of immature-T-cells into immunocompetent mature T-cells, capable of producing lymphokines, IL2, and other immunoeffectors. Lentinan may also directly augment production of CSF (colony stimulating factor) augmenting IL1. A proposed mechanism of a T-cell mediated immunoresponse by lentinan-type (1-3)-ß-glucans is illustrated in Fig. 3.

B. Schizophyllan

Schizophyllan, the extracellular antitumor glucan, is produced by fermentation of *Schizophyllum commune* Fries in a glucose medium containing sodium glutamate and thiamine at 26 - 30°C for 4 - 5 days (yield: 5-6 g/L)(28). Since the native glucan is a large polymer with a molecular weight of approximately six million daltons, its aqueous solutions are highly viscous. Hence, depolymerization was conducted by ultrasonic irradiation (20 KHz) and high speed extrusion to an appropriate molecular size. Following a subsequent deionization step, the depolymerized products were fractionated by acetone precipitation, and finally lyophilized. Scheme 2 shows the outline of the manufacturing process for a clinical grade of schizophyllan, termed SPG, which had molecular weight of 4.5 x 10^5. Also shown are the changes in the molecular weight distribution of schizophyllan resulting from ultrasonic depolymerization of 1% aqueous solutions.

Chemical analysis revealed that native schizophyllan [29] and the ultrasonically treated schizophyllan [30] had a branched (1-3)-ß-glucan structure, similar to scleroglucan and lentinan.

Table 3. THE ANTITUMOR ACTIVITY OF LENTINAN IN MICE (27)

Tumor*	Hosts	Dose**	Timing***	Tumor inhibiiton ratio (%)
Allogeneic:				
Sarcoma-180	CD-1/ICR	1 × 10 (i.p)	1 to 10	100.0
	A/J	5 × 4 (i.p)	1 to 4	96.5
	DBA/2	5 × 4 (i.p)	1 to 4	100.0
Syngeneic:				
A/Ph.MC.S1-fibrosarcoma	A/Ph	1 × 10 (i.p)	1 to 10	100.0
DBA/2.MC.CS-1 fibrosarcoma	DBA/2	1 × 10 (i.p)	1 to 10	79.7
MM-46 carcinoma	C3H/HeN	5 × 2 (i.v)	13, 15	100.0
MM-102 carcinoma	C3H/HeN	10 × 1 (i.v)	7	60.0
L-5178Y lymphoma	DBA/2	10 × 3 (i.v)	7, 14, 21	89.0
P-815 mastocytoma	DBA/2	10 × 4 (i.v)	8, 10, 15, 17	82.0
EL-4 lymphoma	C57BL/6	10 × 1 (i.v)	1	72.0
Lewis lung cancer	C57BL/6	1 × 6 (i.p)	1 to 6	59.3
MH-134 hepatpma	C3H/He	1 × 14 (i.p)	15 to 28	100%
(after surgery)			14 (resect)	(survuival)
Autologous:				
MC-induced primary tumor	DBA/2	1 × 10 (i.p)	1 to 10+	80.5
MC-induced primary tumor	C3H/HeN	1 × 20 (i.p) (with CY: 100×1)	21 to 40+ 1	over 200%
Prevention of Carcinogenesis:				TUMOUR INCIDENCE
Methylcholanthrene induced:	SWM/Ms	1 × 10 (i.p)	21 to 31++	83 → 33%
Methylcholanthrene induced:	DBA/2	1 × 10 (i.p)	14 to 24++	78 → 37%
Adenovirus type-12 induced:	C3H/He	10 × 3 (i.p)	14, 16, 18++	79 → 40%

* All tumors are solid form implanted S.C.
** mg/kg × days.
*** Day of lentinan injection, tumor inoculation is day 0.
\+ 5 mm tumor diameter: day 0
\++ MC or adenovirus inoculation: day 0.

Cultural fluid

↓ Filtration —— Mycelia

Cultural filtrate

↓ (Activated charcoal)

Depolymerization

↓ Ultrasonic irradiation
High speed extrusion

Purification

↓ Ion exchange chromatography
Acetone fractionation
Lyophilization

Clinical type schizophyllan (SPG)
(Molecular mass 4.5 x 10^5)

Irradiation time (min)	M_n x 10^{-4}	M_w x 10^{-4}
0	385	526
3	250	347
6	147	198
10	77.8	102
15	58.9	75.8
20	46.9	57.9
30	33.4	43.7
40	27.4	34.8
60	23.0	27.8
100	19.5	24.1

Scheme 2. Manufacturing process of clinical grade schizophyllan and change in molecular weight distribution in aqueous solution during ultrasonication at 0.10x 10^{-2} g/cm^3.

One out of three glucose residues of the repeat unit was substituted at 0-6 with a single ß-D-glucosyl residue, as shown in Fig. 2. Recently, enzymatic hydrolysis of schizophyllan with endo-ß-glucanase has yielded 6-0-branched tri- and tetra-saccharides, but not laminaritriose. This supports the regular arrangement of ß-glucosyl side chains along the (1-3)-ß-linked backbone. The presence of regularly arranged branches must play a role in the solution stability. The host-mediated antitumor activity of SPG on certain types of tumors in experimental animals has been established [31] (Table 4).

Table 4. Effects of SPG on allogeneic and syngeneic tumors

Tumor	Host	SPG administration route[a]	Effect	Remarks[b]
Mouse allogeneic				
Sarcoma 180	ICR/JCL	i.p.	+	
and others	ddY		+	
Mouse syngeneic				
Lewis lung cancer	C57BL/6	i.p.	+	
MC-induced carcinoma	C57BL/6		+	
Spontaneous leukemia	AKR		+	
Meth-A fibrosarcoma	BALB/c	i.m.	-	5-FU: +
Adenocarcinoma 755	C57B/He	i.m.	+	
L-1210 leukemia	BDG1	i.m.	-	MFC: +
P-388 leukemia	BDG1	i.m.	-	
Rat syngeneic tumor				
BC-47 bladder tumor	AC1/N	i.m.	+	
AMC-60 fibrosarcoma	AC1/N	i.m.	+	

[a] i.p.: intraperitoneal; i.m.: intramuscular.
[b] 5-FU: 5-fluorouracil; MFC: MMC+FU+Ara-C

Table 5. Antitumor activity of synthetic branched cellulose and curdlan

Polysaccharide	Dose mg/kg	Tumor Wt (g)	Tumor Inhibition(%)	Complete Regression
D-Araf-curdlan (S)	10	1.99+1.1	75.2	3/7
D-Araf-curdlan (L)	1	0.33+0.1	99.7	6/7
D-Araf-curdlan (L)	10	0.0	100	6/6
D-Man-curdlan (S)	10	1.39+0.5	82.7	1/7
D-Araf-cellulose(S)	10	5.67+0.4	29.4	0/7
D-Man-cellulose(L)	10	5.46+1.0	32.0	0/7

1. Clinical uses of SPG

The clinical use of SPG and lentinan as immunotherapeutic agents for cancer treatment has been approved in Japan since 1986. These agents are employed in combination with other chemotherapeutics or radiotherapy. Although SPG is currently being used only for immunotherapy of cervical cancers of stage 11 or 111 in combination with radiotherapy, recent clinical studies have demonstrated its beneficial effects in lung and gastric cancer in combined administration with MMC or 5-FU.

In addition to its recognized antitumor activity, the following immunological effects of SPG involving stimulation of the host immune system have been reported [31].

a. Protective effects against bacterial infections in animals.
b. Protective effects against microbial infections, including streptococcal infections in fish (yellow tail).
c. Protective effects against Send virus infections (mice).
d. Anti-HIV activities of partially sulfated schizophyllan.

2. Conformation of (1-3)-ß-glucan chains.

Although the mechanism of tumor-growth inhibition has not yet been fully elucidated, the immunomodulating activities of schizophyllan and other branched (1-3) glucans appear to be closely related to the triple helical structure of their backbone, suggested by viscosity studies [32], as well as X-ray analysis (33-35) (see Fig. 4). The relationships between the molecular sizes of low molecular SPG fractions and their antitumor activities on Sarcoma 180 are shown in Fig. 5. It is apparent that the concentration of triply stranded species fell sharply with decreasing molecular mass. A concomitant loss of the antitumor activity was observed, with essentially no activity being displayed in the random-coil region below a molecular weight of 5 x 104 (36). This confirms that the triple helical conformation in aqueous solutions is essential for the manifestation of antitumor activity. In connection with this, Matsuzaki et al. [37] synthesized branched ß-(1-3)-glucan and -(1-4)-glucan having D-mannose-, D-arabinofuranose, or D-arabinooligosyl side chains using curdlan and cellulose, respectively. D-arabinoyl-, and oligoarabinosyl branched polysaccharides were synthesized by the reaction of 3,4,6-tri-O-acetyl-(1,2-O-ethylorthoacetyl)-ß-mannopyranose, 3,5-d-O-benzoyl-(1,2-O-ethylorthobenzoyl- or 3-O-benzoyl-(1,2,5-O-orthobenzoyl)-ß-D-arabinofuranose with the O-acetylated glucan, respectively. Structural determination and tumor (Sarcoma 180)-inhibiting activities of these branched glucans clearly indicated that branched curdlans, particularly those with D-arabinofuranose- or oligoarabinosyl side chains, exhibit remarkably higher antitumor activity.

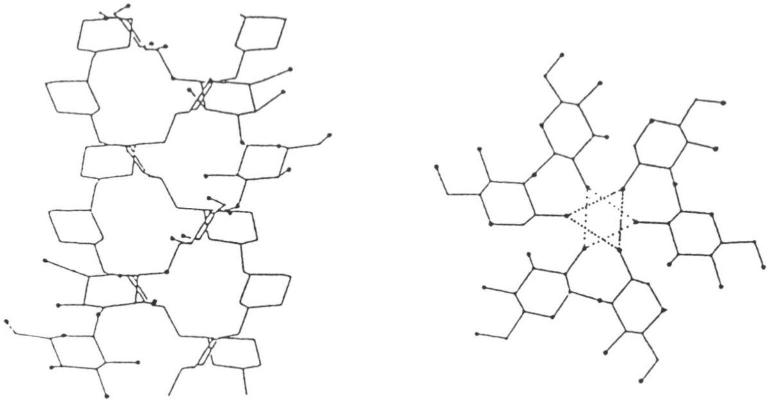

Fig. 4-a Triple helical structure of 6-O-branched (1-3)-ß-D-glucan

a. Projection on ac plane (left), and b. on ac plane (33)

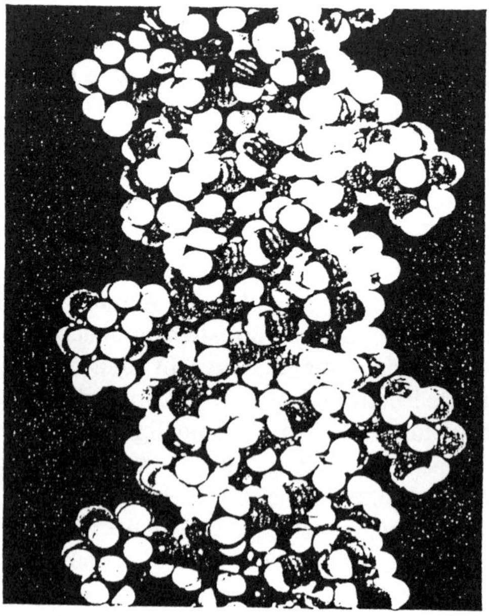

Fig. 4-b Space-filling model of the schizophyllan triple-helix

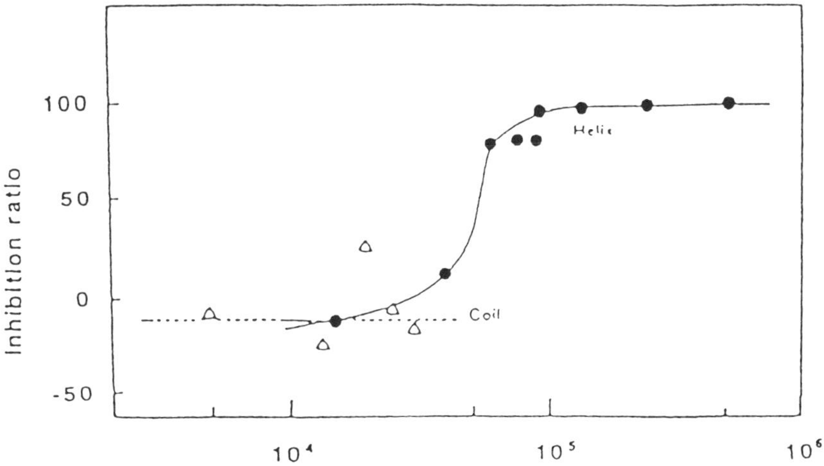

Fig. 5 Molecular weight dependence of the tumor growth inhibition ratio for triple helical species (e) and random coil species (A) of schizophyllan.

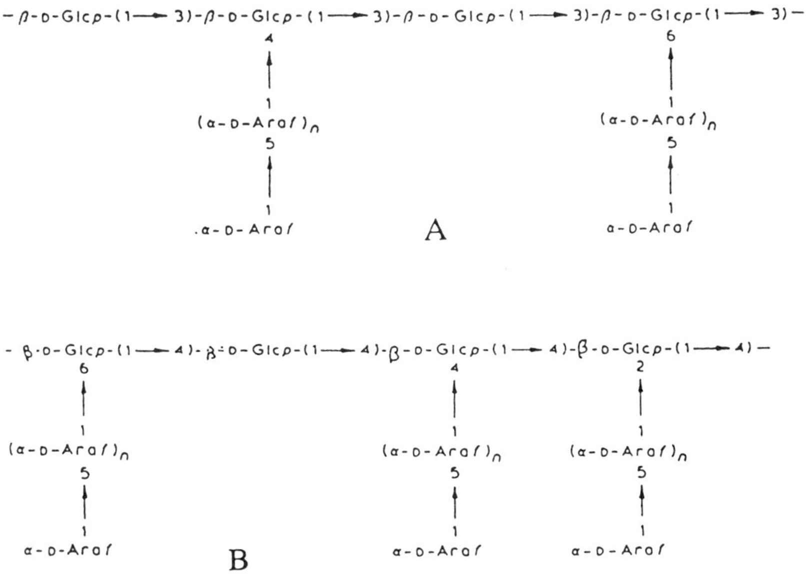

Fig. 6 Synthetic branched (1-3)-, and (1-4)-ß-D-glucan.
A: Curdlan having D-arabinofuranosyl side chains,
B: Cellulose having D-arabinofuranosyl side chains

In contrast, the arabinosyl- or mannosyl-branched celluloses showed very low activity. These results confirm that the ß-linked (1-3)-glucan backbone is essential for antitumor activity (Fig. 6 and Table 5).

Curdlan, extracellularly produced by *Alcaligenes feacalis* var. *myxogenes* , is a linear-(1-3)-glucan (25). This glucan also exhibits antitumor activity. It is a long linear chain with DP > 200 that is not soluble in water but, upon heating a resilient gel is formed. Conformational studies by x-ray analysis indicated that laminaran (DO -35) curdlan generally assumed triple-helix chains [34,35]. On the other hand, ^{13}C-NMR spectroscopic studies suggested that it may take either a single-chain in the anhydrous state, a single-helix in the hydrate state, or the triple-helix form in the gelling state [37]. With regard to the antitumor effect, they suggested that curdlan in the single helix is more effective than in the triple helix form, although such an explanation was not always consistent with X-ray analysis. These conformational studies on the linear and O-6-branched ß-(1-3)-glucans are reminiscent of the unsolved nature of the steric structure of the cell wall glucan of yeast.

IV. STRUCTURAL DIVERSITY OF ANTITUMOR (1-3)-ß-D-GLUCANS

As mentioned before, lentinan, schizophyllan and similar types of fungal glucans have moderately branched structures (db 1/3), and exhibit strong antitumor activity. In continuing studies on the correlation of structure and antitumor activity, we and other workers have isolated various types of branched (1-3)-ß-D-glucans (Table 1). These glucans exhibit different levels of antitumor activity. These isolates include highly branched, water-soluble and alkali-insoluble glucans from Kikurage, the fruiting bodies of *A. auriculajudae* [39,40], less extensively branched glucans from "Reishi" *Ganoderma lucidum* [41], and a highly branched Pestalotan extracellularly produced, by *Pestalotia sp.* 815 [42].

Recently, we also isolated a highly immunoactive, moderately branched glucan by cold alkali extraction of a Chinese edible mushroom (*Volvarielia volvacea)* [43]. In Table 6, the main repeating units of these branched glucans are compared with their antitumor activities against Sarcoma 180 (solid tumor in ICR-JCL mouse, dose: 5-10 mg/kg 10 days). Examination of Table 6 indicates that the presence of antitumor activity is apparently closely related to the mode of branch distributions involving the degree of branching and the glucosyl side chains attached to the (1-3)-ß-linked backbones. Thus, glucans with moderate degrees of branching, i.e., db, 1/5 to 1/3, such as schizophyllan, a cold alkali-extracted *Volvariella* glucan (VVG), and a water-extracted Reishi *Ganoderma* glucan, appear to exhibit the most potent activities, whereas glucans with either extremely high or low degrees of branching are less active.

Table 6. Antitumor activities of glucans of some fungal (1-3)-β-D-glucans

Glucan	db	Dose[a] (mg/kg)	Average tumor weight (g)		% Tumor growth inhibition	Extent of complete regression
			Treated	Control		
Activity against sarcoma 180 solid tumor in ICR-JCL mice						
V. volvacea						
Cold alkali-extracted	1/5	1	2.5	9.5	73.2	3/6
(VVG)		5	0.3	10.2	97.0	4/5
		10	2.0	10.9	81.7	1/5
Hot alkali-extracted	1/2	10	7.2	10.9	33.9	0/5
S. commune						
Schizophyllan	1/3	1	0.3	4.1	95.1	6/10
A. auricula-judae						
Alkali-insolub. glucan	3/4	10	4.7	5.8	18.9	0/4
G. lucidum						
Water-extracted	1/3	10	0.2	9.4	97.7	4/5
Cold alkali-extracted	1/17	10	3.8	9.4	59.6	2/6
Pestalotia sp. 815						
Pestalotan	2/3	5	4.2	9.8	57.3	0/6
Activity against methylcholanthrene fibrosarcoma in DBA/2 mice						
VVG		5	0	2.04	100.0	6/6

[a] Daily intraperitoneal injection for 10 days.

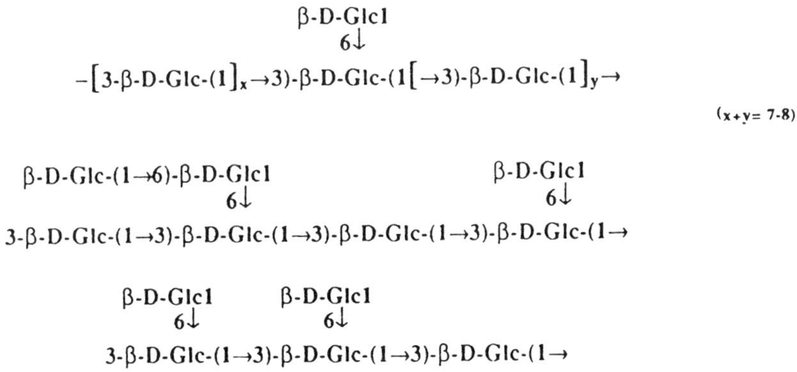

Fig. 7 Structural segements obtained by (1-3)-β-D-glucanase digestion of *V. volvacea* glucan (VVG).

Representatives of the latter two categories include an alkali insoluble kikurage glucan II (db 3/4), Pestalotan, an extracellular glucan (DB 2/3) produced by *Pestalotia* sp. 815, and hot alkali extracted glucans of "Reishi." *Ganoderma lucidum*, (db, 1/17 and 1/23), respectively. The lower activity of some glucans may be attributed, in part, to their insolubility. In this connection it is interesting that the water soluble glucans of Reishi (db, 1/3), and Kikurage (glucan I, db, 2/3) showed strong activity. The importance of solubility is also suggested by the fact that conversion of the glucosyl branch residues into the corresponding, more soluble polyhydroxylated functions resulted in significant activity enhancements of these less active glucans as will be discussed later.

We have recently focused our interest in 0-6-branched antitumor (1-3)-ß-D-glucans on the glucan of a Chinese edible mushroom, "**tsao ru**" in Chinese, or "**Fukurotake**" in Japanese, the fruiting body of *Volvariella volvacea*, in particular the glucan fraction obtained by cold alkali extraction. This moderately branched glucan with a db value of 1/5 was highly purified by ion exchange chromatography (DEAE Toyopearl column). The purified, protein-free glucan ([«] -30_ in M NaOH, Mw 1.5-2 x $1O^6$), was designated as **VVG** (43). It exhibited significantly high antitumor activities, i.e., 81.7% at a dose of 10 mg/kg, 97.7% at 5 mg/kg, and 73.2% at 1 mg/kg, respectively. As summarized in Tables 1 and 7, this glucan was also capable of inhibiting syngeneic tumors, such as Meth-A fibrosarcoma in BALB/c mice and methylcholanthrene-induced fibrosarcoma in DBA/2 mice, when injected intraperitoneally at a dose of 5 mg/kg for 10 days (43, 45). We investigated the structural features of VVG by enzymatic hydrolysis, using exo- and endo-(1-3)-ß-D-glucanase, and performed structural analyses of the oligosaccharide fragments. VVG has predominantly single D-glucosyl groups attached, on the average to every fifth (1-3)-D-glucosyl residue. The results indicated that VVG is structurally heterogeneous with regard to the distribution of glucosyl branches, being composed of sparsely, moderately, and highly branched segments, as shown in Fig. 7. The highly branched segment derived from enzymatic hydrolysis showed a lower level of antitumor activity.

Fig. 8 shows the inhibitory activity of VVG against the growth of Sarcoma 180 solid tumors, when subcutaneously inoculated into ICR-JCL mice. The growth of tumors without VVG treatment shows a linear correlation with time: the average tumor volume of 2.5 mL at day 21, increased exponentially until the average tumor weight reached half of the total body weight after 35 - 42 days.

On intraperitoneal administration at a dose of 5 mg/kg from day 1 to day 10, VVG showed strong antitumor activity and was found to completely eliminate the solid tumors in all mice after 21 days. Intraperitoneal administration of VVG to tumor-bearing mice, at the same dose, starting from days 14 and 21 for 10 days also resulted in growth inhibition of Sarcoma 180, and the mice were

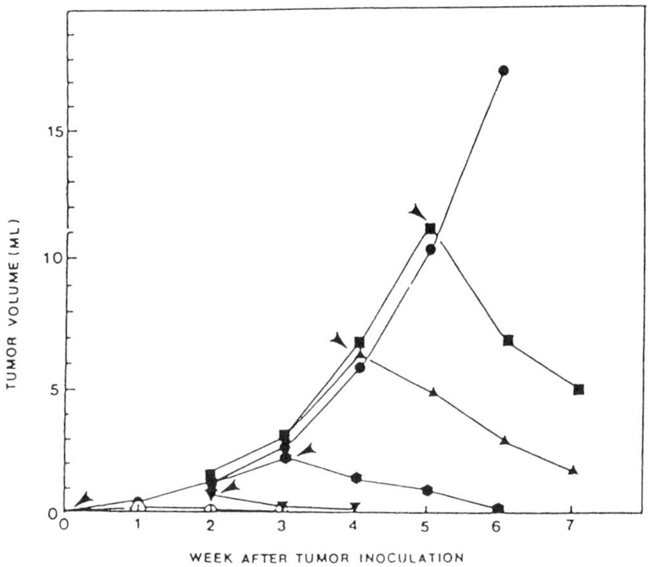

Fig. 8. Growth curves of Sarcoma 180 solid tumors with or without VVG treatment. (●) without treatment (control); (○),(⬢), (▲), and (■) with treatment (5 mg/kg daily for 10 days), arrow (▼) indicate the starting day of VVG injections.

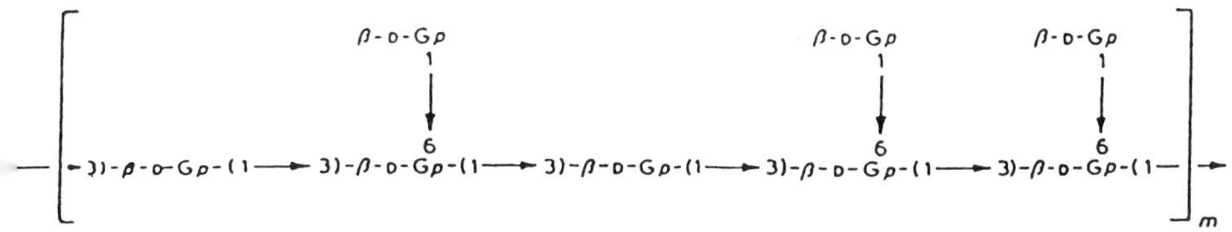

Fig. 9. Possible repeating unit of pestalotan

cured at days 28 and 42, respectively. Administration of VVG into mice at days 28 and 35, when the tumor grew exponentially, also reduced the tumor volume to a great extent, to about one fifth and one half of that in control mice, respectively.

V. EFFECTS OF CHEMICAL MODIFICATIONS OF GLUCAN SIDE CHAINS ON ANTITUMOR ACTIVITY

As described above, **Kikurage** or Tree ear mushroom (*Auricularia auricula judae*) yielded two branched (1-3)-ß-D-glucans, a water-soluble glucan I (db 2/3) and a hot alkali-insoluble glucan II (db 3/4) [46]. The latter was the major constituent of the mushroom and had essentially no antitumor activity (Table 1). The significant difference in the antitumor activities of glucan I and II prompted the removal of part of the side chains in glucan II by a controlled oxidation with a stoichiometric concentration of periodate (0.5-2.0 moles per glucose residue), followed by reduction of the oxidized glucan with borohydride to the corresponding glucan-polyalcohol, and subsequent mild acid hydrolysis (Smith degradation; 0.1 M acid for 24 h at 25°C) [40]. By this procedure periodate-modified glucan fractions with different glucosyl group content should have been anticipated, based on traditional Smith degradations. Most of the partial Smith-degraded glucan fractions exhibited remarkably higher activity than the native glucan II. Methylation analysis and complete acid hydrolysis showed that in these modified glucans the polyhydroxylated groups were still linked to the (1-3)-ß-glucan backbone chain, due to insufficient hydrolysis. This important structure-activity relationship finding revealed that the higher antitumor activity of the periodate-modified glucans was closely related to the content of polyol groups attached to the backbone. When the glucan polyols were hydrolyzed under stronger acidic condition, i.e., with 0.2 M acid for 2 h at 100°C, the acetal linkages were completely cleaved, and the resulting linear (1-3)-glucans lost their original antitumor activity [40,42]. This finding strongly suggested that conversion of D-glucosyl branch residues into the corresponding polyhydroxylated groups should have a pronounced enhancement effect on biological activity. Such enhancements of antitumor activity were observed when this modification procedure was applied to several other branched (1-3)-ß-D-glucans, that were either very highly or sparsely branched, e.g., pestalotan (42), Reishi glucan (41), and even schizophyllan and VVG (44), as shown in Tables 1 and 6. The potent antitumor activity of the glucan-polyols appeared to be directly correlatable with their polyol content [45].

Pestalotan is a highly branched (1-3)-ß-D-glucan, produced in good yields (40 g/10 L of glucose medium) during culture of *Pestalotia* sp 815. This glucan with three ß-glucosyl branches per five (1-3)-linked glucose residues as shown in Fig. 9, becomes water-insoluble after isolation followed by dehydration [42].

The native glucan exhibits a moderate antitumor activity, but after conversion to the corresponding polyol derivative, it showed potent antitumor activities against various mouse-implanted tumors, as summarized in Tables 7 and 8. Also Pestalotan polyol demonstrated augmentation functions, implicating stimulation of RES, activation of macrophages *in vivo,* and stimulation of cell-mediated immunity in tumor (Ehlrich carcinoma)-bearing mice, but not in non-tumor bearing mice.

It must be noted that complete removal of the polyol groups by mild acid hydrolysis resulted in the elimination of the original activity. The glucose branch residues of VVG were also converted to the corresponding 3,6-anhydro derivatives by sulfation, followed by alkali treatment, in order to examine the effects of the molecular shape of the branches on antitumor activity of the 0-6 branched (1-3)-ß-D-glucans. The glucans bearing 3,6–anhydro D-glucosyl branches lost the antitumor activity of the original glucan, or exhibited instead a stimulation effect on the growth of Sarcoma 180 solid tumor cells (inhibition ratio: 63.8%, at a dose of 5 mg/kg). Thus, the removal of the side chains or the introduction of 3,6-anhydro-rings may result in the reduction or elimination of the original antitumor activity. Scheme 3 summarizes the procedures employed for the chemical modification of VVG.

The activity enhancement we observed upon introduction of polyhydroxylated groups may be partly due to increases in water solubility. It is also possible that the presence of numerous polyhydroxylated groups outside of the triple helical assembly of the (1-3)-ß-D-glucans may augment the immunomodulating potency of the host.

To confirm the role of the hydrophilic side chains on the antitumor activity of (1-3)-ß-D-glucans, we examined the effect of chemical derivatization of some linear (1-3)-ß-glucans. For this purpose, curdlan was first partially epoxylated and then glycerated, as shown in Scheme 4. The antitumor activities of the resulting modified curdlans were then compared. Under optimal conditions, a 0.2% solution of curdlan in 0.5 M sodium hydroxide was reacted with 5% epichlorohydrin, affording the epoxylated curdlan (I) with a degree of substitution (DS) of 0.14 (per glucose unit), with substitution occurring most likely at position 0-6. Alkali treatment of I yielded the glyceryl curdlan (II). No epoxy groups were detected in derivative II. These curdlan derivatives were tested for their antitumor activity on Sarcoma 180 solid tumors by intraperitoneal administration of 10 mg/kg for 10 days. In this system, the glyceryl-containing curdlans produced a complete regression of the tumor cells (inhibition ratio: 100%), which was higher than that of the original glucan (inhibition ratio: 72.2%). On the other hand, the epoxylated curdlan derivatives displayed no antitumor activity (inhibition ratio: 8.3%). These results suggest that the introduction of hydrophilic polyol [1-(1,3-dihydroxy-2-propyloxy)-2-hydroxyethyl] or glyceryl groups should result in activity enhancements, whereas the introduction of hydrophobic groups such as epoxy functions may have the opposite effect.

These results provide unambiguous evidence that enhanced antitumor activity may be associated with the triple-helical conformation of the glucan backbone as well as with the presence of polyhydroxylated groups situated outside the triple-stranded backbone that are introduced by substitution at 0-6 of the (1-3)-linked D-glucose residues. In this context it should be noted, as previously discussed, the chemically synthesized, branched (1-3)-ß-glucans with D-arabinofuranosyl or D-mannopyranosyl branches exhibited high antitumor activity, whereas branched (1-4)-ß-glucans with similar side chains had no antitumor activity.

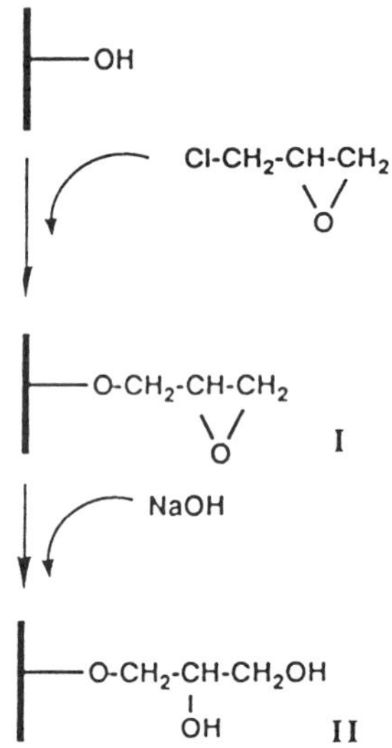

Polysaccharide	Dose[a] (mg/kg)	Average tumor weight (g) Treated	Control	Inhibition ratio (%)	Extent of complete regression
Curdlan	10	2.00	8.20	72.2	2/6
Glyceryl-curdlan	10	0	7.20	100	5/5
Epoxy-curdlan	10	10.46	9.66	-8.3	0/5

[a] Daily intraperitoneal injection for 10 days.

Scheme 4. Derivatizations of curdlan and their antitumor activities against sacroma 180 solid tumor

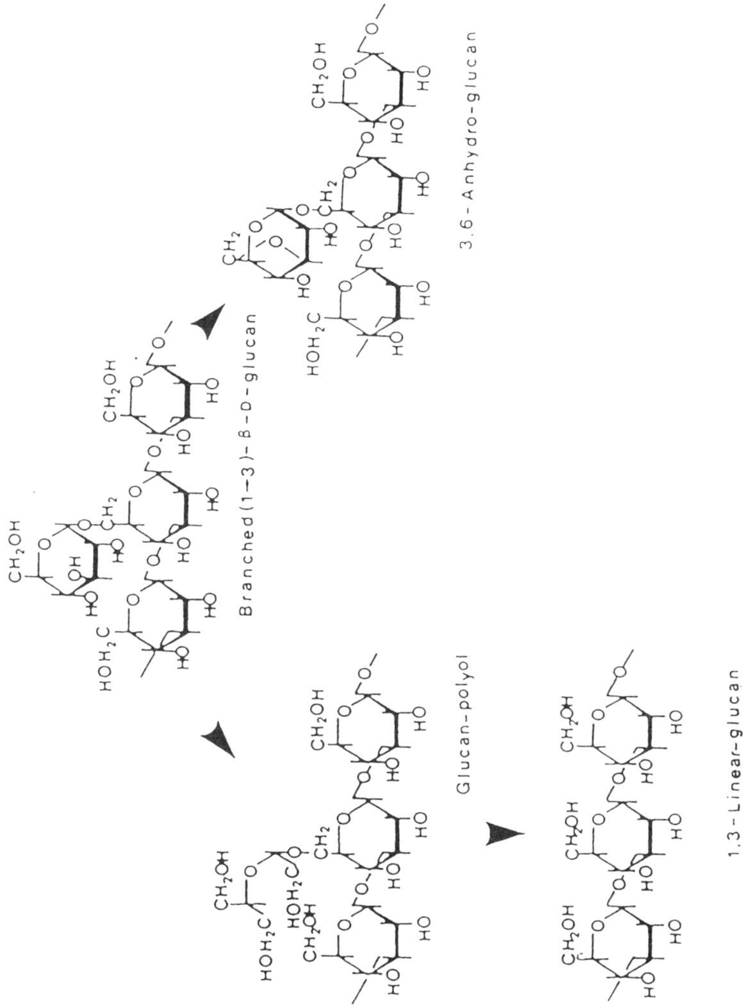

Scheme 3. Chemical modification of (1-3)-β-D-glucan

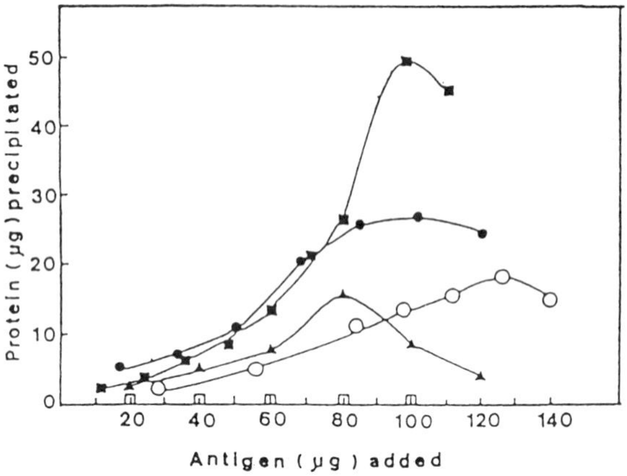

Fig. 10-a. Quantitative precipitation curves of anti-VVG polyol antibody with VVG polyol-BSA (■), VVG polyol (●), VVG (○), BSA (□)

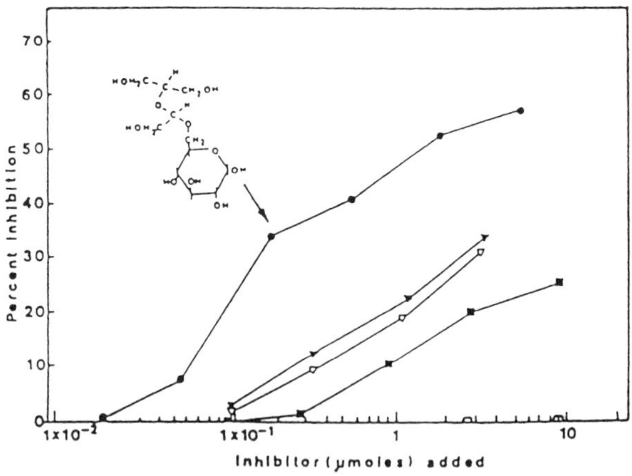

Fig. 10-b. Hapten inhibitions of precipitations with ß-gluco oligosaccharides. Laminaritriose (▽), laminaritetraose (▼), laminaribiose (■), gentiobiose (□), and 6-O-[1-(1,3, dihydroxy-2-propyloxy)-2-hydroxyethyl]-D-glucose (●)

VI. IMMUNOCHEMICAL PROBLEMS OF SOME (1-3)-ß-GLUCANS

Although the mechanism of the antitumor actions of fungal branched (1-3)-ß-D-glucans has not been fully elucidated, immunopharmacological studies especially on lentinan have suggested that a likely explanation may involve the stimulation of the host immunopotentiation, associated with macrophage activation, the promotion of T cell differentiations, the augmentation of NK-activity and others, as previously discussed (27) (cf. Fig. 3). Concerning the antitumor activity of (1-3)-ß-D-glucans, some Japanese workers showed that the mouse serum treated with the glucan displayed enhanced macrophage-mediated cytotoxicity (Sasaki et al. 1982). Similarly, after treatment with CM-TAK (curdlan) by intravenous injection, mouse serum showed a reduction in Sarcoma 180 cells, suggesting formation of a tumor-regressing factor which appeared to be a glycoprotein or protein. Our recent investigations have also shown that when an ^{3}H-labeled VVG-polyol was intraperitoneally administered to the tumor-bearing mice, radioactivity levels appeared in the serum 12-23 hours after the injection, and the serum contained a tumor-inhibiting factor. This observation prompted the preparation of an immunoadsorbent affinity column using antibodies to VVG. Rabbits were immunized with VVG- or VVG polyol-BSA-conjugate. The BSA-conjugate was prepared by a partial periodate oxidation of VVG, followed by reductive amination with bovine serum albumin. Hapten inhibition studies with various (1-6)- and branched (1-3)-linked ß-gluco- oligosaccharides indicated that the purified antibodies recognized the sequence involving the non-reducing terminal groups and possibly the branch points. This was confirmed, using anti-VVG-polyol antibodies, as shown in Fig. 10. When the mouse serum containing tumor-inhibiting factor was passed through the immunoadsorbent column of VVG-antibodies, the tumor-inhibiting factor was, for the most part, retained on the column [47]. This result suggests that the tumor-inhibiting factor is immunochemically similar in nature to the glucan, VVG. This suggests the possible formation of certain glucan-protein conjugates during tissue circulation of the antitumor glucan although isolation and chemical identification of the factor have not yet been achieved. The antibodies or the immunoadsorbent column may provide a useful probe in studies of the immunomodulation mechanism of a specific polysaccharide. In connection with this, it was confirmed that schizophyllan intraperitonally injected into a mouse binds to the macrophages, as immunochemically detected by use of dye-labeled polyclonal antibodies to schizophyllan [48].

VII. SUMMARY

Most fungal ß-D-glucans that constitute the fruiting bodies or are accumulated in cultural fluids, are (1-3)-linked with ß-D-glucosyl branches at 0-6. However, some structural diversity exists in their branch distributions [degree of branching (db) 1/25-1.25]. Lentinan and the sonically-depolymerized schizophyllan (Mw 4.5 x 105) represent two ß-D-glucans with regularly arranged ß-D-glucosyl side chains (db 1/3), that are elaborated by *Schizophyllum commune.* They are currently used as immunotherapeutic agents. Some chemical and immuno-modulating properties of schizophyllan are discussed within this chapter. We also describe the fine structure of VVG, another potent antitumor, branched (1-3)-ß-glucan (db1/5-1/4) of *Volvareella volvacea,* and the use of glucan antibodies as biochemical probes. The antitumor activities of these (1-3) glucans may be based on the triple helical backbone conformation, as well as on the molecular shape and distribution of their side chains, as evidenced by chemical modification of the D-glucosyl branches.

Table 7. Comparison of antitumor activities of some branched (1-3)-β-D-glucans and their derivatives

Polysaccharide	Dose[a] (mg/kg)	Average tumor weight (g)		Inhibition ratio (%)	Extent of complete regression
		Treated	Control		
Activity against sarcoma 180 solid tumor in ICR mice					
V. volvacea					
VVG-polyol	1	0	9.5	100	6/6
	5	0	9.5	100	6/5
	10	2.8	9.5	70.7	3/6
1, 3-Linear-VVG	5	11.6	9.5	-22.5	0/6
3, 6-Anhydro-VVG	5	15.6	9.5	-63.8	0/6
A. auricula-judae					
Glucan	10	8.9	11.0	18.9	0/6
Glucan-polyol	5	0.3	11.0	97.4	6/8
Pestalotia					
Pestalotan	5	4.2	9.8	57.3	3/5
Pestalotan-polyol	1	0	10.2	100	3/5
Activity against meth-A fibrosarcoma in BALB/c mice					
VVG-polyol	5	0.1	3.5	97.1	
Activity against methylcholanthrene-induced fibrosarcoma in DBA/2 mice					
VVG-polyol	5	0.06	2.04	97.1	

[a] Daily intraperitoneal injection for 10 days.

Table 8. Antitumor effects of pestalotan-polyol on allogeneic and syngeneic tumors

Tumor	Host	Effect	Remarks
Mouce allogeneic tumor:			
Sarcoma 180	ICR-JCL	+	
	ddy	+	
	BALB/c	+	
Ehlrich carcinoma	ICR-JCL	+	
Mouse syngeneic tumor:			
L-1210 leukemia	BDF1	+	ILS>23.2%, 1 mg/kg, 10d.
Lewis lung carcinoma	C57 B/6	+	Metastasis
Lewis lung carcinoma	C57 B/6	+	(Radiation therapy)
Meth A fibrosarcoma	BALB/c	+	
Meth A fibrosarcoma (asites)	BALB/c	-	
Meth A fibrosarcoma (asites)	BALB/c	+	5-FU, M.F.C.

References

1. Srivastava, HC, Adams, GA. Can J Biochem. 1962; 40: 1415-14.
2. Creech, HJ., Breuninger, E. Polysaccharide-peptide complex from *Serratia marcescens* cells. Can J Biochem. 1962; 42: 93-603.
3. Whistler RL, Bushway AA, Singh PP, Nakahara W, Tokuzen R. Antitumor polysaccharides. In: Adv Carbohydr Chem 1976; 32: 235-275. Witczak ZJ, Whistler RL. Structure and antitumor activity of ploysaccharides. In: industrial polycsaccharides. The impact of biotechnology and advanced methodologies. Stivalla SS, Crescenzi V, Dea ICM. eds. Gordon and Breach 1987; pp: 157-173.
4. Fitzpatrick FW, Di Carip EJ. Zymosan. Ann NY Acad Sci. 1964; 118: 253-261.
5. Mizuno T. ed. Mushrooms: the versalite fungus-food and medicinal properties. Special issue of Food Reviews International 1995, Marcel Dekker, Inc. New York.

6. Johnson J, Kirkwood S, Misaki A, Nelson TE, Scalleti JV, Smith F. Structure of a new fungal glucan. Chem Ind (London) 1963; 820-822.
7. Misaki A, Kishida E, Kakuta M, Tabata K. Antitumor fungal (1-3)-ß-D-glucans: structural diversity and effects of chemical modification In: Carbohydrates and Carbohydrate Polymers. Yalpani M. ed., 1993: 116-129, ATL Press
8. Di Carlo FJ, Fioore JV. On the composition of zymosan. Science 1958; 127: 756-757.
9. Pilmer L, Bluum L, Lepow LH, Todd EW. The proerdin system and immunity. III. The zymosan assay of prooerdin. J Exp Med. 1956: 103: 1-13.
10. Riggi SJ, Di Luio P. Identification of a reticuloendothelial stimulating agent in zymosan. Am J Physiol .1961; 200: 297-300.
11. Hassid WZ, Joslyn MA, McCready RM. The molecular constitution of an insoluble polysaccahride from yeast, *Saccharomyces cerevisiae*. J Am Chem Soc. 1941; 63: 295-298.
12. Peat S, Whelan WJ, Edwards TE. Polysaccharide of baker's yeast. Part III The presence of 1:6-linkages in yeast glucan J Chem Soc. 1958; 3862-3868.
13. Misaki A, Johnson J, Kirkwood S, Scaletti JV, Smith F. Structuure of the cell wall glucan of yeast. Carbohydr Res. 1968; 6: 150-164.
14. Manners DJ, Masson AJ, Patterson JC, Bjorndal H, Lindberg B. The structure of a ß-(1-6)-D-glucan from yeast cell walls. Biochem J. 1973; 135: 31-36.
15. Manners DJ, Mansson AJ, Patterson. The strucuture of a ß-(1-3)-D-glucan from yeast cell wall. Biochem J. 1973; 135: 19-30.
16. Konishi Y, Misaki A. Fractionation of yeast cell wall and structural elucidation of (1-3)-ß-glucan .1976: Abstr. paper of Jap Agric Chem. Soc.
17. Matsumoto T, Mikami T, Nagase T, Suzuki M, Suzuki S. Relationship between chemical structure and antitumor activity of the mannan of *Saccharomyces cerevisiae* In: *Immunomodulation by microbial products and related synthetic compounds* ed. Yamamura, Y. etal 1982: pp-427, Excerpta Medica
18. Hirase S, Nakai A, Akatsu T, Kobayashi A, Oohara M, Matsunaga K, Fujii M, Kodaira K, Furusho, T, Oomura Y, Wada T, Yoshikumi C, Ueno S, Ohtsuka S. Structural studies on the anti-tumor polysaccharides from *Coroolus versicolor* (Basidiomycetes). I. Fractionation with barium hyroxide. Yakugaku Zasshi (in Japanese). 1976; 96: 413-418.
19. Hirase S, Nakai A, Akatsu T, Kobayashi. A, Oohara M, Matsunaga K, Fujii M, Kodaira K, Furusho T, Oomura Y, Wada T, Yoshikumi C, Ueno S, Ohtsuka S. Structural studies on the anti-tumor polysaccharides from *Coroolus versicolor* (Basidiomycetes). II. Structures of ß-D-glucan

moieties of fractionated polysaccharides. Yakugaku Zasshi (in Japanese). 1976; 96: 419-424.
20. Tsukagoshi S.: In host defense against cancer and its potentiation, edted by Mizuno, D et al: 1975 Univ. of Tokyo Express, p 369 .
21. Chihara G, Maeda Y, Hamuro G, Sasaki T, Fukuoka F. Inhibition of mouse Sarcoma 180 by polysaccharides from *Lentinus edodes* (Berk. Sing. Nature. 1969; 222: 687-688.
22. Chihara G, Hamuro J, Maeda Y, Artai Y, Fukuoka F. Fractionation and purification of the polysaccharides with marked antitumor activiy,especially lentinan, from *Lentinus edodes* (Berk. Sing) an edible mushroom. Cancer Res. 1970; 30: 2776-2781.
23. Sasaki T, Takasukà N. Further study of the structure of lentinan, an antitumor polysaccahride from *Lentinus edodes.* 1976: Carbohydr. Res. 99-104.
24. Singh PP, Whistler RL. Scleroglucan, an antitumor polysaccharide from *Sclerotium glucanicum* . Carbohydr Res. 1974; 37: 245-247.
25. Harada T, Saito H, Misaki A. Curdlan: a bacterial gel-forming ß-1,3-glucan. Arch Biochem Biophys. 1968; 124: 292-298; Saito H, Misaki A, Harada T. A comparison of the structure of curdlan and pachyman. Agric Biol Chem. 1968; 32: 1261-1269.
26. Sasaki T, Takasuka N, Chihara, G, Maeda Y. Antitumor activity of degradaded products of lentinan: its correlation with molecular weight. Gann 1976; 67: 191-195.
27. Chihara G, Immunopharmacology of lentinan and the glucan. J. Immunology and Immunopharmacology .1984; IV: 85-97.
28. Kikumoto S, Miyajima T, Kimura K, Okubo S, Komatsu N. Polysaccharide produced by *Schizophyllum commune* Part II. Chemical structue of an extracellular polysaccharide. J Agric Chem Soc Jpn. 1971; 45: 1747-1753.
29. Kikumoto S, Miyajima T, Yoshizumi S, Fujimoto S, Kimura K. Polysaccharide produced by *Schizophyllum commune* Part I. Formation and some properties of an extracellular polysaccharide. J Agric Chem Soc Jpn. 1970; 44: 337-342.
30. Tabata T, Ito W. Kojima T, Kawabata S, Misaki A. Ultrasonic degradation of schizophyllan. an antitumor polysaccharide produced by *Schizophyllum commune* FRIES. Carbohydr.Res. 1981; 89: 121-135.
31. Matsuo T, Arika M, Mitani M, Komatsu N. Pharmacological and toxicological studies of a new antitumor polysaccharide, schizophyllan. Drug Res. 1982; 32: 595-710.
32. Kashiwagi Y, Norisue T, Fujita H. Triplehelix of *Scizophyllum commne* poysaccharide in dilute solution. Light scattering and viscosity in dilute aqueous sodium hydroxide solution. Macromolecules. 1981; 14: 1220-1255.

33. Deslandes Y, Marchessault RH, Sarko R. Triple-helical structure of (1-3)-ß-D-glucans. Macromolecules. 1980; 13: 1466-1471.
34. Marchessault RH, Deslandes Y, Ogwa K, Sundarajian PR. Can J Chem. 1977; 55: 300-303
35. Kasai N, Harada, T. Ultrastructure of curdlan. In: Fibre diffraction methods French AD, Gardner KH. eds. 1980: 363-383. Am Chem Soc. Synposium Series No. 141
36. Kojima T, Tabata K, Itoh W, Yanaki T. Molecular weight dependence of the antitumor activity of schizophyllan. Agric Biol Chem. 1986; 50: 231-232.
37. Matuzaki, K, Sato T, Enomoto K, Yamamoto I, Oshima R, Hatanaka K, Uryu T, Kaku H, Sone Y, Misaki A. Synthesis of water-soluble, branched polysaccahrides having D-mannopyranose, D-arabinofuranose, or oligo-D-arabinofuranose side-chains and their antitumor activity. Carbohydr Res. 1986; 157: 171-183.
38. Saito H, Ohki, T, Sasaki T. A ^{13}C nuclear magnetic resonance study of gel-forming (1-3)-ß-D-glucans, Evidence of the presence of single-helical conformation in a resilient gel of a curdlan type-polysaccharide from *alcaligenes feacalis var. mxigenes* IFD 1340. Biochemistry. 1977; 16: 908-914.
39. Sone A, Kakuta M, Misaki A. Isolation and characterization of polysaccharides of Kikurage, fruiting body of Auicularia auricula-judae. Agric Biol Chem. 1978; 42: 417-425.
40. Misaki A, Kakuta M, Sasaki T, Miyaji H. Studies on interrelation of structure and antitumor effects of polysaccharides: Antitunor action of periodate-modified , branched (1-3)-ß-D-glucan of *Auricularia auricula-judae*, and other polysaccharides containing (1-3)-glycosidic linkages. Carbohydr Res. 1981; 92: 115-129.
41. Sone Y, Okuda R, Wada N, Kishida E, Misaki A. Strutures and antitumor activities of the polysaccharides isolated from fruiting body and the growing culture of mycelium of *Ganoderma lucidum* . Agric Biol Chem. 1985; 49: 2641-53.
42. Misaki A, Kawaguchi K, Miyaji H, Nagae H, Hokkoku H, Kakuta M., Sasaki T, Structure of Pestalotan, a highly branched (1-3)-ß-D-glucan elaborated by Pestalotia sp 815, and the enhancement of its antitumor activity by polyol modification of the side chains. Carbohydr Res. 1984; 129: 209-227.
43. Misaki A, Nasu M, Sone Y, Kishida E, Kinoshita C. Comparison of the structure and antitumor activity of polysaccharides isolated from Fukurotake, the fruiting body of *Variella volvacea*. Agric Biol Chem. 1986; 50: 2171-2183.

44. Kishida E, Sone Y, Misaki A. Purification of an antitumor-active, branched (1-3)-ß-D-glucan from *Variella volvacea.*, and elucidation of its fine structure. Carbohydr Res. 1989; 193: 227-239.
45. Kishida E, Sone Y, Misaki A. Effects of branch distribution and chemical modification of antitumor (1-3)-ß-D-glucans. Carbohydr Polymers. 1992; 17: 89-95.
46. Misaki A, Kakuta M, Kikurage (Tree-ear. and Shirokikurage (White jelly-leaf.: *Auricularia auricula-judae* and *Tremella fuciformis* In: Mizuno, T. ed. Mushrooms: the versatile fungus-food and medicinal properties. Special issue of Food Reviews International 1995: 211-219, Marcel Dekker, Inc., New York.
47. Kishida E, Sone Y, Shibata S, Misaki A. Preparation and immunochemical characterization of antibodies to branched (1-3)-ß-Glucan from *Variella volvacea.*, and its use in studies of antitumor actions. Agric Biol Chem. 1989; 53: 1849-1859.
48. Tabata K, Itoh W, Hirata A, Sugawara I, Mori S. Preparation of polyclonal antibodies to an antitumor (1-3)-ß-Glucan, schizophyllan. Agri. Biol Chem. 1990; 54: 1953-1959.
49. Shirasugi N, Misaki A. Isolation, characterization, and antitumor activities of the cell wall polysaccharides from *Elsinoe leucospila.* Biosci. Biotech. Biochem.1992; 56: 29-33.
50. Kataoka N, Ikuta J, Kuroshima A, Misaki A. Antitumor activities and immunochemical properties of the cell-wall polysaccharides from *Aureobasidium pullulans.* Bioci Biothech Biochem. 1994; 58: 2145-2151.
51. Sasaki T, Abiko N, Nitta K, Takatsuka N, Sugino Y. Dependence on chain length of antitumor activity of (1-3)-ß-Glucan from *Alcaligenes faecalis var. myxogenes* IFO 13140, and its acid-degraded products. Cancer Res. 1978; 38: 379-383.
52. Chiara G, Hamuro J, Maeda Y, Arai Y, Fukuoka F. Antitumor polysaccharide derived chemically from natural glucan (pachyman.) Nature. 1970; 225: 943-944
53. Usui T, Iwaki Y, Mizuno T, Tanaka M, Shinkai K, Arakawa M. Isolation and characterization of antitumor ß-glucans from the fruit bodies of *Ganoderma applanatum.* Carbohydr Res. 1983; 115: 273-280.
54. Ohno N, Suzuki I, Oikawa S, Sato K, Miyazaki T, Yadomae T. Antitumor activty and structural characterization of glucans extracted from cultured fruit bodies of *Grifola frondosa.* Chem Pharm Bull. (Tokyo). 1984; 32: 1142-1151.

Index